Integrated Circuit Engineering

Design, Fabrication, and Applications

Integrated
Circuit Engineering

Design, Fabrication,
and Applications

ARTHUR B. GLASER
Bell Telephone Laboratories, Inc.
Murray Hill, N. J.

GERALD E. SUBAK-SHARPE
The City College
of the City University of New York

ADDISON-WESLEY PUBLISHING COMPANY
Reading, Massachusetts
Menlo Park, California · London · Amsterdam · Don Mills, Ontario · Sydney

This book was set in Times Roman by Arthur Glaser and Joan McCarthy, using a Graphic Systems phototypesetter driven by a PDP-11/45 running under the UNIX operating system.

Second printing, May 1979

Copyright © 1977 by Bell Telephone Laboratories, Incorporated, and G. Subak-Sharpe. Philippines copyright 1977 by Bell Telephone Laboratories, Incorporated, and G. Subak-Sharpe.

ISBN 0-201-07427-3
HIJKLMNOP-HA-898765432

To the Memory of My Father
Morris Glaser

—

Arthur B. Glaser

With Gratitude to David Peskin

—

Gerald E. Subak-Sharpe

Preface

This book describes the design, fabrication, and applications of silicon, thin-film, and thick-film integrated circuits. Together, these topics constitute what could be called "integrated circuit engineering." The growing use of integrated circuits over the past decade has caused profound changes in the architecture and capabilities of electronic equipment. It is important for every engineer involved in the design of hardware to have some exposure to each of these topics in order to appreciate the advantages and limitations of the various types of integrated circuits.

This is the first truly comprehensive book on the subject. It covers circuit design, layout, and fabrication techniques for all three classes of integrated circuits, some applications of integrated circuits in systems, the modeling of active and passive devices that are used in integrated circuits, assembly techniques, the measurement of various material properties and tests to determine whether completed circuits are operational, and finally, the yield and reliability of integrated circuits. The range of topics that could be included is virtually without limit. For the most part, each of the topics selected for the book is described in some detail.

Since the book is primarily a text for a first-year graduate course in electrical engineering, it is assumed that the reader has had a one-year course in electronic circuits at the level of Millman and Halkias, *Electronic Devices and Circuits* (McGraw-Hill, 1967) and the normal background in circuit and system theory, physics and mathematics. The level of presentation of the topics varies because some topics, such as assembly techniques, can be treated adequately in a qualitative fashion; whereas, circuit design or the physics and chemistry of various processing operations demand a quantitative treatment.

There is ample material for a three-term course so that the instructor can choose topics as desired. The syllabus for a one-semester (three hours per week) senior-level undergraduate course using selected sections of the text is shown in the following table.

One-Term Senior Level Course

Hour	Topic	Section
1	General Survey	1.1-1.3
2,3,4	Basic junction-isolated bipolar process sequence, qualitative description of epitaxy, photoresist process, oxidation, qualitative treatment of diffusion (results of Sections 5.4.1 through 5.4.3).	5.4.1-5.4.3
5	Buried layer, countersunk oxide isolation, chemical vapor deposition of silicon dioxide and silicon nitride	6.1.1, 6.3, 5.6.5, 5.7
6	Dielectric isolation, complementary bipolar IC's	6.5, 6.9
7,8,9	MOS circuits	6.10-6.14
10	Metal gate CMOS process, latch-up in CMOS	6.15
11,12,13	Thin-film circuits	8.4-8.8
14,15,16	Thick-film circuits	9.1-9.10
17,18	Operational amplifier parameters, bipolar difference amplifier, circuit imbalances	12.1-12.3 (without complete derivations)
19,20	Current sources and common-emitter amplifier	12.6, 12.8
21	Level shifters, current mirrors	12.9; first part of 12.10
22	Output stages	12.12
23,24	Voltage references, dc condition of operational amplifier	12.13, 12.14
25	Summary of small-signal and large-signal operational amplifier performance	Summarize results in 12.15 and 12.16
26,27	Signal conditioning	13.1
28	Characterization of logic circuits	14.1
29,30	Ebers-Moll transistor model, emitter-coupled logic, diode-transistor logic	2.10, 14.2.1 14.4.1
31-33	Transistor-transistor logic	Summarize 14.4.2 and 14.4.3; 14.4.5 14.4.6; Summarize 14.4.7
34,35	Integrated-injection logic, start MOS	14.5, 14.6.1
36,37	MOS	14.6.2; Summary of 14.6.3 and 14.6.4
38	CMOS, NOR and NAND gates	14.6.5, 14.6.8

Some instructors might choose to reduce or eliminate the coverage of thin-film and thick-film circuits in order to include more material on system applications from Chapters 13 and 15.

The syllabus for a one-year graduate-level course meeting three hours per week is shown below.

One-Year Graduate Course

Hour	Topic	Sections
1	General Survey	1.1-1.3
2-5	Evaporation and Sputtering	5.1, 5.2
6-7	Epitaxy	5.3
8-11	Diffusion and ion implantation	5.4, 5.5
12-16	Oxidation, nitride deposition, wafer cleaning, photoresist, properties of impurities in silicon	5.7-5.10
17-18	Junction isolated bipolar circuits, buried layer, beam lead technology	6.1, 6.2
19	Countersunk oxide isolation, V-groove isolation	6.3, 6.4
20	Dielectric isolation, inverted chips, CDI, composite masking	6.5-6.8
21	Complementary bipolar IC's, thick-oxide MOS circuits, self-aligned metal gates, composite-insulator gate structures, applications of ion implantation	6.9-6.13
22	Self-aligned polysilicon gate, CMOS	6.14-6.15
23	DMOS, VMOS, silicon-on-sapphire, integrated-injection logic	6.16-6.18
24	In-process measurements	11.1-11.6
25-28	Mask alignment tolerance, design spacings, minimum area transistor, passive elements	7.3-7.5, 4.1 4.5, 4.8, 4.9
29-30	Layout of SIC's	7.6
31	Computer aids for layout	7.7
32-35	Thin-film circuits	8.3-8.8
36-38	Thick-film circuits	9.1-9.10
39	Assembly	10.1-10.7
40-42	Operational amplifier parameters, bipolar difference amplifiers, circuit imbalances	12.1-12.3
43-45	FET difference amplifiers, frequency response, current sources, improving the CMR	12.4-12.7

One-Year Graduate Course

Hour	Topic	Sections
46-48	Single-ended amplifiers, level shifting, current mirrors, superbeta gain stages	12.8-12.11
49-51	Output stages, voltage references, operational amplifiers—dc conditions	12.12-12.14
52-54	Small-signal frequency response and large-signal response of the operational amplifier	12.15, 12.16
55	Operational amplifier macromodels	12.17
56-57	Characterization of logic circuits, Ebers-Moll and Gummel-Poon transistor models	14.1, 2.10, 2.11
58-59	Emitter-coupled logic	14.2
60-65	Transistor-transistor logic	14.4
66-67	Integrated-injection logic	14.5
68-72	MOS logic circuits	3.5, 14.6
73	EFL and threshold logic	14.7, 14.8
74	Flip-flops	15.2
75-77	MSI circuits, memories, PLA's, microprocessors	15.4-15.7

Chapters 2 and 3 are not included in the above syllabi but can be drawn upon when additional background is deemed necessary.

The broad range of topics and the depth of treatment make this book useful as a reference text.

A. B. Glaser would like to thank Bell Laboratories and E. I. Gordon, S. C. Kitsopoulos, D. Koehler, and K. M. Poole for their support of this work. We are indebted to colleagues at Bell Laboratories for many helpful discussions and their comments on various versions of this book. In particular, we thank J. R. Brews, D. B. Fraser, C. F. Gibbon, W. N. Grant, H. K. Gummel, A. S. Jordan, H. P. Lie, A. U. MacRae, D. S. Peck, J. Ruch, J. M. Schoen, H. Seidel, J. Sosniak, S. M. Spitzer, B. A. Unger, and N. C. Wittwer. We thank Edith Coddington who typed earlier drafts of the manuscript and G. Holmfelt of the drafting department who supervised the drawing of a mountain of figures.

J. Gielchinsky deserves special recognition for preparing the solutions manual.

We have taught the material in this book at City College of New York, Bell Laboratories, and the Naval Postgraduate School. We wish to thank Professor S. J. Oh, Chairman of the Department of Electrical Engineering at City College for his support.

This book was typeset using computer and phototypesetting facilities at Bell Laboratories. A. B. Glaser would like to thank B. W. Kernighan, M. E. Lesk, and J. F. Ossanna, Jr., of the Computing Science Research Center for their assistance with the text formatting, and J. Maranzano of the UNIX Support Group for his help with the UNIX system. Finally, the authors acknowledge a deep debt of gratitude to Joan McCarthy for her tireless effort and endless cooperation, so freely given and continually sustained, in typing and retyping of the manuscript and producing the final phototypeset copy.

Murray Hill, N.J.   Arthur B. Glaser
August, 1977   Gerald E. Subak-Sharpe

Contents

1

Prologue

It is barely three decades since Bardeen, Brattain, and Shockley, of Bell Laboratories, invented the bipolar transistor in 1947 and less than two decades since the appearance of the first working MOS field-effect transistor. The silicon integrated circuit, to which this book is largely devoted, was invented by Jack Kilby, of Texas Instruments, in 1958. Thirty years ago, the average family probably owned five active electronic devices, namely, the five tubes in a superheterodyne radio receiver. Today, the average family owns a few hundred transistors in radios and television receivers, stereo phonographs and tape players, and perhaps a calculator or electronic watch. In about one generation, these devices have revolutionized our lives. In the next few years, the average family will own several thousand devices, and it is probable that by the end of the century, each family will own several million transistors.

There was no semiconductor industry until the early 1950's and now it is a $5-billion-a-year industry. Integrated circuits were first marketed in 1962 and about $2.5-billion worth of integrated circuits were sold last year.

This progress was produced by many people in a steady evolutionary stream of technological breakthroughs and improvements. Taken as a whole, these results have produced a global revolution of major proportions, which affects and will affect the lives of people everywhere.

This book is a comprehensive study of processes and techniques of device fabrication, device behavior, digital and analog integrated circuit design, and system applications of integrated circuits. Before starting this study, it is interesting and perhaps amusing to reflect on where we have been and on how the industry emerged, and to conjecture where we might be going.

1.1 INTEGRATED CIRCUITS—ONCE OVER LIGHTLY

An *integrated circuit* is a group of inseparably connected circuit elements fabricated in place on and within a *substrate.* Most often the substrate is silicon and we speak of a silicon integrated circuit, or SIC. Bipolar SIC's are composed of devices that rely on the properties of electrons and holes, whereas unipolar SIC's contain devices, each of which depends on only one type of charge carrier. However, this boundary is becoming less distinct with the introduction of circuits containing both unipolar and bipolar transistors. Silicon integrated circuits are made by the so-called *planar process,* which consists of forming a layer of silicon dioxide on the top surface of the silicon wafer. The oxide layer is then photolithographically patterned to permit the selective introduction of impurities into the bulk of the silicon, thereby altering its electrical properties. In principle, a sequence of three such selective impurity introductions is sufficient to form isolated bipolar transistors. MOS transistors are self-isolating, and only one impurity introduction is required. In both types of circuit, interconnections are made by means of a deposited metal film over a silicon-dioxide insulator on the top surface of the wafer. This film is etched to form the conductor pattern required to connect the circuit elements in the desired way.

The "standard" sequence of operations in which an SIC is made must be sufficient to simultaneously form and interconnect all of its components. This means that it must be possible to make each of the components by some subsequence of the total sequence of operations. If this is not true, it is necessary either to devise a new circuit configuration which can be made using the process sequence, or to modify the process sequence to permit fabrication of the components required by the original circuit. This latter approach is usually tried only if the additional components can be used advantageously in many applications or if it can be demonstrated that the additional components are absolutely essential to obtain the required performance.

A large silicon integrated circuit is typically from 4 to 6 mm on a side, whereas the wafers of silicon are usually 7.6 cm (3 in.) in diameter. Therefore the pattern that defines a particular process step can be replicated many times on the same wafer. The piece of silicon containing one circuit is called a *die* or a *chip.* The nature of the operations is such that it is convenient to process the wafers in lots of between 10 and 25 wafers at a time. Thus, we are talking about a *batch fabrication* technique in which hundreds, or perhaps thousands, of potential circuits are fabricated simultaneously. The batch-fabrication capability is one of the big advantages of integrated circuits because it reduces the labor required per circuit and enables a complicated circuit, such as a four-function calculator chip, to be made for about one dollar per chip if the *chip yield,* that is, the percentage of good chips obtained from the process, is sufficiently high.

After the wafers are fabricated and tested, they are cut apart. Thereafter the individual circuit chips must be handled separately and mounted in packages or in more complex circuit assemblies. At lower levels of integration, the cost of these unit operations may be the dominant component of the overall cost.

It is important to realize that the material characteristics of certain localized volumes of the silicon are chosen so as to provide optimum parameters for the basic transistors. This choice fixes many of the characteristics of the other circuit components such as the diffused resistors. The nature of the processes makes it impossible to repair any of the circuits. Furthermore, the yield of the process is somehow related to the size and complexity of the circuit.

Silicon is not the only substrate material used for integrated circuits, although it has proved to be the most important. Integrated circuits formed on and within the substrate material, as described previously, are often called *monolithic* SIC's. It is possible to produce thin *epitaxial* films (that is, films which bear a relation to the crystal structure of the substrate) of silicon on insulating substrates of *sapphire* or *spinels,* giving rise to a useful class of circuits know as *silicon-on-sapphire* (SOS) integrated circuits.

There is another class of useful integrated circuits known as *film integrated circuits,* whose elements are composed of films formed in place upon an insulating substrate. Thus, SOS circuits could be grouped under this classification, although it is not customary to do so.

Thin-film circuits are composed of films formed by molecular deposition techniques, such as evaporation, sputtering, or electroplating. *Thick films* are deposited as pastes or inks that are suspensions of glass frits (small glass particles), and metal, and metal-oxide powders. These pastes are subsequently fired at a high temperature to impart the final electrical characteristics.

Thin-film IC's are built on an insulating substrate which is usually alumina, although beryllia or glass is sometimes used. In most applications the substrate is inert, the notable exception being microwave circuits where the dielectric properties of the substrate are very important. Metal films are deposited on the substrate by a molecular-deposition process as mentioned above, while dielectric films are either deposited by physical or chemical-vapor-deposition techniques, or formed by anodically oxidizing a previously deposited metal film.

Thin-film components are most often formed by a subtractive process although in some situations the films are selectively deposited in a suitable pattern to form a particular component. In the subtractive method the films are uniformly deposited over the entire surface of the substrate. The desired pattern is then delineated photographically and the excess film is removed by selective chemical etching.

Thin-film circuits are usually passive and contain capacitors, precision resistors, lumped inductors suitable for microwave frequency applications, and interconnections.

Thick-film circuits are made by the selective deposition of the film-forming material on a refractive substrate, such as alumina, which is needed because of the high temperatures required in subsequent process steps. The material that forms the film is called an *ink,* or *paste,* and is a suspension of glass frits, metal and metal-oxide powders, and binders, in a suitable organic vehicle. The ink is deposited by *screening,* which has been widely used for artistic purposes since ancient times. It is then dried and *fired* at a high temperature to form the final thick film. The final properties of the film are very dependent on the temperature and duration of the firing cycle because of the dynamics of the reaction.

The thick-film process is a relatively simple and inexpensive procedure that produces passive circuit elements.

Sometimes different types of integrated circuits are combined in a composite arrangement called a *hybrid integrated circuit,* which usually consists of two or more integrated circuit types, or of an integrated circuit type and discrete elements.

Layout, processing, assembly, and testing of integrated circuits are considered in more detail in Chapters 5 through 11.

1.2 HISTORICAL PERSPECTIVE—HOW DID A NICE COMPUTER LIKE YOU GET STUCK ON A SMALL CHIP LIKE THIS? [1-6†]

Recent past Semiconductors were used in electronics long before the invention of the transistor in 1947. The first radio detectors were point-contact diodes, in which a metal point made contact with galena or silicon. Their use predated the audion tube (Lee de Forest, triode, 1907), but the performance of these early crystal detectors was quite erratic and depended largely on the ability of the operator to find a good rectifying contact. The great advances in radio came about because of the controlled and uniform properties of the vacuum tube (Ambrose Fleming, diode, 1902).

Other semiconductors, principally cuprous oxide and selenium, were used as rectifiers in the 1930's, a time during which great progress was made in the theoretical understanding of semiconductor bulk and surface properties. However, no experimental verification could be obtained because of the lack of suitably pure semiconductor samples.

In the late 1930's and through World War II, much work was done at microwave frequencies and there was great interest, once again, in the

---

† Numbers refer to references at the end of the chapter.

point-contact diode for use as a low-noise, low-capacitance detector. Silicon was chosen as the material by the microwave group at Bell Laboratories. A period of intensive work followed on the properties and the preparation of silicon for electronic applications, which continued well into the sixties. Point-contact microwave detectors, useful at frequencies up to 25 GHz, were developed for radars. The techniques used to make these diodes were applied several years later to the fabrication of point-contact bipolar transistors.

Discrete thin-film passive components have been used in electronic applications at least since the mid-1920's. One of the first thin-film components was the deposited-carbon resistor which was made by the pyrolysis (decomposition at high temperature) of hydrocarbon gases. Platinum metallic-film resistors were made by spraying a platinum compound onto glass rods, which were then heated to decompose the compound, thus producing a deposited film. Capacitors were made by depositing aluminum or zinc films on paper or polymer dielectrics. Higher value capacitors were made with aluminum plates and an anodic aluminum-oxide dielectric.

During World War II, electronics became increasingly important for communication, detection of enemy targets, navigation and fire control systems. Electronic equipment became increasingly complicated; for example, the B-29 bomber required about a thousand vacuum tubes in its electronic apparatus. It therefore became clear that the capability of future weapons systems would be seriously limited by the size, weight, reliability and complexity of their electronic systems. The U. S. National Bureau of Standards had been sponsoring work on methods to simplify the manufacturing process for rugged electronic subassemblies used in proximity fuses. Rubenstein, Ehlers, Sherwood and White of the Centralab Division of Globe-Union, Inc., developed an assembly utilizing a ceramic substrate with metal interconnections and chip capacitors, to which miniature electron tubes were attached. It was the forerunner of hybrid integrated circuits and was probably the first attempt at forming components in place.

After the war, the National Bureau of Standards and Centralab continued their collaboration in this area resulting in the development of high-volume production techniques based on screening. This was the beginning of thick-film technology.

Meanwhile, at Bell Laboratories, the solid-state physics research group headed by Shockley and Morgan, whose members included Bardeen and Brattain, had as one of its projects the study of semiconductor materials. Both Shockley and Brattain had worked on copper-oxide rectifiers and Schottky-barrier theory before the war. Shockley had proposed a solid-state amplifying device, which is now known as the Schottky-gate field-effect transistor. However, it did not work at that time. The solid-state physics group studied the fundamental behavior of semiconductors for almost two and a half years. This culminated in the so-called "magic month" from November 17, 1947 to December 16, 1947, during which the point-contact transistor was invented by Bardeen and Brattain, and important results leading to the invention of the bipolar junction transistor and the junction field-effect transistor were obtained. It is interesting to note, that Bardeen and Brattain were trying to make a good field-effect transistor when they observed minority carrier injection and used it to make a bipolar transistor.

The transistor age   The magic month marked the birth of a new era. It was recognized almost immediately that the new device was very important, but its ultimate revolutionary effect was not foreseen. There was talk of the transistor supplanting the vacuum tube in certain specialized applications, but no serious thought was given to the possibility that the vacuum tube might be an endangered species in less than 25 years.

Colleagues at Bell Laboratories who were there during the first decade of the transistor era say that it was a very exciting time, during which discoveries were made in rapid succession to take transistor electronics from infancy to maturity. Some of the more important milestones are given in Table 1.1.

Progress was also made during the early 1950's on the interconnection problem with the development of printed circuits by Danko and Abrahamson[7] of the Signal Corps. During this time, several companies suggested that evaporated thin films be used in place of the screen deposition techniques of the Centralab assembly technology. They cited as an additional advantage to the use of thin films that, someday in the future, it would also be possible to fabricate thin-film active devices. Much work was done on thin-film active devices without any smashing successes.

Table 1.1 Milestones in transistor electronics

Point-contact transistor	1947
Single-crystal germanium	1950
Junction transistor	1951
Junction field-effect transistor	1951
Zone melting and refining	1952
Single-crystal silicon	1952
Surface-barrier transistor	1953
Oxide masking	1954
Commercial silicon-junction transistor	1954
Diffused-base transistor	1955
Planar transistor	1959
Epitaxial transistor	1960
MOS transistor	1960
Schottky-barrier diode	1960

The possibility of semiconductor integrated circuits was first perceived in 1952 by G. W. A. Dummer of the Royal Radar Establishment in England. In an address to the Electronic Components Conference he said, "With the advent of the transistor and the work in semiconductors generally, it seems now possible to envisage electronics equipment in a solid block with no connecting wires. The block may consist of layers of insulating, conducting, rectifying, and amplifying materials, the electric functions being connected directly by cutting out areas of the various layers."

Probably as a result of this statement by Dummer, the U. S. Air Force began to define, in the mid-1950's, a concept called "molecular electronics," in which new structures were to be devised in order to perform desired functions more directly. The Air Force discussed this topic with Westinghouse in 1957 and 1958, and a contract was awarded in 1959 to examine representative equipment and systematically invent the new devices to perform the necessary functions.

1.2.1 Invention of the Silicon Integrated Circuit

In 1958 the Army was committed to the development of the "Micro-Module," which was an outgrowth of the National Bureau of Standards assembly technique as developed by the Signal Corps. Texas Instruments had an Army contract to work on the Micro-Module approach. Jack Kilby, a recently hired engineer, was assigned to work on an imprecisely defined project dealing with microminiaturization. After trying a scheme to package resistors, capacitors and transistors in a form to rival the Micro-Module, he made the crucial observation that the cost structure of a semiconductor company is such that the only thing it can make in a cost-effective way is a semiconductor. Furthermore, he realized that all of the necessary components—transistors, resistors and capacitors—could be made from a semiconductor and fabricated in place.

The concept of a complete circuit, all of whose elements were made out of semiconducting material, was first demonstrated with a discrete element flip-flop using silicon resistors and *p-n* junction capacitors. The experiment proved that a functioning circuit could be built using only silicon-based components, but it was not integrated. The first integrated circuits were phase-shift oscillators using a distributed-RC network and a flip-flop. The total time from conception of the idea to its reduction to practice was less than three months. The first public announcement by TI of the "solid circuits" was at the IRE convention in March 1959.

At about this same time, Robert Noyce and Gordon Moore of Fairchild Semiconductor were developing the planar process for making diffused transistors with a passivating oxide layer to protect the junctions. The planar process is based on work at Bell Laboratories by Frosch and Derrick[8] who found that silicon dioxide acts as a diffusion barrier against many commonly used dopant impurities in silicon, and by Atalla, Tannenbaum, and Scheibner,[9] also of Bell Laboratories, on oxide passivation of junctions. The development of the planar process was a landmark in that it produced the first modern, diffused transistor, and it has remained the basic process that is in use to this day. In July 1959, shortly after the TI announcement of integrated circuits, Noyce filed for a patent on a silicon integrated circuit, made by the planar process, using reverse-biased p-n junctions to isolate the devices and an adherent metal film on top of the oxide layer to provide electrical interconnection between the circuit elements.

J. T. Wallmark and N. Nelson[15] of RCA were probably the first to use p-n junctions for device isolation. In a patent filed in August 1958 they described a two-dimensional array of JFET's isolated by reverse-biased p-n junctions.

The announcement of the integrated circuit was not hailed as the birth of a new era. Its critics argued that the components were not optimized because the same material is used for transistors, resistors and capacitors. After all, everyone knows that nichrome is a better resistor than silicon and mylar is a better dielectric than silicon dioxide. In addition, since the yield of a circuit with a large number of components is the product of the component yields, it followed that large IC's would be impractical because the yield would be too low. Finally, designs would be expensive and difficult to change and designers would be out of a job. These objections are all true, but perhaps not as serious as the critics envisioned in the early sixties. However, they were strong enough to keep the major companies from getting on the IC bandwagon for several years.

Recall that at that time each branch of the Armed Forces had a program to produce miniature electronic circuits. Once they decided on a course of action, they were not exactly overwhelmed by the semiconductor integrated circuit. The Air Force was committed to molecular electronics and there was a major internal debate with the "purists," who argued that the integrated circuit should not be considered, because *it was a circuit* and no circuits were to be used. Furthermore, they argued, the SIC used resistors which wasted power.

However, the first support for the SIC did come from the Air Force, because a small group there was able to prevail with the argument that this new concept provided an orderly transition to the new era, as it provided a systematic design approach and so would eliminate the need to invent thousands of new devices which would be required in future equipment.

By early 1960 Texas Instruments had a device for commercial evaluation, and in March 1961 Fairchild announced the micrologic family of digital integrated circuits. The SIC was on its way; or was it?

1.2.2 Growth of the SIC

Important milestones in the steady, evolutionary development of integrated-circuit technology since 1961 are shown in Table 1.2. However, the SIC was not an instant success in the marketplace, because users tried to force existing problem solutions into the new framework. Prior to the advent of SIC's, the system builders did their own circuit design and were therefore upset at attempts by the IC companies to usurp this function. After all, what did these silicon growers know about designing circuits?

Attempts were made to duplicate the large variety of existing discrete-component circuits in a technology that is not well suited to producing small quantities of each of a large number of designs. Circuit-design costs were and still are high and are justified only on the basis of wide applicability of the resultant product. So a standoff occurred because the price of an SIC to perform a particular function exceeded the cost of discrete components to perform that same function.

Table 1.2 Evolution of Integrated Circuits

Monolithic integrated circuit	1958
Planar SIC	1959
Commercial monolithic IC (RTL)	1961
Diode-transistor logic	1962
Transistor-transistor logic	1962
Emitter-coupled logic	1962
MOS IC	1962
Complementary MOS	1963
First linear IC's	1964
MOS memory chips	1968
Charge-coupled device (CCD)	1969
MOS calculator chips	1970
Microprocessor	1971
I^2L	1972

At last the semiconductor companies realized that the only way to settle the argument was to produce circuits that performed well-defined system functions and to price these circuits below the cost of equivalent discrete-component circuits. This approach subsequently resolved similar impasses.

Digital systems were easier to attack, because an entire system can be built from a small group of basic elements called *logic gates*. Actually, only one type of gate, either a NAND or a NOR gate, is sufficient. In addition, many useful system functions could be performed by connecting together a small number of gates on a chip. Therefore, it was relatively straightforward to define a family of compatible system building blocks which, if available in SIC form, could be used to implement a wide variety of systems. Many different families of digital circuits have emerged and some have even become extinct by this time. The essential point is that the standardization they provide enables the design cost to be spread over a greater number of units. Thus, the power of the IC technology to produce large quantities of devices can be utilized effectively.

The growth of the worldwide SIC market is shown in Fig. 1.1 where we plot dollar volume from 1962 through 1975. Note, that the average selling price per unit has fallen continuously. Because the complexity of the chips has increased, the price per equivalent gate has decreased from tens of dollars in 1962 to pennies or less in 1976. This tremendous decrease in cost per function has caused the great penetration of electronics into new areas. The growth rate for circuits in the vacuum tube era was about 10% per year, during the transistor era it was 19% per year, during the early years of the IC it was 38% per year and now it is 50-60% per year.

Larger chip sizes become economically feasible as the processing continues to improve. Therefore, it was—and still is—possible to envision more complex functions on a chip. By the late sixties, the gate was accepted as the basic digital circuit element and a replay of an old problem occurred: system designers wanted to get chips containing gates interconnected in a custom manner. Various techniques were proposed to permit the great flood of custom designs to be implemented inexpensively.[10] One method that was developed by Texas Instruments under an Air Force contract, involved producing an array of clusters of gates. These gates would be tested to ascertain which were functioning properly, thus producing a map of good sites. A computer then would take this map and generate a custom metallization pat-

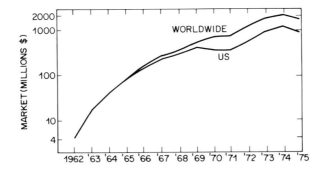

Fig. 1.1 Growth of world SIC markets. (From Kilby.[4])

tern to connect the functioning gates so as to perform a prescribed system function. A computer was actually built using this technique.[11]

What really was needed was a system element whose internal organization is orderly and which has wide applicability so that it can be produced in large quantities.[12] In retrospect the answer is obvious: *memories.*

The answer was not as obvious then, because of the volatility (that is, the loss of memory on power shutdown) of semiconductor memories and their high cost relative to cores. Sixty-four-bit and 256-bit random access memories were introduced without too much success. The introduction in 1970 of the 1-kilobit, fully decoded RAM proved to be the turning point because it took a significantly large function on a single chip and brought the price per bit down to make it competitive with magnetic cores. Since it was expected that the price per bit of semiconductor memory would eventually drop below that of cores, coupled with a significant size and power advantage in semiconductor memory, the system designers suddenly realized that they could live with the volatility of semiconductor memory.

What was truly needed to enable *large-scale integration* (LSI) to be used in a wide variety of applications was a circuit design that could be produced in large quantities and customized by the user to perform a specific task. The answer to this problem is the *microprocessor,* a digital computer on several IC chips, that is customized by the user via *software.*

The microprocessor was first introduced by Intel in 1970 and was accepted rather quickly. In those instances where the microprocessor could be used, the systems designer worried about interfacing the computer to the peripheral devices, and about writing and debugging the program to control the microprocessor. The IC manufacturer, who usurped the circuit design responsibility in the early days of the SIC era, now became involved in writing compilers, assemblers, and software debugging aids. This is a large additional responsibility.

From the introduction of the SIC until 1970 the semiconductor companies were in the circuits business. With the evolutionary development of an LSI capability, and the introduction of microprocessors and such custom functions as calculators and electronic watches, the semiconductor industry has crossed a threshold where it is now supplying complete system functions of an end product on one or several chips. Two of the enormous implications of this development are[13] that lead time and development costs for systems are reduced because custom circuits are not required. Standard circuits are customized by means of specially written software. In addition, new areas of application will open up because a much lower degree of electronic sophistication is required of the user.

This new direction for the semiconductor industry has some other implications, namely: (1) since the chips are more complex there will be fewer designs, but each design will be extremely expensive. Therefore, the effect of a poor choice of part to develop can be disastrous; (2) teamwork between the semiconductor manufacturer and the end user is essential.

Thus, we have seen some of the reasons for the development of the integrated circuit into its present form. This survey is not complete by any means, and the interested reader is urged to pursue the subject further.

1.3 LOOKING INTO A CLOUDY CRYSTAL BALL

We are now in the beginning of an era in which the LSI technology will directly affect our lives, principally through applications of the digital computer for problem solving, process control and entertainment. There are already some examples of products in the latter two categories on the market. Recently introduced programmable, microprocessor-controlled microwave ovens can automatically execute a sequence of operations to produce a tasty meal. There is an ever-expanding variety of video games on the market, and many of the newer models, especially those capable of providing several games, are microprocessor based. Most of the "revolutionary innovations" to be described already exist today, but require hardware that is too expensive to permit significant market penetration and so they are limited to special-purpose environments. It is practical to consider developing systems requiring digital processing for mass use, because of the advent of single-chip 16-kilobit MOS dynamic RAM memories and 16-bit microprocessor chip sets. The almost certain availability of 64-kilobit RAM's and more powerful microprocessors by the end of the decade will provide additional impetus.

The typewriter is giving way to a "word processor," which is a computer-based system, in which the document to be prepared is entered via a keyboard into a file in the computer's memory. The file can be displayed on a video terminal and modified with the help of a text-editing program until the contents are correct. Then a paper copy can be produced by an appropriate output device according to the format specified in a text formatting program. The various typescript versions of this book and the final typeset copy were produced using this "word-processor concept" as implemented on the UNIX system at Bell Laboratories. It should soon be possible to mass produce a stand-alone unit with sufficient memory capacity to provide this kind of capability for letters and short reports. In larger office complexes these word-processing terminals could be linked in a hierarchy of progressively more powerful machines to form a distributed decision and control network incorporating a data-base type filing and retrieval system, computational capability, word processing, and mail distribution. This system also would make real-time accounting with instant operating statements and balance sheets a reality.

Many of these features would also be useful in the home: word processing for producing letters, a data base for maintaining family records and a financial management system, including tax computation. In combination with a telecommunications network, this futuristic computer-based system would permit home shopping and automatic disbursement of funds. Other applications within the home would include security and environmental control systems, programmable and remotely programmable appliances, and a learning and entertainment center.

In the medical field, great advances are possible in diagnostic aids and in prosthetic devices.[14]

The availability of even greater computing power for scientific studies should make possible exciting work on problems such as weather forecasting.

This list of topics is not all-inclusive because of a lack of space and a lack of wisdom. We do not know the form that many of the new gadgets that were listed will take. That will be determined in large part by the ingenuity of you, the readers of this book.

REFERENCES

1. W. F. Pfann. "The semiconductor revolution." Medal Address presented to the *Electrochemical Society*. Chicago, Illinois, May 15, 1973.

2. W. Shockley. "The path to the conception of the junction transistor." *IEEE Trans. on Electron Devices* **ED-23**: 597-620 (July 1976).

3. G. K. Teal. "Single crystals of germanium and silicon—basic to the transistor and integrated circuit." *IEEE Trans. on Electron Devices* **ED-23**: 621-639 (July 1976).

4. J. S. Kilby. "Invention of the integrated circuit." *IEEE Trans. on Electron Devices* **ED-23**: 648-654 (July 1976).

5. D. Kahng. "An historical perspective on the development of MOS transistors and related devices." *IEEE Trans. on Electron Devices* **ED-12**: 655-657 (July 1976).

6. P. E. Haggerty. "Integrated electronics—a perspective." *Proc. IEEE* **52**: 1400-1405 (December 1964).

7. S. F. Danko. "New developments in the auto-sembly technique of circuit fabrication." *Proc. National Electronics Conference*. Chicago, Illinois, 542-550, October 1951.

8. C. J. Frosch and L. Derrick. "Surface protection and selective masking during diffusion in silicon." *J. Electrochem Soc.* **104**: 547-552 (1957).

9. M. M. Atalla, E. Tannenbaum, and E. J. Scheibner. "Stabilization of silicon surfaces by thermally grown oxides." *Bell Sys. Tech. J.* **38**: 749-783 (1959).

10. A. J. Khambata. *Introduction to Large-Scale Integration*. New York: John Wiley & Sons, 1969.

11. M. Canning, R. S. Dunn, and G. Jeansonne. "Active memory calls for discretion." *Electronics* **40**: 143-154 (February 1967).

12. G. E. Moore. "Integrated electronics technology." *Proc. IEEE* **64**: 837-841 (June 1976).

13. C. M. Chang. "The impact of solid-state technology: 1975-1980 overview." *Dataquest Semiconductor Industry Seminar.* Rancho Bernardo, California, October 1976.

14. R. K. Jurgan. "Electronics aids the 'biomedics.'" *IEEE Spectrum* **13**: 76-79 (January 1976).

15. J. T. Wallmark and H. Nelson. *U.S. Patent No. 3,005,937*, filed August 21, 1958; issued October 24, 1961.

2

Diodes and
Bipolar Transistors

The *n-p-n* bipolar transistor is perhaps the most widely used device in integrated-circuit technology. The operation of this transistor is based on the behavior of *p-n* junctions, a topic which is discussed first. This discussion is followed by a description of the behavior of *p-n* diodes and metal-semiconductor (Schottky-barrier) diodes which are finding increased application.

The topic of *n-p-n* bipolar transistors is treated one-dimensionally, from what is believed to be a fresh and rigorous viewpoint, namely according to the Gummel-Poon nonlinear model. The derivation of this model begins with the observation that the emitter and collector currents are approximately equal to a dominant current component, denoted by I_{CC}, which is inversely proportional to the total base charge, denoted by Q_B. This base charge is itself voltage dependent and helps in elegantly explaining the Early effect, high-level injection and other effects, none of which are adequately explained by the older Ebers-Moll transistor model, which may be shown to be a special case of the Gummel-Poon model. Solutions of nonlinear circuits are not always in closed form and numerical methods implemented on a digital computer are usually necessary, even for circuits of moderate complexity. Details of these computer-aided circuit simulation programs, and the algorithms on which they are based, are beyond the scope of this text.[24, 25]

The hybrid-*pi* small-signal transistor model is also briefly discussed because of its wide use in the design of linear circuits. After a description of ion-implanted transistors and two-dimensional transistor analysis, the chapter closes with a short treatment of *p-n-p* lateral transistors.

2.1 DEPLETION WIDTH AND DEPLETION CAPACITANCE [1,2]

The well-known behavior of a *p-n* junction is basic to the operation of integrated circuits and is here briefly recapitulated. Figure 2.1(a) shows a typical doping profile, with N_S the surface *p*-type impurity concentration and N_B the background concentration of the *n*-type material. Under the abrupt junction approximation, shown in Fig. 2.1(b), the depletion layer extends from $x = x_j - x_p$ to $x = x_j + x_n$, where x_j is the location of the metallurgical junction. The depletion-layer width $w = x_p + x_n$ is given by

$$w = \sqrt{2\epsilon_s(N_A+N_D)(V_{bi} \pm V)/qN_A N_D} \tag{2.1}$$

where ϵ_s is the dielectric constant of silicon, V_{bi} is the built-in voltage,

$$V_{bi} = V_T \ln\left[N_A N_D/n_i^2\right] \tag{2.2}$$

and n_i is the intrinsic carrier concentration. The plus sign in Eq. (2.1) is taken for reverse bias, and N_D and N_A are the donor and acceptor concentrations, respectively.

Associated with the depletion layer is the depletion or junction capacitance given by

$$C_j = \epsilon_s/w = \sqrt{\epsilon_s q N_A N_D/2(N_A+N_D)(V_{bi} \pm V)} \tag{2.3}$$

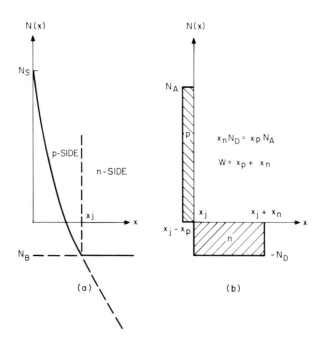

Fig. 2.1 (a) Impurity profile for junction. (b) Abrupt junction approximation.

with its characteristic $V^{-1/2}$ dependence for $|V| \gg |V_{bi}|$. A one-sided abrupt junction occurs when $N_A \gg N_D = N_B$, in which case

$$w = \sqrt{2\epsilon_s(V_{bi} \pm V)/qN_B}, \qquad C_j = \epsilon_s/w = \sqrt{q\epsilon_s N_B/2(V_{bi} \pm V)}.$$

$$(2.4)$$

EXAMPLE 2.1   Find the width of the depletion layer w for a one-sided abrupt junction in silicon. Let the total reverse bias voltage be $V' = V_{bi} + V = 1$ V and let the background concentration $N_B = 10^{15}$ cm^{-3}.

Solution.   It is assumed that $N_A \gg N_D = N_B$ and so Eq. (2.1) becomes $w = (2\epsilon_s V'/qN_B)^{1/2}$, where $q = 1.6 \times 10^{-19}$ C. For silicon $\epsilon_s = 11.8\epsilon_0$, where $\epsilon_0 = 8.85 \times 10^{-14}$ F/cm. Substituting the given values in the equation for w yields $w = 1.14 \times 10^{-4}$ cm or about 1 μm.

EXAMPLE 2.2   Find the capacitance density C, corresponding to the depletion-layer width w given in Example 2.1.

Solution.   From Eq. (2.4), $C = \epsilon_s/w$. Substituting values from Example 2.1 now yields

$$C = \frac{11.8 \times 8.85 \times 10^{-14}}{1.14 \times 10^{-4}} = 9.16 \times 10^3 \quad \text{pF/cm}^2.$$

The change in depletion-layer width as a function of applied reverse voltage is measured by the sensitivity $S_v^w = V\, dw/w\, dV \approx 1/2$ indicating that a one-unit change of width occurs for a half-unit change in voltage. Furthermore, for a one-sided abrupt junction

$$\frac{d}{dV}\left\{\frac{1}{C^2}\right\} = \pm \frac{2}{q\epsilon_s N_B} \qquad (2.5)$$

and so a plot of $1/C^2$ versus V has a slope proportional to the inverse of the doping concentration N_B. This is the basis of a measurement method for N discussed in Chapter 11. Moreover, $1/C^2 = 2(V_{bi} \pm V)/q\epsilon_s N_B$ is a two-segment piecewise linear curve. The contact point with the horizontal axis yields the built-in voltage V_{bi}.

The abrupt-junction approximation is adequate for describing shallow diffused junctions such as the base-emitter junction, but for deep diffused junctions the linearly graded approximation, shown in Fig. 2.2, is more appropriate. The impurity gradient is $a = dN(x)/dx$ atoms/cm^4 and by an analysis similar to that for the abrupt junction, it may be shown that

$$w = [12\epsilon_s(V_{bi} \pm V)/qa]^{1/3}, \qquad C_j = \epsilon_s/w. \qquad (2.6)$$

The general p-n junction is neither abrupt nor linearly graded, but is usually formed by diffusing impurities into silicon material having a constant background concentration N_B. The resulting impurity profile is described by a

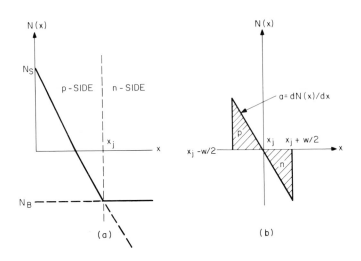

Fig. 2.2 (a) Impurity profile for a graded junction. (b) Linearly graded junction approximation.

Gaussian or complementary error function. Lawrence and Warner[3] have produced a set of curves for both Gaussian and complementary error-function diffusions that show depletion-layer thickness w, depletion capacitance $C_j = \epsilon_s/w$, and penetration of depletion layer a_1/w as a function of normalized applied voltage V/N_B. Each curve is given for a different value of N_B/N_S, where N_S is the surface concentration. The curves shown in Figs. 2.3(a) and 2.3(b) are for $N_B/N_S = 10^{-2}$. Note that at large values of normalized voltage, all curves are asymptotic to a single line which corresponds to the abrupt-junction approximation.

The maximum electric field at a one-sided abrupt junction is given by

$$E_{max} = 2(V_{bi} \pm V)/w \qquad (2.7)$$

and plays a role in p-n junction breakdown, when $V_{bi} + V = V_B$, the breakdown voltage. This subject is discussed in Section 2.5.

EXAMPLE 2.3   Find the value of the maximum electric field, E_{max}, at the one-sided abrupt junction mentioned in Example 2.1.

Solution.   From Eq. (2.7), $E_{max} = 2V'/w$. But from Example 2.1, $V' = 1$ V and $w = 1.14 \times 10^{-4}$ cm. Hence $E_{max} = 2/1.14 \times 10^{-4} = 0.175 \times 10^5$ V/cm. Typical values of breakdown maximum electric fields for cylindrical silicon junctions lie in the range 2×10^5 to 10^6 V/cm and so E_{max} is about one order of magnitude below breakdown.

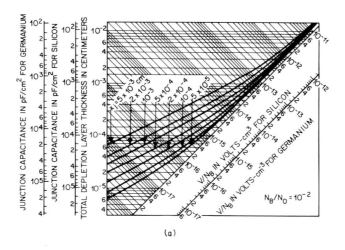

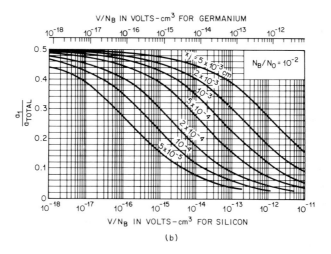

Fig. 2.3 Lawrence-Warner curves for a Gaussian distribution in the range 3×10^{-3} to 3×10^{-2}. (a) Depletion layer width and capacitance versus V/N_B and (b) a_1/w versus V/N_B. (From Lawrence and Warner. Reprinted with permission from *The Bell System Technical Journal,* Copyright 1960, The American Telephone and Telegraph Co.)

2.2 p-n JUNCTION CURRENT-VOLTAGE CHARACTERISTICS

The current-voltage characteristic of a *p-n* junction follows Shockley's law

$$I_D = I_s \lambda(V_D) \qquad \lambda(V_D) = \exp(V_D/V_T) - 1 \qquad (2.8a)$$

where $V_T = kT/q$ is the Boltzmann or thermal voltage and where I_s is the saturation current, whose theoretical value is given by

$$I_s = Aq\left[\frac{D_p p_{n0}}{L_p} + \frac{D_n n_{p0}}{L_n}\right].\tag{2.8b}$$

Here A is the junction area, D stands for diffusion coefficient, L for diffusion length and subscripts p and n denote the type of material. Equilibrium minority-carrier concentrations are indicated by p_{n0} (holes) and n_{p0} (electrons), respectively.

EXAMPLE 2.4   Calculate the theoretical saturation current I_s for an abrupt-junction silicon diode of area $A = 2500\ \mu m^2$ having $N_D = 10^{18}\ cm^{-3}$, $N_A = 10^{16}\ cm^{-3}$, $L_n = 10^{-3}$ cm and $L_p = 10^{-4}$ cm.

Solution. Equation (2.8b) is first modified by means of the Einstein relation $D = \mu V_T$ and the law of mass action, $pn = n_i^2$. Furthermore, $n_{n0} \approx N_D$ and $p_{p0} \approx N_A$ and so

$$I_s \approx Aqn_i^2 V_T\left\{\frac{\mu_p}{L_p N_D} + \frac{\mu_n}{L_n N_A}\right\}.$$

The mobility is a function of impurity concentration.[1] One finds that $\mu_n = 260\ cm^2/V-s$ and $\mu_p = 480\ cm^2/V-s$. Furthermore, $n_i = 1.8 \times 10^{10}\ cm^{-3}$ for silicon at 300°C. Hence,

$$I_s \approx 2.5 \times 10^{-5} \times 1.6 \times 10^{-19} \times 3.64 \times 10^{20} \times 26 \times 10^{-3}\left\{\frac{480}{10^{-4} \times 10^{18}} + \frac{260}{10^{-3} \times 10^{16}}\right\}$$

$$\approx 1.17 \times 10^{-15}\ A.$$

Ideal forward and reverse current characteristics (thick lines) as well as actual measured characteristics (dotted lines) based on Eqs. (2.8a) and (2.8b) are shown in Fig. 2.4 for silicon diodes. Actual diodes deviate from the ideal characteristic for a variety of reasons, such as surface effects, generation and recombination of carriers in the depletion layer, tunneling, high-level injection effects, and bulk material resistance.

A typical curve under reverse bias is shown by curve (e) and is clearly much higher than the ideal reverse-current ratio curve. This is attributed to the existence of a generation current, I_g, which is proportional to depletion-layer width, namely $I_g = I_A(V_{bi} \pm V)^{1/2}$. As a consequence, the actual reverse current I_r is given by

$$I_r = I_s + I_A(V_{bi} \pm V)^{1/2}\tag{2.9a}$$

with the generation term being dominant, $I_r \gg I_s$. The dynamic resistance

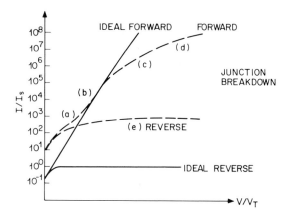

Fig. 2.4 Ideal and actual current-voltage characteristics for a junction diode: (a) recombination current dominates, (b) ideal curve is followed when diffusion current dominates, (c) high-level injection region, (d) bulk series resistance is dominant, and (e) reverse bias characteristic.

of the diode is

$$r_d = \frac{d(V \pm V_{bi})}{dI_r} = \frac{2(V \pm V_{bi})}{I_r - I_s} \approx \frac{2V}{I_r} \,, \quad V \gg V_{bi} \qquad (2.9b)$$

and is very large, because I_r is quite small.

Under forward bias the actual current characteristic deviates from the ideal in a number of well-defined regions. Recombination in the depletion region predominates under low forward bias leading to a

$$\lambda(V/2V_T) \approx \exp(V/2V_T)$$

dependency. As a consequence, the forward current I_f can be taken as the superposition of two current components, namely

$$I_f = I_s\exp(V/V_T) + I_B\exp(V/2V_T). \qquad (2.10)$$

In region (a) of Fig. 2.4, the recombination component, $I_B\exp(V/2V_T)$ dominates. The ideal curve (b) is followed when the diffusion component, $I_s\exp(V/V_T)$, dominates. It will be seen later that Eq. (2.10) is similar to the empirical base current for junction transistors given by Eq. (2.61).

At high current densities, the injected minority carrier density becomes comparable to the majority carrier density and so $p \approx n$, $pn = p^2 = n_i^2\exp(V/V_T)$ or $p = n_i\exp(V/2V_T)$, and the slope of the actual current curve is reduced as shown in region (c). Finally, at very high current densities, the effect of series bulk material resistance dominates, leading to the curve of region (d).

2.3 TEMPERATURE DEPENDENCE OF p-n JUNCTION CURRENTS

Under conditions of approximate charge neutrality $n + N_A = p + N_D$. Using the fundamental relation $pn = n_i^2$ in this equation and solving for an n-type semiconductor yields the equilibrium value

$$2n_{n0} = (N_D - N_A) + \left[(N_D - N_A)^2 + 4n_i^2\right]^{1/2}.$$

Now if $|N_D - N_A| \gg n_i$ and if $N_D \gg N_A$, then

$$n_{n0} \approx N_D; \quad p_{n0} \approx n_i^2/N_D. \tag{2.11a}$$

Similarly, in a p-type semiconductor if $|N_A - N_D| \gg n_i$ and $N_A \gg N_D$, then

$$p_{p0} \approx N_A; \quad n_{p0} \approx n_i^2/N_A. \tag{2.11b}$$

If these results are substituted into the ideal current law, Eq. (2.8b), one obtains

$$I_s = qAn_i^2 \left[\frac{D_p}{L_p N_D} + \frac{D_n}{L_n N_A}\right]. \tag{2.12}$$

The intrinsic carrier concentration $n_i^2 \propto T^3 \exp(-V_{g0}/V_T)$ where V_{g0} is the gap voltage at 0 K (1.21 V for Si). Assuming the bracketed term in Eq. (2.12) is temperature independent, it then follows that the temperature coefficient of the saturation current is given by

$$\frac{T}{I_s}\frac{dI_s}{dT} = 3 + \frac{V_{g0}}{V_T} \approx \frac{V_{g0}}{V_T}. \tag{2.13}$$

However, $V_T \approx 26$ mV at room temperature and therefore the second term of Eq. (2.13) dominates in silicon at normal temperatures.

EXAMPLE 2.5   Compute the temperature sensitivity of the saturation current I_s of a silicon diode at room temperature.

Solution.   From Eq. (2.13) the required sensitivity denoted $S_T^{I_s}$ is given by

$$S_T^{I_s} = \frac{T}{I_s}\frac{dI_s}{dT} = 3 + \frac{V_{g0}}{V_T}$$

$$= 3 + \frac{1.21}{25.9 \times 10^{-3}} = 49.7.$$

Hence, the saturation current increases at a rate that is almost 50 times as fast as the temperature, and so I_s can become very large as T increases.

The temperature coefficient of the ideal forward current I_D for a constant forward bias voltage is, from Eqs. (2.8a) and (2.12),

$$\frac{T}{I_D} \frac{dI_D}{dT}\bigg|_{V_D=\text{const.}} = \frac{V_{g0}-V_D}{V_T} . \tag{2.14}$$

The current increases exponentially with temperature because the right-hand side of this equation is positive for practical forward biases.

For constant diode current I_D, the temperature coefficient of diode voltage is given by

$$\frac{T}{V_D} \frac{dV_D}{dT}\bigg|_{I_D=\text{const.}} = -\left(\frac{V_{g0}}{V_D} - 1\right). \tag{2.15}$$

Hence, the diode voltage decreases with increasing temperature for constant diode current.

2.4 DIFFUSION CAPACITANCE

The depletion capacitance, C_j, is the dominant component of diode capacitance under reverse bias. Under forward bias conditions this is no longer true. If the forward bias V consists of a dc component V_0 and a small signal ac component, such that $V = V_0 + V_1\exp(j\omega t)$ and $V_0 \gg |V_1|$, then it may be shown[1] that the diode admittance $Y(\omega)$ is given by

$$Y(\omega) = \frac{qA}{V_T} \left\{ \frac{D_p p_{n0}}{L_p} \sqrt{1+j\omega\tau_p} + \frac{D_n n_{p0}}{L_n} \sqrt{1+j\omega\tau_n} \right\} \exp\left\{\frac{V_0}{V_T}\right\}$$

$$= G_d(\omega) + j\omega C_d(\omega). \tag{2.16}$$

Note that at low frequencies, when $\omega\tau \ll 1$

$$G_d(0) \approx I_D/V_T,$$

$$C_d(0) \approx (qA/2V_T)[L_p p_{n0}+L_n n_{p0}]\exp(V_0/V_T). \tag{2.17}$$

The first result could have been obtained by directly differentiating Eq. (2.8a). The second result defines a diffusion capacitance $C_d(0)$ whose value increases with forward bias. At high frequencies $\sqrt{1+j\omega\tau} \approx \sqrt{j\omega\tau} = (1+j)(\omega\tau/2)^{1/2}$. Hence, the high-frequency conductance $G_d(\omega)$ is proportional to $(\omega\tau)^{1/2}$ and so it increases, while the high-frequency capacitance $C_d(\omega)$ is proportional to $(\omega\tau)^{-1/2}$ and so it decreases. The equivalent circuits are shown in Figs. 2.5(a) and 2.5(b).

EXAMPLE 2.6 Compute the low-frequency diffusion capacitance for the abrupt junction silicon diode specified in Example 2.4, under a 1 V forward bias.

Solution. From Eq. (2.17), the low-frequency diffusion capacitance is given by

$$C_d(0) = \frac{qAn_i^2}{2V_T}\left[\frac{L_p}{N_D} + \frac{L_n}{N_A}\right]\exp\left(\frac{V_o}{V_T}\right).$$

Substituting appropriate values taken from Example 2.4 yields, at 300 K,

$$C_d \approx \frac{2.5\times10^{-5}\times1.6\times10^{-19}\times3.64\times10^{20}}{5.18\times10^{-3}}\ (10^{-19})\exp\left[\frac{10^3}{25.9}\right]$$

$$\approx 165\times10^{-6}\ \text{F}.$$

Note that this capacitance is very large in comparison with the reverse-bias depletion capacitance of a junction of equal area.

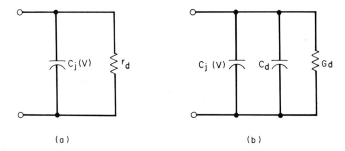

Fig. 2.5   Diode equivalent circuit: (a) under reverse bias and (b) under forward bias.

2.5 p-n JUNCTION BREAKDOWN

The dominant breakdown mechanism in *p-n* junctions is called avalanche multiplication. It occurs when the maximum electric field $E_{\max}$ in the depletion layer exceeds a critical value E_c. This results in impact ionization and the number of carriers leaving the depletion region is multiplied by a factor M. If I_r is the normal reverse current, the avalanche current $I_r{}'$ is now given by

$$I_r{}' = MI_r. \tag{2.18}$$

The avalanche multiplication factor M has been determined empirically and is approximately given by

$$M \approx \frac{1}{1 - (V/V_B)^n} \tag{2.19}$$

where V_B is the breakdown voltage and n is an empirical constant typically between 3 and 6. When avalanching occurs, M approaches infinity. The breakdown voltage V_B is related to a critical electric field in the depletion region, which for a one-sided abrupt junction occurs at the junction such that

$V_B = E_c w/2$. Therefore, with the aid of Eq. (2.4),

$$V_B = \frac{\epsilon_s E_c^2}{2qN} \cdot$$

(2.20)

A typical curve of V_B versus N, the impurity concentration on the lightly doped side is given in Fig. 2.6(a) for several semiconductors. Figure 2.6(b) shows plots of depletion-layer width w at breakdown versus N.

EXAMPLE 2.7   Use Eq. (2.20) to compute the breakdown voltage, V_B, of an abrupt silicon junction, whose background impurity density is $N_B = 10^{16}$ cm^{-3}. The critical field corresponding to this density is $E_c = 4\times10^5$ V/cm. Compare your answer with the empirical formula

$$V_B \approx 60(E_g/1.1)^{3/2}(N_B/10^{16})^{-3/4} \text{ V}.$$

Solution.   From Eq. (2.20), $V_B = \epsilon_s E_c^2/2qN_B$. Substituting the given data,

$$V_B = \frac{11.8\times8.85\times10^{-14}\times16\times10^{10}}{2\times1.6\times10^{-19}\times10^{16}} \approx 52.2 \text{ V}.$$

Use of the empirical formula gives

$$V_B \approx 60(1)^{3/2}(1)^{-3/4} = 60 \text{ V}$$

and so the empirical formula errs somewhat on the high side.

Similar results apply to linearly graded junctions, where it may be shown that the breakdown voltage is given by

$$V_B = \frac{4E_c^{3/2}}{3}\left[\frac{2\epsilon_s}{qa}\right]^{1/2},$$

(2.21)

where a is the impurity gradient. Figures 2.7(a) and (b) show values of V_B and w versus a.

At higher critical fields in heavily doped semiconductors, breakdown occurs as a consequence of tunneling and is referred to as *Zener breakdown*. It is considered that Zener breakdown is dominant at voltages above $6V_g$, where V_g is the gap voltage of the semiconductor material.

2.6 TRANSIENT BEHAVIOR OF p-n JUNCTIONS

When a diode is used as a switch, it is alternatively forward and reverse biased. Figure 2.8(a) shows a circuit arrangement which allows a forward-biased diode, conducting current I_f, to be switched at time $t = 0$ so that it eventually is reverse biased. It is experimentally established that the reverse current, I_r, follows a curve similar to that shown in Fig. 2.8(b). From $t = 0$ to $t = t_1$, the reverse current has a value $I_r = V_R/R_0$ that is solely determined by the external circuit. During this time the diode voltage V_D does not fall below zero, as shown in Fig. 2.8(c), while the minority carrier den-

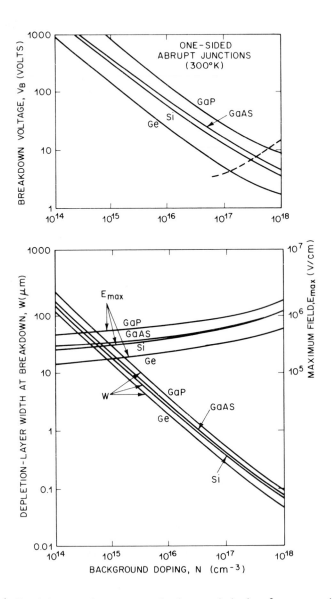

Fig. 2.6 (a) Breakdown voltage versus background doping for a one-sided abrupt junction. (b) Depletion layer width and maximum field versus background doping from a one-sided abrupt junction. (From Sze.[1])

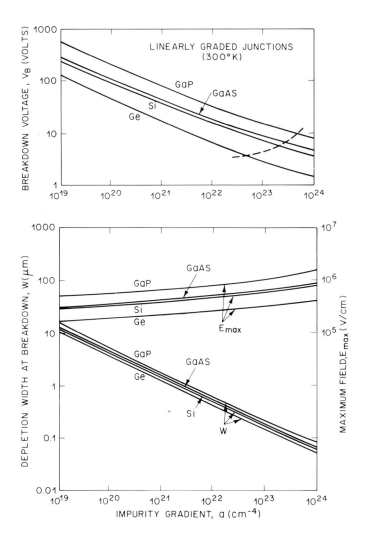

Fig. 2.7 (a) Breakdown voltage versus impurity gradient for a linearly graded junction. (b) Depletion layer width and maximum field at breakdown versus impurity gradient for a linearly graded junction. (From Sze.[1])

sity falls from a value $p_n(0)$ at $t = 0$ to a value p_{n0} at $t = t_1$, as shown in Fig. 2.8(d). At $t = t_1$, called the *storage time* of the diode, the hole density equals p_{n0} at $x = 0$ and the diode voltage goes negative. The decay phase of the diode has now started and lasts until time $t_{off} = t_1 + t_2$, when the reverse current has fallen to a value $I_r/10$. The time t_{off} is called the *turn-off time* or

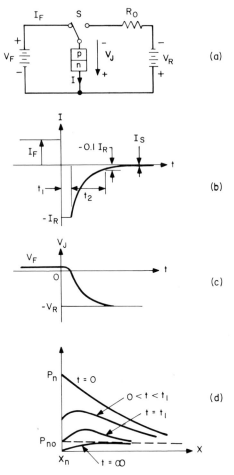

Fig. 2.8   (a) Switching circuit for transient behavior measurement, (b) current response, (c) junction-voltage response, and (d) minority carrier distribution.

recovery time of the diode. Times t_1 and t_2 may be computed by solving the continuity equation for holes under appropriate boundary conditions. However, a simple estimate may be made as follows:

When the diode is forward biased, there is present the charge Q in the n region, due to injected holes. From Fig. 2.8(d)

$$Q \approx \frac{qAp_n(0)X}{2} \tag{2.22a}$$

where X is the base of the triangle. But $I_f = -qAD_p(-p_{n0}/X)$ and so

$$Q = \frac{I_f X^2}{2D_p}. \tag{2.22b}$$

If the decaying portion of the reverse current profile of Fig. 2.8(b) is approximated by a straight line segment which intersects the time axis at $t_1 + t_2$, then the average reverse current $\bar{I}_r$ is given by

$$\bar{I}_r = \frac{I_r\left(t_1 + \dfrac{t_2}{2}\right)}{t_1 + t_2} .$$ (2.23)

The turn-off time $t_{\text{off}} = t_1 + t_2$ is now given by $Q/\bar{I}_R$ and so

$$t_1 + \frac{t_2}{2} \approx \frac{I_f}{I_r} \frac{X^2}{2D_p} .$$ (2.24)

If the length, w, of the n-type material is much greater than the diffusion length L_p, then one may put $X = L_p$ in Eq. (2.24). When $w \ll L_p$, one may put $X = w$. Hence

$$t_1 + \frac{t_2}{2} \approx \begin{cases} \dfrac{I_f}{I_r} \dfrac{\tau_p}{2} & \text{for} \quad w \gg L_p, \\[3mm] \dfrac{I_f}{I_r} \dfrac{w^2}{2D_p} & \text{for} \quad w \ll L_p. \end{cases}$$ (2.25)

Since it may be shown,[1] that the storage time t_1 is theoretically given by

$$\text{erf}\left(\frac{t_1}{\tau_p}\right)^{1/2} = \frac{I_f}{I_f + I_r} ,$$ (2.26)

approximate values for t_1 and t_2 can be obtained. As a general rule, a fast switching p-n junction requires the minority carrier lifetime τ_p to be small. This may be achieved by gold doping to introduce deep impurity levels in the forbidden gap.

EXAMPLE 2.8   Let the ratio of reverse current to forward current, I_r/I_f, be increased from a value of 10 to 100 and compute the percentage reduction in the normalized storage time $x = t_1/\tau_p$.

Solution.   From Eq. (2.26),

$$\text{erf}\sqrt{x_1} = 1/11 = 0.91 \times 10^{-1}$$

and

$$\text{erf}\sqrt{x_2} = 1/101 = 0.99 \times 10^{-2}.$$

It is required to find the value of $100 x_2/x_1$. However, for small values of x,

$$\text{erf } x = \frac{2}{\sqrt{\pi}} \int_x^0 \exp(-t^2)\,dt \approx \frac{2x}{\sqrt{\pi}}$$

and so $2\sqrt{x_1/\pi} \approx 0.91 \times 10^{-1}$ and $2\sqrt{x_2/\pi} \approx 0.99 \times 10^{-2}$. Hence,

$$\left(\frac{x_2}{x_1}\right)^{1/2} \approx \left(\frac{0.99}{0.91}\right) \times 10^{-1} = 1.088 \times 10^{-1}.$$

Therefore, $(x_2/x_1) \approx 1.18 \times 10^{-2}$ and $100 x_2/x_1 \approx 1.2$. We conclude from this that the storage time has been reduced by almost 99% when the current ratio is increased 10 times.

2.7 INTEGRATED CIRCUIT DIODES [3]

The usual bipolar integrated circuit is a four-layer structure corresponding to the emitter, base, and collector regions of the transistor, and also the substrate that holds the circuit. The integrated circuit diode is made by connecting these four layers in one of the six possible ways shown in Fig. 2.9.

The diffusion capacitance is dominant for a forward-biased junction, whereas under reverse bias the depletion capacitance dominates. The substrate is always connected to the most negative point on the power supply and so the collector-substrate junction is always reverse biased resulting in a parasitic substrate-collector depletion capacitance denoted C_{sc}. The other capacitances are denoted C_{be} and C_{bc}, respectively. Capacitance across the diode will be denoted by C_d, whereas capacitance from a diode terminal to ground will be denoted as a parasitic capacitance C_p. Figure 2.9 also includes C_d, C_p, typical turn-off or recovery time t_{off} and also typical breakdown voltage.

When the four-layer structure is connected to form a diode, it is possible to create a parasitic p-n-p transistor. The reverse-biased substrate acts as a collector, the n-type epitaxial layer forms the base and the diffused base is the emitter. Only in case (a), where the base-collector junction is shorted, can there be no p-n-p parasitic transistor action. For all other configurations significant current will be bypassed to the substrate. Typical values of p-n-p current gain are of the order of 1 to 3. Finally, the forward-bias knee-voltage for the various connections differs and may be obtained by measurement or by solving the set of nonlinear transistor equations. These values of current gain and knee-voltage are also indicated on Fig. 2.9.

2.8 SCHOTTKY BARRIER DIODES [1,4]

A rectifying barrier is usually formed when a metal makes contact with a semiconductor. These structures are electrically similar to abrupt one-sided p-n junctions, but several important differences make these diodes, called Schottky diodes, attractive and useful.

First, the diode operates as a *majority carrier* device under low-level injection conditions. As a consequence, storage time—due to storage of minority carriers—is eliminated and an inherently fast response is obtained.

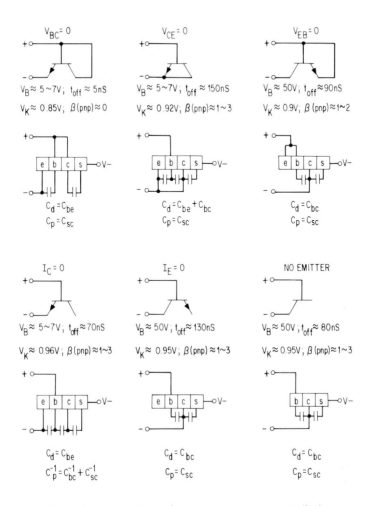

Fig. 2.9 Six methods for connecting a four-layer structure as a diode.

Second, the knee of the forward bias current-voltage curve of the Schottky diodes occurs at a much lower voltage than it does for *p-n* junctions and so Schottky diodes can be used to prevent transistors from going deeply into saturation, as described in detail in Chapter 14. Anticipating this discussion, the effect of the Schottky diode is to shunt excess base current through the diode as majority carrier current, thus reducing the measured saturation time of the combination to as little as 10% of that of the transistor without the Schottky diode.

A brief explanation of Schottky diode operation follows. When a metal is brought into intimate contact with a semiconductor, the conduction and valence bands of the latter form a definite relationship with the Fermi level of the metal as depicted in Fig. 2.10 for three cases; namely thermal equilibrium, forward bias, and reverse bias. The quantity $q\phi_{Bn}$ is the height of the metal-to-semiconductor energy barrier. It represents the energy that must be supplied to an electron in the metal to allow it to escape into the semiconductor. An empirical expression for the barrier height is

$$\phi_{Bn} \approx 0.235\phi_m - 0.352 \tag{2.27}$$

where ϕ_m is the work function of the metal. Since $\phi_{Bn} + \phi_{Bp} = V_g$, where V_g is the gap voltage of the material, ϕ_{Bp} can be evaluated when V_g and ϕ_m are known. Both ϕ_{Bn} and ϕ_{Bp} are determined experimentally, and representative values for several metals are given in Table 2.1. These values have a direct bearing on the saturation current of Schottky diodes as is apparent from Eqs. (2.29a) and (2.29b) below.

Table 2.1 Metal-to-Semiconductor Barrier Heights

Metal	$\phi_{Bn}(V)$	$\phi_{Bp}(V)$
Zr	0.55	0.53~0.55
Mo	0.57~0.59	-
A	0.65	-
W	0.65~0.67	-
Ni	0.66	0.45
Ph	0.68-0.70	0.33
Pd	0.75	-
Au	0.79~0.80	0.25
Pt	0.85~0.87	0.25
PtSi	0.85~0.86	0.20

EXAMPLE 2.9 Calculate the barrier heights for the following types of Schottky diodes: gold to n-type silicon; platinum silicide to n-type silicon and tungsten to n-type silicon. The metal work functions for the aforementioned metals are as follows: gold $\phi_m = 4.8$ V, platinum $\phi_m = 5.3$ V, tungsten $\phi_m = 4.5$ V.

Solution. According to the empirical formula Eq. (2.27), the barrier height

$$\phi_{Bn} = 0.235\phi_m - 0.352.$$

Hence for gold to n-type silicon

$$\phi_{Bn} = 0.235 \times 4.8 - 0.352 \approx 0.78;$$

for PtSi to n-type silicon,

$$\phi_{Bn} = 0.235 \times 5.3 - 0.352 \approx 0.89 \ V;$$

and for tungsten to n-type silicon,

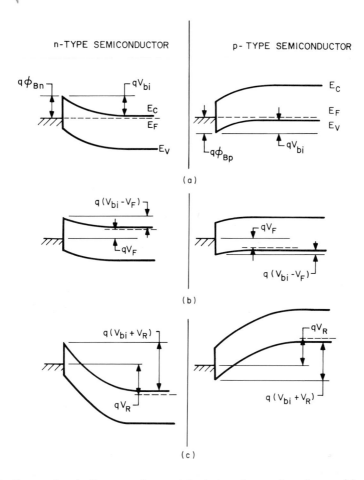

Fig. 2.10 Energy-band diagrams for metal-semiconductor junctions: (a) thermal equilibrium, (b) forward bias, and (c) reverse bias.

$$\phi_{Bn} = 0.235 \times 4.5 - 0.352 \approx 0.71 \ V.$$

Actual values obtained by capacitor-voltage measurement were 0.80, 0.86, and 0.65 V.

A depletion layer of width w extends from the surface of the metal into the semiconductor and so a depletion capacitance $C = \epsilon_s/w$ is formed where

$$w = \left[\frac{2\epsilon_s}{qN_D} \left(V_{bi} \pm V - V_T \right) \right]^{1/2}. \tag{2.28}$$

The positive sign is taken for reverse bias and the other quantities are as defined for Eq. (2.1).

Various theories of current transport in Schottky diodes have been propounded,[1] but are beyond the scope of this text. Suffice to say that the diode current I_D is given by

$$I_D = I_s \lambda(V) \tag{2.29a}$$

$$I_s = AT^2 \exp\left(-\frac{\phi_{Bn}}{V_T}\right) \tag{2.29b}$$

where A is a constant which depends on the choice of theory.

Equation (2.29a) is similar to Shockley's equation for p-n junctions. However, Eq. (2.29b) representing the saturation current is quite different from Eq. (2.8) and is markedly a function of the barrier voltage ϕ_{Bn}. The general conclusion of most theories is that the current transport in silicon Schottky-barrier diodes is due to the emission of majority carriers. The current-voltage characteristic follows theory over many orders of magnitude of the applied voltage. Typical forward characteristics for two types of Schottky diodes and a p-n junction are shown in Fig. 2.11. At sufficiently high reverse voltages, avalanche breakdown occurs, similar to that observed in p-n diodes.

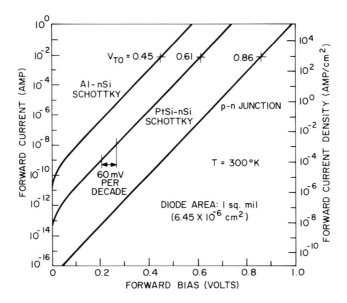

Fig. 2.11 Current-voltage characteristic for Schottky diodes. The breakpoint of the piecewise linear approximation to the characteristic is V_{T0}.

2.9 OHMIC CONTACTS

An ohmic contact should display a linear, symmetrical, current-voltage relationship. No potential barriers—which cause asymmetry—may be present and such a contact will not then add significant parasitic impedance to the structure of which it is part.

Ohmic contacts are approximated in practice by a metal-semiconductor contact, if the semiconductor is heavily doped. Commonly, metal-n^+-n or metal-p^+-p contacts are constructed which are really metal-semiconductor junctions with low or degenerate barriers.

2.10 JUNCTION TRANSISTORS

A bipolar integrated circuit transistor is effectively a four-layer, three-junction device as shown in Fig. 2.12(a) and in the equivalent circuit of Fig. 2.12(b). It incorporates the desired *n-p-n* transistor, but has associated with it an unwanted parasitic *p-n-p* transistor. In its active mode the *n-p-n* transistor has a forward-biased base-emitter junction ($V_{BE} > 0$) and a reverse-biased base-collector junction ($V_{BC} < 0$). If the substrate is given the most negative potential in the circuit, both junctions of the parasitic *p-n-p* will be reverse biased ($V_{BC} < 0$; $V_{SC} < 0$) and so this transistor is in the off mode. According to the Ebers-Moll[5] transistor model, the junction currents I'_{BE}, I'_{BC}, I'_{SC} may be considered linearly related to the exponential junction voltages $\lambda(V_{BE})$, $\lambda(V_{BC})$ and $\lambda(V_{SC})$, defined in Eq. (2.8a),

$$[I'_{BE}, I'_{BC}, I'_{SC}]^t = \mathbf{T}[\lambda(V_{BE}), \lambda(V_{BC}), \lambda(V_{SC})]^t .$$

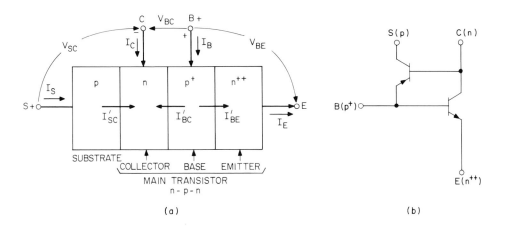

Fig. 2.12 (a) Four-layer, three-junction, bipolar transistor and (b) equivalent circuit.

All junction currents are directed from p to n material with the p material positively biased with respect to the n material. The transmission matrix $\mathbf{T}$ is given by

$$\mathbf{T} = \begin{bmatrix} I_{s1} & -\alpha_I I_{s2} & 0 \\ -\alpha_N I_{s1} & I_{s2} & -\alpha_{SI} I_{s3} \\ 0 & -\alpha_{SN} I_{s2} & I_{s3} \end{bmatrix}$$

where I_{sr}, $(r = 1, 2, 3)$ are saturation currents and the alphas have subscripts that refer to normal and inverse operation, with the additional subscript S denoting the substrate. Junction currents are related to terminal currents by

$$\begin{bmatrix} I_E \\ I_B \\ I_C \\ I_S \end{bmatrix} = \begin{bmatrix} 1 & 0 & 0 \\ 1 & 1 & 0 \\ 0 & -1 & -1 \\ 0 & 0 & 1 \end{bmatrix} \begin{bmatrix} I'_{BE} \\ I'_{BC} \\ I'_{SC} \end{bmatrix}.$$

The terminal behavior of the four-layer structure, obtained from the above equations, is

$$\begin{bmatrix} I_E \\ I_B \\ I_C \\ I_S \end{bmatrix} = \begin{bmatrix} I_{s1} & -\alpha_I I_{s2} & 0 \\ (1-\alpha_N) I_{s1} & (1-\alpha_I) I_{s2} & -\alpha_{SI} I_{s3} \\ \alpha_N I_{s1} & -(1-\alpha_{SN}) I_{s2} & -(1-\alpha_{SI}) I_{s3} \\ 0 & -\alpha_{SN} I_{s2} & I_{s3} \end{bmatrix} \begin{bmatrix} \lambda(V_{BE}) \\ \lambda(V_{BC}) \\ \lambda(V_{SC}) \end{bmatrix}.$$

$$(2.30)$$

The Ebers-Moll model embodies superposition of exponential voltages, $\lambda(V)$, but real transistors easily violate superposition, as is evidenced, for instance, by considering the Early effect.[6] Figure 2.13 shows an experimental

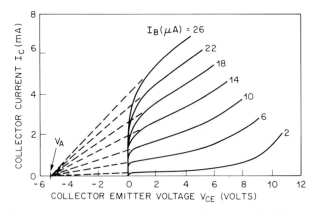

Fig. 2.13 Collector current-voltage characteristic illustrating the Early voltage.

I_C versus V_{CE} characteristic of a transistor. Collector current I_C varies linearly with collector-emitter voltage V_{CE} over a substantial range, and the curves, when projected by straight-line sections, intersect at a voltage point V_A, called the *Early voltage*. If the transistor behavior really obeyed superposition, all straight-line sections of the curves would be approximately parallel to one another. In a real transistor the slope of these lines increases with base current.

Another departure from the ideal relationship of Eq. (2.30) occurs due to carrier injection effects. At low forward voltages the base current I_B shows a $\lambda(V/2)$ dependence, probably due to space-charge recombination, surface recombination in the base, or to both effects, as illustrated by Fig. 2.14. The collector current I_C shows a $\lambda(V)$ dependence in this bias

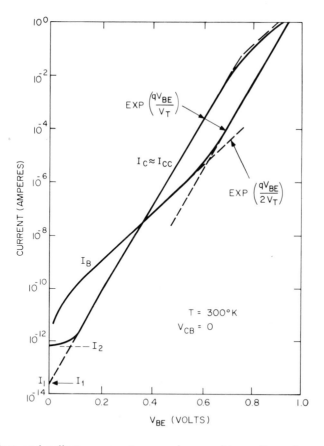

Fig. 2.14 Base and collector current versus base-emitter voltage characteristic illustrating nonideal behavior at both low- and high-current levels.

range. At a bias value of approximately 0.6 V the base current has a "knee-point" and thereafter shows a $\lambda(V)$ dependence. High injection effects occur when the number of minority carriers in the base become comparable with the doping concentration. This affects the collector current I_C, which shows a "knee-point" at $V_{BE} \approx 0.75$ V.

Because of these and similar shortcomings of the Ebers-Moll model, an integral charge control model was proposed by Gummel and Poon[7,8] and will now be described.

2.11 THE GUMMEL-POON MODEL

Figure 2.15 shows a more detailed diagram of the main n-p-n transistor junctions and depletion regions, together with electron and hole densities and currents. A solution for carrier transport is readily obtained under the simple, but unrealistic assumption of zero electric field in the base region. For that case the solution of the continuity equation $\ddot{n} + (n-n_B)/L_B^2 = 0$ yields the electron density in the base, namely

$$n(x) = n_B + \frac{\left[n_1(0)\sinh\left(\dfrac{w-x}{L_B}\right) + n_1(w)\sinh\left(\dfrac{x}{L_B}\right)\right]}{\sinh\left(\dfrac{w}{L_B}\right)} \qquad (2.31)$$

in the range $0 \leqslant x \leqslant w$, where

$$n_1(0) = n_B\lambda(V_{BE}) \qquad n_1(w) = n_B\lambda(V_{BC})$$

and n_B is the equilibrium density of electrons in the base. The electron current in the base region, $I_n(x) = AqD_B\dot{n}$, is now obtained from Eq. (2.31) as

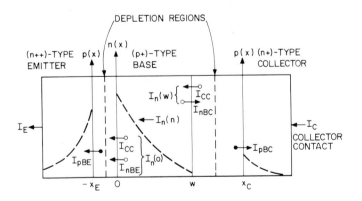

Fig. 2.15 Minority carrier densities and current directions for a bipolar transistor.

$$I_n(x) = -\frac{AqD_B}{L_B} \frac{n_1(0)\cosh\left[\dfrac{w-x}{L_B}\right] - n_1(w)\cosh\left[\dfrac{x}{L_B}\right]}{\sinh\left[\dfrac{w}{L_B}\right]} \qquad (2.32)$$

with $0 \leqslant x \leqslant w$.

The emitter electron current is $I_n(0)$ and the collector electron current is $I_n(w)$. These two currents are both derived from $I_n(x)$, but are not quite equal because of recombination, primarily in the emitter. Nevertheless, it is possible to recognize a dominant, common current component, I_{CC}, by comparing $I_n(0)$ and $I_n(w)$. This component is independent of x and is given by

$$I_{CC} = -\frac{AqD_B n_B}{L_B \sinh\left[\dfrac{w}{L_B}\right]} \{\lambda(V_{BE}) - \lambda(V_{BC})\}. \qquad (2.33a)$$

This value should be compared with one obtained by solving the current equation $I_{CC} = AqD_B(\dot{n} - \dot{\psi}n/V_T)$ which takes into account the existence of an electric field $E(x) = -\dot{\psi}(x)$ in the base region, due to a doping profile $N(x)$ in the base. The result obtained is[7,8]

$$I_{CC} \approx -\frac{A^2 q^2 D_B n_i^2}{Q_B} \{\lambda(V_{BE}) - \lambda(V_{BC})\}. \qquad (2.33b)$$

This formula may also be written in the form

$$I_{CC} \approx -I_s \frac{Q_{B0}}{Q_B} \{\lambda(V_{BE}) - \lambda(V_{BC})\} \qquad (2.33c)$$

where I_s stands for saturation current. Furthermore, Q_B is taken to be the total hole base charge in the case of an n-p-n transistor. On account of charge neutrality in the base $p = n + N(x)$, and so

$$Q_B = qA \int_0^w p(x)\,dx = qA \int_0^w n(x)\,dx + qA \int_0^w N(x)\,dx. \qquad (2.34)$$

Now

$$N_G = \int_0^w N(x)\,dx \qquad (2.35)$$

is defined as the "Gummel number" for the base region and gives the total number of ionized impurities in the base. The total charge corresponding to this number is clearly qAN_G where A is the cross-sectional area of the emitter and base region. The other component, of Eq. (2.34),

$$qA \int_0^w n(x)\,dx$$

may be obtained by integrating Eq. (2.31). This yields an additional zero bias component qAn_Bw due to the equilibrium density and so

$$Q_{B0} = qAn_Bw + qAN_G \approx qAN_G \qquad (2.36)$$

where Q_{B0} stands for the "zero bias" charge.

EXAMPLE 2.10 A transistor has a square emitter 25 μm on a side and zero bias charge $Q_{B0} = 6\times10^{-12}$ C. Calculate the Gummel number N_G.

Solution. From Eq. (2.36), $N_G \approx Q_{B0}/qA$ for $N_G \gg n_Bw$, and so

$$N_G = \frac{6\times10^{-12}}{1.6\times10^{-19}(6.25\times10^{-6})} = 6\times10^{12}.$$

This is a typical value for the number of impurities in the base region. Lower values of N_G tend to cause premature punchthrough, while higher values lead to lower injection efficiency or lower unity-gain frequency f_L.

The other components of

$$qA \int_0^w n(x)\,dx$$

are due to the traverse of electrons across the base region and are defined Q_{nBE}, Q_{nBC}, with the subscripts indicating association with the base-emitter and base-collector junction, respectively. Integrating Eq. (2.31) and comparing with Eq. (2.33) yields

$$Q_{nBE} = -\tau_f I_f, \qquad (2.37)$$

where τ_f is the transit time for electrons in the forward current component I_f. Theoretically

$$\tau_f = \tau_B \left\{ \cosh\left[\frac{w}{2L_B}\right] - 1 \right\} \qquad (2.38a)$$

where $\tau_B = L_B^2/D_B$. For $w \ll L_B$, which is the case for most modern transistors, it is found that

$$\tau_f' = \frac{w^2}{2\eta_f D_B} = \frac{\tau_f}{\eta_f}, \qquad (2.38b)$$

where $\eta_f = 1$ for uniform doping and where typically $2 < \eta_f < 10$ for diffused base transistors.

Current I_f is defined as the forward current component of I_{CC} and so from Eq. (2.33)

$$I_f = -\frac{I_s Q_{B0}}{Q_B}\lambda(V_{BE}). \qquad (2.39)$$

Note that the low-frequency approximation to the unity-gain frequency

f_L is given by

$$f_L = \frac{1}{2\pi\tau_f} . \tag{2.40}$$

EXAMPLE 2.11 An *n-p-n* transistor has the following parameters: base width $w = 0.5\times10^{-4}$ cm, average acceptor concentration in base $N_A = 5\times10^{15}$ cm^{-3}, and actual transit time $\tau_f' = 5\times10^{-10}$ s. Find the constant η_f in Eq. (2.38b).

Solution. From Eq. (2.38b) $\tau_f' = \eta_f\tau_f$ where $\tau_f' = 5\times10^{-10}$ and $\tau_f = w^2/2D_B$. But from Einstein's relation $D_B = \mu_p V_T$ and from tables $\mu_p = 500$ cm^2/V$-$s for $N_A = 5\times10^{15}$. Hence

$$\tau_f = \frac{10^{-8}}{8(26\times10^{-3})500} = 0.96\times10^{-10}.$$

Therefore, $\eta_f = \tau_f'/\tau_f = 5/0.96 = 5.2$.

EXAMPLE 2.12 A transistor has transit time $\tau_f = 5\times10^{-10}$ s. Find its unity-gain frequency, f_L.

Solution. From Eq. (2.40), $f_L = 1/2\pi\tau_f = 1/2\pi(5\times10^{-10}) = 318$ MHz. Actually, f_L is a function of collector current, I_{CC}, which passes through a maximum value and falls sharply at high values of I_{CC}. To a lesser extent, f_L, is a function of V_{BC}, being greater for larger values of V_{BC}.

It may also be shown that

$$Q_{nBC} = -\tau_r I_r \tag{2.41}$$

where τ_r is the reverse transit time for electrons and where I_r is the reverse current component of I_{CC}, that is,

$$I_r = -\frac{I_s Q_{B0}}{Q_B} \lambda(V_{BC}); \quad I_{CC} = I_f - I_r. \tag{2.42}$$

The reverse transit time τ_r can be linked to the forward transit time τ_f by the empirical constant, r, $\tau_r = r\tau_f$. In many practical cases, it is found that r is in the range $10 < r < 500$. This reduces the reverse unity-gain frequency accordingly. The total base charge, Q_B, so far, is given by

$$Q_B = Q_{B0} + Q_{nBE} + Q_{nBC}.$$

However, this equation still lacks the contributions Q_E, Q_C arising from the presence of the base-emitter and base-collector depletion capacitances C_E and C_C. Guided by Eqs. (2.3) and (2.4),

$$C_E = C_{E0}(V_{bi} \pm V_{BE})^{-m} \quad C_C = C_{C0}(V_{bi} \pm V_{BC})^{-m} \tag{2.43}$$

where the positive sign is to be taken for reverse bias and the negative sign for forward bias. These capacitances do not really become infinite when

$V \pm V_{bi} = 0$ and Eqs. (2.43) must, in practice, be modified to allow these capacitances to remain finite.

EXAMPLE 2.13   Let the base-emitter capacitance, C_E, of a transistor be modeled by Eq. (2.43) with parameters $C_{E0} = 3 \times 10^{-12}$ and $m = 1/2$. Find the capacitively stored charge Q_E and discuss the physical significance of the "blowup" of C_E when the forward-bias voltage V approaches the built-in voltage value V_{bi}.

Solution.   From Eq. (2.43)

$$\frac{dQ_E}{dV_{BE}} = C_E = C_{E0}(V_{bi} - V_{BE})^{-m}.$$

Integrating this equation and using the boundary condition $Q_E = 0$ when $V_{BE} = 0$ yields

$$Q_E = \frac{C_{E0} V_{bi}^{1-m}}{1-m} \left[1 - \left(1 - \frac{V_{BE}}{V_{bi}}\right)^{1-m}\right].$$

For a forward bias of $V_{BE} \approx V_{bi}$ one therefore has

$$Q_E \approx \frac{C_{E0} V_{bi}^{1-m}}{1-m}.$$

Inserting the given values yields $Q_E = 6 \times 10^{-12}(0.7)^{1/2} \approx 5 \times 10^{-12}$ C.

Computer simulation[27] shows that the emitter-base depletion capacitance reaches a finite peak value at $V_{BE} \approx V_{bi}$ and then decreases for $V_{BE} > V_{bi}$. Hence, Eq. (2.43) does not model C_E well in this range and should be modified, perhaps to the form

$$C_E = \frac{C_{E0} V_{bi}^{-m}}{(x^2 + b)^{m/2}} \left[1 + \frac{m}{1-m} \frac{b}{x^2 + b}\right]$$

where $x = (V_{BE} - V_{bi})/V_{bi}$ and b is an additional constant.

In any case, the final expression for total base charge is

$$Q_B = Q_{B0} + Q_E + Q_C - \tau_f I_f - \tau_r I_r \tag{2.44}$$

where $dQ_E = dV_{BE} C_E$ and $dQ_C = dV_{BC} C_C$. Equation (2.44) accounts for all holes communicating with the p-type base region in an n-p-n transistor. This charge expression may be normalized with respect to Q_{B0} and divided into two components, a low-level injection component, q_1, where

$$q_1 = 1 + \frac{Q_E + Q_C}{Q_{B0}},$$

and a high-level injection component, q_2, where

$$q_2 = \frac{I_s[\tau_f \lambda(V_{BE}) + \tau_r \lambda(V_{CB})]}{Q_{B0}}.$$

As a result, Eq. (2.44) may be written as a quadratic

$$q_B^2 - q_1 q_B - q_2 = 0; \quad q_B = \frac{Q_B}{Q_{B0}}. \tag{2.45}$$

Under low-level injection $q_2 \ll 1$ and $q_B \approx q_1$ and so the impurities in the base and the depletion capacitances at the junction dominate the base charge profile. As a result, the dominant forward-current component $I_{CC} \propto \lambda(V_{BE})$.

Under high-level injection, $q_2 \gg 1$ and $q_B \approx \sqrt{q_2}$ and so the currents in transit across the base dominate the base charge profile. This implies that the mobile carriers in the base are comparable to, or exceed in number, the impurities there. The onset of high-level injection may theoretically be taken as the knee-voltage $V_{BE} = V_K$ derived by solving the equation $q_2 = 1$. This voltage has a value

$$V_K = V_T \ln\left(\frac{Q_{B0}}{\tau_f I_s}\right) \tag{2.46a}$$

and corresponds to a knee-current

$$I_K = \frac{Q_{B0}}{\tau_f}. \tag{2.46b}$$

EXAMPLE 2.14   An n-p-n transistor has the following parameters: $I_s = 10^{-14}$ A, $Q_{B0} = 2 \times 10^{-11}$ C, $\tau_f = 4.5 \times 10^{-10}$ s. The base-emitter voltage V_{BE} is allowed to range between 0.6 and 0.85 V and the dominant current, I_{CC} measured at these points is -10^{-4} and -10^{-1} A, respectively. Find the total transistor base charge, Q_B, corresponding to these range points. Find also the base voltage and current, V_K and I_K, at the onset of high-level injection.

Solution.   From Eq. (2.33c), $Q_B \approx [-I_s Q_{B0} \exp(V_{BE}/V_T)]/I_{CC}$, which is evaluated at $V_{BE} = 0.6$ V,

$$Q_B = \frac{10^{-14} \times 2 \times 10^{-11} \exp(0.6/26 \times 10^{-3})}{10^{-4}} \approx 2.30 \times 10^{-11} \text{ C}$$

and at $V_{BE} = 0.85$ V,

$$Q_B = \frac{10^{-14} \times 2 \times 10^{-11} \exp(0.85/26 \times 10^{-3})}{10^{-1}} = 3.58 \times 10^{-10} \text{ C}.$$

Hence, while the dominant current component increases by about three orders of magnitude, the total base charge increases by only about one order of magnitude.

From Eq. (2.46b) the knee-current,

$$I_K = \frac{Q_{B0}}{\tau_f} = \frac{2 \times 10^{-11}}{4.5 \times 10^{-10}} = 4.4 \times 10^{-2} \text{ A}.$$

From Eq. (2.46a) the knee-voltage,

$$V_K = V_T\ln\left(\frac{I_K}{I_s}\right) = 26\times10^{-3}\times\ln\left[\frac{0.44\times10^{-1}}{10^{-14}}\right] \approx 0.76 \text{ V}.$$

This voltage marks the onset of high-level injection and for $V > V_K$, $Q_B > Q_{B0}$.

After the knee-point has been reached, the dominant current component $I_{CC} \propto \lambda(V_{BE}/2)$, as shown in Fig. 2.14, because of the relation $q_B = \sqrt{q_2}$. Remarks about the base-current component, also shown in Fig. 2.14, will be made later. The Gummel-Poon model therefore satisfactorily explains normal high-level injection effects. A specific high-level injection condition, called the *Kirk effect*[10, 11] or *base pushout,* may also be accommodated. This effect occurs because the high field region, originally located at the base-collector junction, moves with increasing values of I_{CC} into the epitaxial collector region, thus effectively increasing the base width as shown in Fig. 2.16. This effect can be theoretically accounted for by solving the basic differential equations, with boundary conditions only applied at the

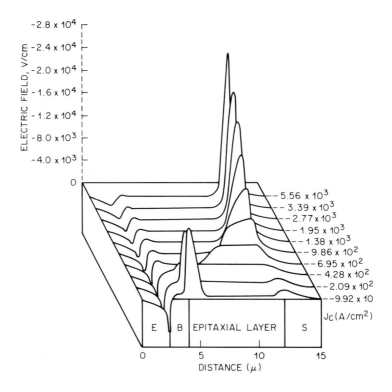

Fig. 2.16 Perspective plot of electric field versus position for various values of collector current density. (From Poon, et al.[11])

actual electric terminals. Such a calculation is beyond the scope of this book, but base pushout may be accommodated into the Gummel-Poon model by modifying the forward transit time τ_f to the value

$$\tau_f' = \tau_f \left(1 + \frac{w_c}{w}\right)^2,$$

where w is the base width and w_c is the width of the epitaxial layer.

Another effect that is elegantly accommodated by the Gummel-Poon model is the Early effect.[6] Here an increase in collector-emitter voltage V_{CE} increases the collector-base depletion-layer width, w_D. Because the effective base width is $w \approx w(V_{CE} = 0) - w_D$, the base width decreases, the minority carrier profile becomes steeper, and I_{CC} increases. In addition, it is observed that the $I_C \approx I_{CC}$ versus V_{CE} characteristic displays a constant and finite slope when I_B is constant, as shown in Fig. 2.13. Moreover, if the straight part of this family of curves is continued, the curves intersect approximately at a point V_A, called the Early voltage, also shown in Fig. 2.13. Starting with Eq. (2.33c) and differentiating with respect to V_{CE} (keeping V_{BE} constant) yields

$$\frac{I_{CC}}{Q_B}\frac{\partial Q_B}{\partial V_{CE}} + \frac{\partial I_{CC}}{\partial V_{CE}} = \frac{I_{CC}\exp(V_{BC}/V_T)}{V_T\lambda(V_{BE})}.$$

But under low-level injection

$$\frac{\partial Q_B}{\partial V_{CE}} = -\frac{\partial Q_C}{\partial V_{BC}}.$$

However, $dQ_C = C_C dV_{BC}$ and so from Eq. (2.43)

$$\frac{\partial Q_B}{\partial V_{CE}} = -C_C, \ |V_{BC}| \gg |V_{bi}|.$$

Hence from the above three equations

$$\frac{\partial I_{CC}}{\partial V_{CE}} \approx (1-m)\frac{I_{CC}Q_C}{V_{BC}Q_B}, \quad \exp(V_{BC}/V_T) \ll 1. \qquad (2.47a)$$

The constant $m = 1/2$ for abrupt and $m = 1/3$ for linearly graded junctions. If one defines a forward current gain β_F such that $I_{CC} \approx \beta_F I_B$ then Eq. (2.47a) may be written

$$\frac{\partial I_{CC}}{\partial V_{CE}} = \frac{I_{CC}C_C}{Q_B}. \qquad (2.47b)$$

The slope therefore depends on the base current I_B, the current gain β_F and approximately inversely on the Gummel number N_G, since $Q_B \approx qAN_G$ for low-level injection. Applying similar triangles to Fig. 2.13, it may be shown that

$$V_A = \frac{I_{CC} - V_{CE}\, \dfrac{\partial I_{CC}}{\partial V_{CE}}}{\dfrac{\partial I_{CC}}{\partial V_{CE}}}, \qquad (2.48)$$

and so

$$V_A \approx \frac{Q_B}{C_C}. \qquad (2.49)$$

Note that V_A is taken positive, and is a constant proportional to Q_B/C_C, while $V_{BC} < 0$ for a reverse-biased base-collector junction. Equation (2.49) allows an immediate estimate to be made of the punchthrough voltage, denoted V_{pt}, which occurs when the base-collector depletion region has become so wide that it touches the base-emitter depletion region. This is equivalent to shorting the emitter and collector of a transistor. Alternatively, it is equivalent to the base-collector depletion charge Q_C eating up all the base charge and so it results when $Q_{B0} + Q_C \approx 0$. When this occurs, I_{CC} of Eq. (2.49) becomes infinite. This condition, from Eq. (2.49), is equivalent to $V_{pt} \approx V_A$ and so one can conclude that punchthrough occurs when the base-collector voltage V_{BC} is of the same order of magnitude as the Early voltage V_A. Note that other transistor voltage limitations are discussed in Section 2.14.

EXAMPLE 2.15 A transistor has a zero bias base charge $Q_{B0} = 2 \times 10^{-11}$ C and a collector capacitance $C_C = 10^{-12}$ F. Find the approximate punchthrough voltage V_{pt}.

Solution. From Eq. (2.49),

$$V_{pt} \approx V_A = Q_{B0}/C_C$$
$$= 2 \times 10^{-11}/10^{-12}$$
$$= 20 \text{ V}.$$

The common current component I_{CC} can be estimated from parameters V_A, V_B, m, and I_s that are all experimentally measurable. These parameters should be of an order given by the theoretical formulae Eqs. (2.48a) and (2.48b). Note, in this connection, that an experimental estimate for the saturation current I_s is given by

$$I_s = I_K \exp(-V_K/V_T) \qquad (2.50)$$

while a theoretical estimate follows from

$$I_s = qAn_i^2 D_B/N_G \tag{2.51}$$

EXAMPLE 2.16   A transistor has knee-voltage $V_K = 0.75$ V and knee-current $I_K = 5 \times 10^{-2}$ A. Calculate the saturation current I_s.

Solution.   From Eq. (2.50),
$$I_s = I_K \exp(-V_K/V_T) = 5 \times 10^{-2} \exp(-0.75/26 \times 10^{-3}) = 1.33 \times 10^{-14} \text{ A.}$$

An important figure of merit for a transistor is the base-transport factor, α_T, defined as the ratio of collected electron current, $I_n(w)$, to emitted electron current $I_n(0)$. Under normal operation $V_{BE} > 0$, $V_{BC} < 0$, one has $I_{CC} \approx I_f$ and so, from Eqs. (2.32) and (2.33), $I_n(w) = I_{CC} + I_{nBC} \approx I_f$ while

$$I_n(0) = I_{CC} + I_{nBE} \approx I_f + \left[\cosh\left(\frac{w}{L_B}\right) - 1\right] I_f = \cosh\left(\frac{w}{L_B}\right) I_f,$$

and so

$$\alpha_T = \frac{I_n(w)}{I_n(0)} \approx \frac{1}{\cosh\left(\dfrac{w}{L_B}\right)} . \tag{2.52}$$

Hence for $\alpha_T \approx 1$, the base width w should be much smaller than the electron-diffusion length in the base, L_B. Note further, that

$$I_{nBE} \approx \left[\cosh\left(\frac{w}{L_B}\right) - 1\right] I_f; \quad I_{nBC} = -\left[\cosh\left(\frac{w}{L_B}\right) - 1\right] I_r. \tag{2.53}$$

Besides the electron current, $I_n(x)$ traversing the base region, there is also a hole-diffusion current contribution I_{pBE}, because of holes that enter the emitter region from the base. The hole density in the emitter is given by

$$p = p_E + p_1(-x_E)\exp\left[\frac{x+x_E}{L_E}\right]$$

and furthermore

$$I_{pBE} = -qAD_E \dot{p}(x)\big|_{x=-x_E}$$

where $x \leqslant -x_E$. Hence

$$I_{pBE} = -\frac{qAD_E p_E}{L_E} \lambda(V_{BE}). \tag{2.54}$$

The emitter efficiency γ is now defined as the ratio of electron current, $I_n(0)$, crossing the base-emitter junction to total emitter current, $I_n(0) + I_{pBE}$. Hence

$$\frac{1}{\gamma} = \frac{I_n(0) + I_{pBE}}{I_n(0)} \approx 1 + \frac{I_{pBE}}{I_{CC}} \, . \tag{2.55}$$

Good emitter efficiency will result when I_{pBE}/I_{CC} is as small as possible. This may be shown to be equivalent to $N_G/N(-x_E)$ being as small as possible and so the emitter region, particularly near the base-emitter junction, should be heavily doped, that is, $N(-x_E) \gg N_G$.

The hole-current I_{pBE}, aids the electron-current $I_n(0)$, and their sum is defined to be the emitter current I_E, that is,

$$I_E = I_{CC} + I_{BE}; \quad I_{BE} = I_{nBE} + I_{pBE}. \tag{2.56}$$

Note that $|I_E| > |I_{CC}|$ and that I_{BE} is the emitter-current component due to the base-emitter junction.

Similarly, there is a hole-diffusion current contribution, I_{pBC}, because of holes that enter the base region from the collector. The hole density in the collector is given by

$$p = p_c + p_1(x_c)\exp\left[-\frac{(x-x_c)}{L_c}\right] \tag{2.57}$$

and furthermore

$$I_{pBC} = -\, qAD_c \dot{p}(x)\big|_{x=x_c}$$

$$= \frac{qAD_c p_c}{L_c} \lambda(V_{BC}) \tag{2.58}$$

where $x \geqslant x_c$. The total collector current, I_C, is

$$I_C = I_{CC} + I_{BC}; \quad I_{BC} = I_{nBC} + I_{pBC}. \tag{2.59}$$

Note that $|I_C| < |I_{CC}|$ and that I_{BC} is the collector-current component associated with the presence of the base-collector junction.

The base current, I_B, is defined as the difference between emitter and collector current

$$I_B = I_E - I_C = I_{BE} - I_{BC}. \tag{2.60}$$

The dominant current component I_{CC}, which was shown to depend inversely on the base charge Q_B, cancels out. In fact, I_B is associated with recombination effects, as exemplified by I_{BE} and I_{BC}.

Equations (2.33b), (2.33c), (2.49), and (2.60) are the basis of the Gummel-Poon model.

Base-current components I_{BE} and I_{BC} appear, in theory, to be computable from Eqs. (2.53) through (2.60). In practice, their value depends on phenomena not included in these equations, such as generation and recombination of carriers in the depletion region, existence of bulk base material, and surface effects. The base current is therefore *best modeled empirically.* One form of modeling uses preexponential factors I_1, I_2, and I_3, together with emission coefficients n_E and n_C. Accordingly, one may write

$$I_{BE} = I_1\lambda(V_{BE}) + I_2\lambda\left|\frac{V_{BE}}{n_E}\right|; \quad I_{BC} = I_3\lambda\left|\frac{V_{BC}}{n_C}\right| + I_4\lambda(V_{BC}). \quad (2.61)$$

EXAMPLE 2.17 An *n-p-n* transistor has the following parameters: $Q_{B0} = 2\times10^{-11}$ C, $C_E = 3\times10^{-12}$ F, $C_C = 10^{-12}$ F, $V_K = 0.76$ V, and $V_{BC} = 5$ V; $I_s = 10^{-14}$ A, $I_1 = -2.7\times10^{-17}$ A, $I_2 = -1.24\times10^{-13}$ A and $n_E = 1.5$. Find the transistor beta at the onset of high-level injection.

Solution. The normalized base charge q_B is given by Eq. (2.45) as $q_B = q_1/2 + [(q_1/2)^2 + q_2]^{1/2}$ where $q_1 = 1 + (Q_E+Q_C)/Q_{B0}$. Hence,

$$q_1 = 1 + \frac{0.76(3\times10^{-12}) - 5(1\times10^{-12})}{2\times10^{-11}} = 0.86$$

while $q_2 = 1$ at onset of high-level injection. Hence,

$$q_B = 0.86/2 + [(0.86/2)^2 + 1]^{1/2} = 1.52$$

and so the main current component can be written, by virtue of Eq. (2.33), as

$$I_{CC} \approx -\frac{I_s}{q_B}\exp\left|\frac{V_{BE}}{V_T}\right| = -\frac{10^{-14}}{1.52}\exp\left|\frac{V_K}{V_T}\right|.$$

The base current component I_{BE} is, from Eq. (2.61),

$$I_{BE} = I_1\exp\left|\frac{V_K}{V_T}\right| + I_2\exp\left|\frac{V_K}{1.5V_T}\right|$$

and so

$$\beta \approx \frac{I_{CC}}{I_{BE}} = \frac{10^{-14}/1.52}{2.7\times10^{-17}+1.24\times10^{-13}\times5.65\times10^{-5}} = 193.$$

Actually β is a function of the dominant current component I_{CC}, with the maximum value of β occurring approximately at $I_{CC}(V_{BE}=V_K)$.

A plot of I_B versus V_{BE} is given in Fig. 2.14. The above remarks about the experimental nature of I_B apply also to the forward-current gain, β_F, and the reverse-current gain, β_R, defined by

$$\frac{1}{\beta_F} = \left|\frac{I_B}{I_C}\right| \approx \left|\frac{I_{BE}}{I_{CC}}\right|, \quad V_{BE} > 0, \; V_{BC} < 0, \quad (2.62)$$

$$\frac{1}{\beta_R} = \left|\frac{I_B}{I_E}\right| \approx \left|\frac{I_{BC}}{I_{CC}}\right|, \quad V_{BC} > 0, \; V_{BE} < 0. \quad (2.63)$$

Both these gains are well known to be current dependent,

$$\beta_F = \beta_F(I_{CC}); \quad \beta_R = \beta_R(I_{CC}) \tag{2.64}$$

.a phenomenon known as the Webster effect.[9] A characteristic curve of β_F versus I_E is shown in Fig. 2.17. Gains β_F and β_R are best obtained by computation using Eq. (2.33) and the empirical Eqs. (2.61). Actual emitter and collector currents can then be obtained from Eqs. (2.56) and (2.59).

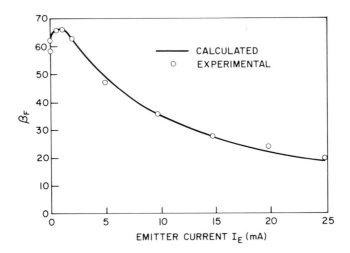

Fig. 2.17 Beta versus emitter current.

Care must be taken as to parameter signs in the Gummel-Poon model and the following remarks are relevant:

p-n-p Transistors. Take $Q_{B0} \approx -qAN_G < 0$; $Q_{B0}/Q_B > 0$; $I_s > 0$; $I_K = -Q_{B0}/\tau_f$;

$$I_{CC} = I_s \frac{Q_{B0}}{Q_B} [\lambda(V_{EB}) - \lambda(V_{EC})];$$

$I_1, I_2 > 0$; $I_3, I_4 < 0$. Wherever voltages occur write V_{EB} and V_{EC}. In the active region $V_{EB} > 0$, $V_{EC} < 0$.

n-p-n Transistors. Take $Q_{B0} \approx qAN_G > 0$; $Q_{B0}/Q_B > 0$; $I_K = Q_{B0}/\tau_f$; $I_s > 0$; $I_1, I_2 < 0$; $I_3, I_4 > 0$. Wherever voltages occur write V_{BE} and V_{BC}. In the active region $V_{BE} > 0$, $V_{BC} < 0$.

Under low-level injection, when $Q_B \approx Q_{B0}$, the Gummel-Poon model goes over into the Ebers-Moll model, which is restated here in a slightly different form than previously given in Eq. (2.30), for a four-layer device:

$$\begin{bmatrix} I_E \\ I_B \\ I_C \\ I_S \end{bmatrix} = \begin{bmatrix} (1+1/\beta_F)I_s & -I_s & 0 \\ (1/\beta_F)I_s & (1/\beta_R)I_s & -I_{ss} \\ I_s & I_{ss}-(1+1/\beta_R)I_s & -(1/\beta_R)I_{ss} \\ 0 & -I_{ss} & (1+1/\beta_R)I_{ss} \end{bmatrix} \begin{bmatrix} \lambda(V_{BE}) \\ \lambda(V_{BC}) \\ \lambda(V_{SC}) \end{bmatrix}.$$

$$(2.65)$$

2.12 BASE-SPREADING RESISTANCE

The base current $I_B = I_B(y)$ is transverse to the current I_{CC}, going from emitter to collector as shown in Fig. 2.18. The base material has a finite resistivity ρ_B and so a voltage drop $v_B(y)$ is developed between the external base contact and the active emitter-base junction. This effect may become serious for large I_B and ρ_B and may result in an actual cutoff of part of the emitter-base junction far removed from the base-contact. The result, known as emitter-crowding, is a reduction of the active emitter region until only the emitter edge facing toward the base contact is active. The effect is distributed throughout the base region and is termed base-spreading resistance. This resistance $r_{bb'}$ is defined as the ratio of the average voltage drop,

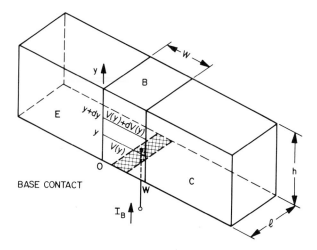

Fig. 2.18 Geometry for base-spreading resistance calculation.

$$\overline{V}_B = \frac{1}{h} \int_0^h v_B(y)\,dy,$$

to the base current I_B producing that drop. That is,

$$r_{bb'} = \frac{\overline{V}_B}{I_B}. \tag{2.66}$$

The value of $\overline{V}_B$ depends on the transistor geometry, the average resistivity of the base material, and the carrier injection level. Consider only the low-level injection case and restrict the discussion temporarily to the rectangular geometry transistor with a uniformly doped base region shown in Fig. 2.18. Then the differential resistance is $dR = \rho_B\,dy/wl$, and the base-current distribution is $I_B(y) = I_B(1 - [y/h])$ and so for any value of y

$$V_B(y) = \frac{I_B\rho_B}{w} \int_0^y \left[1 - \frac{v}{h}\right] dv = \frac{I_B\rho_B}{w}\left[y - \frac{y^2}{2h}\right]. \tag{2.67}$$

This result may now be used to find $\overline{V}_B$ and finally for the rectangular base transistor

$$r_{bb'} = \frac{\rho_b h}{3wl} = \frac{R_{BB'}}{3} \quad \text{(for single-contact stripe).} \tag{2.68}$$

Observe that $R_{BB'} = \rho_b h/wl$ is the resistance of a cross-sectional area w and height h carrying a uniform current density. The factor $1/3$ is due to the distributed nature of this situation. The base-spreading resistance can be reduced by a factor of 4 by placing a second base contact on the opposite side of the base region. This reduces the height of the base to $h/2$ and therefore the base-spreading resistance becomes

$$r_{bb'} = \frac{\rho_b h}{12lw} = \frac{R_{BB'}}{12} \quad \text{(for double-contact stripe).} \tag{2.69}$$

Most transistors have graded base impurity distributions and therefore these results are not directly applicable without some modification. The resistivity of the base region is defined as a function of position by

$$\rho_b(x) = \frac{1}{q\mu N(x)} \tag{2.70}$$

where $N(x)$ is the net impurity concentration in the base as a function of x. In terms of this resistivity, the resistance $R_{BB'}$ becomes

$$R_{BB'} = \frac{h}{q\mu l \int_0^w N(x)\,dx}. \tag{2.71}$$

If the base is uniformly doped so that $N(x) = N$,

$$R_{BB'}(\text{uniform}) = \frac{h}{q\mu w l N} . \tag{2.72}$$

The base-spreading resistance of the graded base transistor normalized to that of the uniform base transistor is the multiplying factor to be applied to the above results to obtain the correct base-spreading resistance. Finally, the mobility is also a function of position and an average value must be used. When this is done, the complete expression for base-spreading resistance becomes

$$R_{BB'} = \frac{K_R h}{qNwl \dfrac{\mu(w) - \mu(0)}{\ln[\mu(w)/\mu(0)]}} \tag{2.73}$$

where $\mu(0)$ is the mobility evaluated at the emitter-base junction and $\mu(w)$ is the mobility at the collector-base junction.

EXAMPLE 2.18   Find the base-spreading resistance, $R_{BB'}$, of a transistor at the onset of high-level injection, when $q_B = 1.5$. Use the empirical formula $R_{BB'} = R_1 + R_2/q_B$ where $R_1 = 10 \ \Omega$ and $R_2 = 100 \ \Omega$.

Solution.   From the above equation

$$R'_{BB} = 10 + \frac{100}{1.5} = 76.7 \ \Omega.$$

The term R_1 represents the constant resistance of the inactive base, while R_2 represents the base resistance of the active transistor. To a first approximation $R_1 + R_2 = R_{BB'}$ of Eq. (2.73).

2.13 THE HYBRID-PI MODEL [12,13]

The hybrid-*pi* model shown in Fig. 2.19 is one of many possible incremental (small-signal linear) models of the bipolar transistor. It gives good agreement with experimental results at high frequencies. The input of the transistor is modeled as one section of a lossy RC transmission line. The feedback elements arise from the depletion layer of the collector-base junction and the resultant base-width modulation. The origin of each of these components will now be considered.

The base-spreading resistance $r_{bb'}$ is the ohmic resistance between the base contact and the active base region underneath the emitter. It was discussed in Section 2.12 and is typically about 100 to 200 Ω for small transistors.

The base-emitter resistance $r_{b'e}$ is due to the dynamic impedance of the emitter-base junction. It is an incremental resistance defined as $|\partial V_{BE}/\partial I_B|$. When the transistor is in the active region, $I_B \approx I_{CC}/\beta_F$, and so

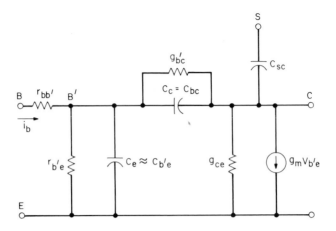

Fig. 2.19 Hybrid-*pi* transistor model.

$$g_{b'e} = \left| \frac{\partial I_B}{\partial V_{BE}} \right| \approx \frac{|I_{CC}|}{\beta_F V_T} = \frac{g_e}{\beta_F} , \quad g_e = \frac{|I_{CC}|}{V_T} . \tag{2.74}$$

Since $r_{b'e} = 1/g_{b'e}$, $r_{b'e} \approx V_T \beta_F/|I_{CC}|$ where $|I_{CC}|$ may be taken approximately equal to $|I_E|$ or $|I_C|$.

The base-emitter capacitance C_e is dominated by the diffusion capacitance $C_{b'e}$ of the forward-biased emitter-base junction. Taking the diffusion term similarly to Eq. (2.17) yields

$$C_{b'e} \approx \tau_f / r_{b'e}. \tag{2.75}$$

Thus the $r_{b'e} C_{b'e}$ time constant is used as an approximation of the transit time for the normal transistor.

Transconductance   When the transistor is in the active region under low-injection conditions, I_{CC} is given by Eq. (2.33). The transconductance g_m is defined as

$$g_m = \left| \frac{\partial I_C}{\partial V_{BE}} \right|_{V_{CE}=\text{const.}} \approx \frac{|I_{CC}|}{V_T} \approx g_e. \tag{2.76}$$

Collector-base conductance   The feedback conductance $g_{b'c}$ is defined as

$$g_{b'c} = \left| \frac{\partial I_B}{\partial V_{CE}} \right| \approx \frac{(1-m)}{\beta_F} \left| \frac{I_{CC} C_C}{qANG} \right| = \frac{\eta}{\beta_F} \frac{|I_{CC}|}{V_T} = \frac{\eta g_e}{\beta_F} , \tag{2.77}$$

where η, the so-called *base-width modulation factor*, is given by

$$\eta = \frac{(1-m)\,V_T C_C}{qAN_G} \tag{2.78}$$

and relates the change in V_{BC} to an equivalent change in V_{BE}. Factor η is typically in the range 10^{-3} to 10^{-5} at useful operating points.

Output conductance g_{ce} is due to the Early effect. It is computed with V_{BE} constant and so $\partial V_{CE} \approx -\partial V_{BC}$. Hence, from Eq. (2.49), one obtains

$$g_{ce} \approx \left| \frac{\partial I_C}{\partial V_{CE}} \right| \approx \left| \frac{\partial I_{CC}}{\partial V_{BC}} \right|$$

$$= \frac{I_s \lambda (V_{BE})}{\left[1 + \dfrac{1}{1-m} \left[\dfrac{V_{BC}}{V_A} + \dfrac{V_{BE}}{V_B} \right] \right]^2 (1-m)\,V_A} \approx \frac{|I_{CC}|}{(1-m)\,V_A}. \tag{2.79}$$

Collector-base capacitance C_C is the depletion capacitance of the junction and is given by Eq. (2.43) as

$$C_C \approx C_{C0}(V + V_{bi})^{-m} \tag{2.80}$$

where m varies from one-third to one-half depending on the type of junction.

2.14 VOLTAGE LIMITATIONS

For a transistor in the common base configuration, breakdown is usually due to avalanching, the phenomenon being similar to that in *p-n* junctions described previously in Section 2.5, and the breakdown voltage is denoted by BV_{BCO}. Empirically, if the reverse current without breakdown is denoted I_{BCO}, then the reverse current with breakdown I'_{BCO} is given by

$$I'_{BCO} = MI_{BCO}, \tag{2.81}$$

similarly to Eq. (2.18). An empirical estimate for M is given by

$$M \approx \frac{1}{1 - (V/BV_{BCO})^n} \tag{2.82}$$

and this equation is similar to Eq. (2.19). Parameter n is approximately between 2 and 6. Avalanche breakdown corresponds to M approaching infinity, that is, V approaching BV_{BCO}. For very narrow base width w in lightly doped base material, breakdown may also occur at a punchthrough voltage, denoted as V_{pt}. It was previously shown that when V_{BC} is of the order of the Early voltage, V_A, punchthrough is liable to occur, that is, $V_{pt} \approx V_A$.

When a voltage is applied to the collector of a transistor in the common emitter configuration with the base lead floating, the base lead will acquire a potential lying somewhere between the collector-emitter potential. Hence V_{BE} will be slightly forward biased and I_C will be the sum of the reverse-biased current of the collector-base junction, $-I_{BC}$ plus the dominant current I_{CC}. Near avalanche breakdown this current is multiplied by a factor M and so

$$M(|I_{CC}| - |I_{BC}|) = |I_E| = |I_{CC}| \left(1 + \frac{1}{\beta_F}\right) \approx \frac{|I_{CC}|}{\alpha_F}$$

or

$$|I_{CC}| = \frac{MI_{BC}}{M - 1 - \left(\dfrac{1}{\beta_F}\right)} . \tag{2.83}$$

When the denominator equals zero, the current I_{CC} is limited solely by external resistances. The breakdown voltage BV_{CEO} is obtained from Eq. (2.82) and the denominator of Eq. (2.83) as

$$BV_{CEO} \approx \frac{BV_{BCO}}{\beta_F^{1/n}} . \tag{2.84}$$

For large values of β_F, the voltage BV_{CEO} is much smaller than BV_{BCO}.

EXAMPLE 2.19 A transistor has a beta of 200 and parameter $n = 4$. Compute the ratio of BV_{BCO}/BV_{CEO}.

Solution. From Eq. (2.84), $BV_{BCO}/BV_{CEO} = \beta^{1/4} = 200^{1/4} = 3.76$.

2.15 ION-IMPLANTED BIPOLAR TRANSISTORS [14,15]

Ion implantation is a technique whereby a desired species of ion, which has been accelerated to high energy, impinges upon the silicon substrate and is buried beneath the surface. Details of the implantation process are described in Section 5.5. The main advantage of ion implantation is the degree of control that it affords over the impurity profile and the total number of impurities that are deposited during an implant. In this section the possible advantages of this process in the fabrication of bipolar transistors are explored. The most interesting possibilities appear to be in connection with the fabrication of the emitter-base region. In a modern transistor this region lies within about 1 μm of the surface, which is within the range of moderate energy implants. In a conventional processing sequence, the base must be formed before the emitter. The high-impurity concentration in the emitter region causes long-range lattice strains and "pushes" the base impurity profile ahead of it and deeper into the collector region. Contrast this with the ion-implantation process in which there is no required order for the formation of the base and emitter impurity profiles. The depth of an implanted profile depends on the particular type of ion and on its energy, which can be precisely controlled. Thus, the emitter can be implanted first, followed by the base, thereby eliminating the emitter push effect.

Another advantage of ion implantation is that it can create impurity distributions that are entirely beneath the surface. This means there is no impurity compensation in the formation of the emitter-base junction. Consequently, there are fewer ionized impurities in the space-charge layer which is also smaller than that of an equivalent diffused transistor. Therefore, the implanted transistor should have a lower space-charge recombination current and the β_F versus collector-current characteristic should be superior at low current levels. Figure 2.20 is a plot of collector current and base current versus base-emitter voltage for a 2 GHz bandwidth ion-implanted transistor. Note that the two plots are practically parallel over many decades of varia-

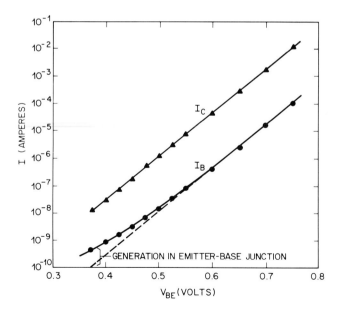

Fig. 2.20 Base and collector current versus base-emitter voltage for ion-implanted transistor.

tion, indicating that emission coefficient $n_L = 0$ in Eq. (2.61). This implies almost constant current gain β_F down to base currents in the order of nanoamperes.

Ion-implanted transistors have very reproducible parameters. Very close control of beta, for transistors with arsenic emitters, is possible because the emitter profile is very sharp. The collector current for a one-dimensional transistor is inversely proportional to the net number of ionized impurities per unit area in the base region. From Eq. (2.33b) with $D/\mu = kT/q$,

$$I_C \approx I_{CC} = -kTn_i^2\left[\frac{\mu A}{N_G}\right]\{\lambda(V_{BE}) - \lambda(V_{BC})\}. \tag{2.85}$$

In this expression it is assumed that the mobility, μ, is constant, and A is the area of the device. Effective values of μ and A can be used for actual devices. The abruptness of the emitter impurity profile makes I_B reproducible, while the implanted base distribution, which is entirely below the surface, allows precise control over N_G and so over I_{CC}. Therefore, precise control of β_F is possible by control of implantation parameters.

A final example of the advantages of ion implantation is control of the match in V_{BE} which is important for balanced circuits such as detectors. From Eq. (2.85), V_{BE} is found to be

$$V_{BE} \approx V_T \ln\left\{\frac{N_G}{\mu A}\left[-\frac{I_C}{kTn_i^2}\right]\right\}. \tag{2.86}$$

Variations in V_{BE} caused by the processing are contained in the term $N_G/\mu A$. There is a slight coupling between μ and N_G. Neglecting the effect of the area on current and junction voltage,

$$\Delta V_{BE} \approx 1.12 V_T\left|\frac{\Delta N_G}{N_G}\right| \tag{2.87}$$

where the factor 1.12 is due to the coupling between μ and N_G. Because of the nature of the implanted base impurity profile, $\Delta N_G/N_G$ is equal to the variation in the impurity dose. A $\pm 1\%$ variation in the impurity dose results in a ± 0.3-mV change in emitter-base voltage.

2.16 TWO-DIMENSIONAL TRANSISTOR ANALYSIS [17-19]

The problem of determining the current distribution in a two-dimensional transistor is extremely difficult. The analysis of one-dimensional transistors is a tractable problem whose solution is a sufficiently good approximation to the behavior of an actual transistor in many practical situations. There are some effects that are inherently two dimensional in nature. For example, the base-spreading resistance and emitter pinch-off effect are due to current in the base region in the direction transverse to the usual transistor current.

In this section the results of a numerical solution of the carrier-transport equations for a two-dimensional transistor are presented. These results are in the form of several computer-generated perspective plots showing the potential distribution and electron-density distribution under several different bias conditions.

Figure 2.21 is a cross-sectional view of the *n-p-n* transistor that was analyzed and all subsequent figures are geometrically oriented according to this figure. It is assumed that a metallic contact is made to the collector region just underneath the epitaxial layer. The doping profile and zero-bias potential distribution are shown in Figs. 2.22(a) and 2.22(b) and should be compared with Fig. 2.21. The doping profile was obtained by solving the diffusion equation, under ideal conditions, for the impurity distributions. The zero-bias potential distribution is obtained from $n = n_i \exp(\Psi/V_T)$.

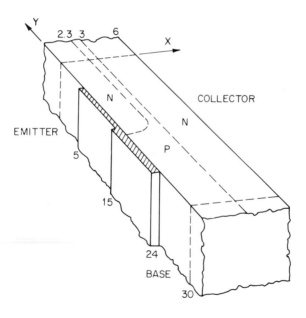

Fig. 2.21 Geometry for two-dimensional transistor analysis.

When the emitter-base junction is forward biased by 450 mV and the collector-base junction is reverse biased by 1 V, the potential distribution and electron-density distribution shown in Figs. 2.23(a) and 2.23(b) are obtained. The potential barrier at the emitter-base junction is lower than that for the zero-bias case and consequently the electron density in the "active" base, i.e., under the emitter, is increased over the zero-bias case which is shown in

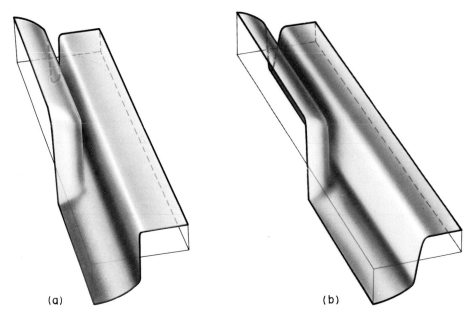

Fig. 2.22 (a) Doping profile and (b) potential distribution for zero bias. (From Heimeier.[17])

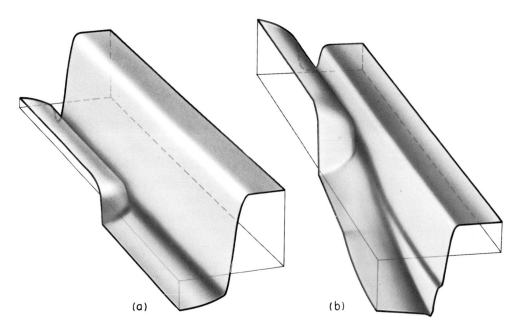

Fig. 2.23 (a) Potential distribution and (b) electron density for 450-mV forward bias, $V_{CB} = 1$ V. (From Heimeier.[17])

Fig. 2.22(a). At the low current density resulting from this bias condition, there is no voltage drop in the collector region underneath the emitter.

As the forward bias on the emitter-base junction is increased, the electron density and the current density increase and, as a result, the canyon in the base region is partially filled in, and emitter-current crowding effects, due to lateral voltage drops in the base region, are evidenced by a gradual slope extending almost to the top of the emitter surface. The conditions for 780 mV forward bias on the emitter-base junction are shown in Figs. 2.24(a) and 2.24(b) for potential distribution and electron density, respectively. There is a 29-mV lateral voltage drop in the base which causes the current density at the leading edge of the emitter to be about twice that of the current density at the center of the emitter. Resistive voltage drops in the collector bulk, evidenced by a depression in the collector surface, are clearly visible. The electron density in the active base is less than an order of magnitude below the electron density in the collector bulk region. There is considerable conductivity modulation in the active base region under these conditions.

The final perspectives for potential distribution and electron-density distribution at 900 mV forward bias on the emitter-base junction are shown in Figs. 2.25(a) and 2.25(b) for potential and electron density, respectively. This case is obviously a high-level injection situation. The electron-density

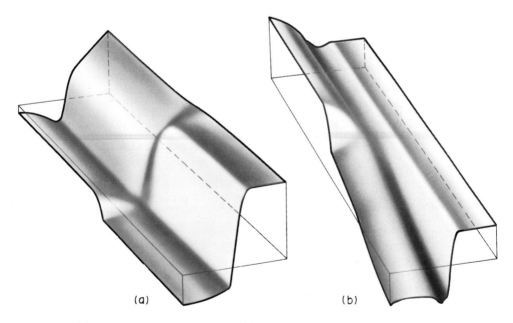

(a)   (b)

Fig. 2.24   (a) Potential distribution and (b) electron density for 780-mV forward bias, $V_{CB} = 1$ V. (From Heimeier.[17])

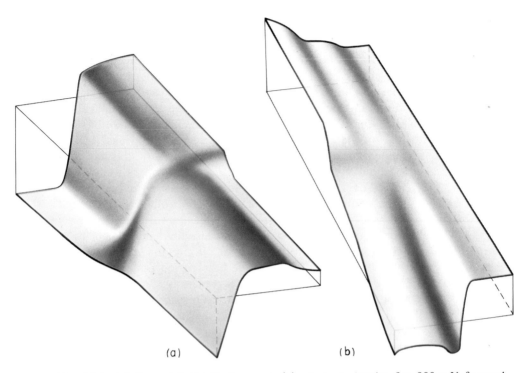

(a)   (b)

Fig. 2.25   (a) Potential distribution and (b) electron density for 900-mV forward bias, $V_{CB} = 1$ V. (From Manck, et al.[18])

distribution in the active base region exceeds that of the collector bulk region, and it decreases almost linearly with distance in the base region. Quasi-charge neutrality is maintained because excess holes tend to neutralize the excess electrons in the base. This results in conductivity modulation and it causes about half the vertical current in the effective base region to be carried by drift and the other half by diffusion. This is the Webster effect.[9]

From these perspectives it is also seen that base widening (Kirk effect) begins to occur at 780 mV forward bias and is quite pronounced at 900 mV forward bias. The Kirk effect is one of the dominant degradation mechanisms under high-injection conditions just as it was in the one-dimensional case. The emitter crowding effect due to lateral base-current IR drops is moderated by conductivity modulation, but it is not a negligible effect.

2.17 LATERAL p-n-p TRANSISTORS [20-23,26]

The lateral *p-n-p* transistor, shown in cross section in Fig. 2.26, consists of two separate, diffused, *p*-type regions which act as emitter and collector,

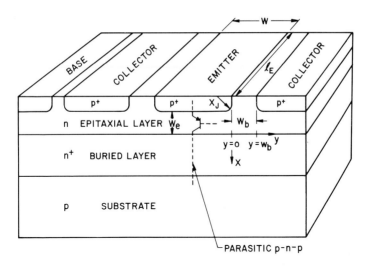

Fig. 2.26 Lateral transistor geometry.

placed in an n-type epitaxial layer. It is called a "lateral transistor" because
most of the controlled current is parallel to the surface, rather than in the
vertical direction as it is in a conventional transistor. This device is important because it can be made by operations that are part of the normal bipolar
silicon-integrated-circuit processing sequence. Thus, it provides, at no
increase in processing complexity, an extra circuit element that is useful in
analog circuits as a dynamic load, current source, and level shifter. A limitation of this device is its poor frequency response.

There are two reasons why the lateral transistor may be a low-gain device. First, the base width, which is a critical parameter in determining both
gain and cutoff frequency, is determined by photolithographic tolerances
rather than as the difference in depth of two impurity distributions. The
base width of a lateral transistor is several microns, whereas that of conventional transistor is often less than a micron. Second, depending upon the
thickness of various layers in the structure, there is either a parasitic diode or
a parasitic vertical p-n-p transistor, shown also in Fig. 2.26, between the base
and emitter of the lateral transistor. This parasitic device increases the
lateral-transistor base current, thus reducing its current gain. Under the
idealizing assumption that all of the lateral transistor current is constrained
between the sidewalls of the emitter stripe and that the bottom face of the
emitter is part of the parasitic device, the current gain of the lateral transistor
will be shown to be a function of the ratio A_L/A_V, where A_L is the emitter
area of the lateral transistor and A_V is the emitter area of the parasitic vertical transistor or diode, which is indicated dotted on Fig. 2.26. A second such

parasitic vertical transistor, originating at the collector, is cut off under normal operation and is therefore not shown. These assumptions are not true for modern lateral devices. A qualitative discussion is presented later.

For simplicity, assume that the emitter has sharp corners rather than rounded ones and is a rectangular stripe of thickness x_j, width w, and length l_E. Then the lateral or sidewall area of the emitter stripe, A_L, is given by

$$A_L = 2x_j(l_E + w). \tag{2.88a}$$

Similarly, the vertical emitter area, A_V, is given by

$$A_V = l_E w. \tag{2.88b}$$

The lateral current, I_L, emanating from area A_L of the emitter and collected by the two striped collectors is given by an equation analogous to Eq. (2.33) as

$$I_L \approx \frac{qA_L D_B p_B}{L_B} \lambda(V_{EB}) \operatorname{csch}\left(\frac{w_b}{L_B}\right). \tag{2.89}$$

Current transport is in terms of holes in the base region, p_B is the equilibrium hole concentration, and D_B, L_B have their usual meaning with respect to minority holes in the base region. Furthermore, the lateral p-n-p transistor is considered in the normal active region with $V_{EB} > 0$ and $V_{CB} < 0$. In addition to this dominant current component there is the hole current, I_{pEB}, associated with the emitter-base junction and, analogously to Eq. (2.53), given as

$$I_{pEB} = \frac{Q_{pEB}}{\tau_B} = \frac{qA_L D_B p_B}{L_B} \lambda(V_{EB}) \tanh\left(\frac{w_b}{2L_B}\right). \tag{2.90}$$

From Eq. (2.90)

$$Q_{pEB} = qA_L L_B p_B \lambda(V_{EB}) \tanh\left(\frac{w_b}{2L_B}\right)$$

and the capacitance, C_{bL}, associated with storage of minority carriers in the base is given by

$$C_{bL} = \frac{\partial Q_{pEB}}{\partial V_{EB}} \approx \frac{qA_L L_B p_B}{V_T} \tanh\left(\frac{w_b}{2L_B}\right) \exp\left(\frac{V_{EB}}{V_T}\right). \tag{2.91}$$

This capacitance is indicated on the hybrid-pi equivalent circuit for the lateral p-n-p given in Fig. 2.27. Note also the transconductance, g_m, shown there is given, by virtue of Eq. (2.89), as

$$g_m = \frac{I_L}{V_T} = \frac{qA_L D_B p_B}{L_B V_T} \lambda(V_{EB}) \operatorname{csch}\left(\frac{w_b}{L_B}\right). \tag{2.92}$$

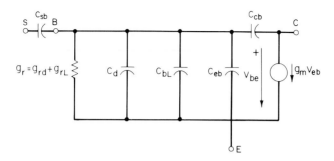

Fig. 2.27 Hybrid-*pi* equivalent circuit for lateral transistor.

In addition to these two parameters, Fig. 2.27 shows three junction capacitances, namely C_{eb}, C_{cb} and C_{sb} whose values are similar to those given in Eqs. (2.43).

The presence of the parasitic, vertical p-n-p, shown dotted in Fig. 2.26, or rather the presence of its forward-biased emitter-base junction, gives rise to a diffusion capacitance C_d, whose value is derivable similar to Eq. (2.17). If hole recombination at the $(n\text{-}n^+)$ interface, or in the n^+ buried layer is neglected, then it may be shown that

$$C_d = C_d(0) \approx \frac{qA_V L_B p_B}{V_T} \lambda(V_{EB}) \tanh\left(\frac{w_e}{L_B}\right). \qquad (2.93)$$

The input conductance, g_r, depends on the base current, I_B, and its uncertainties, and is therefore a function of A_V. It is best established empirically, perhaps from the relation

$$g_r = \left|\frac{\partial I_B}{\partial V_{EB}}\right|_{V_{CB}=\text{const.}} \qquad (2.94)$$

Alternatively it may be obtained from

$$g_r = \frac{g_m}{\beta_F} \qquad (2.95)$$

with β_F being experimentally established. Since g_m is proportional to A_L and g_r is proportional to A_V,

$$\beta_F \propto A_L/A_V. \qquad (2.96)$$

Three basic time constants are associated with the lateral p-n-p, namely:

a)   The transit time, τ_f, of holes across the lateral base region, $\tau_f = w_b^2/2D_B$.

b)   The transit time, τ_e, of holes across the epitaxial layer under the emitter, $\tau_e = w_e^2/2D_B$.

c)   The time constant, τ_r, associated with carrier recombination, $\tau_r = [C_{bL}+C_d+C_{eb}+C_{cb}]/g_r$.

For many practical cases, τ_r, is large in comparison to τ_f and τ_e. The lateral p-n-p cutoff frequency f_T, predicted by the hybrid-*pi* model of Fig. 2.27, is given by

$$\frac{1}{f_T} = \frac{2\pi[C_{bL}+C_d+C_{eb}+C_{cb}]}{g_m}$$

$$\approx \tau_f\left[1 + 2\left(\frac{w_e A_V}{w_b A_L}\right)\right] + \frac{C_{eb}+C_{cb}}{g_m} . \tag{2.97}$$

The term $2w_e A_V/w_b A_L$ is due to charge storage in the epitaxial diode and may be much greater than unity. Because g_m, C_{bL} and C_d are themselves functions of frequency, the hybrid-*pi* model is only adequate when $\omega\tau_e < 1$ or $\omega\tau_f < 1$. A discussion of the frequency behavior of lateral p-n-p's at higher frequencies is beyond the scope of this chapter.

In the analysis of the idealized lateral transistor it was assumed that all holes injected into the epitaxial layer from the bottom face of the emitter contribute to the parasitic current, and only those carriers injected from the side faces of the emitter are collected by the collector. Although marginal, this approximation breaks down when there is a heavily doped n^+-layer (buried layer) under the epitaxial layer acting as a reflector for the minority carriers injected vertically into the epitaxial layer. Actually, if the emitter stripe width is not significantly greater than twice the distance from the bottom of the emitter to the buried layer, most of the holes injected from the bottom face of the emitter will be reflected from the buried layer. If, in addition, the width of the collector region is very wide, most of these holes will be collected. As a rough approximation for use in estimating the dc gain, effective values for areas A_L and A_V are

$$A_L = 2[x_j(w+l_e) + w_e l_e] \tag{2.98a}$$

and

$$A_V = (w-2w_e)l_e \tag{2.98b}$$

where $w > 2w_e$ and it has been assumed that $w/2 - w_e$ from the bottom face, contributes to the lateral current. The ratio A_L/A_V for a lateral transistor with $x_j = 3\ \mu m$, $w_e = 7\ \mu m$, $w = 15\ \mu m$, and $l_e = 20\ \mu m$ is about 12 when calculated using Eqs. (2.98a) and (2.98b), whereas it is about one-third when calculated using Eqs. (2.88a) and (2.88b). This indicates that the inclusion of the buried layer and the use of a narrow emitter stripe and a wide collector, increase the gain by more than an order of magnitude over the simple situation analyzed previously.

The current gain of a modern lateral transistor can be as high as 100. This is in contrast to the very low current gains (about 1) of lateral transistors several years ago.

REFERENCES

1. S. M. Sze. *Physics of Semiconductor Devices.* New York: John Wiley & Sons, 1969.

2. A. S. Grove. *Physics and Technology of Semiconductor Devices.* New York: John Wiley & Sons, 1967.

3. R. M. Warner and J. N. Fordemwalt. *Integrated Circuits, Design Principles and Fabrication.* New York: McGraw-Hill, 1965.

4. W. Schottky. *Naturwissenschaften* **26:** 843 (1938).

5. J. J. Ebers and J. L. Moll. "Large signal behavior of junction transistor." *Proc. IRE* **42:** 1761-1772 (December 1954).

6. J. M. Early. "Effects of space charge layer widening in junction transistors." *Proc. IRE* **40:** 1401-1406 (November 1952).

7. H. K. Gummel. "A charge control relation for bipolar transistors." *Bell Syst. Tech. J.* **49:** 115-120 (January 1970).

8. H. K. Gummel and H. C. Poon. "An integral charge control model of bipolar transistors." *Bell Syst. Tech. J.* **49:** 827-851 (May/June 1970).

9. W. M. Webster. "On the variation of junction-transistor current amplification factor with emitter current." *Proc. IRE* **42:** 914-920 (June 1954).

10. C. T. Kirk. "A theory of transistor cut-off frequency, f_T, falloff at high current density." *IEEE Trans. on Electron Devices* **ED-9:** 164-174 (March 1962).

11. H. C. Poon, H. K. Gummel, and D. L. Scharfetter. "High injection in epitaxial transistors." *IEEE Trans. on Electron Devices* **ED-16:** 455-457 (May 1969).

12. R. L. Pritchard. "Electric-network representations of transistors, a survey." *IRE Trans. Circuit Theory* **CT-3:** 5-21 (March 1956).

13. P. E. Gray and C. L. Searle. *Electronic Principles.* New York: John Wiley and Sons, 1969.

14. R. L. Wadsack, private communication.

15. R. S. Payne, *et al.* "Fully ion-implanted bipolar transistors." *IEEE Trans. on Electron Devices* **ED-21:** 273-278 (April 1974).

16. K. Tada and J. L. R. Laraya. "Reduction of the storage time of a transistor using a Schottky-barrier diode." *Proc. IEEE* **55:** 2064-2065 (November 1967).

17. H. H. Heimeier. "A two-dimensional numerical analysis of a silicon n-p-n transistor." *IEEE Trans. on Electron Devices* **ED-20:** 708-714 (August 1973).

18. O. Manck, H. H. Heimeier, and W. L. Engl. "High injection in a two-dimensional transistor." *IEEE Trans. on Electron Devices* **ED-21**: 403-409 (July 1974).

19. J. W. Slotboom. "Computer-aided two-dimensional analysis of bipolar transistors." *IEEE Trans. on Electron Devices* **ED-20**: 669-679 (August 1973).

20. J. Lindmayer and W. Schneider. "Theory of lateral transistors." *Solid-State Electronics* **10**: 225-234 (1967).

21. S. Chou. "An investigation of lateral transistors—DC characteristics." *Solid-State Electronics* **14**: 811-826 (September 1971).

22. S. Chou. "Small signal characteristics of lateral transistors." *Solid-State Electronics* **15**: 27-28 (January 1972).

23. D. Seitz and I Kidron. "A two-dimensional model for the lateral p-n-p transistor." *IEEE Trans. on Electron Devices* **ED-21**: 587-592 (September 1974).

24. L. O. Chua and Pen-Min Lin. *Computer-Aided Analysis of Electronic Circuits: Algorithms and Computational Techniques.* Englewood Cliffs, N. J.: Prentice-Hall, 1975.

25. L. W. Nagel and D. O. Pederson. "SPICE (simulation program with integrated circuit emphasis)." Berkeley Calif. Univ. of California Electronics Research Laboratory, *Memorandum ERL-M382*, April 1973.

26. D. J. Hamilton and W. G. Howard. *Basic Integrated Circuit Engineering.* New York: McGraw-Hill, 1975.

27. H. C. Poon and H. K. Gummel. "Modeling of Emitter Capacitance." *Proc. of IEEE* **57**: 2181-2182 (December 1969).

PROBLEMS

2.1 The intrinsic carrier concentration n_i is given by

$$n_i = f(\cdot)\, T^{3/2} \exp\left[-\frac{E_g}{2kT}\right]$$

where $f(\cdot)$ is a function of the mass of the electron and the mass of the hole and is assumed to be independent of temperature. Determine the value of $f(\cdot)$ for silicon by plotting log n_i versus $1/T$ using the range $250 < T < 2000$. It is known experimentally that $n_i = 10^{10}$ carriers/cm^3 at 300 K. This result can be used to calculate the product of the effective density of states in the valence band N_V and in the conduction band N_C.

2.2 When holes are injected into an n-type semiconductor under steady-state conditions, the hole density p_n may be shown to obey the differential equation

$$D_p\, \frac{\partial^2 p_n}{\partial x^2} - \frac{(p_n - p_{n0})}{\tau_p} = 0.$$

Solve this equation for a semi-infinite semiconductor sample, using the boundary conditions $p_n(x=0)$ is a constant and $p_n(x=\infty) = p_{n0}$. Note that $x = 0$ represents the surface of the semiconductor sample at which holes are injected.

Solve the differential equation also for a sample of finite length, w, for which $p_n(w) = p_{n0}$. This latter condition supposes that all excess carriers are extracted from the sample at $x = w$. Take the diffusion length $L_p^2 = D_p \tau_p$.

2.3 Plot the depletion layer width w in microns, for a linearly graded silicon junction, against the impurity gradient a cm^{-4} for total applied voltage $V = 0.1$, 1 and 10 V and use the range $10^{18} \leqslant a \leqslant 10^{23}$. Use this result to estimate the value of the built-in voltage V_{bi}.

2.4 Show that when a p-n junction undergoes a small-signal excitation at frequency ω the effective carrier lifetime, τ, must be modified to a value $\tau^* = \tau/(1+j\omega\tau)$ and the effective diffusion length $L^* = L/(1+j\omega\tau)^{1/2}$. Find also an expression for the diffusion capacitance and compare it with Eq. (2.16).

2.5 Describe how the capacitance $C = (q\epsilon_s N_D/2[V_{bi}+V-V_T])^{1/2}$ for a metal to n-type semiconductor diode (Schottky diode) can be used to measure the doping density N_D. How could the measurement be modified if the doping density N_D is not constant?

2.6 Solve the differential equation for the electron density n in the base of a transistor

$$\frac{\partial n}{\partial x} + \frac{nE}{V_T} = \frac{I_{CC}}{qAD_B}$$

where I_{CC} is the dominant current component. Note that the electric field E has been included. Assume I_{CC} is constant for $0 \leqslant x \leqslant w$ and adopt the boundary condition $n(x=w) = 0$. Hence show that

$$n(x=0) \approx -\frac{I_{CC}}{AqD_B} \frac{N_G}{N(x=0)}$$

where

$$N_G = \int_0^w N(x)\,dx$$

is the Gummel number for the base and equals the total number of impurities therein. Furthermore $N(x=0)$ is the value of the acceptor density at the base-emitter junction.

2.7 Let the base-emitter capacitance, C_E, of a transistor be modelled by $C_E = C_{E0}(1-V_{BE}/V_{bi})^{-m}$. Find the capacitively stored charge, Q_E, using the relationship $dQ_E/dV_{BE} = C_E$ and the boundary condition $Q_E = 0$ when $V_{BE} = 0$. What is the physical significance of the "blowup" phenomenon which occurs for C_E when $V_{BE} = V_{bi}$?

2.8 Plot the base charge Q_B versus V_{BE} using the relation

$$Q_B = Q_1/2 + [(Q_1/2)^2 + Q_2]^{1/2},$$
$$Q_1 = Q_{B0} + Q_E + Q_C$$

and

$$Q_2 = I_s[\tau_f\lambda(V_{BE}) + \tau_r\lambda(V_{BC})]$$

where

$$\lambda(a) = \exp(a/V_T) - 1.$$

Take $Q_{B0} = qAN_G = 10^{-11}$ C, $Q_E = V_{BE}C_E$ and $Q_C = V_{BC}C_E$. Take further $C_E = 2\times10^{-12}$ F, $C_C = 10^{-12}$ F, $\tau_f = 5\times10^{-10}$ s, $\tau_r = \tau_f/10$ and $I_s = Q_{B0}\exp(-28)/\tau_f$.

Show that the curve of Q_B versus V_{BE} has a low voltage asymptote $Q_B = Q_1$, a high-voltage asymptote $Q_B = (Q_2)^{1/2}$ and a transition point marked by a knee-voltage $V_{BE} = V_K$, which occurs when $Q_2 = Q_{B0}$. Deduce this voltage from the plot and show that $V_K = V_T\ln(Q_{B0}/\tau_f I_s)$.

2.9 Plot the emitter-base capacitance C_E versus V_{BE} using the "blowup" proof-proposed model

$$C_E = \frac{C_{E0}V_{bi}^{-m}}{(x^2+b)^{m/2}}\left[1 + \frac{m}{1-m}\left(\frac{b}{x^2+b}\right)\right]$$

where $x = (V_{BE}-V_{bi})/V_{bi}$. Take $C_{E0} = 3\times10^{-12}$ F, $b = 1\times10^{-2}$, $m = 0.48$ and $V_{bi} = 0.7$ V. Find the value of peak capacitance.

Find also the capacitively stored charge Q_E using the relations $dQ_E/dV_{BE} = C_E$ and $Q_E = 0$ when $V_{BE} = 0$. Plot Q_E versus V_{BE} and discuss the result.

2.10 Plot $I_C = I_{CC} + I_{BC}$ and $I_B = (I_{BE}-I_{BC})$ versus V_{BE}. From these two plots prepare a third plot of β versus I_{CC} and ascertain the maximum value of β and the value of I_{CC} at which it occurs. Use the Gummel-Poon relations with the following values of model parameters: $I_s = 10^{-14}$ A; $I_1 = -2.8\times10^{-17}$ A; $I_2 = -1.25\times10^{-13}$ A; $I_3 = 9\times10^{-11}$ A; $Q_{B0} = 2\times10^{-11}$ F; $C_E = 3\times10^{-12}$ F; $C_C = 10^{-12}$ F; $\tau_f = 5\times10^{-10}$ s; $\tau_r = 3\times10^{-8}$ s; $n = 1.45$.

3

Field-Effect Devices

3.1 INTRODUCTION

The technology of metal-oxide-semiconductor (MOS) devices has in the past few years expanded into one of the fastest changing fields in electronics. Some reasons for this surge have been the relatively simple fabrication techniques for MOS devices, their high-packing density allowing several thousand devices to be placed on one chip, and the low total-power consumption which can be attained. As a result, large-scale MOS integrated circuits have become available and are finding use in electronic wristwatches, calculators, automotive electronics, and many other applications. A noteworthy application is the 16-kilobit memory fabricated on a single silicon chip.

This chapter introduces the reader to the basic physical principles underlying MOS devices. The aim is to provide simple and tractable results, with reasonable rigor. Junction-field-effect transistors (JFET's) are first discussed, followed by metal-oxide-semiconductor capacitors and transistors. The chapter closes with a discussion of composite insulator MOS devices and of charge-coupled devices (CCD's).

3.2 JUNCTION-FIELD-EFFECT TRANSISTORS

The junction-field-effect transistor (JFET) was first proposed by Shockley[1] and experimentally tested by Dacey and Ross.[2,3] A schematic diagram of the upper portion of a p-channel JFET is shown in Fig. 3.1. It consists of a source terminal at potential V_S and a drain terminal at potential V_D, so that $V_D - V_S$ is negative for a p-channel device. The two metal gates (only one

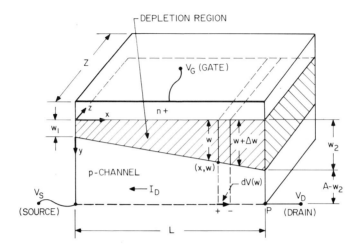

Fig. 3.1 Schematic diagram of a *p*-channel JFET. Note that only the top half of the symmetric structure is shown.

is shown), which contact a shallow layer of n^+ material are shorted together and connected positively with respect to V_S for a *p*-channel device, thus causing a wedge-shaped depletion region as shown. The conducting channel, of length L, is that part of the wedge-shaped *p*-type material outside the depletion region. The channel depth can have a maximum value of *2A* and the channel width is *Z*. Current in the *p*-channel is composed of carriers of only one type, namely, holes that move from source to drain. Hence the JFET is also termed a *unipolar device*. The cross-sectional area available for conducting these holes is modulated in depth by variations in the width of the depletion layer associated with the reverse biased *p-n* junctions. These variations are due to changes in gate-to-source voltage (V_{GS}) and drain-to-source voltage (V_{DS}). When V_{DS} is first increased (with $V_{GS} > 0$) the channel current I_D shown in Fig. 3.2 first increases, although its slope decreases due to the gradual constriction of the channel caused by the widening depletion regions. When voltage V_{DS} reaches a saturation value denoted $V_{DS\ sat}$, the two depletion regions touch at point *P* as shown in Fig. 3.1 and the channel is said to be "pinched off." At this point the current I_D assumes a constant "saturation value," $I_{D\ sat}$, as may be seen in Fig. 3.2. For values of V_{DS} greater than $V_{D\ sat}$, the channel moves beyond pinch-off. The point *P*, denoting coalescence of the depletion regions, moves toward the source, and the current I_D remains constant. In practice, the slope dI_D/dV_{DS} will be slightly positive due to the decreasing channel length *L*. Finally, when V_{DS} is very large, breakdown occurs and the current I_D increases rapidly. Figure 3.2 shows a family of I_D versus V_{DS} curves for various values of V_{GS}.

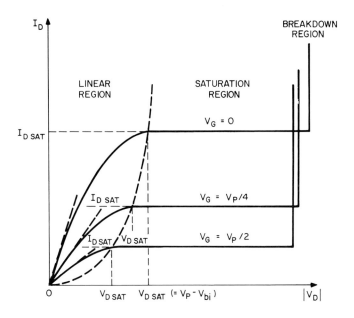

Fig. 3.2 Drain characteristics of a JFET.

To deduce the current-voltage characteristic[4-7] of a p-channel JFET of arbitrary charge distribution, ρ, reference will be made to its upper half portion, shown in Fig. 3.1. Variables will not be a function of z. The junction capacitance per unit area at point (x,w) in Fig. 3.1 is given, by virtue of Eq. (2.3), by

$$C = \frac{\epsilon_s}{w} = \frac{dQ(x,w)}{d\{V(x,0) - V(x,w)\}} = \frac{dQ(x,w)}{dV(w)} \tag{3.1}$$

where the area charge density $Q(x,w)$ is defined as

$$Q(x,w) = \int_0^w \rho(x,y)\,dy. \tag{3.2}$$

Note that $V(x,w)$ is the potential at point (x,w) and $V(x,0) = V_G$ is the gate potential. Hence the voltage across the reverse-biased depletion region of width w, denoted $V(w)$ (and considered positive in magnitude for a p-channel device), is

$$V(w) = V(x,0) - V(x,w) = V_G - V(x,w). \tag{3.3}$$

Now $dQ/dV(w) = [dQ/dw][dw/dV(w)]$, and from Eq. (3.2) $dQ/dw = \rho$ which, when substituted into Eq. (3.1), yields the important result

$$\frac{dV(w)}{dw} = \frac{w\rho(x,w)}{\epsilon_s}. \tag{3.4}$$

This equation implies that the voltage increment dV, needed to increase the depletion width w by dw, is proportional to w and the charge density $\rho(x,w)$ at the boundary.

Consider now the current I_D traversing the channel, as shown in Fig. 3.1. Between x and $x + \Delta x$ the current encounters the elemental resistance dR, where

$$dR = \frac{dx}{2Z\mu_p \int_w^A \rho(x,y)\,dy}. \tag{3.5}$$

The occurrence of the factor 2 is due to the presence of the bottom half-portion of the channel which is not shown in Fig. 3.1. The denominator integral is a function of the channel width. Equation (3.5) can be written as

$$I_D\,dx = -2Z\mu_p \left[\frac{dV(w)}{dw}\right] dw \int_w^A \rho(x,y)\,dy \tag{3.6}$$

because $dV(w) = -I_D\,dR$. Substitution of Eq. (3.4) into this equation, integrating from $x = 0$ to $x = L$, and noting that $w = w_1$ corresponds to $x = 0$ and $w = w_2$ corresponds to $x = L$, yields with the use of Eq. (3.2)

$$I_D = \frac{2Z\mu_p}{\epsilon_s L} \int_{w_1}^{w_2} [Q(x,w) - Q(x,A)]\,w\rho(x,w)\,dw. \tag{3.7}$$

This equation is the basic current equation of the JFET and can be adapted to specific charge distributions in the channel. The depletion width w_2 is related to the voltage $V_{bi} + V_G - V_D$, where $V_D = V(x,w_2)$ and V_{bi} is the built-in voltage defined in Eq. (2.2). Similarly, w_1 is related to the voltage $V_{bi} + V_G$. By virtue of Eq. (3.3), it now follows that $dV(w_2) = -dV_D|_{V_G}$. Hence the dynamic channel conductance, g_D, is given by

$$g_D = \frac{dI_D}{dV_D} = \frac{\partial I_D}{\partial w_2}\frac{dw_2}{dV_D} = -\frac{\partial I_D}{\partial w_2}\frac{dw_2}{dV(w_2)}.$$

Using this equation in conjunction with Eqs. (3.4) and (3.7) now yields

$$g_D = \frac{2Z\mu_p}{L}[Q(x,A) - Q(x,w_2)]. \tag{3.8}$$

Note that this conductance becomes zero when $w_2 = A$. At this point, the two depletion regions coalesce, as aforementioned, thereby pinching off the channel. The corresponding channel pinch-off voltage, V_p, is defined by

$$V_p = V_G - V(x,A) + V_{bi}. \tag{3.9}$$

The drain current I_D, reaches a maximum value, $I_{D\,sat}$, and the drain voltage V_D then has a value

$$V_{D\,sat} = V(x,A), \tag{3.10}$$

as shown on Fig. 3.2. The largest value of g_D, namely g_{D0}, occurs when $V_{DS} = 0$ and therefore $w_2 = w_1$.

$$g_{D0} = \frac{2Z\mu_p}{L} [Q(x,A) - Q(x,w_1)]. \tag{3.11}$$

Also of interest is the transconductance,

$$g_m = \frac{dI_D}{dV_G} = \frac{\partial I_D}{\partial w_2} \frac{dw_2}{dV(w_2)} + \frac{\partial I_D}{\partial w_1} \frac{dw_1}{dV(w_2)} , \tag{3.12}$$

because both w_2 and w_1 are functions of V_G. From Eqs. (3.4), (3.7), and (3.12) it now follows that

$$g_m = \frac{2Z\mu_p}{L} [Q(x,w_2) - Q(x,w_1)]. \tag{3.13}$$

Note that $w_2 = A$ when $V_D \geqslant V_{D\,sat}$, and so $g_m = g_{D0}$.

A special case of interest is uniformly doped p-type semiconductor material, for which $Q(x,w) = qN_A w$. In the case of an abrupt junction, Eq. (2.1) then also gives

$$w_2 = \left[\frac{2\epsilon_s(V_G - V_D + V_{bi})}{qN_A} \right]^{1/2}$$

and

$$w_1 = \left[\frac{2\epsilon_s(V_G + V_{bi})}{qN_A} \right]^{1/2}$$

Substituting these results into Eq. (3.7) yields, after integration and simplification,

$$I_D = - g_{max} \left\{ V_D - \frac{2}{3A} \left[\frac{2\epsilon_s}{qN_A} \right]^{1/2} \left[(V_G - V_D + V_{bi})^{3/2} - (V_G + V_{bi})^{3/2} \right] \right\} \tag{3.14a}$$

where $g_{max} = 2Z\mu_p qN_A A/L$. For a given value of V_G, the maximum current, $I_{D\,sat}$, occurs when $w_2 = A$. It may be obtained from Eq. (3.7) as

$$I_{D\,sat} = - I_p \left\{ 1 - 3 \left[\frac{V_G + V_{bi}}{V_p} \right] + 2 \left[\frac{V_G + V_{bi}}{V_p} \right]^{3/2} \right\}, \tag{3.14b}$$

where

$$I_p = \frac{2Z\mu_p q^2 N_A^2 A^3}{6\epsilon_s L} ,$$

$$V_p = \frac{qN_A A^2}{2\epsilon_s} .$$

Here V_p is the pinch-off voltage defined in Eq. (3.9) and I_p is defined as the pinch-off current.

EXAMPLE 3.1 Compute the pinch-off voltage V_p, the pinch-off current I_p and the value of g_{max}. The JFET has the following parameters: $Z/L = 4$, $A = 10^{-4}$ cm, $N_A = 3.5 \times 10^{15}$ cm^{-3} and $\mu_p = 5.2 \times 10^2$ cm^2/V–s.

Solution. From Eq. (3.14b),

$$V_p = \frac{qN_A A^2}{2\epsilon_s}$$

$$= \frac{(1.6 \times 10^{-19})(3.5 \times 10^{15})10^{-8}}{(2 \times 11.8)(8.85 \times 10^{-14})} \approx 2.7 \text{ V}.$$

Also,

$$I_p = \left[\frac{Z}{L}\right] \frac{\mu_p q^2 A^3}{3\epsilon_s} N_A^2 = \left[\frac{Z}{L}\right] \frac{4}{3} \frac{\mu_p \epsilon_s}{A} V_p^2$$

$$= \frac{4 \times 4 \times 5.2 \times 10^2 \times 11.8 \times 8.85 \times 10^{-14}}{3 \times 10^{-4}} \times 2.68^2 \approx 0.21 \times 10^{-3} \text{ A}.$$

From Eq. (3.14a),

$$g_{max} = \frac{Z}{L} 2\mu_p q N_A A$$

$$= 4 \times 2 \times 5.2 \times 10^2 \times 1.6 \times 10^{-19} \times 3.5 \times 10^{15} \times 10^{-4}$$

$$= 2.33 \times 10^{-4} \text{ mhos}.$$

EXAMPLE 3.2 Discuss the temperature dependence of the pinch-off voltage V_p and pinch-off current I_p on the assumption that the impurity concentration N_A does not vary with temperature over a sufficient region.

Solution. Pinch-off voltage V_p is independent of temperature, but pinch-off current I_p depends on temperature by virtue of its dependence on the mobility μ_p as may be seen from Eq. (3.14b). For low impurity concentrations the mobility decreases with temperature as $T^{-3/2}$. Actual measurements show that for pure materials near room temperature the mobility varies as $T^{-2.5}$ for n-type silicon and as $T^{-2.7}$ for p-type silicon. Actual FET's show an $I_p \propto T^{-2}$ dependence, perhaps because the concentration N_A is not completely temperature independent after all. Hence $T \, dI_p/I_p \, dT = -2$ and so for a 1% change in temperature the pinch-off current decreases by 2%.

Note that the dotted line shown in Fig. 3.2 is the locus of

$$V_{D \text{ sat}} = V_G + V_{bi} - V_p = V_G + V_T \ln\left(\frac{N_A}{n_i}\right) - \frac{qN_A A^2}{2\epsilon_s}$$

which divides the I-V characteristic into two regions, namely: (1) the "triode region" for which $|V_D| < |V_{D\ \text{sat}}|$ and in which Eq. (3.14a) is operative, and (2) the "saturation region" for which $|V_D| > |V_{D\ \text{sat}}|$. As the drain voltage increases, the depletion width increases and the point P, whose potential remains at $V_{DS\ \text{sat}}$, moves toward the source. This causes an effective shortening of the channel length L and, as a result, the saturation conductance $g_D(w{=}A)$ is not zero but has a small finite positive value in practical JFET's. Shown in Fig. 3.2 is the theoretical saturation curve which obeys Eq. (3.14b) and for which $g_D(w{=}A)$ is taken as zero. A third region, the breakdown region, is reached when the gate-to-channel diode suffers avalanche breakdown. This occurs at the drain end of the device, where the reverse voltage is highest and is defined by the locus

$$V_G - V_D = V_B. \tag{3.15}$$

An equivalent circuit of a p-channel JFET in the common-source configuration is shown in Fig. 3.3. The various parameters are obtained as follows:

a)   The transconductance g_m may be obtained from Eq. (3.13) for the case of uniform doping and abrupt junctions.

$$g_m = \frac{2Z\mu_p qN_A}{L}\ (w_2 - w_1)$$

$$= \frac{2Z\mu_p}{L}\ (2\epsilon_s qN_A)^{1/2}\left[\sqrt{V_G - V_D + V_{bi}} - \sqrt{V_G + V_{bi}}\right]. \tag{3.16}$$

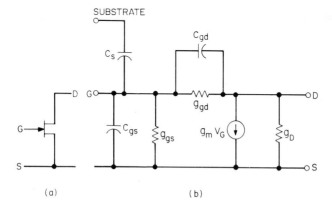

(a)   (b)

Fig. 3.3 Equivalent circuit of a p-channel JFET in common-source configuration.

b)   The capacitance C_{gs} for a symmetrical device equals approximately $ZL\epsilon_s/\bar{w}$, where the average depletion width $\bar{w}$ is $(w_1+w_2)/2$. For a symmetrical device, C_{gd} may be taken equal to C_{gs} and so one has approximately

$$C_{gs} = C_{gd} \approx \frac{2ZL\epsilon_s}{w_1+w_2}.$$ (3.17)

The unity-gain frequency of the device, f_T, is that frequency at which the magnitude of the input current into the capacitor, namely $|j\omega C_{gs} V_G|$, equals the magnitude of the current generated by the controlled current source, namely $|g_m V_G|$,

$$f_T = \frac{g_m}{2\pi C_{gs}} = \frac{qN_A\mu_p(w_2-w_1)^2}{2\pi L^2\epsilon_s}.$$ (3.18)

For high-frequency operation a short channel length L is desirable. Moreover, because $\mu_n > \mu_p$ an n-channel device will achieve a higher operating frequency.

c)   The dynamic conductance g_D for the case of uniform doping may be obtained from Eq. (3.8). The result is

$$g_D = \frac{2Z\mu_p qN_A}{L} (A-w_2)$$

$$= \frac{2Z\mu_p}{L} (2\epsilon_s qN_A)^{1/2}\left[\sqrt{V_p} - \sqrt{V_G-V_D+V_{bi}}\right].$$

d)   To compute the input conductance g_{gs} one may assume a reverse-biased gate-to-source diode, whose current is given by $I_G = I_s\lambda(-V_G/V_T)$, with I_s a saturation current. Hence

$$g_{gs} = \frac{dI_G}{dV_G} = \frac{I_s-I_G}{V_T}.$$ (3.19)

For $I_G \approx 0$ and I_s of the order of less than a nanoampere, the input resistance $1/g_{gs}$ will be of the order of tens or hundreds of megohms. Therefore, the JFET has a very high input resistance. Similar remarks apply to the conductance g_{gd}.

e)   The gate-to-substrate capacitance may be computed geometrically according to the rules given in Section 4.3.

EXAMPLE 3.3   Show that the maximum operating frequency f_T of a JFET can be approximately written as $f_T = \mu V_p/\pi L^2$. Then calculate f_T for an FET with the parameters $\mu_p = 5.2\times10^2$ cm^2/V–s, $V_p = 2.7$ V and $L = 10$ μm.

Solution. From Eq. (3.18)

$$f_T = \frac{qN_A\mu_p(w_2 - w_1)^2}{2\pi L^2 \epsilon_s} .$$

Putting $w_2 - w_1 \approx A$ and substituting $V_p = qN_A A^2/2\epsilon_s$ then yields

$$f_T = \frac{\mu_p V_p}{\pi L^2} = \frac{5.2 \times 10^2 \times 2.7}{3.14 \times 10^{-6}} \approx 4.44 \times 10^8.$$

In practice f_T may be increased by decreasing the channel length L or working with an n-channel device. Note also that $f_T \propto \mu_p \propto T^{-2}$.

A practical integrated-circuit version of a double diffused p-channel JFET is shown in Fig. 3.4a. The p-channel is created by a p-base diffusion into an n-epitaxial layer. The two gates, gate 1, consisting of an n^+ emitter diffusion, and gate 2, consisting of the epitaxial layer, are connected and accessed by means of the gate contact. The drain-gate breakdown voltage is now the same as the emitter-base breakdown voltage, that is, approximately 8 V (see Section 2.7). Note that the impurity distribution in the channel will not be uniform nor will the gate-to-channel junction be strictly abrupt. Nevertheless, for computational purposes it is often convenient to assume a uniformly doped channel and abrupt gate-to-channel junctions.

In most applications JFET's are used in conjunction with bipolar transistors and the processing sequences for these transistors are compatible. Without adding steps to the bipolar process, the JFET pinch-off voltage is predetermined by the design of the bipolar transistors and only the geometric ratio Z/L remains to be specified. This permits control over the maximum transconductance, g_{max}, and hence the maximum current. The addition of one ion-implant step to permit the depth A and impurity concentration N_A of the JFET channel to be specified independently of the bipolar transistors provides control over all JFET parameters. Because the unity-gain frequency f_T is inversely proportional to L, it is customary to select L equal to the minimum linewidth that can be used. This also ensures, for a fixed value of Z/L, that both Z and the device area will be minimized. It is also possible to utilize an n-channel closed geometry of which a combined cross-sectional and top view is shown in Fig. 3.4b. As may be seen, the source and drain consist of n^+ diffusions into n-epitaxial material. The top gate is produced by a p-diffusion, while the bottom gate is the substrate which usually is fixed at the most negative potential. Hence only the top gate can be varied in voltage and receive applied signals. Breakdown of this structure occurs when the base-collector junction avalanches, and V_B is therefore of the order of 60 V (see Section 2.14). For n-channel calculations, polarities of I_D and V_{DS} should be reversed. Furthermore, the mobility μ_n exceeds the mobility μ_p used previously. Otherwise calculations are entirely similar to those used for p-channel devices.

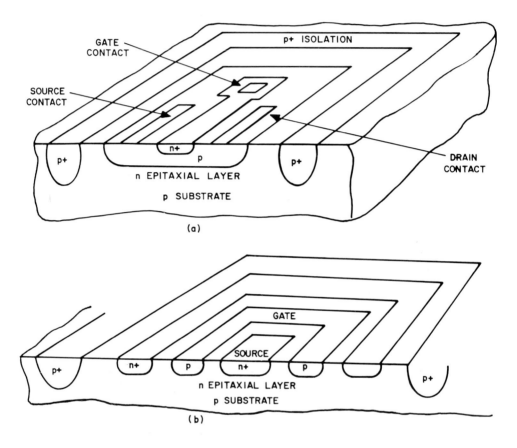

GATE CONTACT

SOURCE CONTACT

p+ ISOLATION

SOURCE CONTACT

n+

p

p+

n EPITAXIAL LAYER

p SUBSTRATE

DRAIN CONTACT

(a)

GATE

SOURCE

p+

n+

p

n+

p

p+

n EPITAXIAL LAYER

p SUBSTRATE

(b)

Fig. 3.4 (a) An integrated circuit, double-diffused p-channel JFET and (b) an n-channel, closed geometry JFET.

3.3 THE MOS CAPACITOR

A cross-sectional view of a simple MOS capacitor is shown at the top of Fig. 3.5. The device consists of a metal gate electrode at potential V_G, a thin insulating layer of silicon dioxide (SiO_2) of thickness x_0 and a substrate of

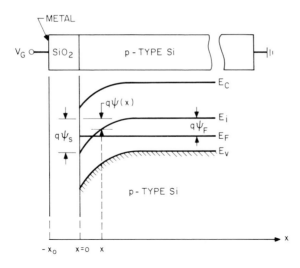

Fig. 3.5 Cross section of an MOS capacitor and its energy band diagram.

p-type silicon (Si), the farthest end of which is at ground potential. The behavior of this structure as a capacitor, as the gate potential V_G is varied, will now be described with the aid of energy-band diagrams. An analogous discussion applies for n-type semiconductor material. Figure 3.5 shows an energy-band diagram appropriate to a metal-insulator-p-type semiconductor device. Shown in this diagram are the conduction and valence bands, as well as the intrinsic Fermi level denoted by E_i and the Fermi level denoted by E_F. The structure shown is in thermal equilibrium because no dc current can flow across the insulator, and therefore the Fermi level is constant. Deep in the semiconductor, the potential $\psi(x=\infty) = 0$ and the carrier densities are given by their thermal equilibrium values, namely $p_{p0} \approx N_A$, $n_{p0} \approx n_i^2/N_A$. Hence,

$$\psi_F = V_T \ln\left(\frac{p_{p0}}{n_i}\right). \tag{3.20}$$

EXAMPLE 3.4   An MOS device has $N_A = 5.4 \times 10^{15}$. Calculate the background potential ψ_F.

Solution.   From Eq. (3.20) and using $n_i \approx 1.8 \times 10^{10}$ cm^{-3} as given by Sze[6][†]

$$\psi_F \approx V_T \ln\left(\frac{N_A}{n_i}\right) = 26 \times 10^{-3} \ln\left(\frac{5.4 \times 10^{15}}{1.8 \times 10^{10}}\right) \approx 0.33 \text{ V}.$$

[†] The exact value of n_i at 300 K differs with different authors. Grove uses 1.4×10^{10} cm^{-3}. Others give values as low as 10^{10} cm^{-3}.

Note that if N_A is assumed independent of temperature, ψ_F is a function of temperature because $n_i = CT^{3/2}\exp(-V_G/2V_T)$.

In Eq. (3.20) ψ_F is the potential difference between the intrinsic Fermi level value E_i/q at $x = \infty$ and the Fermi level E_F/q. As x decreases and the semiconductor surface is approached, the potential ψ, which is a measure of the bending of the intrinsic Fermi level, is no longer zero. At the semiconductor surface $\psi(x=0) = \psi_s$ has its largest value which is called the surface potential. For $\psi > 0$, the energy bands bend down in p-type material and for $\psi < 0$ they bend up. In the absence of any work function difference between metal and semiconductor, the applied gate voltage is given by

$$V_G = V_{ox} + \psi_s \tag{3.21}$$

where V_{ox} is the voltage across the insulator. The potential ψ may be obtained by solving a one-dimensional Poisson equation. Subsequently, the electric field $E = -d\psi/dx$ and the surface charge $Q_s = -\epsilon_s E_s$ can also be obtained. The field at the semiconductor surface, denoted E_s, has been shown to equal[4-7]

$$E_s = \mathrm{sgn}(\psi_s)\,\frac{2V_T}{L_D}\,F\{\psi_s, \psi_F\} \tag{3.22}$$

where $\mathrm{sgn}\,\psi_s = +1$ for $\psi_s > 0$ and $\mathrm{sgn}\,\psi_s = -1$ for $\psi_s < 0$, and L_D is called the extrinsic Debye length for holes

$$L_D = \left[\frac{2\epsilon_s \tilde{V}_T}{qp_{p0}}\right]^{1/2}. \tag{3.23}$$

EXAMPLE 3.5   Calculate the Debye length L_D for an MOS device for which $N_A = 3\times10^{15}$.

Solution.   From Eq. (3.23),

$$L_D = \left[\frac{2\epsilon_s V_T}{qN_A}\right]^{1/2} = \left[\frac{2\times11.8\times8.85\times10^{-14}\times26\times10^{-3}}{1.6\times10^{-19}\times3\times10^{15}}\right]^{1/2}$$

$$\approx 0.106\times10^{-4}\ \mathrm{cm}.$$

The function F is defined by

$$F\{\psi, \psi_F\} = \left\{\lambda(-\psi) + \frac{\psi}{V_T} + \frac{n_{p0}}{p_{p0}}\left[\lambda(\psi) - \frac{\psi}{V_T}\right]\right\}^{1/2} \tag{3.24}$$

where λ was defined in Eq. (2.8a). By Gauss' law, the space charge required to produce this field is then given by $Q_s = -\epsilon_s E_s$, where ϵ_s is the dielectric constant of silicon. Paying proper attention to sign

$$Q_s = -\operatorname{sgn}(\psi_s)\, \frac{2\epsilon_s V_T}{L_D}\, F\{\psi_s, \psi_F\}. \tag{3.25}$$

EXAMPLE 3.6   Calculate the surface charge, Q_s, of an MOS device with $N_A = 3\times10^{15}$ cm^{-3} at the onset of weak inversion, $\psi_s = \psi_F$.

Solution. From Eqs. (3.23) and (3.25), $|Q_s| = \sqrt{2\epsilon_s q N_A V_T}\,|F(\psi_F, \psi_F)|$. Function F is given by Eq. (3.24) with the second term dominating at the onset of weak inversion. Hence $|F(\psi_F, \psi_F)| = (\psi_F/V_T)^{1/2}$ and so

$$\begin{aligned}
|Q_s| &= (2\epsilon_s q N_A \psi_F)^{1/2}\\
&= (2\times11.8\times8.85\times10^{-14}\times1.6\times10^{-19}\times3\times10^{15}\times0.33)^{1/2}\\
&= 1.82\times10^{-8} \ \text{C/cm}^2.
\end{aligned}$$

Note that the value of ψ_F was taken from Example 3.4 and that because $\psi_s > 0$, Q_s is actually negative.

A calculation of the surface charge $|Q_s|$ is shown in Fig. 3.6. Three important characteristic regions can be discerned, namely: accumulation region, depletion and weak inversion region, and strong inversion region.

Accumulation region   This corresponds to negative gate voltages and, because of Eq. (3.21), also to negative values of surface potential. By virtue of Eq. (3.25) the surface charge Q_s is therefore positive and may be shown to have an exponential dependence; that is, $Q_s \propto \exp(|\psi_s|/2V_T)$. Physically the negative charge on the gate induces positive charges at the SiO$_2$-Si interface, thus forming an *accumulation layer* of holes at $x = 0$, as shown in Fig. 3.7(a). The energy bands, which bend upwards, the charge density $\rho \propto \exp(|\psi_s|/2V_T)$, the electric field and the potential are all shown in Fig. 3.7(a). In the accumulation mode the MOS capacitor behaves like an oxide capacitor, and

$$Q_s = C_{ox} V_{ox} = \frac{\epsilon_{ox}}{x_0}\, V_{ox}. \tag{3.26}$$

Depletion and weak inversion region   For moderate positive values of gate voltage, ψ_s will lie in the region $0 < \psi_s < 2\psi_F$. The accumulation layer decreases as V_G is made less negative than in the accumulation region. When $\psi_s = 0$ the so-called *flat-band condition* is reached and the energy bands neither bend upward nor downward. As ψ_s becomes positive, the bands bend downward [Fig. 3.7(b)] and negative charge is induced near the semiconductor surface due to holes being repelled, thus leaving behind uncompensated acceptor ions. Hence a depletion region of width x_d, is

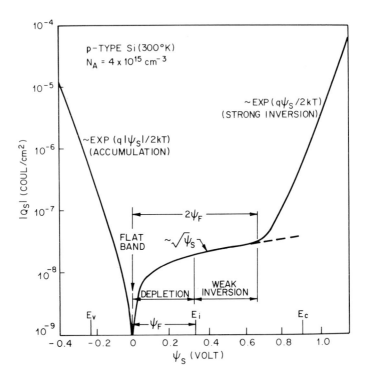

Fig. 3.6 Surface charge as a function of surface potential for an MOS capacitor. (From Sze[6]).

formed, where

$$x_d = \left[\frac{2\epsilon_s \psi_s}{qN_A} \right]^{1/2} \tag{3.27}$$

and may be computed from Eqs. (3.24) and (3.25). The charge density ρ is depicted in Fig. 3.7(b) and the space charge Q_s is now given by virtue of Eq. (3.25) as

$$Q_s = -qN_A x_d; \quad x_d \propto \sqrt{\psi_s} . \tag{3.28}$$

This region of Q_s is shown as the second branch of the curve of Fig. 3.6 for the range $0 < \psi_s < 2\psi_F$. The corresponding differential capacitance, C_D, of the semiconductor space-charge region is given by

$$C_D = \frac{\partial Q_s}{\partial \psi_s} \approx \left[\frac{q\epsilon_s N_A}{2\psi_s} \right]^{1/2}, \quad 0 \leqslant \psi_s < 2\psi_F. \tag{3.29}$$

EXAMPLE 3.7   Show that the differential capacitance of the semiconductor space-charge region is given by

$$C_D = \frac{\partial |Q_s|}{\partial \psi_s} = \frac{\epsilon_s}{L_D} \frac{[-\lambda(-\psi_s) + n_{p0}\lambda(\psi_s)/p_{p0}]}{F(\psi_s, \psi_F)}$$

where $F(\psi_s, \psi_F)$ is defined by Eq. (3.24). Show further that at flat-band condition $(\psi_s = 0)$, $C_D|_{\psi_s=0} = \sqrt{2}\epsilon_s/L_D$ F/cm^2. Finally compute $C_D|_{\psi_s=0}$ for an MOS device having $N_A = 3 \times 10^{15}$ cm^{-3}.

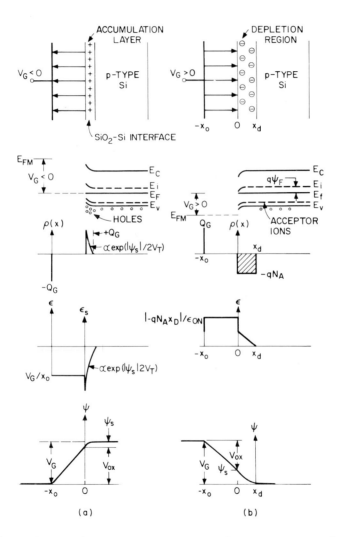

Fig. 3.7 Energy bands, charge distribution, electric field, and potential for an MOS capacitor (a) in accumulation region and (b) in depletion region.

Solution. From Eq. (3.25), $|Q_s| = (2\epsilon_s V_T/L_D) F(\psi_s, \psi_F)$. Hence,

$$C_D = \frac{\partial |Q_s|}{\partial \psi_s} = \frac{\epsilon_s}{L_D} \frac{\left[-\exp\left[-\dfrac{\psi_s}{V_T}\right] + 1 + \dfrac{n_{p0}}{p_{p0}} \left[\exp\left[\dfrac{\psi_s}{V_T}\right] - 1\right]\right]}{F(\psi_s, \psi_F)}$$

which is identical to the above equation. Writing $C_D = (\epsilon_s/L_D) N/D$ where N stands for numerator and D for denominator, one has on expanding N and D as power series

$$N(\psi_s \to 0) = \frac{\psi_s}{V_T}\left[1 + \frac{n_{p0}}{p_{p0}}\right]$$

$$D(\psi_s \to 0) = \frac{\psi_s}{\sqrt{2}\,V_T}\left[1 + \frac{n_{p0}}{p_{p0}}\right]^{1/2}$$

$$\left.\frac{N}{D}\right|_{\psi_s \to 0} = \sqrt{2}\left[\frac{1+n_{p0}}{p_{p0}}\right]^{1/2} \approx \sqrt{2} \quad \text{for} \quad \frac{n_{p0}}{p_{p0}} \ll 1.$$

Hence $C_D|_{\psi_s \to 0} = \sqrt{2}\epsilon_s/L_D$. For $N_A = 3 \times 10^{15}$ cm^{-3}, $L_D = 0.106 \times 10^{-4}$ as shown in Example 3.5. Hence,

$$C_D|_{\psi_s=0} = \frac{1.414 \times 11.8 \times 8.85 \times 10^{-14}}{1.06 \times 10^{-5}} \approx 139.3 \times 10^{-9} \text{ F/cm}^2.$$

At flat-band $\psi_s = 0$, and as shown in the above example $C_D(\psi_s=0) = \sqrt{2}\epsilon_s/L_D$. The electric field and the potential corresponding to the region $\psi_s \geq 0$ are also shown in Fig. 3.7(b). Furthermore, the total capacitance, C, of the MOS capacitor is now the series combination of C_{ox} and C_D. That is,

$$C^{-1} = C_{ox}^{-1} + C_D^{-1}. \tag{3.30}$$

Inversion region   As ψ_s begins to exceed ψ_F and enter the range $\psi_F \leqslant \psi_s \leqslant 2\psi_F$, the intrinsic Fermi level E_i crosses the Fermi level E_F and a very narrow n-type inversion layer (Fig. 3.8) begins to form. At this point the depletion width x_d has almost reached its maximum value $x_{d\ max}$ and any additional negative charge goes into the inversion layer charge Q_n, where

$$Q_s = Q_n - qN_A x_d = Q_n + Q_{sc}. \tag{3.31}$$

Here $Q_{sc} = -qN_A x_d$ may be termed the charge density in the depletion layer. The inversion layer may be thought to extend from the semiconductor surface, at which $x = 0$ and $\psi = \psi_s$, to a point $x = x_i$ in the material at which $\psi = \psi_F$. The total negative charge, Q_n, in the inversion layer is now given by

$$Q_n = -q \int_0^{x_i} n(x)\,dx = q \int_{\psi_F}^{\psi_s} \frac{n(\psi)\,d\psi}{(d\psi/dx)}. \tag{3.32}$$

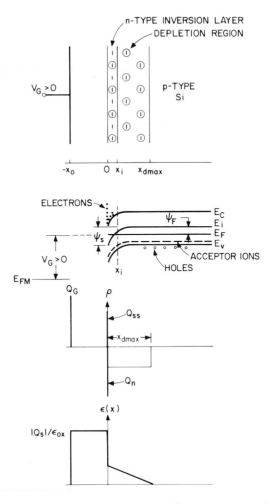

Fig. 3.8 Energy bands, charge distribution, electric field, and potential for an MOS capacitor under inversion.

The above equation can be integrated by computer, noting that

$$n(\psi) = n_i \exp\left[\frac{\psi - \psi_F}{V_T}\right]$$

$$\frac{d\psi}{dx} = -\,\mathrm{sgn}(\psi)\,\frac{2 V_T}{L_D}\,F(\psi, \psi_F). \tag{3.33}$$

The charge density ρ, the electric field E, and the potential ψ are all shown in Fig. 3.8. As Q_n increases, Q_s increases [Eq. (3.31)], and so by virtue of

Eq. (3.29) capacitance C_D increases. Hence the total capacitance C of Eq. (3.30) decreases from its maximum value C_{ox} as is shown in Fig. 3.9.

EXAMPLE 3.8   Compute the value of C_{ox} for an MOS device, having a 1000 Å oxide layer, with a dielectric constant of $3.2\epsilon_0$.

Solution.

$$C_{ox} = \frac{\epsilon_{ox}}{x_0} = \frac{3.2 \times 8.85 \times 10^{-14}}{10^{-5}} = 28.3 \times 10^{-9} \text{ F/cm}^2.$$

Strong inversion   Strong inversion may be considered to start when $\psi_s = 2\psi_F$ and be applicable in the range $\psi_s \geqslant 2\psi_F$. The turn-on or threshold voltage, V_{TH}, for strong inversion is given by Eqs. (3.20) and (3.21) as

$$V_G = V_{TH} = -\frac{Q_s}{C_{ox}} + 2\psi_F = -\frac{Q_s}{C_{ox}} + 2V_T \ln\left[\frac{N_A}{n_i}\right]. \tag{3.34}$$

The corresponding total MOS capacitance C'_{min} is then

$$\frac{1}{C'_{min}} = \frac{x_0}{\epsilon_{ox}} + \frac{x_{d\ max}}{\epsilon_s} \tag{3.35}$$

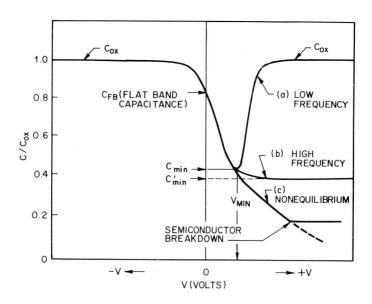

Fig. 3.9   MOS capacitance versus voltage curves: (a) low frequency, (b) high frequency, and (c) nonequilibrium case.

as is indicated on Fig. 3.9. Here $x_{d\ \text{max}}$ is given by

$$x_{d\ \text{max}} \approx \left[\frac{4\epsilon_s \psi_F}{qN_A}\right]^{1/2} = 2\left[\frac{\epsilon_s V_T}{qN_A} \ln\left(\frac{N_A}{n_i}\right)\right]^{1/2}. \qquad (3.36)$$

EXAMPLE 3.9   Compute $x_{d\ \text{max}}$ for an MOS device having $N_A = 5.4\times10^{15}$ cm^{-3}.

Solution.   From Eq. (3.36), $x_{d\ \text{max}} \approx (4\epsilon_s \psi_F/qN_A)^{1/2}$.   But $\psi_F = 0.33$ as shown in Example 3.4 and so

$$x_{d\ \text{max}} \approx \left[\frac{4\times11.8\times8.85\times10^{-14}\times0.33}{1.6\times10^{-19}\times5.4\times10^{15}}\right]^{1/2} \approx 0.4\times10^{-4}\ \text{cm}.$$

Note that $x_{d\ \text{max}}$ would differ for germanium or gallium arsenide, even if N_A is the same because ψ_F is a function of the intrinsic concentration n_i, which differs for different semiconductors.

For $\psi_s > 2\psi_F$ and quasi-static conditions the semiconductor may be considered shielded from further penetration of electric field by the inversion layer containing total charge Q_n. This charge rapidly increases with increasing ψ_s, while the depletion-layer width remains fixed at a value $x_{d\ \text{max}}$. The relationship between $x_{d\ \text{max}}$ and background impurity concentration N_B at 300 K is shown in Fig. 3.10.

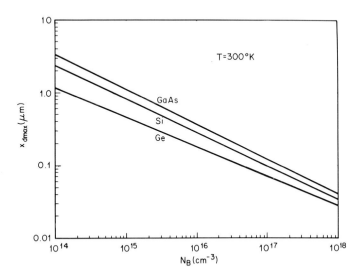

Fig. 3.10 Maximum depletion layer width versus impurity concentration for an MOS capacitor.

As the voltage V_G increases, the capacitance C goes through a minimum C_{min} at $V_G = V_{min}$. The subsequent increase in C, shown in Fig. 3.9, depends on the ability of the electron concentration to follow the applied signal V_G. For the metal-SiO$_2$-Si system it is found experimentally, that this increase is only observed in the frequency range from 5 to 100 Hz. When V_G varies at a higher frequency, MOS capacitor curves do not show this increase and C remains at C'_{min}, where $C'_{min} < C_{min}$ is given by Eq. (3.35). MOS capacitors then behave dynamically as described in Section 3.7.

Under strong inversion the term $\lambda(\psi)$ of Eq. (3.24) is dominant and so

$$F(\psi, \psi_F) \approx \left[\frac{n_{p0}}{p_{p0}}\right]^{1/2} \exp\left(\frac{\psi}{2V_T}\right).$$

From Eqs. (3.22) and (3.23) and the above approximation, one then obtains for $\psi > 0$

$$-\frac{\partial \psi}{\partial x} = \left[\frac{2V_T q n_{p0}}{\epsilon_s}\right]^{1/2} \exp\left(\frac{\psi}{2V_T}\right).$$

Solution of this equation with the boundary condition $x = 0$ when $\psi = \psi_s$ yields

$$x \approx \left[\frac{2V_T \epsilon_s}{q n_{p0}}\right]^{1/2} \left\{\exp\left(-\frac{\psi}{2V_T}\right) - \exp\left(-\frac{\psi_s}{2V_T}\right)\right\}, \qquad (3.37a)$$

from which the value of x_i can be computed when $\psi \approx 2\psi_F$. It may be shown that an upper limit for x_i, the thickness of the inversion layer, is given by

$$x_i \leqslant \left[\frac{2\epsilon_s V_T}{q N_A}\right]^{1/2}. \qquad (3.37b)$$

EXAMPLE 3.10   Compute the upper limit of inversion-layer thickness for an MOS device with $N_A = 5.4 \times 10^{15}$ cm^{-3}.

Solution.   From Eq. (3.37b)

$$x_i \leqslant \left[\frac{2\epsilon_s V_T}{q N_A}\right]^{1/2} = \left[\frac{2 \times 11.8 \times 8.85 \times 10^{-14} \times 26 \times 10^{-3}}{1.6 \times 10^{-19} \times 5.4 \times 10^{15}}\right]^{1/2} = 793 \text{ Å}.$$

Recent computations[8] suggest that this bound is inadequate. The thickness of the inversion layer is probably in the range 20-40 Å.

3.3.1 Effect of Surface Charge

In practice MOS capacitors also have present a fixed positive surface charge Q_{ss}, located in the silicon dioxide, very near the SiO_2-Si interface, as shown in Fig. 3.8. The amount of this charge depends on the silicon crystal orientation, being highest in value for the (111) and least in value for the (100) orientation, and on the oxidation and annealing conditions. As a consequence of charge neutrality at the onset of strong inversion, Eq. (3.31) must now be modified to

$$Q_G = - (Q_{ss}+Q_s); \quad \psi_s = 2\psi_F. \qquad (3.38)$$

Note that Q_G, $Q_{ss} > 0$ and $Q_s < 0$ for $\psi_s > 0$ in p-type material. Further, the threshold or turn-on voltage V_{TH} of Eq. (3.34) now becomes, by virtue of Eq. (3.37)

$$V_{TH} = V_{ox} + 2\psi_F = \frac{Q_G}{C_{ox}} + 2\psi_F = - \frac{Q_{ss}+Q_s}{C_{ox}} + 2\psi_F. \qquad (3.39)$$

Because Q_{ss} is positive while Q_s is negative for $\psi_s > 0$, V_{TH} has been reduced in value somewhat, by the presence of surface charge Q_{ss}. The effect of charge layers on the performance of composite-insulator devices is discussed in Section 3.7.1.

3.3.2 Effect of Work Function Difference

A further modification must be undertaken due to the presence of a nonzero metal-semiconductor work function, denoted ψ_{MS}. Figure 3.11 shows the

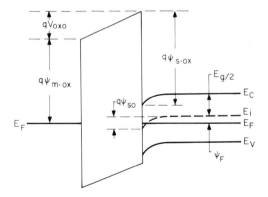

Fig. 3.11 Energy-band diagram for an MOS structure with the gate voltage equal to zero.

energy-band diagram for an MOS structure when the gate voltage, V_G, equals zero. As may be seen, the Fermi level is constant throughout the structure. The following potentials will be referred to:

a) ψ_{m-ox}, metal-silicon-dioxide barrier potential. This potential depends on the choice of gate metal. Table 3.1 presents average values for ψ_{m-ox}.[9]

b) ψ_{s-ox}, silicon-silicon-dioxide barrier potential. This potential has been experimentally found[7] to be approximately 4.35 V, almost independent of crystal orientation and conductivity type.

c) V_{ox0}, voltage across the oxide layer with zero gate voltage ($V_G = 0$).

d) ψ_{s0}, surface potential with zero gate voltage ($V_G = 0$).

e) E_g/q, band-gap voltage which is 1.12 V for silicon at 300 K.

Summing the above-mentioned potentials on either side of the oxide barrier gives

$$\psi_{m-ox} + V_{ox0} = \psi_{s-ox} + \frac{E_g}{2q} - \psi_{s0} + \psi_F. \tag{3.40}$$

The metal-semiconductor work function, ψ_{MS}, may now be identified as

$$\psi_{MS} = \psi_{m-ox} - \psi_{s-ox} - \frac{E_g}{2q} - \psi_F. \tag{3.41}$$

Table 3.1 Work Function for a
Clean Metal Surface in Vacuum

Metal	Ψ_{m-ox} (V)
Mg	3.70
Al	4.30
Ni	4.50
Cu	4.40
Ag	4.30
Pt	5.30
Au	4.80

EXAMPLE 3.11   Compute the metal-semiconductor work function ψ_{MS} for: (a) gold and p-type silicon MOS device; and (b) aluminum and n-type silicon MOS device. Take the impurity concentration $N = 5.4 \times 10^{14}$ cm^{-3} in both cases.

Solution.   From Table 3.1, $\psi_{m-ox} = 4.3$ V for aluminum and $\psi_{m-ox} = 4.8$ V for gold. From remarks in Section 3.2, ψ_{s-ox} will be taken as 4.35 V. The band-gap voltage for silicon is 1.12 V. From Example 3.4, $\psi_F = 0.33$ V for a p-type device and $\psi_F = -0.33$ V for an n-type device of the given impurity concentration. Using Eq. (3.41) now gives:

a)   Gold and p-type silicon

$$\psi_{MS} = 4.8 - 4.35 - 0.56 - 0.33 = -0.44 \ V.$$

b)   Aluminum and n-type silicon

$$\psi_{MS} = 4.3 - 4.35 - 0.56 + 0.33 = -0.28 \ V.$$

The potential ψ_{MS} clearly depends on the metal used for the gate, because of its dependence on ψ_{m-ox}. However, ψ_{MS} also depends on $\psi_F = V_T \ln(N_A/n_i)$. As a result, ψ_{MS} is strongly dependent on doping concentration in both n-type and p-type silicon for aluminum gate electrodes as shown in Fig. 3.12. As may be seen, ψ_{MS} is negative for both n- and p-type silicon, but has a larger magnitude for p-type silicon ($0.6 < |\psi_{MS}| < 1.1$).

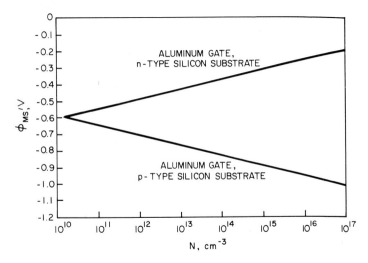

Fig. 3.12 Metal-semiconductor work function for aluminum on n- and p-type silicon.

The basic voltage relationship, Eq. (3.21), must now be modified by reference to the origin $V_G = 0$ as follows

$$V_G = (V_{ox} - V_{ox0}) + (\psi_s - \psi_{s0}).$$

But from Eqs. (3.40) and (3.41), $V_{ox0} = -(\psi_{MS} + \psi_{s0})$ and so V_G may be written as

$$V_G = V_{ox} + \psi_s + \psi_{MS}. \tag{3.42}$$

The modified threshold voltage, V_{TH}, is therefore[10]

$$V_{TH} = -\frac{(Q_s + Q_{ss})}{C_{ox}} + 2\psi_F + \psi_{MS}. \tag{3.43}$$

EXAMPLE 3.12 Compute the threshold voltage V_{TH} for (a) a gold to p-type MOS device and (b) an aluminum to n-type MOS device. Both devices have $N = 5.4 \times 10^{-15}$ cm^{-3}, $C_{ox} = 30 \times 10^{-9}$ F/cm^2. Take $|Q_s| = 4 \times 10^{-8}$ C/cm^2 and $|Q_{ss}| = 5 \times 10^{-8}$ C/cm^2.

Solution. From Eq. (3.43), $V_{TH} = -(Q_s + Q_{ss})/C_{ox} + 2\psi_F + \psi_{MS}$. For gold to p-type silicon, one has from Example 3.11, $\psi_{MS} \approx -0.44$ V and so

$$V_{TH} = -\frac{-4 \times 10^{-8} + 5 \times 10^{-8}}{30 \times 10^{-9}} + 0.66 - 0.44 = -0.11 \text{ V.}$$

In the case $Q_{ss} = 0$, $V_{TH} = 1.55$ V.

For aluminum to n-type silicon, one has from Example 3.11, $\psi_{MS} = -0.28$. Hence

$$V_{TH} = -\frac{4 \times 10^{-8} + 5 \times 10^{-8}}{30 \times 10^{-9}} - 0.66 - 0.28 = -3.94 \text{ V.}$$

In case $Q_{ss} = 0$, $V_{TH} = -2.27$ V. Note particularly the convention as to proper signs and the fact that threshold voltages are somewhat approximate because of the inclusion of Q_{ss} and ψ_{MS} which are experimentally determined quantities.

Care must be taken to give correct signs to all quantities when computing V_{TH}. The correct signs for the relevant quantities for depletion-mode MOS capacitors are given in Table 3.2.

Table 3.2 Signs for Relevant Quantities of a Depleted MOS Capacitor

n-Channel p-Type Bulk		p-Channel n-Type Bulk
$Q_{ss} > 0$		$Q_{ss} > 0$
$Q_s < 0$		$Q_v > 0$
$\psi_F > 0$		$\psi_F < 0$
$\psi_{MS} < 0$		$\psi_{MS} < 0$
$|Q_s, \, n\text{-type}|$	$=$	$|Q_s, \, p\text{-type}|$
$|\psi_F, \, n\text{-type}|$	$=$	$|\psi_F, \, p\text{-type}|$

3.4 THE MOS TRANSISTOR [11-16,36]

When n^+ source and drain regions are diffused adjacent to the MOS capacitor, as shown in Fig. 3.13, an MOS or insulated-gate field-effect transistor (IGFET) is obtained. Current I_D flows from source to drain by virtue of a gate voltage $|V_G| > |V_{TH}|$, which sets up an n-type inversion layer. In the presence of a small applied drain-to-source voltage, $V_{DS} = V_D - V_S$, this layer or channel acts like a resistor and the current I_D is proportional to the applied voltage V_{DS} as shown in Fig. 3.14. As V_{DS} increases, the channel

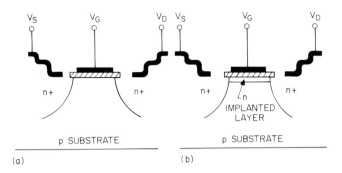

Fig. 3.13 Cross section of an MOS transistor: (a) enhancement mode and (b) depletion mode.

depth x_i at $y = L$ decreases until, at last, $x_i = 0$ when $V_{DS} = V_{DS\,sat}$. When this happens the channel is said to be "pinched off" and the current I_D has reached a saturation value $I_{D\,sat}$. This can be seen also in Fig. 3.14. However, the curves shown have a slight positive slope and true pinch-off, though dramatic, is not supported by detailed numerical calculations.[15] Values of $|V_{DS}| > |V_{DS\,sat}|$ move the pinch-off point $x_i = 0$ towards the source, that is, to a point $y < L$. However, though the channel length is now shortened, the voltage at the pinch-off point remains substantially at $V_{DS\,sat}$ and so $I_{DS\,sat}$ remains essentially the same. In actual short-channel devices, $L < 10\ \mu$m, the saturation current $I_{D\,sat}$ increases somewhat with V_{DS}, for reasons to be discussed later in this section.

The current-voltage characteristics of an IGFET or MOSFET will now be derived, under the following assumptions:

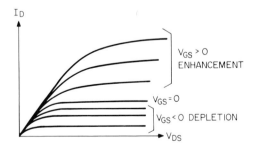

Fig. 3.14 Drain characteristics of MOSFET transistor.

a)   No surface charge Q_{ss} is present.

b)   The carrier mobility μ_n in the inversion layer remains constant.

c)   Reverse leakage current across the depletion layer can be neglected.

d)   Only drift current components contribute to the source-to-drain current I_D.

e)   The electric field E_x in the x-direction is much larger than the electric field E_y in the y-direction. This is the so-called gradual channel approximation introduced by Shockley.[1] A more careful analysis has been given by Pao and Sah.[12]

The total charge per unit area, $Q_s(y)$, induced in the semiconductor at distance y from the source, is given by Eq. (3.34) as $Q_s(y) = -[V_G - \psi_s(y)]C_{ox}$, where $\psi_s(y)$ is the surface potential at the point y. But the charge per unit area in the inversion layer, $Q_n(y)$ is given by Eq. (3.31) as

$$Q_n(y) = -[V_G - \psi_s(y)]C_{ox} + qN_A x_{d\ max}.$$

Assuming that $\psi_s(y) = 2\psi_F + V(y) - V_{ss}$ where V_{ss} is the potential of the substrate and $V(y)$ is the potential at point y in the channel, $(V_S \leqslant V(y) \leqslant V_D)$, gives

$$Q_n(y) = -[V_G - V(y) - 2\psi_F + V_{ss}]C_{ox} + \left[2\epsilon_s qN_A(V(y) + 2\psi_F - V_{ss})\right]^{1/2}.$$

$$(3.44)$$

The differential channel resistance at point y is given by $dR = dy/Z\mu_n|Q_n(y)|$ and so the differential voltage $dV(y)$ is given by

$$dV(y) = I_D dR = \frac{I_D\,dy}{Z\mu_n|Q_n(y)|} \qquad (3.45)$$

where I_D is the drain current, considered to be independent of y. Substitution of Eq. (3.44) into Eq. (3.45) and integration from $y = 0$ to $y = L$ (corresponding to $V(y) = V_S$ and $V(y) = V_D$, respectively) now yields the basic current-voltage relationship of the IGFET, namely,

$$I_D = \frac{\beta}{2}\left\{(V_G - V_S - 2\psi_F)^2 - (V_G - V_D - 2\psi_F)^2\right.$$

$$\left. + \frac{4}{3}K\left[(V_S + 2\psi_F - V_{ss})^{3/2} - (V_D + 2\psi_F - V_{ss})^{3/2}\right]\right\},$$

$$(3.46)$$

where

$$\beta = \frac{Z}{L}\,\mu_n C_{ox}; \quad K = \frac{(2\epsilon_s q N_A)^{1/2}}{C_{ox}}. \tag{3.47}$$

All potentials in this equation are referenced to ground and can take on any value. More particularly, the source potential V_S and the substrate potential V_{ss} need not be equal. Moreover, the above equation is symmetric in terms of V_D and V_S. In particular, $I_D > 0$ when $V_D > V_S$, $I_D = 0$ when $V_D = V_S$ and $I_D < 0$ when $V_D < V_S$. In other words, source and drain regions are interchangeable and designations are a matter of choice. Constant K, by its dependence on the doping concentration, N_A, reflects bulk properties of the device and so the two terms preceded by the K factor are often called *bulk charge terms*.

For small values of drain-to-source voltage

$$I_D \approx \beta\left[(V_D-V_S)(V_G-V_{TH}) - \frac{(V_D^2 - V_S^2)}{2}\right] \tag{3.48}$$

where V_{TH} is given by

$$V_{TH} = 2\psi_F + K\,(|V_S-V_{ss}| + 2\psi_F)^{1/2}. \tag{3.49}$$

EXAMPLE 3.13 Compute β and K for an MOS transistor having parameters $C_{ox} = 30\times10^{-9}$ F/cm^2, $N_A = 5\times10^{15}$ cm^{-3}, $Z/L = 5$ and $\mu_n = 1.4\times10^3$ cm^2/V-s.

Solution. From Eq. (3.47) with $\epsilon_s = 11.8\epsilon_0$,

$$\beta = Z\mu_n C_{ox}/L = 2.1\times10^{-4} \text{ C/cm-s}$$

$$K = \frac{(2\epsilon_s q N_A)^{1/2}}{C_{ox}} = 1.36$$

EXAMPLE 3.14 Compute the turn-on voltage V_{TH} for an MOS transistor using p-type bulk with $N_A = 5\times10^{15}$ cm^{-3} and $V_S = V_{ss}$, and compare it with values obtained from Eq. (3.39) with $Q_{ss} = 0$.

Solution. From Example 3.13 , $K = 1.36$ and from Eq. (3.20), $2\psi_F = 0.64$ V. Hence from Eq. (3.49),

$$V_{TH} = 2\psi_F + K\sqrt{2\psi_F} = 0.64 + 1.36\times0.8 = 1.73 \text{ V}.$$

The exact value obtained from Eq. (3.39) is $V_{TH} \approx 1.55$ as shown in Example 3.12. The value obtained from Eq. (3.49) is an approximation valid for small values of V_{DS}.

Since $-Q_s/C_{ox} = K(V_S-V_{ss}+2\psi_F)^{1/2}$, Eq. (3.49) is compatible with Eq. (3.43). For very small values of (V_D-V_S) the second term of Eq. (3.48) may be neglected and so in this range the IGFET is ohmic and behaves like an ideal linear resistor. Moreover, this equation clearly shows that the gate voltage V_G must exceed the threshold or turn-on voltage V_{TH}, for transistor operation. Actual MOS devices depart from this rule and drain leakage currents are observed in the weak inversion region, for gate voltages above the intrinsic voltage V_I, but below the threshold voltage V_{TH}.

$$V_I(\psi_s = \psi_F) \leqslant V_G \leqslant V_{TH}(\psi_s = 2\psi_F).$$

A detailed discussion of these important effects is beyond the scope of this chapter and the reader is referred to the literature.[17-21] Ignoring the subthreshold conduction effects, the conductance between drain and source, g_D, for very low voltages, V_{DS}, is therefore

$$g_D \approx \beta(V_G - V_{TH}); \quad V_{DS} \to 0 \tag{3.50a}$$

while the transconductance is given by

$$g_m = \frac{\partial I_D}{\partial V_G} = \beta(V_D - V_S). \tag{3.50b}$$

As V_{DS} increases, the slope $\partial I_D / \partial(V_D - V_S)$ decreases, until finally when $V_{DS} = V_{DS\,sat}$ the slope is zero. This may be shown to occur when

$$V_{D\,sat} = V_G - 2\psi_F + G(V_G, K) \tag{3.51a}$$

where

$$G(V_G, K) = \frac{K^2}{2} - K\left[V_G - V_{ss} + \frac{K^2}{2}\right]^{1/2}. \tag{3.51b}$$

The dependence of $G(V_G, K)$ on V_G and K is such that $G(V_G, K)$ is always nonpositive. For values of $K \ll 1$, $G(V_G, K) \approx K\sqrt{V_G - V_{ss}}$ and so

$$V_{D\,sat} = V_G - V'_{TH}; \quad V'_{TH} \approx 2\psi_F + 2K\sqrt{V_G - V_{ss}}. \tag{3.52}$$

The saturation current, $I_{D\,sat}$, may be obtained by substituting Eq. (3.51) into Eq. (3.46). This yields

$$I_{D\,sat} = \frac{\beta}{2}\Bigg\{(V_G - V_S - 2\psi_F)^2 - G^2(V_G, K)$$
$$- \frac{4K}{3}\left[[V_G - V_{ss} + G(V_G, K)]^{3/2} - (V_S + 2\psi_F - V_{ss})^{3/2}\right]\Bigg\}. \tag{3.53a}$$

This equation looks cumbersome, but it can be shown that for $K \ll 1$ and $V_G - V_S - 2\psi_F \gg 1$ one has approximately

$$I_{D\,sat} \approx \frac{\beta}{2}(V_G - V'_{TH})^2 = \frac{\beta}{2}V_{D\,sat}^2. \tag{3.53b}$$

This is the well-known[8] square-law relationship for drain current in the saturation region. The transconductance in this region is given by

$$g_m = \beta(V_G - V'_{TH}). \tag{3.54}$$

It is clear from the simplified models, given in Eqs. (3.48) and (3.53b) that parameters β and V_{TH} play an important role. The more detailed model, Eqs. (3.46) and (3.53a) make use of parameters β, K, ψ_F, and $G(V_G, K)$ all of which are amenable to calculation by digital computer.

The n-channel MOSFET just described is known as an enhancement MOSFET, because an increase in the positive gate-to-source voltage (let $V_{ss} = 0$) increases the conductivity of the inversion layer and so the drain current I_D is increased or enhanced. Consider now a slight modification of the structure of Fig. 3.13(a) and let a shallow n-layer be implanted just below the silicon surface between source and drain. As a consequence, the device will conduct current even for zero gate-to-source voltage. When the gate-to-source voltage is made positive, additional electrons are attracted into the channel, current I_D increases, and the device operates in the enhancement mode. However, when a negative gate-to-source voltage is applied, electrons are depleted from the channel, current I_D decreases, and the device is said to be operating in the depletion mode. The behavior of current I_D, for the two modes of operation is further illustrated in Fig. 3.14. The use of "driver" and "load" MOSFET's belonging to either of these two types is discussed in Chapter 14.

3.4.1 Degradation of Surface Mobility [22,23]

It is found experimentally that the measured surface mobility, $\bar{\mu}_n$, unlike the bulk mobility, μ_n, used in all previous equations, is degraded by the application of a gate voltage $V_G - V_{TH}$. The theoretical reason for this phenomenon is thought to be scattering of charge carriers near the surface. No completely satisfactory theoretical formula exists at present, but an empirical formula is

$$\bar{\mu}'_n = A\mu_n; \quad A = \frac{\ln(1+x)}{x}; \quad x = \frac{V_G - V_{TH}}{V_o} \tag{3.55}$$

where V_o is an empirical constant equal to about 30 volts. Both the current I_D and $I_{D\,sat}$ should therefore be modified to $I'_D = AI_D$ and $I'_{D\,sat} = AI_{D\,sat}$.

EXAMPLE 3.15 Compute the transconductance of an n-channel (p-bulk) MOSFET having $K = 0.1$ and $V_G = 20$. Use the correction Eq. (3.55) for surface mobility. Take $Z/L = 5$, $C_{ox} = 30 \times 10^{-9}$ F/cm^2.

Solution. From Eq. (3.52)

$$V'_{TH} \approx 2\psi_F + 2K\sqrt{V_G} \approx 0.6 + 0.2\sqrt{20} = 1.49 \text{ V}.$$

From Eq. (3.55)

$$x = \frac{V_G - V_{TH}'}{V_o} = \frac{20 - 1.5}{30} \approx 0.62,$$

and so

$$A = \frac{\ln 1.62}{0.62} = 0.78;$$

hence

$$\bar{\mu}_n = A\mu_n = 0.78 \times 1.7 \times 10^3 = 1.33 \times 10^3 \ \text{cm}^2/\text{V-s}.$$

Hence, $\beta = 5 \times 1.33 \times 10^3 \times 30 \times 10^{-9} = 1.99 \times 10^{-4}$ A/V^2 and so from Eq. (3.54)

$$g_m = \beta(V_g - V_{TH}) = 1.99 \times 10^{-4} \times (20 - 1.49) = 3.68 \times 10^{-3} \ \text{mhos}.$$

3.4.2 Effect of Short Channel Length

Modern IGFET's, particularly those used for memory applications where speed and packing density are important, are now fabricated with ever-smaller channel length L. An effect not explained by the "long-channel theory" just outlined is that the threshold voltage V_{TH} is somewhat lowered, while the saturation current $I_{D \ \text{sat}}$ increases somewhat with V_{DS}.[24, 25] It may be shown that these effects can be explained by modifying parameter K of Eq. (3.47) to

$$K' = \frac{K}{1 + 2\xi x_d'/L} = \frac{\sqrt{2\epsilon_s q}}{C_{ox}} \left[\frac{N_A}{(1 + 2\xi x_d'/L)^2} \right]^{1/2}. \tag{3.56}$$

Here ξ is an experimentally determined empirical factor, and x_d' is the depletion width of the drain and source regions, which must be obtained by a two-dimensional analysis,[26]

$$x_d' = \left\{ \left[\frac{2\epsilon_s(|V_S - V_{ss}| + V_{bi})}{qN_A r_0^2} \right]^{0.4} - 0.43 \right\} r_0 \tag{3.57}$$

where r_0 is the depth of the metallurgical junction in the source or drain region. Equation (3.56) shows that the proposed modification is equivalent to replacing the actual impurity density N_A by the effective density, N_e, where

$$N_e = \frac{N_A}{(1 + 2\xi x_d'/L)^2}. \tag{3.58}$$

The new, reduced, threshold voltage is therefore

$$V_{TH}'' = 2\psi_F + \frac{\sqrt{2\epsilon_s q N_A(V_S - V_{ss} + 2\psi_F)}}{C_{ox}(1 + 2\xi x_d'/L)}. \tag{3.59}$$

The basic current equation (3.46) is now modified by replacing K by K'.

The complete expression for I_D in the triode region, due to degraded mobility and short-channel effects, is, using Eqs. (3.46) and (3.55),

$$I_D' = A I_D \big|_{K=K'} \qquad \text{triode region.} \qquad (3.60)$$

In the saturation region the modified equation due to the above two effects is

$$I_D'{}_{\text{sat}} = A B I_{D\ \text{sat}} \qquad (3.61a)$$

where B is called the channel-length modulation term.

$$B = \frac{L}{L - \left[\dfrac{2\epsilon_s(V_{DS}-V_{DS\ \text{sat}})}{qN_A}\right]^{1/2}} . \qquad (3.61b)$$

The finite slope in the saturation region, g_D, for short-channel IGFET's is therefore

$$g_D = A B^2 \left[\frac{\epsilon_s}{2qN_A(V_{DS}-V_{DS\ \text{sat}})}\right]^{1/2} \frac{I_{D\ \text{sat}}}{L} . \qquad (3.2)$$

3.5 EQUIVALENT CIRCUITS [27,28]

A large-signal equivalent circuit MOSFET model suitable for computer-aided circuit design is shown in Fig. 3.15. The parameters for this model are as follows:

1. Voltage-controlled current source I_D has a value given in the triode region by Eqs. (3.46), (3.55), and (3.60), and in the saturation region by Eqs. (3.53), (3.55), and (3.61).

2. Bulk resistances R_s, R_d of the source and drain consist of a bulk term and also a contact term R_c, whose value is discussed in Chapter 4. These resistances are approximately

$$R_s = \frac{R_S L_s}{2Z_s} + R_c; \quad R_d = \frac{R_S L_d}{2Z_d} + R_c \qquad (3.63)$$

   where R_S is the sheet resistance of the diffused source and drain regions, L_s, L_d are the lengths, and Z_s, Z_d are the widths of these diffused areas.

3. Resistances R_{ss}, R_{ds} are the series resistances of the substrate-source and substrate-drain diodes. They are approximately given by

$$R_{ss} = \frac{\rho T_c}{(A_c L_s Z_s)^{1/2}}; \quad R_{ds} = \frac{\rho T_c}{(A_c L_d Z_d)^{1/2}} \qquad (3.64)$$

where T_c, A_c are the chip thickness and area respectively, and ρ is the substrate resistivity.

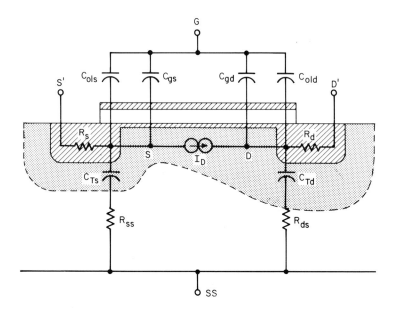

Fig. 3.15 Large signal equivalent circuit of MOSFET.

4. Capacitance C_{ols}, C_{old} are the gate-source, gate-drain overlap capacitances. Their values are

$$C_{ols} = C_{ox} Z l_s; \quad C_{old} = C_{ox} Z l_d \qquad (3.65)$$

where l_s, l_d are the source and drain overlaps respectively. These capacitances are almost zero in self-aligned gate structures.

5. Capacitances C_{gs}, C_{gd} are gate-source and gate-drain capacitances, respectively. Their values depend on the region of operation of the transistor and are approximately as follows:

a) Off region:

$$C_{gs} = C_{gd} = 0.$$

b) Triode region:

$$C_{gs} \approx \frac{C_{ox} ZL}{2} \left[1 + \frac{V_{DS}}{3 V_{DS\,sat}} \right]; \quad C_{gd} \approx \frac{C_{ox} ZL}{2} \left[1 - \frac{V_{DS}}{V_{DS\,sat}} \right].$$

$$(3.66)$$

c)   Saturation region:

$$C_{gs} \approx 2C_{ox}ZL/3; \quad C_{gd} \approx 0. \tag{3.67}$$

d)   Capacitances C_{Ts}, C_{Td} are substrate-source and substrate-drain depletion capacitances. Their values are

$$C_{Ts} \approx \frac{C_{Tso}}{\left[1 + \dfrac{|V_S - V_{ss}|}{V_{bi}}\right]^n} \; ; \quad C_{Td} = \frac{C_{Tdo}}{\left[1 + \dfrac{|V_D - V_{ss}|}{V_{bi}}\right]^n}$$

$$\tag{3.68}$$

where C_{Tso}, C_{Tdo} are constants and n lies in the range $2 \leqslant n \leqslant 3$.

The type of model here described gives reasonable accuracy, without requiring prohibitive computer time.

A small-signal equivalent circuit for an MOSFET, biased in the saturation region, is shown in Fig. 3.16. Parameters are given as follows:

1.   Transconductance g_m is given by Eq. (3.54).

2.   Drain-source conductance g_D is given by Eq. (3.62).

3.   Capacitance

$$C_{GS} = C_{ols} + C_{gs} \approx C_{ols} + 2C_{ox}ZL/3,$$

and

$$C_{GD} \approx C_{old} + C_{gd} \approx C_{old} .$$

4.   Capacitances C_{Ts} and C_{Td} are as previously given.

Of interest is the maximum frequency of operation when the MOSFET is

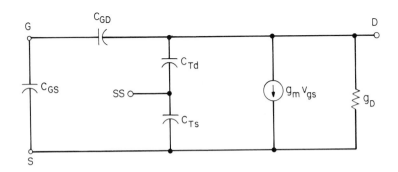

Fig. 3.16 Small signal equivalent circuit of an MOSFET in the saturation region.

biased in the saturation region. For this case the drain-to-source voltage is given, by virtue of Eq. (3.52), as $V_{D\,\text{sat}} - V_S = V_G - V_{TH}' - V_S$. Hence the transit time τ, that is the time taken for electrons to travel from source to drain, is $\tau = L/\mu_n |E_y|$ where the electric field $E_y = L(V_G - V_{TH}' - V_S)$. But the cutoff frequency, f_T, equals $1/2\pi\tau$ and so, from these results, Eq. (3.47), and the definition of g_m,

$$ f_T = \frac{\beta |V_G - V_{TH}' - V_S|}{2\pi L Z C_{ox}} = \frac{g_m}{2\pi L Z C_{ox}} . \tag{3.69}$$

Clearly, a high-frequency MOSFET structure should use a self-aligned gate to minimize parasitic capacitances. Furthermore, it should have a small channel length and use high-mobility substrate material.

Treatment of p-channel devices is entirely similar to that of n-channel devices, just presented. Nevertheless the following changes, chiefly with respect to sign, are necessary:

1. $V_{GS} < V_{TH}$ and $V_{DS} < 0$ for p-channel devices.

2. Replace N_A by N_D everywhere.

3. Take $\psi_F = -V_T \ln(N_D/n_i)$.

4. Take $V_{TH} = 2\psi_F - K(-2\psi_F - V_S + V_{ss})^{1/2}$.

5. Take $V_{D\,\text{sat}} = V_G - 2\psi_F - \dfrac{K^2}{2} + K\left[\dfrac{K^2}{2} - V_G + V_{ss}\right]^{1/2}$.

6. Take

$$ I_D = \frac{\beta}{2} \left\{ -(-V_G + V_S + 2\psi_F)^2 + (-V_G + V_D + 2\psi_F)^2 \right.$$

$$ \left. + \frac{4}{3} K[-(-V_S - 2\psi_F + V_{ss})^{3/2} + (-V_D - 2\psi_F + V_{ss})^{3/2}] \right\}.$$

3.6 COMPOSITE-DIELECTRIC FIELD-EFFECT TRANSISTORS [29]

This class of devices is mainly used in read-only memories (ROM's) and in read-mostly memories described in Chapter 15. A short discussion of MNOS devices and FAMOS devices now follows.

3.6.1 Charge Trapping in Metal-Nitride-Oxide-Silicon (MNOS) Devices

An MNOS device, consisting of layers of aluminum, silicon nitride, silicon dioxide, and silicon, as shown in Fig. 3.17, can be switched between two distinct, more or less permanent, threshold-voltage states by applying a voltage

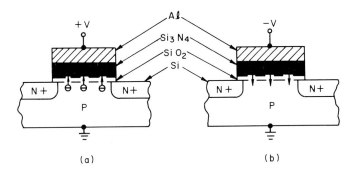

Fig. 3.17 Switching an MNOS memory device (a) to a high threshold state and (b) to a low threshold state.

pulse to the gate. These states are characterized by threshold voltages $V_{TH(\text{high})}$ and $V_{TH(\text{low})}$ as shown in Fig. 3.18. When a positive gate pulse $+V$ is applied, electrons are driven from the substrate into the gate insulator. They are thought to be trapped either at the oxide-nitride interface or in a shallow layer within the nitride. As a result of this addition of an equivalent Q_{ss} [see Eq. (3.43)], the threshold voltage of the device is switched to a high state, $V_{TH(\text{high})}$. The trapped charges are retained in the interface or nitride for long periods of time as shown in the typical retention curve of Fig. 3.18(a). When a negative gate pulse, $-V$, is applied, as shown in Fig. 3.17(b), electrons are driven out of the gate insulator and back into the silicon substrate, thereby changing the threshold voltage to a low value, $V_{TH(\text{low})}$. MNOS devices may be switched between these two threshold values a large number of times, but the nitride becomes degraded after about a million cycles and the threshold values approach each other as shown in Fig. 3.18(b). The device is said to have suffered fatigue. Physical theories explaining charge trapping effects are complicated and not yet completely understood. The p-channel MNOS devices behave in similar fashion. The ability of an MNOS device to adopt either of two distinct threshold-voltage values makes this device useful for read-only memories and electrically alterable read-only memories (EAROM's). The time required to alter V_{TH} is much larger than the time required to read these devices.

3.6.2 Floating Gate Avalanche Injection MOS—FAMOS [30,31]

Another MOS device, called FAMOS, uses a floating-gate structure consisting of polycrystalline silicon embedded in oxide, as shown in Fig. 3.19. The FAMOS structure is made to trap electrons by biasing the drain junction to avalanche breakdown. This procedure causes electrons to be accelerated in

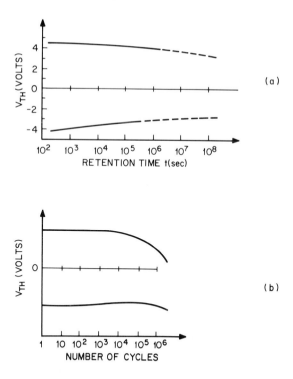

Fig. 3.18 (a) Retention curve of an MNOS device. The solid lines show the measured change of high- and low-threshold states with time. The dashed lines are commonly used extrapolations. (b) Zero-time high- and low-threshold voltages as a function of the number switching cycles.

the depletion region of the reverse-biased drain-substrate junction, and some electrons are thereby injected from the silicon substrate into the silicon dioxide surrounding the floating electrode. Due to the presence of an electric field, because of potential divider action, these electrons then drift to the floating gate, where they are trapped. The "writing" process just described is inefficient and power consuming because only a small fraction of the drain-junction current is injected into the floating gate. On the other hand, the device has excellent trapped-charge retentivity, making removal of electrons from the floating gate rather difficult. Erasure may be accomplished by means of exposing the devices to X-rays or ultraviolet light. FAMOS, unlike MNOS devices, is limited in use to read-only memory applications because of its difficult erasure procedure.

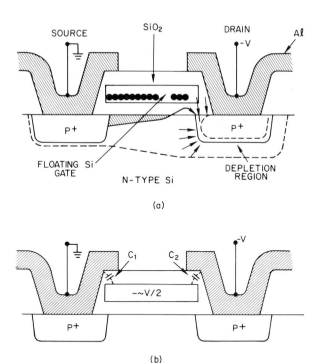

Fig. 3.19 The FAMOS transistor: (a) avalanche injection of electrons into the floating gate and (b) potential of the floating gate due to capacitive voltage divider action.

3.7 POTENTIAL WELLS AND CHARGE-COUPLED DEVICES (CCD's) [32-34]

Consider an MOS capacitor, whose gate voltage, V_{GI}, just causes strong inversion. Then the corresponding inversion-layer charge, Q_{no}, is by virtue of Eqs. (3.31), (3.34), and (3.36) equal to

$$Q_{no} = -V_{GI}C_{ox} + \sqrt{4\epsilon_s qN_A\psi_F} + 2\psi_F C_{ox}. \qquad (3.70)$$

This inversion layer charge, Q_{no}, may be referred to as *the capacity of the potential well,* associated with the MOS capacitor. If thermal equilibrium has existed for some time, this well will be filled with minority carriers, that is, electrons for a *p*-type substrate. These electrons have been supplied to the well by the somewhat slow process of thermal generation and about $t_f = 2\tau N_A/n_i$ seconds are required for the well to fill, where τ is the minority carrier lifetime.

However, if the gate voltage V_{GI} is due to a suddenly applied voltage pulse, then it is observed that the initially empty well of capacity Q_{no} fills by first forming a deep depletion layer $(x_d > x_{d\,max})$ which slowly collapses to depth $x_{d\,max}$ as minority carriers are somehow supplied to the surface to form an inversion layer.

Similarly, if an established gate voltage, V_{GI}, is suddenly reduced in value, it is the majority carriers which respond by moving towards the surface and collapsing the depletion layer. The number of minority carriers forming the inversion layer do not instantaneously change. If the collapse of the depletion layer is not sufficient to produce overall charge neutrality, then majority carriers will accumulate at the surface.

The introduction of potential wells likens charge to a fluid which tends to flow from a high level to a lower one. Hence if MOS capacitors, with potential wells of equal capacity, are placed adjacent to each other, as shown in Fig. 3.20(a) and if one well is partially filled and the other one is empty [see Fig. 3.20(c) for $t = t_2$], then charge will transfer from the filled well to the empty one. This process may be assisted somewhat, by simultaneously reducing the well capacity of the emptying well and increasing the capacity of the filling well by changing gate voltages via a clocking process [see Fig. 3-20(c), for $t = t_3$]. If the charge in the wells, that is, the surface charge under the gates of the capacitors, represents signal information, then this information is usefully transferred and the structure of Fig. 3.20 is called a charge-coupled device.

3.7.1 Charge Transfer

Consider a deep potential well, partially filled with charge representing signal information, which reposes under the phase one, ϕ_1, electrode, when $\phi_1(t_1) = +V$. At time $t_2 > t_1$ phase two electrode is also raised to $\phi_2(t_2) = +V$ and a deep potential well forms under the ϕ_2 electrode. As a consequence, some of the charge diffuses from the ϕ_1 well to the ϕ_2 well. At time $t_3 > t_2$, ϕ_1 is lowered and ϕ_2 is raised in potential, thereby creating an electric field which aids the charge-transfer process. At time $t_4 > t_3$ the charge packet, originally under the ϕ_1 electrode, has been transferred toward the output and now reposes under the ϕ_2 electrode, thereby completing one step in the charge transfer process. For the CCD shown in Fig. 3.20, three clocking phases, ϕ_1, ϕ_2, and ϕ_3, are required because the charge packets must be prevented from flowing backwards toward the input.

Let us now assume an imperfect system[35] such that a fraction $\alpha < 1$ of the charge packet is fed forward during each step, with fraction ϵ of the charge packet left behind. The actual physical causes determining the values of α and ϵ are beyond the scope of this chapter. However, a symbolic model

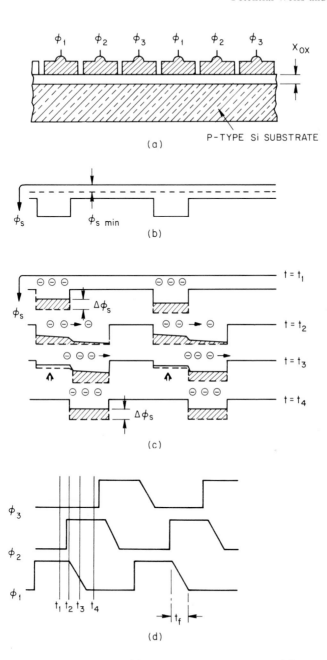

Fig. 3.20 Charge-coupled device: (a) cross section of device, (b) surface potential at beginning of transfer cycle, (c) surface potential during the transfer cycle, and (d) clock waveforms.

for a single-step transfer, as shown in Fig. 3.21, describes the CCD as a linear sampled-data system, with z^{-1} representing the unit-delay operator. From this figure after one transfer step

$$\frac{Q_{out}}{Q_{in}} = \frac{\alpha z^{-1}}{1-\epsilon z^{-1}} \tag{3.71}$$

and after k transfer steps

$$\frac{Q_{out}}{Q_{in}} = H(z^{-1}) = \alpha^k z^{-k}(1-\epsilon z^{-1})^k$$

$$= \alpha^k z^{-k} + k\epsilon\alpha^k z^{-(k+1)} + \cdots + \binom{k+j-1}{j}\epsilon^j\alpha^k z^{-(k+j)} + \cdots$$

$$\tag{3.72}$$

where index j numbers the pulse at the output and

$$\binom{k+j-1}{j} = \frac{(k+j-1)!}{j!(k-1)!} \tag{3.73}$$

is the binomial operator. Note that $H(z^{-1})$ is the transfer function for the CCD. Putting $z^{-1} = \exp(-j\omega T)$ (where $f_s = 1/T$ equals the CCD sampling or clock frequency) yields the frequency response of the CCD. More particularly $|H(\exp[-j\omega T])|$ gives the amplitude-versus-frequency response.

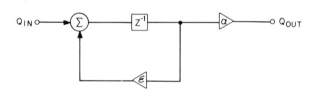

Fig. 3.21 Model for single-charge transfer step.

Finally, Fig. 3.22 shows this response for $k = 24$, that is, 24 transfer elements, with ϵ as a parameter. For practical devices, the *charge residual* ϵ is found presently to lie in the range 0.002 to 0.3. Parameter α is sometimes called the *transfer efficiency*.

The CCD is used as a multistage delay line or shift register in signal processing and storage applications described in Chapters 13 and 15. The input sample is converted into discrete time samples (which are carried as charge packets under the electrodes of the transfer elements) and shifted from one transfer element to the next along the delay line. The imperfec-

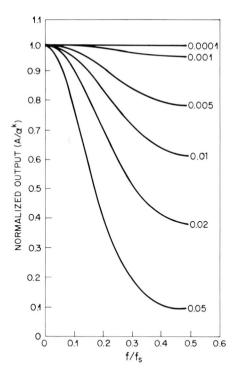

Fig. 3.22 Frequency response of a CCD.

tions of the transfer process are represented by the charge left behind (represented by coefficient ϵ) and by charge lost in transfer (represented by coefficient α).

REFERENCES

1. W. Shockley. "A unipolar 'field-effect' transistor." *Proc. IRE* **40**: 1365-1376 (November 1952).

2. G. C. Dacey and I. M. Ross. "Unipolar field-effect transistor." *Proc. IRE* **41**: 970-979 (August 1953).

3. G. C. Dacey and I. M. Ross. "The field effect transistor." *Bell Syst. Tech. J.*, **34**: 1149-1189 (November 1955).

4. A. S. Grove. *Physics and Technology of Semiconductor Devices.* New York: John Wiley & Sons, 1967.

5. R. S. C. Cobbold. *Theory and Application of Field Effect Transistors.* New York: John Wiley & Sons, 1970.

6. S. M. Sze. *Physics of Semiconductor Devices.* New York: John Wiley & Sons, 1969.

7. P. Richman. *MOS Field-Effect Transistors and Integrated Circuits.* New York: John Wiley & Sons, 1973.

8. F. Stern. "Quantum properties of surface space charge layers." *CRC Critical Review in Solid State Sciences* **5**: 499 (1974).

9. B. E. Deal, E. H. Snow, and C. A. Mead. "Barrier energies in metal-silicon dioxide-silicon structures." *J. of Physics and Chemistry of Solids* **27**: 1873-1879 (1966).

10. F. de la Moneda. "Threshold voltage from numerical solution of the two-dimensional MOS transistor." *IEEE Trans. Circuit Theory* **CT-20**: 666 (1973).

11. C. T. Sah. "Characteristics of the MOS transistors." *IEEE Trans. on Electron Devices* **ED-11**: 324-345 (July 1964).

12. H. C. Pao and C. T. Sah. "Effects of diffusion current on characteristics of MOS transistor." *Solid-State Electronics* **9**: 927 (1966).

13. D. P. Kennedy and P. C. Murley. "Steady state mathematical theory for the insulated gate field-effect transistor." *IBM J. Res. and Develop.* **17**: 2 (1973).

14. D. Vandorpe, J. Borel, G. Merckel, and P. Saintot. "An accurate two-dimensional numerical analysis of the MOS transistor." *Solid-State Electronics* **15**: 547 (1972).

15. M. Heydmann, "Solution numerique bidimensionalle des equations génerales de transport dans les semiconducteurs en régime permanent." *L'onde Electronique* **52**: 185 (1972).

16. J. A. Geurst. "Theory of insulated-gate field effect transistors near and beyond pinch-off." *Solid-State Electronics* **9**: 129 (1966).

17. M. B. Barron. "Low level currents in insulated gate field effect transistors." *Solid State Electronics* **15**: 293-302 (1972).

18. W. Milton Gosney. "Subthreshold drain leakage currents in MOS field-effect transistors." *IEEE Trans. on Electron Devices* **ED-19**: 213-219 (February 1972).

19. R. M. Swanson and J. D. Meindle. "Ion implanted complementary MOS transistors in low-voltage circuits." *IEEE J. of Solid-State Circuits* **SC-7**: (April 1972).

20. R. R. Troutman and S. Chakravarti. "Subthreshold characteristics of insulated-gate field-effect transistors." *IEEE Trans. on Circuit Theory* **CT-20**: 659 (1973).

21. H. Katto and Y. Stoh. "Analytical expressions for the static MOS transistor characteristics based on the gradual channel model." *Solid-State Electronics* **17**: 1283 (1974).

22. A. Popa. "An injection level dependent theory of the MOS transistor in saturation." *IEEE Trans. on Electron Devices* **ED-19**: 774 (1972).

23. J. R. Brews. "Carrier density fluctuations and the IGFET mobility near threshold." *J. Appl. Phys.* **46**: 2193 (1975).

24. H. C. Poon and R. L. Johnston, private communication.

25. L. D. Yau. "A simple theory to predict the threshold voltage of short-channel IGFETS." *Solid-State Electronics* **17**: 1059-1063 (October 1974).

26. M. B. Das. "Physical limitations of MOS structures." *Solid-State Electronics* **12**: 305-336 (May 1969).

27. G. Merckel, J. Borel, and N. Z. Cupcea. "An accurate large-signal MOS transistor model for use in computer-aided circuit design." *IEEE Trans. on Electron Devices* **ED-19**: 681-690 (May 1972).

28. A. Vladimirescu. "Calculator-aided design of MOS integrated circuits." *IEEE J. of Solid-State Circuits* **SC-10**: 151-161 (June 1975).

29. D. Kahng and E. H. Nicollian. "Physics of multilayer-gate IGFET memories." *Applied Solid State Science* **3**, Academic Press (1972).

30. D. Frohman-Bentchkowsky. "A fully decoded 2048-bit electrically programmable FAMOS read-only memory." *IEEE J. of Solid-State Circuits* **SC-6**: 301-306 (October 1971).

31. D. Kahng, W. J. Sundburg, D. M. Boulin and J. R. Ligenza. "Interfacial dopants for dual-dielectric charge-storage cells." *Bell Sys. Tech. J.* **53**: 1723 (1974).

32. W. S. Boyle and G. E. Smith. "Charge coupled semiconductor devices." *Bell Syst. Tech. J.* **49**:, 587-593 (April 1970).

33. S. D. Brotherton. "A theoretical analysis of CCD operation with square clock pulses." *Solid-State Electronics* **17**: 341-348 (April 1974).

34. D. F. Barbe. "The charge-coupled concept." *Report of NRI Progress*, March 1972.

35. G. E. Vanstone, J. B. G. Roberts, and A. E. Long. "The measurement of the charge residual for CCD transfer using impulse and frequency response." *Solid-State Electronics* **17**: 889-895 (1974).

36. D. Kahng and M. M. Atalla. "Silicon-silicon dioxide field induced surface devices," in *IRE-AIEE Solid-State Device Res. Conf.* (Carnegie Inst. of Technol., Pittsburgh, Penn.), 1960.

PROBLEMS

3.1 An IGFET has a gate insulator consisting of thermally grown oxide of thickness x_0 and dielectric constant ϵ_{ox}, covered by a layer of silicon nitride of thickness x_n and dielectric constant ϵ_{ni}. Derive an expression for the threshold voltage V_{TH} of this compound gate insulator transistor and compare it with Eq. (3.43) applicable to the simple oxide gate insulator.

3.2 Five aluminum-gate, n-channel MOS transistors, each with a different oxide thickness $(x_{01}, x_{02}, ..., x_{05})$ are fabricated on the same p-type silicon substrate, having measured impurity concentration N_A. The positive surface charge, Q_{ss}, is known to be constant over the whole wafer. A measurement of the threshold voltage, V_{TH}, is made and yields the values $V_{TH1}, V_{TH2}, ..., V_{TH5}$ versus x_0. Discuss the shape of the curve and its slope, and explain the meaning of the intercept $V_{TH}(x_0=0)$. Determine the value of the surface charge, Q_{ss}, and the metal-semiconductor work function ψ_{MS}.

3.3 Show that the total capacitance C of the MOS capacitor consists of the series combination of C_{ox} and C_D. Show further that C at flat-band is given by

$$C(\text{flat-band}) = \frac{\epsilon_{ox}}{x_0 + \frac{1}{\sqrt{2}} \frac{\epsilon_{ox}}{\epsilon_s} L_D}.$$

3.4 Derive an expression for the flat-band voltage, V_{FB}, of an MOS capacitor. Why is this voltage independent of the charge density in the surface depletion region?

3.5 Two aluminum-gate, n-channel MOS transistors are fabricated on a p-channel ($N_A = 3 \times 10^{15}$ cm^{-3}) substrate. One transistor has $x_{01} = 1000$ Å, the other has $x_{02} = 1500$ Å. If $V_G - V_{FB} = 12$ V, where V_{FB} is the flat-band voltage, calculate the value of drain voltage, V_D, at which $I_{D\,\text{sat}}$ will be observed.

3.6 The threshold voltage V_{TH} in the linear region is given by $V_{TH} = 2\psi_F + \psi_{MS} + (4\epsilon_s q N_A \psi_F)^{1/2}/C_{ox}$ in the absence of Q_{ss}. If ψ_{MS} is considered independent of temperature T, show that the derivative of V_{TH} with respect to temperature is given by

$$\frac{dV_{TH}}{dT} = \frac{d\psi_F}{dT} \left[2 - \frac{(\epsilon_s q N_A/\psi_F)^{1/2}}{C_{ox}} \right]$$

where

$$\frac{d\psi_F}{dT} = \pm \frac{1}{T} \left[\frac{V_{g0}}{2} - |\psi_F(T)| \right].$$

Explain why $dV_{TH}/dT > 0$ for n-channel devices and $dV_{TH}/dT < 0$ for p-channel devices.

3.7 Assume the properties of the depletion region in an MOS transistor are those of a linearly graded junction, such that the channel length modulation term becomes

$$B = \frac{L}{L - \left[\frac{12\epsilon_s (V_{DS} - V_{DS\,\text{sat}})}{qa} \right]^{1/3}}$$

Derive expressions for $I'_{D\,\text{sat}}$ and for g_D and compare with values given in Eqs. (3.61) and (3.62) for abrupt junctions. Comment on the differences.

3.8 Calculate the punchthrough voltage, V_{pt}, of an n-channel MOS transistor fabricated on a p-substrate of impurity concentration N_A cm^{-3}. Assume a channel length of L cm and include the effect of the contact potential, V_{bi}, of the drain-substrate diode.

3.9 The transconductance of a bipolar transistor is shown in Chapter 2 to equal $g_m \approx I_{CC}/V_T$. Compare this transconductance at 300 K with that of an n-channel MOS transistor, using material values and dimensions typical of bipolar and MOS integrated circuit structures.

4

Passive Elements

The purpose of this chapter is to describe the structure of passive elements as realized in integrated circuit form, together with their simple equivalent circuits. The passive elements to be discussed are diffused and thin-film resistors, junction, metal-oxide-semiconductor and thin-film capacitors, and spiral inductors, the latter being useful at very high frequencies.

4.1 RESISTOR GEOMETRY [1,2]

The properties of a resistor are partly geometrical and partly material. The design of an integrated circuit resistor involves specification of a planar shape of resistive material of prescribed thickness, which forms the current path. The resistor value R is given by

$$R = \rho l / A \tag{4.1}$$

where ρ is the resistivity of the slab of material, l is the length of the current path, and A is the cross-sectional area. Note that for a rectangular cross section $A = wt$, where w is the path width and t the effective thickness of the slab. In the design of integrated circuit components, the process is usually specified beforehand and so the resistivity ρ and the thickness t are known. Hence it is convenient to rewrite Eq. (4.1) in the form

$$R = R_S l / w \tag{4.2a}$$

where $R_S = \rho / t$ is called the *sheet resistance*. For materials in which the resistivity is a function of depth, the effective resistivity is used. The dimension of R_S is ohms, but it is common practice to refer to the dimension of sheet resistance as ohms per square, because R_S is the resistance of any square resistor. The measurements of sheet resistance are usually made by means of a four-point probe as described in Chapter 11.

Resistor design begins with the consideration of the straight-line resistor as shown in the plan view of Fig. 4.1(a). The ratio l/w is called the *aspect ratio* or the *number of squares* and is denoted by $n = l/w$. As a consequence Eq. (4.2a) may be written

$$R = R_S n. \tag{4.2b}$$

EXAMPLE 4.1   A semiconductor resistor has sheet resistance $R_S = 200 \ \Omega/\square$ and aspect ratio $n = 10$. Find the resistance R ignoring possible end and edge corrections.

Solution.   From Eq. (4.2b), $R = R_S n = 200 \times 10 = 2 \ k\Omega$.

In silicon integrated circuits it is necessary to have the contacts completely surrounded by resistor material as shown in Fig. 4.1(b). As a consequence, the current flow lines are distorted near the resistor ends, and the total resistance R measured between the contacts is greater than $R_S l_1/w$, where l_1 is the main length of the resistor body. For high-value resistors the simple line geometry would prove too long and so the path is folded as shown in Fig. 4.2.

In general there are two geometric contributions to resistance value: the number of squares in the straight-line segments of the path and corrections due to contacts, bends, and changes of shape. These contributions can be computed by conformal mapping techniques and several situations are shown in Fig. 4.3.

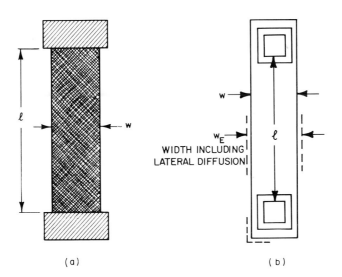

                  (a)   (b)

Fig. 4.1   (a) Plan view of a thin-film straight-line resistor and (b) a straight-line resistor with embedded contacts for silicon integrated circuits.

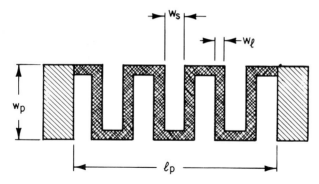

Fig. 4.2 Plan view of a meander resistor.

As an example, the effective number of squares for the meander resistor may be obtained with the aid of Fig. 4.3 and consists of the number of squares in each meander multiplied by the number of meanders.

$$n = \left[\frac{w_p - 2w_l}{w_l} + \frac{w_s}{w_l} + 2k_1 \right] \left(\frac{l_p - w_s}{w_s + w_l} \right) + \frac{w_s}{w_l} + 2k_2. \tag{4.3}$$

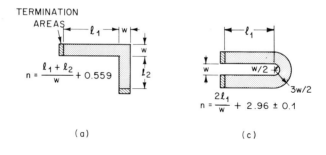

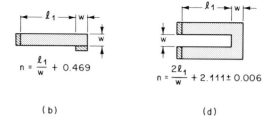

Fig. 4.3 Effective number of squares for various resistor geometries. (From *Thin Film Technology* by R. W. Berry, P. M. Hall, and M. T. Harris, Copyright 1968 by Litton Educational Publishing, Inc. Reprinted by permission of Van Nostrand Reinhold Co.)

Here k_1 is the correction factor for a bend and k_2 accounts for the end contact corrections. One property of the meander shape, which renders it useful at high frequencies, is its low self-inductance.

EXAMPLE 4.2   Calculate the aspect ratio n and the resistance of a 10-meander resistor, with sheet resistance $R_S = 200 \ \Omega/\square$ from Eq. (4.3). Then compare your results with those obtained from the approximate equation

$$n' = \left(\frac{l_p}{w_l + w_s}\right)\left(\frac{w_p}{w_l} + 2k_1\right) + 2k_2.$$

Take $w_s/w_l = 3$, $w_p/w_l = 7$, $k_1 = 0.56$ and $k_2 = 0$.

Solution.   From Eq. (4.3) the number of meanders

$$\frac{l_p - w_s}{w_s + w_l} = 10 = \frac{(l_p/w_l) - 3}{3 + 1}.$$

Hence $l_p/w_l = 43$. From Eq. (4.3)

$$n = (7 - 2 + 3 + 1.12) \times 10 + 3 = 94.2 \approx 94$$

and so the resistance $R = R_S n = 200 \times 94 = 18.8 \ \text{k}\Omega$. From the approximate formula above

$$n' = \left(\frac{43}{1+3}\right)(7 + 1.12) = \frac{43}{4} \times 8.12 \approx 87.3$$

and so $R = 17.45 \ \text{k}\Omega$. Hence use of the approximate formula is accurate to within about 7%.

The power P dissipated in the resistor is given by

$$P = QA_s \tag{4.4}$$

where Q is the power density and A_s is the resistor surface area. For a meander resistor, $A_s = w_p l_p$ is the area of a rectangle which circumscribes the resistor. The power density Q is an important parameter, because it has a direct bearing on resistor aging and reliability.

The maximum power density $Q_{\max}$ is experimentally determined as it depends on the resistor film material, the ambient temperature, the power dissipated on the substrate, the substrate material, the substrate and resistor size and thermal impedance, the required stability of the structure, and possible other factors. For a specified resistor power dissipation P, the minimum resistor surface area $A_{\min}$ is then given by

$$A_{\min} = P/Q_{\max}. \tag{4.5a}$$

For a straight-line resistor, this can be put into the form

$$A_{\min} = (R/R_S) w_{\min}^2 \tag{4.5b}$$

where $w_{\min}$ is the minimum track width as a result of power considerations.

EXAMPLE 4.3   A power density of 5 W/cm^2 and a sheet resistance of 1000 $\Omega/\square$ are specified for a certain resistor material. You are asked to design a 20-kΩ resistor and a 50-kΩ resistor such that each can have a quiescent voltage of 10 V across its terminals. Find the number of squares needed and a suitable track width w, which meets the requirement imposed by the pattern definition process of a minimum linewidth of 50 μm.

Solution.   For the 20-kΩ resistor, $n_1 = R_1/R_S = 20/1 = 20$ and for the 50-kΩ resistor $n_2 = R_2/R_S = 50/1 = 50$. The powers to be dissipated by these resistors are therefore

$$P_1 = \frac{V^2}{R_1} = \frac{100}{20\times10^3} = 5\times10^{-3} \text{ W}$$

and

$$P_2 = \frac{V^2}{R_2} = \frac{100}{50\times10^3} = 2\times10^{-3} \text{ W}.$$

From Eq. (4.5a) minimum areas are therefore

$$A_1 = \frac{P_1}{Q} = \frac{5\times10^{-3}}{5} = 10^{-3} \text{ cm}^2$$

and

$$A_2 = \frac{P_2}{Q} = \frac{2\times10^{-3}}{5} = 0.4\times10^{-3} \text{ cm}^2.$$

The minimum track widths from Eq. (4.5b) are therefore

$$w_1 = (A_1/n)^{1/2} = (10^{-3}/20)^{1/2} \approx 71\times10^{-4} \text{ cm}$$

and

$$w_2 = (0.4\times10^{-3}/50)^{1/2} \approx 28\times10^{-4} \text{ cm}.$$

Hence the 20-kΩ resistor can be designed with $w_1 = 75$ μm, say, but the 50-kΩ resistor must have $w_2 = 50$ μm at least, because of the minimum track-width requirement.

4.2 RESISTOR TOLERANCE

From Eq. (4.2b) it is evident that resistance value depends on sheet resistance R_S and also on the aspect ratio n.

Sheet resistance R_S is a function of manufacturing tolerances and it is also temperature dependent. For semiconductor and thin-film materials the sheet resistance R_S is approximately described by

$$R_S = R_{S0}(1 + \alpha[T-T_0]) \tag{4.6}$$

where T_0 is the ambient temperature, and the working temperature T must be in the range $-55°C$ to $125°C$. Coefficient α is called the temperature coefficient of resistance (TCR) and is given by

$$\text{TCR} = \alpha = \frac{1}{R_{S0}} \frac{dR_S}{dT} .$$

(4.7)

The total resistance variation ΔR due to all causes is:

$$\Delta R = \frac{\partial R}{\partial T} \Delta T + \frac{\partial R}{\partial R_{S0}} \Delta R_{S0} + \frac{\partial R}{\partial n} \Delta n.$$

(4.8)

The first term is due to temperature variations and, from Eqs. (4.2b), (4.6), and (4.7), it is given by

$$\frac{\alpha TR}{1 + \alpha(T-T_0)} \frac{\Delta T}{T} .$$

(4.9a)

The second term is given by

$$R \frac{\Delta R_{S0}}{R_{S0}}$$

(4.9b)

and accounts for changes in resistance value due to the variability in sheet resistance. The third term accounts for the change in resistance caused by variations in the geometric aspect ratio n. For a line resistor $n = l/w$, and when n is large, the variation in n is dominated by changes in w. Hence this term is given by

$$-R \frac{\Delta w}{w} .$$

(4.9c)

Substituting these results into Eq. (4.8) yields

$$\frac{\Delta R}{R} = \frac{\alpha T}{1 + \alpha(T-T_0)} \frac{\Delta T}{T} + \frac{\Delta R_{S0}}{R_{S0}} - \frac{\Delta w}{w} .$$

(4.10)

The first term is deterministic while the other terms are random, because they depend on the control of the manufacturing process. The worst-case bound is the sum of the magnitudes of each of these three terms. It is desirable to design integrated circuits to have characteristics that depend on ratios of component values rather than on absolute values in order to minimize the effect of the variability of R on circuit performance. For two nominally identical resistors R' and R'' in close proximity, one can assume the first two terms in Eq. (4.10) are identical. Hence, assuming $R_0 \approx R_0' \approx R_0''$

$$\frac{R'}{R''} = \frac{R_0'+\Delta R_0'}{R_0''+\Delta R_0''} = \frac{1 + (\Delta R_0'/R_0')}{1 + (\Delta R_0''/R_0'')}$$

$$= \frac{1 - (\Delta w'/w)}{1 - (\Delta w''/w)} \approx 1 - \frac{1}{w} (\Delta w'-\Delta w'').$$

(4.11)

In the worst case the variations oppose one another and so

$$\frac{R'}{R''} \approx 1 \pm \frac{2\Delta w}{w}. \tag{4.12}$$

EXAMPLE 4.4 Determine the absolute accuracy of a film resistor of linewidth $w = 500$ μm, which is to be used in the temperature range $-25°C$ to $75°C$. The TCR for the film is 100 ppm/°C. The resistor material variation is guaranteed to within 3% and the edge uncertainty of the resistor track is 10 μm.

What is the accuracy of the ratio of resistance values for two identical such film resistors in close proximity?

Solution. Assume that the nominal operating temperature is $T_0 = 25°C$. Then from Eq. (4.10), $\Delta R/R \approx |\alpha\Delta T| + |\Delta R_{S0}/R_{S0}| + |\Delta w/w|$. The largest possible temperature deviation $\Delta T = 50°C$ and so the first term equals

$$\alpha\Delta T = 50\times10^{-6}\times100 = 5\times10^{-3}.$$

The second term $\Delta R_{S0}/R_{S0} = 3\times10^{-2}$ and the last term $\Delta w/w = 10/500 = 2\times10^{-2}$. Hence, $\Delta R/R\big|_{max} = (0.5+3+2) \times 10^{-2} = 5.5\times10^{-2}$ or 5.5%. For two such resistors in close proximity, variation in their values R' and R'' is less than $2\Delta w/w = 2\times2\times10^{-2} = 4\times10^{-2}$ or 4%.

Hence the accuracy of the match is determined by the control of the resistor track width, Δw, normalized to the nominal track width w. Since Δw is determined by photolithographic capabilities and is assumed known, the track width w needed for a specific match can be determined from Eq. (4.12). Equation (4.10) does not include the effects of contact resistance or variations in the geometrical correction factors for nonstraight-line resistors. In practical situations a resistor ratio $R_2/R_1 = k$ is usually restricted to integer values. Such ratios are obtained by fabricating $k + 1$ identical resistors, whose values are not precisely known. Resistor R_2 is then formed by the series connection of k of these resistors and resistor R_1 is formed from the remaining $(k+1)$st resistor.

4.3 SEMICONDUCTOR RESISTORS [3-5]

Integrated-circuit resistors can be fabricated in silicon by using either the base material, the emitter material, the collector material, or by pinching the base material with an emitter diffusion. Resistors in base material are by far the most popular, because the sheet resistance is several hundred ohms per square, thus permitting a wide range of useful resistor values. Emitter material is heavily doped with sheet resistance being typically several ohms per square. Resistors made from this material have very low values and may also be used as crossunders. Collector resistors have a large sheet resistance, typically 1000 $\Omega/\square$, and a cross section that differs significantly from that of base and emitter resistors. Furthermore, resistance depends on a carefully controlled isolation diffusion and also on the collector-substrate junction bias. As a result, collector resistors have values that are difficult to control.

Finally, pinched resistors are obtained by constricting the cross section of the base-diffused material with an emitter diffusion. This increases the effective sheet resistance to several thousand ohms per square, but the tolerance on resistance is extremely poor. The pinch resistor is a degenerate junction-field-effect transistor; as a result it is highly nonlinear and has a low breakdown voltage.

Figure 4.4 shows the cross section of a base-diffused resistor together with a simple equivalent circuit showing the parasitic elements. The primary parasitic elements are the distributed voltage-variable capacitance C_2 of the reverse-biased collector-base junction, the capacitance C_1 of the reverse-biased collector-substrate junction and the bulk resistances of the collector and substrate materials. In addition there is a parasitic substrate *p-n-p* transistor, which will be cut off if the *n* epitaxial layer is connected to the most positive voltage point in the circuit. The current gain of this parasitic transistor may also be minimized by putting an n^+ buried layer under the epitaxial layer, because this effectively increases the transistor base width.

Practical resistor values are in the region of 100 Ω to about 10 kΩ, the upper limit imposed by effective utilization of chip area. The TCR for these resistors is in the range of 1000 to 3000 ppm/°C, being higher for higher sheet resistance.

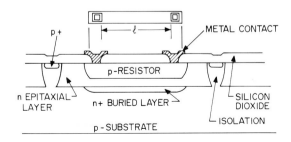

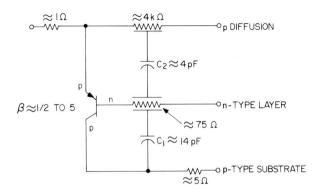

Fig. 4.4 Cross section and equivalent circuit of a base diffused resistor.

The parasitic capacitance C_2 is the capacitance of the the collector-base junction. It can be split into two components, one due to the bottom junction and the other due to the sidewalls. Let A_b be the effective area of the bottom of the junction and let A_s be the effective area of the sidewalls. If C_{bc} is the capacitance density of the bottom of the junction while C_{bcs} is the average capacitance density of the sidewalls, then the total parasitic capacitance C_2 is given by

$$C_2 = C_{bc} A_b + C_{bcs} A_s.$$

For the special case of a bar resistor of length l as shown in Fig. 4.4, $A_b = lw$, and A_s for the cylindrical junction equals $\pi x_j (l+w)$, where x_j is the junction depth.

EXAMPLE 4.5 Compute the parasitic base-collector capacitance of a diffused base resistor of length 300 μm and width 100 μm. The base-collector junction depth is 3 μm, the background concentration $N_B = 10^{16}$ cm^{-3} and the surface concentration $N_s = 10^{19}$ cm^{-3}.

Solution. $N_B/N_s = 10^{-3}$ and for zero bias $(V+V_{bi})/N_B = 0.7 \times 10^{-16}$. Use of the appropriate Lawrence-Warner curve now gives $C_{bc} = 2 \times 10^4$ pF/cm^2. Taking $C_{bcs} = C_{bc}$ gives

$$C_2 = C_{bc}(A_b+A_s) = C_{bc}[lw+\pi x_j(l+w)]$$
$$= 2 \times 10^4 [3 \times 10^{-4} + 3.14 \times 3 \times 4 \times 10^{-6}] = 6.74 \text{ pF}.$$

A similar calculation may be used to determine the parasitic collector-substrate capacitance C_1.

In practice one attempts to place all resistors in a common isolation region. In that case, the parasitic transistor in the equivalent circuit of Fig. 4.4 must be replaced by a multiple emitter device with one emitter for each resistor in the isolation region.

Low value resistors are made using the n^+ emitter material as shown in Fig. 4.5. The n-type emitter region is diffused into the p-type base "pocket," which must be connected to a negative voltage in order to create the required reverse bias, which is necessary to isolate the resistor. For emitter-diffused resistors the TCR is about 100 ppm/°C, which is a ten-to-one reduction from the TCR for base-diffused resistors.

Collector resistors are made from the epitaxial collector material with isolation being provided by the p^+ isolation diffusion. The epitaxial layer is thicker than the base- or emitter-diffused layers, and the resistance values of collector resistors are more difficult to control. In addition they have a higher TCR than base- or emitter-diffused resistors. For these reasons, epitaxial resistors are rarely used.

Pinch resistors are used to realize very high-resistance values in non-critical applications. A cross section of such a resistor is shown in Fig. 4.6. TCR for pinch resistors is typically many thousand parts per million per °C and control of sheet resistance is typical ±50 to 100%.

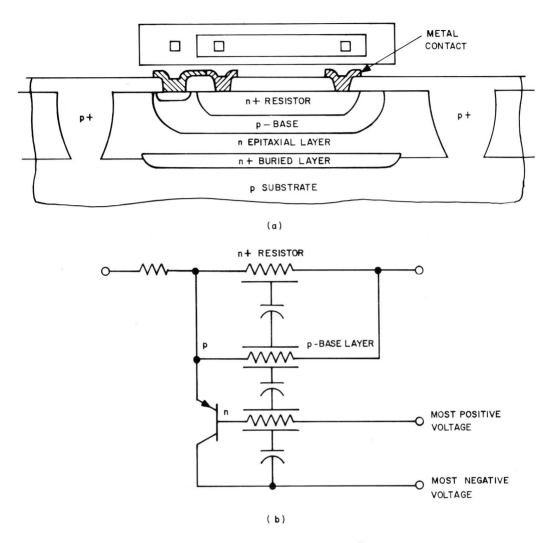

(a)

(b)

Fig. 4.5 Cross section and equivalent circuit of an emitter diffused resistor.

4.4 DESIGN OF DIFFUSED SEMICONDUCTOR RESISTORS [6]

A diffused resistor incorporating two right-angle bends is shown in Fig. 4.7(a). The resistor has track width w and the length between metal contacts measured along a centerline is given by l_c. The metal contacts have the same dimension as the contact oxide windows in the oxide, namely, $w - 2a$.

Figure 4.7(b) shows a cross-sectional, enlarged view of such a resistor. Because diffusion is a three-dimensional phenomenon, the p-diffused layer will extend somewhat under the silicon dioxide window which defined the

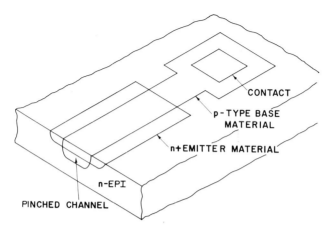

Fig. 4.6   Cross section of a pinched resistor.

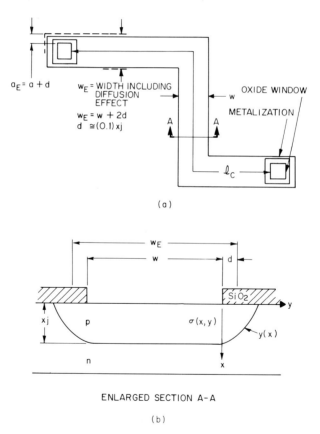

Fig. 4.7   (a) Top view of a resistor with a right-angle bend and (b) cross-sectional view.

resistor track. As a consequence, the track width w is increased to an effective resistor track width w_E, where

$$w_E = w + 2d. \tag{4.13}$$

The track widening factor d, depends on the base-collector junction depth x_j, the junction profile $y(x)$, oxide edges, and the conductivity distribution $\sigma(x,y)$ in the p-base layer. The value of d is typically one-tenth of the base-collector junction depth, x_j, and is usually calculated using a computer. Because of this increased effective width, the aspect ratio n decreases and the principal resistance term becomes

$$R = \frac{R_S l_c}{w_E}. \tag{4.14}$$

However, Eq. (4.14) must be corrected for right-angle bends and also for effects at the contacts. This is done as follows:

a) It was shown in Section 4.1 that bends in the resistor track require corrections to be made in the aspect ratio. More particularly Fig. 4.3(a) shows that the effective resistance of the corner square in a right-angle bend is 0.559 squares. For N bends the correction is

$$R_c = (0.559{-}1)\,NR_S. \tag{4.15}$$

b) The resistor has two end contacts or pads. Far away from the contact there will be laminar flow in the horizontal plane. Near the contacts the current stream lines bend, because the contact is immersed in the semiconductor. This requires a correction term for two contacts of

$$2R_p = R_S\frac{\ln(4)}{\pi}\left[\frac{2(a+d)}{w_E}\right]^2 \tag{4.16}$$

where p stands for "plane."

c) A cross-sectional correction $2R_x$ must also be added because the current passes into the semiconductor from the front edge of the metal contact. The current then spreads out and eventually becomes laminar. The correction is

$$2R_x = \frac{k_x R_S}{w_E} \tag{4.17}$$

where k_x is an empirical constant having a value of about 0.44 for 200 $\Omega/\square$ p-type silicon.

Combining the effects of Eqs. (4.14) and Eq. (4.17) gives

$$R = R_S\left[\frac{l_c}{w_E} - 0.441N + \frac{\ln(4)}{\pi}\left[\frac{2[a+d]}{w_E}\right]^2 + \frac{k_x}{w_E}\right]. \tag{4.18}$$

EXAMPLE 4.6 Design a 3-kΩ base resistor which incorporates two bends. The base material has sheet resistance $R_S = 200\ \Omega/\square$ and the resistor is to have a track width w of 40 μm. The track widening factor d equals approximately $x_{jb}/10$ and the junction depth, x_{jb}, equals 3 μm. If the edge of the oxide window is 6 μm from the edge of the track, find a suitable centerline length l_c, which meets the above specifications. Ignore the pad correction term k_x/w_E in Eq. (4.18).

Solution. From the above information $N = 2$, $w = 40\ \mu$m, $d = x_j/10 = 0.3\ \mu$m, $w_E = w + 2d = 40.6\ \mu$m, and $a + d = 6.3\ \mu$m. The third and fourth terms in Eq. (4.18) prove to be negligible and so $3 = 0.2[(l_c/40.6) - 0.88]$. Hence $l_c \approx 645\ \mu$m.

One last observation on Eq. (4.18) concerns the measurement of the sheet resistance, R_S, of a base-diffused resistor. Sheet resistance is measured immediately after the base-diffusion, but before the end of the process. However, there are high temperature operations after the base diffusion which cause further diffusion of the p-base and further oxidation at the silicon surface, where the current density is highest. Both these effects tend to increase the actual sheet resistance R_{sa} over the measured sheet resistance R_S. Empirically,

$$R_{sa} = k_a R_S \qquad (4.19)$$

where typical values of constant k_a are of the order of 1.06. The sheet resistance R_{sa}, rather than the measured value R_S must be used in design Eq. (4.18).

With care it is possible to obtain resistor ratios of about 1% for resistors in close proximity. Absolute resistance values can be maintained to about 10%.

4.5 THIN-FILM RESISTORS

Thin-film resistors have the same geometric shapes as semiconductor resistors, but their construction is quite different. A typical thin-film line resistor is shown in Fig. 4.8. This resistor consists of a thin resistive film deposited on an insulating substrate. Thin-film resistor materials are metals, metal alloys, and metal compounds as described in Chapter 8. Deposition onto the substrate is by means of vacuum evaporation or sputtering as described in Chapter 5. The nonconducting substrate mechanically supports the film. Aluminum oxides are frequently used as substrate materials and both substrate and film must be compatible with all processing and assembly steps as also described in Chapter 8. Thin-film resistors are being used with increasing frequeny as components in silicon integrated circuits.

Thin-film resistors have sheet-resistance values from about 30 to 1000 $\Omega/\square$, with a $\pm15\%$ initial tolerance. Unlike the semiconductor resis-

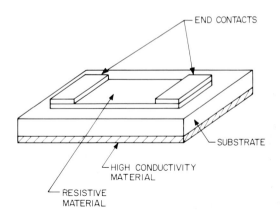

Fig. 4.8 Thin-film line resistor.

tors, thin-film resistors are adjustable by means of a process called "trimming." Final tolerance after trimming is limited by the accuracy of the measuring apparatus and by the amount of time spent in trimming each resistor. Typically 1% is easily obtainable, and 0.1% can be achieved if required. Certain nichrome films (80% nickel, 20% chromium) have a TCR in the +20 to +110 ppm/°C range and so make an almost zero TCR possible. Very precise low TCR resistors can also be fabricated with tantalum.

For most thin-film resistor applications it is necessary to adjust resistance by trimming. This procedure increases resistance, because the aspect ratio n is increased by removal of part of the resistor pattern or because the sheet resistance R_S is increased by decreasing the film thickness. During trimming the resistance is monitored until it reaches the desired value. If R_i stands for initial resistance and R_f for final resistance, then the trim factor T_F is defined

$$T_F = R_f/R_i. \tag{4.20}$$

Thin-film resistors may be trimmed by cutting a notch into the resistor pattern, either abrasively or by laser beam. Another method used primarily for tantalum resistors is anodization, which changes the effective thickness of the film over a wide area. Final resistor accuracy of 0.1% or less is possible.

EXAMPLE 4.7 A thin-film resistor has a sheet resistance $R_S = 1000 \ \Omega/\square$. If the resistor is to be designed with a trim factor of 1.2, find the largest acceptable initial tolerance on the sheet resistance.

Solution.   From Eq. (4.20), the initial resistor value per square is

$$R_i = \frac{R_f}{T_F} = \frac{1}{1.2} = 0.833 \text{ k}\Omega.$$

Hence the initial tolerance on the sheet resistance is $100(1-0.833) = 16.67\%$.

Stability of thin-film resistors is measured by the change of resistance with aging. Catastrophic failures of reliability screened thin-film resistors are extremely rare and so failures are due mostly to parameter changes. Long-term drift of resistance value shows strong dependence on temperature and an accelerated life test yields a linear Arrhenius plot (see Chapter 16). This explains why a maximum power density Q_{max} as per Eq. (4.5a) has to be set and also indicates that most rapid resistance changes occur early in a resistor's life. As a consequence, a thermal stabilization treatment is usually included as a final step in processing some resistor materials. Stability measured as a percentage change in resistance after 1000 hrs at 125°C is less than 0.2% for both nichrome and tantalum nitride films.

For the design of thin-film resistors in the presence of bends, the correction factors of Fig. 4.3 must be applied. No horizontal correction factor for the end contact is needed, because the metal termination extends completely across the resistor track. A vertical end-contact correction factor is needed, because the flow lines crowd to the leading edge of the contact. The empirical correction constant must be evaluated for specific material systems.

4.6 THE DISTRIBUTED RC STRUCTURE [3,7]

A thin-film circuit consisting of resistive material deposited on a substrate of dielectric material, whose underside is coated with high conductivity material (see Fig. 4.8) naturally forms a distributed-RC structure.

Similarly, the diffused resistors shown in Figs. 4.1 and 4.2 with their associated distributed, but voltage-dependent, junction capacitance also form a distributed-RC structure. For these it will be assumed that an average value of capacitance C is taken.

A portion of a one-dimensional resistor model in terms of resistance per unit length r and capacitance per unit length c (with r and c taken constant for simplicity) is shown in Fig. 4.9. Here x is taken along the resistor length, which extends from $x = 0$, to $x = l$ and inductance effects have been ignored.

An infinite number of individual sections of the type shown in Fig. 4.9 constitute a distributed RC transmission line, whose time constant per unit length γ is given by

$$\gamma = rc. \tag{4.21}$$

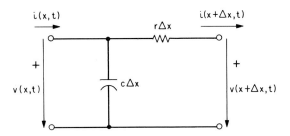

Fig. 4.9 Incremental length of distributed-RC transmission line.

This RC transmission line may be represented by the two-port model depicted in Fig. 4.10(a) with $I_1(s)$, $V_1(s)$, $I_2(s)$, $V_2(s)$ being the terminal variables. The solution of this model is well known and given in terms of the y parameters by

$$\begin{bmatrix} I_1(s) \\ I_2(s) \end{bmatrix} = \frac{\sqrt{s\gamma}}{r} \begin{bmatrix} \coth(\sqrt{s\gamma}) & -\operatorname{csch}(\sqrt{s\gamma}) \\ -\operatorname{csch}(\sqrt{s\gamma}) & \coth(\sqrt{s\gamma}) \end{bmatrix} \begin{bmatrix} V_1(s) \\ V_2(s) \end{bmatrix}.$$

$$(4.22)$$

As may be seen, the two-port is reciprocal with $y_{12}(s) = y_{21}(s)$ and symmetrical with $y_{11}(s) = y_{22}(s)$. The y parameters are transcendental and best studied by computer. Nevertheless, some insight into the behavior of this element may be obtained by studying the following simple but representative transfer functions. Reference will be made to the short-circuit current gain G_i, given by

$$G_i = -\frac{I_2(s)}{I_1(s)}\bigg|_{V_2(s)=0} = -\frac{y_{21}(s)}{y_{11}(s)} = \operatorname{sech}(\sqrt{s\gamma}), \qquad (4.23)$$

and to the short-circuit input impedance Z_{in} given by

$$Z_{\text{in}} = \frac{V_1(s)}{I_1(s)}\bigg|_{V_2(s)=0} = \frac{1}{y_{11}(s)} = \frac{\tanh(\sqrt{s\gamma})}{\sqrt{s\gamma}/r}. \qquad (4.24)$$

Putting $s = j\omega$, $\sqrt{j} = (1+j)/\sqrt{2}$ and $x = l\sqrt{\omega\gamma/2}$ gives

$$1/G_i = \cosh x \cos x + j \sinh x \sin x \qquad (4.25)$$

$$\operatorname{phase}(1/G_i) = \tan^{-1}\{\tanh x \tan x\}. \qquad (4.26)$$

Applying double-angle formulae to Eq. (4.25) yields the magnitude

$$\frac{1}{|G_i|^2} = \frac{\cosh 2x + \cos 2x}{2}. \qquad (4.27)$$

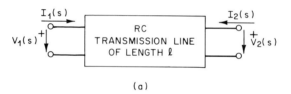

(a)

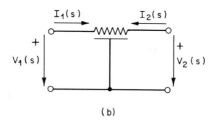

(b)

Fig. 4.10 (a) Definition of terminal variables for two-port model and (b) distributed equivalent circuit.

Because tanh x lies between 0 and 1, and approaches 1 for large values of x, the phase $(1/G_i)$ approaches x for large x and so increases linearly without limit, quite unlike a lumped two-port.

On the other hand, the magnitude $1/|G_i|^2$ is approximately equal to $[\exp(2x)]/2$ for large values of x and so the logarithmic amplitude $10 \log_{10}(1/|G_i|^2)$ increases linearly for large values of x. It may further be shown that the frequency ω_0 at which the magnitude squared $|G_i|^2$ has fallen by 3 dB is given by

$$\omega_0 \approx \frac{2.43}{rcl^2} \tag{4.28}$$

where ω_0 is referred to as the "cutoff" frequency.

EXAMPLE 4.8 Solve the equation $|G_i|^{-2} = (\cosh 2x + \cos 2x)/2$ and show that for $|G_i|^2$ to fall to one-half of its dc value $\omega_0 \approx 2.43/rcl^2$. Note $x = l(\omega\gamma/2)^{1/2}$ and $\gamma = rc$.

Solution. The equation is nonlinear, but $\cosh 2x \gg |\cos 2x|$. Ignoring the second term yields a first approximation, namely, $\cosh 2x = 4$ or $2x = 2.07$. For this value, $\cos 2x = -0.48$ and so this value of $2x$ is too small. Iteration quickly yields a final value of $2x = 2.21$ or $\omega_0 \approx 2.43/rcl^2$ as desired.

A similar treatment for Z_{in} shows that phase (Z_{in}) approaches $-45°$ for large values of x and that $20 \log_{10}|Z_{in}|$ falls off at 3 dB/octave for large values of x.

A symbolic representation of the distributed-RC network is shown in Fig. 4.10(b).

At frequencies up to ω_0 given in Eq. (4.28), the behavior of the distributed network may be approximated by that of a simple lumped-RC network, as follows. Substituting

$$z = l\sqrt{s\gamma} ; \quad r = R; \quad c = C \tag{4.29}$$

into Eqs. (4.22) and then using well-known expansions, yields

$$y_{11}(z) = \frac{z}{R} \coth(z) = \frac{1}{R}\left\{1 + \frac{z^2}{3} - \frac{z^4}{45} + \cdots\right\}$$

or

$$y_{11}(s) = \frac{1}{R} + \frac{sC}{3} - \frac{s^2 RC^2}{45} + \cdots \tag{4.30}$$

$$y_{12}(z) = -\frac{z}{R} \operatorname{csch}(z) = \left(-\frac{1}{R}\right)\left\{1 - \frac{z^2}{6} + \frac{7z^4}{360} - \cdots\right\}$$

or

$$y_{12}(s) = -\frac{1}{R} + \frac{sC}{6} - \frac{7s^2 RC^2}{360} + \cdots . \tag{4.31}$$

Taking only the first two terms of Eqs. (4.30) and (4.31) yields, with a slight further approximation, the π equivalent circuit shown in Fig. 4.11. Note that R and C are given by Eq. (4.29).

For this model, the short-circuit current gain is approximately

$$G_i(s) = \frac{y_{21}(s)}{y_{11}(s)} \approx -\frac{1}{1+j\omega/\omega_0'} \tag{4.32}$$

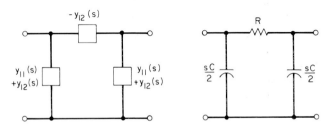

Fig. 4.11 π-equivalent circuit of RC transmission line.

where

$$\omega_0' = 2/RC. \tag{4.33}$$

Hence, the cutoff frequency ω_0' does not differ greatly from ω_0 of Eq. (4.30). At frequencies above ω_0, the lumped equivalent circuit of Fig. 4.11 no longer adequately represents the distributed line.

A T lumped-RC equivalent network obtained in a similar way is shown in Fig. 4.12.

At frequencies above ω_0, lumped modeling may be continued by taking many π (or T) sections in tandem, but this will not be discussed further.

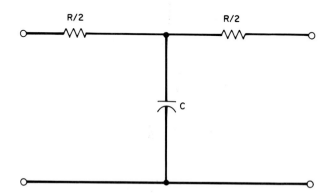

Fig. 4.12 T-equivalent circuit of RC transmission line.

For a thin-film resistor having sheet resistance R_S and capacitance per unit area C_0, also called capacitance density, one has from Eq. (4.28)

$$\omega_0 \approx \frac{2.43 R_S}{C_0 R^2 w^2} . \tag{4.34}$$

For a given resistance value R and selected sheet resistance R_S, plus accompanying parasitic capacitance density C_0, the cutoff frequency depends inversely on the square of the resistor width w. Hence, for high-frequency applications, resistor width should be kept as small as possible, while meeting other desired specifications.

At microwave frequencies, a capacitor is approximated by a short length of transmission line terminated in an open circuit as shown in Fig. 4.13(a). A solution of Eq. (4.22) gives

$$Z_{\text{in}} = \frac{r}{\sqrt{s}\,\gamma\,\tanh(\gamma l)} . \tag{4.35}$$

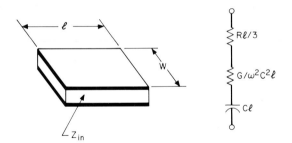

Fig. 4.13 Stripline capacitor and its equivalent circuit.

Expansion of $\tanh(\gamma l)$ and neglecting higher order terms gives

$$Z_{in} \approx \frac{R}{3} + \frac{1}{j\omega C} \,. \tag{4.36}$$

This impedance represents a capacitance of value C in series with resistance $R/3$ due to conductor losses as shown in Fig. 4.13(b). Capacitances of about 1 pF or more can be obtained in the 1 to 10 GHz frequency range. The overall Q, also accounting for possible dielectric losses, is in the range 30 to 3000.

4.7 MICROSTRIP LINE [8,9]

In this section, some properties of microstrip transmission lines are described. These lines find application in microwave integrated circuits. Figure 4.14 is a cross section of a microstrip transmission line. This structure is particularly attractive for fabrication by thin- or thick-film techniques because it consists of a film conductor on a dielectric substrate and a film counterelectrode on the reverse face of the substrate. It is a desirable structure for microwave integrated circuits because it is open and amenable to bonding of discrete semiconductor devices. In addition, the fabrication of microstrip is compatible with the fabrication of inductors, capacitors, and resistors using the film technology.

An additional reason for interest in microstrip is that conductors on printed circuit boards used in high-speed digital circuits are microstrip lines.

A rigorous treatment of microstrip lines is beyond the scope of this book. This section is basically a presentation of some results, due to Wheeler,[9] for calculating the characteristic impedance and some relations for the wavelength λ, the Q, and the attenuation constant α.

The case of primary interest is a conductor of rectangular cross section with width W and thickness t, deposited on a substrate of thickness h and

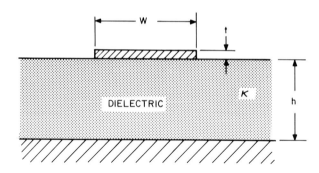

Fig. 4.14 Microstrip transmission line.

relative dielectric constant κ as shown in Fig. 4.14. Typical substrate materials are alumina, beryllia, and sapphire which have dielectric constants between 6 and 12. There is a conducting counterelectrode on the bottom surface of the substrate. Everything above the top surface of the substrate is immersed in a second dielectric of relative dielectric constant κ', which is usually air with $\kappa' = 1$. Such a line will be called a multiple-dielectric microstrip line.

It is convenient to build up the solution for Z_0 in stages. First, consider the case of a single uniform dielectric with relative dielectric constant κ. Wheeler showed that for $(W/h) < 2$

$$Z_0\sqrt{\kappa} = 376.687\left\{\frac{1}{2\pi}\ln\left[\frac{8h}{W_e}\right] + \frac{1}{16\pi}\left[\frac{W_e}{2h}\right]^2 - \cdots\right\} \tag{4.37}$$

and for $(W/h) > 2$

$$Z_0\sqrt{\kappa} = \frac{376.687}{\dfrac{W_e}{4h} + \dfrac{1}{2\pi}\ln\left\{17.08\left[\dfrac{W_e}{2h} + 0.94\right]\right\}} \cdot \tag{4.38}$$

In these equations W_e is the effective width of the conductor which is its geometric width W plus a correction factor to account for the nonzero thickness.

$$W_e = W + \frac{t}{\pi}\left[1 + \ln\frac{2x}{t}\right]. \tag{4.39}$$

The parameter x is given by

$$x = h, \qquad W > (h/2\pi) > 2t,$$

$$x = 2\pi W, \qquad (h/2\pi) > W > 2t. \tag{4.40}$$

The single-dielectric microstrip line with $\kappa = \kappa'$ has few applications because of its high radiation loss. The multiple-dielectric structure is preferable because it concentrates the electric field between the electrodes, thus reducing the loss. Wheeler showed that for the zero-thickness multiple-dielectric microstrip line

$$Z_0 = \frac{376.687}{\pi\sqrt{2(\kappa+1)}} \left\{ \ln\left[\frac{8h}{W}\right] + \frac{1}{32}\left[\frac{W}{h}\right]^2 \right.$$
$$\left. - \frac{1}{2}\left[\frac{\kappa-1}{\kappa+1}\right]\left[\ln\frac{\pi}{2} + \frac{1}{\kappa}\ln\frac{4}{\pi}\right]\right\} \tag{4.41}$$

is valid for narrow strips $(W/h) < 1$, and

$$Z_0 = \frac{376.687}{\sqrt{\kappa/2}} \left\{\frac{W}{2h} + 0.441 + 0.082\left[\frac{\kappa-1}{\kappa^2}\right] \right.$$
$$\left. + \frac{\kappa+1}{2\pi\kappa}\left[1.45 + \ln\left[\frac{W}{2h} + 0.94\right]\right]\right\}^{-1} \tag{4.42}$$

is valid for wide strips, $W/h > 1$. These equations are in good agreement with experiment for very thin strips $(t/h < 0.005)$ when the substrate dielectric constant κ is in the range from 2 to 10 and $0.1 < W/h < 5$.

Finally, the nonzero conductor thickness can best be approximated by replacing W by the effective width W_e defined by Eq. (4.39).

Another parameter, useful for evaluating the behavior of microstrip lines, is the effective dielectric constant of the multiple-dielectric structure. This quantity is used to obtain a uniform-dielectric microstrip line that is electrically equivalent to the multiple-dielectric line under consideration. An accurate empirical formula for κ_e was derived by Schneider.[8]

$$\kappa_e = \frac{\kappa+1}{2} + \frac{\kappa-1}{2}\left[1 + \frac{10h}{W}\right]^{-1/2}. \tag{4.43}$$

Figure 4.15 is a plot of $\sqrt{\kappa_e}$ versus W/h with the dielectric constant of the substrate as a parameter. Observe that it is not a strongly variable function. Using the concept of the effective dielectric constant, Schneider showed that the following relations hold for the important line parameters.

$$Z_0 = \frac{Z_0|_{air}}{\sqrt{\kappa_e}} \tag{4.44}$$

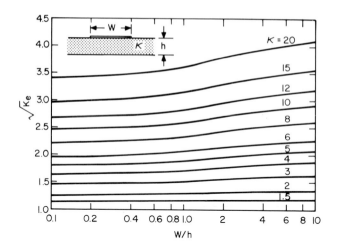

Fig. 4.15 Effective dielectric constant versus *w/h*. (Reprinted with permission from *The Bell System Technical Journal*, Copyright 1969, The American Telephone and Telegraph Co.)

$$\lambda = \frac{2\pi}{\beta} = \frac{\lambda|_{\text{air}}}{\sqrt{\kappa_e}} \tag{4.45}$$

$$\alpha = (\alpha|_{\text{air}})\sqrt{\kappa_e} \tag{4.46}$$

$$Q = \frac{\beta}{2\alpha} = Q_{\text{air}}. \tag{4.47}$$

In these equations the properties of a multiple-dielectric microstrip are related to those of a uniform air-dielectric stripline. The *quality factor Q* defined above may be computed from α and β, the attenuation and phase constant, respectively.

Figure 4.16 is a plot of the characteristic impedance of multiple-dielectric microstrip as a function of the *W/h* ratio for dielectric constants corresponding to several important substrate materials. Note that for the 50-Ω characteristic impedance which is popular in coaxial transmission line, $(W/h) = 1$ on an alumina substrate. A reasonable set of dimensions for the line on an alumina substrate would be $W = h = 20$ mils and $t = 0.25$ mil. The typical attenuation loss as measured at 30 GHz for a 3 in. length of line was 0.78 dB. The calculated loss for this line, assuming uniform current distribution, was 0.69 dB.

4.8 MONOLITHIC CAPACITORS [10]

Two types of monolithic capacitors, both voltage dependent, together with approximate equivalent circuits are shown in Fig. 4.17. A capacitor, with a breakdown voltage of the order of 20-50 V, formed from the reverse-biased

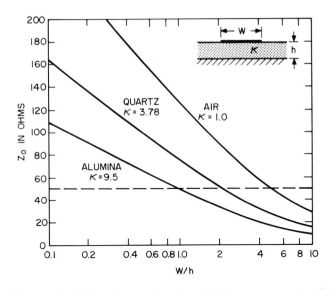

Fig. 4.16 Characteristic impedance of microstrip line versus *w/h*. (Reprinted with permission from *The Bell System Technical Journal*, Copyright 1969, The American Telephone and Telegraph Co.)

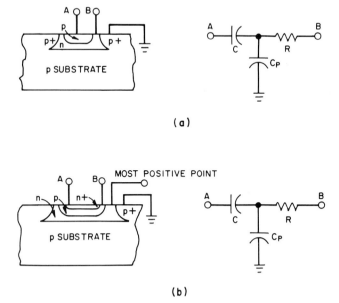

Fig. 4.17 Cross-sectional view and equivalent circuit for two types of monolithic capacitors: (a) collector-base junction and (b) emitter-base junction.

collector-base junction is shown in Fig. 4.17(a). This type of capacitor has a capacitance density C_{bc} in the range of 10^3 to 10^5 pF/cm^3 and with reasonable choices for areas A_b and A_s, total capacitances of the order of 50 pF are possible, without consuming excessive chip area. The capacitor is degraded by a voltage dependent parasitic capacitance C_p, formed by the reverse-biased collector-substrate junction. This parasitic capacitor is identical to the parasitic capacitor C_1 for the base-diffused resistor. Under normal working conditions, the collector-substrate junction is heavily reverse biased, which decreases the value of the parasitic C_p. Nevertheless, the ratio of C_p/C is of the order of 1/10 and it can increase if the reverse bias on the base-collector junction is increased. The capacitor is further degraded by a parasitic resistance R, due to the collector series resistance, that is typically under 100 Ω. This resistance causes the capacitor to have a cutoff frequency ω_0, which is inversely proportional to the RC product. Beyond this cutoff frequency, the structure no longer behaves as a capacitor. This type of capacitor may be used in series or shunt applications, but its voltage dependence causes signal distortion.

Figure 4.17(b) shows another voltage-dependent capacitor that can be used in series or shunt. It is formed from the reverse-biased base-emitter junction. The breakdown voltage of this junction is typically of the order of 5 V. The capacitor is degraded by a parasitic capacitance formed from the reverse-biased collector-base junction and by a parasitic resistance due to the base resistance that is typically about 200 Ω. The ratio of C_p/C is again of the order of 1/10. The capacitor has a cutoff frequency inversely proportional to the RC product.

Both capacitors described have a temperature coefficient of capacitance denoted by

$$TCC = \frac{1}{C} \frac{dC}{dT} \tag{4.48}$$

that is typically of the order of 100 ppm/°C.

These two capacitors have many disadvantages, some of which are overcome by the metal-oxide-semiconductor capacitor, shown in Fig. 4.18. As may be seen, this capacitor utilizes a highly doped n^+ region as its lower electrode and a layer of silicon dioxide having a relative permittivity of about four as its dielectric layer, with a top electrode formed by aluminum metallization.

EXAMPLE 4.9   An MOS capacitor is formed by growing 10^{-5} cm of silicon dioxide over an n^+ layer, whose average concentration $N_D = 2 \times 10^{17}$ cm^{-3}. Show that the oxide-layer capacitance $C_{ox} = \epsilon_{ox}/x_0$ is effectively in series with the depletion-layer capacitance $C_D = \epsilon_s/x_d$, such that the total capacitance $C^{-1} = C_{ox}^{-1} + C_D^{-1}$.

At what applied voltage V is C reduced to $0.80 C_{ox}$ and will an inversion layer have formed already? Take $\epsilon_s/\epsilon_{ox} = 3.69$.

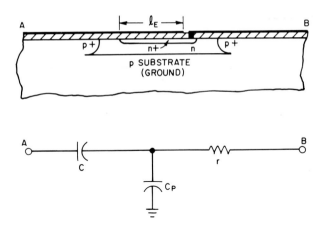

Fig. 4.18 Cross section and equivalent circuit of an MOS capacitor.

Solution. The fact that C_{ox} and C_D are in series is discussed in Section 3.3 [see Eq. (3.30)]. The applied voltage $V = V_{ox} + \psi_s$ as in Eq. (3.21). From $C = 0.8C_{ox}$ one has $C_{ox}^{-1} = 0.8C^{-1} = 0.8(C_{ox}^{-1} + C_D^{-1})$ and so $C_D = 4C_{ox}$. But for two capacitors in series $V_{ox}C_{ox} = C_D\psi_s$ and so $V_{ox} = 4\psi_s$, also. From the former equation $\epsilon_d/x_d = 4\epsilon_{ox}/x_0$ and so

$$x_d = \frac{11.8\times10^{-5}}{4\times3.2} \approx 9.22\times10^{-6}.$$

From Eq. (3.27)

$$2\psi_s = x_d^2 qN_A/\epsilon_s$$
$$= \frac{(85\times10^{-12})(1.6\times10^{-19})(2\times10^{17})}{11.8(8.85\times10^{-14})} = 2.6 \text{ V}.$$

Hence, $\psi_s = 1.3$ V, $V_{ox} = 4\psi_s = 5.2$ V and $V = \psi_s + V_{ox} = 6.5$ V. Onset of strong inversion occurs at

$$\psi_s = 2\psi_F = 2V_T\ln\left(\frac{N_D}{n_i}\right) = 2(26\times10^{-3})\ln\left(\frac{2\times10^{17}}{1.8\times10^{10}}\right) = 0.844 \text{ V}$$

and so an inversion layer has already begun to form by the time $C = 0.8C_{ox}$.

When a positive voltage is applied to the top electrode, positive surface charge forms at the metal-silicon-dioxide interface. This charge is exactly balanced by a negative surface charge which is provided by the n^+ silicon bottom electrode and appears at the silicon silicon-dioxide interface. Since no depletion region is formed, the capacitance is voltage independent. On the other hand, when a negative voltage is applied to the top electrode, a positive charge must be supplied to the silicon silicon-dioxide interface by the n^+ bottom electrode. This can only be accomplished by the creation of a depletion region in the n^+ silicon bottom electrode. Hence, the effective capacitance is partially voltage dependent. This voltage dependence is a function of the oxide thickness and surface impurity concentration of the n^+ layer, and is negligible when compared to the monolithic junction capacitor. Thus, the MOS capacitor introduces very little signal distortion.

The series parasitic resistance, of the order of a few ohms, associated with this capacitor is due to the diffused, highly doped emitter layer. By choosing a suitable area, the value of this capacitor can equal that of the monolithic capacitors, and so the cutoff frequency of the structure is higher.

The parasitic capacitance C_p associated with this capacitor is due to the voltage-dependent reverse-biased collector-substrate junction. The degradation ratio C_p/C is of the same order as before, but is less serious because C is now practically constant. The *TCC* for this structure is of the order of 20 ppm/°C. This capacitor finds wide application both in analog and digital circuits.

4.9 THIN-FILM CAPACITORS [11,12]

Thin-film capacitors usually have a three-layer parallel-plate structure, as shown in Fig. 4.19. The capacitor consists of a bottom or base electrode, a dielectric film and a top or counter electrode. The effective capacitor area A is that of the electrode overlap. The low-frequency capacitance density for this structure is given by $C/A = \epsilon_0 \kappa/d$, where $\epsilon_0 = 8.86 \times 10^{-12}$ F/cm is the permittivity of free space, κ is the relative dielectric constant of the film material and d the dielectric film thickness.

At high frequencies, the effective capacitance density may decrease due to losses in the dielectric, and in the electrodes and leads.

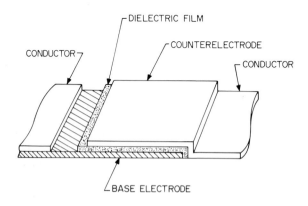

Fig. 4.19 Three-layer, parallel-plate, thin-film capacitor.

The theoretical breakdown voltage, V_D, of the dielectric is related to the dielectric material strength E_d by $V_D = E_d d$, but for dielectric films E_d depends on the types and density of defects and tends to be smaller than the bulk value for the material. As a consequence, one can introduce film dielectric material strength $E_d' < E_d$, where E_d is the strength of the bulk material. Hence $d = V_D/E_d'$ and so d tends to be greater than it would be if computed from the bulk. Because of the complexity and variation of breakdown phenomena, the rated or working capacitor voltage V is given by $V = kV_D$, where constant k has a value lying approximately between 1/10 and 1/2. Hence $d = V/kE_d'$, with E_d' experimentally determined for a particular thin-film dielectric. For SiO_2, the value of d lies in the range from 0.1 to 1 μm. For Ta_2O_5 the range is typically from 0.2 to 0.4 μm. The upper limit for d is usually determined by mechanical stress or by the actual dielectric formation process. The lower limit for d depends on the experimental value of E_d' and sets the upper limit for C/A.

Capacitor losses are due to dissipation in the dielectric, and at high frequencies, to electrodes and leads. This situation may be modelled in terms of the loss tangent δ, where

$$\tan \delta = \tan \delta_0 + RC(\omega - \omega_0) u(\omega - \omega_0). \tag{4.49}$$

Here $\tan \delta_0$ is the dielectric loss tangent, considered constant, while R is the electrode and lead high-frequency resistance. Function $u(\cdot)$ is the unit step function. Resistance R may be reduced by either choosing low-loss electrode materials, a favorable geometric layout, or both.

Other common capacitor materials used for overlap capacitors are Al or Ti-Pd-Au bottom electrodes, a SiO_2 dielectric film and Al or NiCr-Au top electrodes. The approximate capacitance range for this type of capacitor is from 10 to 1000 pF. A related structure having aluminum top and bottom electrodes has a dielectric nearer in composition to something between SiO and SiO_2, and similar properties to SiO_2. Dielectrics of SiO tend to have high-leakage currents, high-loss tangents, and large values of TCC and so do not tend to be used by themselves alone.

4.10 INTERDIGITAL CAPACITORS [13]

To reduce the resistance of lead-in electrodes for the overlap thin-film capacitors discussed in the previous section, the width W of the capacitor may be increased, while reducing the overlap, for a fixed value of capacitance C. However, such a reduction of R is limited as the width may become excessive, while the overlap may become so small as to cause registration errors, with the resultant large variations in capacitance. At this limit, the interdigital capacitor shown in Fig. 4.20(a) is recommended. It consists of two metal terminal strips with interleaving digital fingers, mounted on a substrate and

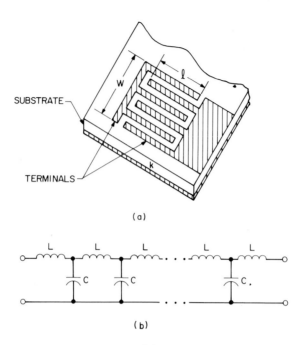

(a)

(b)

Fig. 4.20 (a) Interdigital capacitor and (b) transmission line equivalent circuit.

overlapping a ground plane. There is capacitance from each finger to the ground plane. Moreover, the terminal strip has a small amount of self-inductance between any two adjacent fingers. Therefore, the interdigitated capacitor may be considered to be a capacitively loaded transmission line having the equivalent circuit shown in Fig. 4.20(b). At low frequencies, the parasitic inductors L may be neglected and all capacitors may be considered to be in parallel. The total capacitance is roughly proportional to the number of fingers, and is in fact given by

$$C = \frac{(\kappa+1)\epsilon_0 l}{W} [(N-3)A_1 + A_2] \quad \text{F/length} \tag{4.50}$$

where N is the number of fingers, $\kappa\epsilon_0$ is the relative permittivity, l the finger overlap, and W the capacitor width. Constants A_1 and A_2 represent the contribution of the interior and two exterior fingers, and are tabulated as functions of the ratio of substrate thickness to linewidth of the digital fingers.

At high frequencies, the above formula is no longer adequate and the inductive parasitics must be taken into account. The capacitor must then be treated as a periodic structure with a characteristic impedance that is a function of capacitive loading.

Sometimes such capacitors consist of a bottom floating tantalum or aluminum-tantalum electrode, a tantalum-pentoxide dielectric, and a top

electrode of nickel-chromium-gold as shown in Fig. 4.21. This capacitor may
be considered as a parallel combination of "unit-cells," each composed of
two back-to-back, series-connected overlap capacitors. These unit cells have
low losses because the digital fingers present a short, wide lead-in. The
parallel connection of cells further reduces capacitor loss in proportion to the
number of unit cells. The capacitor is bipolar with an approximate capacitance range of from 100 to 10,000 pF, and an approximate working voltage
of 30 V.

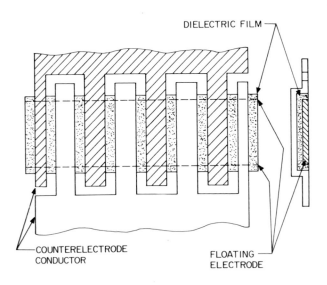

DIELECTRIC FILM

COUNTERELECTRODE
CONDUCTOR

FLOATING
ELECTRODE

Fig. 4.21 Interdigital bipolar capacitor.

4.11 INDUCTORS [14]

Inductance may be realized by a single metallic strip or ribbon of width W
and thickness t. The inductance per unit length of such a single strip is
given by

$$L = 5.08 \times 10^{-3} \left[\ln \left(\frac{l}{W+t} \right) + 1.19 + 0.22W + \frac{t}{l} \right]. \qquad (4.51)$$

The resistance per unit length of such a ribbon is $R = \rho/2(W+t)\delta$
where δ is the skin depth which is proportional to the square root of frequency. For good Q, the ribbon thickness should be several skin depths.
Because of current crowding, the Q is given by

$$Q = \omega L/kR \qquad (4.52)$$

where the coefficient k is an increasing function of w/t and $1.3 < k < 2$.

The size of lumped elements can be greatly reduced because of advances in photolithography. This extends their usefulness to frequencies up to about 10 GHz. In particular, the thin-film spiral inductor shown in Fig. 4.22 provides several nanohenries of inductance, where

$$
L_{\text{spiral}} = 12.57 \, dn^2 \left[2.303 \left(1 + \frac{b^2}{32a^2} + \frac{c^2}{96a^2} \right) \log_{10} \left(\frac{8a}{d} \right) \right.
$$
$$
\left. + \frac{1}{2} + 0.597 \, \frac{c^2}{16a^2} \right] + 12.57 an \, (A_1 + B_1)
$$

$$(4.53)$$

and where b is the film thickness, $D = W + S$ is the pitch, n is the number of turns, $c = nD$, $d = \sqrt{b^2 + c^2}$, $a = a_1 + (n-1)D/2$, W is the width of the path, S is the space between turns, and A_1 and B_1 are constants dependent on b, W, and d.

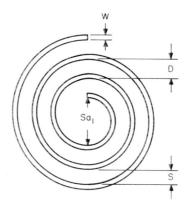

Fig. 4.22 Thin-film spiral inductor.

All dimensions in Eq. (4.53) are in centimeters and L is in nanohenries. One complication with the spiral inductor is that the inner end of the spiral has to be contacted either by means of a wire bond or by means of a film conductor which passes under the spiral. A correction factor must be added to Eq. (4.53) if the wire bond is used. If the film crossunder is used, there will be a small parasitic capacitance between the conductor and each turn of the spiral. This is undesirable because it degrades performance. As a further precaution, when the spiral inductor is fabricated on a microstrip substrate, the grounding-plane metallization immediately underneath the spiral should be removed to reduce the parasitic capacitances to each turn of the spiral.

4.12 COUPLERS [15]

Couplers are multiport devices that are used to transfer energy from one circuit to another. A typical use for a directional coupler is in the measurement of incident and reflected power in order to determine the standing wave ratio.

Coupling takes place whenever two transmission lines are in close proximity. Figure 4.23 shows two coupled microstrip transmission lines. As may

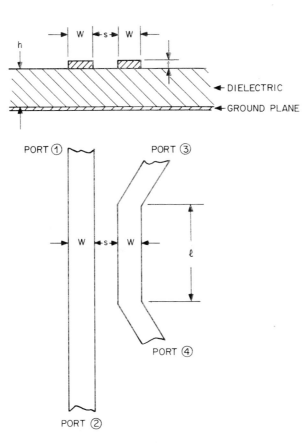

Fig. 4.23 Coupled microstrip transmission lines.

be seen, a length of the lines is a small distance *s* apart. The thickness of the conducting strips is usually several skin depths. When *l* = λ/4, the device is said to be a quarter-wave coupler. Ports 1, 2, 3, and 4 of the microstrip lines are terminated in their characteristic impedances. When there is coupling between ports 1 and 3 while there is no coupling between ports 1 and 4, and ports 2 and 3, then the device is said to be a "backward coupler." Such a coupler is said to have "perfect directivity" from port 1 when port 4 is perfectly isolated from port 1. The "degree of coupling" is a function of the normalized dimensions *W*/*h* and *s*/*h*. When there is coupling between ports 1 and 4, while there is no coupling between ports 1 and 3, and ports 2 and 4, then such a device is called a "forward coupler."

Coupling between parallel microstrip lines may also be through joining branch lines of finite length as shown in Fig. 4.24. Such couplers are well suited to tight coupling. The characteristic impedances of the main and

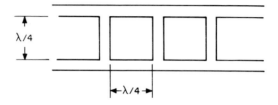

Fig. 4.24 Branch line coupler.

branch lines can be changed from section to section. The length of the branch lines and their spacings are a quarter wavelength at some specified midband frequency. The design equations for couplers are usually very complicated and a digital computer is normally required for coupler design.

The directional coupler is a useful element in microwave circuits, but coupling as just described can occur between adjacent conductors on a printed-circuit wiring board. If one considers a high-speed digital system with pulse rise times of a few nanoseconds, then there is considerable energy content even at frequencies up to several gigahertz. In this situation, printed-circuit wiring boards must be designed to minimize coupling between lines that could adversely interact.

4.13 CIRCULATORS AND ISOLATORS

An ideal *m*-port circulator is defined to be a lossless network in which the signal energy incident on the *i*th port is transferred solely to the (*i*+*l*)st port, while no transmission to the other ports occurs. This effect can be described by the port impedance matrix, which for the case of the ideal loss-less three-port circulator, symbolically shown in Fig. 4.25(a), is given by

$$\mathbf{R} = \begin{bmatrix} 0 & -r & r \\ r & 0 & -r \\ -r & r & 0 \end{bmatrix}$$

where r is called the gyration resistance. The zeros on the diagonal indicate zero port losses and the negative entries are due to the nonreciprocal nature of the circulator.

The scattering matrix $\mathbf{S}$ for the circulator, which describes the transmission of energy between ports and the reflection of energy at ports is derivable from $\mathbf{R}$. It has nonzero elements only in the 12, 23, and 31 positions indicating that all of the energy incident on port 1 is transferred to port 2, all of the energy incident on port 2 is transferred to port 3, and all of the energy incident on port 3 is transferred to port 1. There is no energy reflection at the ports.

A common application of the three-port circulator is in the isolator configuration shown schematically in Fig. 4.25(b). Port 3 is terminated by a matched load and the isolator is now a unilateral two-port with transmission only from port 1 to port 2. There is no reverse transmission from port 2 to port 1. Actually, the signal in the reverse direction is absorbed by the load terminating the third port. This isolator has a power rating determined by the capability of the load rather than by the circulator junction.

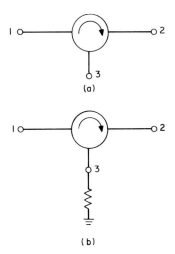

Fig. 4.25 (a) Ideal lossless three-port circulator and (b) an isolator using a three-port circulator.

Figure 4.26 shows a three-port microstrip-line circulator consisting of a transmission line junction of three metal center conductors. The junction is a circular surface S and is sandwiched between two circular ferrite discs made of garnet material. The conductors and ferrite slabs are then placed between

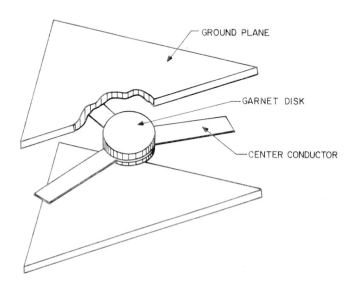

Fig. 4.26 Microstrip three-port circulator.

metal ground planes. A dc magnetic field H_0 is applied perpendicular to the surface of the ferrite discs. By solving Maxwell's equations for the ferrite disc, the electric field and the magnetic field can be calculated. A typical field pattern for the lowest mode excited in the ferrite when the bias field $H_0 = 0$ is shown in Fig. 4.27(a). The magnetic field H is shown by black lines and the electric field by the $\oplus$ for lines coming out of the paper. As may be seen, the electric field is zero along the magnetic-field line across the diameter running from south to north.

When the bias magnetic field H_0 is applied, the field pattern may be made to displace or rotate $30°$ producing the nonreciprocal situation shown in Fig. 4.27(b). Nonreciprocity occurs, because port 3 is situated at a null of the electric field and so transmission occurs between ports 1 and 2 only. This type of circulator is called a field displacement or resonant circulator.

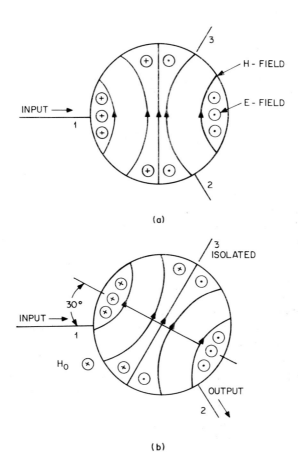

Fig. 4.27 Circulator field patterns: (a) before displacement and (b) after displacement.

REFERENCES

1. R. W. Berry, P. M. Hall, and M. T. Harris. *Thin Film Technology.* Princeton, N. J.: D. Van Nostrand, 1968, Chapters 7 and 11.

2. P. M. Hall. "Resistance calculations for thin film patterns." *Thin Solid Films* **1** (Elsevier, Amsterdam): 277-295 (1967-68).

3. S. K. Ghandi. *The Theory and Practice of Microelectronics.* New York: John Wiley & Sons, 1968.

4. R. M. Warner, Jr., and J. N. Fordenwalt, editors. *Integrated Circuits, Design Principles and Fabrication,* Motorola Series in Solid State Electronics. New York: McGraw-Hill, 1965.

5. C. S. Meyer, D. K. Lynn, and D. J. Hamilton, editors. *Analysis and Design of Integrated Circuits.* New York: McGraw-Hill, 1967.

6. R. D. Brooks, private communication.

7. D. J. Hamilton, F. A. Lindholm, and A. H. Morshak. *Principles and Applications of Semiconductor Device Modeling.* New York: Holt, Rinehart and Winston, 1971, Chapter 1.

8. M. V. Schneider, B. Glance, and W. F. Bodtmann. "Microwave and millimeter-wave hybrid integrated circuits for radio systems." *Bell Syst. Tech. J.* **48**: 1703-1726 (July-August 1969).

9. H. A. Wheeler. "Transmission line properties of parallel strips separated by a dielectric sheet." *IEEE Trans. Microwave Theory and Techniques* **MTT-13**: 172-185 (March 1965).

10. R. M. Warner and J. N. Fordenwalt, editors. *Op. cit.,* Chapter 10.

11. R. W. Berry, P. M. Hall, and M. T. Harris. *Op. cit.,* Chapters 8 and 11.

12. M. Coulton and H. Sobol. "Microwave integrated circuit technology—a survey." *IEEE J. of Solid-State Circuits* **SC-5**: 292-303 (December 1970).

13. G. D. Alley. "Interdigital capacitors for use in lumped element microwave integrated circuits." *G-MTT Symposium Digest,* pp. 7-13 (May 1970).

14. D. A. Daly, S. P. Knight, M. Coulton, and R. Ekholdt. "Lumped elements in microwave integrated circuits." *IEEE Trans. of Microwave Theory* **MTT-15**: 713-721 (December 1967).

15. J. Helszajn. "Nonreciprocal microwave junctions and circulators." New York: Wiley Interscience, 1976.

PROBLEMS

4.1 The TCR of a diffused semiconductor resistor at temperatures in the range 0°C to 50°C is 1000 ppm/°C. The base sheet resistance can be held to within ±8%, while linewidth can be held to within 2 μm.

Calculate the absolute accuracy of a resistor whose linewidth w equals 75 μm.

Neglecting the temperature variation, plot the accuracy as a function of linewidth, for values of w in the range $10 \leqslant W \leqslant 100$ μm.

4.2 It is desired to design a 5-kΩ base resistor which incorporates three bends. The base material has sheet resistance $R_S = 200$ $\Omega/\square$ and the resistor is to be 1000 μm long. If the edge of the oxide window is 6 μm from the edge of the track, find a suitable track width w. Take the track widening factor $d = 0.25$ μm.

4.3 Calculate the parasitic capacitances C_1 and C_2 for the base resistor shown in Fig. 4.4. Ignore the n^+ buried layer and assume a resistor width $w = 20 \ \mu$m and length $l = 400 \ \mu$m and epitaxial-layer dimensions $w = 40 \ \mu$m and $l = 600 \ \mu$m. Take the base-collector junction depth $x_{jb} = 3 \ \mu$m and the collector-substrate junction depth $x_{js} = 8 \ \mu$m. Furthermore take the following capacitance densities at the bottom and sidewall of the junctions:

$$C_{bc} = 200 \text{ pF/mm}^2; \quad C_{bcs} = 200 \text{ pF/mm}^2$$
$$C_s = 100 \text{ pF/mm}^2; \quad C_{ss} = 105 \text{ pF/mm}^2.$$

4.4 Show that the contribution to conductance due to the presence of sidewalls, caused by lateral diffusion, is given by

$$2 G_s L = \int_0^{x_j} r\sigma(r) \, dr \int_{\pi/2}^0 d\theta$$

where G_s is the sidewall conductance per unit length, where r is the radius vector of the sidewalls and θ its angle with the perpendicular.

Take the conductivity $\sigma(r)$ of the p-base region as

$$\sigma(r) = q\bar{\mu} N(r),$$

where $\bar{\mu}$ is the average hole mobility in p-type silicon and where the base-diffusion profile $N(r)$ is given by

$$N(r) = N_s \exp[-(r/L_p)^2].$$

Hence evaluate $2G_s L$. Using the formula $GL = W/R_S$, where R_S is the sheet resistance, show that the track widening factor d (mentioned in Problem 4.2) is exactly given by

$$d = G_s L R_S.$$

Using the following parameters for the p-base diffusion, evaluate d and compare its magnitude with x_j.

$N_s = 4 \times 10^{18} \text{ cm}^{-3};$	$N_B = 10^{15} \text{ cm}^{-3}$
$\bar{\mu} = 265 \text{ cm}^2/\text{V--s};$	$x_j = 3 \times 10^{-4} \text{ cm}$
$R_S = 200 \ \Omega/\square;$	$(x_j/l_p)^2 = \ln(N_s/N_B)$

4.5 A pinch resistor has length l, width w and capacitance C_{be} and C_{bc} per unit area, with reference to the base-emitter and base-collector junction. Show that for a thin pinch resistor the parasitic pinch capacitance equals approximately $C_{pinch} \approx (C_{bc}+C_{be}) wl + 2C_{pad}$, where C_{pad} is the capacitance of each contact-pad region. If a T-bar shaped resistor is taken, with the contact window a square W on a side and registration clearance W, show further that $C_{pad} \approx C_{be} (9W^2+6\pi Wx_j)$ where x_j is the base-collector junction depth.

4.6 Show that the π equivalent circuit parameters (see Fig. 4.11) for a section of a distributed-RC transmission line of length L and width W are approximately given by

$$G = \frac{1}{R} = \frac{\sigma t_1 W}{L} \ ; \quad C = \frac{\epsilon WL}{3t_2} \ .$$

The line consists of resistive material of conductivity σ and thickness t_1, deposited on a substrate of dielectric constant ϵ and thickness t_2.

Show also that the cutoff frequency, ω_0, for this structure is given by

$$\omega_0 = \frac{6t_1 t_2 \sigma}{\epsilon L^2} \ .$$

5

Physics and
Chemistry of Processing

This is the first of five chapters on the fabrication of various types of integrated circuits. It is devoted to a description of the principles of each process required to make either a silicon or a thin-film integrated circuit. The main topics of this chapter are methods of film deposition, doping of silicon, and properties of impurities in silicon. Subsequent chapters describe how these processes are combined to make various types of silicon and thin-film IC's, and how to do integrated circuit layout. Some readers might prefer to become familiar with the material in Chapters 6 and 8 before proceeding with this chapter. The design and fabrication of thick-film circuits is treated in Chapter 9.

The chapter begins with a description of vapor deposition techniques for producing films including resistance, electron-beam, and radio-frequency (RF) evaporation; dc and RF sputtering; sputter, plasma, and ion-beam etching for patterning films; chemical-vapor deposition of silicon dioxide and silicon nitride; and the epitaxial growth of silicon.

5.1 EVAPORATION [1,2]

The conversion of a solid or liquid to the vapor phase requires the addition of energy, often in the form of heat, to increase the energy of the atoms or molecules of the material. The amount of heat required to vaporize a unit quantity of material is called the *latent heat of evaporation, ΔH_e*. The *vapor pressure*, which is the pressure exerted by a saturated vapor on its condensed (solid or liquid) phase, is a function of temperature that is related to the latent heat of evaporation by the Clausius-Clapeyron equation

$$\frac{dp^*}{dT} = \frac{\Delta H_e}{T(V_g - V_c)} \tag{5.1}$$

where p^* is the vapor pressure at temperature T, and V_g and V_c are the volumes of the gaseous and condensed phases, respectively. Since ΔH_e, T, and $(V_g - V_c)$ are always positive, it is seen from Eq. (5.1) that the vapor pressure increases with temperature.

The vapors of metals at low pressure behave like an ideal gas. Neglecting the volume of the condensed phase in comparison with that of the vapor, and invoking the ideal gas law yields

$$V_g = \frac{RT}{p} = \frac{N_{Av}kT}{p}$$

where R is the universal gas constant, k is Boltzmann's constant and N_{Av} is Avogadro's number, 6.0248×10^{23} elementary particles/mole. Substituting these results with $p = p^*$ into Eq. (5.1) and integrating with ΔH_e constant, yields the following important relation for the functional form of the vapor pressure

$$\ln p^* \approx C - \frac{\Delta H_e}{RT} = C - \frac{B}{T} \tag{5.2}$$

where C is a constant of integration which cannot be derived from first principles.[3] This expression is a fairly good representation of the vapor pressure for most materials over a small temperature range at those temperatures for which the vapor pressure is less than 1 torr.[†] Vapor pressures for metals sufficiently high to be useful for evaporation require temperatures above about 600°C.

EXAMPLE 5.1 Aluminum is commonly deposited by evaporation. Its vapor pressure at 815 K is 10^{-11} torr and at 860 K it is 10^{-10} torr. Determine the constants B and C in Eq. (5.2) and use the results to predict the vapor pressure at 1000 K and 1000°C.

Solution. Substituting the given data into Eq. (5.2) yields

$$\ln 10^{-11} = C - (B/815)$$

$$\ln 10^{-10} = C - (B/860).$$

These equations can be solved for B and C with the result $B = 3.586 \times 10^4$ K and $C = 18.677$. At 1000 K the vapor pressure is

$$p^* = \exp\left[18.677 - \frac{3.5864 \times 10^4}{10^3}\right] = 3.4348 \times 10^{-8} \text{ torr}$$

[†] One torr is the pressure of 1 mm of mercury.

which compares very well with the value 4×10^{-8} torr given by Honig.[52] At $1000°C = 1273$ K the calculated vapor pressure is 7.5175×10^{-5} torr while the measured value is 2×10^{-4} torr.

Gas molecules move at very high speeds but they do not travel large distances because of frequent collisions with other molecules. The *mean-free path* λ is the statistical average distance that a particle travels between collisions and it characterizes the ability of a particle to travel undisturbed. It can be shown to be given by

$$\lambda = \frac{kT}{\sqrt{2}\,\pi\sigma^2 p} \tag{5.3}$$

where σ is the molecular diameter. Observe that λ decreases as the pressure increases because there are more particles with which to collide.

EXAMPLE 5.2 The molecular diameter of O_2 is 3.64×10^{-8} cm. Evaluate the mean-free path of oxygen molecules at 300 K at a pressure of 10^{-3} torr and 10^{-6} torr.

Solution. In the *cgs* system, Boltzmann's constant $k = 1.38044\times10^{-16}$ erg/K, and 1 torr $= 1.333\times10^3$ dyn/cm². Substituting the given data into Eq. (5.3) yields $\lambda = 5.278$ cm when $p = 10^{-3}$ torr and $\lambda = 5278$ cm when $p = 10^{-6}$ torr.

The actual path lengths, x, are statistically distributed with an exponential distribution

$$\eta(x) = \frac{1}{\lambda} \exp\left(-\frac{x}{\lambda}\right) \tag{5.4}$$

where $\eta(x)$ is the fraction of particles that has gone at least a distance x without collision. The source-to-substrate distance in the evaporation chamber is a function of the mean-free path because it is desirable for the evaporant molecules to travel in straight lines from the source to the substrate without colliding with other evaporant molecules or residual gas molecules.

It was first observed by Hertz that the *molecular evaporation rate* $dN_e/(A\,dt)$, at a specified vapor pressure, could not exceed an upper limit equal to the impingement rate. The *impingement rate, $dN_i/(A\,dt)$* is the number of molecules in a gas at rest that strike a unit surface in unit time. If the gas obeys Maxwell-Boltzmann statistics and the ideal gas relation, the impingement rate is

$$\frac{1}{A}\frac{dN_i}{dt} = \frac{p^*}{\sqrt{2\pi mkT}} = \frac{1}{A}\frac{dN_e}{dt}\bigg|_{max} \tag{5.5}$$

where N_i is the number of particles striking the surface of area A, p^* is the vapor pressure of the material at the temperature of the surface, and m is the mass of the molecule. The maximum evaporation rate is not achieved because some particles in the vapor phase return to the condensed phase thus forming a return flux. Furthermore, some of the particles in the eva-porant flux may have been reflected back into the gas. The Hertz-Knudsen equation for the evaporation rate R_{ev} is a modification of Eq. (5.5) which accounts for these effects.

$$R_{ev} = \frac{1}{A}\frac{dN_e}{dt} = \frac{\alpha_v(p^*-p)}{\sqrt{2\pi mkT}} \, . \tag{5.6}$$

In this equation α_v is the fraction of evaporant flux that makes the transition from condensed to vapor phase, p is the hydrostatic pressure of the return flux, and N_e is the number of particles evaporating.

The growth rate, G, of a film of constant density, ρ, is related to the evaporation rate by $G = R_{ev}m/\rho$ where m is the mass of an evaporant molecule. From Eq. (5.6)

$$G = \left[\frac{m}{2\pi kT}\right]^{1/2}\frac{\alpha_v(p^*-p)}{\rho} \, . \tag{5.7}$$

EXAMPLE 5.3   The vapor pressure of aluminum at 1830 K is 1 torr. Calculate the evaporation rate R_{ev} and the growth rate of the film under the assumption that the pressure of the return flux is zero and $\alpha_v = 1$.

Solution. The atomic mass of aluminum is 27 and therefore $m = 1.6398\times10^{-29}\times27 = 4.4275\times10^{-28}$ g. Substituting in Eq. (5.6)

$$R_{ev} = \frac{1.333\times10^3}{(2\pi\times4.4275\times10^{-28}\times1.38044\times10^{-16}\times1830)^{1/2}}$$

$$= 5.028\times10^{22} \text{ molecules/s.}$$

The density of aluminum is 2.7 and therefore the growth rate of the film is

$$G = \frac{5.028\times10^{22}\times4.4275\times10^{-28}}{2.7}$$

$$= 8.246\times10^{-6} \text{ cm/s} = 4.947\times10^4 \text{ Å/min.}$$

The growth rate is typically in the 10^4 to 10^5 Å/min range.

An important quantity is the sensitivity of molecular evaporation rate R_{ev} to changes in temperature because it specifies how tightly the process must be controlled. Substitution of Eq. (5.2) for vapor pressure into Eq. (5.6) yields an explicit expression for R_{ev} as a function of temperature. Differentiation of that expression yields

$$\frac{dR_{ev}}{R_{ev}} = \frac{dG}{G} = \left[\frac{B}{T} - \frac{1}{2}\right]\frac{dT}{T} \tag{5.8}$$

where $B = \Delta H_e / R$. The first term in the square bracket is about 20 to 30 for most metals and therefore a small change in temperature causes a large change in evaporation rate. Thus, accurate control of temperature is required to assure a controlled evaporation rate.

EXAMPLE 5.4 Calculate the variation in growth rate of the aluminum film due to a 1% change in temperature. Assume that the value of B obtained in Example 5.1 is valid.

Solution. From Eq. (5.8)

$$\frac{dG}{G} = \left| \frac{3.586 \times 10^4}{1830} - \frac{1}{2} \right| \times 10^{-2} = 0.191,$$

that is, a 1% change in temperature results in a 19% change in growth rate.

A nomogram for use in determining the evaporation rate is given in Fig. 5.1.

Condensation Energy is supplied at the source to convert the evaporant material from a condensed phase to the vapor phase. The evaporant molecules must release energy at the substrate if a film is to form. The minimum amount of energy released at the substrate when the film first forms is related to the vapor pressure of the deposit at the substrate temperature by

$$p^*(T_{\text{substrate}}) = a \, \exp\left(- \frac{L}{R T_{\text{substrate}}} \right)$$

where L has the order of magnitude of a desorption energy term and it depends on the strength of the bond formed between the deposit and the substrate. The thermodynamic condition for condensation of a vapor, on a film or substrate of the same material as the vapor, is that the (partial) pressure p of the evaporant be greater than the vapor pressure of the evaporant at the substrate temperature, that is, $p > p^*(T_{\text{substrate}})$. For atoms condensing on atoms of like material, $L \approx \Delta H_e$. For atoms condensing on a substrate of a different material

$$L = \Delta E_b + \frac{3}{2} R (T_{\text{source}} - T_{\text{substrate}})$$

where ΔE_b is the evaporant-substrate binding energy. This value is appropriate only for the growth of the first monolayer of film, after which the atoms are condensing on atoms of like material.

A high evaporation and deposition rate is desirable because it reduces film contamination from residual gas molecules whose rate of bombardment is constant for constant residual gas pressure. The heat resulting from the

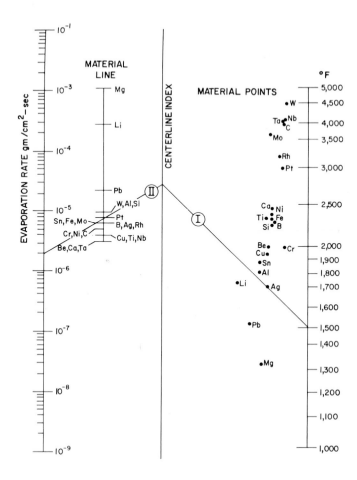

Fig. 5.1 Nomogram for determining the evaporation rate of selected materials in a vacuum. To use this nomogram, draw a line through the desired material point and the temperature at which the evaporation is to be conducted. The intersection of that line with the centerline index determines the index point. Draw a line through the index point and the material on the material line. The intersection of that line with the evaporation scale determines the evaporation rate.

release of energy accompanying condensation must be removed through the substrate holder and associated fixtures, and may place an upper limit on the deposition rate.

Theoretical investigations[4] indicate that incident evaporant molecules will reach thermal equilibrium with the substrate and be adsorbed if the ratio of incident kinetic energy to the energy of desorption is less than 25. Evaporant molecules will continue to be adsorbed without film nucleation until a certain critical population density occurs on the surface corresponding to a steady-state condition, with the flux of re-evaporating molecules balancing that of impinging molecules. However, adsorbed molecules are quite mobile and migrate over the substrate surface to form aggregates. These aggregates are more stable than individual molecules because the molecules are bound to each other by the condensation energy that is released when the aggregate forms. An aggregate whose size is above a critical value will, on the average, continue to grow and form a stable film. Stability of smaller aggregates is not determined solely by the condensation energy because the aggregate has a high ratio of surface area to volume and this means that it has a high surface energy.

Not all atoms impinging on a surface are adsorbed. Some incident atoms are reflected, almost without loss of kinetic energy. After a short dwell time on the surface, some atoms are adsorbed and subsequently re-evaporated. Alternatively, after a short stay on the surface, some atoms react chemically with other atomic species on the surface. The heat of reaction for these chemical reactions is typically 100 kcal/mole. Finally, the incident atoms may associate with evaporant molecules already on the surface.

More details of the nucleation and film-growth processes are found in the references.[4]

Variation of film thickness   Equation (5.6) specifies the evaporation rate without accounting for any directionality of the source. However, directional effects cause the evaporation rate, and therefore the thickness of the deposited film, to be a function of position. Thickness variation of the film is also a function of the orientation of the substrate with respect to the source. It is often necessary to deposit films on a surface that has steps caused, for example, by windows opened in one film layer to permit contacting an underlying layer. The problem of step coverage to produce a continuous film is discussed later in this section.

Consider a point-like source of area dA_e which has spherical symmetry, and a receiving element of area dA_r oriented as shown in Fig. 5.2. It is assumed that the evaporant molecules have a Maxwell-Boltzmann speed distribution at the moment of departure from the source and that all directions of flight are equiprobable. Therefore, the evaporated mass is deposited uniformly on the inside surface of a sphere of radius r. The normal to the receiver area makes an angle ϕ with the line connecting the source and the receiving surface whose subtended area is $dA = r \, d\Omega$ where $d\Omega$ is the subtended solid angle. If the film has a constant density ρ, then its thickness in

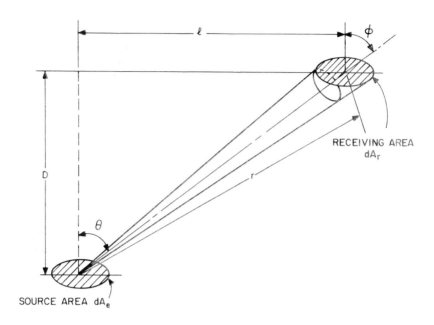

Fig. 5.2 Evaporation from a source of area dA_e onto a plane-parallel receiver.

the differential area dA_r is

$$t = \left[\frac{m}{2\pi kT}\right]^{1/2} \frac{\alpha_v(p^*-p)\cos\phi}{4\pi\rho r^2} . \tag{5.9}$$

Film thickness as a function of position on a finite-area plane-parallel receiver at a perpendicular distance D from the source is found by observing, from Fig. 5.2, that the angle of incidence ϕ equals the angle of emission θ, and substituting for $\cos\phi$ in terms of D and l. The thickness distribution, normalized with respect to the thickness at $l = 0$, is

$$\frac{t}{t(0)} = \frac{1}{\left[1 + (l/D)^2\right]^{3/2}} \tag{5.10a}$$

where $t(0)$ is given by

$$t(0) = \left[\frac{m}{2\pi kT}\right]^{1/2} \frac{\alpha_v(p^*-p)}{4\pi\rho D^2} . \tag{5.10b}$$

   The thickness distribution of a film deposited on a plane-parallel
receiver from a small, flat, plane-surface source can be shown to be

$$\frac{t}{t(0)} = \frac{1}{[1 + (l/D)^2]^2} \,. \tag{5.11}$$

   The plots shown in Fig. 5.3 compare the thickness distribution of films
deposited by each of these sources.

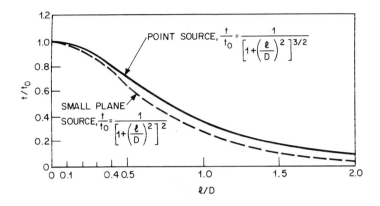

Fig. 5.3 Thickness distribution of films deposited from a point source and a flat
plane source.

EXAMPLE 5.5   (a) A plane substrate is placed at a perpendicular distance of
20 cm from a point source. How large can the substrate be if the film thickness variation is to be kept under 10%? (b) Repeat for a flat plane source.

Solution.   (a) From Eq. (5.10a) with $t/t(0) = 0.9$ and $D = 20$ cm

$$l = D\sqrt{[t(0)/t]^{2/3} - 1} = 20\sqrt{(1/0.9)^{2/3} - 1} = 5.4 \text{ cm}$$

on each side of the perpendicular. Put in another way, the total substrate area over
which a film can be deposited with 10% thickness control is 116 cm^2. (b) From
Eq. (5.11), $l = 4.65$ cm.

Step coverage [5] An important problem encountered in the deposition of
metallization for silicon integrated circuits is the covering of steps. These
steps are caused by windows that are opened in the silicon dioxide layer that
covers the silicon surface (see Chapter 6) to permit the metallization to contact the silicon. Failure to cover these steps adequately can result in non-
functioning or nonreliable devices because of cracks in the metallization.

The problem is geometric in nature as shown in Fig. 5.4. This figure is drawn for the general case of a broad area source such as a sputtering target. (Sputtering is discussed in Section 5.2.) Evaporation sources are small area sources which compound the problem. From this figure it can be seen that part of the source is masked to an observer at point P on the substrate because the substrate has a step of height h and step angle θ. The invisible area of the source is a function of the location of the observer with respect to the step, the step height and angle, h and θ, the dimension of the source L, and the source-to-substrate distance D. Because of their small size, evaporation sources may be totally invisible at some observation points if the source is not centered over the step. From Fig. 5.4 it is seen that point will suffer *slope shadowing* if

$$\theta > \tan^{-1}\left[\frac{D}{(L/2) - x_0}\right] \tag{5.12}$$

where x_0 is the distance from point P to the base of the step. The shadowing, by the top of the step, of a point on the step near its base has been termed *geometric shadowing*. *Self-shadowing* is the shadowing that occurs during film growth due to the interaction of the changing surface geometry with the incoming flux. It is a time-dependent geometric shadowing.

Cracks can occur even with an extended source.[6] The depth of the crack is inversely proportional to the radius of curvature at the bottom of the step. The top edge of the step has little effect. Also, if the step is undercut so that there is a re-entrant portion, a crack occurs pointing toward the bottom of the undercut portion.

The uniformity and step coverage of evaporated films can be improved by moving the substrates in planetary motion during the deposition process. This is not quite the same as having an extended area source because at each particular instant of time, the incident flux is still due to a source of small

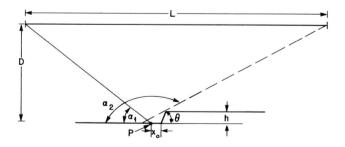

Fig. 5.4 Geometry of the step coverage problem.

angular range $\Delta\alpha$ (see Fig. 5.4) with all of the geometric constraints that that implies. However, the arrangement approximates an extended area source as the ratio of translational or rotational velocity to deposition rate increases.

      Figure 5.5 is a schematic drawing of a planetary motion deposition system. The substrates are held on a substrate holder of diameter A whose center is a distance D from the center of a small area source. The substrate holder is canted at an angle ψ with the horizontal. Commercial evaporators usually have three substrate holders, each about 20 cm in diameter, mounted in the deposition chamber. Each substrate holder rotates about its axis at about 150 rpm, and the entire assembly rotates about the axis through the center of the source at about 150 rpm. The condition for no geometric shadowing to occur for a step at the extreme end of the substrate holder is given by

$$\theta_{max} < \cos^{-1}\left[\frac{b \cos \psi - c \sin \psi + (A/2)}{\sqrt{D^2 + (L/2)^2} + L(b \cos \psi - c \sin \psi)}\right], \qquad (5.13)$$

where θ_{max} is the maximum permissible step angle.

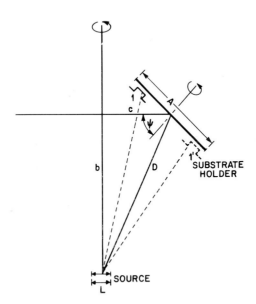

Fig. 5.5 Planetary motion deposition system.

EXAMPLE 5.6   What is the maximum permissible step angle for substrates that are to be coated in an evaporator with $\psi = 45°$, $b = 30$ cm, $c = 10$ cm, $A = 20$ cm, and $L = 2$ cm?

Solution.   From Eq. (5.13) and Fig. 5.5

$$\theta_{max} < \cos^{-1}\left[\frac{30 \cos 45 - 10 \sin 45 + (20/2)}{\sqrt{(30^2+10^2)} + 1 + 2(30 \cos 45 - 10 \sin 45)}\right] = 41°.$$

The maximum step angle is between 30° and 60° depending on the particular deposition apparatus.

Step coverage can be improved by heating the substrates to about 300°C during metal deposition.

5.1.1 Evaporation Techniques [1,7]

Evaporation is conducted in a vacuum environment at a base (initial) pressure in the range 10^{-7} to 10^{-6} torr in an evaporator such as that shown schematically in Fig. 5.6. Large evaporant charges are held in a crucible and are heated either by an electron beam, by RF induction, or by a resistance

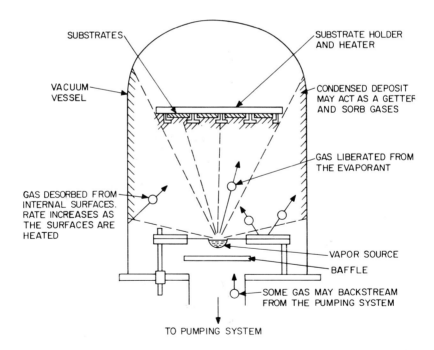

Fig. 5.6 Schematic of an evaporator.

heater. Resistance heating is rarely used in modern integrated circuit production facilities. The first two heat sources have the advantage of supplying at least some energy directly to the charge, which means that the crucible temperature does not have to exceed the vaporization temperature to provide heat flow. An RF-heated source designed specifically for evaporating aluminum is shown in Fig. 5.7(a). The crucible is made of a composite material consisting of boron nitride and tantalum diboride. This composite material is not attacked by aluminum. The crucible is supported on ceramic pillars and surrounded by a water-cooled RF coupling coil that is made of tubing. About two-thirds of the RF energy is absorbed within one skin depth of the surface, and therefore the frequency of the RF supply must decrease as the size of the evaporant charge increases. Frequencies of several hundred kilohertz are sufficient for charges weighing several grams, whereas a charge weighing about a kilogram requires a frequency of about 10 kHz. The power supply typically delivers 500 W. Note that the thickness of the upper portion of the crucible is reduced to avoid the problem of molten aluminum migrating upward and spilling over the brim.

Electron-beam (E-beam) evaporation of materials has advanced from a laboratory art to a production technique in the past few years. The advantages of E-beam evaporation are high evaporation rates, freedom from contamination, precise control, high thermal efficiency, and the possibility of depositing many new and unusual materials. Alloys are usually difficult to evaporate because the constituents have different vapor pressures which often result in films of varying composition unless special precautions are

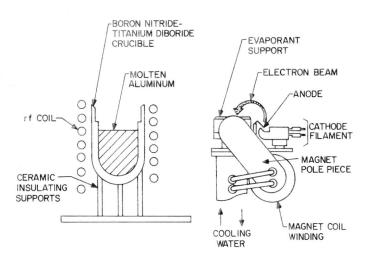

Fig. 5.7 (a) An RF heated source for aluminum evaporation; (b) a 270° bent beam E-gun source.

taken. However, with electron-beam evaporation difficult alloys with vapor-pressure ratios of 100:1 have been processed.

Heating by means of an electron-beam occurs when the beam impinges on the evaporant material and the kinetic energy of the electrons is converted to thermal energy. E-beam guns with about 10 kW capability (5 kV at 2 A) are used for IC fabrication, although guns operating at power levels as high as 1.2 MW have been built. The schematic drawing of a typical 270° bent-beam electron gun is shown in Fig. 5.7(b). Electrons emitted by an incandescent tungsten filament are shaped into a beam by the cathode beam former. The cathode and filament are held at a negative potential and the electrons are attracted towards the grounded anode. The cathode is physically close to the evaporant charge but is offset to protect it from contamination and erosion by ion bombardment. The beam is directed onto the evaporant charge, which is held in a water-cooled crucible or hearth, by means of a magnetic field which effectively bends the beam 270°. Frequently the material to be evaporated is available in rod form which can be continuously fed into the crucible. This is particularly useful for the evaporation of alloys because the rod provides a continuous feed of evaporant, thus producing a steady-state condition so that differences in the vapor pressure of the constituents do not alter the composition of the growing film.

During the deposition process, the pressure in the deposition chamber rises from its base value of 10^{-7} to 10^{-6} torr to about 10^{-5} torr because of outgassing from the various materials in the chamber. It is necessary to continually pump the chamber to maintain the pressure at the desired value. The capacity of the pump, known as the pumping speed, is a function of the gas flow rate in the chamber

$$Q = \frac{d}{dt}(pV) = V\frac{dp}{dt} \tag{5.14}$$

and the conductance of the piping. The last equality in Eq. (5.14) is due to the constant volume of the chamber. The conductance of a pipe is defined as the gas flow rate per unit pressure differential

$$g = \frac{Q}{p_1 - p_2} \tag{5.15}$$

where p_1 is the input pressure and p_2 is the output pressure. The units of g are volume/time.

EXAMPLE 5.7   What is the conductance of (a) two pipes in series and (b) two pipes in parallel?

Solution.   (a) The overall conductance of two pipes in series is found by observing that the pressure drops add. Therefore, from Eq. (5.15)

$$g = \frac{q}{p_1 - p_3} = \frac{q}{(p_1 - p_2) + (p_2 - p_3)} = \frac{1}{\frac{1}{g_1} + \frac{1}{g_2}}$$

where g_1 and g_2 are the conductances of pipes 1 and 2, respectively. (b) For pipes in parallel, the gas flow rates are additive. Hence

$$g = \frac{Q_1 + Q_2}{p_1 - p_2} = g_1 + g_2.$$

When the mean-free path λ of a gas molecule is greater than the diameter D of the pipe or aperature, the flow is said to be *molecular*. The conductance of a long cylindrical pipe in the molecular flow region is

$$g = 12.2 \, \frac{D^3}{\lambda} \quad \text{liters/sec.} \tag{5.16a}$$

For fairly short pipes, Dushman[53] suggests that

$$g = \frac{12.2 D^3}{\lambda[1 + (4D/3\lambda)]} \quad \text{liters/sec.} \tag{5.16b}$$

A quantity related to the conductance is the *pump speed* S_p defined by

$$S_p = \frac{Q}{p} = \frac{1}{p} \frac{d}{dt} (pV) \tag{5.17}$$

where p is the pressure at the point at which the pump speed is measured. The *speed of exhaust* S_E at the outlet of a vacuum chamber can be expressed in terms of the pump speed and conductance of the piping by

$$\frac{1}{S_E} = \frac{1}{S_p} + \frac{1}{g} . \tag{5.18}$$

The pumping speed of a diffusion pump with a liquid nitrogen cold trap of the size used for typical vacuum deposition apparatus is several hundred liters per second in the 10^{-4} to 10^{-9} torr pressure range.

5.2 SPUTTERING [7-11]

Sputtering is a method of physical vapor deposition that involves the removal of material from a solid cathode by bombarding it with positive ions from a rare gas discharge. The cathode may be made of a metal or insulator and, in contrast with thermal evaporation, complex compounds such as pyrex glass can be sputtered with a lesser degree of change in chemical composition.

The simplest way to visualize sputtering is as a process involving the transfer of momentum from impacting ions to surface molecules. If the bombarding ion has mass M_1 and kinetic energy E_1, and if the mass of the stationary target atom is M_2, then application of the laws of conservation of energy and conservation of momentum show that the maximum energy transferred to the stationary atom is

$$E_{max} = \frac{4M_1M_2}{(M_1+M_2)^2} E_1. \tag{5.19}$$

If the energy transferred to the stationary atom is greater than some threshold value, which ranges from about 10 eV to 35 eV, the stationary atom undergoes displacement.

The momentum of the bombarding ions is directed into the surface of the target, and therefore a reversal in its direction is necessary in order to cause the ejection of a target atom. The necessary momentum reversal occurs when the incident ion penetrates into the target where it collides with, and is reflected from, a lower atomic layer in a sequence of collisions as it gives up its energy to the target. If the incident ion has low energy, the mean energy $\bar{E}$ of the struck target atom in the collision is

$$\bar{E} = \frac{(E_{stat}+E_d)}{2} \tag{5.20}$$

where E_d is the threshold energy necessary to displace an atom and $E_{stat} \leqslant E_{max}$ is the kinetic energy of the initially stationary particle after the collision. When the bombarding ion has a very high energy, it buries itself deep in the target lattice and does not cause any sputtering.

An important parameter characterizing the sputtering process is the *sputtering yield, S*, which is defined as the number of ejected atoms per bombarding ion. Calculation of the sputtering yield involves determining the number of atoms displaced by a primary hit. An atom has energy $\bar{E}$ after a collision with a bombarding ion, and makes roughly one collision per interatomic distance. The number of collisions made in slowing down to an energy E_s is

$$N = \frac{\ln(\bar{E}/E_s)}{\ln 2} \tag{5.21}$$

where E_s is the energy binding an atom to the surface of the target. Using a random-walk model, the average number of atomic layers that contribute to sputtering is $1 + N^{1/2}$.

EXAMPLE 5.8   A target material has a binding energy of 10 eV and $\bar{E} = 500$ eV. How many collisions does the bombarding ion undergo in slowing to energy E_s and how many atomic layers contribute to sputtering?

Solution.   From Eq. (5.21) $N = \ln(500/10)/\ln 2 = 5.65$ collisions. The average number of atomic layers involved in sputtering is $1 + N^{1/2} = 1 + 5.65^{1/2} = 3.36$.

The total number of atoms displaced toward the surface per primary collision is $\bar{E}/2E_d$ because only half of the displaced atoms can migrate toward the surface. From these considerations the sputtering yield can be shown to be

$$S = \frac{\bar{E}}{4E_d} \left\{ 1 + \left[\frac{\ln(\bar{E}/E_s)}{\ln 2} \right]^{1/2} \right\} \sigma_p n^{2/3}. \tag{5.22}$$

where n is the number of atoms of the target material per unit volume, and σ_p is the collision cross section which is a function of the energy of the incident ion. In the medium energy range, $\sigma_p = \pi a^2$, where

$$a = \frac{a_0}{\sqrt{Z_1^{2/3} + Z_2^{2/3}}}$$

Z_1 is the atomic number of the bombarding ion, Z_2 is the atomic number of the target atom, and a_0 is the Bohr radius of the atom.

EXAMPLE 5.9   Calculate an approximate value for the sputtering yield of copper bombarded by 1 keV argon atoms. The threshold energy for the argon-copper ion-target pair is 17 eV.

Solution.   Calculation of the sputtering yield requires knowledge of the energies $\bar{E}$ and E_d, and the collision cross section σ_p. In practice the sputtering yield is measured and its value is used in various mathematical models to calculate σ_p. However, we will make several assumptions here to obtain an approximate value for the sputtering yield.

First of all, from Eq. (5.19) the maximum energy that can be transferred from a 1-keV argon ion to a copper atom is

$$E_{\max} = \frac{4(39.94)(63.54)}{(39.94+63.54)^2} \times 1 \text{ keV} = 948 \text{ eV}$$

where $M_1 = 39.94$ is the atomic weight of argon, $M_2 = 63.54$ is the atomic weight of copper, and $E_1 = 1$ keV is the energy of the incident ions. The energy imparted to the copper target atoms is $E_{\text{stat}} \leqslant E_{\max} = 948$ eV. It is assumed here that $E_{\text{stat}} = E_{\max}$, and therefore from Eq. (5.20), $\bar{E} = (948+17)/2 = 482.5$ eV.

A minimum value for the surface binding energy is obtained from the latent heat of vaporization which is $80,070 - 2.53T$ cal/mole for copper. At room temperature this yields a binding energy $E_s = 3.44$ eV/atom. In addition, the atomic volume of copper is 7.09 cm^3/mole, and therefore

$$n = \frac{6.023 \times 10^{23}}{7.09} \frac{\text{atoms/mole}}{\text{cm}^3/\text{mole}} = 8.495 \times 10^{22} \text{ atoms/cm}^3.$$

Finally, we assume that the collision cross section is πa^2 where a is related to the Bohr radius of the target atom. Strictly speaking, this is only true for intermediate values of ion energy. The Bohr radius of copper is $a_0 = 1.1693$ Å, and therefore

$$a = \frac{a_0}{\sqrt{Z_1^{2/3} + Z_2^{2/3}}} = \frac{1.1693 \times 10^{-8}}{\sqrt{18^{2/3} + 29^{2/3}}} = 2.895 \times 10^{-9} \text{ cm}.$$

Hence $\sigma_p = \pi a^2 = 2.634 \times 10^{-17}$ cm^2.

Substituting these values into Eq. (5.22) yields

$$S = \frac{482.5}{4 \times 17} \left\{ 1 + \left[\frac{\ln(482.5/3.44)}{\ln 2} \right]^{1/2} \right\} \times 2.634 \times 10^{-7} \times (8.495 \times 10^{22})^{2/3}$$

$$= 1.312.$$

Measured values for the sputtering yield of copper by argon atoms are between one and eight, depending on ion energy. At 1 keV the measured value is about two.

Gronlund and Moore[11] calculated the sputtering yield using Eq. (5.22) and obtained values that are about half those observed experimentally, but the trend of the yield curve as a function of incident ion energy is correctly predicted. The sputtering yield as a function of incident ion energy is shown in Fig. 5.8 for copper bombarded by argon ions. On the expanded scale it is seen that there is a sputtering *threshold energy* of about 5 to 25 eV for the incident ion, below which sputtering does not occur. This is followed by a small range in which sputtering yield increases linearly with incident ion energy, after which the yield saturates. It is in this region that most of the sputtering is done.

5.2.1 Sputtering Apparatus

Most often the source of ions for bombarding the target is from a glow discharge. A simple sputtering apparatus, shown schematically in Fig. 5.9, is basically a gas-filled diode with plane parallel cathode and anode. The cathode is the target and is made of the material to be sputtered. The substrate upon which the film is to be deposited is placed on the anode. This assembly is enclosed in a bell jar that is filled with a rare gas, usually argon, at a pressure in the order of 10 mtorr, and a sufficient potential difference is established between the cathode and anode to cause the gas to break down, thereby creating a glow discharge. The sputtering apparatus is usually operated in the so-called abnormal glow region of the glow discharge, because in that region the entire cathode is involved and the number of ions is large and controllable. Diode sputtering systems operate with a dc potential of 1 to 5 kV and a current density of 1 to 10 mA/cm^2. Electrodes are typically 5 to 50 cm in diameter with a 1 to 12 cm interelectrode spacing.

Deposition rate is proportional to the ion current I and the sputtering yield S

$$G = CIS \tag{5.23}$$

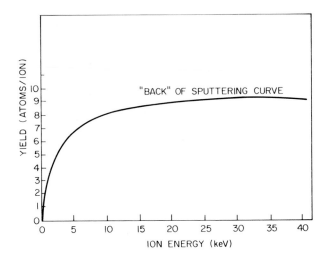

Fig. 5.8 Sputtering yield for copper bombarded by argon ions.

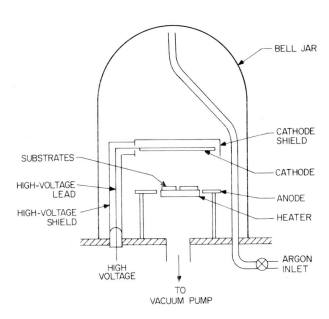

Fig. 5.9 Schematic of a diode sputtering apparatus.

where G is the deposition rate and C is a constant of proportionality that characterizes the particular sputtering apparatus. In addition to the mass of the bombarding ion and its energy, the sputtering yield is a function of gas pressure in the sputtering apparatus, because many of the ejected atoms diffuse back to the cathode when the gas pressure is high. This is important because the power available for sputtering is not unlimited and the only way to increase the ion current without increasing power is to increase the pressure. The practical upper limit on pressure is about 100 mtorr with deposition rates of 100 to 500 Å/min, even with materials such as tantalum which have a low sputtering yield.

Triode sputtering   The advantages of sputtering at low gas pressure include better adhesion of the film to the substrate, because the sputtered atoms have higher energy, and reduced contamination of the film by trapped gas molecules resulting in films of higher density and purity.

A method of maintaining the glow discharge at lower gas pressures than those required for diode sputtering is to inject electrons generated by a thermionic filament into the low-pressure system. The filament is placed in an elbow at the bottom of the triode sputtering apparatus shown in Fig. 5.10 to protect it from sputtered material, and the injected electrons are accelerated through a potential of several hundred volts toward the anode. The plasma is contained in the region between the target and substrate either

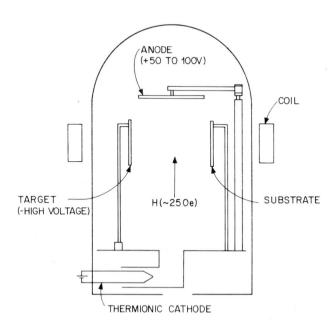

Fig. 5.10   A triode sputtering apparatus.

by a physical "chimney" or by a magnetic field produced by an externally mounted coil. Sputtering occurs when a negative potential of up to several thousand volts is applied to the target, thus attracting positive gas ions from the plasma and repelling electrons. The electron density in the plasma can be controlled by varying the filament current or the accelerating voltage between the filament and anode, while the energy of the bombarding ions is controlled by the target voltage. This independent control of the plasma density provides better control over the film growth rate.

Two disadvantages of triode sputtering are that (1) uniform sputtering from large-area targets is not possible because the ion density is greatest along the axis of the electron beam and at the end closest to the filament, and (2) sputtering in a reactive environment can shorten the filament life.

Reactive sputtering Sputtering is often conducted in the presence of a reactive gas, such as oxygen or nitrogen, to control or modify the property of the deposited film. Usually a relatively small amount of the reactive gas is added to the normal inert atmosphere, although sputtering in a pure reactive gas is possible. Reactive sputtering is one of the most important processes in the manufacture of thin films for electronic circuits because, in many cases, the film properties can be varied from that of a metal through semiconductor all the way to an insulator, simply by changing the proportion of reactive gas in the sputtering atmosphere.

The ratio of the arrival rate of the depositing film to the arrival rate of the reactant gas remains constant at the substrate in order to produce films with a controlled property. This implies that as the deposition rate of the film is increased, the flow rate of the reactant gas must also increase. The ion current, and therefore the flux of sputtered particles, can be increased by increasing the pressure in the sputtering chamber. As the total pressure is increased beyond a certain value, the sputtering yield begins to decrease. When this occurs, the sputtering rate no longer increases linearly with total pressure, and therefore the concentration of the reactant gas in the sputtering atmosphere must decrease.

The exact compound formation mechanism is not known. It is conjectured that at low pressure the reaction takes place at the substrate surface as the film is deposited. At high pressure the reaction is believed to occur at the cathode with the compound being transported to the substrate. A reaction in the vapor phase seems highly unlikely because some means of releasing the energy of formation of the compound and the kinetic energy of the atoms is necessary to prevent spontaneous decomposition.

Radio frequency excited sputtering Excitation by high-frequency radiation is a very important means of producing a discharge in a low-pressure gas. The RF field cannot inherently accelerate ions for bombardment of the target; instead, the sputtering is due to the "automatic" establishment of a dc self-bias by the action of the plasma. A magnetic field parallel to the RF field is useful for containing the electrons, thus improving the efficiency of the process.

A typical RF-diode sputtering apparatus is shown schematically in Fig. 5.11. It consists of a grounded substrate holder and a cathode (target) of the material to be sputtered, enclosed in a low-pressure gas atmosphere. This assembly is connected through an impedance matching network to an RF generator of 1- to 2-kW power output operating at a frequency of 13.56 mHz. An important advantage of RF sputtering is that the target can be either a metal or a *dielectric*.

Electrons have a much higher mobility than do ions, and therefore they are rapidly drawn in large numbers to the target during the time that it is

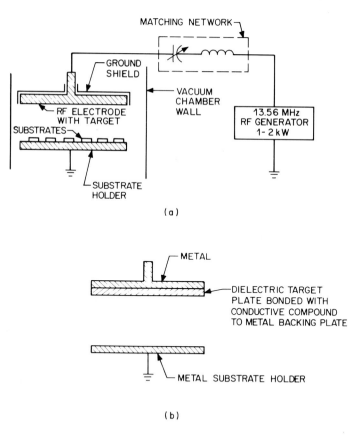

(a)

(b)

Fig. 5.11 (a) An RF diode sputtering apparatus; (b) target assembly for dielectric sputtering.

positive. Most of the current in the external circuit is due to electron collection and therefore the plasma assumes a positive dc potential which varies with applied RF voltage and increases with decreasing pressure. An external blocking capacitor, which may be part of the impedance matching network, causes the cathode to assume a negative potential with respect to the grounded anode. This voltage is of sufficient magnitude to permit sputtering of the target surface by the accelerated ions. Any positive charge developed on the cathode surface during the positive half cycle is neutralized by the collected electrons. This permits the sputtering of dielectric materials which is a great advantage of RF sputtering.

A target assembly for dielectric sputtering is shown in Fig. 5.11(b). The dielectric plate is bonded by means of conducting compounds to a metal backing plate. Note that no external dc blocking capacitor is required when the target is a dielectric.

Magnetron sputtering sources [12] Conventional sputtering sources described in the previous section suffer from two disadvantages; namely, the need for large area cathodes to coat a large substrate area, and excessive substrate heating due to secondary electron bombardment.

The recently developed magnetron-type sputtering source overcomes these disadvantages and promises to replace evaporation as a deposition source in many integrated circuit applications. A magnetron source, shown schematically in Fig. 5.12, consists of a conical annular cathode and a central anode. The anode-cathode assembly is surrounded by an annular magnet and a shield. When a glow discharge is initiated in the region between the central anode and the ground shield, the crossed electric and magnetic fields contain the electrons and force them into long, helical paths thus increasing the probability of an ionizing collision with an argon atom. When the cathode voltage is -700 V, the argon ion current is about 10 A ($\sim$150 mA/cm^2), which is one to two orders of magnitude higher than previous sources. Furthermore, secondary electrons emitted by the cathode due to ion bombardment are bent by the crossed fields and are collected by the ground shields. This eliminates secondary electron bombardment of the substrates which was one of the main sources of substrate heating.

The magnetron sputtering guns are usually used in large chambers with planetary substrate fixturing to provide good step coverage. Even though the flux of sputtered atoms at the cathode is much greater than that of previous sputtering sources, the deposition rate on a substrate is in the hundreds of angstroms-per-minute range (same as quoted previously for sputtering) because previously a small number of substrates (typically seven 7.5 cm diameter silicon wafers) were coated at a time from a large (35 cm diameter) cathode, whereas a large number (seventy-five 7.5 cm wafers) are coated at a time from a single S-gun source. The productivity (wafers per hour) of an

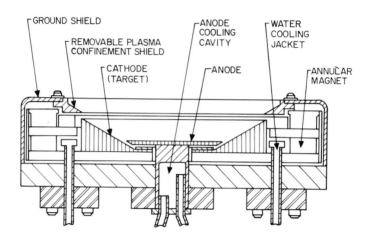

Fig. 5.12 Magnetron-type sputtering source. (Courtesy of Varian Associates.)

S-gun deposition system depositing a tungsten/10% titanium film is three times that of an RF sputtering station with a 35 cm cathode. The deposition rate for pure aluminum is 30 to 50 Å/s by E-beam evaporation in a chamber with a planetary substrate holders with a capacity of seventy-five 7.5 cm silicon wafers. This is greater than the 9 Å/s deposition rate for S-gun sputtering, and so evaporation is still preferred for deposition of pure aluminum. However, many modern silicon integrated circuits use aluminum/4% copper/2% silicon as the metallization. This alloy is more easily deposited with consistent stoichiometry by magnetron sputtering.

The magnetron source can also be used with RF excitation to deposit dielectric materials.

5.2.2 Ion Etching [13,14]

Until fairly recently, chemical etchants were the only means for delineating patterns in films. Since the etchants for polycrystalline films are isotropic, they etch laterally under the edge of a mask feature roughly the same amount as they etch vertically. The result is a wall that is not vertical. The actual profile is determined by a mass transfer process by which the etchant reaches the surface to be etched and the reaction products are removed. The broadening of the pattern is not too troublesome when the horizontal dimensions of the pattern are much greater than the film thickness, but it can be a limiting factor for large-scale integrated (LSI) silicon circuits. In addition, the chemical etchants may leave behind foreign material which could adversely affect the yield of the process.

Ion etching overcomes these difficulties and will be increasingly important in LSI circuit fabrication. The three ion-etching techniques to be described are sputter etching or backsputtering, ion-beam etching, and plasma etching. The apparatus for each of these methods is described and some requirements for the masks that define the patterns are discussed.

Sputter etching   The apparatus for sputter etching is similar to that for RF sputter deposition except that now the substrates are mounted on the cathode (target) instead of the anode. The equipment is relatively simple and the targets are usually quite large (>30 cm) so that many silicon wafers or other substrates can be etched simultaneously. Because of the potential distribution around the target, the etching ions can only impinge on the substrate normal to the surface, thus precluding etching at any other angle. Also, the argon gas pressure is fairly high to produce a large ion current, but it results in mean-free paths λ for the ions that are often smaller than the interelectrode spacing. This can cause considerable back diffusion of sputtered atoms to the cathode surface, thus decreasing the apparent etch rate and producing ditches at the edges of the masked pattern. The effects become more pronounced as the pressure increases.

Ion-beam etching   Figure 5.13 is the schematic representation of an ion-beam etching apparatus. An arc discharge is established between the anode and the filamentary cathode in the upper portion of the chamber. The argon pressure can be low, say 10^{-4} torr, so that the mean-free path of the ions will be large. The discharge is contained and concentrated by the magnetic field produced by the coil. Ions are extracted from the arc by the sieve-like electrode assembly, formed into a beam, accelerated, and fired into the lower portion of the chamber when they impinge upon the substrate holder. The energy and current density of the beam can be varied independently since the arc discharge, the strength of the magnetic field (which affects the ion current density), and the accelerating potential can be adjusted independently. The beam diameter is not much more than 8 cm and therefore the substrates are mounted on a holder that can be moved in two dimensions to permit larger substrate areas to be etched.

Etch rate and masking   The etch rate is a function of the kind of ion, the energy and density of the ions, and the material to be etched. Typical etch rates are in the range 100 to 3000 Å/min for most materials. This poses a problem because of the lack of selectivity between the masking material and the film to be etched. The masking ability of metals such as titanium can be

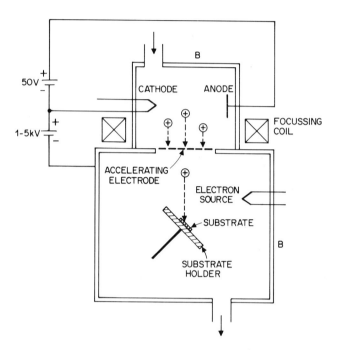

Fig. 5.13 Ion-beam etching apparatus.

improved by adding oxygen to the argon sputtering atmosphere. The titanium oxidizes to form titanium dioxide which is more resistant to ion etching. Figure 5.14 shows a comparison between the etch rates of silicon and titanium for various values of oxygen pressure. From this figure it is seen that the maximum etch rate ratio is about 7 and it occurs when the oxygen pressure is about 10^{-5} torr. Photoresist is also useful as an etch mask in an oxygen-free sputtering atmosphere.

In ion-etching there is no undercutting of the mask but the walls of an etched cut are not necessarily vertical. If the sputtering ions are not incident at right angles to the substrate, a corner is worn off the mask. This has no effect on the resultant pattern until the slope consumes the entire vertical edge of the mask. The cross section of the mask becomes trapezoidal. As the ion bombardment continues the base of the trapazoid decreases and the top face of the etched pattern begins to shrink. In addition the ion bombardment of the vertical wall in the film begins to propagate a sloped edge.

Ion etching can be used to etch a pattern of 1 μm wide lines and spaces in a 1 μm thick film.

Plasma etching   Plasma etching uses basically the same apparatus as sputter etching. However, the etching is done by chemically active ions derived from mixtures of compounds such as CF_4-O_2 instead of inert argon ions. Because of the chemical activity of the ions, they do not have to be accelerated to high energies, and therefore the radiation damage to the substrates is reduced. Photoresist is a suitable mask. The etch rate for SiO_2 is typically 100 Å/min and for Si_3N_4 it is several hundred Å/min.

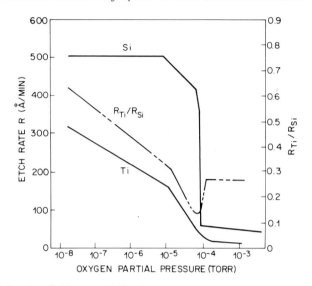

Fig. 5.14   Etch rate of silicon and titanium versus oxygen pressure.

5.3 EPITAXIAL GROWTH

Epitaxy is the ordered growth of a monocrystalline layer that bears a definite relation to the underlying monocrystalline substrate. It is used in the manufacture of integrated circuits to produce the thin layer of silicon in which transistors and other circuit components will ultimately be fabricated. A feature of epitaxy is that the doping of the epitaxial layer is relatively independent of the substrate doping, thus permitting high-quality, lightly doped layers to be grown on heavily doped substrates of either conductivity type. This capability is used in the fabrication of both discrete transistors and bipolar integrated circuits where it is necessary to grow a lightly doped *n*-type layer on n^+-doped material. Some devices, such as microwave *p-i-n* diodes, require multiple epitaxial layers.

   This section consists of a qualitative description of the process and equipment followed by more detailed descriptions of the chemical reactions and reaction kinetics. Finally, the growth of silicon layers on sapphire or spinel substrates is discussed.

5.3.1 Qualitative Description

Epitaxial growth of silicon is usually performed in a horizontal-tube reactor shown schematically in Fig. 5.15. The tube is made of fused silica or quartz and is cooled to prevent silicon from depositing on it and subsequently contaminating the wafers. The susceptor, which holds several wafers, is made of an electrically conductive material such as graphite coated with either boron nitride, silicon carbide, or quartz, and is heated by RF induction to temperatures in excess of 1000°C. Silicon atoms are obtained by the decomposition of a gaseous silicon compound; the most popular reactions being the hydrogen reduction of silicon tetrachloride or trichlorosilane, and the pyrolysis of silane or dichlorosilane.

The simplified overall reaction by which silicon tetrachloride ($SiCl_4$) is reduced to solid silicon with HCl as a reaction byproduct can be written as

$$SiCl_4(gas) + 2H_2(gas) \overset{\sim 1250°C}{\rightleftarrows} Si(solid) + 4HCl(gas). \qquad (5.24)$$

In the reverse reaction, solid silicon is etched by hydrogen chloride. The actual decomposition process is much more complicated. Some additional details are provided later in this section.

Although the hydrogen reduction of silicon tetrachloride is the most popular source of silicon for epitaxy, two disadvantages of this reaction are the high temperatures required for the reaction and the production of HCl. The high temperatures permit considerable out-diffusion of impurities from the substrate (see Section 5.4) thus limiting the abruptness of junctions, whereas the HCl can etch the wafer.

Thus, there is considerable interest in epitaxial growth using the pyrolysis of silane (SiH_4) which yields higher quality films at lower temperatures than does the reduction of silicon tetrachloride. Silane pyrolysis is described by

$$SiH_4(gas) \overset{\sim 1000°C}{\rightarrow} Si(solid) + 2H_2(gas). \qquad (5.25)$$

This nonreversible reaction has no chlorine in the reaction products, thereby excluding the possibility of surface etching. Furthermore, it permits junctions between the epitaxial layer and the substrate to be more abrupt because the diffusion coefficients of the commonly used dopant impurities at 1000°C are more than two orders of magnitude lower than the corresponding values at 1250°C.

The major disadvantages of silane decomposition are:

1.   It is difficult to prevent nucleation of silicon particles in the gas phase.

2.   The deposition rate is much lower than that of the tetrachloride process.

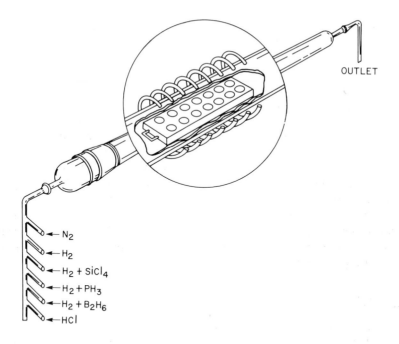

OUTLET

←N₂
←H₂
←H₂ + SiCl₄
←H₂ + PH₃
←H₂ + B₂H₆
←HCl

Fig. 5.15 A horizontal tube epitaxial reactor.

3.   Silane is much more expensive, unstable, and difficult to handle than is
silicon tetrachloride.

Another source of silicon is dichlorosilane (SiH_2Cl_2) which is of
interest because it permits the growth of high quality films at 1150°C with a
high deposition rate. Furthermore, the growth rate is independent of doping
level and is relatively independent of temperature. Pyrolysis is the principal
high-temperature reaction in the decomposition of dichlorosilane:

$$SiH_2Cl_2(\text{gas}) \overset{\sim 1150°C}{\rightleftharpoons} Si(\text{solid}) + 2HCl(\text{gas}). \qquad (5.26)$$

Note that etching of the solid silicon is possible because of the HCl as a
reaction product, but the reverse reaction which results in etching can be
suppressed by processing in excess hydrogen.

Gas-phase epitaxy is a form of vapor deposition because the reaction
constituents are transported in the vapor phase to the vicinity of the substrate surface where the reactions described above occur. However, epitaxial
growth by chemical vapor deposition requires higher temperatures at, and in
the vicinity of, the substrate than for physical vapor deposition so that the
required chemical reactions can proceed. In addition, the silicon atoms that

are deposited on the surface must have sufficient energy and mobility to properly align themselves in the crystal lattice, which is a requirement for the growth of a defect-free, single-crystal film. The deposition rate must be adjusted to give the atoms sufficient time to find a lattice site. A polycrystalline film is formed when the arrival rate of silicon atoms is excessive. Typical industrial growth conditions result in a growth rate of about 1 μm/min.

Referring again to the schematic of the epitaxial reactor, it is seen that there is a rather extensive piping system to control the constituents of the gas entering the reaction chamber. Silicon tetrachloride is introduced into the system by passing hydrogen over the surface of liquid silicon tetrachloride which is maintained at a constant temperature, for example 0°C. The concentration of silicon tetrachloride is diluted to between 1 and 5 mole[†] percent with hydrogen which is used as a carrier gas. Provision is made to introduce dopant gases, such as diborane and phosphane, into the input gas stream to permit the growth of doped layers, and anhydrous HCl is provided for use as a predeposition etchant.

5.3.2 Thermodynamics of the Silicon-Hydrogen-Chlorine System [15-17]

In this section the thermodynamics of the chemical reactions involved in epitaxy by the hydrogen reduction of silicon tetrachloride is studied to determine the chemical composition of the gases in the reactor as a function of the composition of the input gases and of the temperature.

The thermodynamic system to be considered has three components; namely, silicon, chlorine, and hydrogen, and two phases because silicon exists both as an elemental solid and in gaseous compounds. The *number of degrees of freedom v* of such a system is given by the *phase rule* [18]

$$v = 2 + n - f \tag{5.27}$$

where n is the *number of system components* and f is the *number of phases*. The silicon, chlorine, hydrogen system considered here has three degrees of freedom which are the independent variables in our experiment. Obviously temperature is chosen as one independent variable since its influence on the reaction rate and gas composition is of interest. In most practical cases the total pressure is the second independent variable. The final independent variable is the ratio of the total flux of chlorine atoms, J_C, to the total flux of hydrogen atoms, J_H, in the gas phase. In a constant pressure system, J_C and J_H are constant because, by assumption, only silicon exists in both solid and gaseous phases. The ratio J_C/J_H is experimentally meaningful because it is a good indicator of the input gas composition and it is a constant even if the

---

[†] A mole is defined to be that amount of a substance that contains as many elementary units as there are carbon atoms in exactly 0.012 kg of carbon-12. The number of elementary units in a mole is known as Avogadro's number, 6.0248×10²³.

pressure varies because chlorine and hydrogen atoms are not taken from, or added to, the gas phase. Recall that a *flux* is the number of particles crossing a unit area per unit time. Since the chlorine and hydrogen are in the same gas stream, the ratio of the fluxes is equivalent to the ratio of the densities. Hence

$$\phi = \frac{J_C}{J_H} = \frac{\rho_C}{\rho_H} \tag{5.28}$$

where ρ_C and ρ_H are the densities of chlorine and hydrogen, respectively.

The ratio

$$\eta = \frac{J_S}{J_C} = \frac{\rho_S}{\rho_C}, \tag{5.29}$$

where J_S and ρ_S refer to silicon in the gas phase, is chosen as the dependent variable because it is an indicator of the silicon deposition rate since it increases if silicon is dissolving in the gas stream and decreases if silicon is being deposited.

There are many reactions involving solid silicon and gaseous compounds of silicon, chlorine, and hydrogen. Consider, once again, the reaction

$$SiCl_4(gas) + 2H_2(gas) \overset{\sim 1250°C}{\rightleftarrows} Si(solid) + 4HCl(gas). \tag{5.24}$$

According to the *Law of Mass Action*,[18] the rate of the forward reaction is proportional to the partial pressure (or concentration) of the components on the left side of the expression

$$r_f = k_f P_{SiCl_4} P_{H_2}^2 \tag{5.30a}$$

where k_f is the *forward rate constant*. Similarly, the backward reaction is described by

$$r_b = k_b P_{HCl}^4. \tag{5.30b}$$

When the reaction is in dynamic equilibrium $r_f = r_b$ and the *equilibrium constant*

$$K_{SiCl_4} = \frac{k_f}{k_b} = \frac{P_{HCl}^4}{P_{SiCl_4} P_{H_2}^2} \tag{5.30c}$$

is unity.[†] A value of K_{SiCl_4} less than unity means that the reduction of silicon

---

[†] The equilibrium constant can also be written as the ratio of the product of activity coefficients of the products to the activity coefficients of the reactants, each raised to a power given by the number of units of that species in the defining equation for the reaction. Since the activity coefficient of a pure solid is unity, terms involving solid silicon do not appear in the equilibrium constant.

tetrachloride to silicon and hydrochloric acid is favored. One way to achieve this result is to provide excess hydrogen. The conditions for epitaxial growth are such that the reactant partial pressures do not deviate excessively from the equilibrium values because a large flux of reactants to the wafer surface increases the probability of misoriented growth. The other reactions and their equilibrium constants are given in Table 5.1.

Table 5.1 Dissociation Reaction and Equilibrium Constant for Each Species of the Silicon, Hydrogen, Chlorine System

Dissociation	Equilibrium Constant
$SiCl_4 + 2H_2 \rightleftarrows Si + 4HCl$	$K_{SiCl_4} = \dfrac{P_{HCl}^4}{P_{SiCl_4} P_{H_2}^2}$
$SiHCl_3 + H_2 \rightleftarrows Si + 3HCl$	$K_{SiHCl_3} = \dfrac{P_{HCl}^3}{P_{SiHCl_3} P_{H_2}}$
$SiH_2Cl_2 \rightleftarrows Si + 2HCl$	$K_{SiH_2Cl_2} = \dfrac{P_{HCl}^2}{P_{SiH_2Cl_2}}$
$SiH_3Cl \rightleftarrows Si + HCl + H_2$	$K_{SiH_3Cl} = \dfrac{P_{H_2} P_{HCl}}{P_{SiH_3Cl}}$
$SiCl_2 + H_2 \rightleftarrows Si + 2HCl$	$K_{SiCl_2} = \dfrac{P_{HCl}^2}{P_{SiCl_2} P_{H_2}}$
$SiCl + \dfrac{1}{2} H_2 \rightleftarrows Si + HCl$	$K_{SiCl} = \dfrac{P_{HCl}}{P_{SiCl} + P_{H_2}^{1/2}}$

The reaction rates are a function of temperature and the equilibrium constant can be expressed in terms of thermodynamic quantities through the relationship

$$K = \exp(\Delta G / RT) \qquad (5.31)$$

where ΔG is the change in *Gibbs free energy*, R is the universal gas constant, and T is the absolute temperature. For a reaction at constant pressure and temperature, the associated energy change can be written in terms of ΔH, the *heat of formation* of the reaction products, and ΔS, the change in *entropy*, as

$$\Delta G = \Delta H - T\Delta S, \qquad (5.32)$$

where $T\Delta S$ is the amount of work done in a reversible process. Values of ΔH and ΔS at 1500 K for species of the silicon, hydrogen, chlorine system

are given in Table 5.2. Note that a compound with a negative heat of formation is stable because it is produced by an *exothermic* reaction which involves
the release of energy.

Combining Eqs. (5.31) and (5.32) and substituting the value of the gas
constant yields

$$-4.574 \log K = \Delta S - \Delta H / T. \tag{5.33}$$

EXAMPLE 5.10   Calculate the change in Gibbs free energy ΔG and the equilibrium
constants for silicon tetrachloride and the chlorosilanes at 1500 K.

Solution. The   change   in   Gibbs   free   energy   is $\Delta G = \Delta H - T\Delta S$ and
$K = \exp(\Delta G / RT)$ where $R = 1.986$ cal/K·mol. From Table 5.2:

For silicon tetrachloride

$$\Delta G = 59.8 \times 10^3 - (1.5 \times 10^3)(35.9) = 5.950 \text{ kcal/mole}$$

$$K_{SiCl_4} = \exp\left[\frac{5950}{1.986 \times 1500}\right] = \exp(1.997) = 7.37.$$

Therefore, silicon tetrachloride does not dissociate spontaneously even at 1500 K and
about 6 kcal/mole is required to effect the reduction.

For trichlorosilane

$$\Delta G = 49.3 \times 10^3 - (1.5 \times 10^3)(29.9) = 4.450 \text{ kcal/mole}$$

$$K_{SiHCl_3} = \exp(1.494) = 4.45.$$

For dichlorosilane

$$\Delta G = 31.1 \times 10^3 - (1500)(26.0) = -7.90 \text{ kcal/mole}$$

$$K_{SiH_2Cl_2} = \exp(2.652) = 7.052 \times 10^{-2}.$$

Dichlorosilane has a negative Gibbs free energy change at 1500 K and it decomposes
spontaneously with a release of energy.

For chlorosilane

$$\Delta G = 12.9 \times 10^3 - (1500)(23.7) = -22.65 \text{ kcal/mole}$$

$$K_{SiH_3Cl} = \exp(-7.60) = 4.99 \times 10^{-4}.$$

For silane

$$\Delta G = -5.2 \times 10^3 - (1500)(23.6) = -40.6 \text{ kcal/mole}$$

$$K_{SiH_4} = \exp(-13.63) = 1.2 \times 10^{-6}.$$

It is desired to calculate $\eta = \rho_S / \rho_C$ as a function of pressure, temperature, and $\phi = \rho_C / \rho_H$ where the molar density, ρ, of a gaseous component is
assumed to be related to the partial pressure by the ideal gas law $P = \rho RT$.
The total pressure of the gas is the sum of the partial pressures exerted by

Table 5.2 Values of the Change in Entropy and the Change in Enthalphy at 1500 K for the Dissociation Reactions Given in Table 5.1

Species	ΔS_{1500} (cal/mole K)	ΔH_{1500} (kcal/mole)
$SiCl_4$	35.9	59.8
$SiHCl_3$	29.9	49.3
SiH_2Cl_2	26.0	31.1
SiH_3Cl	23.7	12.9
SiH_4	23.6	−5.2
$SiCl_2$	−5.1	−7.7
$SiCl$	−21.5	−61.4

gaseous silicon, hydrogen, and chlorine. A rather tedious calculation involving all possible silicon, hydrogen, and chlorine compounds listed in Table 5.1 yields

$$P = P_{HCl}^2 \Big\{ K_{SiCl_4} X^2 + K_{SiHCl_3} X$$

$$+ K_{SiH_2Cl_2} + K_{SiH_3Cl}/X + K_{SiH_4}/X^2 \Big\}$$

$$+ P_{HCl}\{ K_{SiCl_2} X + 1 + 1/X \} + P_{HCl}^{1/2} K_{SiCl} X^{1/2} \qquad (5.34)$$

where $X = P_{HCl}/P_{H_2}$ is a parameter. In practice, the partial pressure of silicon chloride is negligible, i.e., $P_{SiCl} \ll P$, and, therefore, from the equilibrium constants of Table 5.1 it can be seen that the last term of Eq. (5.34) is negligible. The resulting quadratic equation

$$AP_{HCl}^2 + BP_{HCl} - P = 0$$

is solved for P_{HCl} with X as a parameter. It is then relatively simple to use these results to find all other equilibrium partial pressures, and hence the desired ratios η and ϕ. The actual calculations are omitted.

The results of these calculations are shown in Figs. 5.16 and 5.17. The first of these plots shows the equilibrium composition of the gas phase versus temperature for a total pressure of 1 atm and a chlorine-to-hydrogen ratio of 0.02, which is a typical industrial-growth condition. Also shown in this plot is the equilibrium silicon-to-chlorine ratio. It should be noted that this is probably not the true composition of the gas phase, but rather it is the composition that would occur if the system were allowed to reach equilibrium. If silicon tetrachloride is the source of silicon, the initial value of the silicon/chlorine ratio is 0.25, whereas it is 0.33 if trichlorosilane ($SiHCl_3$) is

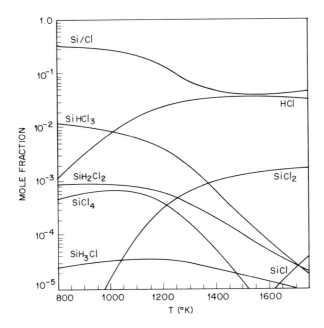

Fig. 5.16 Equilibrium concentration of gas versus temperature for a chlorine/hydrogen ratio of 0.02 and a total pressure of 1 atm. (From Lever.[15] Copyright 1964 by International Business Machines Corp.; reprinted with permission.)

the source. At temperatures below 1050 K the equilibrium Si/Cl ratio is greater than 0.25 indicating that etching of solid silicon could occur with a silicon tetrachloride source.

Epitaxial deposition by the hydrogen reduction of silicon tetrachloride is performed at about 1250°C. It is seen from Fig. 5.16 that trichlorosilane and silicon dichloride are the major silicon-bearing compounds in this temperature range. In addition to the reduction reactions

$$SiCl_4(gas) + 2H_2(gas) \rightleftarrows Si(solid) + 4HCl(gas) \qquad (5.35a)$$

$$SiHCl_3(gas) + H_2(gas) \rightleftarrows Si(solid) + 3HCl(gas) \qquad (5.35b)$$

it is known that the following reaction involving silicon dichloride is important:

$$SiCl_4(gas) + Si(solid) \rightleftarrows 2SiCl_2. \qquad (5.35c)$$

In this reaction silicon transport occurs in the direction from higher to lower temperature.

Figure 5.17 shows the effect of varying the chlorine/hydrogen ratio, ϕ, at a temperature of 1450 K. It is seen that increasing ϕ causes the concen-

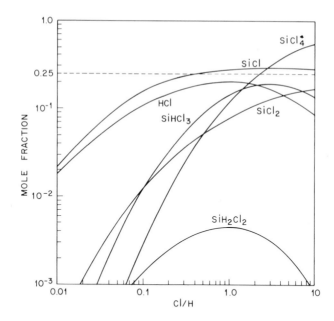

Fig. 5.17 Effect of varying the chlorine-to-hydrogen ratio at 1450 K. (From Lever.[15] Copyright 1964 by International Business Machines Corp.; reprinted with permission.)

tration of silicon chlorides to increase at the expense of hydrogen-containing compounds and the silicon/chlorine ratio, η, to increase. Note that for values of ϕ greater than 0.4, η exceeds 0.25 which means that etching of silicon is possible when the source is silicon tetrachloride.

The growth rate of an epitaxial film as a function of $SiCl_4$ concentration in the input gas stream is shown in Fig. 5.18. Films are usually grown with a low concentration of $SiCl_4$ resulting in a growth rate of about 1 μm/min. The maximum deposition rate occurs at a silicon tetrachloride concentration corresponding to a chlorine/hydrogen ratio of 0.22 which correlates very well with the results presented above.

5.3.3 Kinetics of Epitaxial Growth [19,20]

The thermodynamics of the chemical reaction for the decomposition of silicon tetrachloride was considered in the previous section. In that analysis the chemical system was assumed to be in equilibrium.

The kinetics of epitaxial growth are considered in this section. It is shown that at low temperatures the growth rate is limited by the surface-

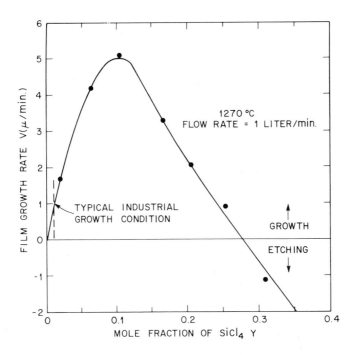

Fig. 5.18 Epitaxial film growth rate versus SiCl$_4$ concentration.

reaction rate coefficient, k_s, and that at high temperatures it is limited by the gas-phase mass transfer coefficient, h_G. The system is assumed to be linear, which means that fluxes are linearly related to driving forces. Furthermore, it is assumed that the reduction of silicon tetrachloride is explained by the single reaction of Eq. (5.24) and the flux of the HCl reaction product is neglected. Therefore,

$$J_1 = h_G(N_G - N_0) \qquad (5.36)$$

where J_1 is the flux of particles from the bulk of the gas stream towards the substrate surface, N_G is the concentration of these particles in the gas stream, N_0 is their concentration at the substrate surface, and h_G is the gas-phase mass-transfer coefficient. Similarly, the reaction at the surface is described by

$$J_2 = k_s N_0 \qquad (5.37)$$

where J_2 is the flux of reactant particles consumed by the reaction on the surface, and k_s is the surface-reaction rate coefficient. In the steady state $J_1 = J_2 = J$, and therefore, from Eqs. (5.36) and (5.37), the concentration at the surface is

$$N_0 = \frac{h_G}{h_G + k_s} N_G \qquad (5.38)$$

and

$$J = \frac{k_s h_G}{k_s + h_G} N_G. \qquad (5.39)$$

This last equation relates the flux to the reactant concentration in the gas stream and the system parameters. Flux is the number of particles per unit area per unit time. If N is the number of silicon atoms per unit volume, then the growth rate of the silicon film is

$$V_y = \frac{dy}{dt} = \frac{J}{N} = \frac{k_s h_G}{k_s + h_G} \frac{N_G}{N}. \qquad (5.40)$$

The temperature dependence of the surface-reaction rate coefficient is expressed in terms of an activation energy E_a by

$$k_s(T) = B \exp(-E_a/RT), \qquad (5.41)$$

where the activation energy for silicon tetrachloride is about 44 kcal/mole which is equivalent to 1.9 eV/atom.

EXAMPLE 5.11   (a) Calculate the growth rate of a silicon film grown from a silicon tetrachloride source at 1250°C. The gas-phase mass-transfer coefficient of the reactor is $h_G = 5$ cm/s, the surface reaction rate coefficient is

$$k_s = 1\times10^7\exp(-1.9 \text{ eV}/kT) \text{ cm/s},$$

$T = 1250°C$, $N_G = 3\times10^{16}$ atoms cm^{-3}. (b) What is the change in growth rate if the reaction temperature is increased by 1%?

Solution.   (a) At $T = 1250 + 273 = 1523$ K,

$$k_s = 1\times10^7\exp\left[-\frac{1.9 \text{ eV}}{8.62\times10^{-5}(\text{eV/K}) \times 1523}\right] = 5.1836 \text{ cm/s}.$$

Substituting in Eq. (5.38) with $N = 5.0\times10^{22}$ atoms cm^{-3} for silicon yields

$$V_y = \frac{dy}{dt} = \frac{5.1836\times5.0}{5.1836+5.0} \frac{3\times10^{16}}{5\times10^{22}} = 1.527\times10^{-6} \text{ cm/s} = 0.916 \ \mu\text{m/min}.$$

(b) Equation (5.38) is a function of T through $k_s(T)$. Differentiating with respect to k_s and collecting terms yields

$$\frac{dV_y}{V_y} = \left[\frac{h_g}{h_g + k_s}\right]\frac{dk_s}{k_s} = 0.491 \frac{dk_s}{k_s}$$

after substitution of the above values for h_g and k_s. Also

$$\frac{dk_s}{dT} = \frac{d}{dT}\left[B \exp\left(\frac{E_a}{kT}\right)\right] = \frac{k_s}{T}\frac{E_a}{kT} .$$

Therefore $(dk_s/k_s) = 14.47(dT/T)$. Substituting this into the previous equation yields the result

$$\frac{dV_y}{V_y} = 0.491 \times 14.47 \frac{dT}{T} = 7.106 \frac{dT}{T} .$$

Thus, if the temperature changes by 1%, the growth rate changes by slightly more than 7%, showing that good control of temperature is essential to the deposition of epitaxial layers of known, reproducible thickness.

The value of the mass-transfer coefficient h_G is affected by the diffusivity D_G of the gas, the geometry of the epitaxial reactor, and the gas-flow rate. These variables interact to determine the boundary layer that is formed in the gas stream at the surface of the substrate. The boundary-layer thickness, $\delta(x)$, is calculated here, as a function of distance, for the simple case of gas (fluid) flow parallel to the surface of a flat plate of length L as illustrated in Fig. 5.19.

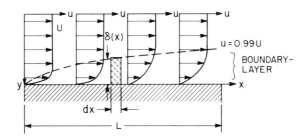

Fig. 5.19 Formation of a boundary layer in epitaxial reactor.

The velocity of the fluid at any point is denoted by $u(x,y)$. It is zero at the surface of the plate and reaches a steady-state value, U, in the gas stream. The boundary-layer thickness is calculated by considering the forces on a volume of unit depth, height $\delta(x)$, and width dx. The frictional force per unit area between two adjacent layers of the fluid is

$$F = \mu \frac{\partial u}{\partial y} \tag{5.42}$$

where μ is the fluid viscosity. Therefore, the frictional force on an elemental area $1 \times dx$ is $F\ dx = ma$, which results in a deceleration of the fluid. Noting that for laminar flow the stream velocity $u = \partial x/\partial t$, and that the mass of the gas is $m = \rho\delta(x)\,dx$ where ρ is the gas density yields

$$F\ dx = \rho\delta(x)u\ \frac{\partial u}{\partial x}\ dx. \tag{5.43}$$

Substituting the expression for frictional force, Eq. (5.42), yields a relation between the partial derivatives of the stream velocity

$$\frac{\partial u}{\partial y} = \frac{\rho}{\mu}\ \delta(x)u\ \frac{\partial u}{\partial x}\ . \tag{5.44}$$

The boundary condition for this equation is found by observing that since velocity of the plate is zero, the velocity of the fluid right next to the plate must also be zero to prevent the occurrence of infinite velocity gradients. An approximate relation for the boundary-layer thickness can be obtained from Eq. (5.44) by approximating the partial derivatives by the quotients of differences, i.e., $\partial u/\partial x \rightarrow U/x$, $\partial u/\partial y \rightarrow U/\delta(x)$, and $u \rightarrow U$. Therefore,

$$\delta(x) \simeq \left[\frac{\mu x}{\rho U}\right]^{1/2} \tag{5.45}$$

is an estimate of the extent of the gas-stream disturbance. This relation for the boundary-layer thickness differs by a multiplicative constant of between two-thirds and five from the results of more precise calculations by Blasius.[21] Finally, the average boundary-layer thickness over the plate is given by

$$\bar{\delta} = \frac{1}{L} \int_0^L \delta(x)\,dx = \frac{2}{3}\ \frac{L}{\sqrt{\mathrm{Re}_L}} \tag{5.46}$$

where Re_L is the *Reynolds number* of the reactor

$$\mathrm{Re}_L = \frac{\rho UL}{\mu}\ . \tag{5.47}$$

The Reynolds number is the ratio of inertial forces to viscous forces and is a very important dimensionless quantity in fluid dynamics. A small Reynolds number indicates that viscous forces predominate, whereas a large Reynolds number indicates that inertial forces predominate. A Reynolds number greater than about 2000 indicates the flow in the tube is turbulent.

The gas-phase mass-transfer coefficient h_G is given in terms of the average boundary-layer thickness $\bar{\delta}$ and the diffusivity D_G by

$$h_G = \frac{D_G}{\bar{\delta}} = \frac{3}{2}\ \frac{D_G}{L}\ \sqrt{\mathrm{Re}_L}\ . \tag{5.48}$$

This result differs from that of a more exact calculation[22] by a factor that is between 0.6 and 0.8 for most gases, practically independent of temperature.

EXAMPLE 5.12   A particular epitaxial reactor has a Reynolds number of 20 and is 10 cm long. What is the gas-phase mass-transfer coefficient if the diffusivity of the gas is $D_G = 8$ cm^2/s.

Solution.   The gas-phase mass-transfer coefficient is given in terms of the Reynolds number, diffusivity, and reactor length by Eq. (5.48)

$$h_g = \frac{3}{2} \times \frac{8}{10} \times \sqrt{20} = 5.3666 \text{ cm/s.}$$

This is typical of reactors such as those studied by Theurer[23] and Shepard[24] in which the stream velocity is about 10 to 30 cm/s. A reasonable range for h_G is 1 to 20 cm/s.

Film growth rate is plotted in Fig. 5.20 as a function of the square root of stream-flow velocity u. At low flow rates, the film growth rate increases proportionally to the square root of U.

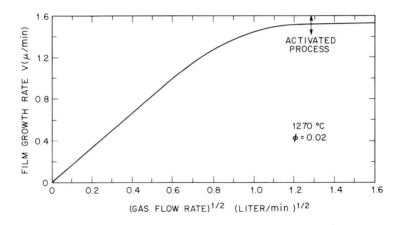

Fig. 5.20   Epitaxial film growth versus gas stream velocity u.

5.3.4 Epitaxial Layers of Silicon on Sapphire or Spinel [25]

The most popular method of electrically isolating integrated circuit components is by forming a reverse-biased p-n junction. This junction has a parasitic capacitance to substrate (ground) which can be substantially eliminated by using an inert material such as sapphire as the substrate instead of silicon.

Sapphire was chosen as an insulating substrate for silicon films because of its mechanical and chemical properties and because a lattice plane could be selected that was a close match to that of silicon. There are difficulties with sapphire because it does not have the same crystal structure as silicon and therefore the lattice match between the two is not good in any plane. However, the long range order in the aluminum-oxygen sites appears to permit epitaxy to take place. The frequently studied orientational relationships are (111) silicon on (0001) sapphire and (100) silicon on ($1\bar{1}02$) sapphire.

Spinels are various mixtures of $MgO:Al_2O_3$ and are closely related to sapphire. Because of their cubic crystal structure they provide a much better crystallographic match to silicon, and it appears that silicon films on spinel exhibit much better crystal structure, higher mobility, and less aluminum autodoping than silicon films on sapphire. The disadvantage of spinel is the lack of high quality crystals of the correct composition.

The quality of thin silicon films on sapphire or spinel is critically dependent on the surface polish of the substrate prior to deposition. The most satisfactory method of surface preparation is to prefire the substrate in hydrogen at 1500°C. The etch rate of sapphire under these conditions is about 0.1 μm/min, and varies directly with temperature.

Epitaxial films on sapphire or spinel are grown either by the pyrolysis of silane or the hydrogen reduction of chlorosilanes or silicon tetrachloride. Growth at low temperature and high rate is desirable to reduce contamination from the substrate, but low temperature deposition often results in poor quality films. Deposition by the pyrolysis of silane at about 1000°C appears to be the lowest temperature at which good quality films can be obtained using these silicon sources. The optimum growth temperature for (111) silicon on (0001) sapphire is about 1200°C whereas for (100) silicon on ($1\bar{1}02$) sapphire it is 1115°C. The growth rates range from 0.1 to 10 μm/min.

The hole mobility of 1.5 to 2.0 μm thick (100) silicon films on sapphire is in the range 120 to 150 cm^2/V-s for hole concentrations in the 10^{16} to 10^{17} cm^{-3} range. This mobility is about 30% of the bulk value for similar doping levels. Electron mobilities of 400 to 550 cm^2/V-s have been reported in 1.5 to 2.0 μm n-type films with electron concentrations in the 10^{16} to 10^{17} cm^{-3} range. These mobilities tend to increase with film thickness as the interface between the silicon film and the substrate becomes less significant. These mobility values are measured by the Hall effect (see Chapter 11) and are overly pessimistic in their effect on the behavior of MOS transistors fabricated in the film. The performance of MOS devices is determined by the characteristics of the 500 Å of the film most remote from the substrate interface. The field-effect mobility (effective mobility in a device) is about 70% that in devices in bulk silicon. Carrier lifetimes are in the low nanosecond range, and therefore these films are not suitable for bipolar devices.

5.4 DIFFUSION [26]

An essential requirement of the planar process is the ability to introduce controlled quantities of dopant impurity atoms into selected regions of the wafer. One method of producing an impurity distribution is by solid-state diffusion. This process also causes the redistribution of impurities whenever a doped crystal is heated. Selectivity is provided by a mask on the top surface of the wafer through which the impurities are introduced.

Solid-state diffusion effects are described by Fick's Laws. The development of the mathematical description begins with the fundamental postulate known as *Fick's first law of diffusion:* Particles diffuse from regions of high concentration to regions of lower concentration. To quantify this statement, let the vector $J(x,y,z,t)$ denote the flux of diffusant particles across the surface S that bounds a region R, and let $N(x,y,z,t)$ be the concentration of diffusant particles. Note that both the flux and concentration are functions of position and time.

The particle flux is

$$J = -D(T,N)\nabla N + \frac{D(T,N)\mathbf{E}}{V_T} N \tag{5.49}$$

where $\mathbf{E}$ is the built-in electric field which may be significant in some applications,

$$\nabla = \frac{\partial}{\partial x}\mathbf{i} + \frac{\partial}{\partial y}\mathbf{j} + \frac{\partial}{\partial z}\mathbf{k}$$

and $D(T,N)$ is the diffusion coefficient which is a function of temperature and possibly concentration for an isotropic material. The diffusion coefficient is also a function of crystal orientation. In general, $D(T,N)$ is a tensor describing the positional dependence of the diffusion process. The minus sign in Eq. (5.49) occurs because particles diffuse toward regions of lower concentration.

Fick's second law of diffusion is obtained by equating the number of particles leaving a small volume, namely $\nabla \cdot J$, with the negative of the accumulation rate, $-\partial N/\partial t$, of particles in that volume. The result is

$$\frac{\partial N}{\partial t} = \nabla \cdot \left\{ D(T,N)\nabla N - \frac{D(T,N)\mathbf{E}}{V_T} N \right\}. \tag{5.50a}$$

When the diffusion coefficient is independent of concentration and drift due to the built-in field is negligible, this equation reduces to

$$\frac{\partial N}{\partial t} = D(T)\nabla^2 N \tag{5.50b}$$

where

$$\nabla^2 = \frac{\partial^2}{\partial x^2} + \frac{\partial^2 y}{\partial y^2} + \frac{\partial^2}{\partial y^2} .$$

In the planar process, impurities are introduced into the crystal from the top surface resulting in a one-dimensional process if effects that occur at the edges of mask features are neglected. Thus, attention is now focused on the solution of the one-dimensional diffusion equation with a concentration independent diffusion coefficient

$$\frac{\partial N(x,t)}{\partial t} = D(T) \frac{\partial^2 N(x,t)}{\partial x^2} \tag{5.51}$$

subject to several important boundary conditions. This partial differential equation is most easily solved by use of the Laplace transformation, which converts it to an ordinary differential equation. The convention used is $LT\{N(x,t)\} = N(x,s)$. Hence

$$\frac{d^2 N(x,s)}{dx^2} - \frac{s}{D(T)} N(x,s) = - \frac{N(x,t=0^-)}{D(T)} . \tag{5.52}$$

Of particular interest is the homogeneous solution of this equation $[N(x,t=0^-) = 0]$, which corresponds to no dopant atoms in the wafer until $t > 0$. For this case

$$N(x,s) = A(s)\exp\left[\left[\frac{s}{D(T)}\right]^{1/2} x\right] + B(s)\exp\left[-\left[\frac{s}{D(T)}\right]^{1/2} x\right]. \tag{5.53a}$$

5.4.1 Constant Surface Concentration

The impurity concentration at the surface of a semi-infinite solid is maintained constant at the value N_0 impurities/unit volume. The appropriate boundary and initial conditions for the diffusion equation are,

$$N(0,t) = N_0 u(t),$$

$$N(x,0^-) = 0. \tag{5.53b}$$

The Laplace transform of the resultant impurity distribution is easily obtained from Eq. (5.53a). The requirement that $N(x,s)$ be finite for all positive values of x implies that $A(s)$ must be identically zero. Also observe that the second of Eqs. (5.53b) has already been applied to Eq. (5.52) to obtain Eq. (5.53a). Taking the Laplace transform of the first boundary condition of Eq. (5.53b), $N(0,s) = N_0/s$, and substituting into Eq. (5.53a) with

$A(s) = 0$ and $x = 0$ yields the value of $B(s)$. Hence,

$$N(x,s) = \frac{N_0}{s} \exp\left[-\left(\frac{s}{D(T)}\right)^{1/2} x\right]. \tag{5.54}$$

The inverse Laplace transform may be obtained from tables. The resultant diffusion profile, written in terms of the error function, is

$$N(x,t) = N_0\{1 - \mathrm{erf}(x/2\sqrt{Dt})\}$$
$$= N_0 \mathrm{erfc}(x/2\sqrt{Dt}). \tag{5.55}$$

The error function is defined to be

$$\mathrm{erf}(y) = \frac{2}{\sqrt{\pi}} \int_0^y e^{-\alpha^2} d\alpha,$$

and clearly $\mathrm{erf}(0) = 0$ and $\mathrm{erf}(\infty) = 1$. Furthermore, the complementary error function $\mathrm{erfc}(y)$ is $1 - \mathrm{erf}(y)$. Error function algebra is summarized in Table 5.3. The total number of impurity particles Q, which have diffused into the slab after a diffusion lasting $t = t_1$ s, carried out at temperature $T_1 K$ is given by

$$Q = A \int_0^\infty N(x,t_1)\, dx = N_0 A \int_0^\infty \mathrm{erfc}(x/2\sqrt{D_1 t_1})\, dx$$

where $D(T_1) = D_1$. From tables of integrals

$$\int_0^\infty \mathrm{erfc}(x)\, dx = \frac{1}{\sqrt{\pi}}$$

and so

$$Q = 2N_0 A \left[\frac{D_t t_1}{\pi}\right]^{1/2} \tag{5.56}$$

where A is the area of the deposition.

If the semi-infinite silicon slab is uniformly doped at a level N_B with a "nondiffusing" impurity of the opposite conductivity type to the impurity being diffused, the solution of the diffusion equation becomes

$$N(x,t) = N_0 \mathrm{erfc}\left[\frac{x}{2\sqrt{Dt}}\right] - N_B. \tag{5.57}$$

Note that in these expressions for the diffusion profiles, the impurity distribution depends upon the product of the diffusion coefficient $D(T)$ and the time.

A p-n junction is formed at that distance x_j into the semiconductor for which the net impurity concentration is zero. Therefore, from Eq. (5.57),

Table 5.3 Error Function Algebra

$$erf \ z = \frac{2}{\sqrt{\pi}} \int_0^z \exp(-\alpha^2) \, d\alpha$$

$$= \frac{2}{\sqrt{\pi}} \left\{ z - \frac{z^3}{3 \cdot 1!} + \frac{z^5}{5 \cdot 2!} + \cdots \right\}$$

$$erfc \ z = 1 - erf \ z = \frac{2}{\sqrt{\pi}} \int_z^\infty \exp(-\alpha^2) \, d\alpha$$

$$erf(0) = 0$$

$$erf(\infty) = 1$$

$$erf(-z) = erf \ z$$

$$\frac{d}{dz}(erf \ z) = \frac{2}{\sqrt{\pi}} \exp(-z^2)$$

$$\frac{d^2}{dz^2}(erf \ z) = -\frac{4}{\sqrt{\pi}} z \exp(-z^2)$$

$$x_j = 2\sqrt{Dt} \ erfc^{-1}\left(\frac{N_B}{N_0}\right). \tag{5.58}$$

The complementary error function profile is plotted in Fig. 5.21 for several different diffusion times. This profile is used whenever shallow, highly doped layers are desired. Typical applications are as emitter diffusions or as the predeposition step in the two-step diffusion to be described shortly.

EXAMPLE 5.13 A p^+-n junction is made by diffusing boron into n-type material with $N_B = 10^{16}$ cm^{-3}. A constant surface concentration $N_0 = 10^{20}$ cm^{-3} is maintained during the diffusion. Calculate the time required to form the junction at a depth of 1.0 μm if the diffusion temperature is 1050°C.

Solution. The diffusion coefficient of boron at 1050°C is 10^{-13} cm^2/s. The junction occurs at that depth for which the impurity concentration in the profile equals the background concentration, i.e.,

$$N_B = N_0 erfc\left(\frac{x_j}{\sqrt{Dt}}\right) = 10^{16} = 10^{20} erfc\left(\frac{10^{-4}}{\sqrt{10^{-13}t}}\right).$$

This equation must be solved for the diffusion time t. The result is $t = 3.306 \times 10^3$ s $\approx$ 55 min.

EXAMPLE 5.14 How long would it take to form the p^+-n junction of the previous example if the diffusion temperature was 950°C?

Solution. The diffusion coefficient of boron at 950°C is about 10^{-14} cm^2/s. Substituting this value for D into the equation of Example 5.13 and solving for the diffusion time yields $t = 3.306 \times 10^4$ sec ≈ 11 hr.

Two methods of performing the error-function diffusion are widely used. In one, the impurity is delivered to the surface in vapor form diluted to the proper concentration with an inert carrier gas such as nitrogen. In the other method, a highly doped silicon oxide, which serves as the source, is deposited (not grown) on the silicon surface. Another technique for p-type diffusions is to place boron nitride wafers between the silicon wafers.

5.4.2 Diffusion from a Limited Source

Consider the case in which a finite number, Q/A, impurities/unit area are somehow deposited on one face of the silicon wafer, where Q is the total number of impurities and A is the area. The wafer is then heated to drive these impurities into the semiconductor. It is assumed that the conditions

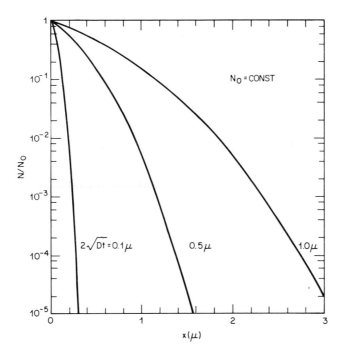

Fig. 5.21 Complementary error function profile.

are such that the impurities cannot evaporate. This condition can be written as

$$N(x,t=0^-) = \frac{Q}{A}\,\delta(x)$$

where $\delta(x)$ is the unit impulse function. Substituting this condition into Eq. (5.52) yields

$$\frac{d^2N(x,s)}{dx^2} - \frac{s}{D(T)}\,N(x,s) = -\frac{Q}{AD(T)}\,\delta(x). \qquad (5.59)$$

For $x > 0$, the solution of this equation is identical to that obtained for the case of constant surface concentration, namely

$$N(x,s) = B(s)\exp\left[-\left(\frac{s}{D(T)}\right)^{1/2}x\right].$$

The new boundary condition to be used in evaluating $B(s)$ is obtained by integrating Eq. (5.59) from $x = 0^-$ to $x = 0^+$ subject to the condition $N(x,s) = 0$ for $x < 0^-$. The result is

$$\left.\frac{dN(x,s)}{dx}\right|_{x=0^+} = -\frac{Q}{AD(T)}. \qquad (5.60)$$

From the previous two equations it now follows that $B(s) = Q/A\sqrt{D(T)s}$, and therefore

$$N(x,s) = \frac{Q}{A\sqrt{D(T)s}}\,\exp\left[-\left(\frac{s}{D(T)}\right)^{1/2}x\right]. \qquad (5.61a)$$

The inverse Laplace transform is

$$N(x,t) = \frac{Q}{A\sqrt{\pi D(T)t}}\,\exp\left(-\frac{x^2}{4D(T)t}\right). \qquad (5.61b)$$

   The solution is exact for a predeposited layer of zero thickness. If the predeposited layer has nonzero thickness h, then the Gaussian distribution is a good approximation to the true impurity profile if $x < Dt/h$ and $t > h^2/D$. A plot of the Gaussian impurity distribution given by Eq. (5.61b) is shown in Fig. 5.22 for three different diffusion times. The Gaussian approximation is generally adequate for deep diffusions.

5.4.3 The Two-Step Diffusion Process

The Gaussian diffusion just described is one example of a two-step diffusion process. In the first step, a complementary error function diffusion is used to deposit a limited number of impurity atoms in a thin layer near the sur-

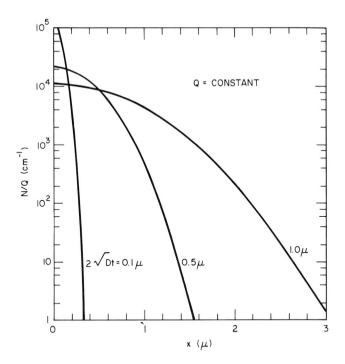

Fig. 5.22 Gaussian impurity profile.

face. The surface concentration, N_0, for this predeposition diffusion is very high, and is determined either by the solid solubility limit of the impurity or by the concentration in a glassy layer that forms on the surface. Any glassy layer that forms during the predeposition is removed and the wafer is placed in a clean furnace for the drive-in diffusion which redistributes the impurities.

Since the predeposition diffusion is done at constant surface concentration, the total number of impurities deposited is given by Eq. (5.56) where t_1 is the time of the predeposition diffusion and D_1 is the diffusion coefficient at the predeposition temperature. Substituting this expression for Q into the Gaussian distribution yields

$$N(x, t_2) = 2N_0 \left[\frac{D_1 t_1}{D_2 t_2} \right]^{1/2} \exp \left(\frac{-x^2}{4 D_2 t_2} \right) \tag{5.62}$$

where D_2 is the appropriate diffusion coefficient for the drive-in and t_2 is the corresponding time. Note that D_1 and D_2 may be quite different because the predeposition and drive-in may be conducted at different temperatures. For

$D_1 t_1 \ll D_2 t_2$, Eq. (5.62) is a good approximation to actual measured profiles.

EXAMPLE 5.15 A two-step boron diffusion is performed into n-type silicon with an impurity concentration $N_B = 10^{16}$ cm^{-3}. The predeposition diffusion is conducted at 950°C with a surface concentration of 10^{21} cm^{-3} for 15 min. The drive-in temperature is 1100°C. (a) How long should the drive-in last to form the junction at $x_j = 3$ μm? (b) What is the impurity concentration at the surface after the drive-in diffusion?

Solution. The diffusion coefficient of boron at 950°C is $D_1 = 10^{-14}$ cm^2/s and at 1100°C it is $D_2 = 2.5 \times 10^{-13}$ cm^2/s. From Eq. (5.62)

$$N(x_j = 3 \ \mu m, t_2) = N_B$$

$$10^{16} = 2 \times 10^{21} \left[\frac{10^{-14} \times 15 \times 60}{2.5 \times 10^{-13} t_2} \right] \exp \left[-\frac{(3 \times 10^{-4})^2}{4 \times 2.5 \times 10^{-13} t_2} \right].$$

a)   Solving this equation for t_2 yields $t_2 = 1.45 \times 10^4$ s ≈ 4.03 hr.

b)   Substituting these values into Eq. (5.62) with $x = 0$ yields $N(0, t_2) = 4.97 \times 10^{18}$ cm^{-3}.

Often the duration of the drive-in diffusion is not much greater than that of the predeposition, or the oxide formed during the predeposition is not stripped prior to the drive-in. Under these conditions, the predeposition cannot be treated as a thin-source layer and Eq. (5.62) no longer gives satisfactory results. The appropriate solution is

$$N(x, t_1, t_2) = \frac{2N_0}{\sqrt{\pi}} \int_{\sqrt{\beta}}^{\infty} e^{-y^2} \mathrm{erf}(\alpha y) \, dy \tag{5.63}$$

where

$$\alpha \triangleq \left[\frac{D_1 t_1}{D_2 t_2} \right]^{1/2}$$

$$\beta \triangleq \frac{x^2}{4(D_1 t_1 + D_2 t_2)}.$$

This integral has been evaluated by Smith[27] for various values of α and β. It is tabulated in Table 5.4 and plotted in Fig. 5.23 with α as a parameter.

Examination of these results show that this two-step process can be used to approximate the complementary error-function profile quite closely without the accumulation of large numbers of dopant atoms on the surface which may damage the semiconductor. The final surface concentration after the drive-in is given by

$$N(0, t_1, t_2) = \frac{2N_0}{\pi} \tan^{-1}\alpha. \tag{5.64}$$

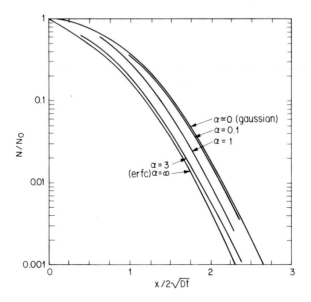

Fig. 5.23 Two-step diffusion profile for several values of α.

The two-step process can also be used to approximate the Gaussian distribution without exposing the semiconductor surface. Finally, the glass may be used as a mask for the next processing step.

5.4.4 Diffusion During Epitaxial Growth

In this section the redistribution of impurities occurring during the growth of an epitaxial layer is studied. In particular, the profile resulting from the growth of a lightly doped n-type layer on an n^+ substrate is derived. This is an important case corresponding to the growth of an epitaxial film over a heavily doped buried layer for bipolar transistors and integrated circuits.

The present problem differs from the previous cases in that the region of interest for the derivation of the diffusion equation is now a function of time. As a consequence, Fick's second law of diffusion must be modified to the following form:

$$\frac{\partial N(x,y,z,t)}{\partial t} = D(T)\nabla^2 N - \nabla N \cdot \mathbf{V} \qquad (5.65)$$

where the velocity vector $\mathbf{V}$ is defined to be

$$\mathbf{V} = \frac{\partial x}{\partial t}\,\mathbf{i} + \frac{\partial y}{\partial t}\,\mathbf{j} + \frac{\partial z}{\partial t}\,\mathbf{k}.$$

Table 5.4 Values of $\sqrt{\pi}\, N(x,t_1,t_2)/2N_0$ as a function of α and β

α \\ β	0.1	0.2	0.3	0.4	0.5	0.6	0.7	0.8	0.9	1.0	1.1	1.2
0.1	0.09015	0.08155	0.07376	0.06672	0.06035	0.05459	0.04938	0.04467	0.04040	0.03655	0.03306	0.02990
0.2	0.17838	0.16119	0.14566	0.13162	0.11894	0.10748	0.09713	0.08777	0.07931	0.07167	0.06477	0.05853
0.3	0.26295	0.23723	0.21403	0.19310	0.17422	0.15719	0.14182	0.12795	0.11545	0.10416	0.09398	0.08479
0.4	0.34254	0.30837	0.27761	0.24993	0.22501	0.20259	0.18240	0.16422	0.14786	0.13314	0.11988	0.10794
0.5	0.41626	0.37374	0.33557	0.30132	0.27058	0.24299	0.21822	0.19599	0.17603	0.15812	0.14203	0.12759
0.6	0.48366	0.43290	0.38751	0.34692	0.31062	0.27814	0.24908	0.22308	0.19982	0.17900	0.16036	0.14368
0.7	0.54464	0.48580	0.43340	0.38673	0.34515	0.30809	0.27505	0.24562	0.21937	0.19596	0.17508	0.15645
0.8	0.59940	0.53264	0.47347	0.42100	0.37447	0.33317	0.29652	0.26398	0.23508	0.20940	0.18657	0.16628
0.9	0.64829	0.57380	0.50812	0.45017	0.39903	0.35385	0.31393	0.27864	0.24742	0.21979	0.19532	0.17365
1.0	0.69176	0.60975	0.53784	0.47475	0.41935	0.37066	0.32783	0.29013	0.25693	0.22765	0.20183	0.17903
1.1	0.73033	0.64100	0.56318	0.49529	0.43600	0.38415	0.33877	0.29900	0.26411	0.23348	0.20655	0.18286
1.2	0.76448	0.66808	0.58465	0.51232	0.44950	0.39486	0.34726	0.30574	0.26946	0.23772	0.20991	0.18553
1.3	0.79470	0.69148	0.60276	0.52634	0.46035	0.40327	0.35377	0.31078	0.27336	0.24074	0.21225	0.18734
1.4	0.82144	0.71164	0.61797	0.53781	0.46901	0.40979	0.35870	0.31449	0.27616	0.24286	0.21385	0.18855
1.5	0.84509	0.72899	0.63069	0.54714	0.47586	0.41482	0.36238	0.31720	0.27815	0.24431	0.21492	0.18933
1.6	0.86601	0.74388	0.64130	0.55469	0.48123	0.41865	0.36511	0.31914	0.27953	0.24530	0.21562	0.18983
1.7	0.88454	0.75666	0.65010	0.56076	0.48542	0.42153	0.36710	0.32051	0.28048	0.24595	0.21607	0.19014
1.8	0.90095	0.76759	0.65739	0.56562	0.48865	0.42369	0.36854	0.32147	0.28112	0.24638	0.21636	0.19033
1.9	0.91549	0.77693	0.66340	0.56948	0.49114	0.42529	0.36956	0.32213	0.28154	0.24665	0.21653	0.19045
2.0	0.92838	0.78491	0.66833	0.57254	0.49303	0.42646	0.37029	0.32258	0.28182	0.24682	0.21664	0.19051
2.5	0.97404	0.81009	0.68228	0.58029	0.49735	0.42887	0.37165	0.32335	0.28225	0.24707	0.21678	0.19059
3.0	0.99920	0.82094	0.68698	0.58234	0.49825	0.42928	0.37183	0.32343	0.28229	0.24708	0.21679	0.19059
∞	1.02843	0.82795	0.68892	0.58291	0.49843	0.42933	0.37184	0.32343	0.28229	0.24709	0.21679	0.19059

Equation (5.65) is the general three-dimensional equation for diffusion within a growing region.

The vector **V** may be associated with the growth of an epitaxial layer, in which case it is the film growth rate. Alternatively, **V** may represent the motion of a boundary between two regions having different diffusion characteristics, such as in diffusion through a growing oxide layer or simultaneous diffusion and etching processes.

In the one-dimensional case to be considered

$$D(T)\,\frac{\partial^2 N}{\partial x^2} = \frac{\partial N}{\partial t} + V_x\,\frac{\partial N}{\partial x} \qquad (5.66)$$

where V_x is the velocity in the x direction.

Growth of intrinsic epitaxial layer on doped substrate   The first set of boundary and initial conditions to be imposed on this equation correspond to the growth of an intrinsic epitaxial layer on a doped substrate. Because impurities from the substrate out diffuse into the epitaxial film during growth, the resulting impurity profile depends upon the diffusion coefficient of the impurity, the growth rate of the film, and the rate at which impurities can escape from the crystal at the gas-film interface.

1.3	1.4	1.5	1.6	1.7	1.8	1.9	2.0	2.5	3.0	4.0	5.0	β / α
0.02705	0.02446	0.02213	0.02002	0.01811	0.01638	0.01481	0.01340	0.00811	0.00491	0.00180	0.00066	0.1
0.05289	0.04779	0.04319	0.03903	0.03527	0.03187	0.02880	0.02603	0.01568	0.00945	0.00343	0.00125	0.2
0.07651	0.06903	0.06228	0.05620	0.05071	0.04575	0.04128	0.03725	0.02228	0.01333	0.00477	0.00171	0.3
0.09720	0.08752	0.07881	0.07097	0.06391	0.05756	0.05183	0.04668	0.02766	0.01640	0.00577	0.00204	0.4
0.11462	0.10297	0.09251	0.08312	0.07468	0.06711	0.06030	0.05419	0.03178	0.01866	0.00645	0.00224	0.5
0.12875	0.11538	0.10340	0.09268	0.08308	0.07448	0.06678	0.05988	0.03475	0.02021	0.00688	0.00236	0.6
0.13982	0.12499	0.11174	0.09992	0.08936	0.07993	0.07150	0.06398	0.03677	0.02120	0.00712	0.00242	0.7
0.14824	0.13219	0.11790	0.10519	0.09387	0.08379	0.07481	0.06680	0.03806	0.02180	0.00724	0.00244	0.8
0.15444	0.13741	0.12230	0.10889	0.09699	0.08642	0.07702	0.06867	0.03885	0.02213	0.00730	0.00245	0.9
0.15889	0.14109	0.12535	0.11141	0.09907	0.08814	0.07844	0.06985	0.03931	0.02231	0.00733	0.00246	1.0
0.16200	0.14361	0.12739	0.11307	0.10041	0.08923	0.07933	0.07056	0.03956	0.02240	0.00734	0.00246	1.1
0.16411	0.14529	0.12872	0.11412	0.10125	0.08989	0.07985	0.07098	0.03969	0.02244	0.00735	0.00246	1.2
0.16552	0.14638	0.12956	0.11478	0.10176	0.09028	0.08016	0.07122	0.03976	0.02246	0.00735	0.00246	1.3
0.16643	0.14706	0.13008	0.11517	0.10205	0.09051	0.08033	0.07134	0.03979	0.02247	0.00735	0.00246	1.4
0.16700	0.14749	0.13039	0.11540	0.10222	0.09063	0.08042	0.07141	0.03980	0.02247	0.00735	0.00246	1.5
0.16736	0.14774	0.13057	0.11552	0.10231	0.09070	0.08046	0.07144	0.03981	0.02247	0.00735	0.00246	1.6
0.16757	0.14789	0.13067	0.11559	0.10236	0.09073	0.08049	0.07146	0.03981	0.02247	0.00735	0.00246	1.7
0.16770	0.14797	0.13073	0.11563	0.10239	0.09075	0.08050	0.07147	0.03982	0.02247	0.00735	0.00246	1.8
0.16777	0.14802	0.13076	0.11565	0.10240	0.09075	0.08050	0.07147	0.03982	0.02247	0.00735	0.00246	1.9
0.16781	0.14804	0.13078	0.11566	0.10240	0.09076	0.08051	0.07147	0.03982	0.02247	0.00735	0.00246	2.0
0.16786	0.14807	0.13079	0.11567	0.10241	0.09076	0.08051	0.07147	......	......	......	......	2.5
0.16786	0.14807	0.13079	0.11567	0.10241	0.09076	0.08051	0.07147	......	......	......	......	3.0
0.16786	0.14807	0.13079	0.11567	0.10241	0.09076	0.08051	0.07147	0.03982	0.02247	0.00735	0.00246	∞

The situation is shown schematically in Fig. 5.24. The gas-film interface is taken as the origin of coordinates which means that the film-substrate interface moves to the right with velocity V_x. The boundary and initial conditions are:

$$N(x,0) = N_1, \tag{5.67a}$$

$$N(\infty,t) = N_1. \tag{5.67b}$$

These two conditions mean that the infinitely thick substrate initially is uniformly doped with N_1 impurities/unit volume. The last condition comes from the continuity equation and the total number of impurities in the system. It is

$$J_x = (h+V_x)N(0,t) \tag{5.67c}$$

where h is a mass-transfer coefficient describing the escape of dopant atoms from the silicon into the gas. If the mass-transfer coefficient $h \ll V_x$ (it usually is for silicon), we can make a transformation of coordinates to a moving coordinate system $x' = x - V_x t$ where x' is the distance from the moving boundary. Substitution into Eq. (5.66) produces the simpler diffusion equation, Eq. (5.51), whose solution has been given previously.

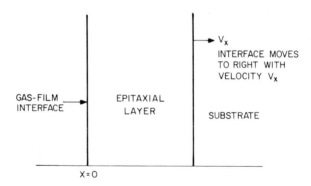

Fig. 5.24 Definition of variables for diffusion during growth of epitaxial film.

Coupled with the boundary conditions of Eq. (5.67), the resultant impurity profile is

$$N(x,t) \approx \frac{N_1}{2} \left\{ 1 + \mathrm{erf}\left[\frac{x - V_x t}{2\sqrt{D_1 t}} \right] \right\}. \tag{5.68}$$

This result shows that the concentration step is smeared in the vicinity of the boundary $x' = 0$.

Growth of doped epitaxial layer on undoped substrate   The second set of boundary and initial conditions of interest correspond to the growth of a doped layer on an initially undoped, infinitely thick substrate. These conditions are

$$N(0,t) = N_2, \tag{5.69a}$$

$$N(\infty, t) = 0, \tag{5.69b}$$

$$N(x, 0) = 0. \tag{5.69c}$$

Once again the gas-film boundary is taken as the origin of coordinates. Equation (5.69a) means the growing film is doped with N_2 impurities per unit volume, whereas Eqs. (5.69b) and (5.69c) say that the substrate is infinitely thick and initially undoped. The solution of Eq. (5.66) with these boundary conditions is

$$N(x,t) = \frac{N_2}{2} \left\{ \mathrm{erfc}\left[\frac{x - V_x t}{2\sqrt{D_2 t}} \right] + \exp\left[\frac{V_x x}{D_2} \right] \mathrm{erfc}\left[\frac{x + V_x t}{2\sqrt{D_2 t}} \right] \right\} \tag{5.70}$$

where D_2 is the diffusion coefficient of the dopant impurity added to the growing epitaxial layer. This result is plotted in Fig. 5.25 as a function of depth, x, normalized to the epitaxial-layer thickness, b. It is virtually impossible to have values of $V_x^2 t/D < 100$ with practical values of diffusion coefficient and epitaxial growth rate.

Growth of doped epitaxial layer on a doped substrate   The final case to be considered is for the deposition of a doped layer on a doped, infinitely thick substrate. The solution to this case is the superposition of the solutions to the two previous cases which, for $h = 0$, is

$$
N(x,t) = \frac{N_1}{2} \left\{ 1 + \mathrm{erf}\left(\frac{x - V_x t}{2\sqrt{D_1 t}} \right) \right\}
$$

$$
+ \frac{N_2}{2} \left\{ \mathrm{erfc}\left(\frac{x - V_x t}{2\sqrt{D_2 t}} \right) + \exp\left(\frac{V_x x}{D_2} \right) \mathrm{erfc}\left(\frac{x + V_x t}{2\sqrt{D_2 t}} \right) \right\},
$$

$$(5.71)$$

where N_1 and D_1 are the concentration and diffusion coefficient of the substrate impurity, and N_2 and D_2 are the concentration and diffusion coefficient of the impurity doping the epitaxial layer.

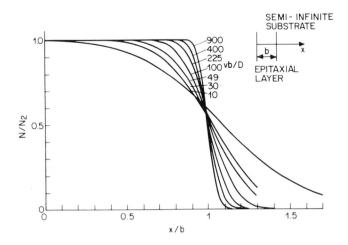

Fig. 5.25 Impurity profile resulting from the growth of a doped epitaxial film on an undoped substrate. (From Rice.[55])

5.4.5 Silicon Dioxide as a Diffusion Mask

Selective diffusions require a mask that is both impervious to the dopant and can withstand the high temperatures necessary for solid-state diffusion. A suitable mask is silicon dioxide[28] which is easily grown or deposited. In addition, it can withstand the high temperatures, is easily patterned using photolithographic techniques, and the diffusion coefficient for the usual silicon dopants is several orders of magnitude lower in silicon dioxide than in silicon.

We now consider the thickness requirements for the masking oxide and, in the next section, diffusion effects at the edge of a mask feature. The growth and properties of silicon dioxide are considered later in this chapter.

Usually there is some growth of oxide during the diffusion process. Anticipating the results of the next section, it is assumed that the oxide thickness as a function of time is given by

$$a(t) = \sqrt{a_0^2 + \sigma^2 t}, \qquad (5.72)$$

where $a(t)$ is the oxide thickness at time t, a_0 is the initial oxide thickness, σ is the oxide growth rate, and t is the growth (diffusion) time. Silicon is consumed as the thermal oxide grows, causing the silicon-silicon-dioxide interface to move. The gas-silicon-dioxide interface also moves because the volume of the silicon dioxide grown is greater than the volume of the silicon consumed. The coordinates for this problem are shown in Fig. 5.26.

The boundary conditions are that the impurity concentration at the gas-oxide interface is constant,

$$N_1(0,t) = N_0, \qquad (5.73a)$$

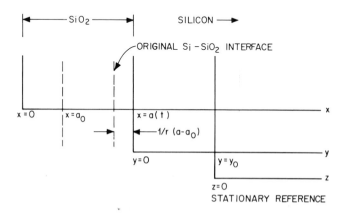

Fig. 5.26 Coordinate system for oxide growth during diffusion.

that the impurity concentration is zero deep in the silicon,

$$N_2(\infty,t) = 0, \tag{5.73b}$$

and that a matching condition is satisfied at the interface between the silicon and silicon dioxide

$$mN_1[x=a(t),t] = N_2(y=0,t). \tag{5.73c}$$

In Eq. (5.73c), m is the *segregation coefficient* defined as the ratio of impurities in the silicon to impurities in the oxide.

Diffusion in the oxide layer is described by

$$D_1 \frac{\partial^2 N_1}{\partial x^2} = \frac{\partial N_1}{\partial t}, \tag{5.74}$$

while in the silicon it is described by the diffusion equation in a region with changing boundaries,

$$D_2 \frac{\partial^2 N_2}{\partial y^2} + V(t) \frac{\partial N_2}{\partial y} = \frac{\partial N_2}{\partial t} \tag{5.75}$$

where

$$V(t) = \frac{\sigma^2}{2ra(t)} \tag{5.76}$$

is the velocity of the silicon-silicon-dioxide interface, obtained with the aid of Eq. (5.72), and r is the volume ratio of silicon dioxide to silicon. Note the similarity of Eq. (5.75) to Eq. (5.66). The solutions to these equations are:

1. For the oxide:

$$N_1(x,t) = N_0 \left\{ 1 - \frac{mDF\ \mathrm{erf}\left(\dfrac{x}{2\sqrt{D_1 t}}\right)}{\left[mDF + \mathrm{erfc}\left(\dfrac{\sigma}{2r\sqrt{D_2}}\right)\right]\mathrm{erf}\left(\dfrac{a(t)}{2\sqrt{D_1 t}}\right)} \right\} \tag{5.77}$$

2. For the silicon:

$$N_2(y,t) = N_0 \frac{m\ \mathrm{erfc}\left(\dfrac{\sigma[(1/r) + (y/a)]}{2\sqrt{D_2}}\right)}{mDF + \mathrm{erfc}\left(\dfrac{\sigma}{2r\sqrt{D_2}}\right)} \tag{5.78}$$

where

$$D \triangleq \left[\frac{D_2}{D_1}\right]^{1/2} \exp\left(-\frac{\sigma^2}{4r^2 D_2}\right) - \frac{\sigma}{2r}\left[\frac{\pi}{D_1}\right]^{1/2} \text{erfc}\left(\frac{\sigma}{2r\sqrt{D_2}}\right) \qquad (5.79a)$$

and

$$F \triangleq \frac{\sigma}{a(t)}\sqrt{t}\, \exp\left[\frac{a^2(t)}{4D_1 t}\right] \text{erf}\left(\frac{a(t)}{2\sqrt{D_1 t}}\right) \qquad (5.79b)$$

where t is the diffusion time.

The number of impurities per unit area absorbed by the oxide layer during a diffusion of t-seconds duration is

$$\int_0^t \int_0^{a(t)} N_1(x,\lambda)\, dx \; d\lambda.$$

Similarly, the number of impurities per unit area diffusing through the oxide layer into the silicon during this time is

$$\int_0^t \int_{a(t)}^{\infty} m N_1(x,\lambda)\, dx \; d\lambda = \int_0^t \int_0^{\infty} N_2(y,\lambda)\, dy \; d\lambda$$

and is a measure of the effectiveness of the oxide layer as a diffusion mask. If the dopant impurity concentration at the silicon surface, $N_2(y=0,t)$, is not to exceed a certain maximum value, N_{max}, then the maximum allowable diffusion time, t_{max}, is given by the solution of $N_2(0,t_{max}) = N_{max}$. These equations are quite complicated and solutions are best obtained by numerical techniques. In addition, the situation is further complicated by the possibility of the formation of glassy layers, and by concentration dependent effects. Empirical curves of the masking properties of SiO_2 for boron and phosphorous diffusions are given in Figs. 5.27 and 5.28.

5.4.6 Impurity Profile Near the Diffusion Mask

The impurity profiles derived in the previous sections are obtained when the diffusion is unmasked so that there are no additional boundaries on the surface of the wafer. They are valid in regions sufficiently far from a mask edge so that the one-dimensional approximation holds. When the wafer surface is partially masked, the one-dimensional diffusion equation no longer describes the process adequately in the vicinity of the mask edge, and either a two- or three-dimensional equation must be used. Solutions of the higher dimensional equations are given by Kennedy and O'Brien[29] for the constant-surface concentration and the instantaneous-source cases and are plotted in Figs. 5.29 and 5.30. From these plots it can be seen that the junctions formed by lateral diffusion under the mask are at about 0.8 as far from the mask edge as the vertical junction is from the surface. Also note that for the

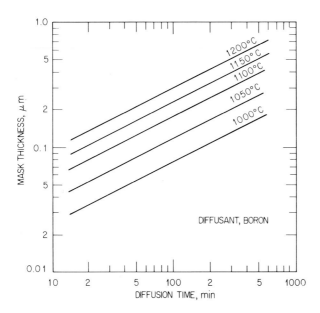

Fig. 5.27 Empirical curves for the effectiveness of SiO$_2$ as a boron diffusion mask.

instantaneous-source diffusion, the doping level in the unmasked region decreases in the vicinity of a mask edge.

5.4.7 Deviations from Simple Diffusion Theory

In this section we consider several important cases in which the results of the previous section, which were based on the diffusion equation with constant diffusivity, are not valid.

Diffusion coefficient is a function of time   Silicon wafers are subjected to many high temperature operations. During each of these operations, there may be a redistribution of impurities. It is possible to calculate an equivalent average diffusion coefficient and diffusion time for this sequence of heat treatments if the diffusion coefficient is only a function of time. The one-dimensional diffusion equation (with no drift field) is

$$\frac{\partial N}{\partial t} = \frac{\partial}{\partial x}\left[D(t)\,\frac{\partial N}{\partial x}\right] = \frac{\partial D(t)}{\partial x}\,\frac{\partial N}{\partial x} + D(t)\,\frac{\partial^2 N}{\partial x^2} \,. \qquad (5.80)$$

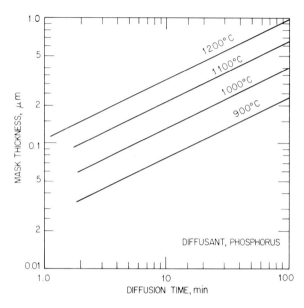

Fig. 5.28 Empirical curves for the effectiveness of SiO_2 as a phosphorous diffusion mask.

If the diffusivity is solely a function of time, the change of variables

$$\Lambda = \int_0^t D(\alpha)\,d\alpha \qquad (5.81)$$

reduces Eq. (5.80) to

$$\frac{\partial N}{\partial \Lambda} = \frac{\partial^2 N}{\partial x^2}.$$

Thus, the previous solutions for constant D can be used to give the concentration in terms of x and Λ. Then Λ can be converted into time t by using Eq. (5.81).

Diffusion with a concentration dependent diffusivity [30-33] The diffusion coefficient can have a complicated functional behavior. It can depend on local electric fields due to impurities in the crystal that are ionized by the thermal emf, the concentration of the diffusant, and the mechanism by which diffusion occurs. The nature of the diffusion coefficient will be discussed in Section 5.8.1.

It was implicitly assumed in applying the simple diffusion equation [Eq. (5.50b)] that the dopant is introduced into an essentially intrinsic crystal, and that the amount of dopant introduced into the crystal is less than the intrinsic carrier concentration at the diffusion temperature. This assump-

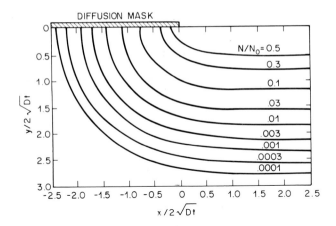

Fig. 5.29 Impurity profile near a mask edge for a constant surface concentration diffusion. (From Kennedy and O'Brien.[29] Copyright 1965 by International Business Machines Corp.; reprinted with permission.)

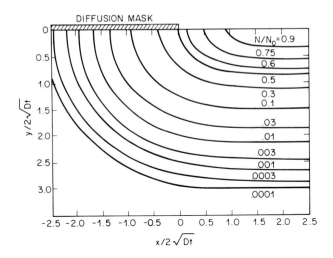

Fig. 5.30 Impurity profile near a mask edge for an instantaneous source diffusion. (From Kennedy and O'Brien.[29] Copyright 1965 by International Business Machines Corp.; reprinted with permission.)

tion is valid for the base diffusion of a bipolar transistor where the crystal is a lightly doped material. The measured emitter diffusion impurity profile differs significantly from that predicted by the simple theory because of the high dopant concentration that is introduced during the emitter diffusion and the presence of a significant dopant profile from the base diffusion.

If the dopant atom does not affect the elastic properties of the host crystal, the concentration dependent diffusivity (diffusion coefficient) can be expressed in terms of the intrinsic diffusivity D_i by

$$D = D_i \left[\frac{1 + \beta f}{1 + \beta} \right] h \tag{5.82a}$$

where

$$f = \frac{N}{2n_i} + \left[\left(\frac{N}{2n_i} \right)^2 + 1 \right]^{1/2}, \tag{5.82b}$$

$$h = 1 + \frac{N}{2n_i} + \left[\left(\frac{N}{2n_i} \right)^2 + 1 \right]^{-1/2} \tag{5.82c}$$

and β is a complicated function whose value at diffusion temperatures in silicon is about 100 for positively charged donor impurites and about 0.01 for negatively charged acceptor impurites. The intrinsic carrier concentration as a function of temperature is given by

$$np = n_i^2 = 1.5 \times 10^{33} T^3 \exp\left(- \frac{1.21}{kT}\right). \tag{5.83}$$

Thus a good approximation to the diffusivity is

$$D = D_i fh \tag{5.84a}$$

for a donor (n-type) impurity, and

$$D = D_i h \tag{5.84b}$$

for an acceptor (p-type) impurity.

The concentration dependent diffusion equation

$$\frac{\partial N}{\partial t} = \frac{\partial}{\partial x} \left[D(N) \frac{\partial N}{\partial x} \right]$$

with $D(N)$ given by Eq. (5.84a) and Eq. (5.84b) can be solved by converting it to an ordinary nonlinear differential equation via Boltzmann's transformation, $y = x/2\sqrt{D_i t}$. The resulting ordinary differential equation is integrated numerically. The solutions are plotted in Fig. 5.31(a) for $\beta = 0$ (p-type impurities) and in Fig. 5.31(b) for $\beta = 100$ (n-type impurities) for diffusions with constant surface concentration. The results for $\beta = 0$ do not differ markedly from the complementary error function profile resulting from a constant surface concentration diffusion with constant diffusivity. The results for $\beta = 100$ show a very pronounced enhancement of the diffusion.

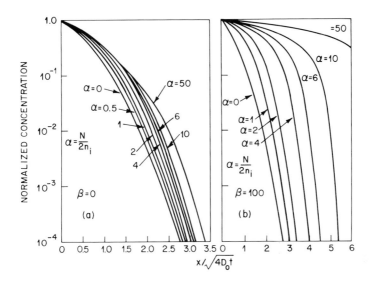

Fig. 5.31 Impurity profile for diffusion with concentration dependent diffusivity (a) for $\beta = 0$, p-type impurities; (b) $\beta = 100$, n-type impurities. (From Hu and Schmidt.[32])

Note also that the concentration gradient $\partial N/\partial x$ evaluated near the surface decreases as the doping increases $(N/2n_i \gg 1)$.

Fair[30] shows excellent experimental verification of this theory using an approximate solution of the concentration dependent diffusion equation in terms of Chebyshev polynomials

$$N = N_0(1.00 - 0.87\,Y - 0.45\,Y^2) \tag{5.85a}$$

where

$$Y = \frac{x}{\sqrt{8N_0 D_i t/n_i}} \tag{5.85b}$$

and, once again, N_0 is the surface concentration. His results for an arsenic diffusion from a doped-oxide source at 1000°C are shown in Fig. 5.32.

The availability of an explicit solution for the dopant profile permits us to write simple expressions for the junction depth and the sheet resistance of the diffused layer. A good estimate of junction depth is obtained by observing that near the surface dN/dY, and therefore the concentration gradient dN/dx, has a negative value that becomes more negative as the depth increases. The concentration gradient can be -10^{25} cm^{-4} in the vicinity of

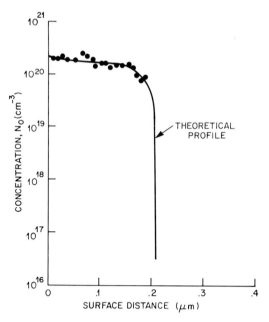

Fig. 5.32 Impurity profile for arsenic diffusion from doped oxide source at 1000°C. (From Fair.[30])

the junction. At 1000°C the intrinsic carrier concentration in silicon is 10^{19} cm^{-3}. An arsenic dopant profile will fall from this value to 10^{16} cm^{-3} in about 100 Å. Therefore, if the junction depth is in excess of 2000 Å, we can use the depth at which $N = n_i$ in place of the true junction depth with less than 5% error. Thus, from Eqs. (5.85a) and (5.85b) with $N_0 \gg n_i$, we find

$$x_j \approx 2.3 \left[\frac{N_0 D_i t}{n_i} \right]^{1/2}. \qquad (5.86)$$

At high doping density ($N > 6 \times 10^{19}$ cm^{-3} for arsenic), the electron mobility becomes independent of concentration ($\mu \approx 75$ cm^2/V−s). Because of the abruptness of the profile, the high concentration value of mobility is valid for all but the deepest 15% of the profile. Therefore, with the use of Eq. (5.85a) and Eq. (5.85b), we find

$$R_S = \frac{1}{q\mu \int_0^{x_j} N \, dx} = \frac{1.56 \times 10^{17}}{N_0 x_j}. \qquad (5.87)$$

Sequential diffusions   The base and emitter of a bipolar transistor are formed by sequential diffusions and the interaction between the dopant profiles must be considered. The description of this interaction requires consideration of a pair of coupled diffusion equations where the flux of each diffusant is a function of the concentration gradient of both dopant profiles. For an *n-p-n* transistor where the subscript 1 refers to the *p*-type base dopant and the subscript 2 refers to the *n*-type emitter dopant

$$
\frac{\partial}{\partial t}\begin{bmatrix} N_1 \\ N_2 \end{bmatrix} = \begin{bmatrix} D_{1i}\dfrac{\partial}{\partial x} & 0 \\ 0 & D_{2i}\dfrac{\partial}{\partial x} \end{bmatrix} \begin{bmatrix} h_1 & -(1-h_1) \\ -f(1-h_2) & fh_2 \end{bmatrix} \begin{bmatrix} \dfrac{\partial N_1}{\partial x} \\ \dfrac{\partial N_2}{\partial x} \end{bmatrix}.
$$

$$(5.88)$$

In this case the f and h factors must be modified to account for the two dopant profiles.

$$
f = \alpha(N_2-N_1) + \sqrt{\alpha^2(N_2-N_1)^2 + 1}
$$

$$
h_j = 1 + \frac{\alpha N_j}{\sqrt{\alpha^2(N_2-N_1)^2 + 1}} \; ; \quad j = 1,2
$$

$$
\alpha = D_{2i}/2n_i
$$

Equation (5.88) must be integrated numerically.

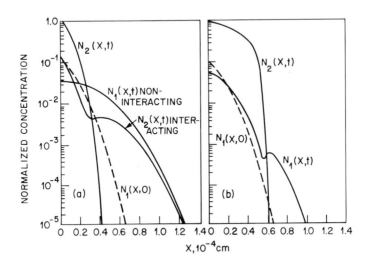

Fig. 5.33 Dopant profiles for sequential diffusion processes. (a) When $f = 1$ the interaction is due solely to local electric fields and (b) when $f \neq 1$. (From Hu and Schmidt.[32])

We will now discuss the results of two specific cases in order to get some feel for the interactive effects. When $f = 1$, the interaction is due solely to the local electric fields within the crystal. Figure 5.33(a) is a comparison between the emitter and base diffusion profiles given by the simple theory (no interaction) and when there is interaction via electric fields. The emitter dopant profile sets up a retarding field that impedes the progress of the base dopant profile, resulting in a dip in the concentration just behind the emitter diffusion front. Examination of these profiles as a function of time shows that the retarded region of the base profile moves at about the same rate as the emitter diffusion front, so that there is no appreciable change in the position of the collector-base junction. The retardation of the collector-base junction is an easily measurable quantity, but it is not a unique measure of the retardation of the overall profile which is affected by other factors such as base width.

The profiles when $f \neq 1$ are shown in Fig. 5.33(b). The p-type impurity in the base is not affected directly when $f \neq 1$ since the diffusivity given by Eq. (5.84b) is not a function of f. However, the profile of the n-type emitter dopant is a function f [see Eq. (5.84a)]. From Eq. (5.88) we note that the flux of the base diffusant is a function of the concentration gradient of the emitter diffusant, and therefore a change in the base profile occurs. Another consequence of the coupling in Eq. (5.88) is a retardation of the emitter profile when the surface concentration of the base is more than 10% that of the emitter.

5.4.8 Gettering

The presence of metal precipitates, even as isolated atoms, can have adverse affects on device performance and should be avoided. For example, gold is often intentionally introduced into the crystal to provide recombination centers that reduce the carrier lifetime. But the quantities of gold, copper, or other metals necessary to have some effect on the carrier lifetime, or to cause soft breakdown of p-n junctions, are so minute that it is impossible to prevent their introduction into the silicon during manufacture, surface preparation, or high temperature operations. It is more advantageous to remove the impurities by using getters.

Gettering can be accomplished by intentionally introducing damage sites on the back surface of the wafer by the use of ion implants[34, 35] or by mechanical abrasion. Heavy metal contaminants will "decorate" these crystal lattice dislocations during a subsequent high temperature anneal. Phosphorus or boron doped glasses are also popular getters as are phosphorus diffusions.[36, 37]

5.5 ION IMPLANTATION [34]

Ion implantation is a technique for introducing a layer of impurities just below the surface of the host material, in this case silicon, by bombarding it with a beam of ions whose energy is in the range of one to several hundred keV. Among the advantages of ion implantation are (1) it offers precise control over the number of impurities introduced into the silicon; (2) it is a low temperature process; (3) impurity layers can be introduced completely below the surface; (4) wide choice of mask material; (5) implanted junctions can be made self-aligned to the mask edge; and (6) impurity layers can be introduced in any order which permits forming the emitter of a bipolar transistor before forming its base region (see Chapter 6 for SIC processing sequences).

The disadvantages are that complex, expensive machinery is required and that the junctions are not automatically passivated.

In this section is described the equipment and technique of ion implantation, the range and straggle of implanted ions and the properties of doped layers, the resultant lattice damage and its annealing, and the phenomena of channeling.

5.5.1 Ion Implantation Apparatus [38]

An ion implantation machine shown pictorially in Fig. 5.34 consists of an ion source, a magnetic mass analyzer that is a filter to separate unwanted species of ions from the beam, an acceleration, focusing, and deflection system, and a target holder. The ion beam is confined to the interior of the system which is a vacuum chamber. The depth to which an ion becomes implanted is proportional to its kinetic energy which is in the 10 keV to 300 keV range for most commercial machines. The beam current, which is at most several milliamperes, is a measure of how many ions are delivered per second and thus is one factor determining the system throughput. The beam diameter is on the order of 1 cm with a divergence of about 0.1 °, and the beam current is in the range from 10 μA to 1 or 2 mA. The ion dose is between 10^{12} and 10^{16} atom/cm^2. For 3 in. diameter wafers, the implantation time is usually a fraction of a minute. Since this is less than the time required to evacuate the chamber, extensive fixturing is employed to permit many wafers to be loaded simultaneously. In one such arrangement,[39] about 50-60 2 in. wafers are mounted on a flat disc which is rotated at about 100 rpm on an axis parallel to the beam. Motion along the second axis is provided by translating the entire disc back and forth perpendicular to the beam. without breaking the vacuum. The primary advantage of ion implantation over diffusion is accurate control over the number and position of the implanted impurities and

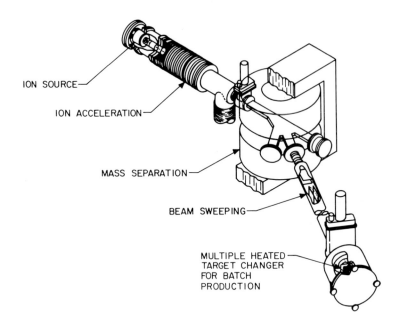

ION SOURCE

ION ACCELERATION

MASS SEPARATION

BEAM SWEEPING

MULTIPLE HEATED
TARGET CHANGER
FOR BATCH
PRODUCTION

Fig. 5.34 An ion-implantation machine.

the reproducibility, especially at high concentrations. In chemical source, solid-state diffusion, control of the impurity distribution relies on control of diffusion time and temperature, gas-flow rates, surface cleanliness and geometry, all of which are variables that are more difficult to control than are current and voltage.

5.5.2 Range and Straggle of Implanted Ions

Heavy charged particles moving through a solid lose energy either via collisions between the ion and nucleii of substrate atoms or by interaction between the ion and electrons in the solid. The particular loss mechanism is a function of the atomic number of the ion, Z_1, and its incident energy, E, with nuclear stopping predominating at low energies and high atomic number, and electronic interaction prevailing at high energy and low Z_1.

The implanted distribution in an amorphous target is roughly a Gaussian distribution and therefore is characterized by a mean, known as the *range,* and a standard deviation, known as the *straggle.* Range is the average distance traveled by an implanted ion while coming to rest in the solid. More important is the *projected range,* R_p, which is the projection of the range R in the direction of incidence, and $\bar{y}$, the mean penetration depth of the implant. When the ion beam is incident approximately normal to the surface $\bar{y} = R_p \cos \theta$ where θ is the angle of incidence. The straggle, ΔR_p, is a measure of the spread of the distribution.

It is customary to assume that nuclear stopping and electronic interaction are independent loss mechanisms, and therefore that the rate at which an incident ion loses energy is given by the sum

$$-\frac{dE}{dx} = N(S_n + S_e) \qquad (5.89)$$

where S_n is the *nuclear stopping power*, S_e is the *electronic stopping power*, and N is the number of substrate atoms per unit volume.

A detailed discussion of energy-loss mechanisms for ions in solids is beyond the scope of this book.[40] Some useful, crude estimates of the range based upon approximate values for S_n and S_e are given along with measured data for range and straggle for several common impurities in silicon.

A useful approximation for the nuclear stopping power, $S_n(E)$, is that it is independent of the incident ion energy and can be written as

$$S_n^0 = 2.8 \times 10^{-15} \frac{Z_1 Z_2}{\sqrt{Z_1^{2/3} + Z_2^{2/3}}} \frac{M_1}{M_1 + M_2} \quad \text{eV·cm}^2 \qquad (5.90)$$

where (Z_1, M_1) and (Z_2, M_2) are the atomic number and mass of the ion and target, respectively, and S_n^0 is the zero-order approximation for the nuclear stopping power. The electronic stopping power is approximated by assuming that the electrons form a free electron gas, and that the stopping power is proportional to projectile velocity as long as that velocity is less than the velocity of an electron having the Fermi energy of the free electron gas. Therefore,

$$S_e(E) = cv_1 = k\sqrt{E} \qquad (5.91)$$

where k depends on both the projectile and target materials, and has the value $k_{Si} \approx 0.2 \times 10^{-15}$ (eV)$^{1/2}$ cm^2 for amorphous silicon. Since S_n^0 is independent of energy and $S_e(E)$ increases with energy, there must be a critical energy, E_c, for which they are equal; namely, $\sqrt{E_c} = S_n^0/k$.

The range is the average distance it takes an ion to give up all its energy. Hence from Eq. (5.89)

$$R = \int_0^R dx = \int_0^E \frac{dE}{N[S_n(E) + S_e(E)]} . \qquad (5.92)$$

Nuclear stopping predominates when $E \ll E_c$, and therefore $-dE/dx \approx NS_n^0$. Thus

$$R = 0.7 \frac{\sqrt{Z_1^{2/3} + Z_2^{2/3}}}{Z_1 Z_2} \frac{M_1 + M_2}{M_1} E_0 \quad \text{Å} \qquad (5.93a)$$

where E_0 is in electron volts. This expression is amazingly accurate for heavy ions, such as arsenic, in silicon, but it overestimates the range of light ions, such as boron, in silicon by a factor of two.

Electronic stopping predominates at high energies and

$$R \approx 20\sqrt{E_0} \ \overset{\circ}{A}. \tag{5.93b}$$

Nuclear stopping involves motion of the substrate atoms which results in radiation damage to the substrate. Radiation damage is most serious for low-energy implants or in the low-energy portion of the track of a high-energy implant.

EXAMPLE 5.16   Calculate the range of 250 keV arsenic atoms implanted in silicon.

Solution.   The atomic number of arsenic is 33 and its mass is 1.2433×10^{-22} gm/atom while for silicon the atomic number is 14 and the mass is 4.662×10^{-23} gm/atom. Therefore, the approximate nuclear stopping power for this ion-target combination is obtained from Eq. (5.90) as

$$S_n^0 = 2.8 \times 10^{-15} \ \frac{14 \times 33}{\sqrt{14^{2/3} + 33^{2/3}}} \ \frac{1.2433 \times 10^{-22}}{1.2433 \times 10^{-22} + 4.662 \times 10^{-23}}$$

$$= 2.345 \times 10^{-13} \ \text{eV·cm}^2.$$

The critical energy E_c at which nuclear and electronic stopping powers are equal is $E_c = (2.345 \times 10^{-13}/0.2 \times 10^{-15})^2 = 1.375 \times 10^6$ eV   and   therefore   nuclear   stopping predominates in this case. From Eq. (5.93) with $E_0 = 250$ keV, the range is $R = 2.09 \times 10^3 \ \overset{\circ}{A}$.

The projected range, R_p, and straggle, ΔR_p, for several common impurities implanted in silicon at 300 K are given in Table 5.5. The distribution of the implanted atoms is approximately Gaussian and is given by

$$N(x) = \frac{\Phi}{\sqrt{2\pi} \ \Delta R_p} \ \exp\left[-\frac{1}{2}\left(\frac{x - R_p}{\Delta R_p}\right)^2\right] \tag{5.94}$$

where $\Phi = \int_0^\infty N(x) \, dx$ is the *dose*. Figure 5.35 shows the distribution of implanted atoms for the case $M_1 < M_2$ and $M_1 > M_2$. The *relative straggle* $\Delta R_p/R_p$ depends primarily on the ratio M_2/M_1.

5.5.3 Channeling

The distribution of ions implanted in a single crystal substrate cannot be predicted theoretically because of a phenomena called *channeling*. It occurs when the incident ion beam is aligned with a low index crystallographic direction and encounters a very open structure with little opportunity for collision. A large fraction of the incident ions penetrate through these channels in the crystal lattice or are guided through the channel by gentle, glancing

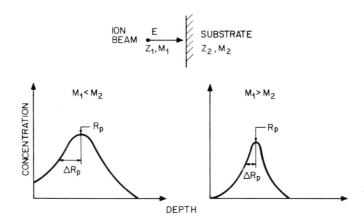

Fig. 5.35 Distribution of implanted ions in an amorphous target for $M_1 < M_2$ and $M_1 > M_2$. To a first approximation the mean depth depends on the ion mass and incident energy; whereas, the relative spread $\Delta R_p / R_p$ depends on M_1 / M_2.

collisions with the strings of atoms which form the channel wall. Channeled ions can penetrate two or three times further than random implants in the same material. Heavier ions penetrate more deeply because electronic stopping decreases with increasing ionic mass. It is also not an efficient process and usually a substantial fraction of the ions are dechanneled into a profile similar to that of random implants, but with peaks occurring at slightly larger than normal values of R_p. As a result, channeled implants have distributions with either a single peak, double peak, or monotonically decreasing characteristics.

Table 5.5 Range and Straggle for Boron, Phosphorus, and Arsenic Ions Implanted in Silicon (From Lee and Mayer.[54])

Energy (keV)		20	30	40	50	70	100	200
					Microns			
B	Range	0.064	0.095	1.25	1.52	2.10	2.80	4.85
	Straggle	0.0285	0.0375	0.0435	0.05	0.059	0.07	0.092
P	Range	0.025	0.036	0.047	0.062	0.086	0.122	0.25
	Straggle	0.0115	0.0165	0.021	0.0255	0.034	0.047	0.078
As	Range				0.031	0.042	0.057	0.11
	Straggle				0.011	0.015	0.02	0.037

In the computation of the maximum range, it is customary to assume that perfectly channeled ions are slowed entirely by electronic interactions and therefore the range is proportional to the square root of the initial ion energy. An estimate of the straggle for perfectly channeled particles is given by $\Delta R \propto \sqrt{d \cdot R}$ where d is the atomic spacing along the channel.

Implants are easily dechanneled by angular disorientation of the ion beam. The critical angle below which channeling occurs is a function of the ion and its energy. For 50 keV arsenic ions in silicon, it is 5.2° with respect to the $\langle 110 \rangle$ axis and 4.0° with respect to the $\langle 100 \rangle$ axis. The critical angle is smaller for boron or phosphorus ions. Furthermore, it decreases with increasing ion energy. An increase in substrate temperature increases the lattice vibrations which, in turn, increases dechanneling. The growth of silicon-dioxide layers on the substrate also tend to dechannel implants. More than 300 Å of silicon dioxide makes a phosphorous implant appear to be random. Similar remarks hold for boron which is less sensitive than phosphorous, and arsenic which is more sensitive than phosphorous.

5.5.4 Lattice Damage and Annealing

It was mentioned above that heavy, low-energy particles are stopped by elastic nuclear collisions that result in displacement of substrate atoms. This lattice damage also occurs in the low velocity portion of high-energy implants. Damage profiles are similar in shape to implantation profiles with the exception that the peak damage occurs slightly closer to the surface.

For light doses, $\Phi < 10^{12}$ ions/cm^2, individual damage tracks are created. Damage builds up linearly with dose until, at some critical dose, the damage tracks overlap and the implanted region becomes amorphous. The critical dose varies with the mass of the implanted species, being smaller for heavier ions.

Apart from substrate damage, implanted ions may not be on the proper substitutional lattice sites and so may not be fully ionized, electrically active donors or acceptors. This is dominated by the presence of many defect centers which trap or compensate implanted carriers. Fortunately, thermal annealing at relatively low temperatures can repair lattice damage in the sense that the implanted ions will be incorporated into electrically active sites, and the residual defects will be reduced to a level where carrier mobilities and lifetimes are usefully large. For example, a 30 min anneal following a room temperature boron implant with a dose of 10^{13} atoms/cm^2 causes the electrical activity to increase monotonically with anneal temperature, reaching the 100% point at 850°C. The annealing behavior is more complex for heavier implant doses, $\Phi > 10^{14}$ ions/cm^2, and displays several distinct regions. At temperatures below 500°C, electrical activity increases monotonically with anneal temperature. The restoration is attributed to the annealing of compensating defects with lightly doped, less damaged regions annealing first. Between 500°C and 600°C it is thought that interstitial silicon atoms replace boron atoms in substitutional sites and, as a consequence, electrical activity decreases with increasing temperature in this range. At temperatures above 600°C, thermal generation and migration of vacancies is thought dominant, and electrical activity again increases monotonically with temperature, reaching the 100% point at 950°C to 1000°C.

Annealing of high-dose implants and severely damaged layers resembles the epitaxial deposition process in some ways. Implanted layers anneal from each side and recrystallize epitaxially onto undamaged portions of the substrate, forming metallurgical imperfections where two regions meet. Permanent surface faults may still remain after the full annealing of shallow implants and amorphous regions that reach the surface. As a general rule, long anneal times correspond to low anneal temperatures and vice-versa.

5.6 OXIDATION OF SILICON

Oxide layers are used during the processing of a planar device as a diffusion mask. They also passivate the device by protecting the silicon surface from most of the harmful contaminants. The dielectric over the channel of an insulated gate field-effect transistor is often a thermally grown oxide layer.

In this section are described the thermal oxidation reactions, oxidation techniques, and reaction kinetics, the redistribution of impurities during thermal oxidation, and oxide deposition.

5.6.1 Oxidation Reactions

The two oxidizing species are oxygen and water vapor. They must diffuse through the growing oxide layer to react with the silicon. In the case of oxygen, the reaction at the silicon surface is quite straightforward.

$$Si + O_2 \rightarrow SiO_2. \tag{5.95}$$

One molecule of oxygen is required to form a molecule of silicon dioxide. The reaction involving water vapor is somewhat more complicated, but the over-all effect is described by

$$Si + 2H_2O \rightarrow SiO_2 + 2H_2. \tag{5.96}$$

In this reaction, two water molecules are required for the formation of each silicon-dioxide molecule. The hydrogen gas formed as a byproduct rapidly diffuses away from the silicon-silicon-dioxide interface. However, the hydrogen molecules react with the silica structure while diffusing through the oxide layer resulting in the replacement of bridging oxygen atoms in the silica structure with nonbridging hydroxyl (OH) groups, thus significantly weakening the oxide and making it more porous.

5.6.2 Oxidation Techniques

The popular oxidation techniques involve the use of dry oxygen, wet oxygen, steam, and high-pressure steam. The first three methods are conducted in an open-tube reactor where the silicon wafers are heated to a temperature in the range 900°C to 1200°C in a stream of water vapor, oxygen, or a mixture of the two oxidants.

A wet-oxygen source is obtained by bubbling oxygen through a constant-temperature, high-purity, water bath whose temperature determines the partial pressure of water vapor in the gas stream for a specified oxygen flow rate.

For steam oxidation, the temperature of the water bath generating the steam provides a measure of the partial pressure and flow rate. Live steam yields poor grades of oxide because of the etching action of excess water.

Properties of the various types of oxides and their etch rates are shown in Table 5.6. Graphs of oxide growth in dry and wet oxygen at various temperatures are given in Figs. 5.36 and 5.37. Note that the oxide resulting from growth in dry oxygen is denser but the growth rate is much slower. It is common practice to initiate and terminate oxide growth in dry oxygen and to use wet oxygen for the intermediate stage. The higher growth rate in wet oxygen is due to the higher solid solubility and larger diffusion coefficient of water vapor in silicon dioxide.

5.6.3 Reaction Kinetics [41]

Thermal oxidation of silicon involves the inward movement of the oxidizing species through the existing oxide layer to react with the silicon at the

Table 5.6 Oxide Properties

Type	Density (gm/cm^3)	Resistivity (Ω-cm)	Dielectric Strength 10^6 V/cm	Etch Rate in Buffered HF ($\overset{\circ}{A}$/s)[†]
Dry O_2	2.24-2.27	3×10^{15}-2×10^{16}	2	6.8
Wet O_2	2.18-2.21			6.7
Steam (1 atm)	2.0-2.20	10^{15}-10^{17}	6.8-9	7.3
SiH_4/CO_2	2.1-2.24			2.06-2.92[††]

[†] Buffered HF etch: 1 part 48% HF, 10 parts NH_4F solution (1 lb NH_4F/680 cc H_2O).

[††] P-etch: 3 parts 48% HF, 2 parts 70% HNO_3, 60 parts H_2O.

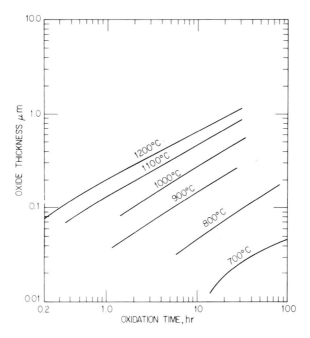

Fig. 5.36 Oxide growth rate in dry oxygen.

silicon-silicon-dioxide interface. The oxidizing agent is introduced into the reactor as a gas and it is brought from the bulk of the gas stream to the wafer surface by a mass-transport process. The flux of oxidant particles incident on the wafer is proportional to the difference in the oxidant concentration in the bulk of the gas stream and at the wafer surface. It is given by

$$J_1 = h_G(N_G - N_o) \tag{5.97}$$

where J_1 is the flux incident on the surface, h_G is the gas phase mass-transfer coefficient, N_G is the concentration of oxidant particles in the gas stream, and N_o is the oxidant concentration at the surface of the wafer. This equation is identical to Eq. (5.36) for the epitaxial reaction.

The oxidant is adsorbed from the gas stream by the existing oxide layer. This adsorption process is described by *Henry's law* which states that in the oxide, the oxidant concentration at the surface is proportional to the oxidant partial pressure at the oxide surface. That is,

$$N_{ox} = Hp_o \tag{5.98a}$$

where H is the Henry's law constant, and p_o is the oxidant partial pressure.

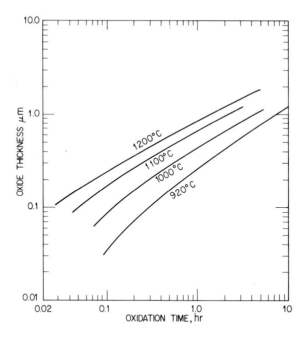

Fig. 5.37 Oxide growth rate in wet oxygen.

The experimentally significant variables are the oxide growth rate and the oxidant concentration in the bulk of the gas stream. It is therefore desirable to express all reactions in terms of the gas stream concentration N_G. Towards that end, denote by N^* the oxidant concentration in the oxide that would be in equilibrium with the oxidant partial pressure in the bulk of the gas stream, i.e.,

$$N^* = Hp_g. \tag{5.98b}$$

The partial pressures, p_o and p_g, are related to the concentrations, N_o and N_G, by the ideal gas law, $p_o = N_o kT$ and $p_g = N_G kT$. Substituting for p_o in Eq. (5.98a) yields a relation between the oxidant concentrations in the gas at the oxide surface and in the oxide at the surface.

$$N_{ox} = HkTN_o \tag{5.99}$$

Similarly, substituting for p_g in Eq. (5.98b) yields a relation between the hypothetical equilibrium concentration N^* and N_G,

$$N^* = HkTN_G. \tag{5.100}$$

These last two expressions permit the incident flux J_1 to be written in terms of concentrations within the oxide as

$$J_1 = \frac{h_G}{HkT} (N^* - N_{ox}). \tag{5.101}$$

Once inside the oxide layer, the oxidizing species diffuses towards the silicon-silicon-dioxide interface. The oxidant flux in the oxide is described by $J_2 = D(\partial N/\partial x)$, which is approximately

$$J_2 = D\left[\frac{N_{ox} - N_i}{a(t)}\right]$$

where $a(t)$ is the thickness of the oxide layer and N_i is the oxidant concentration at the silicon-silicon-dioxide interface. Finally, at the interface the oxidant reacts with the silicon at a rate proportional to the concentration, resulting in a flux

$$J_3 = k_s N_i. \tag{5.102}$$

In the steady state $J_1 = J_2 = J_3 = J$. After some algebra

$$\frac{N_i}{N^*} = \frac{1}{1 + (k_s/h) + (k_s/D)a(t)} \tag{5.103}$$

and therefore

$$J = J_3 = \frac{k_s N^*}{1 + (k_s/h) + (k_s/D)a(t)} \tag{5.104}$$

where $h = h_G/HkT$. Since flux is the number of particles per unit area per unit time, the growth rate is simply the flux divided by the number of particles in a unit volume. Denoting the number of particles per unit volume by N_V yields

$$\frac{da(t)}{dt} = \frac{k_s N^*/N_V}{1 + (k_s/h) + (k_s/D)a(t)} \tag{5.105}$$

as the differential equation describing oxide growth. Separating variables and integrating results in the following expression for oxide growth,

$$a^2(t) + Aa(t) = Bt', \tag{5.106a}$$

where $t' = t + \tau$ is a shifted time scale to account for the presence of an initial oxide layer of thickness a_0 at $t = 0$, and

$$A = 2D\left[\frac{1}{k_s} + \frac{1}{h}\right] \tag{5.106b}$$

$$B = \frac{2DN^*}{N_V} \tag{5.106c}$$

and

$$\tau = \frac{a_0^2 + Aa_0}{B}.$$ (5.106d)

In these expressions, A and B are rate constants.

The two limiting forms of $a(t)$ are of interest. When $t \gg \tau$ and $t \gg A^2/4B$,

$$a^2(t) \approx Bt,$$ (5.107)

and the general growth expression obtained from Eq. (5.106a) reduces to a parabolic growth expression. Thus, B is known as the parabolic rate constant, and by comparison with Eq (5.72), $B = \sigma^2$.

For short oxidation times, $t \ll A^2/4B$, the general expression reduces to a linear relation

$$a(t) \approx \frac{B}{A} t' = \frac{k_s h}{k_s + h} \frac{N^*}{N_V}$$ (5.108)

where B/A is sometimes called the linear rate constant.

Experimental results of Deal and Grove[42] show that for wet oxygen oxidations the apparent initial oxide thickness is zero. The parabolic rate constant B increases from 0.203 $\mu m^2/hr$ at 900°C to 0.720 $\mu m^2/hr$ at 1200°C. The coefficient A is found to vary inversely with temperature, with a value of 0.50 μm at 900°C and 0.05 μm at 1200°C. Furthermore, the activation energy is found to be 16.3 kcal/mole. A plot of oxide thickness versus time for oxidation in dry oxygen indicates that the oxide layer has an initial thickness of 230 ± 30 Å at all processing temperatures. This effect is caused by the initial layer growing at a very fast rate, followed by a strictly linear process that extrapolates back to $a_0 = 210$ Å at $t = 0$. The value of B varies from 0.0011 $\mu m^2/hr$ at 800°C to 0.045 $\mu m^2/hr$ at 1200°C and A varies from 0.370 μm at 800°C to 0.040 μm at 1200°C.

5.6.4 Redistribution of Impurities During Thermal Oxidation [43]

The thermal oxidation of doped silicon causes a change in the impurity concentration profile because when two solid phases (in this case silicon and silicon dioxide) containing the same impurity are in intimate contact, the impurity concentration adjusts itself such that the chemical potential is equal on both sides of the interface. Silicon is continually converted to silicon dioxide during the thermal oxidation process, and the silicon-silicon-dioxide interface advances steadily into the silicon causing a redistribution of the impurities. The three factors that determine the extent and character of the redistribution are the rate of advance of the silicon-oxide interface, the chemical activity coefficients of the impurities, and their diffusion coefficients in oxide. This last factor is important when the impurity tries to escape from the oxide layer.

Equations (5.74) through (5.76) describe the diffusion of impurities in the silicon and in a growing silicon-dioxide film. They were used previously to describe the behavior of SiO_2 as a diffusion mask. Solutions of these equations for oxide films grown in an oxidizing ambient which does not contain any impurity species are described in this section to show the effect of impurity redistribution.

The redistribution effects fall into four classes depending upon the value of segregation coefficient, m, given in Eq. (5.73c)

$$m \triangleq \frac{\text{impurity concentration in the silicon}}{\text{impurity concentration in the oxide}}$$

and the rate of impurity diffusion in the oxide. When $m < 1$, the equilibrium impurity concentration in the oxide is higher than that in the silicon and the oxide tends to deplete impurities from the silicon.

By assumption, the oxidizing ambient does not contain any of the impurity species, and therefore the impurity concentration at the oxide surface must be zero. The situation corresponding to the segregation coefficient $m < 1$ and a slow diffusing impurity (boron) is shown in Fig. 5.38(a). The impurity concentration in the oxide is greater than in the silicon, and both concentrations are relatively high because the slow impurity diffusion through the oxide prevents a loss of impurity atoms. When the impurity diffusion in oxide is fast and $m > 1$, the impurity concentration at the silicon side of the oxide interface can fall almost to zero because the oxide is continually absorbing impurities from the silicon, and these impurities rapidly diffuse through the oxide and escape to the gas stream. This situation is shown in Fig. 5.38(b).

It is interesting to note that there is impurity redistribution even when there is no impurity loss and $m = 1$ because the volume of a silicon dioxide layer is more than twice that of the silicon consumed in growing it. Therefore, if the oxide is to have the same impurity concentration as the silicon, some impurities must be depleted from the silicon to compensate for the volume gained by the oxide growth.

The concentration ratio at the interface is independent of time because both oxide growth and impurity diffusion depend on the square root of time. However, the impurity concentration at the silicon surface does depend upon the segregation coefficient, the ratio of oxidation rate to diffusion rate (B/D), and the ratio of the diffusion constant in oxide to the diffusion constant in silicon ($r = \sqrt{D_{ox}/D}$).

The effect of varying the oxidation-to-diffusion ratio (B/D) on the dopant concentration at the interface is shown graphically in Fig. 5.39 for $m < 1$ (boron) and for $m > 1$ (phosphorus). The change in dopant con-

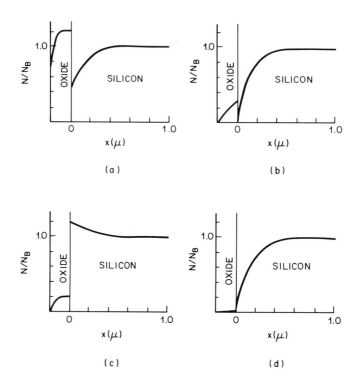

Fig. 5.38 Redistribution of impurities in silicon due to thermal oxidation. (a) Segregation coefficient $m < 1$ and a slow diffusing impurity (boron); (b) $m < 1$ and fast diffusing impurity (boron with H_2 ambient); (c) $m > 1$ and slow diffusing impurity (phosphorus); and (d) $m > 1$ and fast diffusing impurity (gallium).

centration at the surface relative to that deeper in the silicon is more pronounced for wet oxidation because the rate constant for wet oxidation is greater than for dry O_2 oxidation.

5.6.5 Chemical Vapor Deposition of Silicon Dioxide

There are many applications in which it is impossible or impractical to produce a silicon-dioxide film by thermal growth. One such application is the oxide layer that is used as the etching mask for an underlying silicon nitride layer. In these situations, the oxide is deposited by means of a chemical vapor deposition (CVD) reaction, or by sputtering or anodic oxidation. Two methods of chemical vapor deposition are described in this section.

The pyrolysis of silane in oxygen is described by the following reaction:

$$SiH_4(gas) + 2O_2(gas) \rightleftarrows SiO_2(solid) + 2H_2O(gas). \qquad (5.109)$$

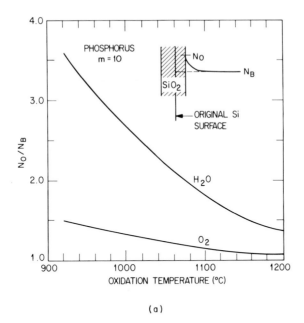

(a)

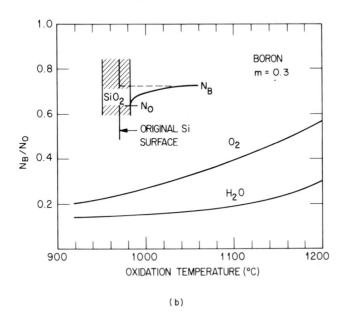

(b)

Fig. 5.39 Surface concentration of impurities in silicon after thermal oxidation (a) phosphorus and (b) boron.

No external heat is required to catalyze this reaction. However, heating the substrates to about 300°C results in films of higher quality and greater uniformity.

The silane, which is diluted with nitrogen, is not allowed to mix with the oxygen until it is in the vicinity of the heated substrates because the reaction occurs at a considerable rate even at room temperature. Typical reaction rates are several hundred angstroms per minute and are about an order of magnitude greater than the oxidation rate.

Boron or phosphorus-doped silicon can be deposited by the addition of B_2H_6 or PH_3 to the silane. Deposited films of silicon dioxide are often not suitable as passivating layers for electronic devices because of their low density. The quality of these films can be improved by a "densification" procedure which typically involves exposing the film to steam at a temperature of 800°C to 900°C for about 15 min.

The density of deposited oxides is inversely proportional to the deposition temperature. Deposited oxides with properties (such as etch rates) similar to that of thermal SiO_2 are deposited at about 1000°C by reactions such as the oxidation of silane with carbon dioxide.[44] Hydrogen is often used as the carrier gas. The deposition rate is independent of the CO_2/silane ratio in the range from 5 to 100. The rate increases with increasing temperature and silane concentration. At 1000°C, the growth rate is 180 Å/min with a carbon dioxide-to-silane ratio of 60.

Doped SiO_2 layers can be deposited by the addition of the halide or oxyhalide of the dopant to the silane.

5.6.6 The Effect of HCl on Thermal Silicon Dioxide [45-48]

The migration of positive ions, particularly sodium ions, in silicon dioxide is a major cause of electrical instabilities in MOS devices. It was observed that the introduction of chlorine in the silicon-oxide interface region tended to cause at least some of the sodium ions incorporated into the growing oxide to be electrically inactive and immobile even under the influence of high electric fields at elevated temperatures. The incorporation of chlorine during oxidation also apparently reduces the density of surface states.

The oxide growth rate is increased during oxidation in the presence of HCl because of the enhanced diffusion of the oxidants, O_2 and H_2O, through the oxide. The HCl also catalyzes the oxidation at the Si-SiO_2 interface, thereby increasing its rate. Finally, HCl contributes H_2O (which diffuses faster than O_2 in silicon dioxide) to the oxidation through the reaction $4HCl + O_2 \rightleftarrows 2H_2O + Cl_2$. The growth rate is a function of the silicon crystal orientation. Without HCl, the parabolic rate constant B for (111) silicon is about 15% larger than that for (100) silicon. The parabolic rate constant for (100) silicon is the larger of the two when the HCl/O_2 ratio is 0.1. At about 2-3% HCl in O_2 (the concentration normally used), the rate constants are equal and almost twice as large as the value without HCl.

The use of HCl during oxidation is not universally accepted. Values of fixed charge less than 10^{11} cm^{-2} and mobile charge less than 10^{10} cm^{-2} have been reported by oxidation with and without HCl. It appears that the use of HCl is a matter of preference and prior experience with a particular oxidation technique.

5.7 DEPOSITION OF SILICON NITRIDE [49]

Silicon-nitride films are used to passivate semiconductor devices because they act as a barrier to the diffusion of metal ions, in particular, sodium ions. The methods of depositing silicon nitride are the nitridation of silane or silicon halides, and by reactive sputtering in nitrogen or ammonia.

One of the most popular methods of producing silicon nitride is by the reaction between silane and ammonia

$$3SiH_4(gas) + 4NH_3(gas) \xrightarrow{\sim 900°C} Si_3N_4 + 12H_2(gas). \qquad (5.110)$$

The reaction temperature is typically 900°C but it can be lowered to 800°C if required. The silane is diluted with nitrogen resulting in a gas mixture which is typically one part silane, 20 parts ammonia, and 40 parts nitrogen. Growth rate of the film is in the 150-220 Å/min range.

Silicon nitride films can be grown over a much wider temperature range using the silicon halides. For example, the overall reaction with silicon tetrachloride is described by

$$3SiCl_4(gas) + 4NH_3(gas) \rightarrow Si_3N_4 + 12HCl(gas) \qquad (5.111)$$

where once again, nitrogen is used as the carrier gas. This reaction has been used for film deposition between 550°C and 1200°C. The deposition rate varies linearly with silicon tetrachloride concentration. Typical rates are in the 10 Å/min to 1000 Å/min range. Also, deposition rate varies exponentially with $-1/T$ over the temperature range of interest.

5.8 CLEANING OF SILICON WAFERS [50]

In the previous sections we have described the main process steps used in the fabrication of silicon and thin film integrated circuits. Processing sequences for various types of silicon integrated circuits are described in the next chapter. The one operation that is often overlooked in a discussion of processing is the cleaning of the wafers. Clean processing is a key to producing devices whose characteristics are stable and reproducible. More than half of the operations in making a silicon IC are for wafer cleaning between processing steps.

Surface contaminants fall into three broad classes; molecular, ionic, or atomic. Molecular contaminants, such as waxes, oils, or resins include oil from fingers or greasy films picked up from the air or from plastic containers. The photoresists used to define patterns on the surface of the wafer are also sources of molecular contaminants as are organic solvent residues. Hydrochloric acid and caustic solutions used as etchants are sources of ionic contaminants. These contaminants adhere to the surface by adsorption or chemisorption, and remain even after washing in deionized water. Atomic contaminants include copper, gold, and other heavy metals.

A suitable wafer cleaning procedure is based on the use of two solutions that contain volatile reagents diluted with pure deionized or quartz distilled water. The first solution contains ammonium hydroxide and hydrogen peroxide (typically five parts by volume H_2O, one part 30% unstabilized H_2O_2, and one part 27% NH_4OH). It removes organic contaminants by the solvating action of the ammonium hydroxide and the oxidizing action of the peroxide.

The second solution is used to remove heavy metals and to prevent their displacement replating from the solution by forming soluble complexes from the resulting ions. It contains hydrogen peroxide and hydrochloric acid (typical composition is six parts H_2O, one part 30% H_2O_2, and one part 37% HCl).

The wafers are placed in the appropriate cleaning solution at 75°C to 85°C for 10 to 20 minutes followed by a rinse in running deionized water. The wafers are then spun dry.

5.9 PROPERTIES OF IMPURITIES IN SILICON

In this section we consider some properties of impurities in silicon, namely, the form of the diffusion coefficient and the solid solubility limits of the common impurities.

5.9.1 The Diffusion Coefficient

The most important modes of diffusion in the situation under consideration here are substitutional diffusion and interstitial diffusion. In substitutional diffusion, the diffusing species jumps from one substitutional site to another. Therefore, the probability of an impurity making a transition is proportional to the probability that the impurity will have enough energy to break the bond at a substitutional site and the probability that an adjacent substitutional site is vacant. Both of these probabilities obey Boltzmann statistics, and therefore the diffusion coefficient is given by

$$D = D_0 \exp\left(-\frac{E_n + E_s}{kT}\right)$$ (5.112)

where $D_0 = 4\nu_0 d^2/6$, ν_0 is the lattice vibration frequency, d is the spacing between crystal planes, E_n is the energy required to break a bond, and E_s is the energy required to form a Schottky defect or lattice vacancy.

Diffusion in silicon is anisotropic because the spacing between crystal planes is a function of crystal orientation

$$d = \frac{a}{\sqrt{k_1^2 + k_2^2 + k_3^2}}.$$ (5.113)

In this equation a is the lattice spacing and k_1, k_2, k_3 are the Miller indices describing the crystal orientation. The lattice spacing for silicon is 5.42 Å, and therefore $d = 5.42$ Å for the (100) orientation, $d = 3.42$ Å for the (110) orientation, and $d = 3.14$ Å for the (111) orientation.

The temperature dependence is dominated by the exponential term, and D_0 is taken to be a constant over the several hundred-degree centigrade range of interest. However, D_{eff}, the effective value of D_0, is a function of concentration because the electric field caused by the ionized impurities in the crystal tends to enhance the diffusion of the slower carriers.

This effect is only important above a critical concentration that is a function of parameters of the host crystal, the impurity, and temperature. The characteristic concentration for boron in silicon is $N_{crit} = 2.5 \times 10^{19}$ cm^{-3} at 900°C and $N_{crit} = 2.5 \times 10^{20}$ cm^{-3} at 1200°C. For phosphorus, it is $N_{crit} = 1 \times 10^{20}$ cm^{-3} at 900°C and $N_{crit} = 1 \times 10^{21}$ at 1200°C. The effective concentration dependent diffusion coefficient of a substitutional impurity is

$$D(N) = D\left[\left(\frac{N}{N_{crit}}\right)^2 + 1\right]\left[1 + \frac{N}{\sqrt{N_2 + 4n_i^2}}\right].$$ (5.114)

Typical substitutional impurities are boron and phosphorous which are the most important p- and n-type dopants, respectively. Figure 5.40 is a plot of the diffusion coefficient of substitutional diffusers in silicon.

Interstitial diffusers, such as gold, move much faster than substitutional diffusers because they move into the interstitial sites which are usually vacant. Thus, it is only necessary for an atom to gain enough energy to squeeze through the lattice constriction in order to move between adjacent interstitial sites. The diffusion coefficient for interstitial diffusers is given by

$$D = D_0 \exp\left(\frac{-E_m}{kT}\right),$$ (5.115)

where E_m is the interaction energy. The diffusion coefficient for interstitial impurities is five or six orders of magnitude greater than the substitutional diffusion coefficient.

5.9.2 Solid Solubility of Impurities in Silicon

The maximum solid solubility is the maximum density of impurity atoms that the semiconductor host crystal can accommodate at a particular temperature. It is a function of the mechanism by which the impurity is incorporated in the crystal and generally increases with temperature until just

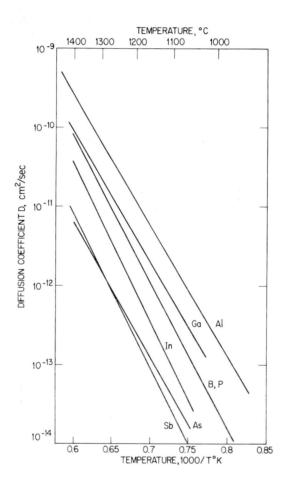

Fig. 5.40 Diffusion coefficient of substitutional impurities in silicon.

before the melting point of the host crystal is reached, at which time there is a steep decline.

The solid solubilities of common impurities in silicon are plotted in Fig. 5.41.

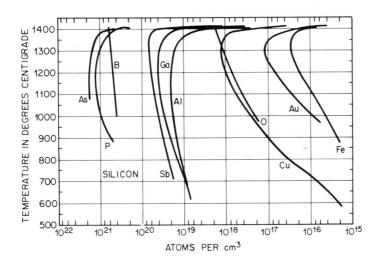

Fig. 5.41 Plot of solid solubilities of some common impurities in silicon. (From Trumbore.[56] Reprinted with permission from The Bell System Technical Journal, Copyright 1960, The American Telephone and Telegraph Co.)

5.10 PHOTORESISTS [51]

Many of the process steps in the fabrication of silicon or thin-film integrated circuits must be applied selectively. For example, a particular dopant is diffused into certain regions of the silicon crystal to form the base region of the transistors. The required selectivity is provided by covering the surface with a material that is correctly patterned. This patterned layer will permit the process to proceed in those regions that are exposed and will inhibit the process in those regions that are covered.

The subject of designing the pattern for these masks is considered in Chapter 7. The actual patterned layer is produced on the substrate by a process called photolithography using a sensitized material called a photoresist. The photolitographic operations are very important steps and often are major yield determining factors. Electron-beams or X-rays are also used to expose the resist and in that case the prefix "photo" should be replaced by E-beam or X-ray. In this section, we briefly describe photoresist materials and processes.

5.10.1 Photoresist Materials

Photoresists are organic compounds whose solubility is affected by exposure to ultraviolet light. There are two classes of resists, namely, negative and positive. In the former, the reaction to ultraviolet light causes the exposed areas to have a lower solubility in the solvent (developer), whereas in a positive resist the exposed areas are more soluble in the developer.

Negative photoresists   The principle components of a negative photoresist are a polymer, a sensitizer, and a solvent system. The most widely used polymers for integrated circuit applications appear to be polyisoprene derivatives. Polyisoprene is the basic polymer in natural latex rubber. It must undergo a process called *cyclization,* during which its molecular weight is reduced in order to be useful as a photoresist. Cyclized rubber is photosensitive as is evidenced by the fact that rubber insulated wire becomes brittle upon prolonged exposure to light. The exposure to light causes cross-linking between molecules which increases the molecular weight of the resultant molecules and makes it more resistant to attack by the solvent. This resin is highly resistant to both acid and alkaline aqueous solutions and oxidizing agents. As a consequence, a given thickness of negative resist film is more resistant than a corresponding thickness of positive resist.

Even though the cyclized rubber resin is photosensitive, a sensitizer is required to shorten the exposure times to a tolerable value. Among requirements for the sensitizer are that it be stable for long periods of time in the presence of the resin prior to exposure, that it remain uniformly dispersed in the solution, and that it be soluble in common solvents. The sensitizers are secret and the reactions that they undergo are quite complex. In general, they are attached to the resin chain and upon exposure to light they form intermediate compounds with an odd number of free electrons called "free radicals" that can enter into cross-linking reactions. The sensitizer may be further polymerized by heat.

Cyclized rubber is a hydrocarbon and therefore any hydrocarbon solvent can be used. Xylene is the most commonly used solvent for polyisoprene.

Positive resists   Among the advantages of positive resists are extreme pattern accuracy (because there is no swelling of the film) with a minimal amount of processing technique and few processing steps. The basic composition of positive resists is similar to that of negative resists; resin, sensitizer, and solvents. However, the materials are different and they produce the desired result in a different way. Positive resists are soluble in strongly alkaline solutions and develop in mildly alkaline solutions. The resins are of low molecular weight to ensure good development in aqueous developing solutions.

Resins used in positive photoresists are usually not photosensitive and have some affinity for water. If the resin has a strong affinity for water it cannot be used in large quantities and still produce a useful image. The purpose of the resin is to improve the adhesion and chemical resistance of the resist film, to prevent the sensitizer from crystallizing out of the coating, and to increase the viscosity of the liquid resists.

In positive photoresists it is the sensitizer that provides the photosensitivity, rather than enhancing it as with negative resists. Sensitizers for positive photoresists do not promote any changes in the resin system. They are usually derivatives of "diazo oxides" or "orthoquinone diazides" which undergo a drastic change in structure upon exposure to light. The reaction products are soluble in alkaline solutions. In addition, the unexposed areas of the resist film can undergo a reaction during development called coupling in which the diazide joins with certain components of the resin system.

The most prevalent solvents are Cellosolve, methyl Cellosolve, or Cellosolve acetate. Aromatic hydrocarbons, such as xylene, are mentioned as diluants.

5.10.2 Typical Photoresist Operation

Each "photoresist operation" in an IC fabrication sequence consists of many steps. A typical photoresist operation consists of (1) substrate cleaning, (2) dehydration bake, (3) final cleaning, (4) resist application, (5) prebake, (6) resist exposure, (7) develop, (8) rinse, (9) dry, and (10) postbake.

The cleaning steps at the beginning of the operation are required to permit the resist film to adhere tightly to the surface. The first step removes organic and inorganic contaminants and particulate matter. The dehydration bake at about 200°C drives off moisture and volatile cleaning residues to permit good resist adhesion. The final clean removes any minor residuals just prior to resist application.

Resists are usually applied by spinning (whirling) at rotational speeds between 3000 rpm and 8000 rpm depending on the viscosity of the resist and the required thickness of the coating. The coating thickness depends on the chemical resistance required of the image and the fineness of the lines and spaces to be resolved. Negative resists can usually resolve about three times the resist thickness. Positive resists are somewhat better. Resist thickness for silicon integrated circuits is about 1 μm.

After coating, the substrates undergo a prebake and dry during which solvents are eliminated from the film to ensure repeatable exposure, the adhesion of the film to the substrate is increased to reduce image damage or distortion during development, and the film is hardened to reduce handling damage. The film is air dried for 10 to 30 min to permit evaporation of the volatile solvents, and then baked for 15 to 30 min at about 80°C to 100°C.

The choice of exposure light source depends on the spectral sensitivity of the resists. Exposure is through a mask by contact printing, off-contact printing, or projection-printing. In contact printing, the mask is in intimate contact with the resist film, whereas in off-contact printing there is a gap of perhaps 20 μm between the surfaces. Projection printing uses a high quality optical system to produce a mask image on the resist film. It requires the most elaborate equipment and the longest exposure time (several minutes) but permits almost unlimited mask life. Contact printing uses the simplest equipment and shortest exposures (fraction of a minute) but the mask life is on the order of 20 to 50 exposures.

After exposure, the image is developed by subjecting the substrates to a spray of the developing solution followed by a water rinse for positive resists. The developing operation is followed by postbaking at about 200°C for about 30 min. This is an essential step to provide films with greater chemical and pinhole resistance, and to improve surface adhesion. Additional polymerization occurs in negative acting resists during the postbake.

REFERENCES

1.   L. I. Maissel and R. Glang. *Handbook of Thin Films.* New York: McGraw-Hill, 1970, Chapter 1.

2.   R. W. Berry, P. M. Hall, and M. T. Harris. *Thin Film Technology.* Princeton: D. Van Nostrand, 1968, Chapter 3.

3.   L. I. Maissel and R. Glang, *op. cit.,* pp. 1-10.

4.   *Ibid,* Chapter 8.

5.   T. C. Tisone and J. B. Bindell. "Step coverage in vacuum deposition." *J. Vac. Sci. Technol.* **11**: 72-76 (January/February 1974).

6.   A. J. Learn. "Evolution and current status of aluminum metallization." *J. of Electrochemical Soc.* **123**: 894-906 (June 1976).

7.   *Physical Vapor Deposition,* Berkeley, California, Airco Temescal, 1976.

8.   L. I. Maissel and R. Glang, *op. cit.,* Chapters 3 and 4.

9.   R. W. Berry, P. M. Hall, and M. T. Harris, *op. cit.,* Chapter 4.

10.   G. Carter and J. S. Colligon, *Ion Bombardment of Solids.* New York: American Elsevier Publishing Company, 1968.

11.   F. Gronlund and W. J. Moore. "Sputtering of Ag by light ions, 2-12 keV." *J. Chem. Phys.* **32**: 1540 (1960).

12.   V. Hoffman. "High rate magnetron sputtering for metallizing semiconductor devices." *Solid-State Technology* **19**: 57-61 (December 1976).

13. H. Dimigen and H. Lüthje. "An investigation of ion etching." *Philips Tech. Rev.* **35**: 199-208 (1975).

14. M. Cantagrel. "Considerations on high resolution patterns engraved by ion etching." *IEEE Trans. on Electron Devices.* **ED-22**: 483-486 (July 1975).

15. R. F. Lever. "The equilibrium behavior of the silicon-hydrogen-chlorine system." *IBM J. of Research and Development* **8**: 460-665 (September 1964).

16. V. S. Ban and S. L. Gilbert. "Chemical processes in vapor deposition of silicon. 1. Deposition from $SiCl_2H_2$ and etching by HCl. *J. of Electrochemical Soc.* **122**: 1382-1388 (October 1975).

17. V. S. Ban and S. L. Gilbert. "Chemical processes in vapor deposition of silicon. 2. Deposition from $SiCl_3H$ and $SiCl_4$." *J. of Electrochemical Soc.* **122**: 1389-1391 (October 1975).

18. W. J. Moore. *Physical Chemistry,* 4th Edition. Englewood Cliffs, N. J.: Prentice-Hall, 1972.

19. A. S. Grove. *Physics and Technology of Semiconductor Devices.* New York: John Wiley & Sons, 1967, Chapter 1.

20. M. S. Bawa. R. C. Goodman, and J. K. Truitt. "Kinetics and mechanism of deposition of silicon by reductions of chlorosilanes with hydrogen." *Chemical Vapor Deposition.* Princeton, N. J.: Electrochemical Society, 1973.

21. H. Blasius, Grenzschichten in Flüssigkeiten mit kleiner Reibung, *Z. Math. v. Phys.* **56** (1908). English translation in *NACA Tech. Memo No. 1256.*

22. H. Schlichting. *Boundary Layer Theory,* 4th Edition. New York: McGraw-Hill, 1960, Chapter 14.

23. H. C. Theuerer. "Epitaxial silicon films by the hydrogen reduction of $SiCl_4$." *J. of Electrochemical Soc.* **108**: 649-653 (July 1961).

24. W. H. Shepherd. "Vapor phase deposition and etching of silicon." *J. of Electrochemical Soc.* **112**: 988-993 (October 1965).

25. G. W. Cullen. "The preparation and properties of chemically vapor deposited silicon on sapphire and spinel." In *Proc. of First International Conf. on Crystal Growth and Epitaxy from the Vapour Phase,* Zurich, Switzerland, pp. 107-125 (September 1970).

26. W. R. Runyan. *Silicon Semiconductor Technology.* New York: McGraw-Hill, 1965, Chapter 7.

27. R. C. T. Smith. "Conduction of heat in the semi-infinite solid with a short table of an important integral." *Australian J. Phys.:* 127-130 (1953).

28. C. J. Frosch and L. Derrick. "Surface protection and selective masking during diffusion in silicon." *J. of Electrochemical Soc.* **104**: 547-552 (1957).

29. D. P. Kennedy and R. R. O'Brien. "Analysis of the impurity atom distribution near the diffusion mask for a planar p-n junction." *IBM J. of Research and Development* **9**: 179-186 (May 1965).

30. R. B. Fair. "Profile estimation of high-concentration arsenic diffusions in silicon." *J. Appl. Phys.* **43**: 1278-1280 (March 1972).

31. H. Wolf. *Semiconductors.* New York: Wiley-Interscience, 1971, Chapter 2.

32. S. M. Hu and S. Schmidt. "Interactions in sequential diffusion processes in semiconductors." *J. of Appl. Phys.* **39**: 4272-4283 (August 1968).

33. J. Crank. *The Mathematics of Diffusion*, 2nd edition. Oxford: Clarendon Press, 1975.

34. K. A. Pickar. "Ion implantation in silicon." In *Applied Solid State Science* **5**, R. Wolfe, editor. New York: Academic Press, 1975, pp. 151-249.

35. T. E. Seidel, R. L. Meek, and A. G. Cullis. "Direct comparison of ion-damage gettering and phosphorus-diffusion gettering of Au in Si." *J. of Appl. Phys.* **46**: 600-609 (February 1975).

36. J. L. Lambert and M. Reese. "The gettering of gold and copper from silicon." *Solid-State Electronics* **11**: 1055-1061 (1968).

37. R. L. Meek, T. E. Seidel, and A. G. Cullis. "Diffusion gettering of Au and Cu in silicon." *J. of Electrochemical Soc.* **122**: 786-796 (June 1975).

38. G. Ryding and A. B. Wittkower. "Industrial ion implant machines." *IEEE Trans. on Manufacturing Technology* **MFT-4**: 21-31 (September 1975).

39. G. I. Robertson. "Rotating scan for ion implantation." *J. of Electrochemical Soc.* **122**: 796-800 (June 1975).

40. J. W. Mayer, L. Eriksson, and J. A. Davis. *Ion Implantation in Semiconductors.* New York: Academic Press, 1970.

41. A. S. Grove, *op. cit.,* Chapter 2.

42. B. E. Deal and A. S. Grove. "General relationship for the thermal oxidation of silicon." *J. Applied Physics*, **36**: 3770-3778 (December 1965).

43. M. M. Atalla and E. Tannenbaum. "Impurity redistribution and junction formation in silicon by thermal oxidation." *Bell Syst. Tech. J.* **39**: 933-946 (July 1960).

44. A. K. Gaind, G. K. Ackermann, V. J. Lucarini, and R. L. Bratter. "Preparation and properties of SiO_2 films from SiH_4-CO_2-H_2." *J. of Electrochemical Soc.* **123**: 111-117 (January 1976).

45. R. J. Kreegler. "The role of HCl in the passivation of MOS transistors." *Thin Solid Films* **13**: 11-14 (1972).

46. R. J. Kriegler, Y. C. Cheng, and D. R. Colton. "The effect of HCl and Cl_2 on the thermal oxidation of silicon." *J. of Electrochemical Soc.* **119:** 388-392 (March 1972).

47. Y. J. van der Meulen, C. M. Osburn, and J. F. Ziegler. "Properties of SiO_2 grown in the presence of HCl or Cl_2." *J. of Electrochemical Soc.* **122**: 284-290 (February 1975).

48. H. Kirabayashi and J. Iwamura. "Kinetics of thermal growth of HCl-O_2 oxides on silicon." *J. of Electrochemical Soc.* **120**: 1595-1601 (November 1973).

49. G. Hass and R. E. Thun. *Physics of Thin Films.* New York: Academic Press, 1969, pp. 293-298.

50. W. Kern. "Cleaning solutions based on hydrogen peroxide for use in silicon semiconductor technology." *RCA Review* **31**: 187-206 (June 1970).

51. W. S. DeForest. *Photoresist.* New York: McGraw-Hill Book Company, 1975.

52. R. E. Honig. "Vapor pressure data for solid and liquid elements." *RCA Review* **23**: 567-586 (December 1962).

53. S. Dushman. *Scientific Foundations of Vacuum Technique*, 2nd edition. New York: John Wiley and Sons, Inc., 1962.

54. D. H. Lee and J. W. Mayer. "Ion-implanted semiconductor devices." *Proc. IEEE* **62**: 1242-1255 (September 1974).

55. W. R. Rice. "Diffusion of impurities during epitaxy." *Proc. IEEE* **52**: 289 (March 1964).

56. F. A. Trumbore. "Solid solubilities of impurity elements in germanium and silicon." *Bell Syst. Tech. J.* **39**: 205-234 (January 1960).

PROBLEMS

5.1 Equation (5.2) for vapor pressure is derived under the assumption that the heat of evaporation is not a function of temperature. An exact expression for vapor pressure is

$$\ln p^* = -\frac{\Delta_e G^0(T)}{RT}$$

where $\Delta_e G^0(T)$ is the Gibbs standard free energy of evaporation. K. K. Kelley in *Bureau of Mines Bulletin 383*, 1935, gives the free energy of evaporation for liquid aluminum in cal/mole as

$$\Delta_e G^0(T) = 65,680 - 43.72T + 4.61 \log T.$$

Calculate the time required to coat a 100 cm^2 substrate with an aluminum film 15,000 Å thick and the amount of energy necessary to vaporize the material from the liquid if the process is conducted at 1900 K. Assume that the pressure of the return flux is zero and that $\alpha_v = 1$.

5.2 The vapor pressure of gold measured at 2040 K is 1 torr. Its density is $\rho = 19.3$ and its mass is 3.2337×10^{-17} gram/mole. Calculate the deposition rate and its sensitivity to small temperature perturbations. Assume that the pressure of the return flux is zero and that $\alpha_v = 1$.

5.3 A film is deposited by evaporation from a point source. The substrate is 10 cm on a side. What is the required source-to-substrate distance if the film thickness uniformity is to be 5%.

5.4 The speed of exhaust is the pumping speed at the output of the vacuum chamber. Show that it can be expressed in terms of the pumping speed of the pump and the conductance of the piping by

$$\frac{1}{S_E} = \frac{1}{S_p} + \frac{1}{g}.$$

What is the speed of exhaust if the pump speed is 50 l/s and the pipe conductance is 10 l/s?

5.5   The sputtering yield of copper in argon was found to be $S = 2.3$ at an ion energy of 600 eV. The measured deposition rate in a particular sputtering apparatus was $G = 200$ Å/min when the external electron current $I_0 = 100$ mA.

a)   Calculate the ion current from Townsend's equation

$$I = \frac{I_0 \exp(\alpha d)}{1 - \gamma[\exp(\alpha d) - 1]}$$

using $\alpha = 0.1$ ions/cm and $\gamma = 0.08$ electron/ion, and the interelectrode spacing $d = 10$ cm.

b)   Compute the constant C that characterizes the sputtering apparatus.

c)   Find the new deposition rate if the ion energy is increased to 1000 eV. $S = 3.2$, $\alpha = 0.5$ ions/cm, and $\gamma = 0.1$ electron/ion.

5.6   Tantalum is bombarded by 2 keV argon ions.

a)   What is the maximum energy that can be transferred to a target atom?

b)   What is the maximum value that the mean energy of a struck target atom can attain if the threshold energy for the argon-tantalum ion-target combination is 26 eV?

c)   What is the average number of atomic layers involved in this sputtering process if the surface binding energy is 10 eV?

The atomic number and mass of tantalum are $Z = 73$, $M = 180.88$. For argon they are $Z = 18$, $M = 39.944$.

5.7

a)   Calculate the change in Gibbs' free energy ΔG and the reaction rate constants for the remaining species in Table 5.2.

b)   Use these values for the rate constants to calculate the equilibrium partial pressures of all species when $X = 10^{-2}$ and 10^{-1}.

5.8   A tubular expitaxial reactor is 3 in. in diameter. The input gas mixture is 2% silicon tetrachloride, 98% H_2 by volume, and its flow rate is 1.5 liters/min. The temperature in the reaction zone is 1250°C and the viscosity of the gas is 3×10^{-4} g/cm-s. Calculate

a)   The gas stream velocity U.

b)   The Reynolds number based on a 3-in. diameter wafer.

c)   The concentration of $SiCl_4$ (molecules /cm^3) in the input gas stream.

d)   The flux of $SiCl_4$ molecules to the surface.

e)   The film growth rate if the reaction is mass transfer limited.

5.9   Calculate the gas phase mass transfer coefficient for a reactor with a Reynolds
       number of 16, a reaction length of 20 cm, and a diffusivity $D_G = 10$ cm^2/sec.

5.10   The temperature dependence of reactions is often expressed in the equivalent
       forms $\exp(-\Delta H/RT)$ and $\exp(-E_a/kT)$. Find the value of ΔH in kcal that is
       equivalent to $E_a = 1$ eV.

5.11   The temperature controller on a diffusion furnace maintains the desired tem-
       perature $\pm 0.1\%$. What is the resultant tolerance on junction depth for an erfc
       diffusion profile at 1000°C?

5.12   Estimate the thickness of silicon dioxide required to mask a boron diffusion at
       1000°C for 1 hr if the solid solubility of boron in silicon dioxide is three times
       that in silicon.

5.13   A two-step boron diffusion is performed into n-type silicon with an impurity
       concentration of 5×10^{15} cm^{-3}. The predeposition is conducted at 1000°C for
       15 min with a surface concentration of 10^{21} cm^{-3}. The drive-in temperature is
       1100°C.

       a)   What is the junction depth if the drive-in diffusion lasts 5 hr?

       b)   What is the surface concentration?

       c)   What is the drive-in time to put the junction depth at 3 μm and the
             resultant surface concentration?

5.14   Calculate the range of 100 keV boron atoms implanted in silicon. The atomic
       number of boron is 5 and its atomic mass is 10.82.

5.15   Select an implant dose of 50 keV boron ions, and a drive-in time and tem-
       perature to result in a 200 $\Omega/\square$ diffused layer and a 1 μm junction depth.
       (Choose the drive-in time and temperature combination to obtain 100% electr-
       ical activity of implanted ions if possible.)

5.16   Show that if a 1 μm layer of silicon is oxidized the thickness of the resulting
       SiO$_2$ layer is 2.22 μm.

5.17   A silicon wafer is covered by an SiO$_2$ film 0.3 μm thick.

       a)   What is the time required to increase the thickness to 0.5 μm by oxida-
             tion in wet oxygen at 1200°C.

       b)   Repeat (a) for oxidation in dry oxygen at 1200°C.

6

Silicon Integrated Circuits

After having given in some detail the chemistry and physics of the processes used in fabricating integrated circuits, we now turn to how these processes are combined to make various types of bipolar and MOS silicon integrated circuits.

6.1 JUNCTION-ISOLATED BIPOLAR CIRCUITS

A silicon integrated circuit consists of several regions, each containing one or more circuit elements, which are interconnected according to a specified network topology. Some method of isolation is necessary to prevent undesired electrical interconnection between regions in the same silicon substrate. For bipolar circuits, the simplest and most popular scheme is known as junction or diode isolation. The isolation is provided by the reverse-biased p-n junction, which is formed between the substrate and the isolation island when the substrate is connected to the most negative potential, as shown in Fig. 6.1. There is a parasitic depletion-layer capacitance and a leakage current associated with the isolating junction.

The substrate for junction-isolated n-p-n bipolar circuits is a lightly doped (5 Ω-cm to 20 Ω-cm) p-type wafer about 125 μm thick. The substrate doping should be light to minimize the parasitic depletion-layer capacitance, but heavy enough to prevent it from inverting to n-type during subsequent processing. Many identical circuits are made simultaneously on each wafer, which is typically 7.6 cm in diameter.

An n-type epitaxial layer that will ultimately be the collector region of the transistors is grown on the substrate. The thickness of this layer is typi-

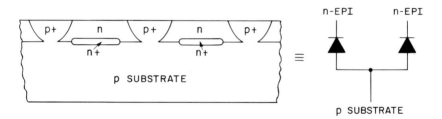

Fig. 6.1 Cross section of the junction-isolated monolithic structure and the equivalent electrical circuit.

cally between 2 μm and 15 μm and its resistivity is in the range from 0.1 Ω-cm to 10 Ω-cm. Both of these quantities are related to the design of the transistor and the maximum collector-base reverse-bias voltage because the collector doping affects both the collector resistance and the collector-base junction capacitance. Thick epitaxial layers are required when the collector region is lightly doped and the transistor operates at a high voltage. Digital circuits usually operate at 5 V and the epi thickness is typically 2 μm to 5 μm. Analog (linear) circuits usually operate at much higher voltages and use epitaxial layers that are 7 μm to 15 μm thick. Immediately after growth, the epitaxial layer is covered by about 5000 Å of thermally grown silicon dioxide, which acts as a protective coating and a mask for selective impurity diffusion in subsequent processing steps as discussed in Section 5.4.5. Figure 6.2(a) shows the fabrication sequence to this point.

The first selective (masked) diffusion is known as the isolation diffusion because it separates the epitaxial layer into isolated islands. It is a high-concentration p-type diffusion (p^+ diffusion), which completely penetrates the epitaxial layer and contacts the underlying p-type substrate as shown in Fig. 6.2(b). A p^+ concentration is desirable for the isolation diffusion to minimize the separation between isolated islands. Also, the isolation diffusion stripe must be wide enough to prevent the depletion layers of the neighboring isolation islands from touching.

Following a regrowth of the oxide, windows are opened for another p-type diffusion that will form the base region of the transistors, as shown in Fig. 6.2(c). The sheet resistance of this diffused layer is typically 200 $\Omega/\square$ and is related to the Gummel number (see Chapter 2), which enters into the expression for current gain. This layer is also commonly used to form diffused resistors. The base diffusion is a two-step process, with the final impurity distribution approximately Gaussian as shown in Section 5.4. In the first step, a controlled number of impurities is predeposited close to the surface in a "thin" layer, either by ion implantation or chemical source diffusion. This is followed by the second step, which is known as the redis-

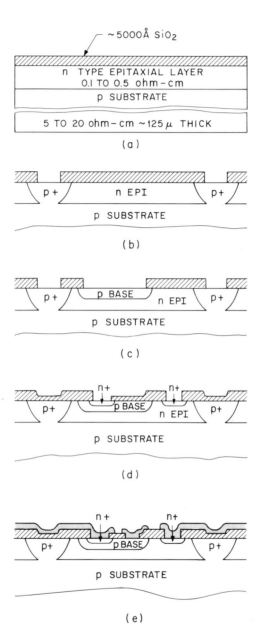

Fig. 6.2 Fabrication sequence for junction-isolated bipolar circuits: (a) Epitaxial layer covered by protective SiO₂ layer; (b) after isolation diffusion; (c) after the base diffusion; (d) after the emitter diffusion; (e) completed device.

tribution or drive-in diffusion. A two-step process is used to obtain an impurity profile with low surface concentration because it provides better control than a single-step diffusion from a constant source. Unlike the isolation diffusion which completely penetrates the epitaxial layer, the depth of the base diffusion is closely controlled, since it determines the position of the collector-base junction. The windows defined by the isolation mask are reopened during the base diffusion to increase the surface concentration in the isolation paths thus reducing the possibility of impurity depletion.

Windows are opened in the interior of the base region as shown in Fig. 6.2(d), and a complementary error function diffusion with a high surface concentration is used to produce the emitter region. The depth of the emitter diffusion must be accurately controlled because it determines the base width which is a critical parameter in transistor design. Note that the base width is the *difference* between the depth at which the base-collector and base-emitter junctions are formed, as shown in Fig. 6.3.

Note also in Fig. 6.2(d) that windows that will ultimately be the collector contacts are opened in the collector region prior to the emitter diffusion in order to have heavily doped material where the silicon is contacted by the aluminum metallization. This is necessary to the formation of an ohmic contact because aluminum is a *p*-type impurity in silicon, and on lightly doped *n*-type material it could cause surface inversion or form a rectifying, Schottky-barrier contact (see Chapter 2).

After completing the diffusion of impurities into the silicon, the next step is to interconnect the circuit components. All contact windows are opened and about 5000 Å to 10,000 Å of aluminum is deposited by evaporating a thin film over the entire surface of the wafer. The desired interconnection pattern is then photolithographically defined and the excess metal is removed by etching to yield the result shown in Fig. 6.2(e).

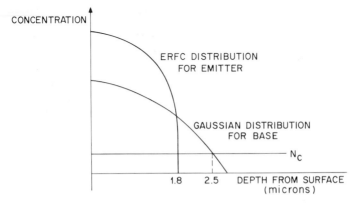

Fig. 6.3 Bipolar transistor impurity profiles.

Finally the circuits are tested to determine which are operative. Then they are separated either by laser scribing or by scribing the wafer with a diamond point and breaking it.

6.1.1 Buried Layer

The basic junction-isolated integrated circuit is a four-layer structure which has two deficiencies that are particularly troublesome in saturating logic circuits. First, a parasitic *p-n-p* transistor is formed by the base, collector, and substrate regions of the four-layer structure. This parasitic transistor is normally cut off because the collector-substrate junction and the collector-base junctions of the four-layer structure are normally reverse biased. When the *n-p-n* transistor saturates, the collector-base junction becomes forward biased and the parasitic *p-n-p* enters its active region. The resultant transistor action diverts current from the collector circuit to the substrate. Second, the desired *n-p-n* transistor is formed in the vertical direction, and hence there is a long path through high-resistance material between the collector contact pad and the active collector region because the collector contact must be on the top surface of the chip. The high series resistance is particularly detrimental in saturating logic circuits because $V_{CE\,\text{sat}}$ is increased. Furthermore, it becomes a function of collector current because of the voltage drop across the series collector resistance.

Both of these undesirable effects can be reduced by introducing an n^+ *buried layer* (often called the *buried collector)* beneath the collector region. The buried layer effectively eliminates the parasitic *p-n-p* transistor by increasing its base width and decreasing its base transport factor. Since the n^+ material has low resistance, it decreases the saturation resistance of the *n-p-n* transistor to alleviate the second effect.

The buried layer is produced by a selective two-step diffusion process before the epitaxial layer is grown. This diffusion approximates a complementary error-function diffusion with a high surface concentration. The process sequence through the base diffusion is shown in Fig. 6.4. Buried layer predeposition may be either by ion implantation[1] or by chemical diffusion. Arsenic or antimony is the dopant because each has a high solid solubility and a small diffusion coefficient. This permits a high doping level while minimizing out-diffusion from the buried layer to the epitaxial layer during subsequent high-temperature processing. Arsenic may be preferred because it fits perfectly into the silicon crystal lattice.

The series collector resistance can be reduced further at the expense of an additional selective diffusion known as the deep-collector diffusion. This diffusion is similar to the isolation diffusion in that it completely penetrates the epitaxial layer as shown in Fig. 6.4(d). It is an n^+ diffusion which serves

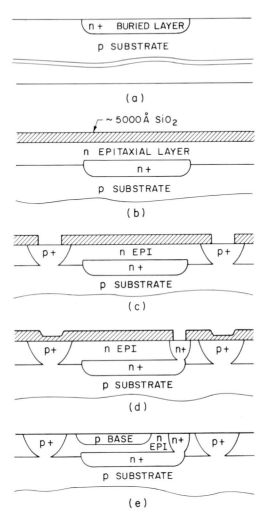

Fig. 6.4 Process sequence for junction-isolated bipolar circuits with buried layers. (a) After selective diffusion of the n^+-buried layer; (b) after epitaxial layer deposition and growth of protective oxide; (c) after isolation diffusion; (d) after deep collector diffusion; (e) after base diffusion.

as a low-resistance contact to the buried layer. A disadvantage of the deep-collector diffusion is that it increases device size because of lateral diffusion and because of tolerances associated with introducing another mask level.

6.2 BEAM-LEAD TECHNOLOGY [2]

The development of silicon integrated circuits was motivated by the need to reduce the number and cost of single-unit operations required to assemble a system. In the last section we described a sequence of operations for batch-fabricating silicon integrated circuits. However, after the wafer is diced to separate the circuits, many single-unit operations are required to mount the die in a suitable protective enclosure and to prepare it for connection to the outside world. A protective enclosure is required because silicon dioxide does not provide sufficient passivation, especially against alkali ion contamination.

If the circuitry on the chip is not complex enough to offset the cost of these single-unit operations, then increasing the complexity of the batch-fabrication sequence to reduce or eliminate single-unit operations is desirable—thus, the incentive to develop the beam-lead, sealed-junction technology.

Beam-lead, sealed-junction circuits are protected by a layer of silicon nitride which is an effective barrier against alkali ions and enables the circuit to withstand hostile environments without further protection. Furthermore, metallic beams cantilevered over the sides of the chip can be mass connected to the external world in one bonding operation. These advantages are obtained at the cost of a more complicated wafer processing sequence.

When applied to junction-isolated bipolar silicon circuits, the process begins after the emitter diffusion. At that point in the processing sequence, all of the contact windows are covered by silicon dioxide. First, contact windows are opened in the oxide and the silicon is lightly reoxidized, as shown in Fig. 6.5(a). The thin-oxide layer is needed in a later step to stop the nitride etch. The wafers are then placed in a reactor, similar to that used for epitaxy, and a layer of silicon nitride approximately 3000 Å thick is grown by a reaction between either silicon tetrachloride or silane with ammonia, as discussed in Section 5.7. A layer of silicon dioxide is deposited on top of the nitride using CVD techniques, because silicon nitride cannot be etched by chemicals that are compatible with photoresist. Using conventional photolithographic techniques, windows slightly larger than those in the lower oxide are opened in the upper oxide layer above the contact windows to the silicon. These steps are as shown in Figs. 6.5(b) and 6.5(c). The upper oxide serves as a mask for the nitride, which is now etched in boiling phosphoric acid, after which the upper oxide layer is removed in a buffered hydrofluoric acid etch. This etch simultaneously opens the windows in the lower oxide layer, thus exposing the silicon surface, as shown in Fig. 6.5(d).

An ohmic contact to the silicon is provided by platinum silicide formed in the contact areas. This is done by sputtering about 500 Å of platinum over the wafer, which is then heated in an inert atmosphere to sinter[†] the

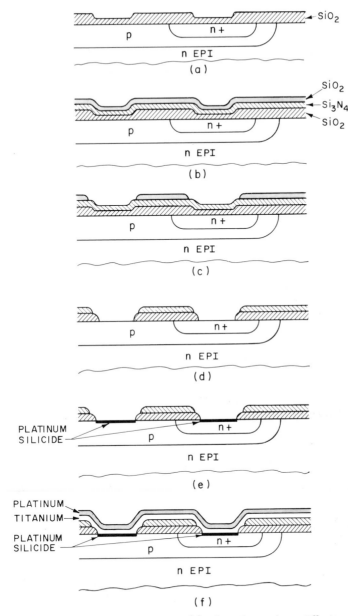

Fig. 6.5 Beam-lead process sequence: (a) after the emitter diffusion; (b) layers of CVD silicon nitride and silicon dioxide are nonselectively deposited; (c) windows opened in upper oxide; (d) after windows are opened in nitride and lower oxide layer and removal of upper oxide; (e) formation of platinum silicide in contact widows; (f) after deposition of titanium and platinum metal films.

platinum with the silicon. Any excess platinum is then removed either by back sputtering or by etching in aqua regia, yielding the result shown in Fig. 6.5(e).

The interconnection metal system must have low electrical resistance and high resistance to corrosion. Furthermore, it must adhere well to both silicon nitride and platinum silicide because the beam leads provide structural support in addition to electrical interconnection. Also, the metal seals the contact windows against contamination. Finally, the metal system must be compatible with the bonding techniques to be discussed in Chapter 10.

No single metal satisfies all of these requirements and therefore a composite metal system is used. Adherence is provided by a layer of titanium, which bonds very well to silicon dioxide, to silicon nitride, and to platinum silicide. This layer, which is 500 Å thick, is sputtered over the surface of the wafer as shown in Fig. 6.5(f). Gold has many desirable properties: high electrical conductivity, excellent corrosion resistance, suitability for thermocompression bonding, and capability of high resolution electroforming. Therefore, it is the choice for the conductive component in the metal system. However, gold can penetrate the titanium adherence layer and this must be prevented because gold radically changes the electrical properties of silicon and can markedly affect the behavior of the circuit. Fortunately, a suitable diffusion barrier is obtained by interposing a layer of platinum or palladium about 1500 Å thick between the titanium and the gold. Figure 6.5(f) shows the process just after the titanium and platinum layers have been deposited.

The metallic interconnection pattern is now defined photolithographically and the excess platinum is removed by either chemical or sputter etching. The result is shown in Figs. 6.6(a) and 6.6(b). The titanium which is not removed at this time, provides an equipotential surface for subsequent electroforming operations during which the gold conductive layer is selectively electroplated onto the platinum to a thickness of about 2 μm. The selectivity is provided by depositing photoresist in all areas that are not to be plated, using a mask that is the negative (though of slightly different dimensions) of the mask that defined the pattern in the platinum.

The beams that will be cantilevered over the side of the chip must be thicker than the intraconnection metal because they provide mechanical support in addition to electrical connection. Therefore, another photolithographic operation is used to expose only the beam pattern to a second electroplating operation, which increases the thickness of the beams to about 12 μm. After the photoresist is removed, the excess titanium is etched in sulphuric acid, leaving the structure shown in Fig. 6.6(c).

The beam-lead circuit separation process involves removing excess silicon from beneath the beam leads, thus making them cantilever over the

---

[†] Sintering is the formation, by heating, of a region in which particles of the various materials, in this case platinum and silicon, are intermixed.

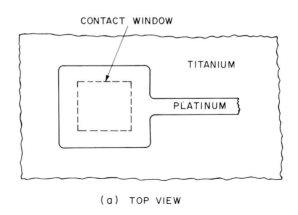

(a) TOP VIEW

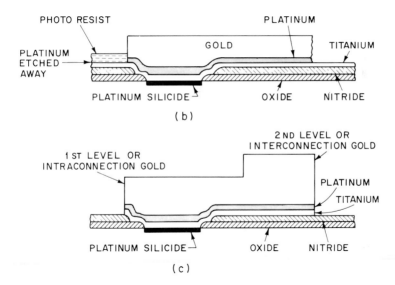

Fig. 6.6 Details of the titanium-platinum-gold metallization system: (a) top view; (b) section prior to electroforming of gold layer; (c) section after electroplating.

edge of the resulting chip. Hence, the chip must be separated by an etching process rather than by scribing. The wafer is mounted face down on, for example, a sapphire disc using a wax as an adhesive. The wafer is then mechanically lapped to a thickness of 50 μm. Both the carrier and silicon wafer are transparent to infrared illumination which is used to align a mask so that the resulting photoresist pattern protects the desired chip areas and exposes the unwanted silicon. The excess silicon is then removed by etching.

6.3 COUNTERSUNK OXIDE ISOLATION [3-5]

The parasitic collector-substrate capacitance of a junction-isolated bipolar circuit has sidewall and bottom components. The LOCOS and ISOPLANAR processes essentially eliminate the sidewall capacitance by replacing p-n junctions at the sidewalls of the collector isolation region with channels of thermally grown silicon dioxide. These processes, which may be thought of as a cross between junction isolation and dielectric isolation (to be described in Section 6.5) are of great interest for high-speed digital circuits because of the reduced collector-substrate capacitance and the smaller device geometry that is achieved by the oxide isolation.

The object of these processes is to produce isolated regions whose borders are defined by a frame of thermally grown silicon dioxide that completely penetrates the epitaxial layer. Since the thickness of an oxide layer is almost twice that of the silicon from which it is grown, and since steps in the oxide are undesirable because they degrade resolution in the subsequent photolithographic operations, it is necessary to remove some silicon from the interisland channels before oxidation to prevent the oxide level from rising above the surface of the wafer. For purposes of illustration, assume that it is desired to isolate bipolar devices in an n-type epitaxial layer. The epitaxial-layer thickness should not exceed 2 μm because of difficulties in growing thick thermal oxide layers.

The starting material is a p-type wafer into which the n^+ buried layers have been diffused and which has a 1.5- to 2-μm n-type epitaxial layer of the desired resistivity. The barrier against the growth of silicon dioxide is silicon nitride. Figure 6.7 shows a comparison of the oxidation rate of silicon and silicon nitride. From this graph, it is seen that the steam oxidation of silicon at 1150°C is approximately 25 times faster than that of silicon nitride. Therefore, for a 2-μm thick isolation oxidation, at least 800 Å of silicon nitride is needed as a mask, so 1500 Å of nitride is used. An oxide layer 200 to 300 Å thick is required below the nitride to stop the nitride etch as in the beam-lead sealed-junction process. Also, an upper oxide layer about 1000 Å thick is required to define the desired pattern in the nitride. The structure thus far described is shown in Fig. 6.8(a). The isolation process is begun by opening windows in the oxide-nitride sandwich to expose all of the silicon surface except the isolated islands. Silicon is thus exposed in the areas which will eventually become the oxide isolation rings. After the pattern has been etched, first in the upper oxide and then in the nitride layer, the upper oxide is removed as shown in Fig. 6.8(b). Trenches are then etched in the silicon to a depth equal to about 55% of the desired oxide thickness. This step is shown in Fig. 6.8(c).

Next the wafers are subjected to a steam oxidation at, for example, 1150°C for 6 hours. The SiO$_2$ that is formed during this oxidation will be level with the original surface of the wafer because the specific volume of sil-

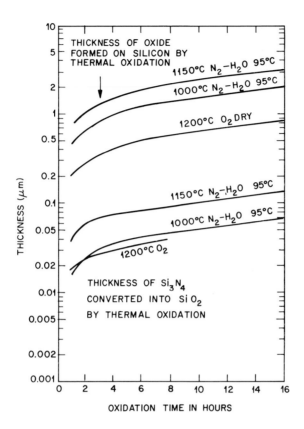

Fig. 6.7 Oxidation rate of silicon and silicon nitride showing the effectiveness of silicon nitride as an oxidation mask.

icon dioxide is 2.2 times that of silicon. Finally the nitride layer is removed. Since it has been oxidized, the upper part of the nitride layer is removed by a buffered hydrofluoric acid etch. This is followed by an etch in boiling phosphoric acid to remove the nitride. The wafer after this step is shown in Fig. 6.8(d).

A small hillock (or bird's beak) is formed at the transition between the recessed oxide region and the silicon nitride mask. The severity of this effect depends on the thickness of the SiO_2 layer beneath the silicon nitride mask, the thickness of the silicon nitride layer, the depth and uniformity of the trench and the undercutting of the oxide, and the thickness of the recessed oxide.[6] The bird's beak becomes more prominent as the thickness of the recessed oxide is increased. It can be suppressed by the use of thicker, more rigid, nitride layers. However, thick nitride layers tend to

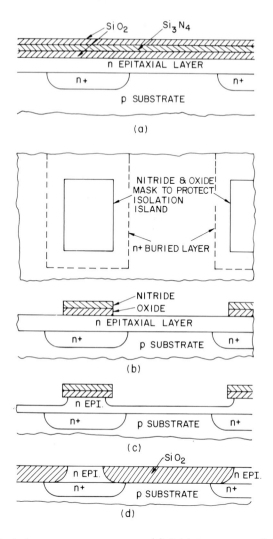

Fig. 6.8 Oxide-isolation process sequence. (a) initial structure; (b) after definition of nitride oxidation mask; (c) after etching of trenches in the epitaxial layer; (d) completed structure.

damage the silicon and degrade the dielectric properties of SiO_2 layers subsequently grown on it.

Transistors can now be formed in the islands by a sequence of two or three selective impurity diffusions. Usually the base width of these double-diffused transistors is very small, resulting in short transit times and good high-speed performance. The high-speed performance is further enhanced by the absence of the sidewall component of collector-substrate capacitance. Since the oxide-isolating ring is electrically inert, it is possible to have the emitter region defined on three sides by the oxide, as shown in Fig. 6.9. The resulting decrease in isolated island area further reduces capacitance by about another factor of two.

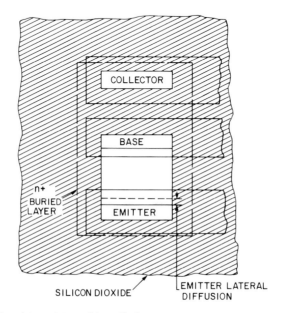

Fig. 6.9 Oxide-isolated transistor with walled contacts.

6.4 V-GROOVE ISOLATION [7,8]

V-groove isolation is another technique for producing small geometry bipolar devices without isolation sidewall capacitance. It uses an anisotropic etch to form trenches around the regions that are to be isolated so they are isolated laterally by air gaps and vertically by a reverse-biased p-n junction between each island and the p-type substrate. In this respect, it is similar to LOCOS isolation described in the previous section.

V-groove isolated circuits are fabricated on (100)-oriented p-type wafers. The first step is the diffusion of an n^+ buried layer, followed by the growth of a thin n-epitaxial layer by the pyrolosis of silane. This method of epitaxial growth is preferred because the reaction occurs at a relatively low temperature, which reduces the out-diffusion of impurities from the buried layer. The thin expitaxial layer is also suitable for the fabrication of high-value epitaxial resistors. A shallow *p*-type base diffusion is performed, yielding the result shown in Fig. 6.10(a).

Isolation is provided by selectively etching V-shaped grooves by means of a preferential etchant that attacks the (100) crystal plane about 30 times as fast as the (111) plane.[9] A suitable etchant is 65% hydrazine, 35% water at

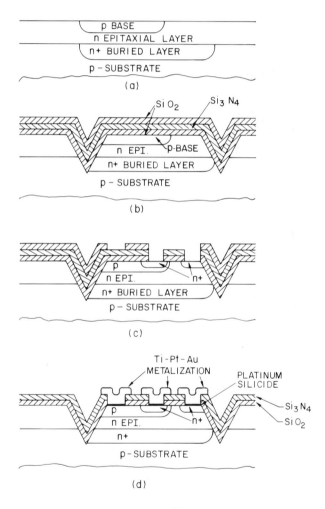

Fig. 6.10 V-groove isolation process sequence. (a) Starting after base diffusion; (b) after etching of V-grooves and deposition of oxide-nitride-oxide sandwich; (c) just after emitter diffusion; (d) after metallization.

$100°C.$[10] The grooves make an angle of 54° with the surface and therefore the width across the top of the groove is solely a function of the epitaxial-layer thickness which the groove must penetrate in order to provide isolation. The isolation grooves consume much less space than the standard isolation diffusion and its associated depletion layers.

After the grooves are etched, the oxide is stripped and an oxide-nitride-oxide sandwich is deposited as shown in Fig. 6.10(b). Windows for all of the contacts are opened in the upper oxide. The emitter and collector windows are opened in the nitride and lower oxide, and an n^+ emitter diffusion is performed, yielding the result shown in Fig. 6.10(c). Finally, the base contact windows are opened and the titanium-platinum-gold metallization used in the beam-lead process is applied and patterned, resulting in the structure shown in Fig. 6.10(d). The composite metallization system was chosen for the V-ATE (Vertical AnisoTropic Etch) process because it permits the close spacing of well-defined metal patterns, that, in turn, result in small device geometries.

Two layers of metallization are used in the V-ATE process to permit the fabrication of high-density circuits. A layer of oxide is deposited above the first level of metallization, followed by the deposition and patterning of the second-level metallization.

A modification and simplification of the V-groove process consists of oxidizing the isolation grooves that are then filled to the wafer surface with polycrystalline silicon. The resultant planar surface permits the use of aluminum metallization.

6.5 DIELECTRIC ISOLATION [11]

Junction isolation and its variants described in the previous section are not suitable for high-voltage applications, such as for linear circuits with supply voltages much in excess of ±30 V because of junction breakdown with reasonable doping levels and thicknesses, or for high nuclear-radiation environments because gamma rays produce photocurrents in p-n junctions. The dielectric-isolation method described in this section produces islands of epitaxially-grown, single-crystal silicon that are completely surrounded by silicon dioxide, thus providing a high-voltage, low-capacitance isolation. The particular dielectric-isolation process to be described is known as the positive-channel-isolation method. It begins with a wafer of n^+ monocrystalline silicon on which a layer of silicon dioxide is grown, as shown in Fig. 6.11(a). Then a layer of polycrystalline silicon is deposited on top of the silicon dioxide. After the polycrystalline silicon is grown, the single-crystal silicon is lapped, polished, and etched to a thickness of 8 to 10 μm, as shown in Fig. 6.11(b). Next, an n-type epitaxial layer of the proper resistivity for the transistor collectors and of suitable thickness is grown on the

n^+ monocrystalline silicon. A silicon-dioxide layer is deposited on top of the epitaxial layer by a pyrolytic reaction. This SiO_2 layer is etched to expose the silicon between what will ultimately be the isolated islands, as shown in Fig. 6.11(c). An etchant that selectively attacks silicon is used to cut through the monocrystalline silicon layers to the underlying silicon-dioxide interface between the monocrystalline and polycrystalline silicon layers. This step is similar to the selective etch used in the V-groove process.

The channel edges are then oxidized and the channels are filled with polycrystalline silicon after which the wafer is polished back to the surface oxide to remove excess polycrystalline silicon from the working areas. A

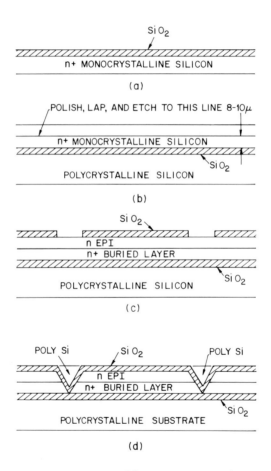

Fig. 6.11 Dielectric isolation process. (a) SiO_2 layer on n^+ monocrystalline silicon; (b) after deposition of polysilicon; (c) after epitaxial layer growth just prior to isolation; (d) after isolation.

final oxidation completes the process yielding the result shown in Fig. 6.11(d). The wafer is now ready for device fabrication.

The positive-channel-isolation method yields devices with good characteristics because the working regions are epitaxially grown and immediately protected against surface damage by the oxide.

Polycrystalline silicon is chosen as the substrate material because its coefficient of thermal expansion, dimensional stability, and rigidity are approximately the same as those of monocrystalline silicon. The thickness of the polycrystalline layer must be sufficient to satisfy the requirement of mechanical strength. However, even the small differences in the mechanical properties of monocrystalline and polycrystalline silicon result in processing problems because the small difference in thermal expansion coefficient causes bowing of the wafer as it cools after the deposition of the polycrystalline layer. After the grinding operation, this effect can result in some isolation islands in which the monocrystalline silicon is too thin or in failure to isolate some islands.

6.6 INVERTED CHIPS [12]

The beam-lead sealed-junction process described in Section 6.2 provides both a mass interconnection capability and protection against contamination, but it is a complicated process. A simpler process which provides the mass interconnections is known as "inverted chips." This process does not seal the junctions with silicon nitride and therefore the inverted chips must be mounted in hermetically sealed packages for maximum environmental protection.

The two major inverted chip interconnection schemes are aluminum bumps, which are ultrasonically bonded to either gold or aluminum mounting pads, and coated copper balls or bumps, which are bonded by either solder reflow or the reflow and subsequent alloying of indium into the copper. We will describe the latter scheme since it has the advantage of providing an initial bond at low temperature (150°C) which can be used for testing circuit operation. After proper operation of the circuit is verified, the indium-copper alloy can be formed. This alloy melts at 700°C, which permits the packages to be hermetically sealed without fear of refloating the chips.

Wafers of inverted chip circuits are processed in the conventional manner, including the aluminum metallization step and deposition of a passivating layer of silicon dioxide. Holes are opened over the contact pad areas, and first aluminum and then copper are deposited over the wafer by evaporation as shown in Fig. 6.12(a). The thickness of the copper layer is increased to between 10 and 15 μm by electroplating. There is about 11,000 Å of aluminum in the bonding pad area. Finally a layer of indium is

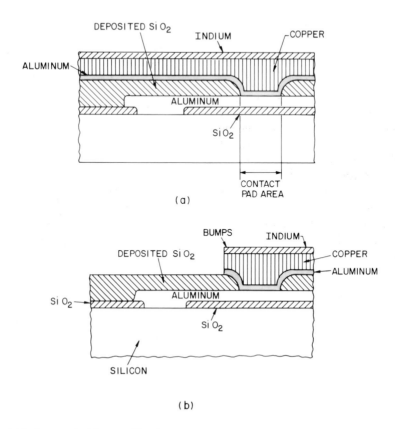

Fig. 6.12 Inverted-chip metallization system.

electroplated over the copper, as also shown in Fig. 6.12(a). Next the contact balls are formed by selectively etching first the indium and then the copper, using a complexed copper nitrate etch which selectively attacks copper in the presence of indium and aluminum. Last, the excess aluminum is removed. The result is indium-coated copper bumps on the aluminum metallization, as shown in Fig. 6.12(b).

The chips are installed by a thermocompression bond of the indium layer. When it is desired to make the final bond, the chips are heated to 450°C for 1 min in a slightly reducing atmosphere followed by 6 min in an oxidizing atmosphere. This alloying of the indium and copper is done simultaneously with the package-sealing operation. The time is limited by the diffusion of $CuAl_2$ through the aluminum layer to the silicon dioxide, with the resultant loss of adherence of metallization to the chip.

6.7 COLLECTOR-DIFFUSION ISOLATION [13]

It was recognized very early in the evolution of bipolar integrated circuits that device size would have to be reduced to obtain high-speed operation, and both device size and processing complexity would have to be reduced to permit complex circuits to be made with high yield. Collector-diffusion isolation (CDI) achieved reductions in both device size and processing complexity by eliminating the isolation diffusion. The resulting structure provides transistors that are suitable for digital logic circuits.

Collector-diffusion-isolated circuits start as wafers of high resistivity p-type material. The first selective diffusion is used to form n^+-type buried layers, which are about 3 to 5 μm deep and have an initial sheet resistance of 15 to 25 $\Omega/\square$. This corresponds to a surface concentration of about 10^{19} cm^{-3} for a complementary error-function distribution. The diffused buried layers form the collector region of the transistors and the resistor isolation regions. A 0.2 Ω-cm, p-type epitaxial layer about 1 to 2 μm thick is then grown. This epitaxial layer will be the base region of the transistors. Isolation is achieved by a deep selective n^+-type diffusion that completely penetrates the epitaxial layer to contact the underlying buried layer. These diffused regions have a sheet resistance of about 6 $\Omega/\square$ and form a collar around every buried collector region. Another selective n^+ diffusion produces the emitter regions to complete the structure shown in Fig. 6.13. An optional, unmasked p-type diffusion over the whole wafer after epitaxial deposition is desirable because it prevents surface inversion, minimizes edge injection from the emitter, provides a built-in field in the base region, and provides control over the composite base-layer resistance.

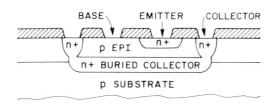

Fig. 6.13 Collector diffusion isolation device structure.

The inverse gain of the CDI transistor is quite high because the emitter region is heavily doped and of comparable size to the collector. This necessitated the development of special circuitry to permit the devices to be used effectively. This process is included mainly for historical reasons since CDI circuits have been replaced by small-geometry junction or isoplanar devices often containing Schottky diodes.

6.8 COMPOSITE MASKING OF BURIED-COLLECTOR JUNCTION-ISOLATED BIPOLAR INTEGRATED CIRCUITS [14]

We have seen in previous sections that in the bipolar SIC process, up to four selective diffusions are made after the growth of the epitaxial layer. The mask that defines the region to be doped by each of the diffusions must be aligned with respect to existing features on the wafer. Allowances must be made in the spacing between features for inaccuracies in mask alignment. This topic is discussed in more detail in the next chapter. However, it is possible to reduce the overall device size by using a technique called *composite masking,* which eliminates the critical mask-alignment steps associated with the isolation, deep collector, and base diffusions.

The basic idea is to use a single composite photomask to pattern a silicon-dioxide-silicon-nitride mask with the features for the isolation, deep collector, and base diffusions, thus eliminating mask-alignment errors. The features of the individual mask levels are separated by using coarse masks to open windows in the oxide using the pattern defined previously in the nitride as the precise mask.

The process begins after the growth of the epitaxial layer. A thin, thermal silicon-dioxide layer is grown followed by the deposition of a silicon-nitride layer about 1500 Å thick. Windows are opened in the nitride layer using a mask whose features are the reverse-tone of the composite of the isolation, deep collector, and base diffusion features. A thick field oxide is then selectively grown using the nitride layer as the oxidation mask, as shown in Fig. 6.14(a). Windows for the isolation diffusion are then opened, using an oversized mask to strip the protective nitride and oxide from the isolation window, as shown in Fig. 6.14(b). After the isolation diffusion and reoxidation, another coarse mask is used to open the deep-collector diffusion windows. After the deep-collector diffusion and reoxidation, the remaining nitride and thin oxide are stripped without benefit of a mask, thus opening windows for the base diffusion.

A conventional masking step is then required for defining the emitter diffusion.

6.9 COMPLEMENTARY BIPOLAR INTEGRATED CIRCUITS [15]

Compatible complementary p-n-p transistors are required to make high-slew rate, low-power operational amplifiers. The ion-implanted structure to be described yields p-n-p transistors with high collector-emitter breakdown voltage and tight control of parameters. The compatible p-n-p/n-p-n structure is shown in Fig. 6.15. It is formed on a 4- to 15-Ω-cm p-type substrate into which an n^+ antimony-doped buried layer is diffused. The buried layer is part of the isolation for the p-n-p transistors. A high resistivity n-type epitax-

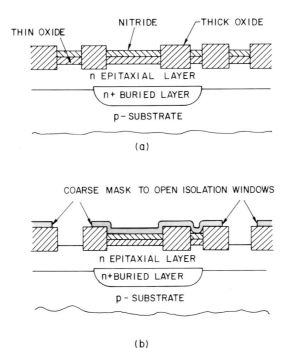

Fig. 6.14 Composite masking of bipolar circuits.　(a) After preparation of thick-oxide composite mask; (b) ready for device isolation.

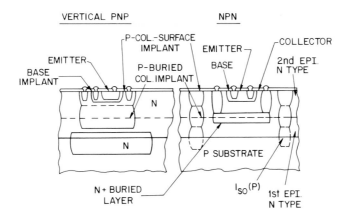

Fig. 6.15 Complementary *p-n-p n-p-n* bipolar transistor structure.

ial layer is now deposited. A boron ion implant is used as the buried layer for the *p-n-p* transistors and as part of the isolation diffusion for the *n-p-n* transistors.

A second *n*-epitaxial layer, of lower resistivity, is now deposited and boron is again implanted in the *p-n-p* transistor regions and into the isolation areas of the *n-p-n* transistors using the same pattern as for the first implant. The boron is diffused until the profiles overlap through the epitaxial layers.

A phosphorus-ion implant is used as a low-temperature predeposition for the base of the *p-n-p* transistor. The base diffusion of the *n-p-n* transistor is adjusted to provide a $200\text{-}\Omega/\square$ sheet resistance for diffused resistors. Then the emitter of the *p-n-p* transistor is diffused, followed by the emitter of the *n-p-n*.

A complementary transistor structure based on a dielectric-isolation scheme in which pits are etched into the *n*-epitaxial layer and subsequently filled with *p*-type epitaxial material prior to the formation of the dielectrically isolated islands has been described.[16] This process is a rather natural extension of the dielectric-isolation technique described above.

6.10 THICK-OXIDE METAL-GATE MOSFET CIRCUITS [17]

The MOSFET is an attractive device for large-scale integrated circuits because it is a self-isolating device.

The first commercial MOSFET process was the thick-oxide, metal-gate process for making *p*-channel enhancement-mode circuits on (111)-oriented silicon. This process was later used with (100)-oriented silicon to reduce the threshold voltage of the enhancement-mode devices. It is the simplest of the IGFET[†] processes. The starting material for *p*-channel enhancement-mode transistors is 3- to 5-Ω-cm *n*-type silicon on which a thermal oxide layer about 5000 Å thick is grown. This oxide is considerably thicker than required to mask the source and drain diffusion, but it is needed as part of the thick field oxide to maintain a sufficient field threshold voltage. Windows are opened to define source and drain regions which are boron diffused and driven to a depth of typically 2 to 3 μm. The field oxide is usually about 1.3 to 1.5 μm thick. It is produced by thickening the existing 5000-Å layer either by thermal growth during and after the drive-in diffusion, or by deposition.

A lower limit on the gate-oxide thickness is imposed by the requirement that the oxide be free of pinholes which would result in shorts. A practical upper limit to the gate-oxide thickness is provided by a tolerable lower limit for the transconductance. Typically, the gate oxide is 1000 Å

[†] The term IGFET for Insulated Gate Field Effect Transistor is the general term for the class of devices that include the MOSFET and is the proper term when the gate insulator is not oxide.

thick and the minimum practical thickness is about 500 Å. The field oxide is stripped from the gate region because it is not possible to remove an oxide layer partially with precise control of thickness. Contact windows are opened in the drain and source regions at the same time. The gate oxide is then thermally grown.

Finally, the source and drain contacts are reopened and the metallization is deposited. The complete process sequence is shown in Fig. 6.16.

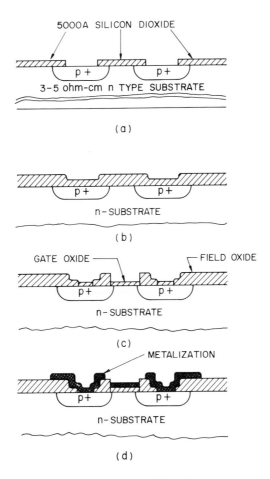

Fig. 6.16 Thick-oxide MOS process sequence. (a) After source-and-drain diffusion; (b) thickening of field oxide; (c) after growth of gate oxide; (d) completed device.

There are large steps in the field oxide in the gate region and at the source and drain contacts that might cause covering problems with the metallization. The countersunk oxide process, which was used previously for bipolar device isolation, can be used to produce the thick field oxide. It may not even be necessary to etch trenches in the silicon prior to oxidation if a step of approximately one-half the total oxide thickness does not cause breaks in the interconnecting metal. A consequence of not forming the trench is that the thick oxide can be grown over the source and drain regions, thus reducing the gate-overlap capacitances. This technique is illustrated in Fig. 6.17.

There is a disadvantage to this process, independent of the choice of crystallographic orientation of the silicon substrate, because the channel and the gate metal are defined by separate masks. This necessitates a gate overlap of from 4 to 7 μm in order to ensure that the entire channel is covered by the gate, and results in an increased value of gate-to-source and gate-to-drain capacitance. The increased gate-to-drain capacitance is particularly troublesome because it is multiplied by the Miller effect when the device is active.

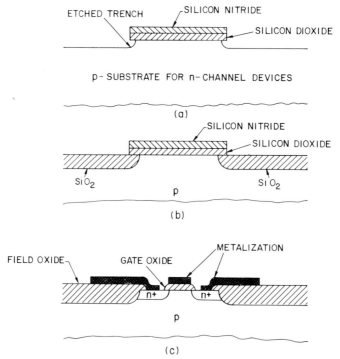

Fig. 6.17 Use of masked oxidation to produce the field oxide: (a) trench etched in those areas not protected by silicon nitride; (b) after growth of the thick, countersunk field oxide; (c) completed device.

When enhancement mode IGFET's are used in logic circuits, the threshold voltage of the transistor becomes the threshold voltage of the gate. Therefore it determines the level at which the gate discriminates between a logic 0 and a logic 1. When (111)-oriented silicon substrates are used, the threshold voltage is in the -3 to -5 V range, while the field-threshold voltage is in the -30 to -50 V range. However, power supply voltages up to about -25 V are required. Because of these reasons, the circuits on (111)-oriented substrates are not T^2L compatible.

Circuits on (100)-oriented substrates are T^2L compatible because the threshold voltage is in the -1.5 to -2.2 V range. However, the field-threshold voltage is in the -10 to -18 V range, which is uncomfortable since a -12 V gate supply is required. In addition, the (100) orientation has lower hole mobility than the (111) orientation requiring a device that is about 20% larger for the same speed.

6.11 COMPOSITE INSULATOR GATE STRUCTURES

The metal-gate process on (111)-oriented substrates has a high field-threshold voltage, which is desirable for self-isolation, but it also has a high gate-threshold voltage which is undesirable. The gate-threshold voltage is a function of gate capacitance. Since the gate-oxide thickness is already at its smallest practical value, the easiest way to increase the capacitance is to change the insulator to one with a higher dielectric constant. This is accomplished by using either a silicon-dioxide-silicon-nitride or silicon-dioxide-alumina dual-dielectric gate structure. Several hundred angstroms of silicon dioxide are required under the nitride or alumina to prevent charge trapping at the silicon surface.

When the gate insulator consists of 200 Å of silicon dioxide and 800 Å of silicon nitride, the effective dielectric constant is 6.8 and the gate-threshold voltage for p-channel enhancement-mode metal-gate transistors is normally -2.4 V. Logic circuits made with these devices can be T^2L compatible. In addition, since the field oxide is about 1.5 μm thick, the field-threshold voltage is in the -30 to -50 V range.

Figure 6.18 is a cross-sectional view of a dual-dielectric IGFET transistor. The processing is essentially the same as described in the last section except that a silicon nitride or alumina film is chemical-vapor-deposited after the relatively thin oxide is grown in the gate region. Therefore, the processing is more complicated than the standard process.

6.12 APPLICATIONS OF ION IMPLANTATION TO SIC PROCESSING [18]

Chemical-source doping has been the traditional method of introducing impurities into a semiconductor. The methods are relatively simple, and the

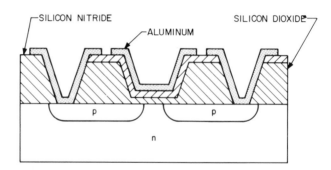

Fig. 6.18 Cross section of a dual-dielectric insulated gate field-effect transistor.

required equipment is comparatively inexpensive. Among its shortcomings are:

1. The protective oxide must be stripped from the region to be doped, thus exposing the bare silicon to impurities other than the desired dopant.

2. Precise control of the number of impurities in the semiconductor is very difficult, if not impossible.

3. The impurity distributions must have their highest concentration near the surface. This means that the total number of impurities is large in regions (such as the emitter) which have been compensated several times, resulting in a degradation of mobility and carrier lifetime.

Ion implantation, described in Section 5.5, is an alternative to chemical-source doping. In ion implantation, the impurities are ionized and shot into the semiconductor using a linear accelerator. This technique allows precise control over the number of impurities that are introduced and their placement in the semiconductor. For the first time, control of impurity concentrations to better than 1% is possible, compared to 5% when the source is doped oxide. Although ion implantation could be used as a substitute for solid-state diffusion for producing impurity profiles, it is presently used mainly to replace the predeposition diffusion in conventional processing sequences. The implanted profile is then redistributed by solid-state diffusion. The precise impurity control permits the fabrication of bipolar transistors with tightly controlled current gain.

Threshold voltage control by ion implantation [17,19] The threshold voltage of an enhancement mode IGFET can be shifted by implanting a thin layer of impurities on the silicon surface. If the layer is very thin, its effect on the threshold voltage is the same as that of charge in the gate insulator. Light ion implants are often used just before deposition of the gate material to adjust the threshold voltage.

This effect can also be used to fabricate depletion-mode IGFET's on the same substrate as enhancement-mode devices by implanting an appropriately doped channel prior to gate material deposition.

Compatible JFET   Another important device is the compatible JFET for use in analog circuits. The major yield-limiting factor for diffused, compatible JFET's is control of the pinch-off voltage, V_p. The ion-implanted JFET is on a 3- to 4-Ω-cm p-type substrate on which an oxide layer is grown. Patterns for the high sheet resistance n-type "tubs" are defined in the oxide and the wafer is subjected to a phosphorus implantation and drive-in. The sheet resistance of these tubs is on the order of 1000 $\Omega/\square$, which is difficult to obtain by conventional techniques. After a reoxidation, the source and drain ohmic contacts are defined and diffused. The one standard-deviation spread of the pinch-off voltage of this ion-implanted JFET is 17% of the median V_p as compared to 67% for the conventional epitaxial and diffused JFET.

6.13 SELF-ALIGNED METAL GATE STRUCTURE [17,19,20]

Reduction of the gate overlap requires either a more accurate mask alignment or some type of self-aligned gate structure. It is also possible to diffuse the drain and source regions from a doped oxide at the same time as the gate oxide is being grown, but that only reduces the overlap capacitance without solving the overlap problem. One method of achieving a self-aligned metal-gate structure is to place the metal gate between the source and drain regions, which are separated by a distance that is larger than the final spacing, as shown in Fig. 6.19. The entire region between the source and drain diffusions is covered by the thin, thermally grown gate oxide. After deposition and patterning of the metal, the wafer is subjected to an ion implantation which extends the source and drain regions to the edge of the gate metal. For the p-channel enhancement-mode device shown, an 80-keV boron implantation easily penetrates the oxide which is about 1000 Å to 1200 Å thick, but is stopped by the metal. This implantation provides a surface concentration of about 10^{14} cm^{-2}. It requires an anneal at 400 to 500°C for about 30 min to remove the radiation damage. Boron diffusion during the anneal is effectively zero so that the gate alignment with the source and drain regions is perfect.

6.14 SELF-ALIGNED POLYSILICON GATE NMOS PROCESS [21]

A process for making self-aligned, silicon-gate, ion-implanted, depletion-mode and enhancement-mode NMOS transistors on one substrate is illustrated in Fig. 6.20 and will now be described. It begins with lightly doped p-type (100)-silicon wafers of about 10 Ω-cm resistivity that have been coated

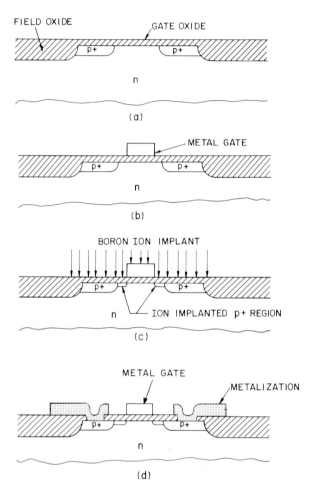

Fig. 6.19 Self-aligned metal gate MOS transistor fabrication. (a) After source-and-drain diffusion and growth of oxide layers; (b) metal gate defined in region between source and drain; (c) boron ion implant to align source and drain regions to gate; (d) final structure.

with 500 Å of thermally grown SiO_2, 400 Å of CVD silicon nitride and about 1000 Å of CVD silicon dioxide. The first mask is used to pattern the upper oxide layer. It is the reverse tone (negative) of the composite image for the gate, source and drain regions of the transistors, so that after the upper oxide layer is etched and the photoresist removed, only those areas of the wafer that will be transistors are still covered by the CVD oxide. A high-energy boron ion implant in those regions not covered by the nitride layer is used to shift the field threshold voltage to a sufficiently high value. This implant is

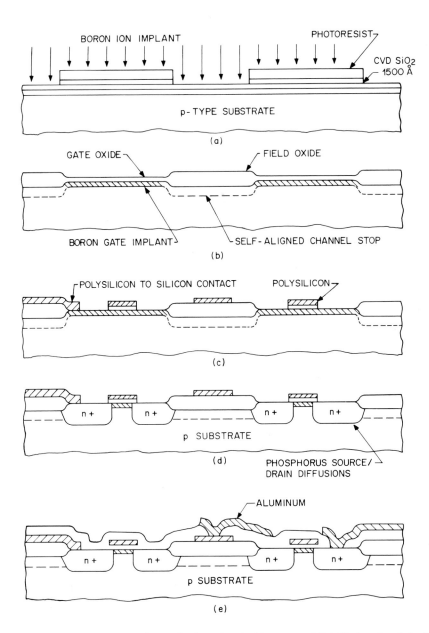

Fig. 6.20 Process sequence for self-aligned polysilicon gate NMOS circuits: (a) after etching of upper oxide layer to expose all regions that will not be transistors; (b) after selective growth of the thick-field oxide and the gate oxide; (c) after deposition and patterning of the polysilicon layer; (d) after source-and-drain diffusion; (e) completed structure.

known as the channel stop and is necessary for closely packed devices. The silicon nitride is now etched using the upper oxide as a mask, and then the upper oxide is stripped.

A selective oxidation is performed to grow a 1.5-μm thick field oxide using the silicon nitride as an oxidation mask. This oxide is partially below the original wafer surface because about 7000 Å of silicon is converted to SiO_2. Thus, the surface discontinuity in the transistor region is only about 8000 Å.

The silicon nitride is now stripped and a gate oxide about 1000 Å thick is grown. A light boron ion implant, for example, 7×10^{11} cm^{-2} at 30 keV, is then used to set the threshold voltage of the enhancement-mode transistors.

If depletion-mode transistors are also required, photoresist is applied and patterned to expose the depletion-mode devices. A phosphorous ion implant, such as 1.4×10^{12} cm^{-2} at 90 keV, is used to create the depletion-mode devices.

Before deposition of the polysilicon gate material, windows are opened wherever it is desired to make a contact between the polysilicon and silicon. Undoped polysilicon is then deposited and patterned to form the transistor gates.

The source and drain contact windows are now opened and the wafer is subjected to a phosphorous diffusion, which simultaneously diffuses the source and drain regions and dopes the polysilicon gate. Alternatively, the doping could have been from an oxide source.

Finally contact windows are opened, and the aluminum metallization is deposited and patterned.

6.15 COMPLEMENTARY MOS STRUCTURES [22,23]

CMOS (Complementary MOS) is the most complex MOSFET process because both n-channel and p-channel devices are made on the same chip. A typical CMOS process begins with n-type (100)-silicon wafers of about 5 Ω-cm resistivity. The n-type material is the background for the p-channel enhancement-mode transistors. The substrate is n-type because the p-type substrate required for n-channel devices must be more heavily doped than the n substrate required for complementary transistors with equal threshold voltages. It is easier to dope the n-type substrate to form p wells than the converse. The first task is to form wells of p-type material of about 3 to 5 Ω-cm resistivity which can be done most easily either by diffusion from a doped oxide source or by an ion implant. Close control over the doping of the wells, especially over the surface concentration, is essential because it affects the threshold voltage of the n-channel devices. The impurity density of 5-Ω-cm n-type material is 10^{15} cm^{-3}. A 5-Ω-cm uniformly doped p-type well has an impurity concentration of 2.5×10^{15} cm^{-3}. The implanted acceptor concentration must be 3.5×10^{15} cm^{-3} since it is necessary to compensate the donor impurities. If the p-well is to be 10 μm deep, the implant dose for 5-Ω-cm resistivity is 3.5×10^{12} cm^{-2}, which is a light implant. The p-well will actually have a Gaussian distribution because it is diffused from a finite, ion-implanted source.

The implantation may be done through a thin-oxide layer that protects the silicon surface. Photoresist is used as the implantation mask. The implant is redistributed by diffusion in an oxidizing ambient, which also increases the field-oxide thickness in what will be the region between transistors. Windows are then etched in the field oxide to define the source and drain diffusions for the p-channel transistors. A p^+ boron diffusion and drive-in in an oxidizing atmosphere forms these regions and further thickens the field oxide. These steps are shown in Figs. 6.21(a) through 6.21(d). Next, windows are opened in the p-well for a high-concentration phosphorous diffusion, which forms the source and drain regions for the n-channel transistors. This diffusion is also followed by a drive-in in an oxidizing atmosphere. The resulting structure is shown in Fig. 6.21(e). The thick oxide is then stripped from the gate region as shown in Fig. 6.21(f), and a thin gate oxide is grown. Then contact windows are opened to the source and drain regions [Fig. 6.21(g)], and the metallization is deposited and patterned by etching. The completed structure is shown in Fig. 6.21(h).

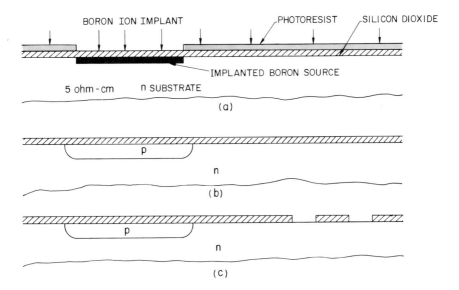

Fig. 6.21 Metal gate CMOS process sequence: (a) Selective boron ion implant to provide dopant for p tubs; (b) p tubs after the redistribution diffusion; (c) opening of windows for the source-and-drain regions of the p-channel transistors.

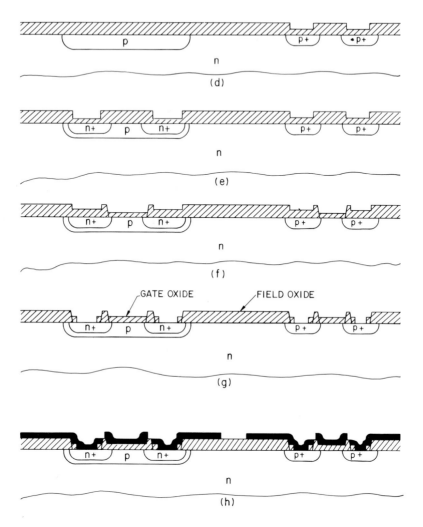

Fig. 6.21 Contd. (d) After p^+ source-and-drain diffusion; (e) after the n^+ source-and-drain diffusion for the n-channel transistors; (f) after growth of the gate oxide; (g) opening of source-and-drain contacts; (h) completed device.

The metal gate CMOS-processing sequence just described does not have self-aligned gates and therefore has larger gate-source and gate-drain capacitances than would be obtained with a self-aligned structure. The major steps in a self-aligned silicon-gate CMOS-processing sequence will now be described.

After the formation of the p-wells as described previously, a field oxide about 0.8 to 1.0 μm thick is grown by thermal oxidation. Windows are then opened in the field oxide wherever there is to be a transistor, and a gate oxide about 1000 Å thick is grown. These steps are shown in Figs. 6.22(a) and 6.22(b). A layer of polysilicon about 0.5 to 0.8 μm thick is deposited. The gate regions are defined, and the polysilicon and underlying gate oxide are etched as shown in Fig. 6.22(c). A layer of boron-doped oxide is deposited over the entire wafer and removed from those regions containing n-channel devices. Next, a layer of phosphorous-doped oxide is deposited over the entire wafer, as shown in Fig. 6.22(d). Next, the circuit is heated to diffuse impurities from the doped oxide into the underlying silicon, thus forming the source and drain regions and simultaneously doping the

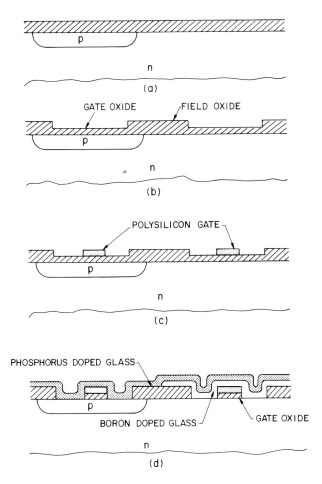

Fig. 6.22 Self-aligned silicon gate CMOS process sequence.

polysilicon gates. Finally, contact windows to the gate, source, and drain regions are opened, and the metallization is deposited and patterned. Channel-stop diffusions are usually required for high-density circuits but have been neglected in this discussion for simplicity.

6.15.1 Latch-up in CMOS Circuits

Examination of the CMOS structure will reveal that a vertical n-p-n bipolar transistor is formed by the n^+ drain/source regions of the n-channel MOS-FET, the p-tub and the n-substrate. Furthermore, there is a lateral p-n-p transistor consisting of the p-tub, n-substrate and p^+ drain/source regions of the p-channel MOSFET. The structure is often complicated further by the inclusion of input protection diodes (p^+n to the substrate and n^+p to the p-well), and n^+ contacts to the substrate and p^+ contacts to the tub. The presence of these bipolar transistors constitute a potential p-n-p-n device that can latch up if certain conditions are met.

A necessary condition for latch-up is that the product of the n-p-n and p-n-p transistor gains exceed unity, $\beta_{n\text{-}p\text{-}n}\beta_{p\text{-}n\text{-}p} \geqslant 1$. In addition, the end junctions of the structure must somehow become forward biased. This condition is not satisfied in normal operation, but it could occur during a transient or in a high radiation environment. The final condition is that the V_{DD} supply and the input circuit be capable of supplying the holding current of the p-n-p-n device.

Methods of reducing or eliminating the latch-up tendency include the use of thick substrates or buried layers, and guard rings around the devices.

6.16 DMOS AND VMOS [24]

Self-registration of the gate was used to reduce the parasitic capacitances, thereby increasing the speed. Another factor that influences speed is the length of the channel. High-speed devices require short channels. In the processing sequences described previously, the channel length was determined by photolithographic considerations, that is, the minimum spacing between features on a mask. The channel length in DMOS and VMOS is determined by the structure and is much smaller than that of previous processes.

6.16.1 DMOS

Double-diffused MOS (DMOS) technology uses two diffusions through the *same* window to produce the source and channel regions. The resultant structure is shown in Fig. 6.23. The substrate can be either n^- or p^-

material. In the first step, windows are opened for the source region and a
p-type diffusion is performed. Next the drain windows are opened and an n^+
diffusion is performed into both the source and drain windows. The
effective channel length is the *difference* between the lateral extent of the p
diffusion and the n^+ diffusion. This is typically in the range 0.4 to 2.0 μm.
Note that the source and drain regions are physically separated by a spacing
that is greater than the channel length. There is a drift region whose length
is the difference between the source-drain spacing and the channel length.
Regardless of the substrate material, this drift region will be n^- because if
the substrate is p^-, the surface will invert. The doping in the drift region
must be at least an order of magnitude less than that in the channel, giving a
sufficiently high breakdown voltage.

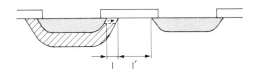

Fig. 6.23 A double-diffused MOS FET.

6.16.2 VMOS

Figure 6.24 is a cutaway view of a VMOS transistor. The structure takes its
name from the V-shaped cross-section of the device. It is similar to DMOS
in that the channel length is determined by the thickness of a p-type layer.
The VMOS transistor is fabricated on the faces of a pyramidal hole etched
into the (100)-oriented silicon substrate. This structure has doping profiles
similar to that of the DMOS transistor, but it has roughly twice the current
carrying capacity for the same surface area.

    The fabrication sequence[25] for n-channel devices begins with n^+
antimony-doped (100) silicon wafers. An unmasked boron diffusion of
about 150 $\Omega/\square$ is driven in until the surface concentration of boron is about
10^{17} cm^{-3}. The boron effectively disappears because of the high concentration of antimony. However, during the subsequent growth of a 3 μm thick
π-type epitaxial layer (5-100 $\Omega/\square$) from a silane source, the boron
outdiffuses from the substrate to form a thin (~1 μm), p-type layer. The
remainder of the epitaxial layer is the drift region. Next, the wafer is
covered by an oxide-nitride-oxide sandwich. The first masking step leaves
the nitride layer to protect the drain region of the transistors during the
growth of a thick thermal field oxide. Channel stoppers are diffused into the

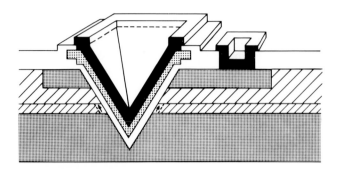

Fig. 6.24 Cross section of a VMOS transistor.

unprotected areas and a thick field oxide is grown. After the field oxidation, the nitride oxidation mask and the thin oxide layer underneath it are removed, and the n^+ phosphorous-doped drain region is diffused and oxidized. The V-groove is now etched through the drain, drift region, channel, and into the source-substrate using an anisotropic etch. The surface is oxidized and the metallization is deposited and defined.

Note that both the channel length and the length of the drift region are independent of photolithographic constraints.

The disadvantages of both DMOS and VMOS are that the transistors are asymmetrical because of the drift region, and therefore source and drain cannot be interchanged. Furthermore, all of the VMOS transistors share a common source terminal.

6.17 SILICON-ON-SAPPHIRE CIRCUITS

The speed of integrated circuits can be increased by reducing the magnitude of parasitic energy-storage elements. This can be done by making devices smaller and on lightly doped substrates, or by devising processes that have inherently small per-unit capacitance from the active region to the substrate or by a combination of both methods. An interesting approach to a process with low inherent parasitics that permits fabrication of complementary IGFET devices involves fabricating the active devices in thin-film silicon, grown on a sapphire substrate.

A lightly doped p-type epitaxial silicon layer 1 to 2 μm thick is grown on the sapphire substrate by the pyrolysis of silane which is preferred to the hydrogen reduction of silicon tetrachloride because the reaction occurs at a lower temperature and severe etching of the sapphire by HCl is avoided. Unfortunately silicon and sapphire do not have the same crystal structure. However, it is possible to grown usable (100)-oriented silicon films on ($1\bar{1}02$)-oriented sapphire.

The epitaxial layer is covered with a thermal oxide and the first mask is used to define the *p* islands which will become *n*-channel enhancement-mode transistors. Another epitaxial deposition and photomasking process is used to define *n*-type islands for the *p*-channel transistors. The masking oxide is stripped and a gate oxide about 1000 Å thick is grown. Polysilicon is then deposited nonselectively and etched to form the gate regions.

Doping of the source-drain regions and of the polysilicon gates is from doped-oxide sources. First a boron-doped oxide is deposited and selectively etched from over the *p* islands. A phosphorous-doped oxide is then deposited. After a single high-temperature drive-in to dope both the *n*- and *p*-channel devices, the doped oxide is stripped and a thin protective oxide is grown. The fifth mask defines the contact windows. Metal is then deposited and patterned by the sixth, and final, mask.

The major advantage of silicon-on-sapphire circuits is that the silicon can be selectively removed from those areas where it is not an active part of the circuit, thus reducing parasitics and increasing speed. Anisotropic etchants are used to remove the silicon because they give sharper edge definition than ordinary etchants.

6.18 SECOND GENERATION I²L PROCESSING

Integrated injection logic (I²L) is a low-power bipolar logic circuit (see Section 14.5). One of the big claims when it was introduced in the early 1970's was that it was merely a circuit technique and that I²L circuits could be fabricated using standard buried collector bipolar technology. Although this is true, it is not desirable because I²L circuits use the buried layer as an emitter and the normal emitter regions as collectors in a multicollector transistor. It became apparent that the performance of I²L circuits could be improved significantly by modifying the structure. It is the purpose of this section to describe some of the modifications to the basic buried collector processes that have been introduced to improve the performance of I²L circuits.

The top view and cross section of a modern I²L structure is shown in Fig. 6.25. It is a multicollector transistor in what is called "stick geometry." Each transistor is a logic gate and gates are packed in an array on the chip. Each gate is surrounded by a guard ring that is either a deep n^+ diffusion to the buried layer[26] (deep collector diffusion) or an oxide collar.[27] The guard ring increases the upward gain of the transistor (gain with the buried layer as the emitter) and reduces crosstalk between adjacent gates by suppressing the parasitic lateral *p-n-p* transistor action.

Ion implantation is used to introduce a precisely specified amount of charge into the intrinsic (active) base region. The dose of this p^- implant determines the Gummel number (active base charge) and thus the gain of the *n-p-n* transistor. Some processes[28] form the intrinsic base region of the

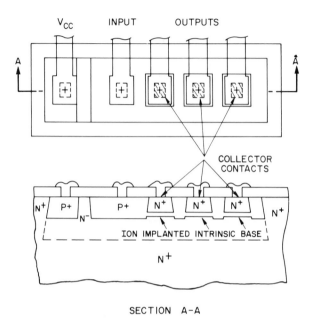

SECTION A-A

Fig. 6.25 Top view and cross section of I²L transistor.

n-p-n transistor by means of out-diffusion of boron from the substrate. This produces an impurity concentration gradient and a drift field in the base region that is opposite to that of an ordinary diffused transistor, thereby increasing the upward gain. Both schemes for producing the intrinsic base are designed to provide n^- doping at the collector side of the collector-base junction. This is necessary to reduce injection from the collector region when the transistor is saturated and to reduce the collector-base capacitance. It is accomplished when the intrinsic base is ion implanted by placing the implanted base dopant distribution sufficiently below the n^+ collector contact diffusion profile by a suitable choice of implant energy. When out-diffusion from the substrate is used, the epitaxial layer is made thick enough to provide an n^- collector region.

A p^+-doped extrinsic base region between the active base region is used to reduce the effect of the parasitic resistance in series with the base. Voltage drops across these parasitic resistors cause the emitter current density to decrease as a function of distance from the base contact, and impose a practical limit of about five on the number of collector contacts per transistor. The reduced current density at remote collector sites can also be combated by the simple layout expedient of increasing the collector area of the more remote sites.[27]

A final processing technique that is sometimes used is to form Schottky barrier diodes in the collector contacts.[28, 29] This reduces the magnitude of the voltage swing at the collector and increases the speed of the gate.

REFERENCES

1. R. A. Moline, R. Lieberman, J. Simpson, and A. U. MacRae. "The growth of high quality epitaxial silicon over ion implanted buried arsenic layers." **J. Electrochem Soc. 121**: 1362-1366 (October 1974).

2. M. P. Lepselter. "Beam lead technology." *Bell Syst. Tech. J.* **45**: 233-253 (February 1966).

3. J. A. Appels, E. Kooi, M. M. Paffen, J. J. H. Schatorj'e, and W. H. C. G. Verkuylen. "Local oxidation of silicon and its application in semiconductor technology." *Philips Research Reports* **25**: 118-131 (April 1970).

4. W. D. Baker, W. H. Herndon, T. A. Longo, and D. A. Peltzer. "Oxide isolation brings high density to production bipolar memories." *Electronics* **46**: 65-70. (March 1973).

5. W. D. Baker, W. H. Herndon, T. A. Longo, and D. A. Peltzer. "The impact of isoplanar technology on high density read/write memories." *1973 IEEE INTERCON Tech. Papers* **6**, paper 30/1.

6. E. Bassous, H. N. Yu, and V. Maniscalco. "Topology of silicon structures with recessed SiO_2." *J. Electrochem Soc.* **123**: 1729-1737 (November 1976).

7. J. Mudge and K. Taft. "V-ATE memory scores a new high in combining speed and bit density." *Electronics* **45**: 65-69 (July 1972).

8. T. J. Sanders, W. R. Morcom, and C. S. Kim. "An improved dielectric-junction combination isolation technique for integrated circuits." *Proc. IEDM,* Washington, D. C., pp. 38-40 (December 1973).

9. R. M. Finne and D. L. Klein. "A water amine complexing agent system for etching silicon." *J. of the Electrochemical Soc.* **114**: 965 (1967).

10. M. J. Declerq, L. Gerzberg, and J. D. Meindl. "Optimization of the hydrazine-water solution for anisotropic etching of silicon in integrated circuit technology." *J. Electrochem Soc.* **122**: 545-552 (April 1975).

11. U. S. Davidsohn and F. Lee. "Dielectric-isolated integrated-circuit substrate processes." *Proc. IEEE* **57**: 1532-1537 (September 1969).

12. A. P. Youmans, R. E. Rose, and W. F. Greenman. "A multichip package utilizing In-Cu flip-chip bonding." *Proc. IEEE* **57**: 1599-1606 (September 1969).

13. B. T. Murphy, V. J. Glinski, P. A. Gary, and R. A. Pederson. "Collector-diffusion isolated integrated circuits." *Proc. IEEE* **57**: 1523-1527 (September 1969).

14. A. E. Cosand. "A very high-speed, low-power, bipolar integrated-circuit process." *Proc. IEDM,* Washington, D. C., pp. 35-37, (December 1973).

15. P. C. Davis and S. F. Moyer. "Ion-implanted, compatible complementary *p-n-p*'s for high slew rate operational amplifiers." *Proc. IEDM,* Washington, D. C., pp. 18-20 (December 1972).

16. J. D. Beasom. "A process for simultaneous fabrication of vertical NPN and PNP's, Nch, and Pch MOS devices." *Proc. IEDM,* Washington, D. C., pp. 41-43 (December 1973).

17. W. N. Carr and J. P. Mize. *MOS/LSI Design and Application.* New York: McGraw-Hill, 1972, Chapter 2.

18. M. P. Lepselter. "Ion implantation—impact on device fabrication." J. Huff and R. Burges, editors. *Semiconductor Silicon, Electrochemical Society,* pp. 842-859, 1973.

19. B. L. Crowder. "Applications of ion implantation for new device concepts." J. of Vacuum Science and Technology. **8**: S71-S75 (May 1971).

20. F. F. Fang and H. S. Rupprecht. "High performance MOS integrated circuit using the ion implantation technique." *IEEE J. of Solid-State Circuits* **SC-10**: 205-211 (August 1975).

21. J. T. Clemens, R. H. Doklan, and J. J. Nolen. "An N-channel Si-gate integrated circuit technology." *Proc. IEDM,* Washington, D. C., pp. 299-302 (December 1975).

22. E. Vittoz, B. Gerber, F. Leuenberger. "Silicon-gate CMOS frequency divider for the electronic wrist watch." *IEEE J. of Solid-State Circuits* **SC-7**: 100-104 (April 1972).

23. W. M. Gosney and L. D. Hall. "The extension of self-registered gate and doped-oxide diffusion technology to the fabrication of complementary MOS transistors." *IEEE Trans. on Electron Devices* **ED-20**: 469-473 (May 1973).

24. N. C. de Troye, "Digital circuits with low dissipation." *Philips Tech. Rev.* **35**: 212-220 (1975).

25. T. J. Rodgers and J. D. Meindl. "VMOS: High speed TTL compatibile MOS logic." *IEEE J. Solid-State Circuits* **SC-9**: 239-250 (October 1974).

26. J. M. Herman III, S. A. Evans, and B. J. Sloan, Jr. "Second Generation I²L/MTL: A 20 ns process/structure." *IEEE Jour. Solid-State Circuits* **SC-12**: 93-101 (April 1977).

27. F. Hennig, H. K. Hingarh, D. O'Brien, and P. W. J. Verhofstadt. "Isoplanar integrated injection logic: A high performance bipolar technology." *IEEE J. Solid-State Circuits* **SC-12**: 101-109 (April 1977).

28. B. B. Roesner and D. J. McGrewy. "A new high speed I²L structure." *IEEE J. Solid-State Circuits* **SC-12**: 114-118 (April 1977).

7

Design of
Silicon Integrated Circuits

7.1 INTRODUCTION

Chapters 5 and 6 were devoted to a study of methods used to process silicon wafers into electronic devices and the consequent fabrication sequences for several types of silicon integrated circuits. In the present chapter, the problem of translating the schematic diagram of a circuit into a set of masks that define the horizontal topography of the silicon integrated circuit is considered.

A flow diagram showing the evolution of a silicon integrated circuit from initial specification through the final layout is shown in Fig. 7.1. The most general case is depicted in this diagram, in which all aspects of the process technology and the circuit configuration can be specified by the design team. In a particular situation, however, the design team has a choice of one of several (standard) processing sequences. Circuit models have been derived for devices made by these process sequences and minimum permissible layout spacings have been calculated for these structures at prescribed operating voltages. The derivation of minimum permissible layout spacings is considered in detail in Section 7.4. For the present, it suffices to say that these spacings are dependent upon the electrical operating conditions of the circuit, the doping levels and profiles of the various doped layers, and the materials and processing techniques used in the fabrication. Therefore, the spacings must be determined jointly by the circuit designers and processors for each group of circuit designs. Thus determined, a set of design spacings is valid only for circuits fabricated by that particular sequence of processing steps and operating under specified dc bias conditions.

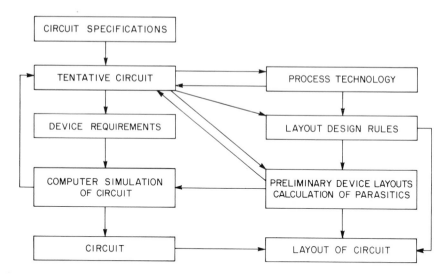

Fig. 7.1 Flowchart for evolution of a silicon integrated circuit.

Photolithography[†] is used during each major processing step to produce a desired pattern on the surface of the wafer. These patterns may be produced directly or via an intermediate mask. Pattern-generation techniques are described in Section 7.2. The statistical dimensional variability of the photolithographic operation and its influence on feature dimensions and placement are considered in Section 7.3.

In Section 7.5, the most elementary layout problem—that of designing a single transistor and a pair of common collector transistors—is considered to illustrate the use of the minimum permissible layout spacings. The layout of a clocked RS flip-flop is detailed in Section 7.6. This example was chosen because it is a nontrivial design that is relatively simple.

The increasing complexity of digital integrated circuits makes it necessary to have computer-based aids for the design and layout of large-scale digital integrated circuits. Computer-based drafting aids and computer-based aids for the placement and interconnection of basic building blocks are described in Section 7.7.

7.2 PATTERN GENERATION

We have already seen that the manufacture of integrated circuits involves the selective deposition or removal of material. The selectivity is provided by

---

[†] The term photolithography is used throughout. Similar considerations, though perhaps with different numerical tolerances, apply to circuits fabricated by electron beam or X-ray lithography.

using a mask to determine upon which areas of the substrate the material is deposited or from which it is removed. In the normal sequence of operations, a master mask is made from the original artwork and working copies, made from the master mask, are contact printed to transfer the pattern to the silicon wafer. The life of a working copy of the mask is limited because of scratches or other defects resulting from the contact printing process. Therefore projection printing is often used as a means of prolonging mask life. Electron-beam lithography modifies the normal sequence of operations in one of two ways because it can be used to produce the photomask or it can write the pattern directly on the silicon wafer.

Photolithography is the primary means of pattern generation for silicon and thin-film circuits. Electron-beam and X-ray lithography have advantages for small-area devices and are becoming increasingly important. Deposition of material for thick-film circuits is always through a screen and that process is explained in Chapter 9.

7.2.1 Initial Artwork

Usually, the initial artwork from which photomasks are made is a highly accurate drawing between 100 and 1000 times final size, which is produced by exposing a photographic film on a computer controlled drafting machine called a coordinatograph. Commercial machines typically have a useful working area about 1 meter on a side, and an accuracy of ± 25 μm, or one part in 40,000. In some instances the coordinatograph cuts Rubylith, which is a sheet of clear mylar laminated with a red overlay. The overlay, which is opaque to green, blue, and ultraviolet light used in subsequent photographic steps, is cut and stripped, leaving a red pattern on a clear background. Regardless of the medium, it is important that the base material be dimensionally stable under varying temperature and humidity conditions.

Another method of generating the initial artwork is the so-called primary pattern generator (PPG). It uses a scanning laser beam to sequentially write the pattern on a glass-based high-resolution photographic plate at either 14 or 35 times actual size.

Electron-beam exposure can be used to write the pattern at its actual size. This technique can be used to produce a photomask, or it can be used to directly expose the resist on each wafer. The latter approach is used when the ultimate resolution is required.

7.2.2 Photoreduction

The production of photographic masks with the final circuit dimensions from the original artwork requires a reduction in size and a replication of the images so that many circuits can be processed simultaneously on a large substrate. These steps are performed photographically.

Basic optical principles often place conflicting demands on the design of a reduction system. Tradeoffs among system parameters are very complex and cannot be covered in any detail in this book. Instead, this section attempts to justify the parameters of typical photoreduction systems by some simple examples.

A simple optical system for producing reduced images consists of a lens of focal length f, which forms a reduced image of an object placed at a distance l_o from the lens on its optical axis. It is easy to show that for a simple thin lens, the lens-to-image distance is

$$l_i = f\left[1 + \frac{1}{R}\right]$$

where $R = l_o/l_i$ is the reduction ratio. For large reduction ratios, l_i approaches f and the object-to-lens distance approaches Rf. We now examine some restrictions on the choice of focal length.

The light-transmission capacity of the lens is often given as a figure of merit called the focal number or F number, where

$$F = \frac{\text{focal length}}{\text{diameter}} .$$

Because of aberrations, a point object will not be imaged as a point but rather as a circle called the *circle of confusion*. The variation in the axial position of the imaging screen so that a theoretical defocused point image falls within the circle of confusion is called the *depth of focus*. Depth of focus increases with increasing F number and the photographic plate should be accurately located and dimensionally stable within this limit. As an example, the depth of focus of an $F2$ lens at $\lambda = 5460$ Å corresponding to green light commonly used for exposing high-resolution photographic plates is about 8.5 μm.

The initial artwork, typically 1 meter on a side, is the object to be reduced. Field size is limited by geometric aberrations which cause off-axis rays to be imaged off of the image plane, resulting in a defocused image. The acceptable deviation of the image from the image plane is related to the depth of focus of the lens. For the same tolerable circle of confusion at the edge of the field, the image size increases as the square of the focal length because the depth of focus is constant for a constant F number. Depth of focus, and therefore image-field size, can be increased by increasing the F number, but this course of action reduces the resolving power of the lens because of diffraction effects. The performance of fast lenses is limited by aberrations which can be reduced by means of complex lens designs to the point where diffraction effects dominate. Unless exposure speed is critical, the lens should not be any faster than necessary to meet the resolution requirements.

EXAMPLE 7.1   Is it possible to make masks for silicon IC's by a single reduction step?

Solution.   MSI complexity circuits are typically 2500 μm on a side, thus requiring a reduction ratio of 400 starting with 1 m artwork. To show that it is impossible to obtain the reduction in one step, assume that an overall reduction ratio of 100 is desired with an image-field size of 2500 μm, and that it is necessary to resolve 1-μm lines and 1-μm spaces. This last requirement implies a spatial frequency response of 500 lines/mm with a reasonably low insertion loss. We simply state that this requirement can be satisfied by an F1.4 lens. The depth of focus of an F1.4 lens at 5460 Å is about 4 μm. Assuming that the image-field size is limited by curvature of field, which is not true for complex lenses, the focal length is 385 mm. The lens-to-object (artwork) distance for a reduction ratio of 100 is 38.5 m, which is an extremely long camera bed. To further complicate matters, it is necessary to maintain the camera in an environment where the temperature is controlled to a small fraction of a degree. Even if the bed is made of gray cast iron, which has a small coefficient of linear expansion, the object distance would vary by 230 μm/°F. A high-quality camera system has corrections for first-order temperature effects and is operated in an environment whose temperature is controlled to ± 0.25°F.

From this example, it is evident that the reduction must be accomplished in stages. To obtain an overall reduction of 400, a suitable combination might be an initial reduction of 40:1 followed by a reduction of 10:1 during the *step-and-repeat* operation, which results in an array of perhaps several hundred identical images. This is accomplished by exposing a small section of the film plate to the image on each exposure. The film plate is then repositioned or "stepped" by a mechanical stage, and the operation is repeated.

EXAMPLE 7.2   What are reasonable parameters for a two-step reduction process?

Solution.   When a two-step reduction process is used, the first reduction image is ten times final size or 25,000 μm, for the previous example, with a requirement for resolving 10 μm lines and spaces. This requirement can be satisfied with an F11 lens which has a depth of focus of 265 μm at a wavelength of 5460 Å. A focal length of 650 mm is required if the image field size is limited by curvature of field. The lenses used in photoreduction cameras are complex, multielement structures that are designed to minimize geometric aberrations and curvature of field. In practice, a complex lens of about one-fourth the above focal length, or 150 mm, is sufficient. The resultant lens-to-object distance is still about 6 meters. The optics for the step-and-repeat camera used for the second-stage reduction must reduce a 25-mm object to a 2.5-mm image with the ability to resolve 1-μm lines and 1-μm spaces. This requires a lens speed of about F1.4. The focal length is typically about 90 mm and thus the resultant object distance is about 90 cm. It is obvious from the above discussion that mechanical and optical constraints impose one strong upper bound on the size of silicon circuits using photolithography.

The variation in linewidth of the image in the photoresist film depends on many factors including the accuracy of the initial artwork, the various reduction factors, the resolution capability of the optical systems, the characteristics of the photosensitive materials, the flatness of the substrates, and the size of the mask. A reasonable number is a few tenths of a micrometer.

7.2.3 Film [1]

The ideal photosensitive material for making masks would have extremely high resolution and high contrast, be capable of reproducing both fine lines and broad light or dark areas with no loss of detail, and it would be on a substrate that is perfectly flat and dimensionally stable. These conflicting demands can only be approached by actual photographic films.

Photographic emulsions are composed of fine grains of silver halides suspended in a gelatinous layer. Since the photosensitive material is a silver halide granule at a discrete point, the medium has only a finite resolution capability. Moreover, the speed (light flux sensitivity), graininess (resolution), and contrast of the film are all related to the grain size with small grains yielding slower, higher resolution emulsions. High-resolution plates used in mask making have emulsion layers that are 5- to 7-μm thick containing silver halide grains 0.01 to 0.1 μm in diameter. These emulsions are capable of resolving about 2000 lines/mm and are most sensitive in the 3000 Å to 5500 Å range. The emulsion is supported on a glass substrate.

An important characteristic of the developed film is its optical density, which is a measure of normalized light transmission. The sharpest images would be obtained if the optical density could change abruptly to zero when going from opaque to transparent areas. This is impossible to achieve because of diffraction effects during exposure and light scattering in the relatively thick emulsion layers. Hence the density transition at a line-edge is gradual and therefore there is an added component to the line-edge uncertainty.

High-resolution plates are delicate, easily damaged, and require careful cleaning in order to avoid contamination. Working plates are subject to a great deal of wear and tear and are usually good for only 20 to 50 contact prints. Projection printing is used in many applications to eliminate the constant need to replace working masks.

7.2.4 Electron Beam and X-Ray Lithography [2-4]

Diffraction-limited resolution is inversely proportional to the focal number of the lens and the wavelength of the light. The fabrication of smaller devices and more densely packed circuits requires increasingly higher resolution.

Therefore electron beams and soft X-rays were investigated as methods of improving image resolution.

According to the de Broglie postulate, the wavelength of a particle is inversely proportional to its momentum. The wavelength of 10-keV electrons is just 0.1225 Å, which is about four orders of magnitude smaller than the wavelength of visible light used in photolithography. Therefore, if an electron optical system with the same parameters as a photo-optical system can be designed, an increase of four orders of magnitude in performance (resolution) would be expected. Unfortunately, this is not the case; in practice the improvement in resolution is only about two orders of magnitude.[5]

Electron optics is similar to conventional optics in many ways and electron lenses are plagued with the same geometric aberrations as conventional lenses. However, a photon lens has sharply defined surfaces at which the index of refraction changes abruptly, which is not the case with electron lenses, and there are some additional aberrations peculiar to electron lenses caused by the existence of both electric and magnetic lenses.

Two factors that place a lower limit on spot size are the transverse thermal velocity of the electrons and spherical aberrations of the lens. To a much lesser extent the beam is affected by chromatic aberration caused by sensitivity of the lens to the spread $e\Delta V$ in particle energy causing a variable focus, and finally by diffraction effects. Beam diameters less than 500 Å can be obtained for beam currents in the 10^{-10} to 10^{-9} A range. It remains to indicate that this current range is useful, and to examine limitations on pattern area.

Electron beams incident on a resist are scattered through large angles by the atoms of the resist. If the beam energy is high enough, backscattered electrons are produced which can also expose the resist. This exposure is roughly spherical in volume with a diameter much larger than the beam-spot diameter. One resist for electron-beam and X-ray lithography is PMM (polymethylmethacrylate). In an example cited by Brewer,[5] a beam of 10 keV, 500 Å electrons expose a PMM resist. A beam current of at least 2.4×10^{-11} A is required to expose the resist; scattering will be minimized or eliminated if the beam current is less than 2.8×10^{-9} A.

The scanning electron-beam machine writes a pattern sequentially rather than simultaneously as in the case of photolithography. A limit on the maximum scan distance is imposed by distortion of the spot caused by the complex electric and magnetic deflecting fields. As the beam deflection angle changes, the beam is brought to a focus at different distances from the source. Therefore, the spot size seen at the target changes. For a 17-cm lens-to-target distance, a 10,000 diameter displacement results in negligible broadening of a 0.15-μm diameter beam. Thus a 2×10^4 spot diameter, edge-to-edge scan can be achieved in a good-quality electron lens system.

Furthermore, 12- to 15-bit D/A converters are available so that digital-computer control of 10,000- to 20,000-spot diameter scans are practical.

Sequential writing permits a pattern that is larger than the scan limit to be written in segments. A disadvantage of sequential writing is the time required to expose a complete wafer. The sensitivity S of an electron resist is defined as the amount of charge required to completely expose a unit area of the surface. Typically the current density, J_y, of the beam is in the range 0.1 to 1 A/cm^2, and S is about 6×10^{-5} C/cm^2 for PMM resist. Therefore, the exposure time is of the order of 60 to 600 μs per spot. To obtain a rough idea of the time to expose a 5-cm diameter silicon wafer assume that the circuits each occupy an area of 3×3 mm and are spaced on 3.5 mm centers. The largest inscribed square is 3.5 cm on a side so that 100 circuits can be put on the wafer. If 25% of the circuit area is to be exposed by a 0.25-μm diameter electron beam, the total number of spots is approximately 4×10^7. At 60-μs exposure per spot the total writing time is 2400 s (40 min).

Exposure times are reduced considerably with newer resists. The sensitivity of a negative resist based on polyglyadyl methacrylate-co-ethyl acrylate[7] (poly for short) is about 100 times greater than PMM. A positive resist[8] based on polybutene-1 sulfone has about the same sensitivity as poly. Using these sensitive resists, exposure times under 20 min per wafer are achieved.

A high-throughput scanning E-beam lithography system using a square beam whose size is equal to that of the smallest feature was recently described by IBM.[6] The use of a shaped beam increases the throughput by a factor of about 25 because round beams of about a fifth the width of the minimum feature are needed to provide adequate resolution.

The rapid generation of high-resolution patterns requires that the entire pattern area be simultaneously irradiated with short-wavelength radiation. A promising technique for accomplishing this is the use of soft X-rays whose wavelength is of the order of 10 Å. There are no efficient X-ray lenses or mirrors. The exposure method involves the use of a beam of X-rays from a "distinct" point source to illuminate the mask so as to produce a shadow on the X-ray resist. A suitable source is an aluminum target, which produces characteristic X-rays at 8.3 Å when bombarded by an electron beam. The master mask is a semitransparent silicon substrate, 3-μm thick, holding a patterned 3000-Å gold film which is highly absorbing to X-rays. This substrate and film combination result in a mask with a contrast ratio of 4:1. PMM resist typically exhibits a dissolution rate that varies as the cube of the exposure so that a 4:1 contrast ratio yields a 64:1 differential dissolution rate of the resist. Thus sharp images can be formed with relatively low contrast masks.

There is a broadening effect because of the nonzero size point source. However, the source diameter, source-to-object spacing and mask-to-surface spacing, can be chosen so that this effect is negligible in comparison with the minimum linewidth to be reproduced.

It should be noted that at the present time, neither electron-beam nor X-ray lithography has replaced photolithography as a production technique. Both techniques require rather large investments in capital equipment which will be resisted until the performance improvements due to reduced device size cannot be obtained in any other way. Electron-beam exposure is being used to make master photomasks and specialized devices.

7.3 MASK-ALIGNMENT TOLERANCE

An important component in each of the design spacings to be derived in the next section is the mask-alignment tolerance, which is a statistical average error between the desired position of a geometric feature in the "ideal chip" coordinate system and its actual position in any of the local chip coordinate systems on the wafer. The numerical values presented in this section and in the section on minimum layout spacings are for illustrative purposes only. They are representative of values that were common in industry about five years ago.

Linewidth control is the statistical variability of any process by which masks can be made. It takes account of the effect of linewidth variations in the original artwork, the resolution of the photographic process used to generate the master mask, and variation of linewidth in the working copies due to variations in exposure and etching. A statistical bound can be placed on each of these effects. The linewidth control is tighter than the summation of the bounds for each of the contributing components because the actual statistics are more optimistic than this worst case. Also, the masks are inspected and those that have excessive feature variation are rejected, which truncates these statistical distributions.

The copying and photolithographic errors, which will be described later, are the description of the statistical uncertainty of a feature edge. The *linewidth tolerance* is twice the sum of the copying and photolithographic errors because edges occur in pairs in nonzero-width lines and spaces.

Placement and registration error is due to the statistical variability in the basic mask-making equipment. Up to a dozen different masks are needed to produce a silicon integrated circuit, and each mask contains the pattern required for one step in the process. The patterns on different masks in the set bear a definite relationship to each other. As mentioned previously, each mask contains many copies of the pattern so that many circuits can be fabricated simultaneously on each wafer. Consider, for the moment, just one copy of the pattern on each mask in the set. Any uncertainty between related features on different mask levels (for example, the positional uncertainty between an emitter-contact window and its corresponding emitter-diffusion edge) is called *registration error*. The pattern is usually replicated by a *step-and-repeat* process, which involves exposing a section of a film plate that is repositioned after each exposure. The difference between where a copy of the pattern should be and its actual position on the mask is the *placement error*.

Copying errors   The master masks are copied to produce submasters, which are then copied to produce the working masks. There are errors associated with this duplication process as a result of random distortions of the pattern caused by surface variations in the photographic plates, imperfect collimation of the light in the printer, and variations in exposure.

Photolithographic errors   Working copies of the masks are used to transfer the patterns to photoresist, which form the resist pattern for an etching operation. Variations in the photoresist exposure and development, as well as variations in the etch rate of the material, are potential sources of error in the photolithographic operation. These are the photolithographic errors.

The final components of the mask-alignment tolerance are tabulated in Table 7.1. The tolerance associated with a feature placement involving two masking operations is the mask-alignment tolerance plus one operator-alignment tolerance.

7.4 MINIMUM LAYOUT SPACINGS

In this section the junction-isolated, bipolar silicon integrated structure is analyzed to determine the minimum permissible spacing required between specified features on the surface of the device, allowing high-yield manufacture of the devices. These minimum spacings are dependent on the material properties, the operating voltages, and the statistical variability of some of the manufacturing operations. For purposes of illustration, the process specified in Table 7.2 is assumed, subject to the ground rules given in Table 7.3.

The dimensions and statements given in the latter table are determined partly by the photolithographic patterning capability, and partly by the requirements imposed upon the design by the techniques used (on the particular processing line), to perform each operation. For example, the minimum feature size on the mask is process dependent because it implies the capability of opening a window of that size in an oxide layer of the thickness required by the particular processing sequence.

Table 7.1 Components of Mask Alignment Tolerance in μm

1. Linewidth Control
    a) Linewidth variation on master mask   ±0.4
    b) Variation in linewidth of chrome copy
       caused by variation in exposure intensity   ±0.2
    c) Variation in chrome copy caused
       by variations in chrome etch   ±0.3

    Linewidth variation on working copy after
    inspection   ±0.50

2. Placement and Registration
    a) Step and repeat tolerance on master   ±0.4
    b) Placement tolerance for features
       between levels on master set   ±0.7

3. Printing of Working Copies   ±0.1

4. Photolithographic Process
    a) Tolerance on photoresist edge
       due to exposure variations   ±0.4
    b) Tolerance on edge of oxide window   ±0.5

    Tolerance on linewidth due only to
    photolithographic processing   ±1.8

5. Operator Alignment Tolerance   ±1.0

    Mask alignment tolerance   ±4.5
    Tolerance for two alignments   ±5.5

It is further assumed that the numerical values of these minimum layout spacings are given for TTL-type digital logic circuits with a 5-V dc supply. It is also assumed that under transient conditions the maximum

Table 7.2 Process Specification for Example (lateral diffusion is assumed to be 80% of the vertical diffusion depth)

Substrate	p-type 10 Ω-cm ±20%, (111) orientation
Buried Layer	40 $\Omega/\square$, 6 μm deep
Epitaxial Layer	0.5 Ω-cm ±10%, 5 ± 0.5 μm thick
Base Diffusion	p-type boron diffusion, nominally 200 $\Omega/\square$, 2 ± 0.5 μm deep
Emitter Diffusion	n-type phosphorus diffusion with n^+ surface concentration, base width is nominally 1 μm
Isolation Diffusion	p^+ boron diffusion, nominally 5 $\Omega/\square$

Table 7.3 Ground Rules

Minimum Feature Size on Mask	$8 \times 8 \ \mu$m
Minimum Width of Metal Line	$10 \ \mu$m
Minimum Metal-to-Metal Spacing	$10 \ \mu$m
for Very Short Runs	$8 \ \mu$m

No metal overlap required on contact
but sufficient allowance must be provided to guarantee a seal.

Emitter contact must be reregistered
in the emitter region.

Minimum Worst-Case Spacing	$1 \ \mu$m

The following quantities were derived
in the previous section:

Linewidth Uncertainty	$\pm 1.8 \ \mu$m
Mask Alignment Tolerance	$\pm 4.5 \ \mu$m
Tolerance for Two Mask Alignments	$\pm 5.5 \ \mu$m

collector-substrate voltage is 10 V, the maximum collector-base voltage is
10 V, and the maximum reverse bias on the emitter-base junction is 5 V.

7.4.1 The Buried Layer

The mechanics of the buried-layer diffusion are examined in this section.
Suitable diffusion times and temperatures are obtained, which yield a buried
layer having the desired thickness and sheet resistance. After a similar look
at the isolation diffusion in the next section, the information at hand will be
applied to the calculation of minimum feature spacings for use in the layout.

Buried layers are made by a two-step diffusion process with the
predeposition being either by thermal diffusion or by ion implantation. The
two-step diffusion process will be assumed to approximate a complementary
error-function diffusion profile (see Section 5.4.3). An impurity with a small
diffusion coefficient is used so that the buried layer will not move appreciably during subsequent high-temperature processing. Either arsenic or
antimony is used as the dopant because both of these elements do not
diffuse rapidly in silicon. Arsenic is often the dopant because it is a perfect
fit in the silicon lattice.

The 10-Ω-cm p-type substrate material is the background for the
buried-layer diffusion. Since the calculation of the extent of depletion layers
is of interest, the impurity concentration corresponding to the high resistivity

limit of 12 Ω-cm is used because it gives the most pessimistic value (i.e., the greatest depletion width). The impurity concentration, N_B, corresponding to 12 Ω-cm is about 10^{15} cm^{-3} as obtained from the curves of Fig. 7.2.

After the buried layer is driven to the desired depth, it is necessary to remove about 0.5 μm from the surface, by an HCl etch in the epitaxial reactor, to provide the clean, defect-free surface required for proper epitaxial growth. The loss of this highly doped layer significantly increases the sheet resistance of the buried layer and must be taken into account.

Irvin[9] has derived a set of curves for the average conductivity of diffused layers with either Gaussian or complementary-error-function profiles in a uniformly doped background of the opposite conductivity type. This was done by integrating the impurity distribution. Two depths denoted by x and x_j are defined, as shown in Fig. 7.3. The ratio x/x_j is a parameter for each set of Irvin curves. In the present case $x/x_j = 0.5/6 = 0.083$. The effective depth of the buried layer after etching is $x_j - x = 5.5$ μm. For the sheet resistance to be 40 $\Omega/\square$ as specified, the average conductivity must be

$$\bar{\sigma} = \frac{1}{R_s(x_j - x)} = \frac{1}{40(5.5 \times 10^{-4})} \approx 45 \ (\Omega\text{-cm})^{-1}. \tag{7.1}$$

The impurity concentration at the surface of the buried layer after the etching is $N_s \approx 10^{19}$ cm^{-3}. This value is obtained from the curves of Fig. 7A.1 in the appendix to this chapter. Hence, the concentration required at the original substrate surface to yield a concentration of 10^{19} cm^{-3} at a

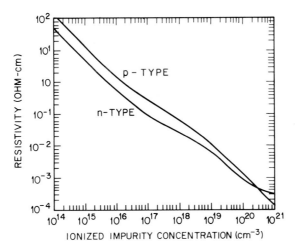

Fig. 7.2 Resistivity of silicon at 300 K as a function of acceptor or donor concentration.

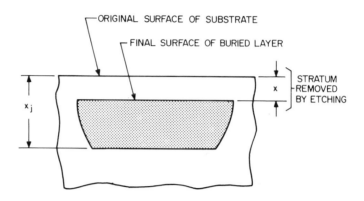

Fig. 7.3 Definition of the variables x and x_j.

depth of $0.5 \, \mu m$ can be easily calculated. Assuming that the two-step diffusion is equivalent to a complementary error-function diffusion with an equivalent diffusion time of 10^4 s yields

$$N(x=0.5 \, \mu m, \quad t=10^4 \text{ s}) = 10^{19} \text{ cm}^{-3}$$

$$= N_0 \text{erfc} \left[\frac{x}{2\sqrt{Dt}} \right]. \qquad (7.2)$$

From Fig. 5.40 it is found that the diffusion coefficient, D, of arsenic in silicon at a drive-in temperature of 1200°C is about 2.5×10^{-13} cm^2/s. Therefore the surface concentration is

$$N_0 = \frac{10^{19}}{\text{erfc}(0.5)} = 2.1 \times 10^{19} \text{ cm}^{-3}.$$

Note that the object of this section has been to determine the surface concentration required for a complementary error-function diffusion such that (1) the junction between the diffused layer and a 10-Ω-cm substrate occurs 6 μm beneath the original surface, and (2) that after a 0.5-μm layer is removed from the surface the resulting diffused layer has a sheet resistance of 40 $\Omega / \square$.

7.4.2 The Isolation Diffusion

The epitaxial layer is grown after the buried layer is diffused, and then the individual isolated regions are defined by a p-type, two-step diffusion process whose impurity profile approximates a Gaussian distribution.

In Section 5.4.6 it was shown that a one-dimensional model of the diffusion process is not valid in the vicinity of a feature edge because the supply of dopant atoms is depleted at the edge of the mask window. The lateral-diffusion distance is typically about 80% of the diffusion distance normal to the surface. It is assumed here that the 80% lateral-diffusion distance is also valid at a depth of 0.5 μm below the original surface, so that the effective buried-layer lateral-diffusion distance in this example is $0.8(6-0.5) = 4.4 \ \mu$m.

Figure 7.4 shows two buried layers, which are to be in different isolated regions, separated by a p-type diffused region whose sheet resistance is typically 5 $\Omega/\square$. It is customary to overdrive the isolation diffusion by about 25% to prevent the depletion layer from extending underneath the isolation diffusion. Therefore, instead of requiring the impurity concentration in the tail of the isolation diffusion profile to equal the epitaxial-layer impurity concentration of $1.1 \times 10^{16} \ \text{cm}^{-3}$ at the worst-case epitaxial thickness of 5.5 μm, it is required to equal the epitaxial-layer concentration at a depth of $1.25 \times 5.5 \approx 6.88 \ \mu$m. The average conductivity

$$\bar{\sigma} = \frac{1}{R_s \times (\text{thickness})} \approx 286 \ (\Omega\text{-cm})^{-1}.$$

The final surface concentration of the p-type isolation region denoted-by N_{02}, is found to be $1.1 \times 10^{20} \ \text{cm}^{-3}$ from the Irvin curves[6] of Figs. 7A.7 and 7A.8. with $x/x_j = 0$ and $N_{BC} = 1.1 \times 10^{16} \ \text{cm}^{-3}$. Although this value is somewhat in error because there is no junction at the bottom of the diffused profile, it is good enough for the present purpose.

For simplicity, assume that both the predeposition and the drive-in diffusions are carried out at a temperature of 1200°C. The final surface concentration for the two-step diffusion process given by Eq. (5.62) is repeated here for convenience.

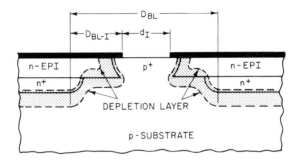

Fig. 7.4 Derivation of the minimum buried layer to isolation-diffusion spacing.

$$N_{02} = N(0,t_1,t_2) = \frac{2N_0}{\pi} \tan^{-1} \left[\frac{D_1 t_1}{D_2 t_2}\right]^{1/2}. \tag{5.62}$$

In this equation N_0 is the impurity concentration that is maintained constant at the surface.

The solid solubility of boron in silicon at the predeposition temperature is chosen as N_0 because of the high final surface concentration requirements. This value is $N_0 = 4 \times 10^{20}$ cm^{-3} as obtained from Fig. 5.41. Since temperatures of predeposition and drive-in are equal, $D_1 = D_2$.

Solving for t_1:

$$\tan^{-1}\alpha = \tan^{-1}\left[\frac{t_1}{t_2}\right]^{1/2} = \frac{\pi}{2}\left[\frac{1.1 \times 10^{20}}{4.0 \times 10^{20}}\right] = 0.432 \text{ radian}$$

$$\alpha = 0.461$$

and

$$t_1 = \alpha^2 t_2 = 0.21 t_2.$$

The two-step diffusion profile has been described by Eq. (5.63)

$$N(x,t_1,t_2) = \frac{2N_0}{\sqrt{\pi}} \int_{\sqrt{\beta}}^{\infty} \exp[-y^2]\text{erf}[\alpha y]\,dy. \tag{5.63}$$

The values of the integral of this equation are given in Table 5.4. The impurity concentration $N(x,t_1,t_2)$ is required to be 1.1×10^{16} cm^{-3} at a depth of about 7×10^{-4} cm and $N_0 = 4 \times 10^{20}$ cm^{-3}. Solving this equation for the integral and substituting the above values yields 2.44×10^{-5}. The parameter β is found to be 9.2 by extrapolating the tabulated values of the integral with α and β as parameters. However, β is related to x, D, t_1, and t_2 by

$$\beta = \left.\frac{x^2}{4D(t_1+t_2)}\right|_{x=7\times10^{-4}\text{cm}}. \tag{7.3}$$

At 1200°C, $D = 2.8 \times 10^{-12}$ cm^2/s, and therefore $t_2 = 3.93 \times 10^3$ s and $t_1 = 8.253 \times 10^2$ s.

It should be noted that the calculations relating to the buried layer and isolation diffusion were done here for the purpose of illustration and are not ordinarily done when computing layout design rules. The process specification sheet given in Table 7.2 completely specifies the process.

7.4.3 The Buried Layer and Isolation Masks

The minimum permissible spacing between features on the buried-layer mask and features on the isolation-diffusion mask can now be developed.

Referring again to Fig. 7.4, it can be seen that the key distance is the buried-layer-to-isolation spacing. The object of the following derivation is to determine the minimum permissible separation between a feature on the buried-layer mask and a corresponding feature on the isolation mask, such that, with high probability, a low-leakage isolating junction is obtained.

The first task is to determine the width of the depletion layer associated with the isolation junction. This is done with the aid of a set of normalized curves of diffused junction characteristics known as the Lawrence-Warner curves.[10] Each family of curves is drawn for a particular value of N_B/N_S where N_B is the background impurity concentration and N_S is the surface concentration. In the present case, the background is the epitaxial layer and therefore $N_B \approx 10^{16}$ cm^{-3}. Furthermore $N_S = N_{02}$ for the isolation region and the value for this quantity was previously found to be 1.1×10^{20} cm^{-3}. Hence $N_B/N_S \approx 10^{16}/1.1 \times 10^{20} = 9.09 \times 10^{-5}$. This ratio will now be used to determine the depletion layer width of the isolation junction. With a maximum reverse-bias voltage of 10 V, the normalized reverse-bias voltage is $V/N_B = 10^{-15}$. From the appropriate family of Lawrence-Warner curves, which is reproduced as Fig. 7.5, the extent of the total depletion region is found to be about 3 μm of which about 1.5 μm is in the epitaxial layer.

The extent of the lateral diffusion associated with the isolation diffusion is about $0.8 \times 7 = 5.6$ μm. The isolation junction will have a low-leakage current if the high-concentration p-type profile does not touch the n^+ buried layer. To obtain this desirable condition, the appropriate design spacing is found, with the aid of Fig. 7.4, to be

D_{BL-I} = buried-layer lateral diffusion

   + isolation lateral diffusion

   + isolation-depletion region in epitaxial layer

   + mask alignment

   + worst-case minimum spacing.   (7.4)

EXAMPLE 7.3 Calculate the buried-layer-to-isolation spacing for the sample process. Assume that the minimum spacing is 1 μm.

Solution. From Eq. (7.5)

$$D_{BL-I} = 5.5 \times 0.8 + 5.6 + 1.5 + 4.5 + 1 = 17 \ \mu m.$$

It is also apparent from Fig. 7.4 that the minimum buried-layer-to-buried-layer separation is determined primarily by the buried-layer-to-isolation spacing, and has the value

$$D_{BL} = 2D_{BL-I} + d_I \qquad (7.5)$$

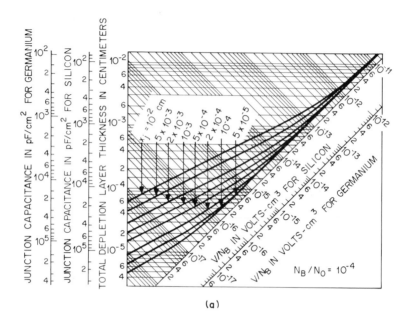

(a)

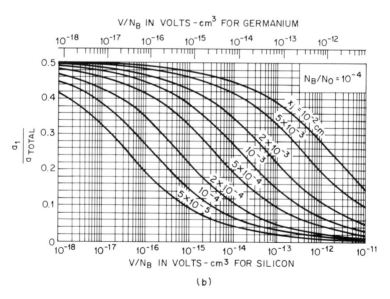

(b)

Fig. 7.5   Lawrence-Warner curve for Gaussian profile with N_B/N_0 in the range 3×10^{-5} to 3×10^{-4}. (a) Total depletion layer width and (b) ratio of depletion layer penetration in side towards the surface to total depletion width versus V/N_B. (From Lawrence and Warner.[10] Reproduced with permission from *The Bell System Technical Journal*, Copyright 1960 The American Telephone and Telegraph Co.)

where d_I is the width of the isolation-diffusion window. If the minimum width of a long window in the masking oxide is taken to be 10 μm, then D_{BL} is 44 μm.

The alignment of the isolation mask with respect to the buried layer poses a bit of a problem because all alignment marks are removed when the surface is etched prior to the growth of the epitaxial layer. This necessitates performing the alignment under infrared illumination because the arsenic-doped buried layer exhibits several known attenuation peaks in the infrared band below 10 μm.[11]

A final point in connection with the isolation-diffusion mask is that the windows opened by this mask are reopened during the base diffusion in order to increase the p-impurity concentration on the surface. This is necessary because it is possible for the surface to become depleted of impurities as a result of impurity redistribution during the oxidations as discussed in Section 5.6.4.

7.4.4 Deep-Collector-to-Isolation Spacing

A deep-collector diffusion is used to reduce the resistance of the conduction path between the external-collector contact and the active-collector region. It is particularly useful for reducing the ON-state output voltage of saturating-type digital logic circuits, and making this voltage less dependent on the load current.

The deep-collector diffusion is another deep diffusion which penetrates the epitaxial layer to contact the underlying buried layer. Its profile is assumed to be Gaussian with a sheet resistance of typically 5 $\Omega/\square$. It is produced by a two-step diffusion process with a long drive-in. The average conductivity corresponding to the 5-$\Omega/\square$ sheet resistance R_s and a 20% overdrive, i.e., a profile that has a impurity concentration equal to that of the epitaxial layer at a depth of 1.2 times the epitaxial layer thickness

$$\overline{\sigma} = \frac{1}{(1.2 \times \text{thickness})R_s} = 303 \ (\Omega\text{-cm})^{-1}.$$

From Fig. 7.6 the surface impurity concentration is found to be $N_s = 6 \times 10^{19}$ cm^{-3}.[†] The appropriate diffusion times are determined by a procedure analogous to that for the isolation diffusion and will not be derived here.

The lateral-diffusion distance associated with the 6.6-μm deep-collector diffusion is $0.8 \times 6.6 \approx 5.3$ μm. In Fig. 7.6 the relationship between the

---

[†] Use of these curves yields results slightly in error because they are derived for a background of the opposite conductivity type which is not the case here.

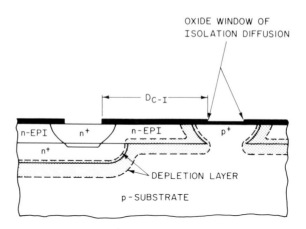

Fig. 7.6 Relationship between deep-collector and isolation diffusions.

deep-collector diffusion and the isolation diffusion is illustrated. The deep-collector diffusion mask is aligned with respect to the oxide window of the isolation diffusion. The minimum spacing between features on the deep-collector diffusion mask and the isolation-diffusion mask must be chosen such that the deep-collector diffusion can make good contact to the buried layer while avoiding the formation of a *p-n* junction between heavily doped layers. Such a junction would have a large leakage current which is undesirable. Note that there is no depletion layer associated with the deep-collector diffusion because it is surrounded by *n*-type epitaxial material. These considerations yield the following expression for D_{C-I}, the minimum deep-collector-to-isolation-mask spacing

$$D_{C-I} = \text{deep-collector lateral diffusion}$$

$$+ \text{ isolation lateral diffusion}$$

$$+ \text{ 10-V depletion layer in 0.5 } \Omega\text{-cm } n\text{-epi}$$

$$+ \text{ mask-alignment tolerance}$$

$$+ \text{ worst-case minimum spacing.} \qquad (7.6)$$

EXAMPLE 7.4   Compute the deep-collector diffusion to isolation-diffusion feature separation for the illustrative process.

Solution.   From Eq. (7.7)

$$D_{C-I} = 5.3 + 5.6 + 1.5 + 4.5 + 1 = 17.9 \ \mu\text{m}.$$

In future examples this value will be rounded up to 18 μm.

7.4.5 Base-Diffusion Spacings

In this section the minimum spacing between features on the base-diffusion mask and the isolation-diffusion and deep-collector diffusion masks are derived. These relationships are illustrated in Fig. 7.7 in terms of distance D_{C-B} and D_{I-B}. The reasoning leading to the design rules governing the base diffusion is essentially the same as in the previous cases. A feature on the base-diffusion mask is positioned with respect to a feature on the isolation-diffusion mask. The deep-collector diffusion features are also positioned with respect to the isolation-diffusion features by an independent process. Therefore the deep-collector-to-base spacing involves the tolerance for two mask alignments.

A 200-$\Omega/\square$ base region that is 2.5 μm thick has an average conductivity $\bar{\sigma} = 1/R_s t = 20 \ (\Omega\text{-cm})^{-1}$, where the nominal depth value of 2.5 μm has been used. The surface impurity concentration, N_S, of a p-type Gaussian diffused layer with this average conductivity in an n-type background with an impurity concentration $N_B \approx 10^{16} \ \text{cm}^{-3}$ is $N_S = 4.0 \times 10^{18} \ \text{cm}^{-3}$. Hence $N_B/N_S = 2.5 \times 10^{-3}$. At a reverse bias of 10 V, $V/N_B \approx 10^{-15}$ and so the total depletion-layer thickness from the Lawrence and Warner curves is 1.4 μm with about a quarter of the depletion layer, that is, about 0.4 μm in the base material. The lateral diffusion is $0.8 \times 2.5 = 2.0 \ \mu$m.

The minimum worst-case mask window spacings are therefore:

$D_{I-B} =$ isolation lateral diffusion

+ depletion layer of isolation junction in epi

+ base lateral diffusion

+ collector-base depletion-layer component in epi

+ mask-alignment tolerance (one mask)

+ minimum worst-case junction spacing.   (7.7)

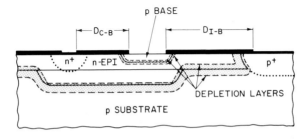

Fig. 7.7 Positioning of the base diffusion with respect to the deep-collector and isolation diffusions.

D_{C-B} = base lateral diffusion

+ collector-base depletion-layer component in epi layer

+ deep-collector lateral diffusion

+ tolerance for two mask alignments

+ minimum worst-case spacing.   (7.8)

The isolation-to-base spacing might be increased above the minimum value calculated using Eq. (7.7) because of pin holes in the isolation mask which could introduce local p-regions that violate the spacing required by this equation. Increasing this spacing by an experimentally determined amount increases the yield.

EXAMPLE 7.5   Calculate the isolation-to-base and deep-collector-to-base spacings for the illustrative process.

Solution.   From Eq. (7.7),

$$D_{I-B} = 5.6 + 1.5 + 2.4 + 1 + 4.5 + 1 = 16 \ \mu m,$$

and from Eq. (7.8),

$$D_{C-B} = 2.4 + 1 + 5.3 + 5.5 + 1 = 15.2 \ \mu m.$$

This latter number will be increased to 16 μm.

The minimum spacing between two windows on the base-diffusion mask corresponding to two base regions on a common buried layer is important in the design of diffused resistors and common collector transistors. It is evident after a little thought that

$D_{B-B} = 2 \times \{$base-lateral diffusion

+ depletion-layer component in epi layer

+ feature-edge uncertainty$\}$

+ minimum separation.   (7.9)

No mask-alignment tolerance is necessary, because all features are on the same mask, but the feature-edge uncertainty must be included.

7.4.6   Emitter Diffusion-to-Base-Diffusion Spacing

The emitter diffusion has a complementary-error-function profile with a nominal depth of 1.5 μm. It is nested inside the base diffusion. The emitter-diffusion mask is aligned with respect to the base-diffusion mask.

Using reasoning analogous to that used in previous cases, the minimum spacing between the emitter-mask window to base-mask window is found

with the aid of Fig. 7.8 to be

D_{E-B} = emitter-lateral diffusion

+ emitter-base depletion layer in p-region

− base-lateral diffusion

+ component of collector base depletion layer in p-region

+ mask-alignment tolerance

+ minimum worst-case junction spacing.   (7.10)

Using a background concentration of about 10^{17} cm^{-3} at the emitter-base junction, $V/N_B \approx 5 \times 10^{-17}$ for this junction. The depletion-layer width is about a 0.25 μm. Hence the calculated minimum emitter-base spacing is 5.2 μm ($\approx$6 μm).

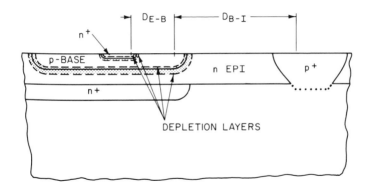

Fig. 7.8 Illustrating the relationship between the emitter and base-mask diffusion features.

7.4.7 The Emitter Contact

Reregistration of the emitter contact means that one mask is used to define the windows for the emitter diffusion and another mask is used to open contact windows that must always be within the emitter region. This contact mask also opens the contacts to the base and collector regions. If the final, minimum worst-case separation between the emitter-contact window and the edge of the emitter diffusion is taken to be 1 μm, then with reference to Fig. 7.9 and on the assumption that the contact mask is aligned with respect to the emitter-diffusion oxide feature, the smallest emitter-to-contact-window spacing guaranteeing the nonshorting contact is

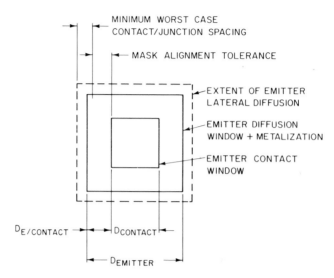

Fig. 7.9 Placement of the emitter-contact window.

$$D_{E/\text{contact}} = \text{minimum worst-case contact-junction spacing}$$

$$+ \text{ mask-alignment tolerance}$$

$$- \text{ emitter-lateral diffusion.} \qquad (7.11)$$

The minimum emitter-to-contact spacing is 4.3 μm ($\approx$5 μm).

7.4.8 Placement of the Base Contact

Figure 7.10 is a top and side view of the base region of a transistor showing the relationship between the base contact, the base diffusion, and the edge of the emitter diffusion. The minimum (worst-case) distance between the base-contact feature on the contact mask and the edge of the base-diffusion-mask feature will now be calculated. All of the contact windows are on the same mask, which is assumed to be aligned with respect to the emitter-diffusion features. Therefore the desired result obtained by inspection of Fig. 7.10 is

$$D_{\text{base/base contact}} = \text{minimum contact-junction spacing}$$

$$+ \text{ component of collector-base depletion layer in base}$$

$$- \text{ base-lateral diffusion}$$

$$+ \text{ tolerance for two mask alignments.} \qquad (7.12)$$

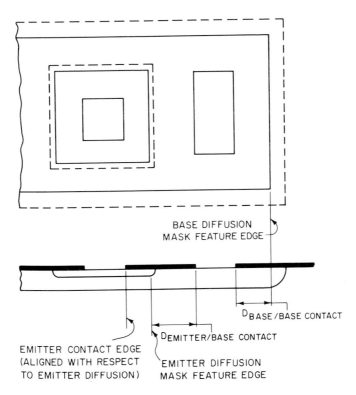

Fig. 7.10 Illustrating the relationship between the base contact and the base and emitter diffusions.

From this figure the following expression for $D_{\text{emitter/base contact}}$, the minimum worst-case distance between the base contact feature on the contact mask and the edge of the emitter-diffusion mask feature, can be derived

$D_{\text{emitter/base contact}}$ = emitter-lateral diffusion

+ component of emitter-base-depletion layer in p—region

+ mask-alignment tolerance

+ minimum contact-junction spacing.   (7.13)

These equations specify the position of the base contact. Using the values appropriate to the illustrative example, these spacings are $D_{\text{base/base contact}} = 5\ \mu\text{m}$, and $D_{\text{emitter/base contact}} = 7\ \mu\text{m}$.

7.4.9 Contacts in the Collector Region

A window is opened in the collector-contact region prior to the emitter diffusion in order to increase the impurity concentration at the surface. The minimum spacing between the collector-contact window on the emitter-diffusion mask and the base-diffusion-mask feature is less than the deep-collector-to-base-diffusion spacing because the emitter diffusion is so shallow. Therefore the deep-collector diffusion-mask features can be reopened with the emitter-diffusion mask with no further consideration of possible consequence. Finally, note that because of the large deep-collector lateral-diffusion distance and the small lateral-diffusion distance associated with the emitter diffusion, the collector-contact window on the emitter-diffusion mask can be 1 or 2 μm larger than the deep-collector diffusion window.

The results of this section are summarized in Table 7.4.

Table 7.4 Minimum Design Spacings

	Microns
Buried Layer to Isolation	17
Deep Collector to Isolation	18
Base Diffusion to Isolation	16
Deep-Collector-to-Base Diffusion	16
Emitter Diffusion to Base Diuffsion	6
Emitter Diffusion to Emitter Contact	5
Emitter Diffusion to Base Contact	7
Base Diffusion to Base Contact	5
Minimum Metal Linewidth	10
Minimum Metal-to-Metal Spacing for Short Runs	8
Minimum Feature Size	8×8
Emitter-Emitter Base-Base	10

7.5 THE MINIMUM-AREA TRANSISTOR

The minimum-layout spacings derived in the preceding section are now used in the layout of a minimum-area transistor and two transistors having a common collector terminal. The integrated-circuit transistor consists of a sequence of nested features, the smallest being the emitter contact. The emitter-contact dimensions are chosen to be the minimum obtainable mask-feature size of 8 μm by 8 μm so as to achieve a minimum-size transistor.

An emitter contact of this size, which results in an 18-μm by 18-μm emitter, will yield a transistor with a peak in the β-versus-collector-current characteristic at about 1 to 2 mA.

The process-specification sheet given in Table 7.2 specifies that the emitter contact must be reregistered inside the emitter diffusion. According to the derived design rules this requires a frame 5 μm wide around the contact. Hence the emitter-diffusion-mask feature is 18 μm by 18 μm. No metal overlap is required over contact windows. However, this does not mean that the metallization for the emitter connection can be 8 μm wide because there is a requirement that all contacts be sealed, i.e., that bare silicon should not be exposed. The mask-alignment tolerance is 4.5 μm, and it is necessary to provide a border of metal of at least this width around the contact to account for any mask misalignment between the emitter-contact mask and the metal mask. The border width in this design is 5 μm.

The emitter is diffused into the base and must not contact the collector. In order to guarantee this condition, the base-diffusion-mask window forms a frame at least 6 μm wide around the emitter-diffusion-mask window. The frame is enlarged on the fourth side so that contact can be made to the base region. It is easy to see that the placement of the base contact is metal limited. A minimum metal-to-metal spacing of 8 μm is permitted for short distances. The inside edge of the base-contact-mask feature must be a minimum of 13 μm from the edge of the emitter-diffusion-mask feature so as to provide the 5-μm metal border to guarantee sealing the base contact, and the 8-μm metal-to-metal spacing. The base contact is an 8-μm wide strip that is 5 μm away from the edge of the base diffusion. These features are clearly shown in Fig. 7.11.

The collector region is contacted by means of a deep-collector diffusion to the buried layer. It can be seen from the design rules of Table 7.2 that the contact placement is determined by the deep-collector-to-base spacing, and is not metal limited.

Finally, the isolation diffusion is placed around the transistor at the minimum distance of 17 μm from the buried layer. The transistor area, exclusive of the isolation band, is now 64 μm by 109 μm.

EXAMPLE 7.6   Perform the layout of two minimum-area transistors on a common collector.

Solution.   Both transistors are on the same buried layer and in the same isolated region because of the common collector. The base-emitter regions are identical to that of the minimum-area transistor which was just designed. The minimum spacing between the base-diffusion-mask features is determined by the base-lateral diffusion, collector-base depletion layer, and a minimum worst-case spacing.

Figure 7.12 is the resultant layout from which it can be seen that the area of this transistor pair is about 1.5 times that of the single transistor.

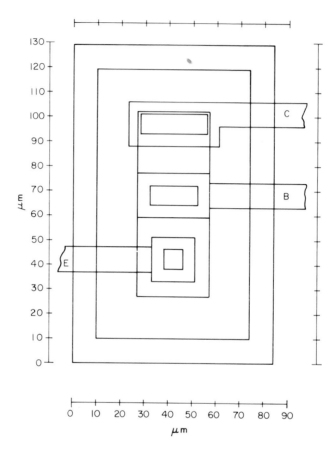

Fig. 7.11 Layout of a minimum-area transistor according to the design rules of Table 7.2.

7.6 LAYOUT OF SILICON INTEGRATED CIRCUITS

Integrated-circuit layout is at least as much art as it is science, and there is no unique solution to any layout problem. In general, one tries to find a topological arrangement that occupies the smallest area, has a favorable aspect ratio, requires the smallest number of diffused crossunders, and takes advantage of any symmetries in the circuit.

The set-reset flip-flop circuit shown in the circuit diagram of Fig. 7.13 will be used to illustrate some particulars of the layout problem. This circuit is complex enough so that a good layout is not immediately apparent, yet it

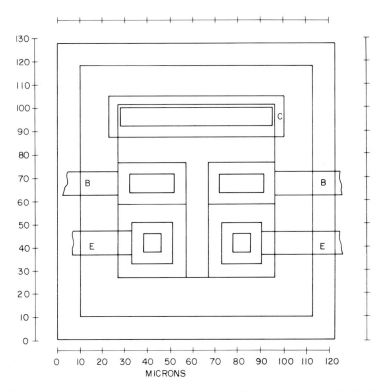

Fig. 7.12 Layout of two minimum-area transistors sharing a common buried layer.

is simple enough to be a tractable example. This example consists of a series of sketches and a written explanation that was generated while actually performing the layout, followed by a scale drawing of the final result.

The design rules of Table 7.2 are used for this layout, and it is assumed that the current levels are such that minimum emitter-area transistors can be used.

7.6.1 Estimate of the Chip Area

Before doing any layout, it is possible to get an idea of the area required for the integrated circuit. The circuit requires two single transistors, two common collector-transistor pairs, two double-emitter transistors, and 44 kΩ of resistors. In addition, there are seven terminals that must be brought out to bonding pads at the periphery of the chip. It is entirely possible that with such a small circuit the chip size may be determined by the requirements for the bonding pads, i.e., the chip size may be terminal limited.

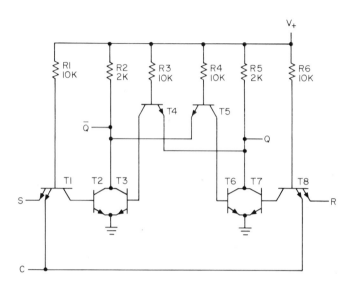

Fig. 7.13 Circuit diagram of a clocked set-reset flip-flop.

Assume initially that the components can be optimally packed so that there is no wasted space. From Figs. 7.11 and 7.12 it is found that each single-emitter transistor, including half of the isolation band, requires 74×119 μm and the common collector transistors require 114×119 μm. The double-emitter transistor will be the same width as the single-emitter transistor but its length will be increased by 25 μm to account for the extra emitter and the metal-to-metal space. Hence, the total area required by the transistors is 66,000 μm^2. The resistors are made from 200-$\Omega/\square$ base material. Slightly less than 220 squares are required because the contacts add some resistance. If 210 squares are used and the stripes are 10-μm wide with 10-μm spaces, an area of 42,000 μm^2 are consumed. The minimum area for a contact is 18 μm by 18 μm. Therefore the minimum active area for this circuit is 112,000 μm^2, which does not include isolation for the resistor region or make allowances for any metallization requirements.

In a large circuit, the active area tends to be about 30% to 40% above this minimum. For this small circuit, the active area is probably 60% to 70% above the minimum. Therefore, if the circuit is square, it would be about 425-μm on a side.

It must now be determined if seven bonding pads can be arrayed outside of a 425- by 425-μm active rectangle. Bonding pads are typically 125 μm on a side and about 100-μm separation is required between pads. From Fig. 7.14, it can be seen that eight bonding pads can fit comfortably around a 425-μm by 425-μm active area. Probably two more pads could be placed easily before the chip would have to be enlarged. Thus, simple circuits like multiple-input gates are terminal limited rather than active-area limited.

7.6.2 Topological Arrangement of Circuits

The object of this phase of the design is to redraw the circuit diagram of Fig. 7.13 in a manner that will place the actual devices and their physical placement in better perspective. At this stage of the design, the physical sizes of the various components are of no consequence.

A circuit of any reasonable complexity inevitably requires crossovers because it is nonplanar. In the absence of two-level metallization, the crossover is accomplished, if possible, by utilizing a resistor or the base or collec-

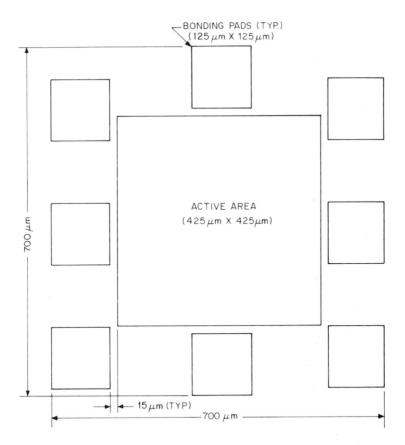

Fig. 7.14 The active area and array of bonding pads for illustrative layout example.

tor of a transistor which happens to be in series with one of the leads so that one of the crossing elements is under the silicon surface. Sometimes it is necessary to use an emitter-diffusion crossunder, i.e., it is necessary to introduce a low-valued resistor in series with one of the leads to take the conduction path beneath the surface.

Figure 7.15(a) is the original circuit, redrawn to group elements that share common isolated regions. Transistors with common collectors are in the same isolated region. Similarly, all resistors are placed on a single buried layer in an isolated region. Note that the crosscoupling of the transistors

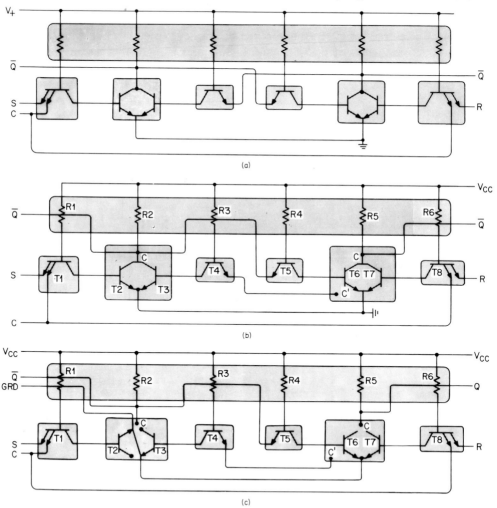

Fig. 7.15 (a) Circuit of RS flip-flop redrawn to show grouping of circuit elements. (b) Alleviating the cross-coupling problem. (c) Opening a path for the trapped ground line.

must be redrawn to make it planar and the externally accessible points of the circuit must be brought to the periphery of the chip. This presents several crossover situations which must be resolved.

In Fig. 7.15(b) the crosscoupling situation is resolved by noting that the lead from the collectors of T_2, T_3 to the emitter of T_5 can go over the body of resistor R_3, and the lead from the emitter of T_4 to the collectors of T_6, T_7 can be connected to any part of the buried layer. Therefore, the buried layer of this transistor pair can be extended to accommodate a second deep-collector diffusion and contact. Note also that outputs Q and $\overline{Q}$ can be brought to the chip periphery by crossing the bodies of resistors R_1 and R_6. However, the ground line is trapped inside the clock line.

Figure 7.15(c) shows a way to bring the ground line out over the body of resistor R_1 by recalling that access to the active collector region of the transistor pair T_2, T_3 is through the buried layer which, for this purpose, can be regarded as an equipotential surface. Therefore, transistor T_2 can be rotated by 180° to open a path for the ground lead.

This design looks promising enough to begin putting in more details of the various devices. The features of the transistors and isolation are sketched but the metallization is represented as single lines. The double-emitter transistors have the emitters side by side facing the base contact. The sketch of the tentative layout is shown in Fig. 7.15(d). Some of the

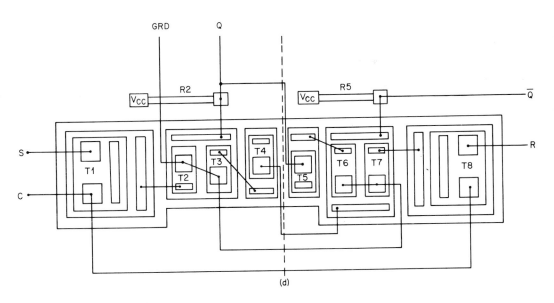

Fig. 7.15 Contd. (d) Tentative layout

resistors are also included, but this approach was abandoned when a quick size computation based on numbers obtained from Figs. 7.11 and 7.12 showed that the width of this layout would be over 600 μm and its height would be about 300 μm. A 2:1 aspect ratio would be desirable if two flip-flops were being put on one chip, but in the present example a nearly square design is preferred.

This layout is not a total waste of effort because it does show that all of the connections can be made without introducing any diffused crossunders. A more desirable layout might be obtained by observing that (1) the layout is almost symmetrical about a vertical centerline between the two small, single-emitter transistors, and (2) there are long metal runs for the clock inputs and the ground line. If the layout is cleaved along this vertical center-line and the left-hand section is rotated underneath the right-hand section, the transistor package is more nearly square and the clock terminals are adjacent. The resistors are on a common buried layer in an isolated region that surrounds this transistor package. The resultant layout shown in Fig. 7.15(e) appears to be almost square.

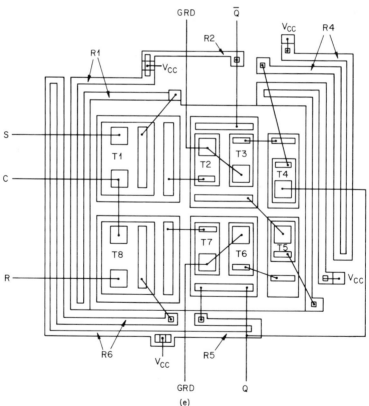

Fig. 7.15 Contd. (e) Revised layout with square aspect ratio.

The problems with this arrangement are that the resistor buried layer is quite narrow in places which can lead to excessive voltage drops in the buried layer if the current is large, and that there are two ground lines which must be connected via a ring of metallization around the periphery of the chip.

We begin the new topological arrangement by again placing the multiemitter transistors T_1 and T_8 in adjacent positions to minimize the length of the common-clock line. The R, S, and C leads can easily be brought to bonding pads at the bottom of the chip. The base of each of these transistors is connected to a resistor on the common resistor buried layer. This buried layer will be above transistors T_1 and T_8 because the region below the transistors is devoted to metallic connections to the outside world as shown in Fig. 7.15(f). Transistors T_4 and T_5 are also connected to base resistors in the common resistor region, and like transistors T_1 and T_8 are the input transistors for a TTL NAND gate. We will therefore try to place these transistors on either side of, and in the same orientation as, T_1 and T_8. The common-collector transistor pairs T_2, T_3 and T_6, T_7 will be placed above the input transistors. The base contacts for these common-collector transistors are on the bottom to facilitate connection to the collector regions of the input transistors.

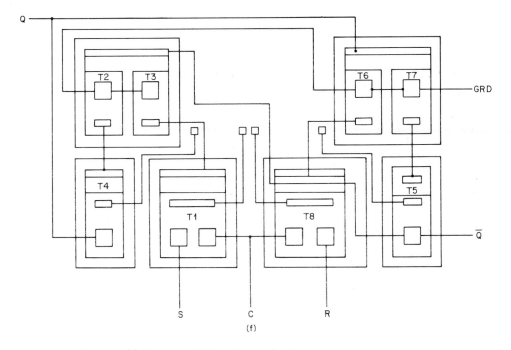

Fig. 7.15 Contd. (f) Layout with broad sea of resistors.

The complete drawing for this topological arrangement, including metallization, is shown in Fig. 7.16. Note that the resistors are used as diffused crossunders in several situations. In particular, the cross-coupling is accomplished by taking the lead from T_2, T_3 across several resistors and over an extended region between the base-emitter region of T_8 and its collector contact. The cross-coupling lead from T_6, T_7 is taken around the outside of the chip. The output connections can be made anywhere along these cross-coupling leads. Particularly convenient places appear to be the upper left-hand corner of the chip for a connection to the lead from T_6, T_7 which is the Q output, and the emitter terminal of T_5 for the $\overline{Q}$ output. The ground lead from T_2, T_3 is taken through several resistors including the collector-load resistor R_2 to connect with the ground at T_6, T_7.

The transistor dimensions given in Fig. 7.11, will be used together with the following: (a) an 8-μm emitter-to-emitter spacing for the multiemitter transistors; (b) the metal rules given in Table 7.3; and (c) a 10-μm wide band of isolation around the entire active area of the chip. It then follows that the minimum width of the active region is 372 μm (Problem 7.4). When the active region has this minimum width, it is also easily seen that the width of the resistor buried layer is 110 μm. Similarly, the minimum height for the active region as dictated by the transistors is 263 μm. However, this only provides a 110-μm by 75-μm buried layer for the resistor region which is far below the area required for optimally packed resistors. This means that the resistor region will expand upward into a T-shaped region.

7.7 COMPUTER AIDS FOR INTEGRATED-CIRCUIT LAYOUT

Computer aids for integrated-circuit layout can be placed in two broad categories: namely, drafting aids and automated design aids. It was recognized long ago that the generation by hand of the final artwork for the master masks, and cutting and stripping of a laminated mylar material known as Rubylith, would not be practical for patterns of the complexity required for large-scale silicon integrated circuits. Computer-controlled drafting machines quickly appeared in order to expedite this job. A powerful, user-oriented design language called XYMASK was written at Bell Laboratories. It enables the designer to use simple, English-like statements to specify geometric features. The XYMASK input data is converted by the program into the machine-language instructions required to drive the drafting machine. The use of this type of preprocessor program has the added advantage of buffering the circuit designer from the programming peculiarities of the particular artwork generator.

Part of the XYMASK coding for the flip-flop designed in the last section is shown in Table 7.5. Each statement specifies a geometric shape, the

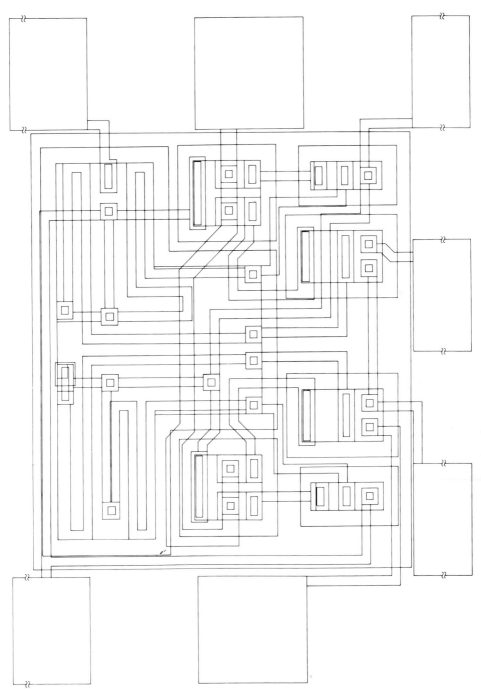

Fig. 7.16 Final drawing for layout of RS flip-flop.

Table 7.5 Section of XYMASK Code for Flip-Flop

EMIT	CLUMP	0,0	
RECT	N8	0,0	18,18
RECT	N21	5,5	13,13
RECT	N19	0,0	18,18
END	CLUMP		
TRANS	CLUMP	0,0	
RECT	N3	90,59	120,134
INC	EMIT	96,65	
RECT	N2	90,59	120,109
RECT	N21	95,96	115,104
RECT	N19	90,91	120,109
RECT	N5	95,125	115,133
RECT	N8	93,123	117,135
RECT	N21	95,125	115,133
RECT	N19	90,120	120,139
END	CLUMP		
ISOLATION	CLUMP	0,0	
BLOB	N10	28,27 73,27 73,151 137,42	
ETC	73,42 73,27 147,27 147,166 237,166		
ETC	237,42 147,42 147,27 277,27		
ETC	277,176 28,176		
END	CLUMP		
FLIPFLOP	CLUMP	0,0	
INC	ISOLATION	0,0	
REPEAT	TRANS	0,0,2,344,0	
PATH	N10	10 63,290 172,290 172,176	
BLOB	N10	28,295 43,295 43,437 481,437 481,285	
ETC	483,285 483,176 497,176 496,452 28,452		

mask on which it is to appear, and point coordinate pairs to describe its location. Thus the word *RECT* denotes a rectangular shape and the designation *N2* assigns this feature to the base-diffusion mask. The coordinate points (0,0) and (30,50) are located at diagonally opposite points of the rectangular feature. A feature which constitutes a path is denoted by the word *PATH*. The first numerical entry is the path width followed by coordinate pairs denoting the breakpoints of the centerline of the path. A *CLUMP* is an associated group of features. It is given a name to make subsequent reference possible. For example, the single-emitter transistor is called *TRAN* in this coding. This group of features can be placed as an entity in a more complex layout, by means of the include statement *INC.* This hierarchical structure clearly makes this language very powerful.

The next step in the evolution towards a completely automated layout capability is a computer-based, interactive design and drafting aid. In principle, the designer never has to make any drawings or do any coding by hand, although it is advisable to do all of the topological arrangement steps described in the previous section before beginning an interactive session with the machine. This type of machine accepts inputs from a keyboard and a digitizer-plotter (which is a drafting machine that contains digital-position servos so that it can be precisely positioned by the computer) and displays outputs either on a CRT screen or on the digitizer-plotter. In addition, it has an analog-to-digital converter which is used to report the pen position to the computer when the digitizer-plotter is used as an input device.

Geometric information, generated either from the keyboard or by digitizing points with the digitizer-plotter, is stored in the computer data base. It can be manipulated to move and/or modify the features. Data listings for the designer are available in a language such as XYMASK. When a satisfactory design is obtained, the information in the data base is used to generate the instructions for the appropriate artwork generator.

The ultimate goal is to have a computer-based machine that will accept as input the circuit diagram and specifications from which it will produce a verified layout. The furthest advances in design automation have come in the layout of digital MOS integrated circuits. This automated design capability has made it possible to utilize the LSI potential of MOS processes for custom digital-logic applications, where the quantities involved would not have been sufficient to justify the design costs without design automation. Automation has enabled many manufacturers to use custom MOS logic in their products.

A computer-aided design (CAD) system for MOS artwork was described by Mattison.[12] It is typical of CAD aids for MOS layout.

Custom digital logic is made from a set of basic building blocks known as polycells. These polycells range in complexity from NAND and NOR gates, to exclusive-OR gates, flip-flops, adders, shift registers, and counters. Each circuit has been characterized and its layout is coded and stored in the data base of the computer. Computer programs are available to simulate the behavior of a custom circuit made by interconnecting polycells, and to verify the sufficiency of the tests used to ascertain that the finished chip is functioning properly.

All polycells in the set have the same height, and their width is an integral number of unit widths. The cell layout is such that power and ground leads are carried through the cell on common spacings. The three operations required for automated layout using these polycells are the determination of relative placement of the cells, assignment of actual cell positions which is known as chip building, and the interconnection of the polycells. The arrangement of a polycell chip is shown in Fig. 7.17. The cells are arrayed, back-to-back, in rows. Between pairs of cell rows are spaces called interconnect channels. Conductors on two separate levels are required to permit arbitrary placement of vertical and horizontal conductor segments. The impedance level of MOS circuits is very high and therefore a reasonable amount of lead resistance can be tolerated. This permits the use of diffused conductors in the vertical direction and the metallization as horizontal conductors.

The placement algorithm analyzes the connections between polycells and the sizes of the polycells to determine the number of rows of cells required to yield a square chip, the placement of cells within these rows, and the orientation of the rows to form the interconnect channels. Relative cell-placement assignments are converted into actual placements and a cell maze is constructed for the wiring router according to the layout design rules during the chip-building phase. The circuit designer may intervene at this stage to manually optimize the placement. A major consumer of space in the interconnect channels are leads that must go from one channel to another around the outside of a row of cells. Thus, it is important to place the cells such that the cells within a given row are heavily connected and with only sparse connection required to other rows. When connections between rows are required, special feedthrough cells are inserted in the intervening row to break the connection net into two subnets each of which is confined to only one channel. It is important that the chip-building routine be programmed

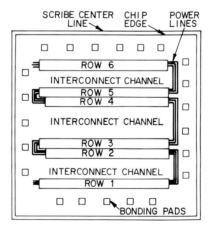

Fig. 7.17 The arrangement of polycells in back-to-back rows separated by interconnect channels.

to recognize that when assigning terminals to nets, the inputs for logic gates are interchangeable and that the mirror image of a polycell is allowed. This reduces the number of crossovers required and saves space.

The interconnection routine is based on a line-routing algorithm due to Hightower[13] which does not guarantee to find the shortest path available or, in fact, to find any path at all. It is extremely fast, with an accuracy limited only by the capability to store small numbers, and the operating area is bounded solely by the largest number that the computer can store. This algorithm finds a path between two points A and B as shown in Fig. 7.18 by first constructing perpendicular line segments through point A extending in each direction from boundary to boundary. An escape point is sought such that a line constructed through this escape point will extend beyond the previous boundaries of point A. If such an escape point is found, it defines a new point A and the process is repeated.

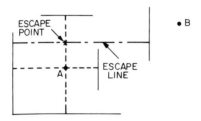

Fig. 7.18 Illustrating the method used by the Hightower algorithm to find a path from point A to point B.

Comparison of automatic versus manual polycell layout for eight circuits, as given by Mattison, is shown in Table 7.5. The automated layout typically requires about 10% more area than a corresponding manual layout.

Table 7.6 Comparison of Automatic and Manual Layout Techniques

	Number of Cells	Number of Point Pairs	Dimensions (mils)		Number of Contacts	
			Manual	Auto	Manual	Auto
A	142	359	108×126	110×123	379	361
B	68	166	91×117	109×117	284	242
C	117	289	149×160	171×151	451	414
D	142	275	145×145	128×141	346	381
E	131	329	157×167	161×174	603	491

However, automation does reduce layout time and design costs considerably. A 10% area penalty is a small price to pay on a custom-designed chip produced in small quantities. If, and when, production quantities warrant it, the design can be modified by hand to reduce the chip area.

REFERENCES

1. L. I. Maissel and R. Glang. *Handbook of Thin Film Technology.* New York: McGraw-Hill, 1970, Chapter 7.

2. G. R. Brewer. "Electron and ion beams in microelectronic fabrication process, Part 1." *Solid-State Technology* **15** (July 1972).

3. G. R. Brewer. "Electron and ion beams in microelectronic fabrication process, Part 2." *Solid-State Technology* **15** (August 1972).

4. A. B. El-Kareh and J. C. V. El-Kareh. *Electron Beams, Lenses and Optics.* New York: Academic Press, 1970, Volumes 1 and 2.

5. G. R. Brewer. "The application of electron/ion beam technology to microelectronics." *IEEE Spectrum* **8**: 23 (1971).

6. H. S. Yourke and E. V. Weber. "A high-throughput scanning-electron-beam lithography system, EL1, for semiconductor manufacture: general description." *IEEE International Electron Devices Meeting*, Washington, D. C., December 6-8, 1976.

7. L. F. Thompson, J. P. Ballantyne, and E. D. Feit. "Molecular parameters and lithographic performance of poly(glycrdyl methacrylate-co-ethyl acrylate): A negative electron resist." *J. Vac. Sci. Technol.* **12**: 1280-1283 (November/December 1975).

8. M. J. Bowden, L. F. Thompson, and J. P. Ballantyne. "Poly(butene-1 sulfone)—a highly sensitive positive resist. *J. Vac. Sci. Technol.* **12**: 1294-1296 (November/December 1975).

9. J. C. Irvin. "Resistivity of bulk silicon and of diffused layers in silicon." *Bell Syst. Tech. J.* **41**: 387-410 (March 1962).

10. H. Lawrence and R. M. Warner, Jr. "Diffused junction depletion layer calculations." *Bell Syst. Tech. J.* **41**: 389-404 (March 1960).

11. W. Spitzer and H. Y. Fan. "Infrared absorption in n-type silicon." *Phys. Rev.* **108**: 268-271 (1957).

12. R. L. Mattison. "Design automation of MOS artwork." *Computer* **7**: 21-27 (January 1974).

13. D. W. Hightower. "The interconnection problem: a tutorial." *Computer* **7**: 18-32 (April 1974).

PROBLEMS

7.1 Calculate the minimum design spacings for the illustrative process mentioned in Section 7.4 if the maximum collector-base voltage is 20 V and the maximum reverse bias on the emitter-base junction is 10 V.

7.2 Calculate minimum design spacings for the following bipolar process that is to be used for linear circuits operating with ± 15-V power supplies.

Substrate: p-type, 10-Ω-cm $\pm$ 20% (111) orientation.

Buried layer: 40-$\Omega/\square$, 6-μm deep.

Epitaxial layer: 5-Ω-cm $\pm$ 10%, 10 $\pm$ 1-μm thick.

Base diffusion: p-type boron diffusion, nominally 200-$\Omega/\square$, 4 $\pm$ 0.5-μm deep.

Emitter diffusion: n-type phosphorus diffusion with n^+ surface concentration, base width is nominally 1 μm.

Isolation diffusion: p^+ boron diffusion, nominally 5-$\Omega/\square$.

7.3 Pass a vertical cutting plane through the middle of the minimum area transistor of Fig. 7.11 and draw the corresponding cross section.

7.4 Calculate the dimensions of the active region of Fig. 7.15(g).

7.5 Do the layout of the AND-OR-INVERT gate shown schematically in Fig. 14.15.

7.6 Do the layout of the J-K flip-flop shown schematically in Fig. 15.6.

Chapter 7 Appendix

This appendix contains two groups of charts that are very useful in integrated circuit design. Figures 7A.1 through 7A.8 are curves of surface concentration versus average concentration for diffused layers in silicon. They are from Irvin[9] and are reprinted with permission from *The Bell System Technical Journal*, Copyright 1962, The American Telephone and Telegraph Co. Figures 7A.9 through 7A.14 are curves of depletion-layer thickness and capacitance per unit area for diffused junctions. They are from Lawrence and Warner[10] and are reprinted with permission from *The Bell System Technical Journal*, Copyright 1960, The American Telephone and Telegraph Co.

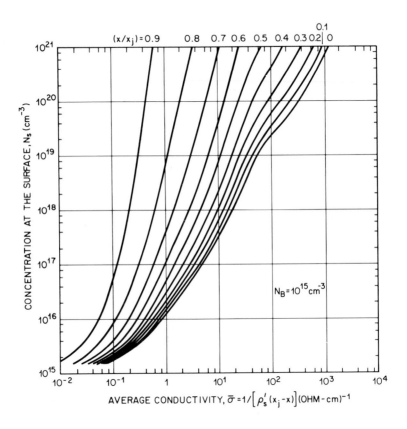

Fig. 7A.1   Surface concentration as a function of average conductivity for n-type Gaussian-diffused layers in silicon for $N_B = 10^{15}$ cm^{-3}.

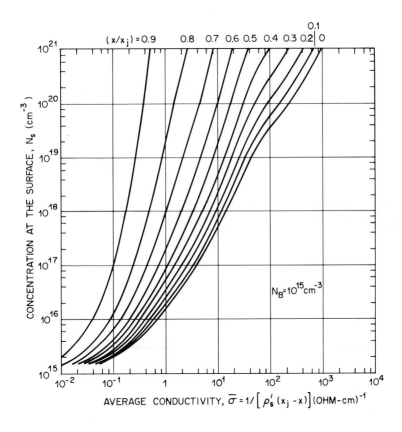

Fig. 7A.2 Surface concentration as a function of average conductivity for n-type complementary error function diffused layers in silicon for $N_B = 10^{15}$ cm^{-3}.

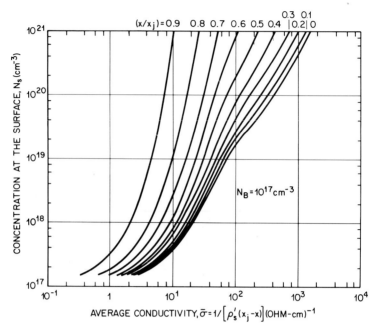

Fig. 7A.3 Surface concentration as a function of average conductivity for n-type Gaussian-diffused layers in silicon for $N_B = 10^{17}$ cm^{-3}.

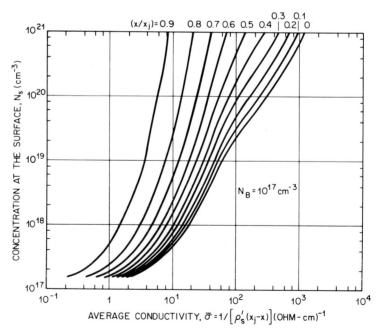

Fig. 7A.4 Surface concentration as a function of average conductivity for n-type complementary error function diffused layers in silicon for $N_B = 10^{17}$ cm^{-3}.

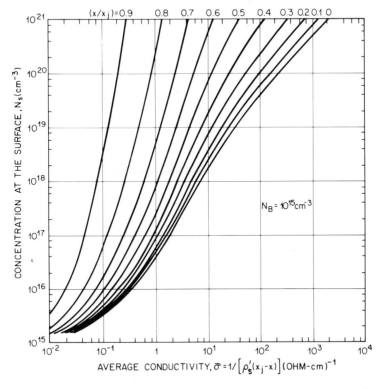

Fig. 7A.5 Surface concentration as a function of average conductivity for p-type Gaussian-diffused layers in silicon for $N_B = 10^{15}$ cm^{-3}.

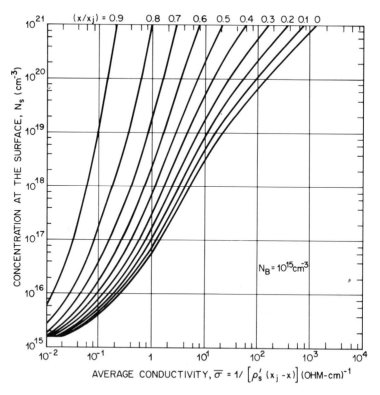

Fig. 7A.6 Surface concentration as a function of average conductivity for p-type complementary error function diffused layers in silicon for $N_B = 10^{15}$ cm^{-3}.

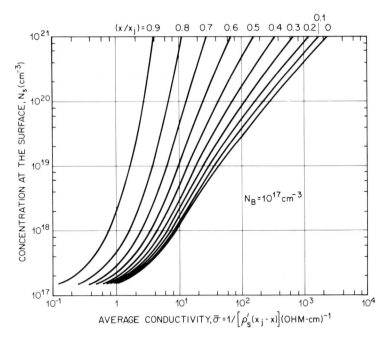

Fig. 7A.7 Surface concentration as a function of average conductivity for p-type Gaussian-diffused layers in silicon for $N_B = 10^{17}$ cm^{-3}.

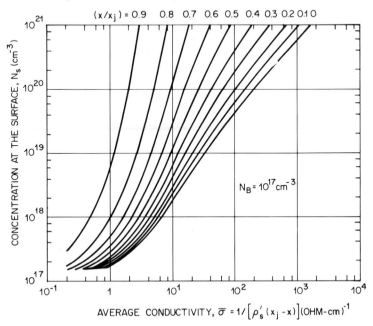

Fig. 7A.8 Surface concentration as a function of average conductivity for p-type complementary error function diffused layers in silicon for $N_B = 10^{17}$ cm^{-3}.

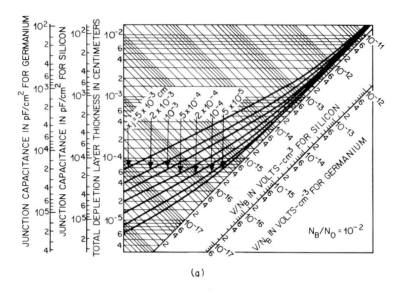

(a)

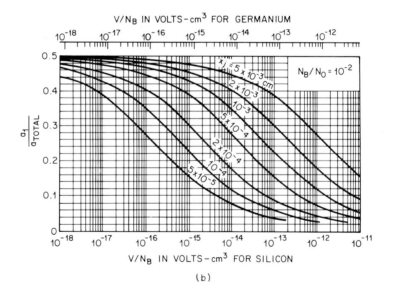

(b)

Fig. 7A.9 (a) Total depletion-layer thickness and capacitance per unit area versus V/N_B with junction depth as a parameter for Gaussian-diffused junctions with N_B/N_0 in the range 3×10^{-3} to 3×10^{-2}. (b) Ratio of depletion layer penetration on the side toward the surface to total depletion thickness as a function of V/N_B.

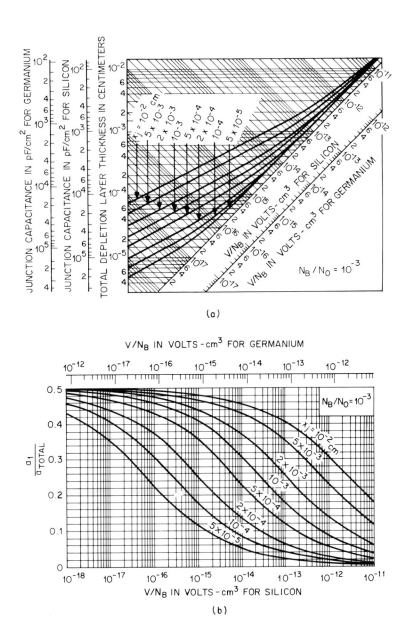

(a)

(b)

Fig. 7A.10 (a) Total depletion-layer thickness and capacitance per unit area versus V/N_B with junction depth as a parameter for Gaussian-diffused junctions with N_B/N_0 in the range 3×10^{-4} to 3×10^{-3}. (b) Ratio of depletion layer penetration on the side toward the surface to total depletion thickness as a function of V/N_B.

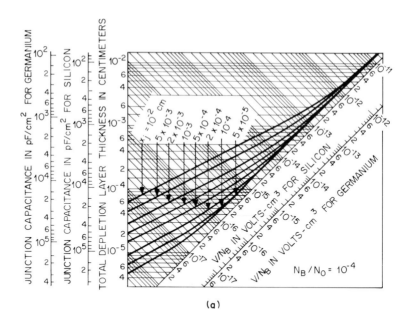

(a)

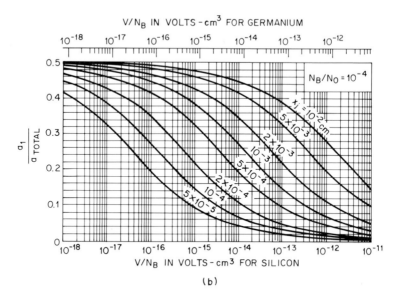

(b)

Fig. 7A.11 (a) Total depletion-layer thickness and capacitance per unit area versus V/N_B with junction depth as a parameter for Gaussian-diffused junctions with N_B/N_0 in the range 3×10^{-5} to 3×10^{-4}. (b) Ratio of depletion layer penetration on the side toward the surface to total depletion thickness as a function of V/N_B.

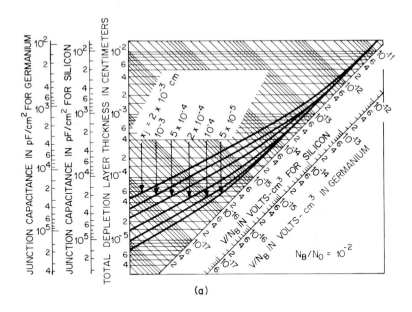

(a)

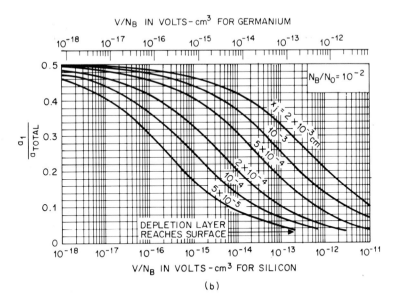

(b)

Fig. 7A.12 (a) Total depletion-layer thickness and capacitance per unit area versus V/N_B with junction depth as a parameter for complementary error function diffused junctions with N_B/N_0 in the range 3×10^{-3} to 3×10^{-2}. (b) Ratio of depletion layer penetration on the side toward the surface to total depletion thickness as a function of V/N_B.

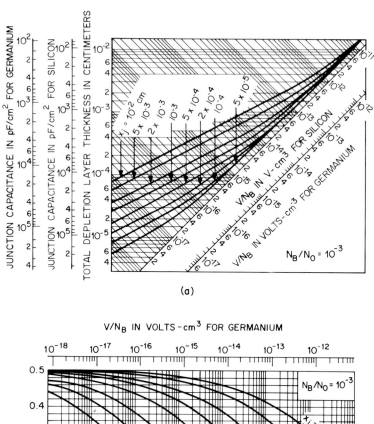

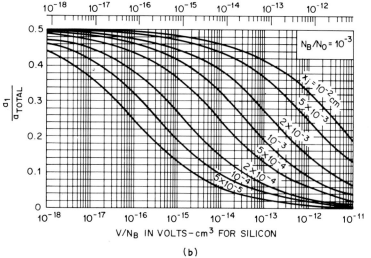

Fig. 7A.13 (a) Total depletion-layer thickness and capacitance per unit area versus V/N_B with junction depth as a parameter for complementary error function diffused junctions with N_B/N_0 in the range 3×10^{-4} to 3×10^{-3}. (b) Ratio of depletion layer penetration on the side toward the surface to total depletion thickness as a function of V/N_B.

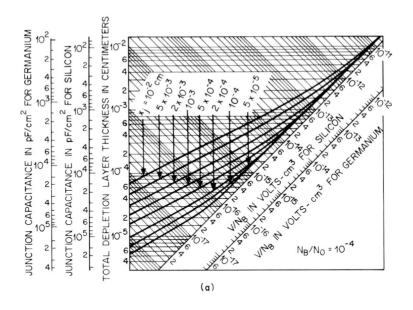

(a)

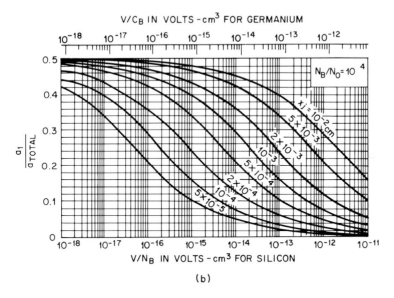

(b)

Fig. 7A.14 (a) Total depletion-layer thickness and capacitance per unit area versus V/N_B with junction depth as a parameter for complementary error function diffused junctions with N_B/N_0 in the range 3×10^{-5} to 3×10^{-4}. (b) Ratio of depletion layer penetration on the side toward the surface to total depletion thickness as a function of V/N_B.

8

Thin-Film Circuits

8.1 INTRODUCTION

This is the first of two chapters on the design of film circuits. Practical film circuits contain only resistors, capacitors, and conductor patterns[†] which is in contrast with silicon integrated circuits considered in the previous chapters. Film circuits are often supplemented by silicon transistors and/or integrated circuits, other semiconductor devices, and appliqued discrete components to form hybrid integrated circuits. In this chapter we consider the design of thin-film circuits. Thick-film circuits are considered in the next chapter.

Thin films are deposited on inert substrates by chemical or physical vapor-deposition techniques, as described in Chapter 5. These films are usually hundreds of angstroms to several microns thick. Conduction in thin films is treated briefly in the next section.

The requirements for the inert substrate are described in Section 8.3. Usually it is of fine-grained, 99+% alumina. Beryllia is sometimes used when increased thermal conductivity is desirable. The lowest-cost practical substrate material is glass.

Section 8.4 is a description of the fabrication sequence for thin-film circuits. The design of resistors and capacitors is described in Section 8.5 and Section 8.6. Conductor materials are considered in Section 8.7. Finally, miscellaneous circuit-design considerations are discussed in Section 8.8.

[†] Inductors with values of several nanohenries which are useful at frequencies of the order of 1 GHz are considered in Chapter 4. Further, silicon circuits on sapphire and spinel substrates are really thin-film circuits but we are classifying them as silicon integrated circuits.

8.2 CONDUCTION IN METAL FILMS [1]

Metal, metal-alloy or conducting transition metal compound films are used for resistors. Single or multilayer metal films are used for conductors. Metals have finite conductivity because the conduction electrons are continually interacting with the crystal lattice. The resistivity of a metal can be expressed in terms of the electron mean-free path λ, which is the average distance traveled by an electron between collisions, as

$$\rho = \frac{h}{2q^2}\left[\frac{3}{\pi n^2}\right]^{1/3}\left(\frac{1}{\lambda}\right), \tag{8.1}$$

where h is Planck's constant, n is the density of conduction electrons, $\lambda = v_f\tau$, where τ is the relaxation time and v_f is the speed of colliding electrons on the Fermi surface. If the crystal lattice was perfectly periodic and there were no crystal defects, the mean-free path would be infinite and the resistivity would be zero. At room temperature, the mean-free path for most metals is about 300 Å because the lattice periodicity is disturbed by phonon scattering, i.e., interactions between electrons and the thermal vibration of the crystal lattice, impurity and defect scattering, and scattering from the surface of a thin sample.

Phonon scattering is the dominant scattering mechanism in pure bulk material. It is due to thermal vibration of the crystal lattice and is therefore temperature dependent. The phonon resistivity is proportional to temperature T when the temperature is above a critical value which is called the Debye temperature. The Debye temperature is less than 300 K for most metals. Therefore, the temperature coefficient of resistivity, defined by

$$\alpha = \frac{1}{\rho}\frac{d\rho}{dT}, \tag{8.2}$$

is simply $\alpha_{ph} = 1/T$ above the Debye temperature, T_D.

The periodicity of the lattice can be disturbed by an impurity or by a defect such as a grain boundary, vacancy, or dislocation. Assuming that the electron scattering due to these phenomena is independent of phonon scattering, and using the relation

$$v_f = \frac{h}{2m}\left[\frac{3n}{\pi}\right]^{1/3},$$

where m is the mass of the electron, the resistivities can be added yielding

$$\rho = \rho_{ph}(T) + \rho_r = \frac{m}{nq^2}\left[\frac{1}{\tau_{ph}} + \frac{1}{\tau_r}\right]. \tag{8.3}$$

The residual resistivity ρ_r due to impurity and defect scattering is almost independent of temperature and therefore $d\rho/dT$ is independent of impurity content, a phenomenon known as Matthiessen's rule. Thus

$$\lim_{T \to 0} \rho = \rho_r.$$

If the sample is extremely thin, the effect of scattering from the boundaries must also be considered. Sondheimer developed a theory for scattering from the boundaries of a thin film by considering an electron with a specified velocity at an arbitrary point in the film, and integrating over all directions for the velocity and all points in the film. He obtained a rather complicated function relating the resistivity ρ of a film of thickness d to the resistivity ρ_∞ of a film of infinite thickness. The two limiting cases for this expression are

$$\rho = \rho_\infty \left[1 + \frac{3\lambda_\infty}{8d} \right] = \frac{mv_f}{nq^2} \left[\frac{1}{\lambda_\infty} + \frac{3}{8d} \right] \tag{8.4a}$$

which occurs when $(d/\lambda_\infty) > 0.1$, and

$$\frac{\rho}{\rho_\infty} = \frac{4\lambda}{3d \left[\ln(\lambda_\infty/d) + 0.4228 \right]} \tag{8.4b}$$

for $(d/\lambda_\infty) < 0.1$, where λ_∞ is the electron mean-free path in a sample of infinite thickness.

The so-called "thin-film" limit for the Sondheimer relation, given in Eq. (8.4b), is not useful in most practical applications because the defect density increases as film thickness decreases, causing a reduction in the mean-free path. The electron mean-free path for phonon scattering in metals is of the order of 100 to 600 Å at room temperature. Therefore, for films whose thickness exceeds about 50 Å, the "thick-film" limit of the Sondheimer expression is appropriate. Most practical films are more than 50 Å thick and are thus not "thin films" in the physicist's sense of the term. In integrated circuit terminology thin films are those films deposited by molecular deposition techniques such as evaporation, sputtering, or electroplating. In the next chapter we consider thick films which are formed by screening the constituents onto the substrate and firing these materials to form the film.

Let ρ_T be the resistivity of a thin film at T K and let ρ_r be the resistivity at 0 K (low temperatures). Then from Eq. (8.1), $\rho_T = C\lambda_{ph}^{-1} + \rho_r$ and from Eqs. (8.3) and (8.4), $\rho_r = C\lambda_0^{-1}(1+3\lambda_0/8d)$. Note that λ_0 is the mean-free path due only to temperature-dependent effects such as impurity and defect scattering. Hence,

$$\lambda_0 = \frac{d}{\left[\dfrac{d\rho_r}{\lambda_{ph}(\rho_T - \rho_r)} - \dfrac{3}{8} \right]} \tag{8.5}$$

and so the "residual" mean-free path λ_0 can be calculated from the film thickness d, the phonon-scattering dependent mean-free path λ_{ph} and from knowledge of the film resistance at $T = 0$ and temperature T. Values of λ_0 may be much greater than the film thickness $(\lambda_0 \gg d)$ implying that the film is of high purity and large crystallite size, that is, composed of plate-like crystallites whose diameter is much larger than the film thickness.

8.3 SUBSTRATES [2,3]

The substrate for most thin-film circuits should be inert, acting simply as a carrier for the film components. A notable exception is the microstrip transmission line, described in Chapter 4, in which the substrate is an integral part of the circuit. The properties of practical substrates do influence the performance and cost of the resultant circuits.

Ideally a substrate should have:

a)   A perfectly smooth surface to permit the growth of thin, defect-free films. This is especially vital for capacitor dielectric films so as to eliminate points at which the electric-field strength is exceptionally high.

b)   A temperature coefficient of expansion that is similar to that of the film so as to minimize mechanical stresses in the films.

c)   High mechanical strength and thermal shock resistance to enable the substrate to withstand the rigors of processing.

d)   High thermal conductivity to efficiently remove heat, thus permitting the fabrication of high-component-density circuits operating at high power density.

e)   Inertness to chemicals used in circuit processing.

f)   Zero porosity to prevent the entrapment of gases which might contaminate the films during deposition.

g)   High electrical resistance and low dissipation factor.

h)   Low cost.

i)   Uniform physical properties.

Commonly used substrate materials include alumina, beryllia, silicon, sapphire and spinels, and various glasses. Silicon must be coated with a dielectric layer which is usually 6000 to 10,000 Å of thermally grown SiO_2. Alumina is probably the most popular substrate material because of its high thermal conductivity, good resistance to chemical attack, and high compressive strength for lead bonding. Being a ceramic, it is composed of sintered granules. Therefore, its surface finish and porosity are not as good as those of glass or the single-crystal substrates. The thermal conductivity of various substrate materials are plotted in Fig. 8.1. Alumina substrates for thin-film circuits can be polished to improve the surface finish. Capacitors usually require a smoother surface than that of as-fired ceramic. This finish is obtained by selectively glazing the substrate. The glazed surface has the finish and low porosity of glass and, because it is thin, a thermal conductivity approaching that of the ceramic.

Figure 8.2(a) is a stylus tracing of a substrate surface. The fluctuations can be described in terms of three components, called the roughness, waviness, and flatness, as shown in Figs. 8.2(b), 8.2(c), and 8.2(d).

Roughness is a short-range random fluctuation which may be characterized numerically by the average deviation of the trace from some arbitrary mean value. However, any use of statistical averages tends to obscure the occurrence of deep scratches. Thus, it might be useful to also specify the maximum peak-to-valley deviation.

Waviness is a periodic variation, characterized by a peak-to-peak amplitude h_2 and a repetition interval λ, and flatness is the long-range random fluctuation characterized by the peak-to-peak variation h_3. The surface roughness and flatness for several common substrate materials is given in Table 8.1.

The substrate surface affects the circuit in several ways. Patterns must be defined in the films to form the various electrical components. This is

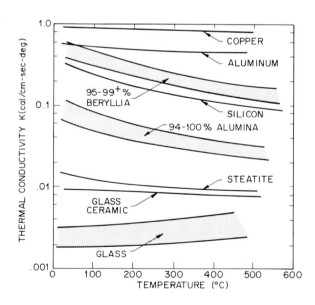

Fig. 8.1 Thermal conductivity of various substrate materials.

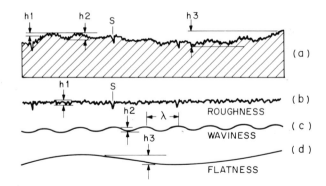

Fig. 8.2 Substrate surface finish. (a) Stylus tracing of surface; (b) roughness; (c) waviness; (d) flatness.

Table 8.1 Substrate Roughness and Flatness

Substrate Material	Roughness	Flatness
Soda-Lime Glass	<250 Å	20 μm/cm
Corning Code 7059 Glass	<250 Å	40 μm/cm
Polished Sapphire	<250 Å	<1 μm/cm
Polished 99.94% Alumina (ultrafine grain)	≈250 Å	<0.1 μm/cm
As Fired 99.5% Alumina (fine grain)	≈10,000 Å	50 μm/cm
Glazed Alumina	<250 Å	50 μm/cm

done photolithographically, using the same techniques as for silicon integrated circuits. The substrate flatness determines how intimately the mask is in contact with the surface. Therefore, it influences the minimum linewidth and spacing. Resistor films are relatively tolerant of surface roughness, until that roughness is comparable to the thickness of the film. Resistive films on rough substrates exhibit a higher sheet resistance than a similar film on a smooth substrate because of the apparent increase in length between two points and because of thin spots in the film. For films of nominally equal thickness on smooth and rough substrates, those on the rough substrate exhibit a more negative temperature coefficient and are less stable during thermal aging than those on smooth substrates. The performance of capacitors is critically affected by surface roughness because of the region of high electric-field strength in the vicinity of an isolated defect protruding from an otherwise smooth surface. This defect could cause premature break-down of the dielectric film.

Ceramic materials are superior to glasses because they are more resis-tant to thermal shock, have better mechanical and chemical stability at higher temperatures, are not attacked by hydrofluoric acid solutions and ano-dizing solutions that are used in processing thin-film circuits.

8.4 FABRICATION SEQUENCES FOR THIN-FILM CIRCUITS

A thin-film circuit consists of circuit elements patterned from several thin films, of different materials, that are deposited on a suitable substrate. It is possible to directly form the components by selectively depositing the films through masks. However, the most popular technique for producing thin-film circuits is the so-called *subtractive process* in which a multilayer film is deposited uniformly over the substrate, followed by the selective removal of material from the individual layers to form the desired components. In this section the subtractive process is described in some detail.

8.4.1 Fabrication of Resistor-Conductor Circuits

Many different materials have been used to form thin-film resistors. The properties of some of these resistive films are considered in the next section. In order to talk about specific materials in the description of the subtractive process, we consider a structure consisting of a tantalum-nitride resistive film and a conductor structure of titanium, palladium and gold. Tantalum nitride is chosen for the resistive layer because it is a stable film which produces a good anodic oxide that passivates the resistor. Films of β-tantalum are used in the fabrication of capacitors. The conductor films are discussed in more details in Section 8.7.

Substrate preparation   The subtractive process begins with a substrate uni-formly coated with resistive and conductive films. If the ceramic is glazed or if the substrate is glass, a film of tantalum oxide must be placed between the substrate and the resistive film as an etch barrier, because glass and glazes are attacked by hydrofluoric acid used in the tantalum etchant. This tantalum-oxide film is formed by sputtering a thin layer of tantalum which is then oxidized. A tantalum-nitride resistive film typically in the range of 300 Å to 1500 Å thick is sputtered onto the ceramic substrate. This is fol-lowed by the successive evaporation of about 1000 to 2000 Å of titanium, 2000 Å of palladium, and 13,000 to 18,000 Å of gold. High conductivity cir-cuits use a 50,000 Å gold layer which is formed by electroplating. The titanium is used to provide good adherence of the palladium-gold conductor film to the tantalum-nitride resistive film. The palladium film is used as a diffusion barrier to separate the titanium and gold. Another material which is used to provide adherence for the conductor film is nichrome. However, the disadvantage of nichrome is that chromium diffuses through the composite structure at high temperatures and forms an oxide on the surface, which is difficult to remove.

The single-etch method   The single-etch method is so called because the pattern to be etched in each of the film layers is defined by its own photoresist layer. Either positive or negative photoresist can be used. The first step of the single-etch process is to apply photoresist and define the conductor pattern. The exposed conductor material is then removed by selective etching as shown in Fig. 8.3(a). Next, the photoresist is stripped and a new

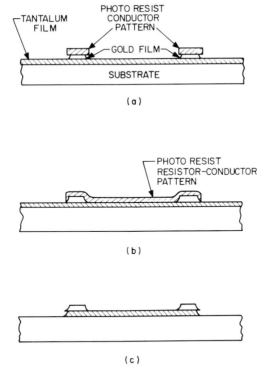

Fig. 8.3 The single-etch process. (a) After etching of conductor pattern; (b) photoresist layer to define combined resistor-conductor pattern; (c) after etching of resistive layer.

layer of photoresist is applied and exposed to define the combined resistor-conductor pattern. The resistive film is then etched and the photoresist is stripped. These steps are illustrated in Figs. 8.3(b) and 8.3(c).

The single-etch method has replaced the older *dual-etch* method in which a single application of positive photoresist is first used to define the composite resistor-conductor pattern and then re-exposed through the conductor mask to define the conductor pattern.

Selective etching of metal films   The subtractive process described above is predicated on the ability to selectively remove layers of the composite thin film. This is done either by immersion or spray etching with chemical etchants, or by sputter etching. The chemicals used to etch films are basically the same as those used to dissolve the bulk material. Etchants for thin films are usually more dilute than those for bulk material resulting in a reduced etch rate, and minimizing attack on the photoresist and the film underlying it.

Etchants for some common films are given in Table 8.2. In general, etchants based on reactions in which metal ions change valency states are preferred to oxidizers which evolve gas. Some form of agitation is required to remove bubbles from the surface if gas evolution occurs.

Anodization[4]   Anodization is the electrochemical oxidation of a metal anode in an electrolytic cell. It is used to form an adherent oxide film on certain metals called valve metals. Among these metals are tantalum, titanium, tungsten, and aluminum. The anodic process has been used to trim resistors to value by creating an oxide at the expense of the parent film. However, in recent years laser trimming has been displacing the anodic method. The anodic film is also used as the dielectric film for thin-film capacitors.

The anodization reaction can be described by the following equations

$$kM + nH_2O \rightarrow M_kO_n + 2nH^+ + 2ne \quad \text{at the anode}$$

$$(8.8)$$

$$2ne + 2nH_2O \rightarrow nH_2 + 2nOH^- \qquad \text{at the cathode}$$

where M denotes the metal atom and k and n denote the number of ions and oxygen atoms required to form the oxide. Water is provided by the aqueous electrolyte. The anode is oxidized and hydrogen is liberated at the cathode. A list of some valve metals, their anodization rate, and properties of the anodic oxide is given in Table 8.3.

The growth rate of the anodic oxide film is proportional to the current density in the electrolytic cell, and is given by Faraday's law

$$\frac{d}{dt}[d] = \frac{JG}{96,500D} \tag{8.8}$$

Table 8.2 Etchants for Common Thin Film Materials

Film	Etchant
Aluminum	Hydrochloric acid, a mixture of phosphoric acid and nitric acid, or sodium hydroxide.
Nichrome	Hydrochloric acid, a mixture of hydrochloric acid and copper chloride, or a mixture of hydrochloric acid and nitric acid.
Titanium	Dilute hydrochloric acid.
Tantalum	A mixture of nitric acid and hydorfluoric acid.
Tantalum Pentoxide	Hydrofluoric acid.
Gold, Palladium	Aqua regia or a mixture of potassium iodide and iodine.
Platinum	Aqua regia or mixtures of alkali cyanides and hydrogen peroxide.
Silicon Monoxide	Mixture of ammonium fluoride and ammonium hydroxide.
Silicon Dioxide	Buffered hydrofluoric acid.
Manganese Oxide	Hydrochloric acid or hydrogen peroxide.
Copper	Hydrochloric acid, nitric acid, and ferric chloride.

where d is the thickness (in cm) of the anodic oxide film, G is the equivalent weight of the oxide, J is the current density (in A/cm^2), 96,500 (in A-s/equivalent) is Faraday's constant and D (in gm/cm^3) is the oxide density.

Table 8.3 Properties of Anodic Oxides

Metal	Anodic Oxide	Dielectric Constant of Oxide	Attainable Thickness, μ	Maximum Thickness/Volt Angstrom/V
Aluminum	Al_2O_3	10	1.5	3.5
Tantalum	Ta_2O_5	22	1.1	16
Titanium	TiO_2	40	15	
Tungsten	WO_3	42		
Zirconium	ZrO_2	12	>1	12-30

EXAMPLE 8.1   The parameters for tantalum pentoxide are $D = 8$ gm/cm^3 and $G = 44.18$ gm/equivalent. Calculate the growth rate of a tantalum-pentoxide film at a current density of 2 mA/cm^2.

Solution.   From Faraday's law

$$\frac{d}{dt}\,[d] = \frac{2\times10^{-3}\times44.18}{96,500\times8} = 11.4\times10^{-8} \text{ cm/s} = 11.4 \text{ Å/s.}$$

Tantalum has an anodization current efficiency of 100%, that is, an oxide film is formed during anodization to the exclusion of all other reactions. If a constant current is maintained in the electrolytic cell, the anodic-oxide will continue to grow at the constant rate, and the voltage across the oxide will continue to increase until the oxide breaks down. Oxide films of tantalum and aluminum break down by arcing.

EXAMPLE 8.2   The differential field strength, E_d, at constant anodization current density J is defined as

$$E_d = \frac{dV/dt}{\dfrac{d}{dt}\,[d]}$$

where dV/dt is an experimentally measured rate of voltage rise and $d[d]/dt$ is the growth rate mentioned in Example 8.1.

If a tantalum-pentoxide film, growing at a constant current density of 1 mA/cm^2, has a measured dV/dt of 0.33 V/s, find the differential field strength E_d.

Solution.   The growth rate at 1 mA/cm^2 for Ta$_2$O$_5$ is found from Faraday's law to be 5.7 Å/s (see Example 8.1). Hence $E_d = 0.33/5.7\times10^{-8} = 5.8\times10^6$ V/cm.

When a constant voltage is maintained across the electrolytic cell, the growth rate of the oxide decreases rapidly as the voltage across the oxide film approaches the applied voltage causing the anodization current to decrease, thus yielding a film of reproducible thickness. However, the thickness of the anodic-oxide film does not approach a limiting value.

Trim anodization of resistors   Tantalum resistors are anodized to minimize the effect of electrocorrosion and to adjust the resistors to their final value. The minimum anodization voltage for tantalum resistors is 30 V to protect them against corrosion. Tantalum oxide grows at the rate of 16 Å/V while consuming 6.3 Å of tantalum per volt. Nitrogen-doped tantalum films oxidize at the rate of 4.5 Å/V. The oxide film grows at the rate of 16.7 Å/V.

The anodization is conducted at constant current until the correct resistance value is measured. The resistor is anodized at one value of current until the resistance is close to the final value. Then the current is reduced to a lower value to reduce the anodization rate. The resistor is anodized for a few milliseconds and then its value is remeasured. This cycle is repeated until the resistor is within the desired tolerance.

Anodization for forming capacitor dielectric   Capacitor dielectric films are anodized initially at constant current. When the voltage across the dielectric film reaches the desired value, that voltage is maintained across the cell until the current drops to a small fraction of its initial value. Anodization and soaking at constant voltage minimizes the oxide-metal interface states which contribute to capacitor loss and permits more precise control of the final dielectric film thickness than is possible by constant current anodization. An example of film growth versus time for a tantalum film anodized at a constant voltage of 120 V is shown in Fig. 8.4. After 30 min of anodization, a 4-min error in the anodization time results in a nominal thickness error of 0.05%. However, the actual thickness error of 9.6 Å is comparable to the thickness of a monolayer and the roughness of the film. Thus the film thickness cannot be reproducibly controlled with this precision.

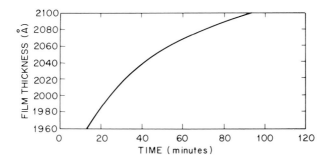

Fig. 8.4   Anodic-oxide film growth versus time for a tantalum film anodized at a constant voltage of 120 V. (From *Thin Film Technology* by R. W. Berry, P. M. Hall, and M. T. Harris, Copyright 1968 by Litton Educational Publishing, Inc. Reprinted by permission of Van Nostrand Reinhold Co.)

Laser trimming of thin-film resistors   Laser machining can be used to alter the geometry of a thin-film resistor so as to increase the total number of squares of resistance. Whereas trim anodization effected the change in resistance by reducing the effective thickness of the film, thereby increasing its sheet resistance, laser trimming leaves the film properties substantially unchanged and varies the geometry. Figure 8.5 shows several typical resistor designs that are trimmed by laser machining. They include the bar, top hat, ladder, and loop.

Fabrication of capacitors   A thin-film capacitor is a parallel-plate structure consisting of a thin-film electrode, dielectric, and counterelectrode on a glazed substrate. The cross section of a tantalum-tantalum-oxide thin-film capacitor is shown in Fig. 8.6. Fabrication of tantalum capacitors begins with

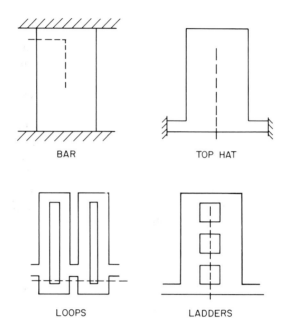

Fig. 8.5 Bar, top-hat, ladder and loop resistor designs for laser trimming.

the sputtering of a very thin tantalum layer and its subsequent thermal oxidation to form a tantalum-oxide film that protects the glaze during subsequent etching operations. This is followed by the sputtering of a tantalum film about 4000 Å thick. The capacitor dielectric is tantalum pentoxide formed by anodizing the surface of this tantalum film. The anodization is conducted at a constant current density of 0.1 to 1.0 mA/cm² until the desired anodization voltage is reached. Then that voltage is maintained for about 30 min to reduce the metal-oxide surface interface states and heal the oxide. After thorough cleaning to remove all traces of the electrolyte, the conductive counterelectrode is deposited by evaporation.

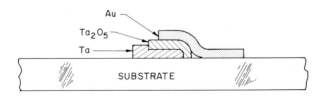

Fig. 8.6 Tantalum-tantalum-oxide thin-film capacitor.

8.4.2 Fabrication of RC Circuits on a Single Substrate

Figure 8.7 shows the process sequence for the fabrication of tantalum thin-film circuits containing resistors and capacitors on the same substrate. The process is somewhat more complicated than the subtractive process for resistor-conductor circuits described earlier in this section. This is partly because a metal deposition for the capacitor counterelectrode is required after the dielectric film formation, and partly because tantalum nitride is used for stable resistors whereas β-tantalum is used for the capacitors.

The sequence begins with the selective glazing of the ceramic substrate in those areas that will ultimately be capacitors. A thin layer of tantalum is deposited and thermally oxidized to form an underlay that acts as an etch stop to protect the glaze during the tantalum etch. Next, about 4000 Å of β-tantalum is deposited by sputtering, resulting in the uniformly coated substrate shown in Fig. 8.7(a). The first patterning operation is to remove the β-tantalum film from the resistor areas and to outline the base electrode region of the capacitors as shown in Fig. 8.7(b). Next, the capacitors are preanodized, followed by the sputtering of the tantalum-nitride resistive film and the evaporation of a titanium-palladium film to form the contact areas. These steps are shown in Figs. 8.7(c), 8.7(d), and 8.7(e). The function of the titanium-palladium film is to ensure the integrity of the low-resistance contact between the tantalum layers and the conductor film to be deposited later. The titanium-palladium layers are then etched to form contact areas for the resistors and capacitors, as shown in Fig. 8.7(f). The next step is to selectively etch the resistor tracks and to remove the resistive film from the capacitor areas. The anodic-oxide film over the β-tantalum in the capacitor areas acts as an etch stop during the etching of the tantalum-nitride film. The resistors thus formed are then thermally stabilized either by thermal oxidation during a high-temperature bake or by anodization. This is followed by a reanodization of the capacitor dielectric film and evaporation of a nichrome-gold conductor film for the counterelectrode of the capacitors and the interconnection wiring. The nichrome-gold film is removed from the resistor areas and the interconnection pattern is generated by selective etching. Finally, the resistors are adjusted to their final value either by laser trimming or by anodization. These steps are illustrated in the remaining sections of Fig. 8.7.

8.5 DESIGN OF RESISTORS

The design of thin-film resistors involves the choice of a suitable film, the determination of the film thickness to yield the desired sheet resistance, the choice of a suitable pattern with the required number of squares, and the selection of the linewidth and spacing that results in a power density that is

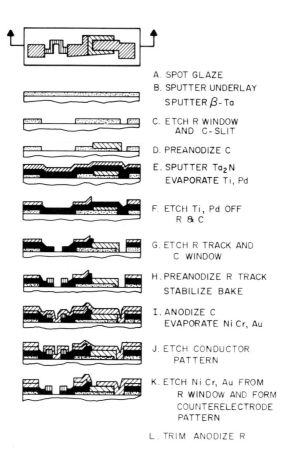

A. SPOT GLAZE

B. SPUTTER UNDERLAY
   SPUTTER β-Ta

C. ETCH R WINDOW
   AND C-SLIT

D. PREANODIZE C

E. SPUTTER Ta_2N
   EVAPORATE Ti, Pd

F. ETCH Ti, Pd OFF
   R & C

G. ETCH R TRACK AND
   C WINDOW

H. PREANODIZE R TRACK
   STABILIZE BAKE

I. ANODIZE C
   EVAPORATE Ni Cr, Au

J. ETCH CONDUCTOR
   PATTERN

K. ETCH Ni Cr, Au FROM
   R WINDOW AND FORM
   COUNTERELECTRODE
   PATTERN

L. TRIM ANODIZE R

Fig. 8.7 Fabrication of tantalum thin-film circuits containing resistors and capacitors on a single substrate.

within acceptable limits imposed by resistor stability requirements. A complete discussion of resistor design is beyond the scope of this text. We present here a qualitative discussion of these design variables, some quantitative relationships, and examples based on the tantalum thin-film technology.

The choice of film material is based on considerations such as the temperature and voltage coefficients of resistance, the noise coefficient, stability of resistance with time, the allowable power density, and the method of film deposition. One commonly used resistive film is the 80%-nickel, 20%-chromium alloy known as Nichrome. This material has a bulk resistivity of 108 $\mu \Omega$-cm and a nominal TCR of 110 ppm/°C according to Berry, et al.,[6] and a bulk resistivity of 110 $\mu \Omega$-cm and a TCR of 85 ppm/°C as cited by Maissel.[7] The resistivity of Nichrome films as thin as about 50 Å remains at the bulk value, provided the film has the same 80%-nickel, 20%-chromium composition as the bulk material.[8] Nichrome films with a TCR approximately equal to zero at room temperature can be prepared by adding small amounts of copper or aluminum to the film. Nichrome can be deposited by vacuum evaporation although it may be difficult to obtain a film of uniform composition unless special techniques are used, because of the difference in vapor pressure of chromium and nickel. Nichrome films are not stable in high humidity, especially under dc load, because of electrolysis, and are often protected by a thick silicon monoxide overcoating. Nichrome films can be made 1 μm thick yielding a sheet resistance of about 1 $\Omega/\square$. Films having sheet resistances of several tens of ohms per square become quite thin, but Nichrome resistors with sheet resistance up to 500 $\Omega/\square$ are commercially available.

Tantalum is a so-called refractive metal because of its high melting point. One of its prime advantages as a resistor material is stability with time. Tantalum has a high recrystallization temperature and its films do not anneal below several hundred degrees centigrade. It forms an adherent oxide layer which protects the surface. Bulk tantalum has a resistivity of 14 $\mu \Omega$-cm, and films having resistivities of 20 $\mu \Omega$-cm and the same temperature properties as bulk material have been prepared. Thin films of tantalum are usually deposited by sputtering. Undoped, sputtered tantalum films are usually in a tetragonal crystal structure known as *beta phase* which has a resistivity that is much higher than that of bulk material. The resistivity of β-tantalum is 180 $\mu \Omega$-cm with a TCR between -100 and $+100$ ppm/°C. Nitrogen-doped tantalum films have resistivities in the range of 280 to 400 $\mu \Omega$-cm and TCR's between -100 and -200 ppm/°C. In addition, they exhbit much better stability on exposure to air than do films of β-tantalum. The sheet resistance of tantalum-nitride films is typically 50 $\Omega/\square$ or 100 $\Omega/\square$, although 300 $\Omega/\square$ films are sometimes used. This range of sheet resistances corresponds to films that are nominally 250 to 1000 Å thick.

The permissible power density for any resistor material is a function of the resistor material, the substrate, any conductive films on the substrate, and external heat sinks. The resistor material itself affects the power density in that it imposes an upper limit on temperature of the film because of stability considerations. The temperature of the resistive film is given by

$$T = T_A + Q/h \qquad (8.9)$$

where T_A is the ambient temperature, Q is the power density, and h is the thermal conductivity of the substrate to an infinite heat sink. For glazed ceramic substrates 0.76 mm (30 mils) thick, h is about 31 mW/cm² °C.

EXAMPLE 8.6 The temperature of a thin-film resistor was measured to be 120°C when it was dissipating 0.51 W, and 220°C when it was dissipating 1.1 W. Find estimates for the thermal conductivity, $\hat{h}$, and for the ambient temperature, T_A.

Solution. From two pairs of values (T_1, Q_1), (T_2, Q_2)

$$\hat{h} = \frac{Q_2 - Q_1}{T_2 - T_1} = \frac{1.1 - 0.51}{220 - 120} = 5.9 \times 10^{-3} \text{ W/cm}^2$$

$$\hat{T}_A = \frac{T_2 Q_1 - T_1 Q_2}{Q_1 - Q_2}$$

$$= \frac{220 \times 0.51 - 120 \times 1.1}{0.51 - 1.1} = 33.5°C.$$

The resistance change during aging is a statistical process that is a function of time and temperature. It is given by

$$\frac{\Delta R}{R} = \frac{AR_S}{50} \left(\frac{t}{t_0} \right)^n \exp\left(-\frac{T_0}{T} \right) \qquad (8.10)$$

where t is the time in hours, t_0 is a characteristic time that depends on the film material, substrate material and surface, and the film temperature range, T_0 is a characteristic temperature for the materials combination indicative of an activation energy, and T is the film temperature. In addition, R_S is the sheet resistance, A is a constant determined by the method of anodization and the exponent n is determined by the temperature range. Typical values for tantalum-nitride films on unglazed alumina substrates are $t_0 = 25$ hr, $T_0 = 3200$ K and $n = 0.3$ for $T \leqslant 150°C$ and $t_0 = 10^{-7}$ hr, $T_0 = 7900$ K and $n = 0.5$ for T between 150°C and 250°C. The constant A is approximately unity. Equation (8.10) is the mean value of a statistical distribution. This distribution is rather broad with values of standard deviation often equaling the mean. The results on resistor aging should be viewed in that light. Equations (8.9) and (8.10) together specify a maximum power density for the tantalum-nitride resistive film,

$$Q \leqslant h \left\{ \frac{T_0}{\ln(\Delta R_S/50) + n \ln(t/t_0) - \ln(\Delta R/R)} - T_A \right\}, \quad (8.11)$$

in terms of the lifetime and resistance stability requirements. Curves of maximum expected resistance change versus time for various power densities are given in Fig. 8.8.

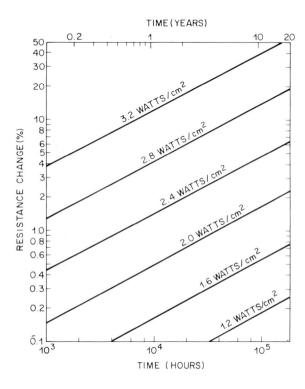

Fig. 8.8 Expected resistance change versus time for 50 $\Omega/\square$ tantalum-nitride films on 1.25-mm thick Corning 7059 glass substrates.

The allowable maximum power density is used, together with the power-dissipation requirement of the resistor, to place one limit on the physical size of the resistor. The outline of a simple bar resistor and a meander resistor are shown in Fig. 8.9. The area of the bar resistor is simply the width of the resistor track multiplied by its length. The effective area of the meander resistor is the area of the rectangle enclosing the resistor pattern, that is,

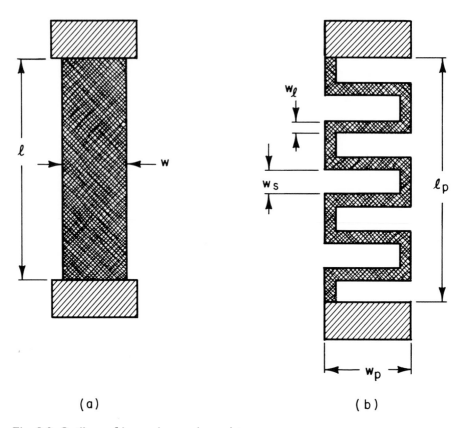

(a)   (b)

Fig. 8.9 Outlines of bar and meander resistors.

$$A_{MR} = w_p l_p$$

where w_p is the pattern width and l_p is the pattern length. Knowing the power density and the required power dissipation,

$$A_R = P/Q. \tag{8.12}$$

Knowledge of the resistance value and sheet resistance of the film specifies the number of squares of resistor material, $n = R/R_s$. For a bar resistor,

$$n = \frac{l_b}{w_b} = \frac{A_B}{w_b^2} = \frac{A_R}{w_b^2}. \tag{8.13}$$

This last equation relates the total resistance to the linewidth and total area of the resistor. The approximate number of squares for a meander resistor, neglecting the bend correction factor for each meander, is the number of meanders multiplied by the number of squares per meander. Thus, for a

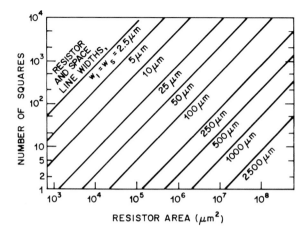

Fig. 8.10 Number of squares for a meander resistor with equal width lines and spaces.

meander resistor,

$$n = \frac{A_R}{w^2(1+w_s/w_l)} \, . \tag{8.14}$$

This equation is plotted in Fig. 8.10 for $w_s = w_l$.

Resistor precision sometimes imposes a lower limit on the linewidth because variations in the geometric pattern during the process of its definition in the resistive film will affect the final value of the resistor. The fractional variation of resistor value due to variations in the sheet resistance of the film and dimensional tolerances on the pattern is given by

$$\frac{\Delta R}{R} = \frac{\Delta R_s}{R_s} + \frac{\Delta l}{l} - \frac{\Delta w}{w} \, . \tag{8.15}$$

There is parasitic series inductance and shunt capacitance associated with thin-film resistors. We will not present any quantitative results about these parasitics. The following are conclusions about parasitic inductance and capacitance associated with thin-film resistors from Berry, *et al.*[5]

1.   Lead inductance can be reduced by placing components close together.

2.   The meander pattern is basically a noninductive pattern. The bar resistor has a higher inductance per unit length of path. The series inductance of thin-film resistors is small.

3.   Series inductance of a thin-film resistor is directly proportional to the linear dimension of the pattern.

4.   Changing the sheet resistance of the film affects the resistance value without changing the series inductance. Scaling the pattern changes the series inductance without affecting the resistance.

5.   Increasing the spacewidth-to-linewidth ratio of a meander resistor, while keeping the length constant, increases the series inductance.

6.   The substrate thickness and dielectric constant affect the parasitic shunt capacitance. This capacitance is reduced by using a thin substrate with a low dielectric constant.

7.   The meander resistor has more shunt capacitance than the bar resistor. The capacitance between branches of the meander pattern are in series. Therefore a pattern with many short meanders has a smaller shunt capacitance than one with a few long meanders.

8.   Decreasing the size of the pattern reduces the shunt capacitance without changing the resistor value.

9.   Higher sheet resistance implies a smaller pattern for the same resistor values and, therefore, it implies a smaller shunt capacitance.

10.   The shunt capacitance is increased by decreasing the spacewidth-to-linewidth ratio.

11.   Finally, the high-frequency performance of the resistor is limited either by the parasitic inductance or capacitance depending, to a large extent, on the resistance value.

8.6 DESIGN OF CAPACITORS [5]

The thin-film structure described in Section 8.4 is a parallel-plate capacitor whose capacitance is given by

$$C = \kappa \epsilon_0 \frac{A}{d} \qquad (8.16)$$

where ϵ_0 is the permittivity of free space, κ is the dielectric constant of the dielectric material, d is the thickness of the dielectric film, and A is the effective area of the capacitor, that is, the area of the dielectric film that has both an upper and a lower plate. Fringing effects are neglected because the thickness of the dielectric film in a thin-film capacitor is much less than the area of the plates. A characteristic of the dielectric film is the capacitance density

$$\frac{C}{A} = \frac{\kappa \epsilon_0}{d} \; . \qquad (8.17)$$

The dielectric film thickness, breakdown electric-field strength, E_B, and the breakdown voltage, V_B, are related by

$$E_B = \frac{V_B}{d} \, . \tag{8.18a}$$

The working voltage and field strength in an actual capacitor is some fraction of that given by Eq. (8.18a),

$$E_R = KE_B = \frac{KV_B}{d} \tag{8.18b}$$

where the constant K is in the range 0 to 1. This last equation places a lower limit on the thickness of the dielectric film. A very useful film material property which determines the area of a thin-film capacitor is the capacitor charge storage factor. It is the product of the capacitance density and the working voltage. From Eqs. (8.17) and (8.18b),

$$V_R \frac{C}{A} = \kappa \epsilon_0 K E_B, \tag{8.19}$$

where $V_R = KV_B$ is the rated voltage of the capacitor. The capacitor charge storage factor for a tantalum-oxide film is 3.1 $\mu C/cm^2$. The value chosen for K is determined by considerations of capacitor reliability, and not yield. This means that if a capacitor designed according to Eq. (8.19) passes its initial test, it has a low probability of failing during its lifetime. Dielectric films that are very thin or very thick have higher defect densities than films of moderate thickness. The optimum thickness range for tantalum-oxide films on ceramic is found to be 3000 to 3500 Å.

Thin-film capacitors are not ideal energy storage elements. The sources of loss in a thin-film capacitor are dielectric loss, and resistance in the capacitor electrodes and leads. The dielectric loss for thin-film dielectrics is independent of frequency to a first approximation. It is convenient to represent the actual thin-film capacitor as the series connection of an ideal capacitor and an equivalent resistor which accounts for the loss. The driving-point impedance of the capacitor using that equivalent circuit is

$$Z = \frac{\tan \delta'}{\omega C} + R' - \frac{j}{\omega C} \, , \tag{8.20}$$

where $\tan \delta'$ is the dielectric loss and R' is the equivalent electrode and series lead resistance. A typical value for $\tan \delta'$ for a tantalum-oxide dielectric is 0.003. Note from this expression that the dielectric loss is the dominant loss mechanism at low frequencies, whereas at high frequencies it is the resistor losses that dominate. Considering the distributed nature of the thin-film capacitor leads to a transcendental expression for the driving-point impedance. At low frequencies, if the loss is small enough, the effective resistance in series with the dielectric is

$$R' = R_L + \frac{R_E}{3} \tag{8.21}$$

where R_E is the electrode resistance and R_L is the lead resistance. This result is valid only for the crossed electrode pattern shown in Fig. 8.11(a).

Unlike resistors which can be trimmed to final value either by anodization, etching, abrasion, or laser adjustment of the pattern, the values of a capacitor cannot be adjusted in any continuous fashion. Control of capacitor value rests in the ability to control the capacitance density of the structure and the overlap area of the electrodes. We have already seen that the thickness of an anodic oxide is a function of the anodization voltage. The capacitance density of a tantalum-oxide structure can be controlled to within several percent. With careful monitoring of the deposition process, the control of capacitance density for a deposited dielectric structure can be almost as good.

The most elementary electrode configuration is the crossed-electrode pattern shown in Fig. 8.11(a). The dimensions of each of the electrodes is dependent on the photolithographic line-definition capability in the thin film. This is on the order of several microns in current thin-film technology. The overlap electrode area, which is the determining factor for capacitance, depends upon the control of the width of each of the electrodes and the rota-

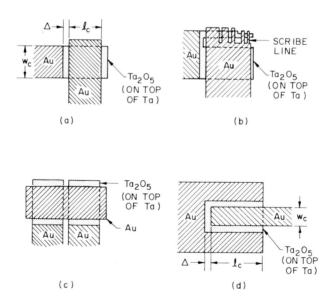

Fig. 8.11   Capacitor patterns. (a) Crossed electrodes; (b) main capacitor with binary-weighted trimming capacitors; (c) nonpolar capacitor; (d) minimum resistance pattern.

tional misalignment which increases the area by

$$\frac{\Delta A_c}{A_c} = \sec(\Delta\theta) - 1. \tag{8.22}$$

The overlap area is independent of translational misalignments. A one-degree rotational misalignment, which is not difficult to achieve, corresponds to an area increase of less than 0.016%. This implies that the dimensional control of the electrode width is the dominant factor in determining the capacitor area.

The capacitance value can be adjusted in discrete steps using the pattern shown in Fig. 8.11(b). It consists of a main capacitor in parallel with a set of binary weighted trimming capacitors which can be selectively removed by laser or mechanical scribing. This type of pattern can provide an adjustment range of 10% of the design value. The final value can be within 0.2% of the design value.

Tantalum capacitors are considered to be polar devices. It is important that the tantalum electrode be made positive with respect to the counterelectrode to prevent dielectric breakdown under large reverse bias, although small-signal bias variations do not pose any problem. A nonpolar capacitor can be constructed, when required, by placing two capacitors in series back to back, as shown in Fig. 8.11(c). This arrangement requires four times the area of a polar capacitor of comparable value because of the series connection.

Finally we show, in Fig. 8.11(d), the minimum resistance pattern. The lead and electrode resistance can be reduced substantially by contacting the base electrode material on three sides. If R_s is the sheet resistance of the tantalum film and Δ is the minimum metal-to-metal spacing, it can be shown that the maximum resistance in series with the dielectric is

$$R' < R_s \left[\frac{\Delta}{2l_c} + \frac{w_c}{12l_c} \right]. \tag{8.23}$$

8.7 CONDUCTOR FILMS

Conductor films are usually multilayer composite films of several metals because no one material possesses all of the desired properties. The ideal conductor material should have:

1. low resistivity;

2. good adherence to the substrate and to other films; and

3. a top surface that is resistant to corrosion and is easily bondable.

Furthermore, the conductor film must be compatible with other films, chemicals, and processes used in the fabrication of thin-film circuits.

The bulk resistivity of several metals and the sheet resistance for a film 1000 Å thick are given in Table 8.4. Good conductivity for the composite conductor film is provided by a film of gold or copper that is typically from 1000 Å to 10,000 Å thick. Note that a 10,000-Å-thick gold layer has a sheet resistance of about 0.03 $\Omega/\square$. Below 1000 Å, the sheet resistance increases at a rate greater than the reciprocal film thickness because the bulk resistivity value is no longer appropriate. It becomes difficult to evaporate gold films thicker than 10,000 Å because of a limited amount of material in the eva-porator source charge, the length of time required, and heating of the sub-strate and fixtures. Gold films can be plated to a thickness of about 250,000 Å before they become mechanically unsound.

Table 8.4 Sheet Resistance of Metal Films

Metal	Bulk Resistivity ($\mu\Omega$-cm)	Sheet Resistance For 1000 Å ($\Omega/\square$) Film
Copper	1.7	0.20
Gold	2.4	0.27
Aluminum	2.8	0.33
Palladium	11	1.3
Palladium-Gold 50-50 Alloy	21	3
Titanium	55	10
Nichrome	100	15

Unfortunately, metals with good conductivity usually exhibit poor adherence to substrate materials. The adherent properties of the composite conductor film are provided by a layer of titanium, chromium, or nichrome under the low resistivity film.

The composite conductor film must be corrosion resistant and its top surface must be bondable to permit the attachment of appliqued com-ponents, to form a hybrid IC, and leads to connect this circuit to the outside world. Copper films are subject to oxidation and are therefore protected by a film of a more noble metal such as gold.

In addition, gold is well suited to thermocompression bonding. How-ever, it is not suitable for solder connections because it forms a brittle alloy with solder. This problem can be alleviated by using a selectively deposited buried rhodium layer as a solder barrier.

Another consideration in choosing conductor materials is compatibility with the rest of the processing. Aluminum is an interesting conductor material which is easy to deposit by evaporation, has a low resistivity, and forms a protective anodic oxide. Its one disadvantage is that it is not compatible with conventional soldering or welding techniques. However, it is amenable to ultrasonic bonding methods. Methods of lead and component attachment are discussed in Chapter 10 on "Assembly Techniques."

8.8 LAYOUT OF THIN-FILM CIRCUITS

In a general way, the layout problem for thin-film circuits is the same as that for silicon integrated circuits, namely to convert the circuit schematic to a set of masks that are used to make the circuit. It differs in detail because of the differences in the materials and the physical sizes of the features. Recall that in silicon integrated-circuit technology the minimum feature size was in the range from several microns up to about 10 μm. Minimum feature sizes in the thin-film technology are about an order of magnitude larger. Typically the minimum conductor linewidth or line-to-line spacing is about 50 μm as compared to 10 μm in the illustrative SIC design rules of Chapter 7.

The first step in the design of a thin-film circuit which is to be used in conjunction with SIC's and discrete components, as part of a hybrid integrated circuit, is to partition the circuit components among the various technologies. This step requires careful consideration of the properties of the circuit elements as realized in each of the technologies. The choice of materials for the film structure might be made at this step. More commonly, the film circuits are restricted to be made of a certain class of materials for which the processing capability exists at a given facility. The partitioning problem is not considered, other than to note that it exists and must be solved because it plays a major role in determining cost.

The size of each component on the substrate of the HIC must now be determined. The sizes of SIC's and other appliqued components are known as a result of having designed those components, or from catalogs. The minimum area for each resistor is determined by using conditions such as Eqs. (8.11) through (8.15) to relate the geometric pattern, resistance value, tolerance, power dissipation, and linewidth. All resistors should be designed to have the same power density so that they all age in a similar manner. This is not always possible because, for example, the power dissipation for some resistors is so low that the small area necessary to achieve constant power density would require a linewidth smaller than the minimum attainable value which is typically about 50 to 100 μm. The absolute minimum for resistor track width is 50 μm. The active area of each resistor is bordered by a protective region, typically 400 μm wide, when individual adjustment of the resistance value is by anodization. No protective border is necessary when the resistors are laser trimmed. Figure 8.12 shows several typical resistor configurations, including the protective area. Those portions of the resistor that are outside of the anodized area are sometimes called dumbbells. When the resistors are placed on the substrate they should be aligned parallel to the edges of the substrate to facilitate the use of automated trimming equipment.

The size of the film capacitors is based on the capacitance density or charge storage factor, and the choice of a pattern based on constraints such as the loss, adjustment capability, or nonpolarity. Capacitors also require a protective border of about the same width as that required for resistor anodization.

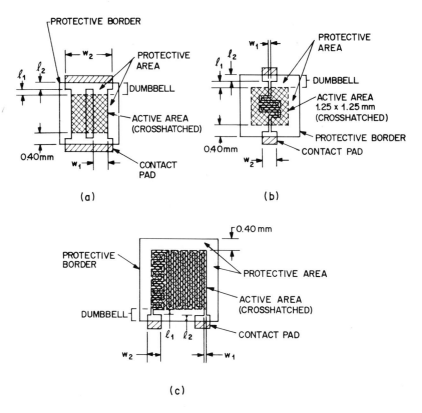

Fig. 8.12 Plan view of resistors designed to be trim anodized showing protective border surrounding the active area: (a) parallel bar resistors; (b) meander resistor smaller than minimum active area; (c) meander resistor larger than minimum active area. In this figure $l_1 = l_2 = 0.2$ mm and $w_2 = 0.4$ mm minimum.

The lines in the conductor pattern must have a minimum linewidth of 100 μm and the minimum spacing between conductors should be 100 μm. However, in the vicinity of a beam-lead SIC, the conductor pattern must have smaller dimensions and a minimum linewidth of 50 μm and a minimum space of 50 μm is used.

Crossovers are often required in circuits of any reasonable complexity. There are methods for batch fabricating these crossovers *in situ* by multiple plating operations, but these methods will not be described here because they are complex and are not often used. For our purposes, crossovers are made by the thermocompression bonding of wire or ribbon between bonding pads. These bonding pads are typically 125 μm by 200 μm and are at least 200 μm from other conductors. Figure 8.13 shows the details of a crossover.

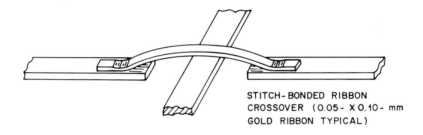

STITCH-BONDED RIBBON
CROSSOVER (0.05- X 0.10- mm
GOLD RIBBON TYPICAL)

Fig. 8.13 Ribbon crossover.

Bonding pads must also be provided for the attachment of appliqued components. In general, the bonding pads must be larger than the terminals of the component. Two typical termination arrangements are shown in Fig. 8.14. Note that if the appliqued component is large enough, conductors can be brought between the terminals and the appliqued component also serves as a crossover. A protective border of about 500 μm should be included around each appliqued component.

Bonding pads for the attachment of external leads are situated about the periphery of the substrate. They are larger than internal pads to permit the use of a lead frame if desired. A preferred dimension is about 1.5 mm on a side. Additional points should be brought out to bonding pads to permit the circuit to be tested. If this is not possible or practical, internal test

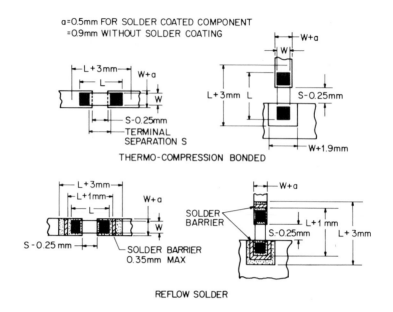

Fig. 8.14 Terminal arrangements for appliqued components.

pads should be employed to permit testing of the thin-film circuit before assembly of the hybrid circuit.

REFERENCES

1. R. W. Berry, P. M. Hall, and M. T. Harris. *Thin Film Technology.* Princeton: D. Van Nostrand, 1968, Chapter 6.

2. *Ibid.,* Chapter 9.

3. L. I. Maissel and R. Glang, editors. *Handbook of Thin Film Technology.* New York: McGraw-Hill, 1970, Chapter 6.

4. R. W. Berry, P. M. Hall, and M. T. Harris. *Op. cit.,* pp. 271-285.

5. *Ibid.,* Chapter 11.

6. *Ibid.,* p. 337.

7. L. I. Maissel and R. Glang, editors. *Op. cit.,* pg. 18-8.

8. R. H. Alderson and F. Ashworth. "Vacuum-deposited films of nickel-chromium alloy." *British J. of Applied Physics* : 8205-210 (1957).

PROBLEMS

8.1 A film 6000 Å thick has a phonon mean-free path, $\lambda_{ph} = 360$ Å at 300 K. It is found from measurements at 300 K and at low temperature (near 0 K) that the ratio of film resistivities is $\rho_T/\rho_r = 34$. Calculate the residual mean-free path, λ_0, of the film.

8.2 The growth of an anodic oxide film is given by Faraday's law as

$$\frac{d}{dt}[d] = \frac{JG}{96,500D} ,$$

where d is the oxide film thickness in centimeters, J is the current density in A/cm^2, G is the equivalent weight of the oxide in gm/equivalent and D is its density in gm/cm^3. The value 96,500 A-s/equivalent is Faraday's constant. From empirical considerations it appears that the current density, J, obeys the relation $J = \alpha \exp(\beta E)$, where E is the electric field in volts/cm and α and β are constants.

Show that the oxide film growth rate at constant voltage may be expressed as

$$\frac{d}{dt}[d] = \frac{Bd^2}{t}$$

where B is a suitable constant $(B = 0.372 \times 10^4$ cm$^{-1})$.

8.3 Anodization to form a capacitor dielectric is first performed at a constant current of 1 mA/cm^2, while the voltage rises linearly, reaching 120 V after 7 min. Anodization is then continued at a constant voltage of 120 V, until a total time of 60 min has elapsed from the beginning of the anodization process. If the film is tantalum oxide having $G = 44.2$ gm/equivalent, $D = 8$ gm/cm^3 and $B = 5.8 \times 10^6$ V, find the total film thickness d attained.

What is the error in thickness if the total process had been 65 min instead of 60?

8.4 A tantalum-nitride thin-film resistor was aged for 1000 hr at a power level of 3/4 W. The mean value of the normalized resistance change $\Delta R/R$ was 0.45%.

Find the film temperature T and the thermal conductivity h of the substrate. Refer to Eqs. (8.9) and (8.10) and take the ambient temperature $T_S = 300$ K, $T_0 = 3200$ K, $t_0 = 25$ hr and $n = 0.3$ for $T \leqslant 150°C$ and $T_0 = 7900$ K, $t_0 = 10^{-7}$ hr and $n = 0.5$ for T between 150°C and 250°C.

8.5 To minimize the effect of resistance variability on circuit performance, two nominally identical resistors R' and R'' are placed in close proximity and the circuit characteristic is made to depend only on the ratio R'/R''. Show that if R' changes from R_0' by $\Delta R_0'$ and if R'' changes from R_0'' by $\Delta R_0''$ $(R_0' \approx R_0'')$, then in the worst possible case, the variation leads to

$$\frac{R'}{R''} = 1 \pm 2\left(\left|\frac{\Delta l}{l}\right| + \left|\frac{\Delta w}{w}\right|\right).$$

9

Thick-Film Circuits

9.1 INTRODUCTION [1-7]

This chapter covers the properties and design of thick-film circuits which are used in many applications ranging from consumer electronics to sophisticated digital computers. A thick-film circuit consists of passive components which are formed in place on the substrates by the selective application of materials in the form of pastes. These pastes are subsequently *fired* at high temperatures to form the desired films. Often discrete components, such as capacitors and silicon integrated circuits, are added to form a thick-film hybrid IC.

Thick-film circuits are made in a continuous process. The vertical and horizontal dimensions of the thick-film pattern are two to five times larger than the dimensions of thin-film or silicon integrated circuits. Thus, the equipment and manufacturing process is much simpler than thin-film or SIC equipment and processes. Hence, both the capital investment for a thick-film production facility and the cost of masks for a particular circuit are less than for a thin-film circuit. Moreover, the manufacturing process is simple and design changes can be easily made. Tooling for a new circuit is relatively inexpensive and small production runs can easily be accommodated. Production rates are high and capital investment for production equipment is relatively low. As a consequence, it is possible to consider using a thick-film circuit in applications where a thin-film film circuit would not be feasible or economical.

The chief passive components formed from thick films are conductive interconnections, high-precision resistor networks, and sometimes capacitors which are usually limited to values of 2000 pF. However, it is often more economical to use discrete capacitors, which are applique mounted to the thick-film circuits as are active silicon integrated chips containing transistors, diodes and diffused resistors. An important feature of the thick-film technology is a robust crossover which is not available in thin films.

A brief qualitative description of the thick-film process is given in the next section, followed in Sections 9.3 through 9.5 by descriptions of the properties of inks used to form conductive, resistive, and capacitive elements. Sections 9.6 and 9.7 discuss the design and layout of thick-film resistors and capacitors, Section 9.8 presents some guidelines for partitioning complex circuits, while Section 9.9 treats the substrate. The remaining section deals with the additional mounting of active chips or discrete components.

9.2 QUALITATIVE DESCRIPTION OF THE THICK-FILM PROCESS

The materials that will ultimately form the film are prepared as a suspension or emulsion of particles in a suitable organic vehicle of solvents. The suspended materials are then forced through a patterned screen by means of a moving squeegee to form the desired geometrical pattern. By analogy to the conventional screen printing process, these suspensions of particles are called inks. Thick-film inks typically consist of some or all of the following constituents:

1.   Finely divided metal or alloy powders which give the film cohesion or *joinability.* For the case of conductive films these metals determine the electrical conductivity.

2.   Metallic oxides, which in the case of resistive films determine the film resistivity.

3.   Finely divided glass frits, which chiefly determine the *adherence* of the film to the substrate and of different metals in the film to each other. Sometimes additional chemical bonding agents are added.

4.   An organic vehicle to provide basic flow (rheological) properties. Among the constituent agents are solvents, which dilute the paste and can be later evaporated, and surfactants (surface active agents), which permit solid particles to be wetted by the vehicle and be properly dispersed within it.

Two critical paste properties that control print quality are viscosity and surface tension. The paste or ink must not flow through the screen until forced to do so by the wiping action of the squeegee. Moreover, the paste must then adhere to the substrate and hold its form without running, when the screen breaks away.

After the ink is screened onto the substrate, usually through a stainless steel mesh, it is air dried at about 20°C for 5 to 15 min to permit leveling of the mesh pattern and to protect line definition. This is followed by a bake at about 125°C to 150°C to dry out the volatile solvents. The ink is then subjected to a firing sequence. In the initial phase of the sequence the organic vehicle must burn off completely without contaminating the film. During the peak of the firing cycle, which takes place either in a reducing or an oxidizing atmosphere depending on the particular film material, the glass frits and the bonding materials melt and/or react with the substrate to form an adhesion layer for the film. At the same time the metallic particles are sintered, thus forming the film. The film must be maintained at the peak temperature which ranges from 500°C to 1000°C for several minutes during which the film formation occurs. This is followed by a cooling period. The entire firing cycle takes about 45 minutes.

In the usual fabrication sequence for conductive and resistive films, the conductive pattern is first screen printed and fired. Then the resistor pattern is deposited over the conductive terminations and fired at the same, or a lower, peak temperature than the conductive pattern. Although it is possible to fire conductor and resistor patterns simultaneously, this approach is not recommended for best results, because the peak temperature will then not be optimum for each of the different compositions.

The recommended fired film thickness of many commercially available thick films is approximately 20 μm, which should be controlled to $\pm 10\%$ to ensure reproducible fired properties. Preferred resistor width is 1000 to 1250 μm (40 to 50 mils) or greater. Since the sheet resistance is $R_s = \rho/t$, it is clear that resistance depends inversely on the *fired* film thickness. After firing, a printed resistor will be within ± 5 to $\pm 50\%$ of the final desired value. Resistors are intentionally designed to be about 30% lower than the desired value because pattern changes after firing only increase the resistance. Resistors are then abrasively or laser trimmed to the final value. After trimming, a hybrid thick-film circuit is completed by attaching appliqued components and wire bonding the necessary interconnections. Properties of commonly used inks and factors influencing their choice will now be briefly elaborated.

9.3 CONDUCTIVE INKS

Conductive inks are based on noble metals such as silver, gold, platinum, and palladium. These metals are chosen because of their good electrical conductivity and joinability, when in the form of alloys. It is frequently advisable to use binary or ternary alloys of these metals, such as Pt-Au, Pt-Ag, Pd-Ag, or Au-Pt-Pd to obtain particular benefits such as better adhesion of film to substrate, improved film density, better reproducibility, or lower cost.

When selecting a conductor composition, several factors must be considered.

Solderability Fired conductor films should be suitable for soldering to allow the attachment of external leads or appliqued components. Sometimes a reduction of circuit interconnection resistance may be obtained by immersing the substrate in molten solder, thereby tinning only the solderable conductive thick film.

As a figure of merit, one may speak of the *tinnability,* that is, the ability of certain materials to be wet by solder. Table 9.1 describes this ability for a variety of conductive films. As may be seen from the comments in this table, a problem encountered in soldering is the phenomenon of *leaching,* or *erosion,* which is the dissolution of thick-film metal components such as silver and gold, when brought into contact with molten solder. When leaching occurs, a conductive film will *dewet* the substrate, resulting in circuit failure. This very undesirable phenomenon is measured in terms of the time (usually in seconds) taken for the solder to dewet. Leaching is temperature dependent and the time to dewet decreases with increasing solder temperature.

Tin- and lead-based solders are mostly used for bonding and lead attachment, while gold- and tin-based solders are usually used for hermetic sealing of packages.

Soldering of external leads often requires the screening of a dielectric solder dam to prevent any gold conductor patterns from coming into contact with solder. Such a dam is deposited at the same time as resistive overglazes, which are discussed later in this section. Figure 9.1 shows noncon-

Table 9.1 Unplated Tinnability (10 Sn:90 Pb at 625°F)

Material	Classification	Main Problem
Cu	Fair-Poor	Oxidation
Ag	Fair-Good	Erosion
Au	Fair-Good	Erosion
Ag-Pd	Good	-
Ag-Au-Pd	Good	-
Ag-Pt	Good	-
Au-Pt	Good	-
Au-Pd	Good	Some erosion
Au-Pt-Pd	Good	-
Pt	Fair-Good	Erratic
Pd	Fair	Oxidation
Mo-Mn	Poor	Requires plating
Mo	Poor	Requires plating
W	Poor	Requires plating
Ti-Zr	Poor	Requires plating

tinuous solder dams which are preferred because the parasitic capacitive or resistive effect of the glaze is minimized.

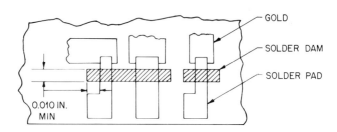

Fig. 9.1 Noncontinuous solder dams to prevent solder from contacting gold conductor pattern.

Fired film adhesion[8] The adhesion of a conductive thick film to the substrate depends on processing conditions and particularly on the amount of flux, such as Bi_2O_3 with glass, used in the ink. This glassy flux can improve the adhesion of the fired film, but it can also degrade its tinnability. Fluxes in the form of low-melting-point borosilicate glasses seem to give the best results. The adhesion of a binary or ternary alloy thick film also depends on the composition of metals and on the firing temperature.

Fritless or reactively bonded gold-based pastes are used when thermocompression bonding of beam-lead silicon integrated circuits or of lead frames is contemplated. Pastes containing glass frits are not recommended here because although the device may be reliably attached to the conductor, it cannot, in practice, be removed and replaced if found to be defective. Fritless pastes contain small amounts of Cu_2O in place of the glass frit. Adhesion of the film takes place by chemical reaction at the film-substrate interface, where copper aluminate compounds are formed.

Film adhesion is often measured by means of a tensile strength test in which a nail-head bond placed on a thick-film pad is pulled upward and away at 90° from the substrate. Table 9.2 displays the results of such a test for various film materials. As may be seen, a tensile strength of about 180 kg/cm² (2500 psi) forms a lower bound and proves adequate, while a figure of 720 kg/cm² (10,000 psi) represents excellent adhesion.

Film compatibility Compatibility of conductive inks with resistive or dielectric inks is very important when making a thick-film circuit. If films

Table 9.2 Results of Tensile Adhesion Test
(Adhesion to 90-96% Alumina)

Material	Classification	Typical Tensile Strength (kg/cm)
Cu	Fair	170
Ag	Fair	-
Au	Poor	-
Ag-Pd	Good	210
Ag-Au-Pd	Fair	204
Ag-Pt	Good	140-320
Au-Pt	Good	210
Au-Pd	Fair	140
Au-Pt-Pd	Good	320
Pt	Fair-Good	140-350
Pd	Poor-Fair	-
Mo-Mn	Excellent	700
Mo	Fair	-
W	Excellent	700
Ti-Zr	Excellent	700

are not compatible, interaction between the glassy phases of the two films may occur at the firing temperature. This can lead to deterioration of component characteristics, to deformation of layers, and to a possible loss of film adhesion. Manufacturers of inks will generally ensure that a particular range of their suggested products is fully compatible. Some manufacturers have arranged their suggested processing sequence so that conductive inks are first printed and fired at the highest temperature, with resistive or other layers printed and fired sequentially, at lower temperatures.

Line resolution   The printing of fine lines ranging in width from 50 to 150 μm (2 to 6 mil), requires special screens and precision techniques. Generally, the attainable linewidth depends also on the firing process. Fine-line, gold-based conductor compositions (Au, Pt-Au, and Pd-Au), having a high solids content, generally permit line resolution of about 150 μm.

The minimum conductor width for normal applications is 250 μm, which is equivalent to two openings of a 200-mesh screen. Larger widths are preferred and the spacing between conductors, conductive pads, or resistors should be at least 250 μm (10 mil), as shown in Fig. 9.2. Resistor-conductor circuit design and device attachment is discussed in Sections 9.6 and 9.9.

Screened electrical film resistance   The resistance of a fired thick-film conductor should be as low as possible. Experimental values obtained for 250-

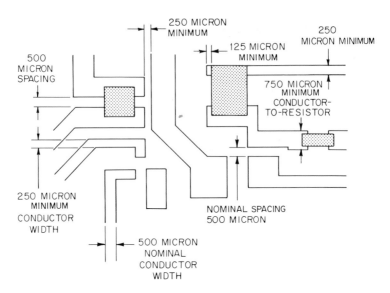

Fig. 9.2 Minimum spacing for conductor patterns.

μm wide films screened with 200 mesh are presented in Table 9.3. Because film resistivity depends on film thickness, which in turn depends on linewidth, all films measured were 250 μm wide. The measured values for film resistivity can often be considerably reduced by tinning the conductor patterns with solder.

Film stability[9]   Film stability is the ability of a thick film to retain its properties under conditions of storage and service. For instance, it is well

Table 9.3 Experimental Screened Film Resistance Values for 250 μm (10-mil) Wide Films

Material	Classification	Experimental Resistance (Ω/cm)
Cu	Good	11.8
Ag	Good	7.9
Au	Good	23.6
Pt	Fair	78.7
Ag-Pd(80:20)	Fair	59
Ag-Pt(80:20)	Fair	78.7
Au-Pt(80:20)	Poor	256
Au-Pd(80:20)	Poor	394
Ag-Au-Pd(20:55:25)	Poor	181
Au-Pt-Pd(60:20:20)	Poor	394

known that some thick-film circuits suffer loss of substrate adhesion under high-temperature storage conditions. Another well-known effect leading to loss of stability is caused by silver migration under high-voltage and high-humidity conditions. Such migration occurs at resistor-conductor terminations and between parallel conductors, when silver compositions are employed. To guard against loss of stability, fired films are often protected by coating them with a protective glaze overlayer which is fired at an elevated temperature. Such glazes protect the film from humidity, from organic solvents present in the atmosphere, and from other contamination. Factors also affecting film stability are film size and operating environment.

Some good and bad points in the behavior of conductive films are briefly summarized below:

Single metal films   Pure gold conductors are not solderable by means of lead-tin alloys because of leaching, although many gold-bearing conductors are solderable. On the other hand, any of the gold-bearing conductors may be used for eutectic die bonding. *Fritless* gold-bearing conductors must be used for thermocompression bonding of beam-lead silicon integrated circuits, particularly if bond repair is contemplated. No cracking of glass film-substrate bonds can occur with fritless gold conductors as a consequence of bonding loads. Upper conductors of crossovers may be made of gold or gold-bearing materials. Gold films are expensive.

Platinum conductors are costly and sometimes suffer from reproducibility problems. They are generally solderable, die bondable, and thermocompression bondable.

Palladium films have adhesion problems and may give rise to oxide formation.

Copper films are markedly dependent on paste particles and have to be fired in inert or reducing atomospheres.

Binary alloys[10, 11]   Pt-Au films are most expensive, but they have excellent erosion or leach resistance and are recommended for circuits where discrete components may have to be replaced. Pt-Au films may be used for eutectic die bonding and they are thermocompression bondable.

Pd-Ag films are economical and have many desirable properties such as good adhesion, good solderability, and preform die bondability. Pd-Ag is not thermocompression bondable and could be subject to silver migration under conditions of high stress and high humidity. Pd-Ag films are frequently used to make pads for soldered leads and components, as well as for conductor patterns incorporating simple resistor networks not requiring crossovers.

Pt-Ag films are economical, solderable, die bondable with preforms, but not thermocompression bondable. Pt-Ag films are not stable in high-humidity conditions and under electrical bias.

Pd-Au films may prove to be porous and highly resistive. However, they show good solderability, are reasonably economical, and are sometimes used in place of Pd-Ag films for crossovers to eliminate silver migration problems.

Ternary alloys   Ag-Au-Pd films do not have silver migration problems, but may suffer from adhesion problems.

Au-Pt-Pd films are more resistant to solder erosion than are Au-Pt films. Furthermore the ternary alloy improves the fired density of lands and improves the adherence to the ceramic substrate. However, Au-Pt-Pd films are costly and may have unduly high resistance.

9.4 RESISTIVE INKS [6,12,15]

A resistive ink has three main components, namely, a conducting material, a flux or frit to provide film adhesion to the substrate, and an organic vehicle to provide desirable flow properties and screenability. The conducting material might be a metallic element or its oxide, an alloy, or compounds of the elements and oxides. The most common conducting material in the constituents of resistor ink is ruthenium.

An important class of resistor inks use ruthenium dioxide as the conducting material. Ruthenium dioxide is stable up to 1000°C under normal atmospheric conditions. Like all of the oxides of ruthenium it is a semiconductor, but its electrical properties are those of a metal. It has a resistivity of about 50 $\mu\Omega$-cm and a TCR of 3000 ppm/°C. The TCR can be reduced by doping the ruthenium dioxide with niobium pentoxide.

A typical ruthenium dioxide resistor ink[15] is fired at a peak temperature of 760°C. The firing operation consists simply of the burning off of organic constituents and the fusion of the glass component. During the firing operation, the glass and conductive components remain chemically inactive and therefore the sheet resistance is effectively independent of the peak firing temperature as shown in Fig. 9.3.

A popular family of resistor pastes marketed by DuPont under the trade name Birox, uses a ruthenate as the conductive component. The recommended peak firing temperature is 850°C, and the sheet resistance of the fired film is relatively independent of firing temperature and length of the firing cycle as shown in Figs. 9.4(a) and 9.4(b). However, the TCR does change markedly as the peak firing temperature deviates from the recommended value as shown in Fig. 9.5.

Another class of pastes uses mixtures of silver and palladium as the conductive component. The conductivity of the fired film is thought to be mainly controlled by the presence of palladium oxide, with the palladium

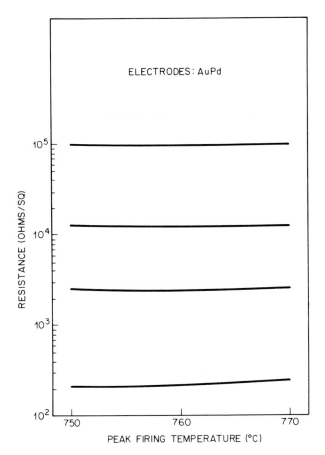

Fig. 9.3 Sheet resistance versus peak firing temperature for several ruthenium dioxide resistor pastes.

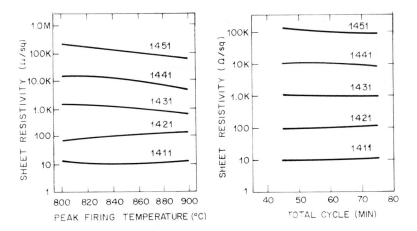

Fig. 9.4 Sheet resistance of Birox 1400 Series resistor films as a function of (a) the peak firing temperature and (b) the length of the firing cycle. (Courtesy of E. I. Du Pont De Nemours & Company.)

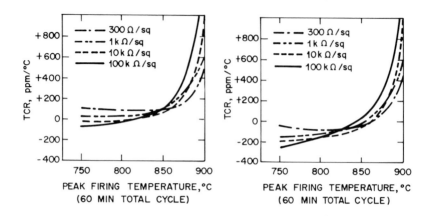

Fig. 9.5 Effect of peak firing temperature on TCR of Birox 1400 Series resistor films: (a) from −55°C to 25°C and (b) from 25°C to 125°C. (Courtesy of E. I. Du Pont De Nemours & Company.)

silver providing better joinability, film density, and uniformity. Palladium oxide and palladium silver are formed during the firing process, but neither reaction proceeds to completion. The resultant composition depends critically on the firing profile and particularly on the peak firing temperature and time at that temperature. The resistance of a palladium-silver film increases with an increase of palladium, indicating a greater palladium-oxide formation. Because such a fired film is, to some extent, susceptible to further oxidation at elevated operating temperatures, it is called a *reactive, dynamic,* or *nonequilibrium*-resistive film. These films are normally protected by a glaze fired at about 500°C to prevent further change in resistance due to oxidation.

The glass flux must be compatible with the conducting material, the substrate, and the conductor films. The flux, consisting, for example, of Bi_2O_3 and glass frit, is dispersed throughout the ink. Good adhesion and tinnability may be obtained by the use of low melting point lead borosilicate glasses as the flux. Generally, it is preferable to match the expansion coefficient of the frit to that of the metals (rather than to that of the substrate) in order to achieve good film adhesion.

During the firing process, the particles of the conducting material and the flux sinter and fuse forming coherent deposits, known as *lands,* which adhere to the ceramic substrate. Shrinkage of the deposited volume occurs as the ink sinters during firing, and it is important that cracks and fissures do not occur.

The role of the organic vehicle is fourfold. The vehicle consists of a resin, a flow-control agent, a solvent, and a surfactant. The resin provides the desired viscosity for the inks, while preventing undue formation of gels. The purpose of a flow-control agent is to eliminate *secondary flow* and maintain fine-line definition. Secondary flow is defined as a further flow of the ink which may take place after screen printing has been completed in normal ambient conditions.

The solvent—which has by far the largest percentage weight among components of the vehicle—is used to effect mixture of metal powders and flux with adequate dispersion and breakup of pigment agglutinates. It is desirable that the solvent have low volatility so that inks do not dry out rapidly during screening. The surfactant makes the ink incompatible with water so that the rheological properties of the composition are not affected by humidity variations. Finally, inks should be stable, with low volatility and good flow properties, and capable of passing through fine-line mesh screens used for pattern definition.

A secondary characteristic of a resistive film is the temperature coefficient of resistance (TCR). This is defined as

$$\text{TCR} = \frac{1}{R}\,\frac{dR}{dT} \times 10^6$$

and is expressed in parts per million per degree centigrade. TCR may be established by measuring film resistance at two temperatures. Depending on processing, sheet resistance, and temperature range, TCR may be positive, zero, or negative. For some resistive films it is possible by means of processing techniques to restrict TCR to within ± 50 ppm/°C over the -55°C to 125°C temperature range.

Thick-film resistors are also subject to change of resistance value for different applied dc levels and this secondary effect seems to be a function of sheet resistance. Voltage coefficient of resistance (VCR) is defined as

$$\frac{1}{R(V_1)}\,\frac{R(V_2) - R(V_1)}{V_2 - V_1} \times 10^6$$

where V_2 and V_1 are voltages developed across the same film and leading to resistance values R_2 and R_1, respectively. Good thick-film resistors should have VCR values smaller than 50 ppm/V.

Another important parameter for thick-film resistors is noise. Noise measurements usually compare the noise output of a thick-film resistor against that of a standard wirewound resistor. This measurement is restricted to a specified frequency range and the measured value is usually expressed in decibels as 10 log (noise power of sample/noise power of standard). Measurement of typical thick-film resistors yield values from -30 dB to $+30$ dB. As an example, a resistor with a noise output of $+20$ dB has a noise level

100 times greater than that of a specific standard resistor, over a particular range of frequency.

9.5 DIELECTRIC INKS

Dielectric inks are screened and fired to produce dielectric films for use in parallel-plate capacitors, crossovers, protective overglazes, and hermetic seals. These different structures will now be examined.

Dielectric parallel-plate capacitors   Typical parallel-plate capacitor structures, shown in Fig. 9.6, consist of a base or bottom electrode, a dielectric layer and an upper electrode. Usually the base electrode is first printed and fired followed by the printing, drying, and simultaneous firing of two separate dielectric layers. This ensures freedom from pinholes because one dielectric layer will fill the pinholes of the other. The structure is completed by printing and firing the upper electrode layer.

Normally, silver-bearing electrodes are undesirable in such a capacitor structure because the stress of high dc voltage might produce silver migration. High-permittivity dielectrics ($\kappa \approx 1000$) are frequently based on the ferroelectric ceramic material barium titanate. Low-permittivity dielectrics are nonferroelectric materials, such as magnesium titanate ($\kappa \approx 14$), zinc titanate ($\kappa \approx 19$), titanium oxide ($\kappa \approx 95$), or calcium titanate ($\kappa \approx 160$).

All dielectric-material properties are temperature dependent, with the temperature coefficient of capacitance (TCC) varying from a few ppm/°C for low dielectric constant material, to 30% or more for high dielectric constant materials for temperatures in the range from $-55°C$ to $+125°C$. A typical curve of capacitance change versus temperature for a high dielectric constant capacitor is shown in Fig. 9.7. Low-permittivity capacitors possess a higher

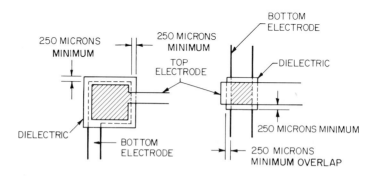

Fig. 9.6 Thick-film, parallel-plate capacitor structures.

degree of temperature stability, with no more than a few percent linear change of capacitance between −55°C and 125°C.

High dielectric constant capacitors are also dc voltage and frequency-dependent as shown in Fig. 9.8. These curves are characteristic of the ferroelectric material employed, and therefore, reflect changes encountered also in conventional ceramic capacitors. Dissipation factors for such capacitors, also shown in Fig. 9.8, are of the order of 3-5% with breakdown voltages in the region of 4 to 20 VDC/μm. The maximum capacitance density obtainable using high dielectric constant inks is at present about 1200 pF/cm^2/μm.

Crossover   As the complexity of thick-film circuits increases, it may become necessary to incorporate conductor crossovers. An advantage of the thick-film technology is the robust crossover. A typical crossover structure is shown in Fig. 9.9. A thick-film circuit requires a minimum of three screenings to form the lower conductor, dielectric layer, and upper conductor of the crossover structure.

Three basic types of dielectric inks are in use for conductor crossovers, namely, glasses, recrystallizing glasses, and glass-ceramic mixtures. The dielectric layer is usually double printed to minimize the occurrence of pinholes which could cause shorts between the conductors.

Crossover materials must be compatible with conductors and have matching thermal expansion coefficients. Glass dielectric compositions must be gradually fired, with the first conductor layer being fired at the higher temperature and the glass layer being fired some 100°C lower. This avoids excessive flow of glass leading to movement of the top conductor, an effect

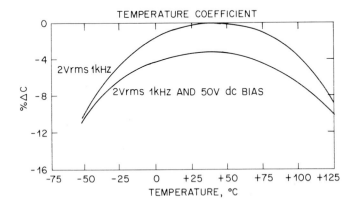

Fig. 9.7 Capacitance variation versus temperature for a high dielectric constant capacitor.

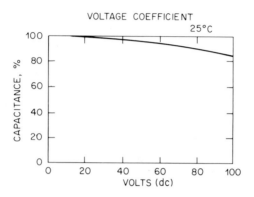

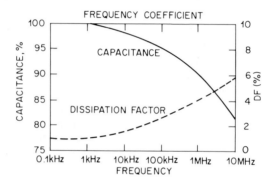

Fig. 9.8 Voltage and frequency dependence of a high dielectric constant capacitor.

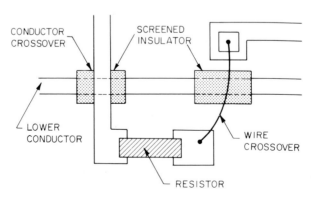

Fig. 9.9 Thick-film conductor crossover.

known as swimming, and necessitates careful control of firing procedures. Interconnection between conductor layers is made by etching windows into the glass insulation layer, through which the next conductor layer may be printed. Values of breakdown strength of the order of 4 to 20 VDC/μm are common, with insulation resistance figures of 10^{10} Ω/cm^2 measured in the 10- to 100-V range.

A recrystallizing glass dielectric exhibits no reflow subsequent to the first firing, because the material loses its vitreous nature and develops a crystalline structure. This absence of flow during later firings permits good dimensional registration to be maintained without requiring a gradual firing procedure. Typical insulation-resistance values are 4×10^{11} Ω/cm^2/μm, with dielectric breakdown strength in excess of 500 V for thick crossovers.

Glass-ceramic mixtures are designed for high viscosity at normal firing temperatures so that very little flow takes place. Moreover, gradual solution of the ceramic component on subsequent refiring leads to a further increase in viscosity.

Some applications combine both thick and thin films on the same substrate. A hybrid crossover would consist of a thick-film lower conductor and dielectric layer, and a thin-film upper conductor. Typical crossover dielectric properties are lised in Table 9.4.

Table 9.4 Crossover Dielectric Properties

Property	Unit	Thick Film	Thick/Thin Film	Remarks
Dielectric Constant	-	9-12	6-9	1 kHz
Dissipation Factor	%	0.05-0.25	0.5-1.0	1 kHz
TCC	ppm/°C	<75	<100	-
Simulation Resistance	ohms	>10^{12}	9	>10^{10}, 100 VDC
Breakdown Volts	volts	>400	>500	-
Firing Temperature	°C	850-900	850-900	-
Thermal Expansion Coefficient	-	4.6×10^{-6}	6.3×10^{-6}	0–300 °C
Capacitance Density	pF/cm^2	200-300	125-140	-

Protective overglaze and hermetic seals These overglazes provide protection for thick-film resistors and capacitors against environmental conditions such as humidity, reactive contaminants, or reducing atmospheres.

Glazes are normally composed of low melting point vitreous glasses which are fired at temperatures of about 500°C. Such glazes are low in cost and provide good electrical insulation, but they have the disadvantage of being brittle and poor conductors of heat.

9.6 DESIGN AND LAYOUT OF THICK-FILM RESISTORS

Many aspects of resistor design were considered in Chapter 4. Some of the results from Chapter 4 are specialized in this section to the design of thick-film resistors.

Once again, it is convenient to work with the sheet resistance $R_S = \rho/t$, where t is now the thickness of the *fired* film. Because fired film thickness t is considered fixed, the design of a thick-film resistor to have resistance value R consists of choosing a suitable ink having sheet resistance R_S and then providing the necessary aspect ratio n in the resistor body. Inks are available in decade values of R_S from $1 \, \Omega/\square$ to $10 \, \text{M}\Omega/\square$. Nevertheless, such design is constrained by power considerations, for it is found that thick-film resistors on 96% alumina substrates should have a maximum power density Q_{max} of about $8 \, \text{W/cm}^2$. The minimum surface area A_{min} required for a straight line (bar) resistor of value R dissipating power P is given by Eq. (4.5b) as $A_{min} = P/Q_{max} = (R/R_S) w_{min}^2$, where w_{min} is the minimum track width imposed by the power constraint.

Typical thick-film resistor layouts and minimum dimensions are shown in Fig. 9.10. Short resistors have $l < w$ and should have an aspect ratio $1/3 < n < 1$. Long resistors $(l > w)$ should have an aspect ratio $1 < n < 10$. Note that the resistor length l is the length between conductors and not the full length of the resistor pattern. If the long resistor is made too long $(n > 10)$, it is liable to develop hot spots, which ultimately may lead to breakdown. Long resistors are also difficult to trim. If a resistor is made too short, its final value becomes increasingly difficult to predict accurately because of termination effects.

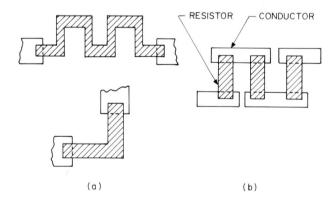

(a)   (b)

Fig. 9.10 Thick-film resistor patterns.

Top-hat resistors are employed for aspect ratios greater than three, even though they take up much substrate area. A large resistor can be made with a low resistivity ink, thus eliminating a screening operation.

Meander or zig-zag resistors, which are used in silicon and some thin-film designs, should be avoided here because of the danger of hot-spot formation at corners and also because they are difficult to trim abrasively. Figure 9.10 shows parallel resistors in meander form which are acceptable.

As a general rule, a resistor should be made as large as the available area will allow. Resistor width should be 1000 to 1250 μm, and for safe registration tolerance the resistor should overlap the conductor by 250 μm or more. Moreover, the conductor terminations should be wider than the resistor track by at least 250 μm.

Resistor values are adjusted by abrasive or laser trimming which involves removing part of the resistive film as shown in Figs. 9.10 and 9.11. Consequently, the final resistance increases, and for good design the value of the untrimmed resistor should be about 70-80% of the final trimmed resistance.

As an example, suppose a 20-kΩ resistor is desired and one designs to 80% of final value. Hence the untrimmed resistor should have a value $R = 20 \times 0.8 = 16\ k\Omega$. If an ink with sheet resistivity $R_S = 8\ k\Omega/\square$ is available, then the aspect ratio n of the untrimmed resistor is $n = R/R_S = 2$. If the resistor width w is chosen to be 1250 μm, then the resistor length l will be 2500 μm and the resistor area $A = w \times l$ will be 3.125×10^{-2} cm^2. If the resistive film has a rated power density $Q_{max} = 3.2$ W/cm^2, then the resistor can dissipate a rated power of $P = Q_{max} \times A = 100$ mW. Hence, it is essential that the final resistor does not encounter a voltage or current stress in excess of this value.

Abrasive trimming is simplified if all bar resistors are aligned in either the X or the Y direction and if possible along only one axis. It is necessary to leave a trimming space of about 1250 μm between resistors as shown in Fig. 9.11.

Finally, because each ink requires a separate screen printing, the number of different resistor inks used on each substrate should be kept to a minimum and should not exceed three in number. In general the size of the substrate, the inks available, and the power requirements will determine the best layout and ink combination. The following is a brief summary of recommended design practices.

1. Conductive termination pads for all resistors must be screened by the same screen and fired before the resistor material is deposited.

2. Resistors must be screened on bare ceramic and must be oriented with sides parallel to the substrate edges. No resistor is to be located closer than 500 μm from the substrate edge.

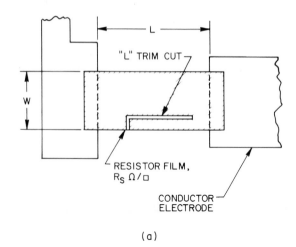

(a)

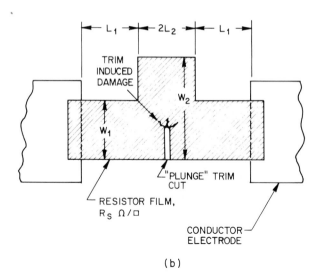

(b)

Fig. 9.11 Resistor dimensions and trim cuts.

3.   Large dimensional differences should be avoided for resistors printed during the same screening so that variations in print thickness and as-fired tolerance are minimized.

4.   Ordinary as-fired tolerances for resistors is 20-30%. Rectangular resistors should be designed for 0.7 of final value and top-hat resistors for 0.5 of final value to allow for trimming.

5. Circuit layout must allow for probing and trimming. It is preferable to probe resistors at external lead bonding pads as shown in Fig. 9.12. If a resistor is to be probed at one or both of its terminals, a probe pad at least 500 μm on a side must be provided with 750 μm being preferred. This is also shown in Fig. 9.12.

6. For optimum TCR tracking, resistors should have the same screening direction, be screened with the same ink, and have the same width.

7. Low-value resistors may be made by paralleling two or more low-aspect ratio resistors.

    Large-value resistors may be made by connecting two or more high-aspect ratio resistors in series. A meander arrangement utilizing bar resistors connected by straight-line conductor shorting bars occupies a relatively small area.

    Closed resistor patterns, with a break in the conductor that can be closed by a wire bond, should be avoided as this is an expensive procedure.

8. Resistors which are to be air-abrasively trimmed should be at least 1000 μm from the nearest conductor line, as shown in Fig. 9.11. For laser trimming, 500 μm suffices.

9. For good design, 1250-μm clearance should be provided between resistors and crossovers. Conductor width should have a preferred value of about 500 μm. Wide conductors are recommended because they minimize circuit resistance. Spacings between conductors, resistors, and pads should have a preferred minimum value of 50 μm.

10. Many circuits are nonplanar, that is, they cannot be drawn on a plane without crossovers. Figure 9.9 shows both a screened and a wire crossover. Screen crossovers require extra processing steps, while wire crossovers require additional bonding operations. Both components increase the cost of the circuit. Moreover, crossovers always pose a potential danger of shorts and invariably have an associated parasitic capacitance, often amounting to several picofarads. Crossovers should be avoided in any design if at all possible.

11. Sometimes a crossover can be avoided by running a conductor between the connection pads of a mounted chip capacitor, as shown in Fig. 9.13. For safety, the conductor passing under the mounted capacitor chip is given a protective insulating coating.

9.7 DESIGN AND LAYOUT OF THICK-FILM CAPACITORS [13]

By analogy with resistor design, where it was found convenient to define the sheet resistance R_S, it is convenient to define the capacitance density C_s,

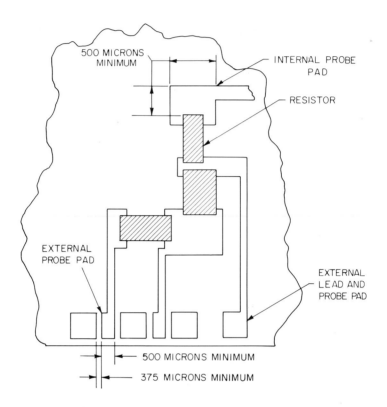

Fig. 9.12 Placement and dimensions of bonding pads.

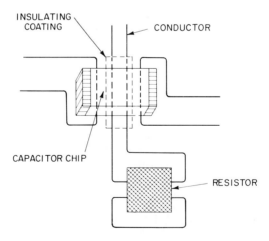

Fig. 9.13 Use of appliqued capacitor chip to facilitate a crossover.

where

$$C_s = \kappa \epsilon_0 / t. \qquad (9.1)$$

Here κ is the dielectric constant, ϵ_0 is the permittivity of free space and t is the dielectric thickness. The actual capacitance of a parallel-plate capacitor is given by

$$C = \kappa \epsilon_0 A / t \qquad (9.2)$$

where A is the capacitor area and therefore the capacitance density C_s is given by

$$C_s = C / A. \qquad (9.3)$$

Because the capacitance density varies inversely with dielectric thickness, there is an upper limit for this density set by how small t can be made realibly. Dielectric thickness t in turn is determined by the dielectric strength of the material and by the necessity to avoid pinholes.

If, for example, $\kappa = 100$ and t for a printed dielectric is 500 μm, then $C_s = 1850$ pF/cm^2. For a capacitor area of 0.313 cm^2, the capacitance C would then be 560 pF.

A possible capacitor structure was previously shown in Fig. 9.6. As a general rule, the printed dielectric film should overlap the bottom electrode by at least 250 μm for each edge and the top or counter electrode must fall completely within the bottom electrode area. The plate overlap area then represents the useful capacitor area.

Capacitor trimming is carried out by removing a portion of the top electrode and dielectric. As in resistor design, if the capacitor is to be adjusted, room must be left on one side of the capacitor for trimming. Different capacitor configurations are possible, but in any case bottom and top electrodes must be properly insulated from each other.

9.8 CIRCUIT PARTITIONING

When a circuit is too complex to be manufacturable on a single substrate, either because of economic or reliability considerations, it must be partitioned into smaller, functional entities. There are many criteria to be applied in the partitioning procedure such as thermal constraints, cost of alternative packages, reliability, and producibility. The following guidelines provide a reasonable compromise between all design objectives.

1. Make each subcircuit perform a distinct function to simplify the specification and testing of the individual modules. Try to keep all elements required to perform a specific function in the same module.

2. Keep the component count and component density uniform to permit the use of standard processing and packaging techniques.

3. Distribute the high-power dissipating elements uniformly to maintain a reasonably uniform power density on each substrate and approximately equal power dissipation in each package. Make sure that the power dissipated in each module is within the capability of the package.

4. Leave physically large components outside the package when that is possible.

5. Use silicon IC chips whenever possible to take advantage of the high component density.

6. Yields associated with various assembly operations impose a practical upper limit on the number of SIC's or appliqued components that should be mounted on one substrate. The number of these components in a particular module should be limited to an amount commensurate with these yields.

9.9 THE SUBSTRATE [14]

The major requirements for a thick-film substrate are as follows:

1. Mechanical strength, good thermal conductivity, and high bulk and surface resistivity.

2. Smooth surface texture (consistent with good film adhesion) and chemical and physical compatibility with the fired thick film.

3. Low camber, that is, small deviation from surface flatness.

4. Ability to withstand normal film firing temperatures between 500°C and 1000°C.

5. Close tolerances on overall dimensions and positioned holes.

6. Low cost.

One ceramic material, namely alumina (Al_2O_3) displays a good combination of the above requirements. Table 9.5 lists characteristics for 96%, and 99.5% alumina and for beryllium oxide (BeO) materials. A widely used thick-film ceramic is 96% alumina which is available in a variety of shapes and sizes at reasonable cost. Although beryllium oxide (BeO) has about ten times the thermal conductivity of alumina, it is toxic when finely powdered and also more costly.

Most thick-film screening is done on 5×5-cm by 0.625-mm thick 96% alumina substrates. For combined thick- and thin-film circuits, 99.5% alumina substrates are used as these give more stringent tolerance on substrate flatness and a better surface for the thin films.

One frequently used method of substrate manufacture is ribbon casting of the ceramic prior to firing. In this process alumina powder of fine particle size is suspended in a liquid carrier to form what is called a *ceramic slip*. This

Table 9.5 Properties of Ceramic Substrates

	94% Al_2O_3	96% Al_2O_3	99.5% BeO
Specific Gravity	3.62	3.70	2.88
Safe Temperature at Continuous Heat (°C)	1500	1550	1500
Tensile Strength (kg/cm^2)	1400	1760	-
Compressive Strength (kg/cm^2)	22,150	26,360	>13,000
Flexural Strength (kg/cm^2)	3090	3230	1760
Thermal Conductivity (cal/cm-s-°C)	0.034	0.041	0.28
Dielectric Strength at 60 Hz kV/mm	8.3	8.3	8.7
Volume Resistivity (Ω-cm):			
25°C	$>10^{14}$	$>10^{14}$	$>10^{14}$
100°C	7.0×10^{13}	2.0×10^{13}	$>10^{14}$
300°C	4.4×10^{10}	1.1×10^{10}	$>10^{14}$

slip is placed on a glass support and the thickness of the slip is regulated by a device called a *doctor blade*. After oven drying, the slip is called *green* and is sufficiently plastic to have the substrate configuration, including holes, stamped. Next, it is sintered by firing in a furnace between 1500°C and 1800°C. After sintering, the ceramic is unaffected in structure, size, and surface finish by any subsequent firing of the thick-film inks.

Substrate thickness normally is between 0.5 and 1.0 mm, with tolerances ranging from 25 μm to 125 μm. Tolerances on the positioning of holes range also from 50 μm to 125 μm. Substrate areas are typically between 3.5 cm^2 to 35 cm^2 and it is important that the substrate not be bowed by more than 40 μm/cm because a large camber causes a variation in substrate-screen separation and thereby degrades screen printing quality. As a result of camber considerations, substrates that are to be screened uniformly are limited to 6-cm by 6-cm square. At increased cost, 96% alumina substrates can be supplied flatter than 40 μm/cm. Extremely flat substrates are also more resistant to stresses introduced during thermocompression bonding.

The diameter of any hole in the substrate should be more than two-thirds of the substrate thickness, with a minimum diameter of approximately 0.5 mm. If a hole has any electrical feedthrough function, then a conductor pad larger than the hole size is required on each side of the substrate to provide at least 0.375 mm of conductor material on all sides of the hole. In all cases metallized holes should be about 125 μm to 200 μm larger than the lead wires.

Conductor contact areas or pads should be from 1.25 mm to 2.5 mm in length. A minimum clearance of 0.75 mm should be provided between any resistor or conductor and the edge of the substrate. If a cover is to be hermetically sealed to the substrate, a glassified sealing bond of about 1.25 mm should be allowed. Furthermore, conductors should not be run around the edge of the substrate.

9.10 MOUNTING OF CHIPS AND OTHER DISCRETE COMPONENTS

The full utility of thick-film circuits is realized when they are incorporated in a hybrid circuit with appliqued components such as capacitor chips and silicon integrated circuits.

Silicon chips are mounted on a conducting pad by die bonding as described in Chapter 10. Semiconductor chips should not be mounted close to the edge of the substrate. Figure 9.14 shows an active chip mounted on a bonding pad. Most transistor and diode dies are of the order of 0.625 mm by 0.625 mm or larger. The mounting pad should provide a border of preferably 250 μm on each side of the chip. The length of wire interconnections, shown in Fig. 9.14 should be kept to a minimum and should be between 1 and 2 mm in length. Both 25-μm gold and 25-μm aluminum

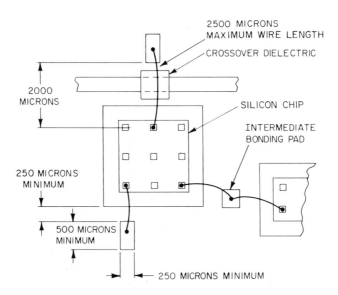

Fig. 9.14 Mounting and interconnection of silicon chips on thick-film circuits.

wires are commonly used for such interconnections. Discrete components usually should not be mounted across resistors or conductors, with the capacitor chip shown in Fig. 9.13 being an exception to this rule. Wires should not cross over exposed conductor lines or semiconductor dice. When a crossover is necessary, the conductor line must be overglazed at the crossover point to prevent contact. In addition, wires should never cross other wires. When making thermal compression or ultrasonic bonds between semiconductor dice, an intermediate bonding pad should be used, as also shown in Fig. 9.14. One should never bond directly from semiconductor chip to semiconductor chip.

REFERENCES

1. M. L. Topfer. *Thick Film Microelectronics.* New York: Van Nostrand-Reinhold, 1971.

2. A. V. Planer and L. S. Phillips. *Thick Film Circuits.* New York: Crane Russak Co., 1972.

3. D. W. Hamer and J. V. Biggers. *Thick Film Hybrid Microcircuit Technology.* New York: John Wiley & Sons, 1972.

4. J. Agnew. *Thick Film Technology.* Rochelle Park, N. J.: Hayden, 1973.

5. R. A. Rickoski. *Hybrid Microelectronic Circuits—The Thick Film.* New York: John Wiley & Sons, 1973.

6. C. A. Harper, editor. *Handbook of Thick Film Microelectronics.* New York: McGraw-Hill, 1974.

7. A. E. Zinnes, private communication.

8. W. A. Crossland and L. Haule. "Thick film conductor adhesion reliability." *Solid-State Technology* **14**: 42-47 (February 1971).

9. P. W. Polinski. "Stability of thick film resistors under high electromagnetic stress." *Solid-State Technology* **16**: 31-35 (May 1973).

10. M. Hansen. *Constitution of Binary Alloys.* New York: McGraw-Hill, 1958.

11. R. P. Elliott. *Constitution of Binary Alloys,* First Supplement. New York: McGraw-Hill, 1965.

12. R. E. Trease and R. L. Dietz. "Rheology of pastes in thick film printing." *Solid State Technology* **15**: 39-43 (January 1972).

13. C. W. Young. "A thick film capacitor design nomograph." *Electronic Packaging and Prod.* **11**: 71-72 (July 1971).

14. W. D. Kingery. *Introduction to Ceramics,* 2nd Edition. New York: John Wiley & Sons, 1976.

15. T. H. Lemon. "Thick-film ruthenium resistor pastes." *Platinum Metals Review* **17**: 14-20 (January 1973).

PROBLEMS

9.1   The circuit of an emitter-coupled astable multivibrator is shown in Fig. P9.1.
Assume that the transistor die size is 1500 μm by 1500 μm and that a suitable
0.1 μF capacitor chip is 8 mm by 6 mm.

   a)   How many screening operations are required to produce this circuit?

   b)   Design a suitable thick-film circuit layout. Make reasonable engineering
assumptions for anything that is not specified.

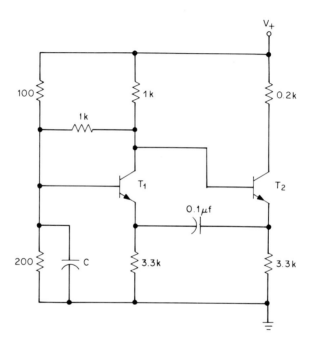

Fig. P9.1

9.2   An active-RC filter using four operational amplifiers and capable of realizing
any biquadratic transfer function is described in Section 13.2. Figure 13.7 is
reproduced here for convenience as Fig. P9.2. The element values are:
$R_1 = 25$ kΩ, $R_5 = 2.67$ kΩ, $R_6 = 1.33$ kΩ, $r = 50$ kΩ, $C_1 = C_2 = 0.5$ μF
and all other resistors are 1 kΩ.

   a)   Design a suitable thick-film circuit layout assuming that dual op-amp
chips and applique capacitors are used. The capacitor chips are 10 mm by
12 mm. The op-amp chips are 2500 μm by 2500 μm and have the terminals for each amplifier on opposite edges of the chip. Make any other
necessary assumptions.

   b)   How many screening operations are required? Can the number of
screening operations be reduced by changing the design of the resistors?
What are the advantages and disadvantages of this redesign?

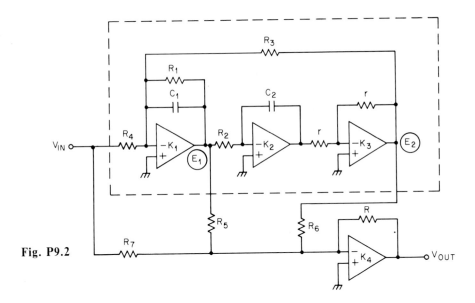

Fig. P9.2

9.3 (a) Design a thick-film circuit layout for the twin-tee network shown below. No appliqued components are permitted. (b) How many screening operations are required and what is the size of the resulting circuit? (c) Prepare a sketch to scale for the pattern that defines each screening operation.

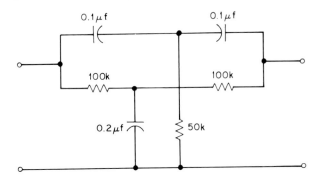

Fig. P9.3

10

Assembly Techniques

The various techniques used in the assembly and packaging of integrated circuits will now be discussed.[1-3] The operations which must be performed in assembling a circuit include mounting chips to the substrate or package base (Section 10.1) and bonding of leads to connect the chips to film circuits or package terminals (Section 10.2). There is also a description of several common types of packages.

In recent years inflation and consequent escalating costs have increased the cost of packaging integrated circuits, and this has forced manufacturers to have the assembly operations performed offshore—that is, outside the continental United States—where labor costs are a small fraction of the cost in the United States. Assembly is usually done in places like Taiwan, Hong Kong, and Korea. However, labor costs are constantly rising even offshore, assemblies are becoming more complex and therefore more difficult to perform without error, and economic contraints, such as tariffs, are subject to change. These factors have prompted consideration of methods for reducing assembly costs. The methods fall into two classes: (1) mass interconnection of leads and (2) automated bonding operations. The mass interconnection schemes to be described are beam leads (Section 10.3), tape-carrier packages (Section 10.4), and inverted chips (Section 10.5). Programmable, computer-controlled, automatic wire bonders are an example of the second class of methods and are described in Section 10.2.

10.1 CHIP BONDING

The electrical connection to most SIC chips is made via bonding pads on the top surface or face of the chip. Usually the chip or die is bonded to the substrate of a film circuit or the base of a package by means of a eutectic, solder, or epoxy bond.

10.1.1 Eutectic Die Bonding

Eutectic materials, such as gold-silicon, gold-tin or aluminum-germanium are bimetallic systems that are characterized by a phase diagram[4] such as Fig. 10.1 for the gold-silicon system. The material composition is plotted along the horizontal axis with pure gold on the left and pure silicon on the right. Temperature is plotted on the vertical axis. Each curve defines the boundary of a region at which the nature of the metallic system changes. For example, below the horizontal line the material is solid, while in the region at the left bounded by the curve and horizontal line there is a gold-silicon liquid in which all of the available gold cannot dissolve. Note that at

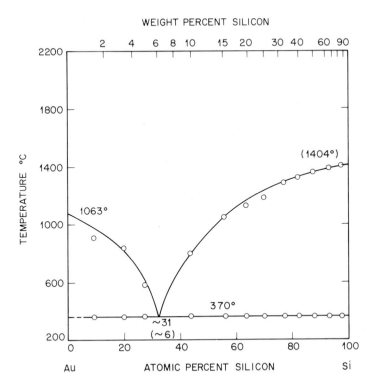

Fig. 10.1 Phase diagram for the gold-silicon system. (From Hansen.[4])

31 atomic percent silicon there is one point at which the solid exists in equilibrium with the liquid. This special point is known as the *eutectic point* and it defines the lowest temperature at which solid and liquid can coexist in this bimetallic system. The gold-silicon eutectic temperature is 370°C, but if there is an excess of either element there will be crystalline silicon or gold in equilibrium with the melt until the temperature is raised sufficiently.

There are several common methods of forming eutectic bonds such as heating a gold-plated silicon chip and gold-plated substrate that are in firm contact, or by placing thin sheets of the eutectic alloy between the chip and substrate. It is essential that sufficient gold be present to ensure that the melt remains on the left branch of the phase diagram as melting continues until the supply of gold is exhausted or the temperature falls below the eutectic temperature. Besides gold-silicon there are whole families of gold-based preforms that find application.

The gold-tin eutectic, consisting of 10% gold and 90% tin, is an alternative material that is popular for thin-film hybrid circuits because the eutectic temperature is 217°C. However, temperatures encountered in any subsequent package sealing process must be limited to prevent refloating the chips. The aluminum-germanium eutectic which is made at about 420°C is used in an all-aluminum system that is not subject to the formation of intermetallic compounds at high temperatures. It also provides the best radiation hardness and good thermal impedance, although the latter is also dependent on a variety of other factors.

10.1.2 Solder Die Bonding

Solder die bonding consists of heating a plated substrate and a plated silicon chip, pretinned with a suitable solder compound. Solder melting temperatures, even for low-temperature solders, are usually above the processing temperatures that the chip will experience after having been die bonded. Solder die bonds provide low thermal impedance between die and substrate. However, solders usually require flux, which must be removed by a flux removal step.

10.1.3 Epoxy Die Bonding

The die may also be bonded to the ceramic substrate by means of an epoxy cement. The poor thermal resistance associated with epoxy material may be completely circumvented by the use of aluminum or gold-filled epoxy compounds. However, a disadvantage of the epoxy bond is "outgassing," that is, the subsequent release of gas, usually while experiencing thermal stress. Outgassing introduces unknown contaminants into the hermetic package, usually with an adverse effect on reliability and lifetime.

10.1.4 Appliqued Chips

Hybrid IC's often require various passive chips that are not silicon based. Typical of these chips is the ceramic capacitor with metallized ends. These chips are usually bonded by solder-reflow techniques in which the chip is pretinned with a "fluxless" tin-lead solder which wets the substrate pad areas when the chip is heated to about 260°C.

10.2 WIRE BONDING

Conventional faceup SIC chips are electrically connected to film circuits or package terminals by means of thin wire leads which are attached either by thermocompression (TC) or ultrasonic bonding with temperature assist (US TC). In both of these techniques intimate contact must be obtained between the two metallic surfaces during which contact plastic deformation must occur so that unsaturated atomic bonds on both surfaces can react. Surface irregularities, oxide formation, and absorbed moisture layers all act to prevent intimate metal-to-metal contact, thereby complicating the bonding process. Types of wire bonding are: (1) ball-wedge or (2) wedge-wedge to be described subsequently. In most cases, the interconnecting media is either gold or aluminum wire ranging in diameter from 25 μm to 100 μm. Generally gold wire is TC or US-TC bonded and the type of bond is ball-wedge. Aluminum wire is US-TC bonded and the type of bond is wedge-wedge. Bonding pads for both ball-wedge and wedge-wedge are approximately 125 μm on a side.

10.2.1 Thermocompression Bonding

Thermocompression bonding occurs at the interface of two metallic surfaces in intimate contact, where plastic deformation and diffusion occurs during a controlled time, temperature and pressure cycle. Plastic deformation at the interface results in intimate surface contact, in an increase of the bonding area, and in the destruction of any interface films. Controlled conditions and clean materials are necessary for the creation of strong bonds. Typical bonding temperatures are between 250°C and 350°C which are well below the temperature required for interface melting. It is desirable to use ductile bonding wire to prevent elastic springback after pressure is removed. A typical bonding cycle, excluding mechanical positioning, lasts a fraction of a second and the required temperature is achieved either by mounting the substrate and package on a heated column, by a heated bonding tool, or by a combination of both of these methods.

It is generally found that low bonding temperatures yield weak bonds because of occurrence of surface contamination and that high bonding temperatures for sufficiently long times result in fragile bonds because of alloying and intermetallic compound formation.

One type of thermocompression bonding apparatus feeds the wire through a heated tungsten carbide capillary. One or both sides of the face of the capillary are used as the bonding tool. Gold wire may be cut by an electric discharge leaving a ball at the end whereas aluminum wire does not leave a ball when severed by a flame or discharge. Both types of wire can be mechanically cut.

Gold wire may be used to form a ball or nail-head bond, details of which are shown in Fig. 10.2. The work to be bonded is precisely positioned under the heated capillary. This can be performed manually with the aid of micromanipulators and magnifying optics, or automatically under computer control. The capillary is lowered into position causing deformation of the gold ball under pressure and temperature to give a characteristic "nail-head bond," which is made on the silicon chip. The capillary is then raised and the work moved into position on the package for the second bond, which is made by deforming the wire with one edge of the capillary, and results in a wedge bond. After completing the wedge bond, the tool is withdrawn, the wire is severed either mechanically or by a capacitive discharge, and the ball

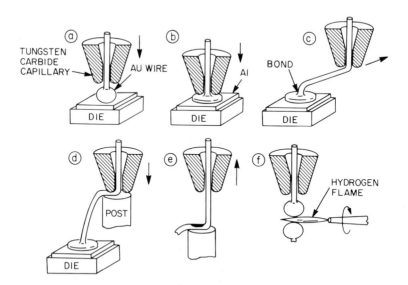

Fig. 10.2 Steps in performing a nail-head bond. (a) Initial positioning of work; (b) deformation of ball to form nail-head bond; (c) positioning of capillary for second bond; (d) formation of wedge bond with capillary edge; (e) withdrawal of capillary from wedge bond; (f) flame cutting of wire leaving ball.

needed for subsequent bonding is formed. During bonding, the work area is bathed in nitrogen gas to prevent oxidation. The first bond need not be a ball bond. In particular, both bonds are wedge bonds when aluminum wire is used, although aluminum wire still (1977) presents handling and control difficulties and is not currently used on automatic wire bonders. Figure 10.3 shows the sequence of operations for forming a wedge bond on wire. In this case the wire is initially bent around the capillary edge so that the first bond is made with the capillary acting as a wedge. After repositioning, the second bond is also made using the edge of the capillary. At this point the wire is severed either by automatically controlled scissors or by vibrating the capillary with the wire in tension. This latter technique invariably causes the wire to break at its weakest point, which is immediately after the bond. In either case the wire ends up bent around the capillary edge and is prepared for the next bonding sequence. To prevent damage to the second bond during the wire cutting or breaking operations a "double-wedge bond" is often used.

10.2.2 Ultrasonic Wire Bonding with Temperature Assist

The source of energy for ultrasonic bonding is mechanical rather than thermal. An elastic vibration in the frequency range 20 kHz to 60 kHz is created by an ultrasonic transducer. This vibration is coupled to the bonding tip through a mechanical transformer matched to provide maximum transfer of

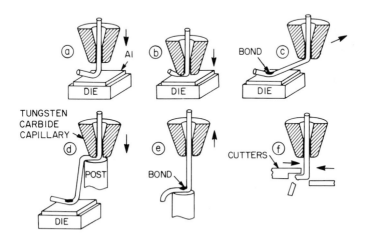

Fig. 10.3 Steps in performing a wedge bond. (a) Initial positioning of work; (b) formation of wedge bond with edge of capillary; (c) repositioning of capillary for second bond; (d) formation of second bond; (e) withdrawal of capillary from bond; (f) severing of wire with mechanical cutters.

energy. The bonding tip vibrates parallel to the bond interface and probably induces a shear mode of vibration resulting in a localized temperature rise of approximately 300°C. As a result ultrasonic bonds can be made immediately adjacent to temperature-sensitive materials without adverse effects. In principle, the bond can be made using ultrasonic energy as the sole heat source. However, the quality of the bond can be improved by heating the substrate to about 120°C. Either gold or aluminum wire can be used and bonding pad size is of the same order as for TC bonding.

10.2.3 Automatic Wire Bonders [7,8]

The present trend is towards automated chip and wire bonding machines. These machines rely on a pattern recognition system to align—either manually or automatically—four points in the coordinate system of the chip and four points in the coordinate system of the package. This information, together with the ideal locations of bonding pads on the chip and on the package, is then stored in the computer memory and is both necessary and sufficient to determine the actual location of every bonding pad both on the chip and on the package. The automatic bonding machine then carries out the bonding operation at a rate of two wire bonds per second (early 1977), with rates of four wire bonds per second being projected for the near future.

Only gold-wire ball-wedge bond systems are presently (early 1977) possible with automatic machines. The problem of handling and controlling aluminum wire in automatic bonders has not been solved at the time of writing.

It is present opinion (early 1977) that automatic bonding machines by their flexibility, relative low cost, and smaller volume handling capacity will successfully compete and coexist with tape-carrier machines, which, though capable of handling huge volumes, are less flexible and more expensive.

10.3 BEAM LEADS [5]

The processing of beam-lead chips was described in Section 6.2. It will now be shown how these chips are mounted and electrically connected in just two operations.

The beam leads are thermocompression bonded either by means of a heated hard tool or by a compliant bonding apparatus which is shown in Figs. 10.4(a) and 10.4(b). Bonding force and heat are, for the latter case, applied to the "beam leads" through a compliant member, made of a frame of thin aluminum sheet, thereby compensating for variations in lead thickness, substrate surface flatness, and bonding equipment. The beam-lead bonding operation requires that the chip be precisely positioned so that the beams are correctly aligned with the fine-line conductor pattern on the sub-

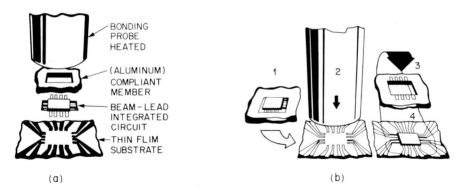

Fig. 10.4 Bonding of beam lead chips. (a) Exploded view of bonding setup; (b) sequence of operations.

strate. Since there are many closely spaced beam leads, this operation is critical. A suitable method involves aligning the chip to a reference grid internal to the microscope reticle, followed by a separate alignment of the substrate pattern and the compliant member to the same grid. The advantages of this technique are that a simple, flat-faced, bonding tool can be used and that lead deformation, and therefore bond integrity, is more consistent. The major disadvantages are that additional operations are required to fabricate the compliant members and that alignment is further complicated by the need for registration of three separate parts.

   Beam-lead devices can be bonded by: (1) TC-compliant tape bonding; (2) wobble-tip TC bonding, (3) hard-tip compensating lead TC bonding, and (4) hard-tip TC bonding.

   In addition to bonding, the assembly of "beam-lead sealed junction" chips requires some form of packaging. The junction seal is provided by a silicon-nitride layer, but the chip, including the thin metallization, must be completely protected by an encapsulant from electrolysis due to excessive moisture collecting between closely spaced leads, and from mechanical abuse. This protection is provided by a thin coating of silicone RTV rubber. The beam leads provide the necessary medium for thermal transfer from chip to substrate.

10.4 TAPE-CARRIER PACKAGING [9-11]

Several methods have been devised for reducing the number of bonding operations required to connect an SIC. Among these techniques are beam leads, and inverted chips, to be mentioned in Section 10.5, which reduce the number of assembly operations at the expense of more complex chip processing. The tape-carrier packaging technique to be described may be considered a development stemming from the G.E. minimod technique, and will be an increasingly important assembly method as labor costs increase, because it permits mass interconnection of all the leads on a chip without unduly complicating the chip processing.

The technique consists of the repetitive manufacture of miniature printed circuits on a metal tape which resembles an ordinary film and comes either as a laminate or as a gold-plated copper foil. This tape is moved by rollers whose teeth engage in sprocket holes at its edges. The tape carries laminated copper conductors which are cantilevered free of the tape at the inner periphery, where they will subsequently be bonded in a single operation to an integrated-circuit chip. The conductors are also cantilevered free of the tape at the outer periphery, where they will subsequently be bonded in another single operation to a lead frame, ceramic substrate, flexible printed circuit or printed-circuit board.

Figure 10.5 shows the process of inner lead bonding. The tape, moving on rollers, is positioned over the chip and the inner tape leads are accurately positioned over the chip pads. The downward motion of a die bonding head attaches all inner leads in a single operation. The simultaneous (gang) bonds are made to gold bonding pads on the chip. Recall that the metallization on a conventional chip is aluminum, and therefore the chip processing must be complicated slightly to produce gold bonding pads about 25 μm high. A diffusion barrier, such as titanium, is required under the gold to

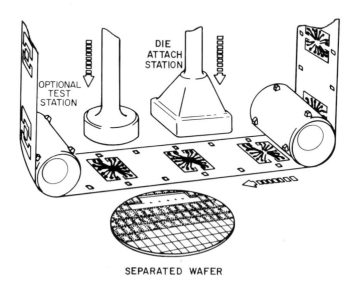

SEPARATED WAFER

Fig. 10.5 Bonding of chips to the tape carrier.

keep the gold from coming in contact with the aluminum or silicon. Not all the bumps are exactly the same height, but their ductility allows the higher ones to be compressed as the bonding tool forces the inner leads against the chip during the simultaneous bonding operation. After bonding, the chip is supported on the tape by the inner cantilevered leads. At this point, the center portion of the assembly can be insulated by means of a drop of resin. To make outer bonds to a ceramic substrate or printed circuit board, the inner portion of the tape, which is supporting the chip, is stamped or cut out. This is done by carefully positioning it over the lead frame, after which downward movement of the lead-frame bonding tool stamps out the tape, and bonds the outer cantilevered leads to bonding pads on the ceramic substrate in a single operation as shown in Fig. 10.6.

Final encapsulation in a conventional DIP package is carried out in the usual molding operation.

10.5 INVERTED CHIPS [6,12-15]

Another mass interconnection scheme is that of inverted chips originally proposed by IBM. The chip processing for one possible type of inverted chip metallization is described in Section 6.6. However, there are two methods of forming the joint. One uses solderable bumps or balls and the other uses

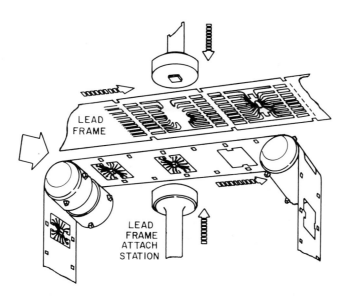

Fig. 10.6 Bonding of chip in tape carrier to the lead frame.

nonsolderable bumps, the attachment being made by thermocompression or ultrasonic methods.

A solder system utilizes the controlled collapse of solder bumps as shown in Fig. 10.7. Solder is deposited in and over the contact holes (a) in the glassy SiO_2 protective layer of the chip. When these solder deposits are heated, a solder ball (b) forms in the shape of a spherical lump or bump. The chip is now placed face down (c) and aligned with pretinned areas on the substrate. Heating takes place, the solder bumps melt and combine forming molten columns of solder (d). Surface tension forces of the molten solder columns pull the unit into proper location and suitably placed untinned areas on the chip prevent the molten solder from shorting adjacent bumps.

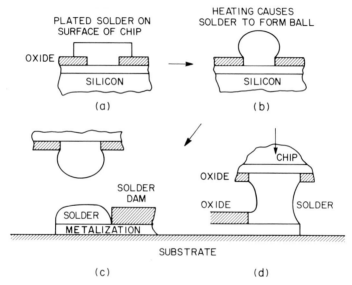

Fig. 10.7 Inverted chip system using the controlled collapse of solder bumps.

Some difficulties in using inverted chips are that bonds are made by applying forces directly to the chip and that they are not inspectable. When welding techniques are used for attachment, it is difficult to replace defective chips because the bonds may part at any of several interfaces during disassembly leaving a rough surface.

10.6 PACKAGES

Four types of packages are used for silicon integrated circuits, namely transistor type metal cans, molded plastic or ceramic dual-in-line (DIP) packages, ceramic flat packs, and hermetic or plastic chip carriers. The ceramic and metal packages are hermetically sealed. Flat packs are also used to package hybrid circuits.

The metal can package commonly used is similar to that first used to package transistors and is shown in Fig. 10.8. It is made of Kovar, an iron-nickel-cobalt alloy with a thermal expansion coefficient which closely matches the coefficient of certain sealing glasses. The gold-plated package base or header comes with gold-plated leads glassed into place. The chip is eutectically bonded to the header and the connecting leads are wire bonded to the terminals. Following an extensive cleaning procedure and bakeout to remove surface moisture, packages are sealed in an inert atmosphere by welding a Kovar cap to the header. The welding process is rapid, and results in a high yield of hermetically sealed packages. This package is often used for linear circuits where the shielding provided by the metal can is useful.

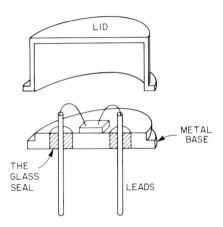

Fig. 10.8 Exploded cross section of metal can package.

The dual-in-line package was developed primarily for ease of equipment assembly and is by far the most popular package. It is available in sizes ranging from 8 pins to more than 40 pins. A typical DIP is shown in Fig. 10.9. The leads are arranged in two rows on 0.100 in. centers and are sufficiently rugged to withstand automatic insertion on printed circuit boards.

A variety of ceramic dual-in-line packages are in existence. One version consists of a ceramic base and top, and a Kovar lead frame which is separately mounted to the base with a glass frit. The lead frame is shaped for best access to a single centrally located monolithic chip. The chip is bonded, either eutectically or by metal-filled epoxy, to a gold-plated

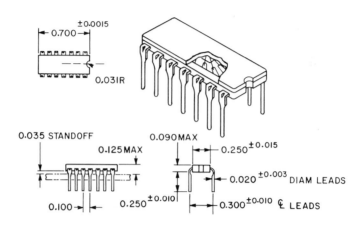

Fig. 10.9 Orthographic projection and cutaway view of a typical DIP package.

molymanganese pad on the package base. The package is furnace sealed using a glass frit between base and top. The advantage of this package is, of course, ease in subsequent assembly.

In the plastic DIP the monolithic chip is connected to the Kovar lead frame by the bonding wires. Subsequently it is encapsulated by injection molding in an epoxy plastic which results in a rugged inexpensive package.

Figure 10.10 is an exploded view of a typical small ceramic flat pack. Ceramic flat packs are used for hybrid IC's as well as SIC's. In most of the

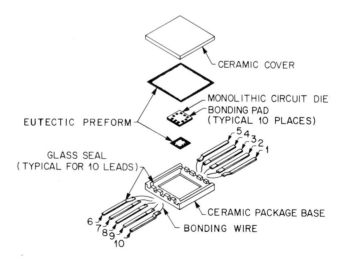

Fig. 10.10 Exploded view of a ceramic flat package.

larger flat packages, the package base is ceramic with lead frames bonded in place using a high-temperature glass. The package seal is made eutectically with a gold-germanium or gold-tin preform sandwiched between a gold-plated Kovar (or ceramic) lid and a gold-plated ceramic "window frame" on the package base. The frame is mounted to the base with glass at the same time that the lead frames are mounted.

When the flat packs are used exclusively for SIC chips, the chips are eutectically or metal-epoxy bonded to the package base as in the previous cases. In the sealing operation the package is purged with nitrogen, evacuated and refilled with a nitrogen-helium mixture, and finally the lid is attached by heating the preform to its eutectic temperature. The rise, duration, and fall time, and the peak temperature of the heating cycle must be carefully controlled to ensure good sealing yields. Typically, the sealing cycle lasts 1 to 2 min with lid temperatures reaching 450°C. The package base and internal components can be kept cool by appropriate heat sinking in the event that this is necessary. As package size increases, the sealing yield decreases because of lid warpage and because of pinholes caused by gases given off from components inside the package.

Chip carriers, shown in Fig. 10.11 are quite similar to ceramic flat packs, but the leads, instead of being brought out at the ends (see

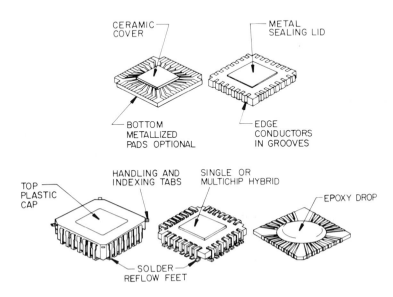

Fig. 10.11 LSI package standards for chip-carrier assemblies.

Fig. 10.10) are either brought out in the form of a film on top of the chip-carrier package (see leadless Type A) or they are brought to the edge (leadless Type B) or to the edge and underneath the chip-carrier package (leaded Type A). Chip-carrier packages are used for high-density circuits such as memories. The chip carriers are mounted to thin- or thick-film circuits or special chip-carrier mounting pads, either via an interconnecting element, or by soldering, or by a solder-reflow technique. Since several chip-carrier assemblies may be mounted on one substrate or special "mother board," the technique lends itself to hybrid-circuit assembly.

10.7 THERMAL CONSIDERATIONS

Thermal considerations should not be overlooked in microcircuit assembly. It is shown in Chapter 16 that the useful operating life of a semiconductor circuit is strongly affected by junction temperature. Consequently, it is very important to keep the temperature of the active components below some maximum level. This necessitates providing a low thermal impedance path between the active elements and the ambient. As a typical example of a thermal impedance path between a chip and the ambient consider Fig. 10.12(a), which shows a chip eutectically bonded to a ceramic substrate,

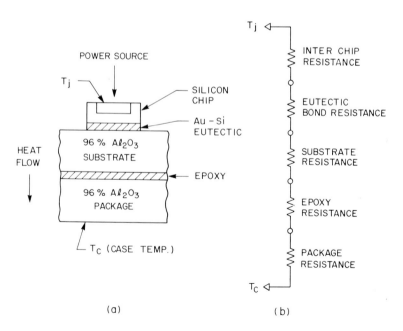

(a)   (b)

Fig. 10.12 (a) Cross-sectional view of a silicon chip mounted in a package; (b) electrical equivalent of the thermal resistance path.

which in turn is epoxy bonded to the base of a ceramic flat pack. Using a thermal-electrical analogy for a heat transfer, with applied power corresponding to current and temperature rise corresponding to voltage, it is seen from Fig. 10.12(b) that there are five thermal resistances from the heat source to the case exterior. The overall temperature rise from junction to case $(T_j - T_c)$ will be directly proportional to the thermal resistance R_{JC}. Because glass frit bonds have very low thermal conductivity, emphasis in any but very low-power circuits is on the use of eutectic die bonds and metal-filled epoxy cements to bond directly to either a ceramic package base or to a metal header. Typically the thermal resistance from junction to case may be of the order of 40 to 60°C/W.

Concepts such as these are also extended to the physical design and determine the minimum spacing between circuit boards in a rack. The allowable density of components depends on the method of heat removal. A more detailed discussion of thermal considerations is beyond the scope of this book.

REFERENCES

1. C. A. Harper, editor. *Handbook of Electronic Packaging.* New York: McGraw-Hill, 1969, Chapter 9.

2. D. W. Hamer and J. V. Biggers. *Thick Film Hybrid Microcircuit Technology.* New York: Wiley-Interscience, 1972, Chapter 11.

3. S. K. Ghandi. *Theory and Practice of Microelectronics.* New York: John Wiley & Sons, 1969, Chapter 9.

4. M. Hansen. *Constitution of Binary Alloys.* New York: McGraw-Hill, 1958.

5. M. P. Eleftherion. "Assembling beam-lead sealed-junction integrated-circuit packages." *Western Electric Engineer* **11**: 16-26 (December 1967).

6. E. M. Davis, W. E. Harding, R. S. Schwartz, and J. J. Corning. "Solid logic technology: versatile, high-performance microelectronics." *IBM J. of Research and Development* **8**: 102-114 (April 1964).

7. S. Kulicke, private communication.

8. B. H. McGahey, private communication.

9. A. van der Drift, W. G. Gelling and A. Rademakers. "Integrated Circuits with Leads on Flexible Tape." *Philips Technical Review* **34**: 85-95 (1974).

10. W. Patstone. "Tape carrier packaging boasts almost unlimited Potential." *EDN* **19**: 30-37 (1974).

11. S. E. Grossman. "Film carrier technique automates the packaging of IC chips." *Electronics* **47**: 89-95 (May 1974).

12. P. A. Totta and R. P. Sopher. "SLT device metallurgy and its monolithic extension." *IBM J. of Research and Development* **13**: 226-238 (May 1969).

13. L. F. Miller. "Controlled collapse reflow chip joining." *IBM J. of Research and Development* **13**: 239-250 (May 1969).

14. L. S. Goldmann. "Geometric optimization of controlled collapse interconnections." *IBM J. Research and Development* **13**: 251-265 (May 1969).

15. K. C. Norris and A. H. Landzberg. "The reliability of controlled collapse interconnections." *IBM J. of Research and Development* **13**: 266-271 (May 1969).

11

In-Process Measurement and Testing of Integrated Circuits

11.1 INTRODUCTION

In any complex manufacturing process one must evaluate the properties of the product at various stages during its fabrication. This is to ensure that the process is under control and that the final product has a good chance of performing properly. A number of nondestructive tests useful for in-process monitoring in the fabrication of silicon and thin-film integrated circuits are presented in this chapter. Of particular interest is the measurement of thickness and resistivity of both semiconductor and thin-film layers, the surface impurity concentration of semiconductors, and the thickness and dielectric constant of dielectric films. Some properties can only be determined by destructive measurements which are used for process characterization rather than for routine in-process evaluation.

The in-process measurements monitor the progress of the fabrication process and weed out those wafers whose properties are sufficiently far from the norm such that they would not provide satisfactory working circuits. Unfortunately, not all of the circuits on a good wafer are satisfactory, and therefore some testing must be performed on the completed circuits as discussed in the latter part of the chapter.

11.2 FOUR-POINT PROBE MEASUREMENT OF RESISTIVITY [1]

The four-point probe shown in Fig. 11.1 is widely used for measuring resistivity on thin slices or wafers. The spacing between adjacent probes is fixed at a centimeters. A reference current I is introduced between the outer

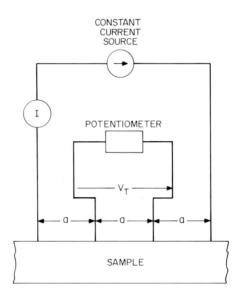

Fig. 11.1 Schematic diagram for a four-point probe measuring setup.

probes and the resulting voltage V_R is measured between the inner probes. Care must be taken to ensure that all four probes make good ohmic contact with the material. When the sample is a semiconductor, measurements should be made at several current levels and in both polarities to be sure that the contact is truly ohmic.

   If the sample is thick, and if the probes are sufficiently far away from the edge, it can be shown that the resistivity ρ is given by

$$\rho = 2\pi a \, \frac{V_R}{I} \qquad\qquad (11.1\text{a})$$

while for thin samples

$$\rho = C(t/a) \, \frac{V_R}{I} \qquad\qquad (11.1\text{b})$$

where $C(t/a)$ is a correction factor whose value depends on the ratio of a sample thickness t to probe separation a and $I = V_G/R_G$ where V_G is a voltage measured across a known standard resistance R_G as shown in Fig. 11.1. These correction factors are given in Table 11.1.

EXAMPLE 11.1 A four-point probe measuring set has a probe separation $a = 0.127$ cm. Measured   values   for   a   palladium-silver   film   were: $V_R = 12.5\times10^{-3}$ V, $I = 10\times10^{-3}$ amps. Calculate the resistivity ρ of the film.

Solution.   From Eq. (11.1a),

$$\rho = 2\pi a V_R / I = 6.28 \times 0.127 \times 12.5 \times 10^{-3} / 10 \times 10^{-3} \approx 1 \ \Omega\text{-cm}.$$

This method can be used to measure resistivity in the range 0.001 Ω-cm to 300 Ω-cm. Extending it for higher resistivity samples is dependent on probe design, and may be limited by problems of ohmic contacts and surface leakage of current.

Table 11.1 Four Point Probe Corrections for Finite Thickness of Wafer

$\dfrac{t}{a}$	$\dfrac{C(t/a)}{t}$	$\dfrac{t}{a}$	$\dfrac{C(t/a)}{t}$
0.4	4.5301	1.0	4.1761
0.5	4.5206	1.1111	4.0370
0.5555	4.5088	1.25	3.8480
0.6250	4.4861	1.4286	3.5978
0.7143	4.4408	1.6666	3.2746
0.8333	4.3511	2.0	2.8717

11.2.1 Thickness of Films by Resistivity Measurement

The thickness of a film may be determined indirectly by measuring the sheet resistance. When this method is used for metal films, it is necessary to know the bulk resistivity ρ of the metal, and to restrict the film thickness to be greater than the mean-free path of electrons in the material so that the actual resistivity of the film is close to the bulk resistivity. For metals at room temperature the film thickness must be greater than several hundred angstroms. The average resistivity of diffused layers having Gaussian or complementary error-function impurity profiles has been calculated.[2]

The sheet resistance is usually measured with a four-point probe having equally spaced, in-line points. It can be shown that the sheet resistance is given by

$$R_s = \frac{\rho}{t} = \frac{\pi}{\ln 2} \frac{V_R}{I} \qquad (11.2)$$

when the measurement is carried out on an infinite sheet. As a consequence

$$t = \rho \frac{\ln 2}{\pi R_G} \left(\frac{V_G}{V_R} \right). \qquad (11.3)$$

The accuracy of this measurement is about $\pm 10\%$. For epitaxial layers, which may be assumed uniformily doped, this method will measure thicknesses from 1 μm upwards. For nonuniformily heavily doped layers, with heavy surface concentrations, this method should not be employed. For lightly, nonuniformily doped layers, this method will measure thicknesses in the range 2000 to 10,000 Å. Correction factors must be incorporated into Eq. (11.3) when the sheet is of finite extent, but this subject will not be pursued further here.

11.3 RESISTIVITY AND CARRIER MOBILITY [3]

The resistivity of, and carrier mobility in, samples in wafer form can be measured by an ingeneous method due to van der Pauw. This method actually permits measurement of the Hall effect coefficient[4] and the resistivity from which the carrier mobility can be calculated. With reference to Fig. 11.2, a transresistance $R(ab|dc)$ is defined by

$$R(ab|dc) = \frac{V_{dc}}{I_{ab}} \tag{11.4}$$

where $V_{dc} = V_d - V_c$ and I_{ab} is the current entering terminal a and leaving terminal b. Current I_{ab} is measured by observing the voltage V_G developed across the standard resistance R_G, and therefore

$$I_{ab} = \frac{V_G}{R_G} . \tag{11.5}$$

A voltage measurement across terminals d and c then yields V_{dc} and establishes transresistance $R(ab|dc)$. The standard resistor, R_G, is chosen to be about ten times as large as $R(ab|ab)$, the input resistance looking into terminals a and b.

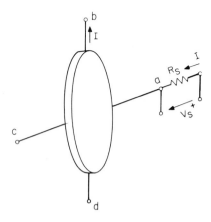

Fig. 11.2 Sample geometry for van der Pauw measurement.

Van der Pauw showed that the resistivity, ρ, of a wafer whose thickness is t may be computed from

$$\rho = \frac{\pi t}{2 \ln 2} \{R\,(ab\,|\,dc) + R\,(bc\,|\,ad)\}\cdot f\left\{\frac{R\,(ab\,|\,dc)}{R\,(bc\,|\,ad)}\right\} \qquad (11.6)$$

where $R\,(bc\,|\,ad) \triangleq V_{ad}/I_{bc}$. The van der Pauw's function, $f(\cdot)$, depends only on the ratio of the transresistances and is plotted in Fig. 11.3. From this figure it is seen that $f(1) = 1$, $f(\infty) = 0$, and that $0 \leqslant f \leqslant 1$.

Van der Pauw's method may be used to measure sheet resistances in the range $0.5\ \Omega/\square$ to $5000\ \Omega/\square$.

The establishment of ρ then reduces to a thickness measurement to determine t, and several voltage measurements to determine the two transresistances.

Van der Pauw also showed that the carrier mobility μ could be expressed in terms of a "cross-resistance" using the Hall effect.

$$\mu = \frac{t}{B}\ \frac{\delta R\,(bd\,|\,ca)}{\rho}\ . \qquad (11.7)$$

In this equation $\delta R\,(bd\,|\,ca)$ represents the change in "cross-resistance" due to the magnetic field B.

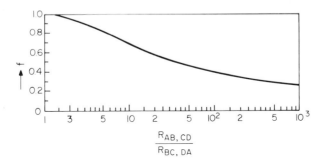

Fig. 11.3 Van der Pauw's function.

11.4 FILM THICKNESS—INTERFERENCE TECHNIQUES

Film thickness can be determined quite accurately by interference techniques. Two cases are considered, namely, (1) a "transparent" film on a reflective substrate, and (2) a nontransparent film on a possibly nonreflecting substrate. The first case corresponds to an epitaxial layer of silicon deposited over a highly doped substrate, whereas the second case occurs in the thin-film technology.

11.4.1   Interference Measurement of Dielectric Film Thickness [5]

The situation to be considered is shown in Fig. 11.4. A beam of mono-chromatic light of wavelength λ is incident on the surface of the film of thickness t at an angle Φ with the normal to the surface. The transparent, nonabsorbing film material has refractive index $n_1 = \sqrt{\mu_1 \epsilon_1}$, where μ is the permeability of the material and ϵ is its permittivity. At the air-film interface part of the energy of the incident wave is transmitted and part is reflected because of the impedance mismatch between the film and air. The direction of propagation of the wave is altered as it enters the film because of the increased optical density. This change of direction is given by Snell's law,

$$n \sin \Phi = n_1 \sin \Phi'$$

where n is the refractive index of the medium above the film. The optical path length in the film is $n_1(AB+BC)$, and the difference δ in path length between a phase of the reflected wave and a corresponding phase of the transmitted component that is emerging at point C is

$$\delta = n_1(AB+BC) - AD$$

$$= 2n_1 t \cos \Phi'. \tag{11.8}$$

Here it was assumed that air $(n = 1)$ is above the film and from geometrical considerations, $AC = 2AB \sin \Phi'$, $AD = AC \sin \Phi$, $t = AB \cos \Phi'$.

At the air-film interface the reflected wave undergoes a phase change of π radians, or an equivalent change in path length of $\lambda/2$, because the epitaxial film is optically denser than air. It is assumed that the measurement is made at a frequency such that the refractive index of the film is greater than

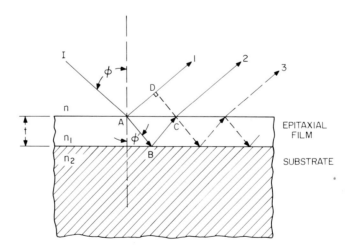

Fig. 11.4 Reflection and refraction at a dielectric film interface.

that of the substrate, and that the extinction coefficient of the substrate is negligible, so that there is no phase change accompanying reflection at the substrate.[†] For some wavelength λ_o for which this is true, the path difference δ is $\lambda_o/2 = 2n_1 t \cos \Phi'$ resulting in destructive interference and the occurrence of interference fringes. At this wavelength the phase of the reflected ray is advanced by $\lambda_o/2$, and the phase of the ray that traversed the film is retarded by $\lambda_o/2$. The zero-order (first) fringe occurs at λ_o, the first-order fringe occurs when $\delta = \lambda_1 + (\lambda_1/2)$, and the mth-order fringe occurs when

$$\delta = \left(m + \frac{1}{2} \right) \lambda_m. \tag{11.9}$$

The actual measurement procedure involves plotting fringe intensity versus wavelength. Since it is hard to determine the actual value of m, it is desirable to obtain a relation for thickness in terms of successive fringe maxima. From Eqs. (11.8) and (11.9)

$$t = \frac{[m + (1/2)]\lambda_m}{2n_1 \cos \Phi'}. \tag{11.10}$$

Consider two distinct fringe maxima $\delta_m = [m + (1/2)]\lambda_m$ and $\delta_{m+x} = [m + x + (1/2)]\lambda_{m+x}$, and solve for $[m + (1/2)]$. Substituting this result into Eq. (11.10) yields

$$t = \frac{x\lambda_{m+x}}{\lambda_m - \lambda_{m+x}} \frac{\lambda_m}{2n_1 \cos \Phi'} \tag{11.11}$$

where λ_m and λ_{m+x} are the wavelengths corresponding to the mth and $(m + x)$th fringe maxima, and x is the *difference* in the fringe order.

EXAMPLE 11.2 A silicon-dioxide film of refractive index $n_1 = 1.46$ produced reflectivity maxima at $\lambda_m = 620 \times 10^{-7}$ cm and at $\lambda_{m+5} = 370 \times 10^{-7}$ cm. The angle of incidence $\Phi' = 30°$. Find the film thickness t.

Solution. From Eq. (11.11),

$$t = \frac{5 \times 370 \times 620 \times 10^{-14}}{250 \times 10^{-7} \times 2 \times 1.46 \times 0.866} = 1.82 \times 10^{-4} \text{ cm.}$$

A requirement of this procedure is that the optical properties of the material, in particular the refractive index, be known. The refractive index may be obtained by a technique involving the measurement of the Brewster angle.[6] Additional constraints are that the epitaxial layer resistivity must be greater than about 1 Ω-cm to provide satisfactory transmission, and the film thickness must be at least 2 μm so that a sufficient number of fringes can be obtained.

[†] If this condition is not satisfied, there will be a phase shift of between 0 and π radians associated with the reflection from the substrate.

11.4.2 Ellipsometry [6,20]

The two major disadvantages of the interference method for measuring film thickness are that the optical properties of the film must be known and that the film must be thick enough for several fringes to occur. These deficiencies are inherent in the method because only the amplitudes of the incident and reflected waves are considered.

Before describing the ellipsometry technique, a few definitions from electromagnetic theory are useful. A plane electromagnetic wave consists of mutually orthogonal electric and magnetic fields. If the amplitude of these fields is a real constant independent of time, the electromagnetic wave is said to be *linearly polarized* or *plane* polarized. It is customary to use the direction of the electric field as the direction of polarization. A *circularly polarized* wave is composed of two linearly polarized waves of equal amplitude at right angles to each other. In this case the composite electric- or magnetic-field vectors appear to describe a circle. The most general polarization is *elliptical*. It corresponds to two linearly polarized waves of unequal amplitude at right angles.

Ellipsometry is a more complex technique than interference, and is based upon changes in polarization that occur when a wave is reflected. If the film is nonabsorbing, and if the optical properties of the substrate are known, then both the film thickness and its indices of refraction can be determined simultaneously by ellipsometry. The optical properties of the substrate must be known for both ellipsometry and interference methods.

The diagram of Fig. 11.4 illustrates that there are two interfaces between the film and adjacent media. The boundary conditions at any interface require the tangential components of both the electric and magnetic fields to be continuous, which results in a set of so-called Fresnel reflection coefficients at each interface. The Fresnel coefficients for the film-substrate interface depend upon the properties of the substrate material in a rather complex manner. However, when the optical properties of the substrate and the angle of incidence are known, the Fresnel coefficients at both the air-film and film-substrate interfaces become functions of the angle of refraction at the air-film interface and the index of refraction of the film.

The total amplitude of the reflected wave is the sum of the reflected ray (1) and the rays (2), ..., (*n*) that have undergone multiple reflections at both interfaces. The basic equation of ellipsometry relates the amplitude of the reflected wave polarized parallel to the plane of incidence to the amplitude of the perpendicularly polarized reflected wave. These amplitudes are obtained by performing the required summations separately for each component polarization. Denoting the complex-valued reflected wave amplitudes by $R_\parallel$ and $R_\perp$ and taking the ratio yields the form of the basic ellipsometry equation

$$\frac{R_{\parallel}}{R_{\perp}} = (\tan \Psi)\exp(j\Delta) \tag{11.12}$$

where Ψ and Δ are complicated functions of the index of refraction and the film thickness. However, each point in the (Ψ, Δ) plane corresponds to a particular pair of values for the film index of refraction, n_1, and film thickness t. A typical set of plots of Eq. (11.12), generated by computer, is shown in Fig. 11.5 for transparent films of silicon with $n_2 = 4.05$, $k_2 = 0.028$, $\lambda = 5461$ Å, and $\Phi = 70°$. The parameter for each curve is the index of refraction, n_1, which is underlined, and the curves are marked at intervals of 20° with values of β, where the film thickness, t, is given by

$$t = \frac{(180N+\beta)\lambda}{360(n_1^2 - \sin^2\Phi)^{1/2}} .$$

Note that N takes on integer values, depending on the thickness of the film. Arrows on the curve indicate the direction of increasing β or thickness t. Each set of curves is generated for a particular set of substrate properties, n_2, k_2, and particular values for the angle of incidence, Φ, and the wavelength λ of the incident radiation. Once n_1 and β have been established

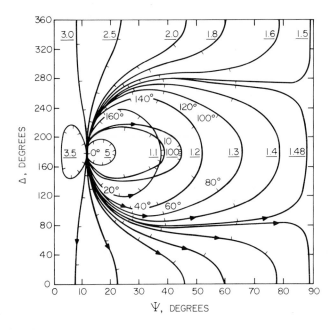

Fig. 11.5 Typical plots from ellipsometry measurement.

from the set of curves (corresponding to a value ψ, Δ) then thickness t may be obtained from the equation.

A typical ellipsometer uses a helium neon laser source, polarizer, quarter-wave compensator, and detector. The polarizer produces a linearly polarized wave which is then converted to elliptic polarization by the quarter-wave compensator. This elliptically polarized radiation is incident on the sample and the reflected wave is passed through an analyzer. When the analyzer is adjusted to null the detector, the ellipticity produced by the analyzer, polarizer, and compensator is the opposite of that caused by the reflection. Thus, $\tan \Psi$ and Δ can be determined.

The ellipsometer is exact in principle and is widely used for measurement of film thickness in the range 1 to 1000 Å. It is not too accurate near the origin of the (Ψ, Δ) plane where the curves for different refractive indices converge. For a detailed mathematical description of this technique and particulars of computer programs the reader is referred to the literature.[20]

11.4.3 Interference Measurement of Opaque Film Thickness [7]

The simplest method of measuring the thickness of nontransmitting films is by multiple-beam interferometry using Fizeau fringes. This situation is similar to that for transparent films, except in this case the multiple reflections are between the film-coated substrate and an auxiliary optically flat, partially transmitting, partially reflecting surface, as shown in Fig. 11.6. The angle of incidence, Φ, is almost 0 radians so that all of the reflected waves almost coincide spatially. Assuming that there is a phase change of π radians at each reflection, the change in phase during a path traversal is

$$\alpha = \frac{4T\pi}{\lambda} + 2\pi = 2N\pi, \quad \text{or} \quad T = \frac{(N-1)\lambda}{2} \tag{11.13}$$

where T is the separation between the film and the reflecting surface, and λ is the wavelength of the incident radiation. The condition for constructive interference is that the phase change be an integer multiple of 2π radians and the distance between the maxima of successive fringes corresponds to a distance $T = \lambda/2$.

A small angle between the film-coated substrate and the reflecting plate translates changes in the vertical dimension to horizontal displacements of the fringes. The fringe spacing is essentially $\lambda/2$ when the incident illumination is almost normal to the surface and the wedge angle between the film and reflector is very small. Film thickness, t, is measured by introducing a step into the film which is then coated with a thin layer of highly reflecting material, as also shown in Fig. 11.6. This step abruptly changes the distance T and causes an abrupt fringe displacement, d. The thickness is measured by

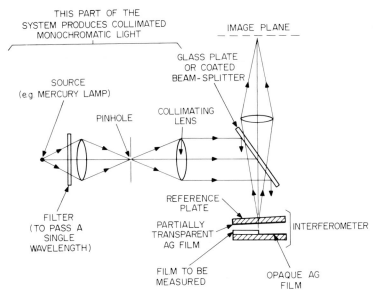

Fig. 11.6 Experimental setup for interference measurement of opaque films.

comparing the fringe displacement to the fringe spacing. Thus

$$t = \frac{d}{D} \cdot \frac{\lambda}{2} \tag{11.14}$$

where D is the fringe separation and d is the fringe displacement due to the step.

The resolution of this method is about $\lambda/400$ (that is, about 14 Å for $\lambda = 5461$ Å) because for highly reflecting surfaces the width of a fringe is about $D/40$, and displacements of about one-fifth of a fringe width can be measured.

11.4.4 Thickness Measurement Using Electromechanical Transducers

Electromechanical measuring instruments are used to measure the surface roughness of films and substrates. They can also be used to measure film thickness by measuring the mechanical movement of a stylus as it traverses a film-substrate step.

A stylus tip, having a radius of the order of 2.5 μm, traces variation in the surface that is traversed. The transducer attached to the stylus translates the motion into an electrical signal which is amplified and recorded on a conventional strip recorder. If the gain of the system is known, then displacement on the graph can be correlated directly with actual dimensional variations.

Machines of this type which are available, such as Talystep, Talysurf and Dectac, can measure displacements from perhaps 40 Å to about 10,000 Å with an accuracy of several percent.

11.5 MEASUREMENT OF JUNCTION DEPTH [21]

Silicon integrated circuits are often made by a sequence of nested diffusions as described in Chapter 6. An important characterization of the process is the depth of the junctions. Because the junctions are formed below the silicon surface, the direct methods of determining junction depth involve sectioning the device which is a destructive process.

The conventional method of sectioning a device is to lap it at an angle, θ, of several degrees to the surface, as shown in Fig. 11.7, to enlarge the apparent vertical dimension by a factor of $1/\sin \theta$. Contrast between the n- and p-doped layers is provided by a selective chemical stain or plating, whereby chemical reactions proceed at different rates on n-type and p-type silicon.

Staining, that is the reaction of silicon with an acid stain to form a film, is a tricky business that is perhaps best learned experimentally. This difficult art essentially consists of exposing the lapped, beveled chip surface to an acid solution, such as 0.1% nitric acid (HNO_3) in concentrated hydrofluoric acid (HF). The action of the stain is to darken the p-type region—at a rate depending on the resistivity of the p-type material—while leaving the n-type region essentially unchanged. The reactions occurring during staining are not completely known, but apparently a film of silicon monoxide, silicon hydride, H_2SiF_6, or some similar substance is created. Activation of this film growth or staining action seems to require the presence of some impurity in the hydrofluoric acid. Best results are obtained if the resistivity of the p-type material is in the range 0.1 to 0.25 Ω-cm because in this range the film grown on the p-type material then appears black. Regions of higher resis-

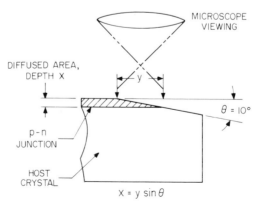

Fig. 11.7 Measurement of junction depth by lapping.

tivity are more difficult to stain. It appears that staining takes place up to the edge of the depletion region. A refined version of this experiment, involving reverse- and forward-biased junctions, has been used to measure values of depletion layer width. Once the junction has been effectively stained, the depth can be measured with a microscope having a calibrated reticle or it may be measured by interference methods. For further details the reader is referred to the literature.[21]

EXAMPLE 11.3　The horizontal measurement obtained by the angle-lap technique was 17 μm. The lap angle ϕ was 10°. Find the junction depth t.

Solution.　$t = y \tan \phi = 17 \tan 10° \approx 3 \ \mu$m.

An alternate method which requires very sophisticated instrumentation is to cut most of the chip with a laser, fracture the junction area of interest, and view the junctions with an electron microscope after they have been stained. This method is useful on smaller geometry devices with multiple junctions.

11.6 MEASUREMENT OF IMPURITY PROFILES

An accurate method for determining impurity profiles is important for the characterization of fabrication sequences. Several successive impurity diffusions and other high-temperature operations are required to make a bipolar transistor, and during each of these operations there is impurity redistribution which affects the electrical characteristics of the completed devices. It is necessary to be able to measure impurity profiles over variations in the impurity concentration of many orders of magnitude. In this section the spreading-resistance technique and the junction-capacitance method for measuring impurity profiles are described.

11.6.1 Measurement of Spreading Resistance [8,9]

The spreading-resistance measurement involves contacting the surface of a mechanically chemically polished sample with two probes, separated by a distance d. The radius of the probe tip, denoted by a, is about 1 μm and is much smaller than the probe separation. The probes are pressed against the surface by a force of several grams. A voltage V of several millivolts is applied to the sample via the probes and the resultant current I is measured. The *spreading resistance, $R_{sp} = V/I$.* It can be shown that for two probes contacting a very thick, semi-infinite slab of uniform resistivity material,

$$R_{sp} = \rho/2a, \tag{11.15}$$

where ρ is the resistivity.

The local resistivity of a doped semiconductor $\rho(x)$ may be related to the carrier concentration function $N(x)$ by

$$N(x) = \frac{1}{q\rho(x)\mu[N(x)]} \tag{11.16}$$

where the carrier mobility μ is itself a function of carrier concentration. When the resistivity is not uniform, the expression for spreading resistance is modified to include a complicated correction factor f, which includes the normalized probe spacing $d/2a$, and the values of resistivity that are measured in many, very thin layers.

$$R_{sp} = \frac{\rho_1}{2a} f[d/2a; \rho_1, \rho_2, \cdots, \rho_n] . \tag{11.17}$$

In this equation, f is the correction factor, ρ_1 is the resistivity of the surface layer, and ρ_i is the resistivity of the ith layer.

The apparatus is calibrated by measuring specially prepared samples of known impurity type and resistivity. Calibration curves for n- and p-type silicon samples are shown in Fig. 11.8. Notice that these curves are for a specific ratio of normalized probe spacing $2d/a$ and probe force g.

Before a transistor can be profiled, it must be beveled and polished in a manner similar to that of the bevel and stain method described in Section 11.5. The spreading resistance is plotted versus position on the bevel

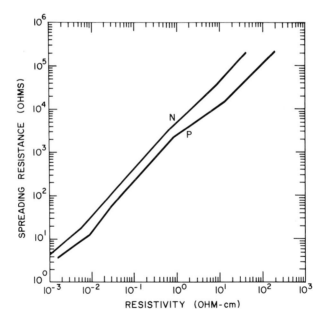

Fig. 11.8 Calibration curves for p- and n-type polished silicon samples. Probe spacing is 100 μm and probe load is 20 g. (After Gruber and Pfeifer.[9])

edge $n \cdot s$ where s is the step size and n is the number of the step. The layer depth, $t = ns \sin \theta$ is a function of the bevel angle θ which is usually less than 15°. Uncorrected spreading-resistance profiles may be "corrected" to account for the close proximity of a conducting layer (such as a highly doped region) or an isolating layer (such as a p-n junction). This corrected value of resistivity, denoted ρ_{corr}, may by virtue of Eq. (11.17) be written as

$$\frac{\rho_{corr}}{2a} = \frac{R_{sp}}{f[2d/a; \rho_1, \rho_2]} \tag{11.18}$$

for a two-layer structure. When $\rho_2 = 0$ there is a conducting boundary and when $\rho_2 = \infty$ there is an isolating boundary. It is often necessary to resort to iterative techniques to determine the correction factor; a procedure that is beyond the scope of this book.

Both corrected and uncorrected data are then converted from resistivity to impurity concentration via the Irvin resistivity versus impurity concentration curves.

EXAMPLE 11.4 An n-type epitaxial layer on p-type material had a depth $d = 5 \ \mu m$. The spreading resistance R_{sp} was measured with probe tips of radius $a = 2 \ \mu m$ and was found to be $R_{sp} = 2500 \ \Omega$ for a ratio of $\rho_2/\rho_1 = 1$. The correction factor $f[2d/a; \rho_1, \rho_2]$ obtained from tables equaled 1.25. Find the corrected resistivity $\rho_{1 \ corr}$.

Solution.   From Eq. (11.18),

$$\rho_{1 \ corr} = \frac{2aR_{sp}}{f} = \frac{2 \times 2 \times 10^{-4} \times 2500}{1.25} = 8 \times 10^{-1} \ \Omega\text{-cm}.$$

This value, when looked up on the Irvin curve[2] of ρ versus concentration produces an impurity concentration of the epitaxial layer of approximately 8×10^{15}. In a practical measurement, a sharp dip in this value is observed as the p-n junction depth is approached.

The spreading resistance method may require the use of a digital computer to process the raw measured data and produce finished plots of carrier concentration versus depth. This method is useful for carrier concentrations from 10^{13} to 10^{20} carriers/cm³. A typical impurity profile obtained by this method is shown in Fig. 11.9. Programs are applicable for both n-type and p-type conductivities and (100) and (111) wafer orientations. Accuracy of the method is about 10%.

11.6.2 Impurity Concentration from Capacitance-Voltage Measurement [10]

A reverse-biased p-n junction exhibits a voltage-dependent depletion-layer capacitance. It will now be shown that the impurity concentration can be inferred from a knowledge of the capacitance variation with changes in voltage.

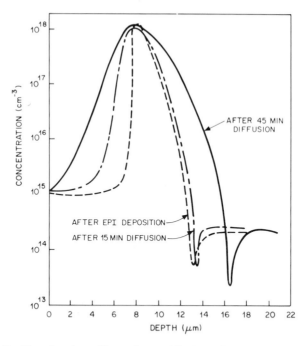

Fig. 11.9 Profiles showing effect of out-diffusion of buried layer dopant during epitaxial deposition and subsequent high-temperature processing. These curves are raw spreading resistance measurement data. (After Schroen.[8])

Recall from Chapter 2 that the depletion-layer capacitance per unit area for an abrupt *p-n* junction is given by

$$C = \frac{\epsilon_s}{w} = \left[\frac{q\epsilon_s}{2} \frac{N_A N_D}{N_A + N_D} \frac{1}{V + V_{bi}} \right]^{1/2} , \qquad (11.19)$$

where V is the applied voltage and V_{bi} is the built-in voltage. Differentiating this equation with respect to applied voltage yields

$$\frac{d}{dV} \left[\frac{1}{C^2} \right] = \frac{2}{q\epsilon_s} \frac{N_A N_D}{N_A + N_D} . \qquad (11.20)$$

If the *n* side of the junction is much more heavily doped than the *p* side (one-sided abrupt junction) $N_D \gg N_A = N(w)$, where $N(w)$ is the acceptor concentration in the *p*-material at the edge of the depletion layer, then from Eqs. (11.19) and (11.20)

$$N(w) = \left[\frac{q\epsilon_s}{2} \frac{d}{dV} \left[\frac{1}{C^2} \right] \right]^{-1} \qquad (11.21)$$

$$w = \frac{\epsilon_s}{C} .$$

A plot of $1/C^2$ versus V obtained from measured data is shown in Fig. 11.10. A plot of $N(w)$ versus w where w is the distance from the metallurgical junction to the edge of the depletion region prepared from Fig. 11.10 and from Eqs. (11.21) is shown in Fig. 11.11.

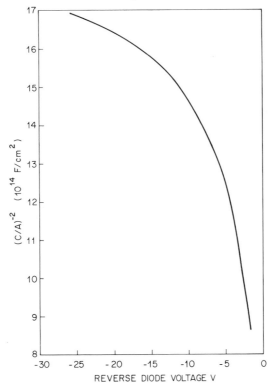

Fig. 11.10 Plot of $1/C^2$ versus V for a p-n junction.

11.6.3 MOS Capacitance Method [11-13]

The MOS capacitor was analyzed in Section 3.3. We will now show how measurement of the C-V variation of the MOS capacitor permits determination of surface concentration.

When a moderate gate bias of the correct polarity is applied to the MOS capacitor, it divides between the oxide and space-charge layer to satisfy the constraints of a capacitive voltage divider consisting of an oxide capacitance and a depletion-layer capacitance. The oxide layer is of thickness x_o and therefore the oxide capacitance per unit area is $C_{ox} = \epsilon_{ox}/x_o$. A differential change in applied gate voltage V_G results in voltage changes across both capacitors such that

$$dV_G = dV_{ox} + dV_s = \frac{dQ_s}{C_{ox}} + \frac{dQ_s}{C_s}, \qquad (11.22)$$

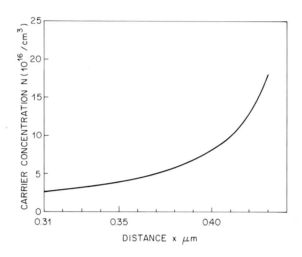

Fig. 11.11 Plot of $N(w)$ versus w.

where Q_s is the space-charge density per unit area, $C_s = \epsilon_s/w$ is the depletion-layer capacitance per unit area (which is voltage dependent) and w is the depletion-layer width which now is the distance from the surface of the semiconductor to the edge of the depletion layer.

$$dQ_s = C \; dV_G = qN(w)\,dw. \tag{11.23}$$

In this equation, $N(w)$ is the doping concentration at the edge of the space-charge layer and $C = (C_{ox}^{-1} + C_s^{-1})^{-1}$ is the total measured MOS capacitance per unit area. From the definition of C_s we find that

$$dw = \epsilon_s d(C_s^{-1}) = \epsilon_s d(C^{-1}), \tag{11.24}$$

and thus from the last two equations

$$N(w) = \left[\frac{q\,\epsilon_s}{2}\; \frac{d}{dV_G}\left[\frac{1}{C^2}\right]\right]^{-1}. \tag{11.25}$$

This result shows that the doping density at distance w from the surface of the semiconductor to the edge of the space-charge layer can be calculated from the slope of the $1/C^2$ versus V_G curve. Note the similarity between this equation and Eq. (11.21).

EXAMPLE 11.5   A pulsed MOS C-V measurement on a silicon wafer yielded the following data:

$$V_{G1} = -4 \text{ V}; \quad C_1^{-2} = 92\times10^{15} \text{ cm}^2/\text{F}^2$$

$$V_{G2} = -4.5 \text{ V}; \quad C_2^{-2} = 108\times10^{15} \text{ cm}^2/\text{F}^2$$

It was also ascertained that $C_{ox} = 2.1\times10^{-8}$ F/cm^2. Calculate the impurity density $N(w)$ and the depletion width w.

Solution.   Because $C^{-1} = C_{ox}^{-1} + w/\epsilon_s$,   therefore $w = \epsilon_s\{(C_{av}^{-2})^{1/2} - C_{ox}^{-1}\}$ where $C_{av}^{-2} = (C_1^{-2} + C_2^{-2})/2$ and so represents the average value. For the above data $C_{av}^{-2} = (92+108) \times 10^{15}/2 = 10\times10^{16}$ and so $C_{av}^{-1} = 3.16\times10^8$. But $C_{ox}^{-1} = 0.48\times10^8$ and so

$$w = 11.8\times8.854\times10^{-14}\{3.16-0.48\} \times 10^8 = 2.8\times10^{-4} \text{ cm.}$$

To find $N(w)$ we approximate

$$\frac{dC^{-2}}{dV_G} \approx \frac{C_2^{-2} - C_1^{-2}}{V_{G2}-V_{G1}}$$

$$= \frac{(108-92) \times 10^{15}}{4.5-4} = 32\times10^{15}.$$

Hence from Eq. (11.25)

$$N(w) \approx \frac{2}{q\epsilon_s}\left[\frac{dC^{-2}}{dV_G}\right]^{-1}$$

$$= \frac{2}{1.6\times10^{-19}\times11.8\times8.854\times10^{-14}\times32\times10^{15}} = 3.739\times10^{18}.$$

In a practical measurement, many values of C^{-2} taken at applied voltages V_G would be plotted and values of w could then be computed by taking the average values of adjacent points on the curve. The slope of the curve provides the average value of $N(w)$ for points through which a least square straight line can be fitted. $N(w)$ may also be obtained from values of adjacent points, as illustrated above.

If the positive voltage V_G applied to the gate of a p-type bulk MOS capacitor is increased, the depletion region width w will increase until the surface inverts as described in Chapter 3. This inversion is accompanied by a sharp increase in the number of electrons, Q_n, at the oxide-silicon interface. After the occurrence of inversion, the depletion width w stays at a maximum value of w_{max} which corresponds to a space-charge voltage $V_{s\ inv}$. The charge density at the surface then becomes

$$Q_s = Q_n - qN_A w_{max} \tag{11.26}$$

where the first term Q_n is the charge in the inversion layer and the second term $-qN_A w_{max}$, which is due to the space-charge layer, remains constant.

Furthermore,

$$w_{max} = \left[\frac{2\epsilon_s V_{s\ inv}}{qN_A} \right]^{1/2}. \tag{11.27}$$

After inversion the MOS capacitance C no longer decreases with the square root of gate voltage, because the surface is no longer being depleted. This condition is clearly shown in Fig. 11.10 and the onset of inversion corresponds to a gate turn-on voltage, V_{GT},

$$V_{GT} = \frac{qN_A w_{max}}{C_{ox}} + V_{s\ inv}. \tag{11.28}$$

The capacitance C remains constant for values of gate voltage V_G exceeding V_{GT}.

The capacitance-voltage characteristic of an MOS capacitor is a function of frequency with a low frequency limiting value, C_{LF}, that approaches the oxide capacitance, C_{ox}, at the onset of strong inversion. This frequency dependence leads to a modified form of Eq. (11.25) for the carrier concentration in which C_{HF} is the high frequency value of the MOS capacitance.

$$N(w) = \frac{2}{q\epsilon_s} \left[\frac{1 - (C_{LF}/C_{ox})}{1 - (C_{HF}/C_{ox})} \right] \left[\frac{d}{dV_G} \left(\frac{1}{C_{HF}^2} \right) \right]^{-1}. \tag{11.29}$$

This equation is valid for both depletion and inversion so the distance over which the profile can be determined has been increased.

EXAMPLE 11.6   Prove that

$$\frac{C_{LF}}{C_{ox}} = 1 - qwN(w) \frac{dC_{HF}^{-1}}{dV}$$

Hence determine C_{LF}/C_{ox} given the following data

$$w_1(\mu m) = 0.4, \quad N(w_1) = 1.11 \times 10^{15}\ cm^{-3};$$

$$w_2(\mu m) = 0.5, \quad N(w_2) = 1.03 \times 10^{15}\ cm^{-3};$$

$C_{ox} = 1.9 \times 10^{-8}$ F/cm.

Solution.   From Eq. (11.29) one has

$$\frac{C_{LF}}{C_{ox}} = 1 - \frac{q\epsilon_s N(w)}{2} \left[\frac{1 - C_{HF}}{C_{ox}} \right] \frac{d(C_{HF}^{-1})^2}{dV}.$$

But

$$d(C_{HF}^{-1})^2 = 2C_{HF}^{-1}d(C_{HF}^{-1}) \quad \text{and} \quad C_{HF}^{-1} - C_{ox}^{-1} = w/\epsilon_s.$$

Hence $C_{LF}/C_{ox} = 1 - qwN(w)d(C_{HF}^{-1})/dV$ as was required to prove.

To prove the second part, we note that

$$\frac{d(C_{HF}^{-1})}{dV} \approx \frac{C_{HF2}^{-1} - C_{HF1}^{-1}}{V_2 - V_1}.$$

Now

$$C_{HF1}^{-1} = \frac{w_1}{\epsilon_s} + C_{ox}^{-1}$$

$$= \frac{0.4 \times 10^{-4}}{11.8 \times 8.854 \times 10^{-14}} + 0.527 \times 10^8 = 0.91 \times 10^8$$

and corresponds to

$$V_1 = \frac{q w_1 N(w_1) C_{HF1}^{-1}}{2}$$

$$= \frac{1.6 \times 10^{-19} \times 0.4 \times 10^{-4} \times 1.11 \times 10^{15} \times 0.91 \times 10^8}{2} = 0.323 \text{ V}.$$

Similarly,

$$C_{HF2}^{-1} = \frac{w_2}{\epsilon_s} + C_{ox}^{-1}$$

$$= \frac{0.5 \times 10^{-4}}{11.8 \times 8.854 \times 10^{-14}} + 0.527 \times 10^8 = 1.01 \times 10^8$$

and corresponds to

$$V_2 = \frac{q w_2 N(W_2) C_{HF2}^{-1}}{2}$$

$$= \frac{1.6 \times 10^{-19} \times 0.5 \times 10^{-4} \times 1.03 \times 10^{15} \times 1.01 \times 10^8}{2} = 0.416 \text{ V}.$$

Hence

$$\frac{d(C_{HF}^{-1})}{dV} \approx \frac{(1.01 - 0.91) \times 10^8}{0.46 - 0.323} = 1.075 \times 10^8.$$

Taking average values for w and $N(w)$ one has

$$w = \frac{w_1 + w_2}{2} = 0.45 \times 10^{-4}$$

and

$$N(w) = \frac{N(w_1) + N(w_2)}{2} = 1.07 \times 10^{15} \text{ cm}^{-3}.$$

Hence

$$C_{LF}/C_{ox} = 1 - 1.6 \times 10^{-19} \times 0.45 \times 10^{-4} \times 1.07 \times 10^{15} \times 1.075 \times 10^8$$

$$= 1 - 0.828 \approx 0.17.$$

This value is too low, but inaccuracies of 50% or more may have been introduced through computations of slope.

The low-frequency capacitance is usually measured at a few hertz whereas the high-frequency capacitance is measured at frequencies in the megahertz range. A theoretical doping density curve obtained from Eq. (11.29) is shown in the solid curve of Fig. 11.11 while the dashed curve corresponds to uncorrected data obtained from Eq. (11.25). In the absence of inversion, both curves would coincide. A 5% accuracy range of measurement is indicated and extends from

$$3L_B \leqslant w \leqslant 2L_B\sqrt{u_B} \qquad (11.30)$$

where, $L_B = \sqrt{V_T\epsilon_s/qN}$ is the bulk Debye length and $u_B = \ln(N/n_i)$, with the intrinsic carrier density n_i being approximately 1.45×10^{10} cm^{-3} at 300 K for silicon.

When the impurity concentration is almost uniform, or changes very little over a distance of w_{max}, then a single $C(V)$ curve is insufficient to compute the impurity concentration. In that case it is necessary to prepare a series of $C(V)$ curves and to calculate w_{max} for each curve. Each curve is produced after removing a thin layer of silicon, much smaller than w_{max} and it is assumed that the impurity concentration is constant over the thickness of the removed layer.

It may be shown that the absolute accuracy of the MOS capacitance measurement technique is within about 10%, but the method is only applicable for thin, lightly doped layers (concentrations below 10^{18} carriers/cm^3).

11.7   INTRODUCTION TO TESTING

A prime advantage of integrated circuits, aside from their small size, is that the batch fabrication processes by which they are made result in a low cost per circuit function. Unfortunately, not all of the potential circuits in a batch work, and therefore it is essential to have some means of separating the good product. The complexity of modern LSI circuits is such that it may not be possible, or economically practical, to ascertain with 100% certainty that a particular circuit is good. It is essential to perform enough tests to be confident that the product is good. Many of the tests to be described might also be performed by the customer for lot acceptance.

The manufacturer has several motives for testing completed devices. For instance, when a new product is first put into production, it is necessary to thoroughly *characterize* its behavior so that performance limits can be set for subsequent large-scale production tests. This information is also used in the preparation of specification sheets. Tests of a diagnostic nature on the final product are also used to provide feedback to adjust the processing controls.

Performance verification tests are a subset of the above characterization tests which are used in large scale production to determine whether a device meets a minimum standard.

Often there is a family of devices which are derived from the same design by *sorting* according to desired parameter values from the total population.

Reliability tests are used to ensure that a device will operate under given conditions for a desired time interval. Such tests are described in detail in Chapter 16.

Some or all of the following electrical tests are used to achieve the above objectives:

DC tests may be performed on analog or digital systems. The device under test is excited by a dc voltage or current and the desired physical variable is measured after all transients have died out.

AC steady-state tests are primarily used for analog systems. The device is excited by a sinusoidal or other periodic waveform and the response is also periodic.

Functional tests on a digital circuit are used to verify its logical operation, for example, the truth table for a combinational circuit.

Dynamic tests are pulse measurements to determine transient response.

A test is termed *operational* when the device is loaded to simulate the intended system environment.

11.8 TESTING OF DIGITAL CIRCUITS [14,15]

The design of digital integrated circuits is considered in Chapter 14. Testing of digital integrated-circuit chips can be divided into two broad categories:

1. Quantitative measurements of dc terminal and transfer characteristic. Such tests would also include power consumption and verification that all terminals are operative.

    Measurements of dynamic quantities such as propagation, delay, rise and fall time and characteristics of transient response.

    Measurement techniques are similar to laboratory techniques, but must include provisions to probe chips when unpackaged circuits are being tested. Because of the large number of measurements that must be made, equipment is automated and test results are communicated directly to a computer.

2. Qualitative tests are used to establish that the circuit is performing the logical function for which it was intended.

Many difficulties exist with respect to testing large, complex logic circuits. A number of these difficulties are listed.

1. For a combinational logic circuit having N inputs, the total number of possible logic tests is 2^N. This number may be so large that it becomes impractical to carry out the tests in a sufficiently short time. Fortunately there exists a minimum set of tests, smaller by possibly several

orders of magnitude, that is sufficient to locate any single fault observable at the system output. Computer aids are usually necessary to establish this set of tests. An automated test set may also be required, because the tests can still involve a very large number of measurements. Faults encountered are usually *stuck-at* faults. A node is said to be stuck at zero (or one) if it is permanently unable to change its zero logic (one logic) state.

2.   In testing combinational integrated circuits, determination of basic stuck-at faults may be insufficient. Integrated circuits may suffer from other faults, such as pin shorts, intermittent faults, or even the presence of spurious components. Many present-day test procedures do not include these types of faults in their repertoire, nor do they consider the possible presence of multiple stuck-at faults.

3.   A sequential logic circuit is a digital circuit with memory, whose output depends on previous input and output states. Hence, a sequential logic circuit must be brought to a known initial state from which testing can commence. This can be done either by providing auxiliary inputs, which force the memory elements into a correct state, or by a sequence of primary input vectors (i.e., excitation of all primary inputs) known as a *homing sequence.*

4.   It is essential that the initial circuit design makes adequate provision for the subsequent testing of the circuit. This might mean that the circuit includes more than the minimum number of elements necessary to realize the desired system function. When testing digital integrated circuit chips it is sufficient to detect faults and usually unnecessary to locate them, because the chip is nonrepairable. An exception occurs in the testing of a new design when the fault must be located and corrected. However, new designs are more easily debugged on a simulator.

5.   When testing digital systems assembled from many chips, the test algorithm is only required to isolate a fault to within a collection of elements on a given individual chip, since only the chip is replaceable.

11.9 TESTING MEMORIES [16,17,22]

Random access memories (RAM's) will be discussed in some detail in Chapter 15 and those unfamiliar with memories might want to read that chapter before continuing with this section. Three types of tests are required for such memories, the first of which is a parametric test. Here one measures dc characteristics, such as leakage current, breakdown voltage, power consumption, and minimum and maximum output states as a function of output sink or source currents. The second type of test is functional in nature and includes pattern tests at rated cycle rates and truth table tests.

Finally, dynamic tests include measurement of access time, setup or hold times, strobe width and cycle times.

Faults can occur in the memory array, in the address register and decoders, in the output buffers and drivers, in the read/write circuitry, in the data-in/data-out circuitry and in the sense amplifiers. These faults can result in functional disorders, such as the inability to read or write. They can also lead to parametric disorders, such as an unacceptable output level and dynamic difficulties, such as slow access times.

A RAM is completely functional if every cell of the memory is capable of storing a zero or a one without any interaction between cells. The cell addressing circuits must correctly access the desired cell and the sense amplifiers must operate correctly so that zeroes and ones are properly indicated. In the case of dynamic memories, the cells must retain information for a specified period without being refreshed.

Test patterns are used to discover functional faults. These patterns must be able to locate memory cells whose behavior varies depending upon the contents of neighboring cells. For example, writing a one into a particular cell may cause another memory cell to change state, or timing relations may change when a repetitive pattern is replaced by a nonrepetitive pattern.

Many examples of test patterns could be given, but only a few are mentioned here:

"Walking ones" or "march" Every memory cell is initially set to zero and the first cell is read to check the zero. Then a one is written into the first cell and it is read to verify it is still in the one state. The test bit moves to the second cell and the process is repeated. The procedure is continued until each cell has in turn been set to one. The memory is now fully scanned in a read mode, to verify that it is full of 1's. The process is then repeated from the minimum address to the maximum address by changing 1's to 0's, and then fully scanning the memory to ensure that it is full of 0's. The test length for this procedure is proportional to N, the number of cells in the memory array.

This test pattern proves every cell can be set to zero and to one. It also verifies that all cells are being addressed correctly and so this pattern constitutes a very thorough test, particularly of decoder problems.

"Galloping" or "ping-pong" A program first writes data into every cell of the memory. The first cell is then read, after which it is written with the complement of its previous data. The program then continues by reading cell one, reading cell two, reading cell one, reading cell three, ..., reading cell one, reading the last cell. The data at cell one is complemented and the cell rewritten, and the test continues, starting with cell two. After n passes through the memory array the last cell is reached. Then this procedure is completely repeated, starting with the memory filled with the complement of the original data.

This procedure requires a test time approximately proportional to N^2. The ping-pong test is a very thorough test, particularly of pattern sensitivity. The test checks that operating on a cell in one location has no effect on the contents of a cell in another location. Galloping or ping-pong tests are very time consuming, typically requiring many seconds on a computer-controlled test set. This makes the test less meaningful for large memories, namely 4 kilobits or larger, particularly during the production stage.

In order to minimize pattern testing time and taking account of memory structure, one may take the view that a cell is most closely *electrically related* to another cell, when both happen to be in the same row or column. Moreover, two cells may be considered most closely *geometrically related* to each other, when they are in the same row, column or diagonal. Hence, using the row and column of the test bit as the field—rather than the whole memory as in the previous test—one may do a ping-pong read-only in the row and column of the test bit. Moreover, the test bits may be placed on a diagonal to allow for geometrical relationships. This type of test is less lengthy than the previous test and will now be described.

"Shifted Diagonal"   Every cell in the memory is initially set to zero and then the main diagonal, offset by one (that is, the superdiagonal consisting of cells 0,1; 1,2; 2,3; and so on), is written with 1's. The complete memory is read doing a row-column ping-pong. That is, cell (0,0) is read, then cell (0,1) is read, and so on, for the rest of the row. The column of the test bit is then similarly exercised. This is repeated for each test bit in the diagonal. The offset of the diagonal is now increased by one so that the new diagonal lies on locations (0,2; 1,3; 2,4; ...) the read mode is now repeated for each test bit in this diagonal, after which the offset of the diagonal is again increased by one. For a 4-kilobit memory, 64 diagonals must be so exercised, for instance. The test bits are then replaced by 0's in a field of 1's and the inverse pattern repeated. This test pattern checks proper access times and also determines that all cells in the memory exist and that there are no sense amplifier or data loss problems. The time required for this test procedure is proportional to $N^{3/2}$ and so is less time consuming than the ping-pong test. The test is therefore more useful in exercising large memories, even though it is not quite as complete as the ping-pong test, previously mentioned.

Refresh tests   Dynamic MOS RAM's must be tested to see whether they store information accurately for a specified time period because of decay of stored charge. After the whole matrix is written, a period of time is allowed to elapse and then the matrix is read at high speed. The memory is sometimes disturbed during such a test to simulate actual conditions in a memory system.

Even when memories have passed parametric and functional tests, they may still be practically useless, unless they can respond within some reasonable time specification. Particularly important are chip select access, read access, and write access time measurements. Also important are timing relationships such as for data, address setup, and so on. A more detailed discussion of dynamic testing however is beyond the scope of this chapter.

11.10 TESTING OF ANALOG CIRCUITS

Because there exists a wide variety of analog systems, with widely differing requirements, the following remarks are confined to tests used to characterize a differential input operational amplifier. (See Chapters 12 and 13 for a description of operational amplifiers and their applications.) Such tests are illustrative of the type of tests used to characterize many analog systems. A differential input, single-ended output operational amplifier has inverting and noninverting input terminals, an output terminal, a real or implied ground, and two power supply terminals. Most of the quantities that characterize the performance are complex functions of frequency. These include the differential mode gain, A_{dd}, which is typically 10^5, the common mode gain, A_{cc}, and the common mode rejection ratio 20 log $|A_{dd}/2A_{cc}|$ which may be of the order of 100 dB. Also of interest are the output impedance and the three impedances measurable between the input terminals and from each input terminal to ground.

Clearly the test set must have a large dynamic range and must be capable of making measurements out to at least the unity gain frequency of the amplifier, which is typically several megahertz. These ac steady-state tests are different in character from the functional tests for digital circuits. Fewer tests are performed and fewer excitation signals are required, but the test is conducted over a wide frequency range.

Transient performance is characterized by the maximum rate of change of the output voltage, known as the slewing rate and also in terms of the step input response.

Other measurements of interest are carried out with zero input voltage or current. Among these are noise measurements and offset voltage and current measurements.

For high volume components such as operational amplifiers, special automated test sets have to be designed.

11.11 FUNCTIONAL TESTING OF A HYBRID ANALOG CIRCUIT [18]

The tests so far described merely indicate whether a circuit is good or bad. The test now to be described on the TOUCH-TONE® oscillator results in the adjustment of the circuit to meet required specifications. It is an example of a total engineering solution, in that the circuit was designed to be functionally tested and adjusted by an automatic test set.

A TOUCH-TONE® telephone transmits dialing information to the central office by means of audio frequency tones which represent the called number. A TOUCH-TONE® telephone contains two oscillators, each capable of producing four audio frequencies. The frequencies are in two groups, the low frequency group being 698, 771, 853.5, and 942.5 Hz and the high frequency group consisting of 1211, 1338, 1479, and 1636 Hz. These frequencies were carefully selected to avoid any harmonic relationships or duplication which might occur due to the caller's voice. Each push button selects one frequency from each group; thus there are sixteen possible combinations of which only ten are used for dialing at the present time, although facility for twelve is provided.

The initial approach to meeting the specifications with an IC is almost obvious—a notch-type RC audio oscillator using a twin-tee filter as the frequency selective feedback network. A direct-coupled amplifier provides the loop gain required to maintain oscillation, while the variable-gain buffer amplifier serves the dual function of isolating the oscillator from the line and providing a constant output signal level.

Telephone signaling requirements impose a tolerance of ±0.1% on the oscillator frequencies and the notch depth of −30 dB must be maintained to within ±0.5 dB for linearity requirements. If an attempt to meet these specifications is made by imposing close tolerances on the individual resistor and capacitor values, these tolerances would be so stringent as to render the device uneconomical, if not impossible, to produce. This fact can be better appreciated by doing the following exercise. Assume that the null frequency is inversely proportional to some RC time constant. Imposing the frequency stability requirement and considering only first-order variation terms in R and C yields

$$\frac{1}{f_0(1\pm0.001)} = T(1\pm0.001) = (R\pm\Delta R)(C\pm\Delta C).$$

Thus

$$0.001 = \frac{\Delta C}{C} + \frac{\Delta R}{R},$$

which means that the tolerance on both R and C must be better than 0.1%. It is possible to adjust the value of a tantalum thin film resistor to about 0.1%, but it is impractical to even approach this tolerance with capacitors. However, to satisfy the frequency criteria, it is not necessary for the resistors or capacitors to have tightly controlled values so long as certain RC products are tightly controlled. The notch depth is set by adjusting resistors only.

A complete TOUCH-TONE® oscillator consists of two substrates, one containing all the tantalum thin-film capacitors and the second containing the tantalum resistors together with the conductive interconnection pattern. Two beam-lead sealed junction silicon chips and two thermistors are mounted on the resistive substrate. One of these silicon chips contains diodes used in the amplitude-limiting circuit of the oscillators. The second silicon chip contains all the active elements. The thermistors are used in the output stage to render the oscillator amplitude independent of temperature changes.

A special test set was designed to test and adjust a circuit while it is operating. The adjustment procedure consists of first adjusting the notch depth and then adjusting the frequency of oscillation. It can be shown[18] that the notch depth can be adjusted in either direction, while the frequency of oscillation can only be lowered. The test set measures resistance and frequency, stores the data, and then computes correction factors and new resistance values. Hence tantalum thin-film and silicon-monolithic technology have been effectively combined to produce a hybrid integrated circuit for the telephone system by engineering the circuit so that notch depth and frequency of oscillation are independently adjustable, and designing a special test set to perform the adjustment. The oscillator has been designed so that the null frequency of the feedback network is changed by varying one resistor in the series output arm in one of the parallel tees. Each resistor value is selected by any of four push-button switches.

11.12 AUTOMATIC TEST EQUIPMENT [19]

The basic elements of a computer controlled system are shown in Fig. 11.12. The hardware part of the system consists of automatic circuit test equipment which is activated by a special purpose control computer. The basic elements of the system are as follows:

a)   A data base of circuit descriptions for the device or system under test. Such a data base incorporates details of integrated circuit types, wiring diagrams, pin-loading requirements, and so on. An effective data base is basic to creating a simulatable model.

b)   Circuit or test simulation so as to create a fault-free model, which forms the standard against which actual products are judged. An actual circuit is then exercised by means of a set of stimuli called an *input vector*. The resultant set of responses, called the *output vector*, is then compared with that of the fault-free model and "good" products are thereby distinguished from failing ones. The set of input-test vectors should be sufficient to allow for adequate fault resolution in failing products. It is often more efficient to excite the circuit under test by the input stimulus vector and to compare the observed-response vector to

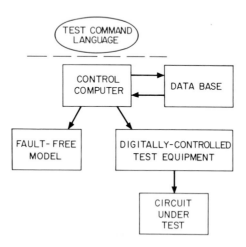

Fig. 11.12 Block diagram of a computer-controlled automatic test set.

the expected-response vector, which is obtained by prior simulation and stored in the computer.

c)   Test generation is the specification of a set of input-stimuli vectors, which are regarded as sufficient to completely test the digital circuit. For certain classes of circuits there are algorithmic methods for generating these input vectors. In other cases the methods are heuristic. For nonrepairable circuits it is only necessary to detect faults. For repairable circuits these vectors should be sufficient to detect and also locate faults.

d)   Each computer-controlled automatic test set has an associated test command language. This is a convenience for the engineer because it enables the test program to be written in English-like statements, which are then compiled into machine-language instructions for execution by the control computer. Results of tests are displayed on a suitable printout and stored in a data base. Statistical analysis of this data is then possible. A system of this type might cost several hundred thousand dollars.

11.13 ECONOMICS OF INTEGRATED CIRCUIT TESTING

The costs of automatic test systems is considerable with price tags of several hundred thousand dollars being quite common. The price of a test system depends on its flexibility, the particular type of measurements it can perform, its speed, and its data-handling capability. This sophisticated equipment is required because perhaps as many as several hundred thousand tests may be necessary to ascertain the performance of a complex integrated circuit, and there are many of these circuits on each wafer. The cost of testing a particular integrated circuit design may be a significant fraction of the total production cost. This may be justified or even desirable if the circuit performs an indispensable function in the system application for which it is intended.

EXAMPLE 11.7 The cost of testing each integrated circuit depends upon the number of tests required, the rate at which the test set can perform the tests, the cost of labor to operate and maintain the test set, and depreciation of the equipment. Make some reasonable assumptions and thereby compute an estimate of the cost of testing circuits.

Solution. Assume that an automated test set has an initial price of $500,000 and that it is to be fully depreciated in 10 yr. Furthermore, it is assumed that the maintenance cost is $50,000 per year, and that labor cost to operate the machine is also $50,000 per year. If these are the only costs associated with this machine, then it costs $150,000 annually to operate this test set.

The machine runs 10 hr each day and 300 days per year, i.e., about 10^7 operating seconds/year. Hence, machine time costs about 1.5 cents per second. A single measurement and storage of the result is assumed to require 100 μs.

It is not uncommon for a complex integrated circuit to require in excess of 100,000 measurements for the complete test. Thus it costs about 15 cents to test each circuit on the wafer using this machine. However, this is not the test cost per *good* circuit which depends on the circuit yield and may be several times the chip test cost.

The cost of testing can be expressed in a more mathematical way by assuming that a wafer is processed at a cost of W dollars and produces N "potential" circuits after separation. Each chip site on the wafer is given a preliminary test at a cost of D dollars and $M \leqslant N$ chips are found to be good. Hence the total expenditure is $(W+ND)$ dollars, and the cost of each good chip is $(W+ND)/M$ dollars. Note that the yield at this preliminary test stage is M/N.

Each good chip is then packaged at a cost of P dollars per chip and subjected to a final dynamic and logic test at a cost of L dollars per chip. Hence the total cost per chip before dynamic testing is $(W/M) + (DN/M) + P$ and the total cost after logic testing is $(W/M) + (DN/M) + P + L$. Assume further that only G chips $(G \leqslant M)$ are good after final test. Hence the total cost C per good tested chip is

$$C = \frac{[(W+ND)/M] + P + L}{G/M}$$

$$= \frac{(W+ND) + (P+L)M}{G} . \qquad (11.31)$$

The total testing cost, $(ND+LM)$ dollars, must be borne by the G good chips. Therefore the true test cost per saleable circuit is

$$T = \frac{ND+LM}{G} . \qquad (11.32)$$

Hence the ratio of test cost per chip T to the total cost per chip C is

$$\frac{T}{C} = \frac{1}{1 + (W+PM)/(ND+LM)} . \qquad (11.33)$$

This value is clearly of importance in any integrated circuit pricing study. However, it is difficult, if not impossible, to set an upper limit on this ratio without additional information about the application intended for this circuit.

EXAMPLE 11.8   Consider that a preliminary test and inspection adds a cost of 10¢ to each chip. Further, let the cost of a wafer be $50 and let there be 150 chip sites. Show that for the preliminary test and inspection to be economical $Y/Y' > 1.3$, where $Y = M/N$ is the chip yield with preliminary test and inspection, while $Y' = M'/N$ is the yield when such tests are omitted.

Solution.   From Eq. (11.31) and the above data $D = 0.1$, $W = 50$ and $N = 150$. The processing cost per good chip, C', without test and inspection is $C' = (W/N)/Y'$, while the processing cost per good chip, C, with preliminary test and inspection is $C = [(W+ND)/N]/Y$. For the preliminary test and inspection to be economical $C/C' < 1$, or $Y/Y_1 > [1 + (DN/W)] = 1.3$.

REFERENCES

1.   P. F. Kane and G. B. Larrabee. *Characterization of Semiconductor Materials.* New York: McGraw-Hill, 1970, p. 91.

2.   J. C. Irvin. "Resistivity of bulk silicon and diffused layers in silicon." *Bell Syst. Tech. J.* **41**: 387 (1962).

3.   S. M. Sze. *Physics of Semiconductor Devices.* New York: Wiley-Interscience, 1969, p. 42.

4.   L. J. van der Pauw. "A method of measuring specific resistivity and Hall effect of disc or arbitrary shape." *Philips Research Reports* **13**: 1-9 (February 1958).

5.   P. F. Kane and G. B. Larrabee, *ibid.,* pp. 226-228.

6.   P. F. Kane and G. B. Larrabee, *ibid.,* pp. 316-317.

7.   P. F. Kane and G. B. Larrabee, *ibid.,* pp. 229-231.

8. W. H. Schroen, G. A. Lee, and F. W. Voltmer. "Comparison of the spreading resistance probe with other silicon characterization techniques." *National Bureau of Standards Special Publication 400-10*, Spreading Resistance Symposium, NBS, Gaithersburg, Maryland, June 1974.

9. G. A. Gruber and R. F. Pfeifer. "The evaluation of thin silicon layers by spreading resistance measurements." *National Bureau of Standards Special Publication 400-10*, Spreading Resistance Symposium, NBS, Gaithersburg, Maryland, June 1974.

10. R. P. Donovan and R. A. Evans. "Incremental sheet resistance technique for determining diffusion profiles." *National Bureau of Standards Special Publication 337*, Washington, D. C., 1970, pp. 123-131.

11. S. M. Sze, *ibid.,* p. 90.

12. W. van Gelder and E. H. Nicollian. "Silicon impurity distribution as revealed by pulsed MOS C-V measurements." *J. Electrochemical Soc., Solid-State Science* **118**: 138-141 (January 1971).

13. J. R. Brews. "Correcting interface-state errors in MOS doping profile determinations." *J. Appl. Phys.* **44**: 3228-3231 (July 1973).

14. S. M. Reddy. "Easily testable realizations for logic functions." *IEEE Trans. Computers* **C-21**: 1183-1188 (November 1972).

15. A. D. Friedman. *Fault Detection in Digital Circuits.* Englewood Cliffs, New Jersey: Prentice Hall, 1971.

16. R. E. Huston. "Testing semiconductor memories." *Digest of Papers*, IEEE Symposium on Semiconductor Memory Testing, Cherry Hill, New Jersey, October 1973.

17. K. Jackson and B. Sear. "The need for time testing in memories and LSI circuits." IEEE Semiconductor Memory Testing Symposium, Cherry Hill, New Jersey, November 1974, pp. 7-32

18. L. A. Priolo and W. B. Reichard. "Thin-film technology enters a new era." *The Western Electric Engineer* **II**: 44-51 (December 1967).

19. K. To and R. E. Tulloss. "Automatic test systems." *IEEE Spectrum* **11**: 44-51 (September 1974).

20. E. Passaglia, R. R. Stromberg and J. Kruger, editors. "Ellipsometry in the Measurement of Surfaces and Thin films." *NBS Misc. Publ. 256*, Government Printing Office, Washington, D. C., 1964.

21. R. M. Burger and R. P. Donovan. *Fundamentals of silicon integrated device technology.* Englewood Cliffs, N. J.: Prentice-Hall Inc., Vol. I, 1968.

22. *Digest of 1974 Semiconductor Memory Testing Symposium*, Cherry Hill, N. J. (November 1974).

PROBLEMS

11.1 An *n*-type buried layer measured at a depth of 8 μm had a resistivity of 7.5×10^{-3} Ω-cm. The buried layer to *p*-type substrate junction was at a depth of 14.3 μm, with a ratio of ρ_2/ρ_1 (at 8 μm) $\approx$ 267. The probe tip radius

$a = 2$ μm. If the correction factor $f[d/a; \rho_2/\rho_1 = 267]$ is given by the following table

d/a	1	2	3	4	5
f	3.3	2	1.6	1.45	1.3

find the original spreading resistance measurement R_{sp}. What is the expected value of R_{sp} if the probe tip radius "a" is changed to 1.5 μm?

11.2 An MOS C-V measurement yielded the following data of C^{-2} versus applied voltage V_G.

$-V_G$	3	3.5	4	4.5	5	5.5	6	6.5	7	7.5	8
C^{-2}	28	35	45	53	62	72	80	90	98	105	114

where V_G is in volts and C^{-2} is in cm$^2 \times 10^{15}/F^2$. Plot C^{-2} versus V_G and thus deduce the average impurity density $N(w)$ using Eq. (11.25).

If $C_{0x} = 2.5 \times 10^{-8} F/cm^2$, find also the depletion width w at each voltage measurement. Do this by correcting C^{-1} for C_{0x}^{-1} and by averaging adjacent values of C^{-2}.

11.3 A corrected doping profile measured by the MOS capacitor method [see Eq. (11.29)] yielded the following data:

w (μm)	0.3	0.35	0.4	0.45	0.5	0.55	0.6	0.65
$N(w)$ $10^{15}/cm^3$	1.25	1.17	1.11	1.08	1.03	1.02	1.01	1

Capacitor C_{0x} consisted of a thermal oxide (grown at 1100°C in dry O_2), 1450 Å thick (as determined by ellipsometry, described in Section 11.4.2), having a relative dielectric constant of 3.2.

Determine and plot C_{LF}/C_{0x} versus V.

Note: First determine, C_{HF}^{-1}, using $C_{HF}^{-1} = C_{0x}^{-1} + w/\epsilon_s$. Since voltage $V = qwN(w)C_{HF}^{-1}/2$ a plot of C_{HF}^{-1} versus V can now be undertaken. From this plot, the slope dC_{HF}^{-1}/dV may be obtained and so C_{LF}/C_{0x} can be determined from a modified version of Eq. (11.29) by using

$$C_{LF}/C_{0x} = 1 - qwN(w) \frac{dC_{HF}^{-1}}{dV} .$$

11.4 The truth table for a simple AND/OR circuit is given by

X_3	X_2	X_1	f
0	0	0	1
0	0	1	0
0	1	0	1
0	1	1	0
1	0	0	1
1	0	1	0
1	1	0	0
1	1	1	0

Express this truth table in terms of the expansion

$$f = C_0 + C_1 X + C_2 X_2 + C_3 X_2 X_1 + C_4 X_3$$
$$+ C_5 X_3 X_1 + C_6 X_3 X_2 + C_7 X_3 X_2 X_1$$

and thereby determine the eight expansion coefficients C. Draw also a realization of the logic circuit that satisfies the truth table and this expansion.

Show further that a general test schedule for the circuit is provided by

C_0	X_1	X_2	X_3	
0	0	0	0	
0	1	1	1	
1	0	0	0	T_1
1	1	1	1	
d	0	1	1	
d	1	0	1	T_2
d	1	1	0	

where the d's stand for "don't care." Discuss why T_1, consisting always of four tests (even for n variables X_1, X_2, ..., X_n), tests the OR gates for both stuck-at-one and stuck-at-zero faults, besides testing the input and output of AND gates for stuck-at-zero faults. Finally, show that test T_2 (always consisting of n tests for n variables X_1, X_2, ..., X_n) tests the AND gate leads which are stuck at one.

Note that the test schedule is independent of the test coefficients C_1, C_2, ..., C_{2^n-1} (for variables) and therefore is independent of the function realized.

11.5 The total cost, C, per good chip after tests and inspection is by virtue of Eq. (11.31) given by

$$C = \frac{C_c}{Y} + \frac{C_a}{Y} = \frac{(W+PM)/N}{G/N} + \frac{(ND+LM)/N}{G/N} .$$

Note that the chip yield with test and inspection procedures is $Y + G/N$. Note further, that the added cost, C_a, per chip due to test and inspection is $C_a = (ND+LM)/N$. Finally, the processing cost per chip, C_c, without test and inspection costs is $C_c = (W+PM)/N$.

Consider now test and inspection procedures discontinued and the consequent chip yield Y'. Hence the total chip cost, C', without test and inspection is given by $C' = C_c/Y'$.

Prove that for test and inspection procedures to be economical, it is necessary that

$$\frac{Y}{Y'} > 1 + \frac{ND+LM}{W+PM} .$$

12

Analog Integrated Circuits

This chapter is the first of four chapters on the design and application of integrated circuits. Analog systems are considered in the next chapter. Digital circuits and systems are considered in the two succeeding chapters.

The most popular analog integrated circuit is the direct-coupled, high-gain, differential input operational amplifier because it can be used in a wide variety of signal processing applications. Part one of this chapter is concerned with the analysis of each component of the operational amplifier. The performance of complete operational amplifiers is considered in the second part of the chapter.

We begin with a discussion of some fundamental operational amplifier parameters.

12.1 OPERATIONAL AMPLIFIER PARAMETERS

The prototype of an operational amplifier is a voltage-controlled voltage source that produces an output voltage proportional to the difference between two input signals.

$$v_o = A\,(v_{s1} - v_{s2}) = A v_d \qquad (12.1)$$

In this equation $v_d = v_{s1} - v_{s2}$ is the *differential* or *difference* input voltage and A, a real positive constant, is the gain. An ideal operational amplifier has infinite gain. In actual amplifiers, A is finite and a complex-valued function of frequency. Operational amplifiers are used in feedback loops whose *closed-loop* gain differs from the ideal value by an error term which goes to zero as A approaches infinity. Furthermore, the frequency dependence of the amplifier gain can result in stability problems unless the gain-frequency characteristic is properly controlled. Proper control of the frequency response is a distinguishing factor between an operational amplifier and an ordinary high-gain amplifier.

Information is conveyed by a differential signal in many applications such as in a high-noise environment, and therefore the characterization of the terminal behavior of a differential-input, differential-output amplifier is important. The first stage of an operational amplifier is a differential amplifier, and the complete operational amplifier is a differential-input, single-ended output amplifier which is a special case.

Characterization of a Differential Amplifier [1-3] A differential-input, differential-output amplifier produces a pair of related output voltages v_{o1} and v_{o2}, in response to the difference $v_{s1} - v_{s2}$, between the two input signals. An equivalent situation is obtained when v_{s1} and v_{s2} are replaced by three sources as shown in Fig. 12.1. The average value of the input sources v_{s1} and v_{s2} is called the *common-mode* signal,

$$v_{cm} = \left(\frac{v_{s1} + v_{s2}}{2} \right), \tag{12.2a}$$

while the information-bearing signal is the differential signal

$$v_d = v_{s1} - v_{s2}. \tag{12.2b}$$

An ideal, balanced differential amplifier would respond only to the difference voltage v_d, and would completely reject the common-mode voltage v_{cm}. Actual differential amplifiers respond to both difference and common-mode inputs, and produce both difference and common-mode outputs as described by

$$\begin{bmatrix} v_{od} \\ v_{oc} \end{bmatrix} = \begin{bmatrix} A_{dd} & A_{dc} \\ A_{cd} & A_{cc} \end{bmatrix} \begin{bmatrix} v_{id} \\ v_{ic} \end{bmatrix} \tag{12.3}$$

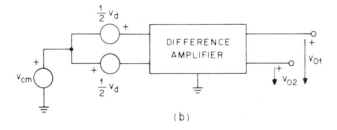

(b)

Fig. 12.1 Excitation of difference amplifier using common-mode and difference-mode sources.

where v_{od}, and v_{oc} are the differential and common-mode components of the output voltage, v_{id}, and v_{ic} are the differential and common-mode components of the input voltage, and the elements of the gain matrix relate each component of the output voltage to a component of the input voltage. For example, A_{dc} is the ratio of the differential output voltage to common-mode input voltage when the difference-mode input voltage is zero. The diagonal terms of the gain matrix are the more important because they relate the differential output to differential input, and common-mode output to common-mode input. The off-diagonal terms only occur in unbalanced amplifiers and they result in *mode conversion,* that is, they produce a common-mode output component from a differential input, or *vice versa.*

There are several figures of merit describing the performance of an actual differential amplifier relative to that of an ideal difference amplifier. The output of a differential amplifier, taken in differential mode, is

$$v_{od} = A_{dd} v_{id} + A_{dc} v_{ic}.$$

The *differential input-output figure of merit,* also called the *double-ended rejection ratio,* is denoted by Γ_d and defined as

$$\Gamma_d = \left| \frac{A_{dd}}{A_{dc}} \right|. \tag{12.4}$$

In an ideal, balanced differential amplifier, Γ_d is infinite because $A_{dc} = 0$.

There are many situations in which the output of a differential amplifier is observed in a single-ended fashion. The output voltage in those cases is

$$v_o = \frac{v_{od}}{2} + v_{oc} = \left[\frac{A_{dd}}{2} + A_{cd} \right] v_{id} + \left[\frac{A_{dc}}{2} + A_{cc} \right] v_{ic}.$$

The output component due to differential excitation is $[(A_{dd}/2) + A_{cd}]v_{id}$ and the output component due to common-mode excitation is $[(A_{dc}/2) + A_{cc}]v_{ic}$. The ratio of the gains defined by these terms is called the *single-ended rejection ratio* and denoted by Γ_s,

$$\Gamma_s = \left| \frac{(A_{dd}/2) + A_{cd}}{(A_{dc}/2) + A_{cc}} \right|. \tag{12.5}$$

When the amplifier is perfectly balanced $A_{cd} = 0$, and $A_{dc} = 0$, and Eq. (12.5) reduces to the more familiar *common-mode rejection ratio,*

$$CMR = \left| \frac{A_{dd}}{2 A_{cc}} \right|. \tag{12.6}$$

The common-mode rejection ratio of integrated circuit operational amplifiers is typically in the range from 2000 to 100,000.

The rejection ratios Γ_d and CMR are related by

$$\Gamma_d = \frac{CMR}{|\Delta|} \tag{12.7}$$

where

$$\Delta = \left| \frac{A_{dc}}{2A_{cc}} \right| \tag{12.8}$$

is the *differential error factor.*

The equivalent circuits which determine the gain factors, A_{ij}, are different and so each gain is a different function of frequency. Hence, the various rejection ratios are also functions of frequency.

Input offset voltage   The ideal difference amplifier is symmetric and balanced so that when the differential input voltage is zero, the differential output voltage is also zero. In actuality, a temperature dependent, nonzero differential input voltage, known as the *input offset voltage,* is required to null the output voltage. The input offset voltage for a differential-output amplifier is defined as

$$V_{os} = (v_{i2} - v_{i1}) \Big|_{v_{o2} = v_{o1}}. \tag{12.9a}$$

Operational amplifiers with single-ended output are designed such that the output voltage is zero when $v_{i2} = v_{i1}$. Therefore, the offset voltage for a single-ended output operational amplifier is

$$V_{os} = (v_{i2} - v_{i1}) \Big|_{v_o = 0} \tag{12.9b}$$

The input offset voltage is a measure of component mismatch. The input offset voltage for an operational amplifier with a bipolar transistor differential amplifier input stage is typically less than 5 mV with a temperature coefficient between 5 and 50 μV/$^\circ$C. Expressions for input offset voltage and its temperature coefficient are derived for particular amplifier circuits in the following sections.

Input current   A differential amplifier using bipolar transistors requires an *input bias current* which is proportional to the collector current and inversely proportional to the current gain of the transistors. This current is usually considered to be the average value of the current into the two inputs.

$$I_b = \frac{i_{i1} + i_{i2}}{2}.$$

Imbalance in circuit components gives rise to slightly different bias currents, which can be described by an *input offset current* I_{os} defined as the difference between the two input currents.

$$I_{os} = i_{i1} - i_{i2} \; .$$

In many applications, the input bias current imposes limitations on the impedance level of the input circuit.

Common-mode input range and latch-up Real operational amplifiers only accept a finite range of common-mode input signals because they contain transistors (or other active devices) whose operating range is restricted. In a bipolar transistor operational amplifier, the positive limit of the common-mode range is set, for example, by the saturation of the input stage transistors as the input signal approaches the positive supply voltage. The negative range limit is imposed by cutoff of the input stage transistors as the input approaches the negative supply voltage. The finite common-mode signal range means that high-voltage operational amplifiers are required when large common-mode signals are encountered. This is one example of how the operating environment for the device affects the choice of device technology.

The operational amplifier is used in a feedback loop. When the common-mode limit is exceeded, positive feedback paths are created that result in a *latch-up* condition that will persist even after the excessive common-mode excitation is removed.

Slew-rate limitations Operational amplifiers have a maximum output current capability that is determined by the design of the output stage. Assuming that there are no capacitors at internal nodes, the maximum rate of change of the output voltage when the amplifier is capacitively loaded is

$$S_R = \left. \frac{dv_o}{dt} \right|_{max} = \frac{i_{o\ max}}{C_L} \tag{12.10}$$

where C_L is the load capacitor and S_R is known as the *slew rate*.

Real operational amplifiers have capacitance at various internal nodes. These capacitors cause slew-rate limiting even when the amplifier output is not capacitively loaded.

12.2 BIPOLAR DIFFERENTIAL AMPLIFIER

Expressions for the differential and common-mode gain are developed in this section for the bipolar transistor differential amplifier shown in Fig. 12.2(a). It is assumed initially that the circuit is symmetric, that is, $R_{E1} = R_{E2} = R_E$, $R_{S1} = R_{S2} = R_S$, $R_{C1} = R_{C2} = R_C$, and that transistors T_1 and T_2 are identical. The effect of asymmetries in the circuit are considered later in the section. The equivalent circuit of this amplifier is shown in Fig. 12.2(b).

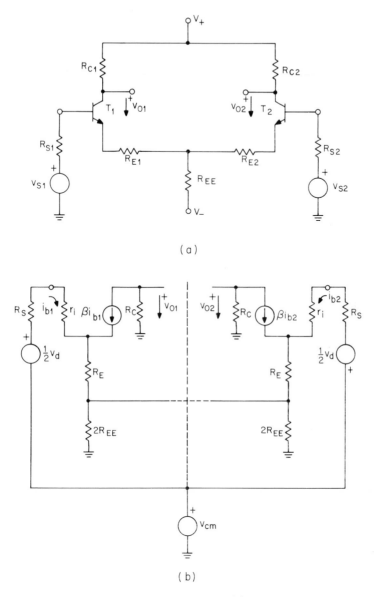

Fig. 12.2 Bipolar transistor differential amplifier. (a) Circuit diagram; (b) equivalent circuit for analysis.

Difference mode gain When the common-mode excitation is identically zero, i.e., $v_{cm} = 0$, a differential input voltage v_d drives one side of the difference amplifier up while the other side is driven down by an equal amount. This situation is analogous to the seesaw with equal-length arms.

Therefore, according to Bartlett's bisection theorem, a vertical ground plane can be passed through the center of the circuit, grounding all horizontal lines that are cut. The equivalent circuit for obtaining the difference-mode gain is shown in Fig. 12.3. Note that $v_o = Av_d/2$ and the symmetric output from the opposite side will be $v_o' = -Av_d/2$. Thus the differential output voltage is $v_{od} = v_o - v_o' = Av_d$, and therefore the difference mode gain is $A_{dd} = v_{od}/v_d = A$.

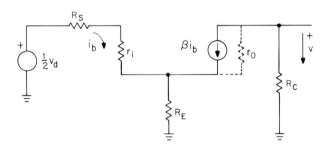

Fig. 12.3 Equivalent circuit for calculating the difference-mode gain.

The analysis is simplified considerably by considering the effect of a resistor in series with the common terminal of an active three-terminal two-port network as shown in Fig. 12.4. The current and voltage gains of the network are defined to be

$$A_I = \frac{i_o}{i_i} \tag{12.11a}$$

and

$$A_V = \frac{v_o}{v_i} = \frac{i_o R_L}{i_i R_{in}} = A_I \frac{R_L}{R_{in}}. \tag{12.11b}$$

Kirchhoff's voltage law applied to the loop containing the input port yields

$$v_i = v' + (i_i - i_o)R = v' + i_i(1 - A_I)R. \tag{12.12a}$$

A similar result for the output loop is

$$v_o = i_o R_L = v'' + (i_i - i_o)R = v'' - i_o \frac{(A_I - 1)R}{A_I}. \tag{12.12b}$$

The important result obtained from Eqs. (12.12) is that a two-port network with a resistor R in series with its common terminal is equivalent to a net-

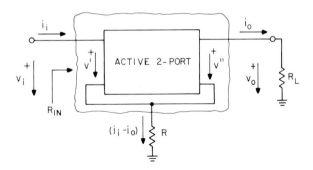

Fig. 12.4 Effect of resistor in common terminal of active two-port.

work with a grounded common terminal that has a resistor of value $R(1-A_I)$ in series with the input port and a resistor of value $R[1-(1/A_I)]$ in series with the output port.

Applying these results to the difference-mode equivalent circuit of Fig. 12.3 yields

$$A_I = \frac{R_E - \beta r_o}{R_E + R_C + r_o} .$$

(12.13)

The input resistance R_{in} is given by virtue of Eq. (12.12a) as

$$R_{in} = R_S + r_i + (1 - A_I)R_E.$$

(12.14a)

Substituting Eq. (12.13) for A_I in this equation and realizing that the differential input resistance R_{id} of the differential amplifier is twice the value R_{in} of one section, yields

$$R_{id} = 2R_{in} = 2\left\{ R_S + r_i + R_E \frac{[R_C + (1+\beta)r_o]}{R_E + R_C + r_o} \right\}.$$

(12.14b)

In many practical cases, $r_o \gg R_C, R_E$ and therefore this equation reduces to

$$R_{id} \approx 2[R_S + r_i + (1+\beta)R_E].$$

(12.14c)

When $R_E = 1\ k\Omega$, $\beta = 100$ and $I_c = 10\ \mu A$, $R_{id} \approx 350\ k\Omega$.

The difference-mode voltage gain A_{dd} is obtained by substituting the current gain and differential-mode input impedance expressions Eq. (12.13) and Eq. (12.14b) into Eq. (12.11) $[A_V = A_I(R_L/R_{in})]$

$$A_{dd} = -\frac{(\beta r_o - R_E)R_C}{(R_S + r_i)(R_E + R_C + r_o) + R_E[R_C + (1+\beta)r_o]} .$$

(12.15a)

For $r_o \gg R_C, R_E$, the difference-mode gain becomes

$$A_{dd} \approx - \frac{\beta R_C}{R_S + r_i + (1+\beta) R_E} = - \frac{2\beta R_C}{R_{id}} . \qquad (12.15b)$$

To determine the output impedance R_o recall that the open-circuit voltage $v_o(OC)$ is given by

$$v_o(OC) = \left[\lim_{R_C \to \infty} A_V \right] v_i. \qquad (12.16a)$$

According to Thevenin's theorem, when R_C equals the output resistance R_o, the output voltage v_o will be half its open circuit value. Therefore

$$\frac{v_o(OC)}{2} = A_V \Big|_{R_C = R_o} v_i. \qquad (12.16b)$$

Taking the ratio of Eq. (12.16a) and Eq. (12.16b), using Eq. (12.15a) for the voltage gain, and solving for R_o, yields

$$R_o = r_o + \frac{R_E(R_S + r_i + \beta r_o)}{R_S + r_i + R_E} . \qquad (12.17)$$

From this equation it is apparent that $r_o \leqslant R_o < (1+\beta) r_o$ as R_E varies from zero to very large values.

Common-mode gain   When the difference amplifier is excited by a common-mode signal, the current in any of the horizontal links crossing the vertical cutting plane shown in Fig. 12.2(b) is identically zero. Therefore, it is sufficient to consider only one side of this circuit. The equivalent circuit for common-mode analysis is shown in Fig. 12.5. A large value for R_{EE} is desirable so as to minimize the variation in emitter current, and the associated variation in transistor parameters, as a function of common-mode level. If an ideal current source were used in place of R_{EE}, the instantaneous emitter current of the transistors in the differential pair would be independent of common-mode signal level, and therefore the common-mode gain would be zero.

The common-mode equivalent circuit is structurally identical to the difference-mode equivalent circuit. Hence the gain, input impedance, and output impedance can be written by inspection of Eq. (12.14), Eq. (12.15), and Eq. (12.17).

$$A_{cc} = - \frac{[\beta r_o - (R_E + 2R_{EE})] R_C}{(R_S + r_i)(R_C + r_o) + (R_E + 2R_{EE})[R_S + r_i + R_C + (1+\beta) r_o]} .$$

$$(12.18a)$$

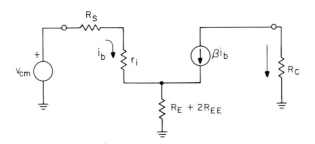

Fig. 12.5 Equivalent circuit for common-mode gain.

In many practical cases, $R_{EE} \gg R_E$, $r_o \gg R_E, R_C$, the transistor current gain is large and $\beta r_o \gg 2R_{EE}$ so that the common-mode gain is approximately

$$A_{cc} \approx -\frac{R_C}{2R_{EE}}. \qquad (12.18b)$$

To find the input impedance we note from Fig. 12.2 that the two halves of the amplifier are effectively in parallel as far as common mode voltage v_{cm} is concerned. Hence from Fig. 12.5 and using Eq. (12.15a)

$$R_{ic} = \frac{v_{cm}}{i_{in}} = \frac{1}{2}\left\{ R_S + r_i + \frac{(R_E+2R_{EE})[R_C + (1+\beta)r_o]}{R_E+2R_{EE}+R_C+r_o} \right\}. \qquad (12.19a)$$

Under the same assumptions used to simplify the previous equation,

$$R_{ic} \approx \frac{1}{2}\,[r_i + (1+\beta)(2R_{EE}\|r_o)]. \qquad (12.19b)$$

The output impedance is given by

$$R_{oc} = \frac{1}{2}\left\{ \frac{(R_E+2R_{EE})(R_S+r_i+\beta r_o)}{R_S+r_i+R_E+2R_{EE}} \right\} \qquad (12.20a)$$

$$\approx \frac{\beta r_o}{2 + (r_i/R_{EE})}.$$

The common-mode rejection ratio for this bipolar difference amplifier with a single-ended output is found by using Eq. (12.15a) and Eq. (12.18) for the gains, yielding

$$CMR = \left| \frac{A_{dd}}{2 A_{cc}} \right|$$

$$= \frac{1}{2} \left| \frac{\beta r_o - R_E}{\beta r_o - (R_E + 2R_{EE})} \right.$$

$$\left. \times \frac{(R_S + r_i)(R_C + r_o) + (R_E + 2R_{EE})[R_S + r_i + R_C + (1+\beta) r_o]}{(R_S + r_i)(R_C + r_o) + R_E[R_S + r_i + R_C + (1+\beta) r_o]} \right|.$$

$$(12.21a)$$

Using the approximate forms of the gain equations, the common-mode rejection ratio reduces to

$$CMR \approx \frac{\beta R_{EE}}{R_S + r_i + (1+\beta) R_E} . \qquad (12.21b)$$

EXAMPLE 12.1 Compute the difference-mode gain A_{dd}, the differential- and common-mode input impedances, and the common-mode rejection ratio CMR for the differential amplifier of Fig. 12.2(a) with $R_{C1} = R_{C2} = 25\ k\Omega$, $R_{S1} = R_{S2} = 500\ \Omega$, $R_{E1} = R_{E2} = 500\ \Omega$, and $R_{EE} = 100\ k\Omega$. The transistor parameters are $\beta = 100$, $r_i = 25\ k\Omega$, and $r_o = 5\ M\Omega$.

Solution. Substituting the given values in Eq. (12.15b) yields $A_{dd} = -33$. From Eq. (12.14b) the differential mode input impedance is found to be $R_{id} = 151.5\ k\Omega$. Finally, from Eq. (12.21b) the common-mode rejection ratio is $CMR = 332$.

12.3 EFFECT OF CIRCUIT IMBALANCES

Mismatches in the transistor parameters and other circuit elements inevitably occur in practical differential amplifiers and result in error voltages. In this section the different causes of imbalance are considered separately to determine their first-order effect. Higher-order effects and interactions between first-order effects are neglected. The computations would be intractable if all of the effects were considered simultaneously. The total error output is obtained by summing all of these first-order contributions.

Figure 12.6 shows a differential amplifier whose base-emitter circuit is unbalanced by letting $R_{E1} = R_E + \delta R_E$, $R_{E2} = R_E$, and $r_{i1} = r_i + \delta r_i$, $r_{i2} = r_i$. The differential output voltage due to the common-mode excitation is

$$v_{od} = v_{o1} - v_{o2} = \beta R_C (i_{b2} - i_{b1}). \qquad (12.22)$$

Applying Kirchhoff's voltage law around the loop containing the base and

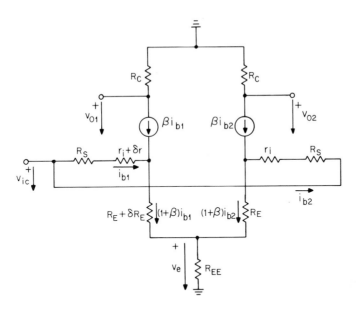

Fig. 12.6 Equivalent circuit for base and emitter resistor imbalance.

emitter of both transistors yields

$$i_{b2} = \left\{ 1 + \frac{\delta r_i + (1+\beta)\delta R_E}{r_i + R_S + (1+\beta) R_E} \right\} i_{b1}. \tag{12.23}$$

The output voltage is found by substituting this expression for base current into Eq. (12.22)

$$v_{od} = \beta R_C \frac{\delta r_i + (1+\beta)\delta R_E}{R_S + r_i + (1+\beta) R_E} i_{b1}. \tag{12.24}$$

      To relate this last result to the common-mode gain A_{cc} we need a relation between v_{ic} and i_{b1}. This relation is obtained by writing the KVL equation for the input loop of transistor T_2, and substituting for i_{b2} from Eq. (12.23).

$$v_{ic} \approx (1+\beta) R_{EE} \left\{ 2 + \frac{\delta r_i + (1+\beta)\delta R_E}{r_i + R_S + (1+\beta) R_E} \right\} i_{b1}. \tag{12.25}$$

Usually R_{EE} is so much larger than any of the other resistances that the base current is not appreciably changed by the small resistive imbalance, and for $\beta \gg 1$, $v_{ic} \approx 2\beta R_{EE} i_{b1}$. Forming the ratio of v_{od} to v_{ic}, using Eq. (12.14c) for the difference-mode input resistance and Eq. (12.18b) for the approximate value of A_{cc} yields

$$\frac{v_{od}}{v_{ic}} \approx -\frac{2A_{cc}}{R_{id}} [\delta r_i + (1+\beta)\delta R_E]. \tag{12.26}$$

Equation (12.26) is the first-order contribution to the differential-mode error signal due to an imbalance in r_i and R_E. Since it can be either positive or negative, the magnitude of Eq. (12.26) is the important information.

To consider the effect of unequal source resistance let $R_{S1} = R_S + \delta R_S$, and $R_{S2} = R_S$. Since R_S and r_i are in series, R_S is always paired with r_i in the expressions describing the effects of imbalanced r_i and R_E. It therefore follows immediately by inspection of those results that

$$\frac{v_{od}}{v_{ic}} \approx -\frac{2A_{cc}}{R_{id}} \delta R_S. \tag{12.27}$$

The effect of mismatch of β, R_C, and r_o are derived in a similar manner, and the results are given in Table 12.1. Figure 12.7 aids the computation of the contribution of δr_o.

The total differential-mode error due to a common-mode input signal is, to a first order, given by the sum of the individual error terms.

$$A_{dc} \approx -A_{cc} \left[\frac{\delta r_i + \delta R_S + (1+\beta)\delta R_E - (R_S+r_i)\delta\beta}{(R_{id}/2)} - \frac{\delta R_C}{R_C} - \frac{R_C \delta r_o}{(R_C+r_o)^2} \right] \tag{12.28}$$

Each term in this equation can be either positive or negative and thus these terms can add or subtract from each other. Recall that the differential error factor is defined in Eq. (12.8) as $\Delta = A_{dc}/A_{cc}$ with $v_{id} = 0$.

The differential error factor is the bracketed term in Eq. (12.28) and is the ratio of differential-mode output signal to common-mode output signal due to a common-mode excitation and should be as small as possible.

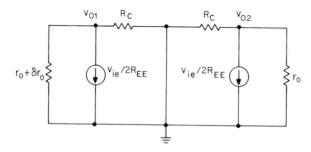

Fig. 12.7 Equivalent circuit for output impedance imbalance.

Table 12.1 Effect of Mismatched Component on Output of Difference Amplifier

Component	Equation
Emitter resistor and input resistance	$\dfrac{v_{od}}{v_{ic}} \approx -\dfrac{2A_{cc}}{R_{id}}\,[\delta r_i + (1+\beta)\delta R_E]$
Source resistance	$\dfrac{v_{od}}{v_{ic}} \approx -\dfrac{2A_{cc}}{R_{id}}\,\delta R_S$
Current gain, β	$\dfrac{v_{od}}{v_{ic}} \approx \dfrac{2A_{cc}}{R_{id}}\,(R_S + r_i)\delta\beta$
Collector load resistor	$\dfrac{v_{od}}{v_{ic}} \approx A_{cc}\,\dfrac{\delta R_C}{R_C}$
Output resistance	$\dfrac{v_{od}}{v_{ic}} \approx A_{cc}\,\dfrac{R_C\delta r_o}{(R_C + r_o)^2}$

EXAMPLE 12.2 Evaluate the worst-case error factor Δ for the differential amplifier of Example 12.1 with $\delta R_S = 10\ \Omega$, $\delta R_C = \delta r_i = 500\ \Omega$, $\delta R_E = 1\ \Omega$, $\delta r_o = 5\ k\Omega$, and $\delta\beta = 2$.

Solution. Substituting these values and the element values from Example 12.1 into Eq. (12.28) yields $\Delta = 1.694$, which means that every volt change in common-mode output voltage is accompanied by a change of 1.694 V in the differential-mode output.

Input offset voltage   The small-signal behavior of the bipolar transistor difference amplifier and the effect of circuit imbalances were considered in the previous section. The effect of dc imbalance and its temperature variation is considered in this section. The circuit to be analyzed is shown in Fig. 12.8.

The quantitative description of the effect begins with the Kirchoff voltage law equations for the base-emitter loops

$$v_{i1} - V_{BE1} = R_1 I_{E1} + R_{EE}I + V_-$$

$$v_{i2} - V_{BE2} = R_{EE}I + R_2 I_{E2} + V_- . \qquad (12.29)$$

where the total current $I = I_{E1} + I_{E2}$, $R_1 = R_{E1} + [R_{S1}/(1+\beta)]$ and $R_2 = R_{E2} + [R_{S2}/(1+\beta)]$. It is assumed that the current gains of the transistors are identical since the gain of modern transistors is large and small differences have a minor effect on the resistors R_1 and R_2. Furthermore, let

$$\frac{I_{E1}}{I_{E2}} = x, \qquad (12.30a)$$

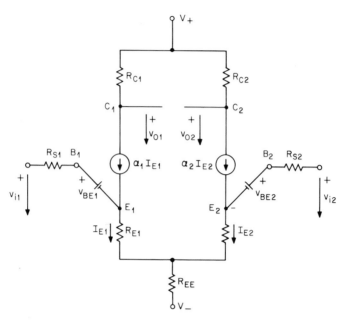

Fig. 12.8 Equivalent circuit for determining input offset voltage.

and from the requirement that the differential-mode output voltage be zero, $\alpha_1 R_{C1} I_{E1} = \alpha_2 R_{C2} I_{E2}$,

$$x = \frac{\alpha_2 R_{C2}}{\alpha_1 R_{C1}} . \tag{12.30b}$$

Using Eq. (12.30a) and Eq. (12.30b) in Eqs. (12.29) and solving for the offset voltage yields

$$V_{os} = v_{i1} - v_{i2} = \Delta V_{BE} + \frac{R_1 x - R_2}{1+x} I. \tag{12.31}$$

When the input voltages are approximately zero, and $-V_- > V_{BE}$, $I = -V_-/R_{EE}$ and therefore

$$V_{os} = \Delta V_{BE} - \frac{R_1 x - R_2}{1+x} \frac{V_-}{R_{EE}} . \tag{12.32a}$$

It is desirable to express the second term of this equation in terms of component variations. In Eq. (12.30b), let $R_{C2} = R_{C1} + \delta R_C$ and $\alpha_2 = \alpha_1 + \delta\alpha$, and considering only first-order variations,

$$x \approx 1 + \frac{\delta\alpha}{\alpha} + \frac{\delta R_C}{R_C} = 1 + \frac{1}{1+\beta} \frac{d\beta}{\beta} + \frac{\delta R_C}{R_C} .$$

Furthermore, if $R_2 = R_1 + \delta R$

$$V_{os} \approx \Delta V_{BE} - \frac{R_1 V_-}{2R_{EE}}\left[\frac{1}{1+\beta}\frac{d\beta}{\beta} + \frac{\delta R_C}{R_C} - \frac{\delta R}{R_1}\right]. \qquad (12.32b)$$

The first term is the V_{BE} match of the transistors and is the dominant component of offset voltage. The second term is the effect of imbalances in other circuit components and can have either sign depending on the direction of the imbalance.

An important quantity is the variation in offset voltage with temperature, the *input offset voltage drift*. It can be determined by a straightforward evaluation of the derivative of V_{os}. The expression for V_{os} is quite complicated and the derivation of a general expression for dV_{os}/dT is tedious and not very instructive. Since the ΔV_{BE} term is dominant for reasonable parameter values, we will consider the temperature variation of that term. The emitter current of a transistor is approximately $I_E \approx I_s \exp(V_{BE}/V_T)$, and therefore from Eq. (12.30) and Eq. (12.32)

$$V_{os} \approx \Delta V_{BE} = V_T\left[\ln x - \ln\left(\frac{I_{s1}}{I_{s2}}\right)\right]. \qquad (12.33)$$

If the transistors have identical current gains and the collector load resistors are matched, the first term of this equation is zero because $x = 1$, and any offset is due to a mismatch in the junction leakage currents caused primarily by small differences in the emitter areas. We will consider the ratio of leakage currents and the ratio of current gains to be independent of temperature, and therefore, from Eq. (12.33),

$$\frac{dV_{os}}{dT} = \frac{\Delta V_{BE}}{T} + V_T\left[\frac{1}{R_{C2}}\frac{dR_{C2}}{dT} - \frac{1}{R_{C1}}\frac{dR_{C1}}{dT}\right]. \qquad (12.34a)$$

When both resistors have the same temperature coefficient of resistance, and $R_{C1} = R_C$, $R_{C2} = R_C + \delta R_C$, this equation becomes

$$\frac{dV_{os}}{dT} = \frac{\Delta V_{BE}}{T} - \frac{V_T}{T}\frac{\delta R_C}{R_C}\left[\frac{dR_C/R_C}{dT/T}\right]. \qquad (12.34b)$$

EXAMPLE 12.3 (a) Determine the input offset voltage for a differential amplifier with $R_{C2}/R_{C1} = 1.05$, $\alpha_2/\alpha_1 = 1.01$, $\alpha_2 = 0.99$, $\Delta I_s/I_s = \pm 0.05$, $R_S = 500\ \Omega$, $R_E = 50\ \Omega$, $R_{EE} = 10^5\ \Omega$, $\delta R/R = 0.05$ (i.e., 5% tolerance resistors) and $V_- = -15$ V. (b) Find the offset voltage drift if the *TCR* is 2500 ppm/°C and if the V_{BE} match of the transistors is 3 mV.

Solution.

a)   The factor x describing collector circuit balance is $x = 1.05 \times 1.02 = 1.071$. The input offset voltage is given by Eq. (12.32b) and Eq. (12.33) (evaluated at room temperature)

$$V_{os} = V_T\left[\ln x - \ln\left(\frac{I_{s1}}{I_{s2}}\right)\right] - \frac{R_1 V_-}{2R_{EE}}\left[\frac{1}{1+\beta}\frac{d\beta}{\beta} + \frac{\delta R_C}{R_C} - \frac{\delta R}{R_1}\right]$$

$$= V_T\left[\ln x - \ln\left(\frac{I_{s1}}{I_{s2}}\right)\right] - \frac{R_1 V_-}{2R_{EE}}\left[\frac{\delta \alpha}{\alpha} + \frac{\delta R_C}{R_C} - \frac{\delta R}{R_1}\right]$$

$$= 0.026[\ln 1.071 - \ln(1.05)] - \frac{[(1-0.98)500 + 50]\times 15}{2\times 10^5}[0.01+0.05-0.05]$$

$$= 3.052\times 10^{-3} - 4.55\times 10^{-5} = 3.007 \text{ mV}.$$

Note that in the worst case, the components of the first term are additive because they are randomly varying quantities. However, the derivation of these components suggests a method of offset voltage compensation, namely, to intentionally change x by means of an adjustable resistor to reduce the offset voltage.

b)   From Eq. (12.34b)

$$\frac{dV_{os}}{dT} = \frac{3\times 10^{-3}}{300} - 0.026(0.05)(2500\times 10^{-6}) = 6.75 \ \mu\text{V}/^\circ\text{C}.$$

Input current   The *input bias current* for a bipolar difference amplifier is the average base current

$$I_{\text{bias}} = \frac{1}{2}\left\{\frac{I_{C1}}{\beta_1} + \frac{I_{C2}}{\beta_2}\right\} \tag{12.35a}$$

and the *input offset current* is the difference between the base currents when $v_{o2} = v_{o1}$,

$$I_{os} = \Delta I_B\Big|_{v_{o2}=v_{o1}} = I_{C2}\left\{\frac{1}{\beta_2} - \frac{R_{C2}}{R_{C1}}\frac{1}{\beta_1}\right\}, \tag{12.35b}$$

where β is the dc common emitter current gain, and once again we used the result $(I_{C1}/I_{C2}) = (R_{C2}/R_{C1})$. The magnitude of the error can be written in terms of the parameter mismatch as

$$I_{os} = \frac{I_{C2}}{\beta}\left\{\pm \frac{\delta\beta}{\beta} \pm \frac{\delta R_C}{R_C}\right\},$$

from which it is seen that the current gains of the differential transistor pair must be well matched to have a low offset current.

The *error current drift* is the derivative of ΔI_B with respect to temperature. On the assumption that the resistor ratio R_{C2}/R_{C1} is independent of temperature to a first approximation, and that the temperature coefficient $k_\beta = (1/\beta)(d\beta/dT)$ is the same for β_1 and β_2, the drift in offset current becomes, after some algebra,

$$\frac{d}{dT} I_{os} = \left\{ \frac{1}{I_{C2}} \frac{dI_{C2}}{dT} - k_\beta \right\} I_{os}. \tag{12.36}$$

Thus, the drift in input offset current is proportional to the offset current. The first term in the braces is the temperature coefficient of the collector current and is given approximately by

$$\frac{d}{dT} (\ln I_{C2}) \approx \frac{m}{T} + \frac{V_{g0}}{\eta T V_T}$$

where $m = 1.5$, $\eta = 2$, and $V_{g0} = 1.21$ V for silicon. At room temperature this temperature coefficient is approximately $0.08/°C$. Unfortunately the temperature coefficient of beta is not constant. However, it can be approximated by

$$k_\beta = \frac{1}{\beta} \frac{d\beta}{dT} = \begin{cases} -0.005/°C & \text{for} \quad T > 25°C \\ -0.015/°C & \text{for} \quad T < 25°C. \end{cases}$$

Hence the terms in the braces of Eq. (12.36) are additive. Furthermore, since these terms depend upon fixed parameters nothing can be done do reduce them. Drift is minimized by reducing the offset current through the use of special input circuitry.

12.4 FET DIFFERENTIAL AMPLIFIER

The junction field-effect transistor is used for the input differential amplifier stage of an operational amplifier when extremely high-input impedance and low-leakage current is required. An FET input stage on the same chip as a bipolar operational amplifier requires a more complex processing sequence, usually involving ion implantation, in order to optimize the parameters of both the FET's and bipolar transistors.

An FET differential amplifier is shown in Fig. 12.9. The equivalent circuit of the FET is a voltage controlled current source with finite input resistance, r_{gs}, and output resistance, r_{ds}, as shown in Fig. 12.10. Using the same type of analysis as for the bipolar difference amplifier, it is easily seen that the difference-mode current gain is

$$A_I = \frac{R_S - g_m r_{gs} r_{ds}}{R_D + r_{ds} + R_S}$$

$$\approx - \frac{g_m r_{gs} r_{ds}}{R_D + r_{ds}} \tag{12.37}$$

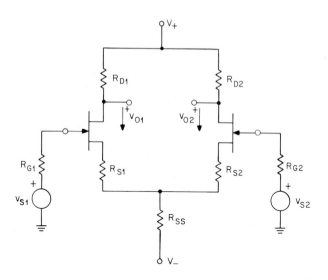

Fig. 12.9 FET differential amplifier.

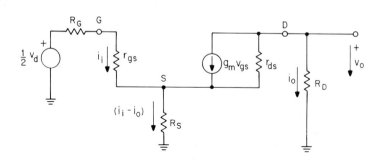

Fig. 12.10 Equivalent circuit of FET differential amplifier for difference-mode gain.

where the approximation is valid for $r_{ds} \gg R_S$, and $g_m r_{gs} r_{ds} \gg R_S$.

The input resistance of the equivalent circuit of Fig. 12.10 is $R_G + r_{gs} + (1-A_I)R_S$. Substituting the approximate form of Eq. (12.37) for

A_I in this expression yields

$$R_{\text{in}} \approx R_G + R_S + r_{gs}\left[1 + \frac{g_m r_{ds} R_S}{r_D + r_{ds}}\right]. \qquad (12.38a)$$

Usually $r_{gs} \gg R_S$, R_G, and therefore Eq. (12.46a) reduces to

$$R_{\text{in}} \approx r_{gs}\left[1 + \frac{g_m r_{ds} R_S}{R_D + r_{ds}}\right] = \frac{1}{2} R_{id}, \qquad (12.38b)$$

because for the difference amplifier $R_{id} = 2R_{in}$. Typical values for FET circuit parameters are $r_{gs} \approx 10^{11}$ Ω, $r_{ds} \approx 3 \times 10^5$ Ω, and $g_m \approx 1 \times 10^{-3}$ mho, and therefore R_{id} is more than 10^{11} Ω.

To obtain A_{dd} one uses the relation $A_{dd} = A_I R_D / R_{in}$ together with Eqs. (12.37) and (12.38a) to obtain

$$A_{dd} = -\frac{g_m R_D r_{ds}}{R_D + r_{ds}} \frac{1}{1 + \dfrac{R_G + R_S}{r_{gs}} + g_m \dfrac{R_S r_{ds}}{R_D + r_{ds}}}. \qquad (12.39a)$$

In most practical situations the external resistances in the gate-drain circuit are negligible in comparison with r_{gs}, $(R_S, R_G \ll r_{gs})$, and the drain load resistor R_D is usually small in comparison with r_{ds}. When these assumptions are true, A_{dd} reduces to

$$A_{dd} = -\frac{g_m R_D}{1 + g_m R_S}. \qquad (12.39b)$$

For this case the differential input resistance becomes

$$R_{id} \approx 2r_{gs}(1 + g_m R_S). \qquad (12.40)$$

The output resistance is obtained by using Thevenin's theorem, forming the ratio of A_{dd} when $R_D \rightarrow \infty$ to A_{dd} when $R_D = R_{\text{out}}$, and observing that this ratio has the value two. Solving for $R_{\text{outd}} = 2R_{\text{out}}$ yields

$$R_{\text{outd}} = 2R_{\text{out}} = 2\left\{r_{ds} + R_S \frac{R_G + r_{gs}(1 + g_m r_{ds})}{R_G + R_S + r_{gs}}\right\}. \qquad (12.41a)$$

When R_S and R_G are much less than R_{gs}, $(R_S, R_G \ll r_{ds})$, this expression simplifies to

$$R_{\text{outd}} \approx 2[r_{ds} + R_S(1 + g_m r_{ds})], \quad R_S, R_G \ll r_{gs}. \qquad (12.41b)$$

If, in addition, $g_m r_{ds} \gg 1$, then $R_{\text{outd}} \approx 2r_{ds}(1 + g_m R_S)$.

The approximate common-mode behavior of the FET difference amplifier is given by the following expressions for the common-mode gain A_{cc}, and the common-mode input and output impedances.

$$A_{cc} \approx -\frac{R_D}{2R_{SS}} \tag{12.42a}$$

$$R_{oc} \approx \frac{r_{ds}}{2}\,(1+g_m R_{SS}) \tag{12.42b}$$

$$R_{ic} \approx \frac{r_{gs}}{2}\,(1+g_m R_{SS}). \tag{12.42c}$$

The common-mode rejection ratio for a balanced FET differential amplifier with single-ended output is from Eq. (12.39a), Eq. (12.39b), and Eq. (12.42a)

$$CMR = \left|\frac{A_{dd}}{2A_{cc}}\right| = \frac{g_m R_{SS}}{1+g_m R_S}\,. \tag{12.43}$$

Effect of imbalance in an FET difference amplifier   Imbalance in the drain loads, source resistors, generator resistors, and transconductances of the active devices introduce error terms. These errors are manifested as difference outputs from common-mode excitation and *vice versa* just as in the bipolar case. Using a procedure analogous to that for the bipolar case it can be shown that

$$\frac{v_{od}}{v_{ic}} = A_{dc} = \pm A_{cc}\left[-\frac{1}{R_S + (1/g_m)}\left[\delta R_S + \frac{1}{\delta g_m}\right] \pm \frac{\delta R_D}{R_D} \pm \frac{\delta r_{ds}}{r_{ds}}\right]$$

$$\pm \frac{A_{dd}}{\sqrt{v_{ic}}}\,[I_B \delta R_G + I_{os} R_G] \tag{12.44}$$

where I_B is the input bias current or static gate leakage of the FET, and I_{os} is the input offset current. The differential error factor A_{dc}/A_{cc} is also of interest for the FET difference amplifier. From the above equation, Eq. (12.43) and Eq. (12.8)

$$\Delta = \frac{A_{dc}}{A_{cc}} = \pm \frac{1}{R_S + (1/g_m)}\left[\delta R_S + \frac{1}{\delta g_m}\right] \pm \frac{\delta R_D}{R_D} \pm \frac{\delta r_{ds}}{r_{ds}}$$

$$\pm \frac{2CMR}{\sqrt{v_{ic}}}\,[I_B \delta R_G + I_{os} R_G] \tag{12.45}$$

12.5 FREQUENCY RESPONSE OF A DIFFERENTIAL AMPLIFIER

The frequency response of a balanced bipolar difference amplifier is considered in this section. It will be shown later that the results to be derived are also valid for FET amplifiers by making an appropriate change of symbols.

     The differential mode equivalent circuit of Fig. 12.3 is redrawn as Fig. 12.11(a) using the hybrid-pi transistor model which is described in Chapter 2. The effect of feedback in the emitter resistor can be represented in a more convenient form by placing an equivalent resistor, as given by Eq. (12.12), in series with the input and output ports. This is shown in Fig. 12.11(b).

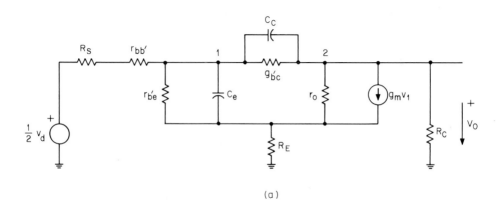

(a)

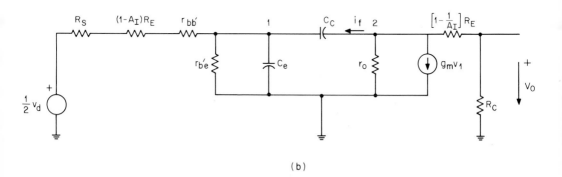

(b)

Fig. 12.11 (a) Equivalent circuit for difference-mode frequency response analysis; (b) equivalent circuit with grounded emitter neglecting $g_{b'c}$.

There is feedback from the internal collector node to the internal base node through the collector-base capacitance, C_c. The current, i_f, through the feedback element, Y_f, is

$$i_f = Y_f(v_2 - v_1) = Y_f(K-1)v_1$$

$$= Y_f \left[1 - \frac{1}{K} \right] v_2 \qquad (12.46)$$

where the voltage gain $K = v_2/v_1$, and in this case, $Y_f = sC_c$. Thus, the feedback capacitor can be replaced by a capacitor $C_1 = (1-K)C_c$ across $r_{b'e}$, and another capacitor $C_2 = [1 - (1/K)]C_c$ across r_o. This is known as the *Miller effect*, and Eq. (12.46) is sometimes known as *Miller's theorem*.[2]

In order to minimize the algebraic complications in the following results, it is assumed that $\beta \gg 1$ so that the current gain A_I is large. The impedance, Z_2, across the controlled current source $g_m v_1$ is C_2 in parallel with $R_2 = (R_C + R_E) \| r_o$,

$$Z_2 = \frac{1/C_2}{s + (1/R_2 C_2)} = \frac{R_2}{1 + sR_2 C_2} . \qquad (12.47)$$

The RC time constant of this pole in the output circuit is much smaller than that of the pole in the input circuit which involves the feedback capacitance multiplied by $(1 - K)$. Therefore the pole in the output circuit can be considered to be independent of the gain K. Because voltage $v_2 = -\beta Z_2 i_B = -\beta Z_2(v_1/r_{b'e})$, one may write

$$K = -\frac{\beta R_2}{r_{b'e}} \frac{1}{1 + sR_2 C_2} = \frac{K(0)}{1 + sR_2 C_2} \qquad (12.48)$$

where $K(0)$ is the zero frequency value of the voltage gain K.

Figure 12.12 is the equivalent circuit used to calculate the overall differential-mode voltage gain $A_v(s)$. The voltage transfer function from the input to node 1 is

$$\frac{v_1}{(v_d/2)} = \frac{r_{b'e}}{R_S + R_{in} + (1+\beta)R_E} \frac{1}{1 + s\tau_{in}} \qquad (12.49a)$$

where $R_{in} = r_{bb'} + r_{b'e}$ and the time constant of the input circuit is

$$\tau_{in} = \frac{r_{b'e}[R_S + r_{bb'} + (1+\beta)R_E]}{R_S + R_{in} + (1+\beta)R_E} [C_e + (1-K)C_c]. \qquad (12.49b)$$

The Miller-effect capacitance $(1-K)C_c$ dominates the base-emitter capacitance C_e in determining the input time constant since $|K| \gg 1$ for reasonable values of β, R_C, R_E, and $r_{b'e}$. There is a voltage divider consisting of

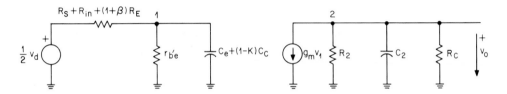

Fig. 12.12 Equivalent circuit for difference-mode gain.

R_E and R_C between node 2 and the output node. Thus

$$A_v(s) = \frac{v_o}{(v_d/2)} = \left(\frac{R_C}{R_C+R_E}\right)\left(\frac{v_2}{v_1}\right)\frac{v_1}{(v_d/2)}$$

and $A_{dd} = 2A_v$ is

$$A_{dd} = \left(\frac{R_C}{R_C+R_E}\right)\left(\frac{r_{b'e} + (1+\beta)R_E}{R_S + R_{in} + (1+\beta)R_E}\right)\left(\frac{K}{1+s\tau_{in}}\right). \qquad (12.50)$$

This is an important result that will be used throughout the remainder of the chapter. Note that $K \approx K(0)$ out to frequencies far beyond that of the input circuit pole, because $\tau_{in} \gg R_2C_2$, and therefore the voltage gain is a single time-constant function.

The common-mode equivalent circuit is shown in Fig. 12.13. It is structurally equivalent to the differential-mode circuits. However, because of the low common-mode gain and the large common-mode input voltage, there is a significant feedthrough component from input to output through the collector-base capacitance. Following the same line of reasoning as used in deriving the Miller theorem

$$i_1 = sC_c(v_o-v_1) = sC_c(K'-1)v_1, \qquad (12.51)$$

which is here interpreted as a capacitor of value $(1-K')C_c$ from node 1 to ground and the parallel combination of a capacitor C_c and ideal current

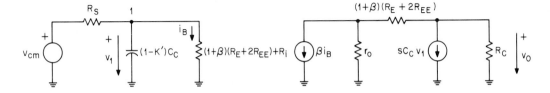

Fig. 12.13 Equivalent circuit for common-mode gain.

source $sC_c v_1$ from node 2 to ground. The voltage gain K' is very small, but even in this case C_c is the dominant capacitance in the input circuit because it bypasses $(1+\beta)(R_E+2R_{EE}) + R_{in}$ whereas C_c bypasses only $r_{b'c}$. Analysis of the input voltage divider yields

$$\frac{v_1}{v_{cm}} = \frac{(1+\beta)(R_E+2R_{EE}) + R_{in}}{(1+\beta)(R_E+2R_{EE}) + R_{in} + R_S} \frac{1}{1+s\tau_{in}'} \qquad (12.52a)$$

where

$$\tau_{in}' = \frac{R_S[R_{in} + (1+\beta)(R_E+2R_{EE})]}{(1+\beta)(R_E+2R_{EE}) + R_{in} + R_S} (1-K')C_c. \qquad (12.52b)$$

Making the approximation that $(1+\beta)(R_E+2R_{EE})$ is much greater than the impedance from node 2 to ground so that the current from the βi_B generator divides in the resistive current divider, the common-mode voltage gain is found, after some algebra, to be

$$A_{cc}(s) = \frac{v_o}{v_{cm}} \approx -\left\{ \frac{r_o R_C}{[r_o + (1+\beta)(R_E+2R_{EE})][R_{in} + (1+\beta)(R_E+2R_{EE})]} \right\}$$

$$\times \frac{1+s\tau_z}{(1+s\tau_{in}')(1+sR_C C_c)}$$

$$(12.53a)$$

where the zero time constant is

$$\tau_z = [R_{in} + (1+\beta)(R_E+2R_{EE})]C_c. \qquad (12.53b)$$

For reasonable component values $\tau_z > \tau_{in}' > R_C C_c$, and therefore as the frequency is increased the common-mode gain will be constant until $1/\tau_z$, after which it increases at 6 dB/octave because of the feedthrough effect. The pole at $1/\tau_{in}'$ causes the common-mode gain to again be constant. Finally, the pole at $1/R_C C_c$ causes it to fall at the rate of 6 dB/octave. The time constant associated with the difference-mode pole is also larger than τ_{in}' and $R_C C_c$. Therefore, the common-mode rejection ratio will deteriorate at a maximum rate of 12 dB/octave before it again becomes a constant at very high frequencies.

12.6 CURRENT SOURCES

A current source is a two-terminal circuit element that can maintain a prescribed terminal current regardless of the external load. It therefore follows that a constant current source has an infinite output impedance. Current sources are used as biasing elements and dynamic load elements.

In this section we are concerned with both the large-signal nonlinear behavior and the incremental small-signal behavior of some popular bipolar transistor current sources.

The collector characteristics of a common emitter transistor approximate the desired constant-current behavior. However, the small-signal output impedance of a grounded emitter transistor with a collector current of 1 mA is of the order of 50 kΩ. The primary problem is to establish the desired operating point and to maintain it over a wide temperature range. If the current gain of the transistor is sufficently large, $I_C \approx I_E$ and is given by

$$I_E = I_s \exp(V_{BE}/V_T). \tag{12.54}$$

Thus, one way of establishing a constant collector current is to maintain the exponent constant. A popular method of accomplishing this is to use a diode to bias the current source transistor.

12.6.1 Diode-Compensated Current Source

Figure 12.14 shows a diode-compensated transistor current source consisting of diode-connected transistor T_1, which temperature compensates transistor T_2. Resistor R_1 is used to establish the reference current I_1 and current $I_{C2} = I$ is the controlled source current.

Applying Kirchhoff's voltage law to the loop that includes R_1 and T_1 yields

$$I_1 = \frac{V_+ - V_{BE1}(T)}{R_1(T)}. \tag{12.55a}$$

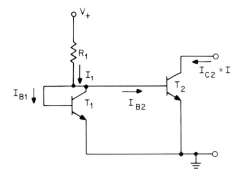

Fig. 12.14 Diode compensated current source.

In many applications the supply voltage greatly exceeds both the diode voltage $V_{BE1}(T)$ and its temperature variations which are of the order of 200 mV over the usual temperature range. Therefore $V_{BE1}(T)$ in Eq. (12.55a) can be replaced by $V_{BE1}(T_0) \triangleq \bar{V}_1$, where T_0 is the nominal ambient temperature. This substitution yields an approximate expression for reference current

$$I_1 \approx \frac{V_+ - \bar{V}_1}{R_1(T)} . \tag{12.55b}$$

Even this approximate expression for the reference current is a function of temperature because of the variability of R_1. In general, $I_1 = I_1[V_{BE1}(T), R_1(T)]$ has the following temperature variation

$$\frac{dI_1}{dT} = \frac{\partial I_1}{\partial V_{BE1}} \frac{dV_{BE1}}{dT} + \frac{\partial I_1}{\partial R_1} \frac{dR_1}{dT} .$$

With the aid of Eq. (12.55a) and Eq. (12.55b) one finds

$$\frac{1}{I_1} \frac{dI_1}{dT} \approx -\frac{1}{V_+ - \bar{V}_1} \frac{dV_{BE1}}{dT} - \frac{1}{R_1} \frac{dR_1}{dT} . \tag{12.56}$$

The first term on the right side of this equation is approximately

$$-\frac{1}{V_+ - \bar{V}_1} \frac{dV_{BE1}}{dT}\bigg|_{I_{C1}=\text{constant}} \approx \frac{V_{g0} - \bar{V}_1}{V_+ - \bar{V}_1} \frac{1}{T} . \tag{12.57}$$

The second term can be evaluated by assuming that the resistance has an approximately linear variation with temperature, that is, $R_1 \approx R_1(T_0)[1 + \alpha(T - T_0)]$. Hence

$$\frac{1}{R_1} \frac{dR_1}{dT} = \frac{\alpha}{1 + \alpha(T - T_0)} . \tag{12.58}$$

For a 200 $\Omega/\square$ base diffusion α is typically about 2.5×10^{-3} parts/°C. On the other hand, if the reference supply $V_+ = 30$ V, $V_{g0} = 1.21$ V for silicon and $\bar{V}_1 \approx 0.7$ V, we have, from Eq. (12.57), a variation of 5.8×10^{-5} parts/°C. Therefore the variation in reference current is due mainly to the temperature variability of the resistor R_1.

Kirchhoff's current law applied to the collector node of T_1 yields $I_1 = I_{B1} + I_{B2} + I_{C1}$. On the assumption that both transistors are identical and that the dc betas exceed 40 or so, $I_1 \approx I_{C1} \approx I_{C2}$ because of Eq. (12.54).

Assuming that the transistors are initially matched, this simple bipolar-transistor current source is compensated for variations in V_{BE}, but the output current is still a function of temperature because of the temperature sensitivity of R_1. The controlled current I_{C2} will differ from I_{C1} by a factor $\exp(\Delta V_{BE}/V_T)$ if the transistors are not matched.

12.6.2 The Logarithmic Current Source

The addition of a resistor in the emitter of the source transistor converts the diode-compensated current source into the so-called logarithmic current source shown in Fig. 12.15. This source can be made to have a positive, zero, or negative temperature coefficient of source current. It derives its name from the fact that $\ln(I_1/I) = \Delta V/V_T$ where $\Delta V \triangleq V_{BE1} - V_{BE2}$. The voltage ΔV can be related to the collector currents by solving Eq. (12.54) for V_{BE} and substituting in the definition.

$$\Delta V = V_T \ln\left[\frac{I_{C1}}{A_{E1}} \cdot \frac{A_{E2}}{I_{C2}}\right]. \tag{12.59}$$

In this equation A_{E1} and A_{E2} are the emitter areas, and the result that the saturation current is proportional to emitter area was used in the derivation. Another expression for ΔV is obtained by applying KVL around the loop encompassing T_1, T_2, and R_2.

$$\Delta V = (I_{B2} + I_{C2})R_2 \approx I_{C2}R_2. \tag{12.60}$$

From these last two equations it is seen that the controlled current is never greater than the reference current when the transistors have the same emitter areas, and that ΔV is a function of I_{C1}, I_{C2}, and temperature. It was shown previously that the reference current varied with temperature. If we want $dI_2/dT = 0$, the ratio I_{C1}/I_{C2} must also vary with temperature.

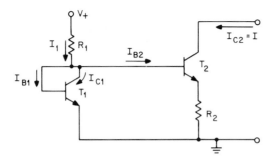

Fig. 12.15 Logarithmic current source.

Differentiation of Eq. (12.59) and Eq. (12.60) yields

$$\frac{d\Delta V}{dT} = \frac{\partial \Delta V}{\partial V_T}\frac{dV_T}{dT} + \frac{\partial \Delta V}{\partial I_{C1}}\frac{dI_{C1}}{dT} + \frac{\partial \Delta V_1}{\partial I_{C2}}\frac{dI_{C2}}{dT}$$

$$= \frac{\Delta V}{T} + \frac{V_T}{I_{C1}}\frac{dI_{C1}}{dT} - \frac{V_T}{I_{C2}}\frac{dI_{C2}}{dT} \qquad (12.61a)$$

$$= R_2\frac{dI_{C2}}{dT} + I_{C2}\frac{dR_2}{dT}. \qquad (12.61b)$$

Combining these results with $\Delta V = I_{C2}R_2$ and Eq. (12.56) yields

$$\frac{1}{I_{C2}}\frac{dI_{C2}}{dT} = \frac{1}{T}\frac{\delta v}{1+\delta v} - \frac{1}{R_2}\frac{dR_2}{dT}\frac{\delta v}{1+\delta v} - \frac{1}{R_1}\frac{dR_1}{dT}\frac{1}{1+\delta v}$$

$$(12.62)$$

where $\delta v = \Delta V/V_T$ is the base-emitter voltage difference normalized to the thermal voltage.

This equation describes the temperature behavior of the logarithmic current source. Since there is one positive term and two negative terms, the temperature coefficient of the current can be positive, negative or zero. The zero temperature coefficient case is usually of greatest interest.

An integrated circuit processing sequence is designed primarily to achieve specific transistor parameters and the circuit designer must accept the other parameters that result. Under the assumption that both resistors are fabricated by the same impurity deposition, they will have identical temperature coefficients dR/dT which will be specified *a priori*. Therefore Eq. (12.62) cannot be equal to zero for an arbitrary choice of δv. This leads to a bound on Eq. (12.62) if full temperature compensation is to be achieved.

$$\frac{dR}{dT} \geqslant \frac{\delta v}{1+\delta v}\frac{1}{T}. \qquad (12.63)$$

The properties of the material layer used to form the resistors are restricted by this inequality.

For ease in notation, the following temperature sensitivities are defined:

$$S_T^{I_{C2}} = \frac{dI_{C2}/I_{C2}}{dT/T} \qquad (12.64a)$$

$$S_T^{R_1} = \frac{dR_1/R_1}{dT/T} \qquad (12.64b)$$

$$S_T^{R_2} = \frac{dR_2/R_2}{dT/T}. \qquad (12.64c)$$

Using these sensitivity factors, Eq. (12.62) becomes

$$S_T^{I_{C2}} = \left(1 - S_T^{R_2}\right) \frac{\delta v}{1+\delta v} - S_T^{R_1} \frac{1}{1+\delta v} . \qquad (12.65)$$

For prescribed resistor sensitivities this equation will be zero if

$$\delta v = \frac{\Delta V}{V_T} = \frac{S_T^{R_1}}{1 - S_T^{R_2}} . \qquad (12.66)$$

To determine the effect of reference voltage supply variations on the output current, Eq. (12.59) and Eq. (12.60) are rewritten in the following manner to explicitly show the dependence on V_+.

$$\delta v = \frac{I_{C2}R_2}{V_T} = \ln\left(\frac{I_{C1}}{I_{C2}}\right) = \ln\left(\frac{V_+}{R_1 I_{C2}}\right). \qquad (12.67)$$

Differentiation of this equation with respect to V_+ yields

$$S_{V_+}^{I_{C2}} = \frac{dI_{C2}/I_{C2}}{dV_+/V_+} = \frac{1}{1+\delta v} . \qquad (12.68)$$

In many practical designs $\delta v > 1$ and so $S_{V_+}^{I_{C2}} < 1/2$. Note that the controlled current is never independent of supply voltage.

EXAMPLE 12.4 Design a temperature-independent logarithmic current source to supply a current of 500 μA using equal-area junctions and a supply voltage of 15 V. Calculate the sensitivity to supply voltage. The *TCR* of both resistors is 2500 ppm/°C, and the ambient temperature is 25° C = 298 K.

Solution. Both resistors have the same *TCR* and therefore have the same resistor sensitivity factor. From Eq. (12.64),

$$S_T^R = \frac{T}{R}\frac{dR}{dT} = (2.5\times10^3 \text{ ppm/°C})(298 \text{ K})(10^{-6}) = 0.745.$$

Note that although the *TCR*, which is an incremental quantity, is expressed in ppm/°C, the absolute temperature is used in computing the sensitivity factor because all activation processes are functions of absolute temperature.

The design of the logarithmic current source begins with the determination of δv from Eq. (12.66). $\delta v = 0.745/(1-0.745) = 2.922$, and therefore $\Delta V = 0.026\times2.922 \approx 76$ mV. However, $\Delta V = I_{C2}R_2$ where I_{C2} is the 500 μA source current. Therefore, $R_2 = 152$ Ω. The reference current $I_1 \approx I_{C1}$ is found by combining Eqs. (12.59) and (12.66) to yield

$$I_{C1} = I_{C2}\left(\frac{A_{E1}}{A_{E2}}\right)\exp\left[\frac{S_T^{R_1}}{1-S_T^{R_2}}\right]. \qquad (12.69)$$

Thus $I_{C1} = 5\times10^{-4}\exp(2.922) \approx 9.3$ mA, and therefore, from Eq. (12.55b),

$R_1 \approx (15-0.75)/9.3 \times 10^{-3} = 1.53 \text{ k}\Omega$. Note that with only two adjustable parameters, namely R_1 and R_2, we can only specify the source current and its temperature dependence and must then solve for the required reference current.

It should also be noted that the current source is only approximately temperature independent because higher-order temperature-dependent terms are neglected and it was assumed that the junctions were identical. This latter assumption may not be valid in some applications and more complex circuits have to be used.

Finally, from Eq. (12.69) the supply voltage sensitivity is

$$S_{V_+}^{I_{C2}} = \frac{1}{1+2.922} = 0.255$$

which means that if the supply voltage doubles $\Delta V_+/V_+ = 1$ and $\Delta I_{C2}/I_{C2} = 0.255$, i.e., a 25% change.

12.6.3 Small-Signal Behavior of Current Sources

An ideal constant current source has an infinite dynamic output impedance. The output impedance of the diode-compensated current source presented in Section 12.6.1 is that of a common emitter-connected transistor operated at the source current. This is typically about 50 kΩ for a transistor operating at a current of 1 mA.

Feedback can be used to modify the impedance level. The simplest example in this context is the logarithmic current source of the preceding section. The resistance in series with the emitter of the source transistor reduces the ratio of output current to reference current as we have seen. It also introduces degeneration which increases the output impedance. The small-signal equivalent circuit of this current source is shown in Fig. 12.16, where r_d is the dynamic resistance of the diode biasing element which is usually negligible in comparison with r_i. When r_o is much greater than r_i and r_d, the output impedance is easily shown to be

$$R_{out} = \frac{(r_d+r_i)(r_o+R_2) + (1+\beta)R_2 r_o}{r_d+r_i+R_2} . \qquad (12.70\text{a})$$

Usually $r_d = V_T/I_E$ is much smaller than the other resistors, and $R_2 < r_o$. Therefore Eq. (12.70a) can be simplified somewhat to

$$R_{out} = \frac{r_i + (1+\beta)R_2}{r_i+R_2} r_o. \qquad (12.70\text{b})$$

Due to the negative feedback in R_2, R_{out} can be many times larger than r_o, if r_i and R_2 are comparable in value and $\beta \gg 1$. The value of R_{out} is somewhere between the common-emitter and the common-base output resistance of transistor T_2 depending on the value of R_2. It is left as an exercise to show that several related controlled current sources can be obtained by placing additional transistors across the reference diode of Fig. 12.14.

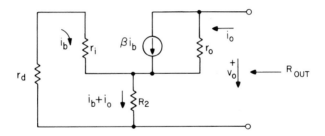

Fig. 12.16 Small-signal equivalent circuit of logarithmic current source.

It is also left as an exercise to show that the three-transistor current source shown in Fig. 12.17 has an output resistance

$$R_{\text{out}} = \frac{\beta_2}{2} r_o \qquad (12.71)$$

where r_o is the common-emitter output resistance of transistor T_2 and β_2 is its gain.

12.6.4 Emitter Degenerated Current Source

It was assumed above that the two transistors in the current source were matched so that $\Delta V = 0$ if $I_{C1} = I_{C2}$. In practice $\Delta V \neq 0$ because of unavoidable differences in emitter area. The effect of these area mismatches can be reduced by means of degeneration in a resistor in the emitter of the refer-

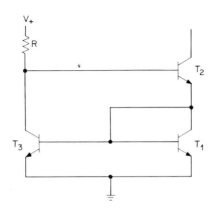

Fig. 12.17 Three-transistor current source with high-output impedance

ence transistor, as shown in Fig. 12.18(a). The Thevenin equivalent circuit for T_1 and bias resistors R_1 and R_3 is a voltage source

$$V_R = \frac{R_3}{R_1+R_3}\, V_+ + \frac{R_1}{R_1+R_3}\, V_{BE1}(T) \tag{12.72}$$

in series with a resistance $R_{eq} \approx R_1 \| R_3$ which leads to the simplified circuit of Fig. 12.18(b). Writing Kirchhoff's voltage law around the base-emitter loop of T_2 and solving for the base current I_B yields

$$I_B = \frac{R_3(V_+ - V_{BE1})}{R_1R_3 + (1+\beta_2)(R_1+R_3)R_2} + \frac{(R_1+R_3)\Delta V}{R_1R_3 + (1+\beta_2)(R_1+R_3)R_2} \,. \tag{12.73}$$

The resistors and junction voltages are functions of temperature. The output current I_{C2} is easily obtained from Eq. (12.73) by multiplying by β_2. Note the similarity of the first term of the resultant equation to Eq. (12.55a). Once again, if $V_+ \gg V_{BE1}(T)$ and if temperature variations in V_{BE} are small, $V_{BE1}(T)$ can be replaced by the constant $\overline{V}_1 = V_{BE1}(T_0)$, where T_0 is the normal ambient temperature. With the indicated substitutions

$$I_{C2} = \frac{\beta_2}{1+\beta_2} \left\{ \frac{R_3(V_+ - \overline{V}_1)}{[R_1R_3/(1+\beta_2)] + R_2(R_1+R_3)} \right.$$

$$\left. + \frac{(R_1+R_3)\Delta V}{[R_1R_3/(1+\beta_2)] + R_2(R_1+R_3)} \right\}. \tag{12.74a}$$

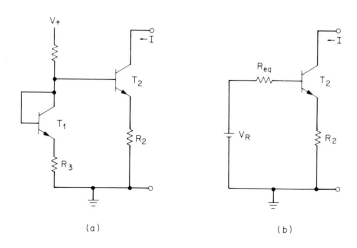

(a)   (b)

Fig. 12.18 (a) Emitter degenerated current source; (b) simplified circuit.

If $\beta_2 \gg 1$,

$$I_{C2} = \frac{R_3}{R_2} \frac{V_+ - \bar{V}_1}{R_1 + R_3} + \frac{\Delta V}{R_2}. \tag{12.74b}$$

Usually $R_2 = R_3$ and therefore the first term is approximately I_{C1}, the collector current of the reference transistor. It is the dominant term of this equation. Hence, $I_{C2} = I_{C1} + (\Delta V/R_2)$ and from Eq. (12.59) it follows that $\Delta V \approx V_T \ln(A_{E2}/A_{E1})$.

Note that in this case the perturbation in output current due to the mismatch is *linearly* related to ΔV.

EXAMPLE 12.5   Determine the output current $I = I_{C2}$ for an emitter-degenerated current source with $R_1 = 50 \ k\Omega$ and $R_2 = R_3 = 5 \ k\Omega$. The supply voltage is 20 V and the base-emitter junction voltages are matched to ± 2 mV. The leakage current for the transistor is $I_s = 2 \times 10^{-16}$ A.

Solution.   The output current is given by Eq. (12.74b) when $\bar{V}_1$ is known. Using an iterative procedure, consider the dominant first term of Eq. (12.74b) with $\bar{V}_1 = 0$ to find a good estimate of I_{C1} from which $\bar{V}_1$ can be found. The result is $I_{C1} = I_{C2} = 364 \ \mu A \approx I_s \exp(V_{BE}/V_T)$, from which we find $V_{BE} = 0.734$ V. Repeating this procedure with $\bar{V}_1$ equal to 0.734 V yields a new value of 0.733 V for V_{BE} which is used as the value for $\bar{V}_1$. From Eq. (12.74b), $I_{C2} = 350.3 \pm 0.4 \ \mu A$.

12.7 IMPROVING THE COMMON-MODE REJECTION RATIO OF DIFFERENCE AMPLIFIERS [3,4]

Current source biasing   It was shown in Eq. (12.21b) that the common mode rejection ratio was proportional to R_{EE}, the impedance of the emitter circuit biasing element. It is desirable for $R_{EE} > R_E/2$ as shown in the derivation of Eq. (12.21). However, diffused resistors do not often exceed 20 kΩ in value. The dynamic impedance of a current source is often used as the biasing element to provide a large effective resistance.

Consider, for example, the difference amplifier with an uncompensated current source shown in Fig. 12.19. The dynamic resistance R_{EE} is given by virtue of Eq. (12.70), with appropriate substitution of variables, as

$$R_{EE} \approx \left[1 + \frac{\beta R_3}{(R_1 \| R_2) + r_i + R_3} \right] r_o.$$

With a reasonable value for β and R_3 the dynamic impedance can be brought into the megohm range.

It is easy to show that the bias current is

$$I = \frac{1}{R_3} \left\{ \frac{V_+ R_2 + V_- R_1}{R_1 + R_2} - V_{BE3} \right\} \left\{ \frac{\beta R_3}{(R_1 \| R_2) + (1 + \beta) R_3} \right\}. \tag{12.75a}$$

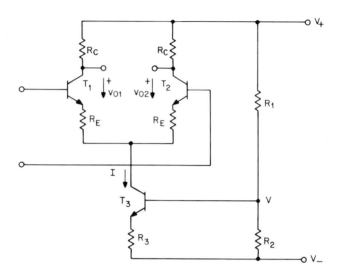

Fig. 12.19 Difference amplifier with current source biasing.

When $\beta \gg 1$ and $\beta R_3 \gg R_1 \| R_2$, this equation reduces to

$$I \approx \frac{1}{R_3}\left\{\frac{V_+ R_2 + V_- R_1}{R_1 + R_2} - V_{BE3}\right\}. \qquad (12.75b)$$

EXAMPLE 12.6 Calculate the common-mode rejection ratio for a difference amplifier with $R_{C1} = R_{C2} = 25\ k\Omega$, $R_{S1} = R_{S2} = 500\ \Omega$, $R_{E1} = R_{E2} = 50\ \Omega$, $R_1 = 20\ k\Omega$, $R_2 = 1\ k\Omega$, $R_3 = 4\ k\Omega$. The transistor parameters are $r_i = 25\ k\Omega$, $r_o = 5\ M\Omega$, and $\beta = 100$.

Solution. The dynamic resistance of the current source is

$$R_{EE} = 5\times10^6\left[\frac{20\times10^3/21 + 25\times10^3 + (101)4\times10^3}{20\times10^3/21 + 25\times10^3 + 4\times10^3}\right] = 7.178\times10^7.$$

Using this value for R_{EE} in Eq. (12.21b) yields

$$CMR = \frac{10^2\times7.178\times10^7}{500+2.5\times10^4 + (101)\times50} = 2.39\times10^5.$$

This value of *CMR* should be compared with the result of Example 12.1 where we found the *CMR* to be 332. This high rejection ratio is only obtained with a perfectly balanced amplifier. However, single-ended rejection ratios Γ_s of about 2000:1 can be obtained with this simple current source biased difference amplifier. Rejection ratios Γ_d of 50,000:1 for double-ended output are commonly possible.

Common-mode feedback biasing   Further improvement in the single-ended rejection ratio may be obtained by using a circuit with common-mode feedback, as shown in Fig. 12.20. This configuration reduces the common-mode gain by negative feedback and thereby enhances the rejection ratio. The common emitter-junction voltage of the second stage is approximately equal to the common-mode output signal of the first stage, $(v_{o1} + v_{o2})/2$. This common-mode voltage is used to determine the bias of the current source transistor T_5. The variation in common-mode voltage at the base of the current source transistor causes a variation in bias current for the first stage that opposes the change in bias due to the change in common-mode input voltage.

A reduction in common-mode gain A_{cc} by a factor of three to five is obtained with this type of circuit. Note that the output of the second stage is single-ended. To compute the overall single-ended rejection ratio of this circuit, denote the outputs of the first stage by

$$v_{od1} = A_{dd1}v_{id} + A_{dc1}v_{ic},$$

$$(12.76)$$

$$v_{oc1} = A_{cd1}v_{id} + A_{cc1}v_{ic}.$$

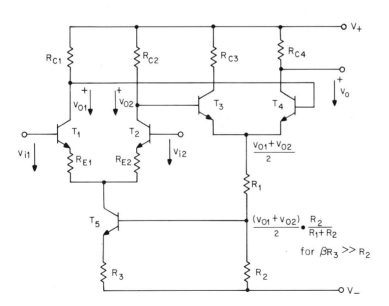

Fig. 12.20   Common-mode feedback biasing of a difference amplifier.

These voltages are the inputs to the second stage, whose output is

$$v_o = \frac{v_{od2}}{2} + v_{oc2}$$

$$= \left(\frac{A_{dd2}}{2} + A_{cd2}\right) v_{od1} + \left(\frac{A_{dc2}}{2} + A_{cc2}\right) v_{oc1}. \tag{12.77}$$

Substitution of Eqs. (12.76) into Eq. (12.77) yields an expression for the second-stage output in terms of the first-stage inputs.

$$v_o \approx \left\{\frac{A_{dd1}A_{dd2}}{2}\right\} \left\{v_{id} + \left[\frac{A_{dc1}}{A_{dd1}} + 2\,\frac{A_{cc1}A_{cc2}}{A_{dd1}A_{dd2}}\right] v_{ic}\right\}. \tag{12.78}$$

The approximation involves neglecting some terms because $|A_{dc2}| \ll A_{cc2}$ and $|A_{cd2}| \ll |A_{dd2}|$. The second term in the square brackets, $2A_{cc1}A_{cc2}/A_{dd1}A_{dd2}$, is always much smaller than A_{dc1}/A_{dd1}. Hence the single-ended rejection ratio of the overall amplifier, Γ_s, is approximately

$$\Gamma_s \approx \left|\frac{A_{dd1}}{A_{dc1}}\right| = \Gamma_{d1} \tag{12.79}$$

which is the double-ended rejection ratio for the first stage. Note that in this amplifier A_{cc1} is greatly reduced by the common-mode feedback, and so rejection ratio $CMR_1 = |A_{dd1}/2A_{cc1}|$ is also greatly increased.

12.8 SINGLE-ENDED AMPLIFIERS

Single-ended amplifier stages are widely used in linear integrated circuits to provide the high gain that is desirable in an operational amplifier. In this section we shall present a method for biasing a common-emitter stage that is applicable to integrated circuit implementation, and analyze both the Darlington emitter follower and the cascode amplifier.

12.8.1 The Common-Emitter Stage

The common-emitter amplifier can provide both voltage and current gain with moderate input and output impedance levels. In discrete circuit implementations, the operating point is stabilized with large amounts of dc feedback and high ac gain is obtained by heavy capacitive bypassing of the emitter resistor. This method is obviously useless for an integrated circuit amplifier because very large capacitors cannot be made on an IC chip. It is now shown that the inherent parameter match between integrated transistors can be exploited in the biasing of a common-emitter amplifier.

The basis of a simple constant current source is to set the operating point of a common-emitter amplifier by means of a diode operating at the same current density as the emitter-base junction of the transistor. The diode and transistor have identical temperature characteristics, and therefore the transistor current is equal to the diode current independent of temperature.

This same basic principle can be used to bias the common-emitter amplifier stage. A modification is required to permit the input signal to be introduced at the base terminal of the amplifier. Kirchhoff's current law applied at the collector node of T_1, in the circuit of Fig. 12.21, yields the relation

$$I_1 = I_{C1} + I_{B1} + I_{B2} \approx (\beta_1 + 2) I_{B1}$$

where the approximation is valid when β_1 is large. Using this relation, together with the following expression for base current I_{B1},

$$I_{B1} = \frac{V_+ - I_1 R_1 - V_{BE1}(T)}{R_2} \tag{12.80a}$$

and observing that usually V_+ is sufficiently larger than both $V_{BE1}(T)$ and its temperature variation, yields

$$I_{B1} = \frac{V_+ - \overline{V}_1}{R_2 + (\beta_1 + 2) R_1} \tag{12.80b}$$

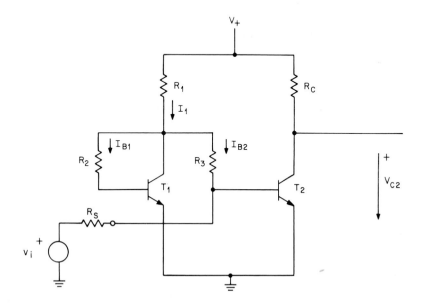

Fig. 12.21 Biasing of common-emitter amplifier.

where $\bar{V}_1 = V_{BE1}(T_0) \approx V_{BE1}(T)$. An equation similar to Eq. (12.80a) can be obtained for I_{B2}. Using that equation, Eq. (12.80a), and $V_{BE2} = V_{BE1} + \Delta V_{BE}$,

$$I_{B2} = \frac{R_2}{R_3} I_{B1} - \frac{\Delta V_{BE}}{R_3}. \tag{12.81}$$

Thus I_{B2} is related to I_{B1} by a resistor ratio and the V_{BE} match of the transistors. Typically ΔV_{BE} is several millivolts and its temperature variation is several microvolts per degree centigrade.

The quiescent output voltage $V_{C2} = V_+ - I_{C2} R_C$ is found, with the aid of Eq. (12.80b) and Eq. (12.81), to be

$$V_{C2} = \frac{\left\{ R_2 + (\beta_1+2)R_1 - \beta_2 \dfrac{R_2 R_C}{R_3} \right\} V_+ + \beta_2 \dfrac{R_2 R_C}{R_3} \bar{V}_1}{R_2 + (\beta_1+2)R_1} + \frac{\beta_2 R_C}{R_3} \Delta V_{BE} \tag{12.82}$$

which is approximately $V_+/2$ when $\beta_1 = \beta_2 = \beta \gg 1$, is large, $R_C = R_1/2$, and $R_2 = R_3$.

The small-signal equivalent circuit of this common-emitter stage is shown in Fig. 12.22. The resistance seen looking back from R_3 across R_1 is

$$R = \frac{R_1(R_2+r_i)}{R_2 + r_i + (\beta+1)R_1} \approx \frac{R_2+r_i}{\beta+1} \tag{12.83}$$

which is usually much smaller than R_3. Therefore the input impedance of this amplifier is approximately r_i in parallel with R_3. The current and voltage gains are

$$A_I = \frac{i_o}{i_{in}} = -\beta \frac{R_3 r_o}{(R_3+r_i)(R_C+r_o)} \tag{12.84a}$$

and

$$A_v = A_I \frac{R_C}{R_{in}} = -\beta \frac{R_C r_o}{r_i(R_C+r_o)}. \tag{12.84b}$$

The overall voltage gain $A_{vi} = v_o/v_i$ which takes account of the generator impedance is

$$A_{vi}(0) = -\beta \frac{R_C r_o}{[r_i + R_S + (r_i R_S/R_3)](R_C+r_o)} \tag{12.85}$$

where we have explicitly noted that this is the gain at zero frequency.

The input impedance can be increased and the gain stabilized at a smaller value by placing a resistor in series with the emitter of each transis-

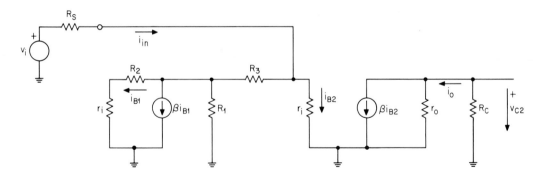

Fig. 12.22 Small-signal equivalent of common-emitter amplifier.

tor. The results of the previous analysis are still valid in this case, provided the transistor parameters are modified to account for the emitter degeneration.

Figure 12.23(a) is the ac small-signal equivalent circuit of the common-emitter amplifier, which is redrawn in Fig. 12.23(b) by applying Miller's theorem to the feedback capacitor. This circuit is identical to the difference mode equivalent circuit of the difference amplifier, Fig. 12.11, except for component values. Therefore, it immediately follows from Eq. (12.49a), Eq. (12.49b), and Eq. (12.50) with $R_E = 0$, $R_{eq} = R_C \| r_o$, and $K(0) = -\beta R_C r_o / r_{b'e}(R_C + r_o)$, that

$$A_{vi}(s) = \frac{A_{vi}(0)}{1 + s\tau_{in}}, \qquad (12.86a)$$

where $A_{vi}(0)$ is given by Eq. (12.85) and

$$\tau_{in} = \frac{r_{b'e}(R_S + r_{bb'})}{R_S + R_{in}}\left[1 + \frac{\beta R_C r_o}{r_{b'e}(R_C + r_o)}\right]C_{b'e}. \qquad (12.86b)$$

12.8.2 Cascode Amplifier

A cascode amplifier is a series connection of a common-emitter amplifier and a common-base amplifier as shown in Fig. 12.24(a). From this figure it is evident that the emitter current of T_2 is equal to the collector current of T_1. The complete equivalent circuit using the hybrid-pi model is shown in Fig. 12.24(b). It will be shown that the coupling network between the transistors is essentially resistive and that the Miller-effect capacitance at the input is substantially reduced over that of a common-emitter stage.

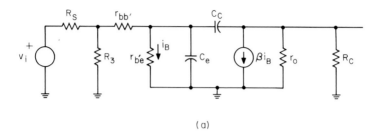

(a)

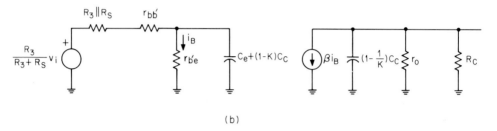

(b)

Fig. 12.23 (a) Small-signal equivalent of common-emitter amplifier for frequency response analysis; (b) small-signal equivalent circuit after application of Miller's theorem.

The Miller-effect voltage gains are defined as $K_1 = v_2/v_1$ and $K_2 = v_o/v_3$. Figure 12.25 is the cascode-amplifier equivalent circuit with the feedback capacitors replaced by the equivalent Miller-effect capacitors to ground. For the present, assume that nodes 2 and 2' are disconnected. Kirchhoff's current law applied at node 2' yields

$$i_T = (1+\beta_2) i_{b2} + \frac{1}{r_{o2}} (v_o - v_2).$$

Analysis of the shunt arm shows that

$$v_2 = v_3 - i_{b2} R_2' = \left[1 + (1+s\tau) \frac{R_2'}{R_2}\right] v_3$$

where $\tau = (R_2' \| R)(1 - K_2) C_{c2}$. From these two results, we find that

$$i_T = -\frac{[1 + \beta_2 + (R_{i2}/r_{o2})]}{R_{i2}} \cdot \frac{1 + \left[\dfrac{1 + \beta_2 + (R_2'/r_{o2})}{1 + \beta_2 + (R_{i2}/r_{o2})}\right] s\tau}{1 + (R_2'/R_{i2}) s\tau} v_2 + \frac{v_o}{r_{o2}}.$$

(12.87)

where $R_{i2} = R_2' + R_2$. However, $R_2' \ll r_{o2}$, $R_{i2} \ll r_{o2}$, and $R_2'/R_{i2} \approx 1$.

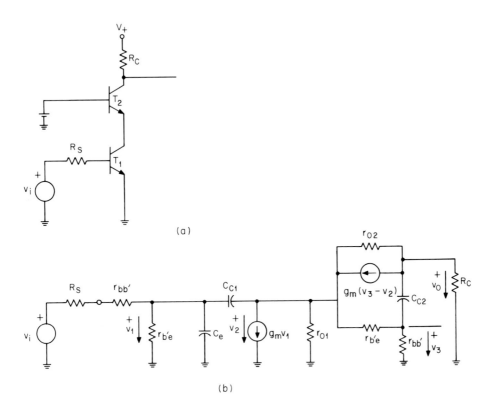

Fig. 12.24 (a) Circuit diagram for cascode amplifier; (b) equivalent circuit for cascode amplifier.

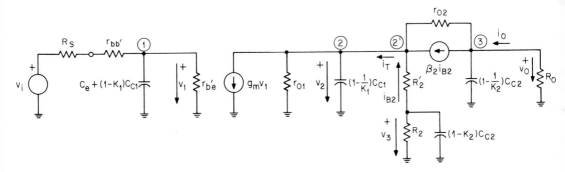

Fig. 12.25 Equivalent circuit for cascode amplifier redrawn using Miller's theorem.

Therefore the multipliers on the time constants in this equation are each approximately unity and this equation reduces to

$$i_T \approx -\frac{1+\beta_2}{R_{i2}} v_2 + \frac{v_o}{r_{o2}} \, . \tag{12.88}$$

The input impedance of the second stage is $R = -v_2/i_T$. From Eq. (12.88) with $K' = v_o/v_2$,

$$R = \frac{R_{i2}}{(1+\beta_2) - K'(R_{i2}/r_{o2})} \, . \tag{12.89}$$

To get an idea of the magnitude of K', temporarily assume that r_{o2} is infinite so that all of the second stage current goes through the load resistor, R_C. Then $v_o = -\beta_2 i_{b2} R_C$, $i_{b2} = -v_2/R_{i2}$, and $K' = v_o/v_2 = \beta_2 R_C/R_{i2}$. If $\beta \gg 1$, the denominator of Eq. (12.89) becomes $\beta_2[1 - (R_C/r_{o2})]$ which is typically about $0.9\beta_2$. For simplicity, we will use $R = R_{i2}/\beta_2 \approx V_T/I_E$.

The interstage coupling network consists of R in parallel with r_{o1} and $[1 - (1/K_1)]C_{c1}$, and is essentially resistive because the capacitor is, at most, several picofarads.

We next consider the output resistance of the cascode which is $r_o' = v_o/i_o$ when the input voltage $v_i = 0$. Straightforward analysis of the equivalent circuit shown in Fig. 12.26 yields

$$r_o' = R_{i2} + (1+\beta_2)r_{o2} \tag{12.90}$$

which is much higher than the output resistance of a common-emitter stage.

Consider now the reduced equivalent circuit shown in Fig. 12.27. The voltage ratio $K_1 = v_2/v_1$ is

$$K_1 = \frac{\beta_1}{1+\beta_2} \frac{R}{R_1} \, . \tag{12.91a}$$

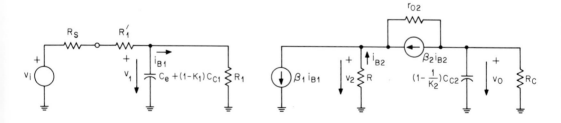

Fig. 12.26 Equivalent circuit for output resistance

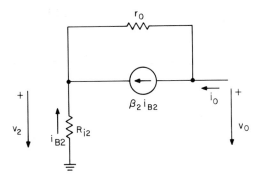

Fig. 12.27 Reduced equivalent circuit.

Using the approximate value of R determined above,

$$K_1 \approx \frac{\beta_1}{(1+\beta_2)^2}\frac{R_{i2}}{R_1} \approx \frac{1}{\beta} \qquad (12.91\text{b})$$

because $R_1 \approx R_{i2}$ and it is assumed that $\beta_1 = \beta_2 = \beta$. Thus K_1 is almost zero which practically eliminates the Miller-effect capacitance multiplication.

The overall voltage gain of the cascode amplifier is

$$A_v(s) = \frac{A_v(0)}{(1+s\tau_{\text{in}})(1+s\tau_{\text{out}})} \qquad (12.92)$$

where

$$A_v(0) = \frac{v_o}{v_i} = -\frac{\beta_1\beta_2 R_C}{(1+\beta_2)(R_S+R_{i1})[1 + (R_C/r_{o2}')]}, \qquad (12.93\text{a})$$

and the time constants are

$$\tau_{\text{in}} = \frac{R_1(R_1'+R_S)}{R_{i1}+R_S}\,(C_{c1}+C_{b'e}) \qquad (12.93\text{b})$$

and

$$\tau_{\text{out}} = \frac{R_C r_{o2}'}{R_C+r_{o2}'}\,C_{c2}. \qquad (12.93\text{c})$$

Both of the time constants are small and of about the same size. Hence, the cascode amplifier is a wideband configuration, and its frequency response is described by a two-pole function.

EXAMPLE 12.7 Compute the voltage gain and time constants for a cascode amplifier with $\beta_1 = \beta_2 = 100$, $R_C = 5\text{ k}\Omega$, $R_{i1} = R_{i2} = 2\text{ k}\Omega$, $R_S = 1\text{ k}\Omega$, $r_o = 50\text{ k}\Omega$, $C_{C1} = C_{C2} = 0.3\text{ pF}$, $C_{b'e} = 0.2\text{ pF}$.

Solution. From Eq. (12.90) $r_o' = 2 \times 10^3 + (101) \times 5 \times 10^4 = 5.052$ MΩ. The dc gain is evaluated from Eq. (12.93a),

$$A_v(0) = - \frac{10^2 \times 10^2 \times 5 \times 10^3}{(1 + 10^2)(1 \times 10^3 + 2 \times 10^3)[1 + (5 \times 10^3/5.052 \times 10^6)]} = -1.649 \times 10^2.$$

The time constants are found from Eqs. (12.93b) and (12.93c),

$$\tau_{in} = \frac{2 \times 10^3 (2 \times 10^3 + 1 \times 10^3)}{2 \times 10^3 + 1 \times 10^3} (3 \times 10^{-13} + 2 \times 10^{-13}) = 1.0 \times 10^{-9} \text{ s},$$

$$\tau_{out} = \frac{5 \times 10^3 \times 5.052 \times 10^6}{5 \times 10^3 + 5.052 \times 10^6} (3 \times 10^{-13}) = 1.499 \times 10^{-9} \text{ s}.$$

Therefore,

$$A_v(s) = - \frac{164.9}{(1 + 1 \times 10^{-9} s)(1 + 1.499 \times 10^{-9} s)}.$$

12.9 DC-LEVEL SHIFTING

DC-level shifting is required in an operational amplifier because the input and output must be at the same dc level and *p-n-p* transistors with good high-frequency response are difficult to obtain. The dc shift can be accomplished by virtue of the voltage drop across R_1 in the circuit of Fig. 12.28(a). Transistors T_1 and T_3 are emitter followers while T_2, R_2, and V_R form a rudimentary current source that draws a current I_1 through resistor R_1. This circuit provides a level shift

$$V_{shift} = V_{BE1} + V_{BE3} + I_1 R_1 \qquad (12.94)$$

where V_{BE1} is a function of I_1, and V_{BE3} depends upon the value of R_3. The small-signal gain of the level shifter would be the gain of two cascaded emitter followers if T_2 were an ideal current source with infinite source impedance.

$$A_V \approx 1 - \frac{r_{i1}}{(1 + \beta_1)[R_1 + r_{i3} + (1 + \beta_3) R_3]}. \qquad (12.95)$$

In actuality the source impedance of T_2 is several hundred kilohms which is comparable to the input impedance of T_3. This can impose a restriction on the value of R_1 because for high gain R_1 must be much smaller than the driving point impedance at node 1.

A simple modification shown in Fig. 12.28(b) introduces *positive* feedback to increase the voltage gain. A useful gain in excess of unity can be obtained. It is evident that the dc conditions can still be satisfied with the current in R_2 equaling the sum of the emitter currents in transistors T_2 and T_3. The small-signal equivalent circuit is shown in Fig. 12.29. It is assumed that $R_2 \gg r_2/(1 + \beta_2)$.

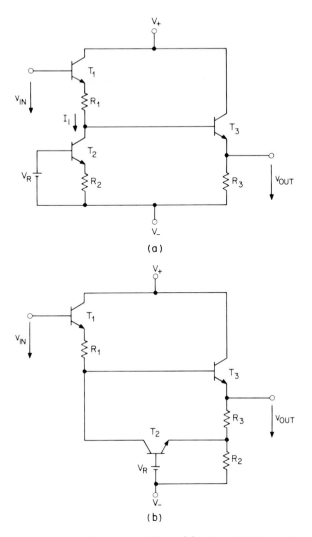

Fig. 12.28 (a) A simple resistive level shifter; (b) a level shifter with positive feed-back.

The details of the analysis, which involves applying Kirchhoff's current law at nodes A and B and writing expressions for the voltage drop across resistors R_1 and R_2, are omitted. The resultant expression for voltage gain is

$$A_v = \frac{(1+\beta_1)(1+\beta_2)[r_3 + (1+\beta_3)R_2]}{(1+\beta_1)(1+\beta_2)\left[r_3+(1+\beta_3)\left\{R_2+\dfrac{r_2}{(1+\beta_2)}\right\}\right] + (1-\beta_2\beta_3)[r_1+(1+\beta_1)R_1]}.$$

$$(12.96a)$$

If all transistors have the same small-signal parameters and $\beta \gg 1$ the voltage gain becomes

$$A_v \approx \frac{1}{1 - (R_1/R_2)} \,. \tag{12.96b}$$

We would expect that the input impedance may be negative because of the positive feedback. The expression for input impedance $Z_{in} = v_{in}/i_1$ is

$$Z_{in} = r_1 + (1+\beta_1)R_1 + \frac{(1+\beta_1)(1+\beta_2)}{1-\beta_2\beta_3} \left\{ r_3 + (1+\beta_3) \left[R_2 + \frac{r_2}{(1+\beta_2)} \right] \right\}. \tag{12.97a}$$

When the transistors have identical small-signal parameters,

$$Z_{in} \approx (1+\beta)(R_1 - R_2). \tag{12.97b}$$

The input impedance is negative when $R_2 > R_1$ which is the condition for positive voltage gain.

The circuit will be stable when driven from a source whose internal impedance is

$$R_s \ll \frac{\beta_1 R_2}{A_v} \,. \tag{12.98}$$

Finally, the output impedance of the level shifter is approximately

$$Z_{out} \approx \frac{R_1 A_v}{\beta_2} \,. \tag{12.99}$$

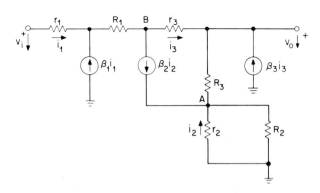

Fig. 12.29 Small-signal equivalent circuit of level shifter.

EXAMPLE 12.8   Design a positive feedback level shifter to have a gain of 3 and a voltage shift of 10 V. Compute the input and output impedances of this level shifter. Assume V_R is chosen such that the collector current of T_2 is 1 mA, and that $\beta_1 = \beta_2 = \beta_3 = 100$. The base-emitter leakage current for the transistors is 2×10^{-16} A.

Solution. Assuming identical transistors and $\beta \gg 1$, the gain is given by Eq. (12.96b)

$$A_v = 3 \approx \frac{1}{1 - (R_1/R_2)} .$$

Therefore, $R_2 = 1.5 R_1$. The value of R_1 is determined from the required voltage shift. The base-emitter voltage drop of the transistors at 1 mA is 0.76 V at room temperature. This is not a rapidly varying function and therefore V_{BE3} is assumed equal to V_{BE1} in Eq. (12.94). Hence,

$$V_{\text{shift}} = 2 \times 0.76 + 1 \times 10^{-3} R_1 = 10$$

and therefore $R_1 = 8.48$ kΩ, $R_2 = 12.72$ kΩ. From Eq. (12.97b)

$$Z_{\text{in}} \approx 101 (8.48 - 12.72) \times 10^3 = -427.8 \text{ k}\Omega$$

and from Eq. (12.99)

$$Z_{\text{out}} \approx 8.48 \times 10^3 \times 3/100 = 254.4 \ \Omega .$$

The value of the remaining resistor R_3 is determined from the negative-supply voltage and the current from the output-emitter follower which are not specified.

12.10 DIFFERENTIAL TO SINGLE-ENDED CONVERSION—CURRENT MIRRORS

The purpose of the circuits described in this section is to convert the full differential output voltage available between the collectors of the differential pair into a single-ended signal referenced to ground. The basis for these circuits is the diode-compensated current source described in Section 12.6. The "reference current" for the current source comprised of p-n-p transistors T_3 and T_4 in Fig. 12.30 is provided by transistor T_1, and the controlled source current drives the same node as transistor T_2. We are interested in the small-signal behavior about the quiescent operating point.

When T_1 draws an incremental current out of node 1, T_2 delivers an identical incremental current into node 2. The incremental current drawn by T_1 through reference transistor T_3 increases the emitter-base voltage of transistors T_3 and T_4 equally. This causes T_4 to produce an (approximately) equal increment of current into node 2. Thus, the current source "mirrors" the current in the left side of the amplifier and delivers it to the output node. The output current i_L is effectively twice the change in current through

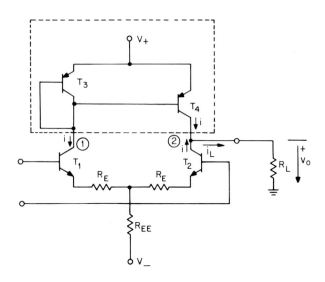

Fig. 12.30 Differential amplifier with current mirror load.

either side of the difference amplifier. Therefore, the full differential voltage is developed across R_L.

The equivalent circuit of the current mirror is shown in Fig. 12.31 after the application of Miller's theorem. The current transfer relation between i_{b4} and i_{in} is found by applying Kirchhoff's current law to the input node

$$i_{in} = i_{b4} + (1+\beta_3)\,i_{b3} + v/r_o$$

and noting that $i_{b4} = v_1/r$ and that $r_{in} = r' + r$. Analysis of the input circuit yields

$$i_{b4} = \cfrac{i_{in}}{(2+\beta_3) + s\tau_{in}\left[\cfrac{(1+\beta_3)\,r' + r_{in}}{r_{in}}\right]} \ , \qquad (12.100)$$

where

$$\tau_{in} = rC_c(1-K) \qquad (12.101)$$

and $K = v_o/v_1$. The output current is

$$i_o = \frac{\beta_4 r_o}{R_L+r_o} \ \frac{1}{1+s\tau_{out}} \ i_{b4} \qquad (12.102)$$

where

$$\tau_{out} = \frac{r_o R_L}{R_L+r_o} \ C_c\left(1 - \frac{1}{K}\right). \qquad (12.103)$$

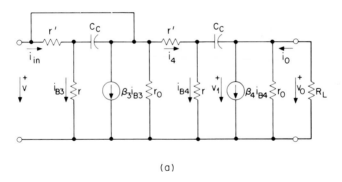

(a)

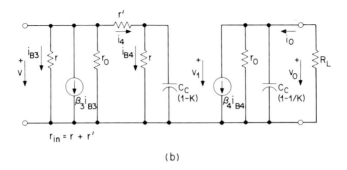

(b)

Fig. 12.31 (a) Equivalent circuit of current mirror; (b) equivalent circuit after application of Miller's theorem.

The output voltage $v_o = -i_o R_L$ and $i_{b4} = v_1/r$. Therefore, from Eq. (12.102), the low frequency value of K is found to be

$$K = -\frac{r_o R_L}{R_L + r_o}\frac{\beta_4}{r} \approx -\beta_4\frac{R_L}{r}.$$
(12.104)

Thus $\tau_{\text{out}} \approx R_L' C_c \ll \tau_{\text{in}}$ because $|K| \gg 1$, where R_L' is R_L in parallel with r_o, and therefore the-current mirror is a single time-constant circuit.

Finally, the current transfer ratio between the input and output is

$$\frac{i_o}{i_{\text{in}}} = \frac{\beta_4}{(2+\beta_3)[(1 + (R_L/r_o)]}\frac{1}{1 + s\tau_{\text{in}}\left[\dfrac{(1+\beta_3)r' + r_{in}}{(2+\beta_3)r_{\text{in}}}\right]}.$$
(12.105)

At low frequencies, this transfer ratio approaches unity when $R_L \ll r_o$ and β_3, β_4 are large.

The main disadvantages of this simple current mirror are that r_o, the output resistance of a common emitter stage, is relatively small and it unduly shunts the output and that the differential amplifier is asymmetrically loaded. These differences will be alleviated shortly.

It is also useful to note that the incremental output resistance of a current source with a constant reference current is very useful as a dynamic load. It is a means of providing a high resistance load which does not require as much space on an integrated circuit die as an equivalent passive resistor, and in addition, it has a much smaller dc-voltage drop.

A simple way to increase the output resistance of the current mirror is to add resistor R_E in series with the emitters of transistors T_3 and T_4 thus making them appear to be more like common-base transistors, and increasing the effective value of r_o. The new output impedance, R_o, becomes

$$R_o = r_o + R_E \frac{(r_i + R_E + \beta r_o)}{r_i + 2R_E}$$

$$\approx \left[1 + \frac{\beta}{2}\right] r_o. \tag{12.106}$$

Modified single-ended converter   A single-ended converter that loads the differential amplifier in a symmetrical fashion can be obtained by the addition of an emitter follower to the previous circuit as shown in Fig. 12.32. The emitter follower provides voltage coupling between the collector and base of T_3 while preserving a high impedance level at the collector of T_3.

Assuming that the voltage gain of the emitter follower is unity yields the equivalent circuit of Fig. 12.33 where R', the resistance shunting the

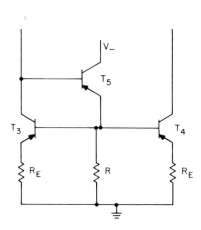

Fig. 12.32 Current mirror modified to provide symmetric loading.

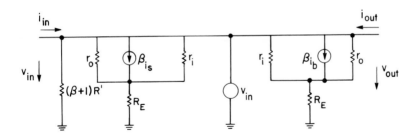

Fig. 12.33 Equivalent circuit of modified current mirror.

input terminals, is the emitter follower load resistor R in parallel with the input impedance of T_3 and T_4,

$$R' = R \| \frac{1}{2} [r_i + (1+\beta) R_E].$$

When the input current is zero, the input of transistor T_4 is effectively at ground and the output impedance of transistor T_4 is

$$R_o = \frac{v_{out}}{i_{out}} = r_o + R_E \frac{\beta r_o + r_i}{R_E + r_i}$$

$$\approx (1+\beta) r_o. \qquad (12.107)$$

The impedance looking into the input terminals is the output impedance looking into the collector of T_3 in parallel with $(1+\beta) R'$. Using the approximate form of Eq. (12.107), assuming that $R = R_E$, and recalling that our assumption that the gain of the emitter follower was close to unity implies that $\beta \gg 1$ yields

$$R_I = \frac{v_{in}}{i_{in}} \approx \frac{(3+\beta) R_E}{2r_o + (1+\beta) R_E} (1+\beta) r_o \propto (1+\beta) r_o. \qquad (12.108)$$

Therefore, the driving point impedance at either port is much larger than r_o.

The overall current gain of the modified single-ended converter is $A_{Ic} = i_{out}/i_{in} \approx 1$. The transconductance from the input of one side of the difference amplifier to the output of the current mirror is

$$\frac{i_{out}}{(v_d/2)} \approx \frac{\beta}{R_S + r_i + (1+\beta) R_E} A_{Ic} \qquad (12.109a)$$

where A_{Ic} is the current gain of the single-ended converter and Eq. (12.15b) was used. The transconductance of the other side of the difference amplifier is

$$\frac{i}{(-v_d/2)} \approx - \frac{\beta}{R_S + r_i + (1+\beta)R_E} . \qquad (12.109b)$$

The current through the load resistor R_L' is $i + i_{out}$, and since $A_{Ic} \approx 1$ for the modified single-ended converter, the output voltage becomes

$$v_o = -i_L R_L' = - \frac{\beta R_L'}{R_S + r_i + (1+\beta)R_E} v_d, \qquad (12.110)$$

where R_L' is the load resistor R_L in parallel with the output impedance look-ing into the collector circuits of the differential amplifier and the single-ended converter. Comparison of this result with Eq. (12.15) shows that the single-ended voltage developed across the load resistor is the full differential voltage swing when the difference amplifier is subjected to the same load.

12.11 SUPERBETA   GAIN   STAGES   AND   CASCODE   DIFFERENTIAL AMPLIFIERS [5,6]

Superbeta transistors are diffused transistors whose emitters are driven very deeply into the base region so that the effective base width is very narrow. This has the effect of greatly increasing the current gain over that of a nor-mal transistor while reducing the collector-base breakdown voltage. Typical superbeta transistors have current gains of 2000 to 5000 at a collector current of about 1 μA and a collector-emitter breakdown voltage of about 4 V.

These transistors are most useful as the input transistors in an opera-tional amplifier because they permit very low input bias currents with small offset voltage and current. Because of the low breakdown voltage, the superbeta transistor is usually used in a compound connection with a normal transistor to form an ultrahigh beta transistor that can withstand normal operating voltages. A cascode amplifier using a superbeta transistor is shown in Fig. 12.34. The forward-biased diode string consists of a diode-connected superbeta transistor in series with a diode-connected normal transistor, and is used to keep the collector-base voltage of the superbeta input transistor at 0 V. This prevents punchthrough breakdown of the input transistor.

A useful configuration is the superbeta cascode differential amplifier shown in Fig. 12.35. The superbeta input transistors are operated with $V_{CB} = 0$ V by bootstrapping the bases of the standard transistors T_3, T_4 to the common-mode voltage. This prevents punchthrough of the superbeta transistors. This cascode differential amplifier stage has several advantages over FET or bipolar Darlington input stages. Matched pairs of superbeta transistors exhibit a typical offset voltage of 0.5 to 1.0 mV with temperature drifts of 2 μV/°C. The drift for FET's is 40 to 50 μV/°C. Standard transis-tors in the Darlington configuration can meet low bias current specifications,

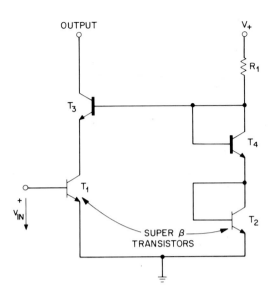

Fig. 12.34 Cascode amplifier with superbeta transistors.

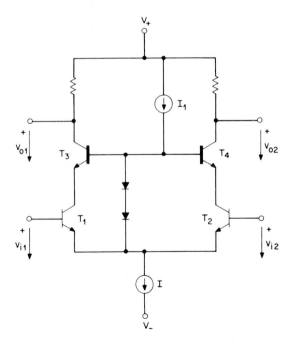

Fig. 12.35 Superbeta cascode differential amplifier. Superbeta transistors are shown with thinner base.

but the offset voltage is a function of the β mismatch; with a 10% β mismatch resulting in 2.5 mV offset. Furthermore, Darlington input differential amplifiers have higher noise and lower common-mode rejection than simple (or cascode) differential amplifiers.

It can be shown that the differential-mode gain, common-mode gain, and differential-mode input resistance of a cascode differential amplifier are equivalent to that of a standard differential amplifier.[6] The cascode differential amplifier does provide a significant increase in common-mode input resistance and differential input-differential output figure of merit, Γ_d. In addition, the cascode configuration offers wider bandwidth than the conventional differential pair.

12.12 OUTPUT STAGES

Operational amplifiers and many other linear amplifiers are voltage-to-voltage transducers; that is, they deliver an output voltage in response to an input voltage excitation. A good voltage source must have a low output impedance.

The output stage of an operational amplifier provides current gain and impedance-level transformation to isolate the load from the input and intermediate stages. The output stage usually is a variant of the emitter follower which has a low output impedance, good current gain, and nominal-unity voltage gain.

Class A emitter follower   Consider the Class A emitter follower, shown in Fig. 12.36, which consists of a single n-p-n transistor driving a resistive load. When the output voltage v_o is zero, there is a quiescent bias current $I_{BQ} = -V_-/R_E$ and the load current is zero. When the input voltage rises above its quiescent value the transistor conducts more heavily to supply the load current i_o and an increased current in the emitter resistor R_E. The increased current during positive input excursions comes from the positive supply via a low-impedance path. During negative input excursions, the load current is supplied by the negative power supply through the fixed resistor R_E. If the peak value of the output voltage is $\hat{V}_o < |V_-|$, and $\hat{I}_o$ is the magnitude of the load current due to the most negative voltage $V_o = -\hat{V}_o$, then it can be shown that the quiescent bias current must be

$$I_{BQ} \geqslant \frac{V_-}{\hat{V}_o + V_-} \hat{I}_o. \qquad (12.111)$$

Since V_- is negative and $\hat{V}_o < |V_-|$ this relation indicates that the quiescent bias current I_{BQ} must be chosen considerably larger than the magnitude of the load current $\hat{I}_o$ due to the most negative-going output voltage

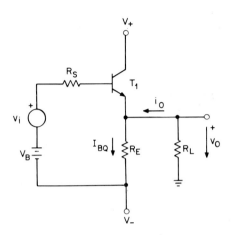

Fig. 12.36 Class A emitter follower.

$V_o = -\hat{V}_o$. The Class A emitter follower is a simple circuit but it has the disadvantage of high quiescent power dissipation and different output impedances for positive- and negative-going waveforms. These disadvantages can be overcome quite easily by using a Class AB output stage.

Class AB emitter follower output stage   The Class AB output stage shown in Fig. 12.37 utilizes an n-p-n and a p-n-p transistor. It provides bidirectional output swings with low values of quiescent current, I_{BQ}, and therefore it is the usual choice for an operational amplifier output stage.

The biasing diodes D_1 and D_2 are always conducting and the quiescent bias current I_{BQ} of transistors T_1 and T_2, in the absence of load current ($V_o = 0$), is given by

$$I_{BQ} = \frac{(V_{D1} + V_{D2}) - (V_{BE1} + V_{BE2})}{2R_E}. \qquad (12.112)$$

It depends on the sum of diode voltages, the sum of transistor base-emitter voltages, and the value of emitter resistor R_E. In any case the quiescent bias current may be made small if the numerator of Eq. (12.112) is almost zero. A positive input signal will then turn on T_1 and turn off T_2, and *vice versa* for a negative input signal. For a positive input signal transistor T_1 is conducting and transistor T_2 is off. The power dissipation in transistor T_1 is $P_1 = I_o(V_+ - V_o - I_o R_E)$. Because $V_o = I_o R_L$ this equation may be written as

$$P_1 = I_o[V_+ - I_o(R_L + R_E)]. \qquad (12.113)$$

The output current $\hat{I}_o$ resulting in maximum power dissipation $\hat{P}_1$ in T_1 may be obtained by differentiation of Eq. (12.113). It is given by

$$\hat{I}_o = \frac{V_+}{2(R_L + R_E)} \tag{12.114}$$

and may be much larger than the quiescent bias current I_{BQ} given in Eq. (12.112). Substituting Eq. (12.114) into Eq. (12.113) yields the maximum power dissipation $\hat{P}_1 = V_+\hat{I}_o/2$. Current $\hat{I}_o$ usually represents the maximum output current the stage can handle without exceeding the safe power dissipation ratio $\hat{P}_1$.

Short circuit protection   Current $\hat{I}_o$ may, however, be exceeded due to a short circuit to ground or to one of the power supplies. If a short circuit to ground occurs, the full positive supply voltage V_+ is effectively across the ON transistor T_1 if R_E is small. In that case the output current must be limited to the value I_{oL}, where

$$I_{oL} = \frac{\hat{P}_1}{V_+} \tag{12.115a}$$

if $\hat{P}_1$ is not to be exceeded. On the other hand, if a short to a supply voltage occurs, then the voltage across the output transistor could reach the value

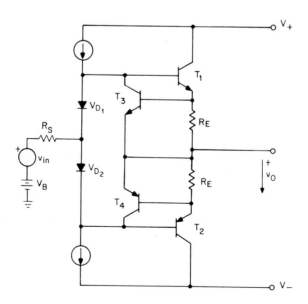

Fig. 12.37 Class AB output stage with short-circuit protection.

$V_+ - V_-$ and the output current limit would have to be set to

$$I'_{oL} = \frac{\hat{P}_1}{V_+ - V_-} . \tag{12.115b}$$

An optional protective circuit using transistors T_3, T_4 as "clamps" is also shown in Fig. 12.37. Transistors T_3 and T_4 are normally off and only turn on when the voltage $I_o R_E$ due to the output current exceeds the cut-in voltage of the clamp transistor. In that case T_3 (T_4) turns on and its collector current shunts current away from the base of T_1 (or T_2) thus preventing the output current I_o from increasing beyond a safe current limit, which is given by $I''_{oL} = V_{BE3}/R_E$ or V_{BE4}/R_E.

Clearly this current limit is temperature sensitive because of its dependence on V_{BE3} (or V_{BE4}). However, the temperature coefficient dV_{BE}/dT is negative and dR/dT is positive, and so an increase in temperature decreases V_{BE} and therefore decreases the limiting current I''_{oL}, thereby controlling the transistor junction temperature and the maximum power dissipation $\hat{P}_1$.

12.13 VOLTAGE REFERENCES [13,14]

There are many applications in which a stable, constant voltage reference is needed. These include regulated power supplies, comparator circuits, and some types of logic circuits. The circuit of Fig. 12.38 provides a stable,

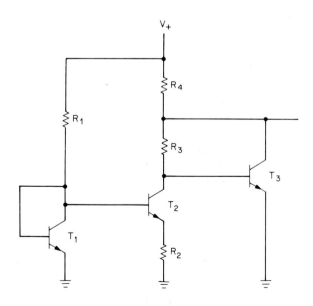

Fig. 12.38 Stable voltage reference.

temperature-independent, output voltage approximately equal to the bandgap voltage of silicon, yet it requires only standard bipolar transistors and resistors.

Identical transistors T_1 and T_2, and resistors R_1 and R_2 form a logarithmic current source in which the current density in the emitter of T_2 is less than that of T_1 because of the voltage developed across R_2. Recall from Eq. (12.62) that the temperature variation of the collector current of T_2 can be made positive, zero, or negative by proper choice of R_1 and R_2. Transistor T_3 senses the temperature-dependent voltage developed across R_3 and provides an output voltage

$$V_o = V_{BE3} + I_{C2}R_3$$

$$= V_{BE3} + \frac{R_3 \Delta V_{BE}}{R_2} , \qquad (12.116)$$

where $\Delta V_{BE} = V_{BE1} - V_{BE2}$ results in different current densities in T_1 and T_2. It was shown previously that $\Delta V_{BE} = V_T \ln(J_1/J_2)$. An accurate expression for the base-emitter voltage of a transistor is[12]

$$V_{BE} = V_{g0}\left[1 - \frac{T}{T_0}\right] + V_{BE0}\left[\frac{T}{T_0}\right] + nV_T \ln\left[\frac{T_0}{T}\right] + V_T \ln\left[\frac{I_c}{I_{co}}\right], \qquad (12.117)$$

where V_{g0} is the bandgap voltage extrapolated to absolute zero, and V_{BE0} is the base-emitter voltage at T_0 and I_{co}. The last two terms of this equation are small and of the same order as nonidealities of the transistors, and are therefore ignored. From Eq. (12.116) and Eq. (12.117)

$$V_o = V_{g0}\left[1 - \frac{T}{T_0}\right] + V_{BE0}\left[\frac{T}{T_0}\right] + \frac{R_3}{R_2} V_T \ln\left[\frac{J_1}{J_2}\right]. \qquad (12.118)$$

For zero temperature drift of output voltage $\partial V_o/\partial T = 0$,

$$\frac{\partial V_o}{\partial T} = -\frac{V_{g0}}{T_0} + \frac{V_{BE0}}{T_0} + \frac{R_3}{R_2}\frac{V_T}{T} \ln\left[\frac{J_1}{J_2}\right], \qquad (12.119)$$

and therefore

$$V_{BE0} + \left[\frac{T_0}{T}\right]\left[\frac{R_3}{R_2}\right]\Delta V_{BE} = V_{g0}. \qquad (12.120)$$

In this equation, the first term is the base-emitter voltage of transistor T_1 at the initial temperature T_0 and the second term is proportional to the difference in the base-emitter voltages of T_1 and T_2.

Substituting Eq. (12.120) into Eq. (12.118) shows that when $V_o = V_{g0}$, the output voltage is perfectly temperature compensated. In practice it turns out that V_o should be slightly larger than V_{g0} to account for the small terms that were neglected.

12.14 THE OPERATIONAL AMPLIFIER—DC CONDITIONS [7,9]

In this section it is shown that the low-frequency gain of an integrated operational amplifier is limited by thermal feedback from the output circuit to the input differential amplifier. This is a general result that applies to any integrated circuit which has both high gain and a high-power output stage.

The simplified circuit diagram of a representative integrated circuit operational amplifier is shown in Fig. 12.39. It consists of a differential amplifier input stage with a current mirror differential to single-ended converter, a high-gain, single-ended amplifier, and a broadband, low-gain output stage. The gain of an operational amplifier is typically between 10,000 and 100,000 and its frequency response (see Section 12.15) is such that it is stable even when the output terminal is connected directly to the inverting input.

We first compute an approximate value for the electronic voltage gain of this amplifier. The output stage is a Class AB emitter follower. Therefore its voltage gain is approximately one, and it transforms a load impedance R_L

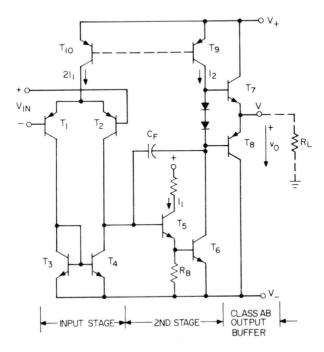

Fig. 12.39 Simplified schematic of operational amplifier.

at the output terminal into a load $\beta_7 R_L$ on the intermediate stage. When R_L is at its lowest permissible value, β_7 will also be low because transistor T_7 is operating at a high current density. Under these conditions the load on the intermediate stage is at its minimum value, and therefore the output resistance of the intermediate stage transistors can be neglected. Hence, the voltage gain is

$$A_v = \frac{v_o}{v_{in}} \approx \frac{g_{m1}\beta_5\beta_6\beta_7 R_L}{1 + (r_{i2}/r'_{o1})} \tag{12.121}$$

where r'_{o1} is the equivalent output resistance of the differential amplifier and current mirror, and it was assumed that $g_{m1} = g_{m2}$, and $\beta_7 = \beta_8$.

The parameter values for typical amplifiers[8] are $g_m = 192\ \mu$mhos, $\beta_5 = \beta_6 = 150$, $\beta_7 = 50$, $r'_{o1} = 5.7\ M\Omega$, and $R_L = 2000\ \Omega$. Using these values in Eq. (12.121) yields $A_v \approx 312,000$, but the gain measured on an actual integrated circuit is much less than this value because of thermal feedback.

The important elements of the thermal feedback problem are illustrated in Fig. 12.40. The output stage of an operational amplifier can deliver 50 to 100 mW to a load, while internally dissipating a similar amount of power, causing the temperature of the chip and package to rise. In spite of the good thermal conductivity of the chip and package, there are small thermal gradients ranging from several tenths of a degree to several degrees across the length of the chip.

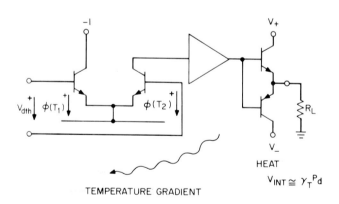

Fig. 12.40 Thermal feedback problem.

Recall that junction voltage is a function of temperature and therefore the temperature rise of the chip as a whole causes the junction voltages to decrease if the currents are constant. However, the thermal gradient across the chip induces a small differential mode voltage, in the first stage, which is proportional to the power dissipated in the output stage.

The power dissipated in a Class AB output stage is the power delivered by the supply minus the power delivered to the load,

$$P_d = \frac{V_+ V_o - V_o^2}{R_L} \quad \text{for} \quad V_o > 0$$

(12.122)

$$P_d = \frac{-V_- V_o - V_o^2}{R_L} \quad \text{for} \quad V_o < 0.$$

This function exhibits peaks at $V_o = V_+/2$ and $-V_-/2$, and zeros of power dissipation at $V_o = 0$, V_+, and $-V_-$.

At frequencies below the reciprocal of the thermal time constant, the temperature difference between a pair of matched components in close proximity is, to a first approximation,

$$\Delta T \approx \pm K_T P_d,$$

(12.123)

where ΔT is the temperature differential in degrees centigrade, and K_T is a proportionality constant with dimensions of degrees centigrade per watt corresponding to the thermal resistance between the two matched components. When the dominant components of the input circuit are the differential transistor pair, the thermally induced input voltage is

$$V_{dth} \approx \pm \frac{dV_{BE}}{dT} K_T P_d = \pm \gamma_T P_d$$

(12.124)

where $dV_{BE}/dT \approx -2$ mV/°C for a silicon junction. The $\pm$ signs are required because the direction of the thermal gradient is unknown.

The value of K_T depends upon the layout of the integrated circuit chip and the thermal characteristics of the package. In a good thermal design, the high-power transistors approximate either a point source or a line source of heat, and the input stage components are placed symmetrically on the resulting isothermal lines, as shown in Fig. 12.41. Typical values for K_T are about 0.3°C/W in a TO-5-type metal can package. Therefore, the thermal feedback voltage constant is about 0.6 mV/W.

Substituting Eq. (12.122) into Eq. (12.124) yields an expression for the thermal feedback voltage in terms of the output voltage which is plotted in Fig. 12.42. The electrical gain which is a straight line and the apparent thermally controlled, open-loop gain are also plotted in this figure. The thermal-feedback factor is found [from Eq. (12.122) and Eq. (12.124)] to be

$$\beta_T = \frac{\partial V_{dth}}{\partial V_o} = \pm \frac{\gamma_T}{R_L} (V_s - 2V_o)$$

(12.125)

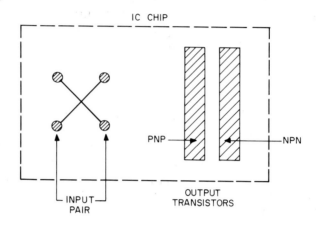

Fig. 12.41 Simplified chip layout for amplifier showing a quad of cross-connected input transistors positioned to reduce thermal feedback.

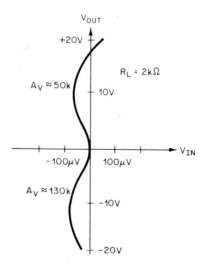

Fig. 12.42 Effect of thermal feedback on voltage gain.

where $V_s = V_+$ for $V_o > 0$ and $V_s = -V_-$ for $V_o < 0$. This is the electrical equivalent for the thermal-feedback network.

The apparent open-loop gain is obtained by applying this feedback factor to the electrical gain μ

$$A_v(0) = \frac{\mu}{1 \pm \dfrac{\gamma_T \mu}{R_L} (V_s - 2V_o)} \tag{12.126}$$

which can be either positive or negative, as can be seen from Fig. 12.42. From Eq. (12.125) it is observed that the incremental feedback is greatest at $V_o = 0$, V_+, and $-V_-$ which are the zeros of dissipated power, and it is zero at the power peaks $V_o = V_+/2$ and $-V_-/2$. The *maximum usable gain* is the gain when the thermal feedback is a maximum. Thus

$$A_v(0) \Big|_{max} = \frac{R_L}{\gamma_T V_s} \approx \frac{1}{\gamma_T I_{max}} \tag{12.127}$$

where $I_{max} = V_s/R_L$ is the maximum value of load current if the amplifier output could swing all the way to the supplies.

The phase of the open-loop gain can vary as a function of output voltage due to the thermal feedback. Therefore it is necessary to apply sufficient electrical feedback, as shown in Fig. 12.43, so as to reduce the amplifier gain below the maximum usable gain in order to obtain predictable

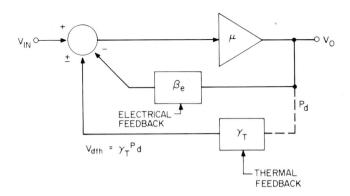

Fig. 12.43 Block diagram showing effect of electrical and thermal feedback.

characteristics. The closed-loop gain is then given by

$$A_v(0) = \frac{\mu}{1 + \mu(\beta_E \pm \beta_T)} \qquad (12.128)$$

where β_E is the electrical feedback factor. The most commonly observed effect of thermal feedback is low-frequency distortion due to the nonlinear transfer characteristic. Higher frequencies are unaffected because the differential thermal coupling falls off at an initial rate of 6 dB/octave starting at about 100 to 200 Hz.

12.15 SMALL-SIGNAL FREQUENCY RESPONSE OF THE OPERATIONAL AMPLIFIER [7]

A good approximation to the small-signal frequency response of the operational amplifier of Fig. 12.39 can be obtained from the equivalent circuit of Fig. 12.44. The intermediate stage is modeled as a high-gain, frequency-independent amplifier with capacitive feedback due to the compensating capacitor, C_F. The feedback reduces the input impedance of this stage far below the output impedance of the first stage over the frequency range of interest. Therefore, the input differential amplifier and current mirror are replaced by a frequency-independent, voltage-controlled current source, $i_{in} \approx g_{m1}v_d$. The output stage is an emitter follower which has unity gain and is neglected.

Then $v_o \approx -i_{in}/sC_F$ and the magnitude of the voltage gain is

$$|A_v(\omega)| = \left|\frac{v_o}{v_d} \; (s=j\omega)\right| = \frac{g_{m1}}{\omega C_F} . \qquad (12.129)$$

Upon setting $|A_v(\omega)| = 1$, we obtain the *open-loop unity gain cutoff frequency*, ω_u ,

$$\omega_u = \frac{g_{m1}}{C_F} , \qquad (12.130)$$

which is also the gain-bandwidth product of the closed loop amplifier. A typical value for the feedback capacitor C_F is 30 pF and $g_{m1} = 192 \; \mu$mhos. The unity gain frequency is $f_u = \omega_u/2\pi = 1.02$ MHz which is typical of that measured on integrated operational amplifiers.

This result is valid if the feedback capacitor C_F is the dominant frequency-response determining element. However, there are poles in the differential amplifier and current mirror transfer functions that can degrade the frequency response. There are many operational amplifier applications in which the inverting input is excited by both the input signal and the feedback signal, with the noninverting input terminated in a resistor. In these applications the frequency response may be limited by a pole determined by the stray capacitance at the common emitter node. This can be seen by analyzing the circuit of Fig. 12.45(a). The formulae of Section 12.5 can be modified to account for the stray capacitance C_s, and the response computed as the superposition of the responses due to a common-mode excitation $v_1/2$ and a difference-mode excitation v_1. Another method will be used here.

Observe that when $v_2 = 0$, transistor T_2 is a common-base amplifier driven by an emitter follower. Figure 12.45(b) is the equivalent circuit obtained by looking back into the emitter of the emitter follower T_1 and into the emitter circuit of T_2. In this figure R_{EE} is the output resistance of the current source or the value of the biasing resistor, and C_s is the value of the

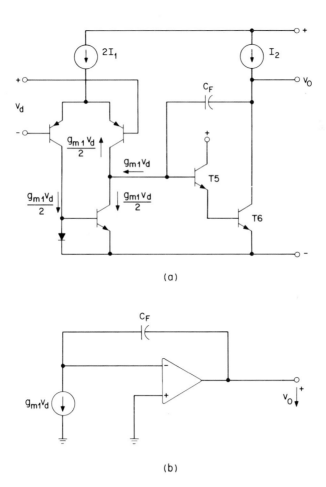

(a)

(b)

Fig. 12.44 (a) Simplified schematic of operational amplifier for high-frequency analysis; (b) simpler equivalent circuit.

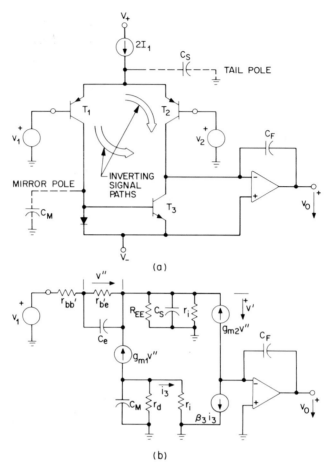

Fig. 12.45 (a) Effect of tail and mirror poles; (b) equivalent circuit for analysis.

stray capacitance from the common emitter terminal to ground. The voltage transfer ratio relating the common-base amplifier input voltage v' to the excitation v_1 is

$$\frac{v'}{v_1} \approx \frac{1}{2} \frac{1 + r_{b'e}C_e s}{1 + \left[r_{b'e}C_e + \dfrac{C_s}{2g_{m1}} \right] s} \tag{12.131}$$

where $g_{m1} = \beta_1/r_{b'e}$. When the stray capacitance is negligible, the coupling is wideband because the pole and zero cancel. Using reasonable values for stray capacitance, it is possible for the $C_s/2g_{m1}$ term to become commensurate with the $r_{b'e}C_e$ term at low current levels when the transconductance is low. When this happens, the interstage coupling-network transfer function rolls off at 6 dB/octave starting at the pole frequency. It flattens out again at frequencies above the zero frequency.

In Section 12.10 it was shown that the current mirror is described by a single-pole transfer function. Because of the feedback around the second stage there is a virtual ground at the output of the mirror, and therefore its pole time constant is approximately

$$\tau_m \approx r(C_e+C_m) \tag{12.132}$$

where C_e is the emitter transition capacitance, C_m is the stray capacitance at the mirror-input node, and r is the dynamic resistance of the mirror-input diode, $r = V_T/I$. At low current levels r becomes large and the mirror pole frequency might become a significant consideration. At $I = 1 \mu A$, $1/2\pi\tau_m \approx 0.9$ MHz which would introduce significant phase shift and degrade the phase margin of an operational amplifier whose cutoff frequency $\omega_u = 1$ MHz.

Thus, there is a conflict between the desirability of operating the first stage at low current to reduce the input bias and offset currents and (as we shall see in the next section), improve the slew rate, and the disadvantage of reducing the current to a level at which the frequency response characteristics of the input stage degrade the overall phase response of the amplifier.[10, 11]

We now consider the behavior of the intermediate stage amplifier which was previously assumed to be an ideal integrator. This stage is an emitter-follower, common-emitter amplifier cascade with capacitive feedback. The emitter follower will be ignored to simplify the analysis. The input stage is represented by its Norton's equivalent of a current source in shunt with R_i and C_i. Similarly the load of the output stage is R_L in shunt with C_L. The reduced circuit is shown in Fig. 12.46(a) and its equivalent circuit is shown in Fig. 12.46(b). In the equivalent circuit $r_{bb'}$ is neglected and the RC networks loading the ports are absorbed into the impedances at the ports of the transistor model.

The transresistance of the amplifier can be shown to be

$$\frac{v_o}{i_s} = \frac{N(s)}{D(s)} = -\frac{g_m R_1 R_2[1 - (sC_p/g_m)]}{1+sA+s^2B} , \tag{12.133}$$

where $C_p = C_F + C_c$ is the total feedback capacitance

$$A = R_1(C_1+C_p) + R_2(C_2+C_p) + g_m R_1 R_2 C_p$$

and

$$B = R_1 R_2[C_1 C_2 + C_p(C_1+C_2)].$$

Under normal operating conditions the poles are widely separated and there-

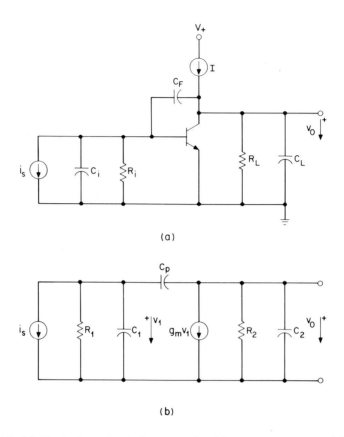

Fig. 12.46 (a) Equivalent circuit for analysis of intermediate stage; (b) equivalent circuit with transistor and external parameters lumped together.

fore the denominator of Eq. (12.133) can be approximately factored as $D(s) \approx 1 + s/p_1 + s^2/p_1 p_2$ where p_1 is the dominant pole, i.e., $p_1 \ll p_2$. Therefore, $p_1 \approx 1/A$ and under normal operating conditions the last term of A is much larger than the other two terms so that

$$p_1 \approx \frac{1}{g_m R_1 R_2 C_p} \tag{12.134a}$$

and

$$p_2 \approx \frac{g_m C_p}{C_1 C_2 + C_p (C_1 + C_2)}. \tag{12.134b}$$

Observe that p_1 is inversely proportional to g_m whereas p_2 is directly proportional to g_m. Thus as g_m increases, the poles are split apart with p_1 driven toward the origin and p_2 driven toward infinity.

The effect of pole splitting in the μA741 operational amplifier is shown in Fig. 12.47. Values used are $g_m = 1150\ \mu$mhos, $C_1 = C_2 = 10$ pF, $R_1 = 1.7\ M\Omega$, and $R_2 = 100\ k\Omega$. When $C_p = 0$ both poles are in the kilohertz range. When the compensating capacitor $C_p = 30$ pF, the poles are $p_1 \approx 2.5$ Hz and $p_2 \approx 66$ MHz, and so the amplifier is effectively a single-time constant circuit over a wide frequency range.

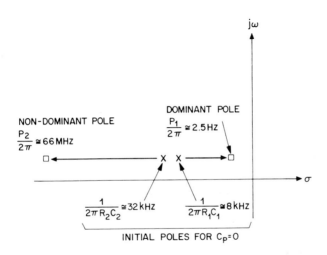

Fig. 12.47 Root locus of poles for μA 741.

12.16 LARGE-SIGNAL RESPONSE OF THE OPERATIONAL AMPLIFIER [7]

When an operational amplifier is overdriven by a step excitation, its output is a ramp whose slope is determined by internal current and capacitances. The slope of the resulting output ramp is known as the *slew rate*.

Figure 12.48 is the model used to calculate the slew rate of the illustrative operational amplifier. As an idea of the magnitude of the input signal required to overdrive this circuit, a 60 mV signal is sufficient to switch 90% of the current to one branch of the simple bipolar difference amplifier. The slew rate is the maximum rate of change of the output voltage, $dv_o/dt\,|_{max}$. In this circuit the input stage is a current source that drives a capacitive load. The maximum output current of the first stage is $2I$ which is the total

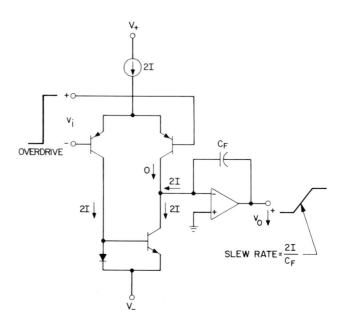

Fig. 12.48 Slew rate of operational amplifier.

current available from the current source in the emitter circuit. Therefore

$$\left.\frac{dv_o}{dt}\right|_{max} = \frac{2I}{C_F} = S_R \tag{12.135}$$

and the slew rate is limited by the current level of the input stage and the feedback capacitor if the current levels in the intermediate stage are greater than $2I$. Recalling that the unity gain open-loop cutoff frequency is $\omega_u = g_{m1}/C_F$ [Eq. (12.130)] permits the slew rate to be related to the frequency response of the amplifier.

$$\left.S_R\right|_{max} = \frac{\omega_u}{g_{m1}} 2I \tag{12.136a}$$

However, for a simple bipolar difference amplifier $g_{m1} = I/V_T$, and therefore

$$\left.S_R\right|_{max} = 2\omega_u V_T \tag{12.136b}$$

where V_T is the thermal voltage. This means that for the simple bipolar difference-amplifier input stage, the slew rate and the unity-gain cutoff frequency are related by a fundamental constant.

EXAMPLE 12.9   Determine the slew rate of an operational amplifier with a simple bipolar difference-amplifier input stage with $I = 1 \ \mu A$ and $\omega_u = 2\pi \times 10^6$ rad/s.

Solution.   Assuming that the slew rate is limited by the current capability of the input stage, the slew rate is given by Eq. (12.136b) as

$$S_R \Big|_{max} = 2\omega_u V_T = 2 \times 2\pi \times 10^6 \times 26 \times 10^{-3} = 3.27 \times 10^5 \ V/s.$$

A simple way to improve the slew rate is to put a resistor in series with each emitter of the difference amplifier. This reduces the g_m of the input-stage transistors to

$$g_m' = \frac{g_m}{1 + g_m R_E} \tag{12.137}$$

where g_m' is the effective transconductance. The transconductance decreases rapidly as R_E increases beyond $1/g_m$. If $g_m R_E = 19$, the slew rate is increased by a factor of 20. Recall that the offset voltage will increase if the emitter resistors are not well matched.

EXAMPLE 12.10   What is the slew rate of the operational amplifier of Example 12.9 when a 5-kΩ resistor is placed in series with the emitter of each difference-amplifier transistor?

Solution.   At a current of 1 μA the transconductance is

$$g_m = 10^{-6}/26 \times 10^{-3} = 38.46 \ \mu mho.$$

When a 5-kΩ resistor is in series with the emitter, the effective g_m is found from Eq. (12.137):

$$g_m' = \frac{38.46 \times 10^{-6}}{1 + 5 \times 10^3 \times 38.46 \times 10^{-6}} = 32.36 \ \mu mho.$$

From Eq. (12.136a), the slew rate is

$$S_R \Big|_{max} = \frac{2 \times 2\pi \times 10^6 \times 10^{-6}}{3.226 \times 10^{-5}} = 3.896 \times 10^5 \ V/s$$

which is a 19% increase over the case without the resistors.

The slew rate imposes an upper limit on the frequency of a sinusoidal excitation which an operational amplifier can faithfully reproduce with full output swing. The slew rate required to reproduce an output voltage $v_o = V_m \sin \omega t$ is

$$S_R \Big|_{max} = \omega V_m. \tag{12.138a}$$

Hence, for a specified slew rate the maximum excitation frequency

$$\omega_{\max} = \frac{1}{V_m} S_R \Big|_{\max} \qquad (12.138b)$$

The frequency band from dc to $\omega_{\max}$ is called the *power bandwidth* of the amplifier. Amplifier performance does not degrade gracefully when the power bandwidth is exceeded, rather the output waveform abruptly becomes triangular.

The power bandwidth of operational amplifiers is often quite modest considering their small-signal high-frequency capability. For example, the μA741 has a $V_m = 10$ V power bandwidth of 10.7 kHz corresponding to its slew rate of 0.67 V/μs.

As a final point, we now show that stray capacitance at the common-emitter node of the input difference amplifier affects the slewing behavior of the operational amplifier. The worst case occurs when the operational amplifier is used as a voltage follower because the output is tied directly to the input. This situation is illustrated in Fig. 12.49. A positive step input of sufficient magnitude cuts off transistor T_1 because its emitter voltage is initially held constant by the stray capacitance C_S. The collector of T_4 is at virtual ground because of the feedback around amplifier A. Hence, the output voltage appears across the feedback capacitor C_F and, with the exception of

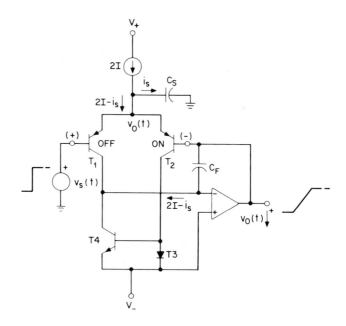

Fig. 12.49 Slew rate degradation due to stray capacitance at emitter node of difference amplifier.

the small emitter-base voltage drop, it also appears across the stray capacitance C_S. Transistor T_2 is an emitter follower that tracks the output voltage. Its collector current is $2I - i_s$ which is mirrored by the current mirror and charges the feedback capacitor C_F. Thus, the degraded slew rate S_R^* is

$$S_R^* = \frac{2I - i_s}{C_F} \approx \frac{i_s}{C_S} \approx \frac{2I}{C_S + C_F} . \tag{12.139}$$

EXAMPLE 12.11   For the μA741 operational amplifier $I = 10 \; \mu$A, $C_S \approx 4$ pF, and $C_F \approx 30$ pF. Calculate the slew rate without C_S and the degradation due to C_S.

Solution.   The slew rate without C_S is given by Eq. (12.137) as

$$S_R = \frac{2I}{C_F} = (2 \times 10^{-5})(30 \times 10^{-12}) = 6.67 \times 10^5 \text{ V/s}.$$

When C_S is taken into account, the degraded slew rate given by Eq. (12.139) is

$$S_R^* \approx \frac{2I}{(C_S + C_F)} = \frac{2 \times 10^{-5}}{34 \times 10^{-12}} = 5.88 \times 10^5 \text{ V/s},$$

which is a degradation of 11.8%.

A negative step cuts off transistor T_2 and transistor T_1 conducts the sum of the source current and the stray-capacitor charging current. Since T_2 is OFF, the current mirror is inactive, and the total current into the integrating capacitor C_F is the current coming from the collector of T_1. Therefore

$$v_o(t) = -\frac{1}{C_F} \int_0^t \left[2I + C_s \frac{dv_i}{dt} \right] d\tau$$

$$= -\frac{2I}{C_F} t - \frac{C_S}{C_F} V_i u(t) \tag{12.140}$$

where the input voltage is $v_i(t) = V_i u(t)$. The first term is the normal slewing response of the amplifier without any stray capacitance C_S and the second term is a step voltage that enhances the output response. It is proportional to the stray capacitance.

12.17 MACROMODELS OF OPERATIONAL AMPLIFIERS [12]

In the next chapter we present some system applications of operational amplifiers. Often several operational amplifiers are required, even in these simple examples. As we have already seen, the integrated circuit operational amplifier is a rather complicated circuit. One or two operational amplifiers described at the device level can severely tax the capability of large circuit simulation programs run on a digital computer. Therefore, it is desirable to have a reasonably simple model that accurately describes the terminal behavior of the operational amplifier so as to facilitate a detailed system analysis with the aid of a digital computer. At the very least, this saves time, and (perhaps huge) expense of simulating the complete system using device-level models. More importantly, these macromodels may make the analysis of complex systems possible.

The operational-amplifier macromodel shown in Fig. 12.50 accurately describes the input and output characteristics, differential- and common-mode gain and frequency response, quiescent dc conditions, offset behavior, and large-signal characteristics of the operational amplifier. The number of branches and nodes in the macromodel is about one-sixth that of the device level model of the operational amplifier. Further, the eight p-n junctions of the macromodel compares with 60 to 80 junctions in the actual amplifier. Computation time required to simulate amplifiers, timers, and filters is reduced by a factor of 6 to 10.

The macromodel has a unity voltage-gain difference amplifier as the input stage. This stage uses ideal, dissimilar transistors, a constant current

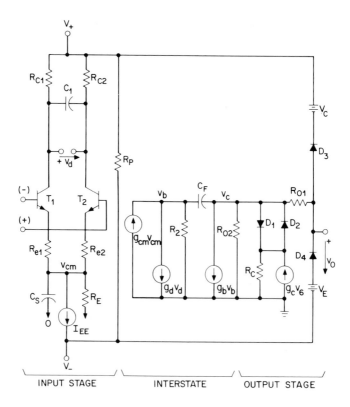

Fig. 12.50 Operational amplifier macromodel.

bias source, and resistors and capacitors. It is designed to provide the desired differential and common-mode input characteristics and the required voltage and current offsets. Capacitors C_S and C_1 affect the slew rate and phase response, respectively. The differential and common-mode gains are provided by voltage-controlled current sources and resistors described by g_{cm}, g_d, R_2, and R_{o2}. Feedback capacitor C_F provides the dominant time constant. Diodes are used in the output stage to produce the desired maximum short-circuit current and maximum voltage excursions.

The specification of the input-stage parameters begins with the determination of the collector current necessary to meet the slew rate requirements. Under quiescent conditions $I_{C1} = I_{C2} \approx I$. Equation (12.135) is used to define the negative-going slew rate,

$$I_{C1} = \frac{1}{2} C_F S_R^-. \tag{12.141}$$

If $S_R^+ > S_R^-$, n-p-n transistors should be substituted for the p-n-p input transistors. Recall from the previous section that the positive slew rate performance of an amplifier with a p-n-p transistor input stage is degraded by stray capacitance C_S. Therefore, given the positive slew rate requirement, Eq. (12.139) is solved for C_S:

$$C_S = \frac{2I}{S_R^+} - C_F. \tag{12.142}$$

The input current requirements are satisfied by choosing different values of current gain for each input transistor. In addition, the input voltage offset V_{os} is obtained by making the quiescent collector currents unequal according to the relation

$$I_{s1}^{(2)} = I_{s1}^{(1)} \exp\left(\frac{V_{os}}{V_T}\right) \approx I_{s1}^{(1)}\left[1 + \frac{V_{os}}{V_T}\right], \tag{12.143}$$

where $I_{s1}^{(2)}$ refers to the emitter-base leakage current of transistor 2 and $I_{s1}^{(1)}$ similarly refers to transistor 1. Recall that the input bias current is the average base current, and the input offset current is the difference between the base currents. Hence

$$I_{B1} = I_{\text{bias}} + \frac{I_{os}}{2},$$

$$\tag{12.144}$$

$$I_{B2} = I_{\text{bias}} - \frac{I_{os}}{2}.$$

Using this equation and Eq. (12.141) yields $\beta_1 = I_{C1}/I_{B1}$. From Eq. (12.143), $I_{C2} = I_{C1}\exp(V_{os}/V_T)$, and from Eq. (12.144) we find $\beta_2 = I_{C2}/I_{B2}$.

The unity gain open-loop cutoff frequency $\omega_u \approx A_d \omega_{3dB}$. At low frequencies, $A_d = g_d g_b R_2 R_{o2}$, and g_b is chosen equal to $1/R_{C1}$ in order to obtain

$$\omega_u = \frac{1}{R_{C1} C_F} \tag{12.145}$$

which is similar to Eq. (12.130) and must be solved for the appropriate value of R_{C1}.

Once the collector currents and current gains are known, I is found as

$$I = \left[\frac{\beta_1 + 1}{\beta_1} + \frac{\beta_2 + 1}{\beta_2} \right] I_{C1} \tag{12.146a}$$

and

$$R_E \approx \frac{V_A}{I}, \tag{12.146b}$$

where V_A is the Early voltage which is typically 200 V for the transistors used in operational amplifiers.

If $I_{C1} = I_{C2}$, then $g_{m1} = g_{m2}$. In addition, if $R_{C1} = R_{C2}$ and $R_{e1} = R_{e2}$, then

$$R_{e1} = \frac{\beta_1 + \beta_2}{2 + \beta_1 + \beta_2} \left[R_{C1} - \frac{1}{g_{m1}} \right]. \tag{12.147}$$

The excess phase, $\Delta\Phi$, at the unity gain frequency is modeled by the time constant involving capacitor C_1.

$$\Delta\Phi = \arctan\left| \frac{2C_1}{C_F} \right| \tag{12.148}$$

If phase margin $\Phi_m = 90° - \Delta\Phi$, then

$$C_1 = \frac{C_F}{2} \tan(90° - \Phi_m). \tag{12.149}$$

Analysis of the intermediate stage shows that the transconductance

$$g_{cm} = \frac{1}{CMR\ R_{C1}}, \tag{12.150}$$

where CMR is the common-mode rejection ratio.

The output stage provides low dc and ac output impedance and the output voltage and current limitations. When the amplifier is in the quiescent state at low frequency, the output resistance is

$$R_{out} = R_{o-ac} + R_{o2} \tag{12.151}$$

where R_{o-ac} is the high-frequency output resistance. The corner frequency above which the total output impedance is R_{o-ac} is

$$\omega_c = \frac{1}{R_{o2}C_F(1+g_bR_2)} \ , \tag{12.152}$$

where g_b is obtained from the low-frequency differential gain $A_d = g_d g_b R_2 R_{o2}$ as

$$g_b = \frac{A_d R_{C1}}{R_2 R_{o2}} \ . \tag{12.153}$$

Finally, the output voltage and current limitations are provided by the diode clamps.

Diodes D_1 and D_2 limit the short-circuit output current to I_{sc}. The maximum current that the voltage-controlled current source $g_b v_b$ can supply is $2I_{C1}R_2g_b$ which is about 100 A with typical component values. This current supply capability is limited to the desired short-circuit current of the operational amplifier by a proper choice of the clamp-diode leakage current, namely

$$I_{sD1} = I_{sD2} = (2I_{C1}R_2g_b - I_{sc})\exp\left[-\frac{R_{o1}I_{sc}}{V_T}\right]. \tag{12.154}$$

The controlled-current source $g_c v_o$ in shunt with R_c form a voltage reference that determines the onset of output current limiting. For proper operation, the product $R_c g_c$ should be unity and R_c should be small so that the voltage drop across it is no more than about 1% of the diode voltage V_{D1} or V_{D2}. In Eq. (12.154) the diode voltage is identified as $R_{o1}I_{sc}$. Therefore, from Eq. (12.154),

$$R_c = \frac{V_T}{100R_{o1}I_{sc}} \ln\left[\frac{2I_{C1}R_2g_b - I_{sc}}{I_{sD1}}\right]. \tag{12.155}$$

The output voltage is constrained by clamp diodes D_3 and D_4 to be in the range

$$-V_- + V_E - V_{D4} \leqslant v_o \leqslant V_+ - V_C + V_{D3}.$$

Therefore, the bias voltage V_C and V_E are given by

$$V_C = V_+ - V_{out}^+ + V_T\ln\left(\frac{I_{sc}^+}{I_{sD3}}\right) \tag{12.156}$$

and

$$V_E = V_- + V_{out}^- + V_T\ln\left(\frac{I_{sc}^-}{I_{sD4}}\right).$$

We will now compute the parameter values for an operational amplifier macromodel and compare the computed response of the model to that of the amplifier.

12.17.1 Experimental Verification of the Model

The characteristics of a representative operational amplifier are given in Table 12.2.

Table 12.2 Operational Amplifier Characteristics

C_F	30pF	$\Delta\psi$	16.8°
S_R^+	0.9 V/μs	CMR	106 dB
S_R^-	0.72 V/μs	R_{out}	566 Ω
I_{bias}	256 nA	R_{o-ac}	76.8 Ω
I_{os}	0.7 nA	I_{sc}^+	25.9 mA
V_{os}	0.299 mV	I_{sc}^-	25.9 mA
$A_{VD}(0)$	4.17×10^5	V_{out}^+	14.2 V
$A_{VD}(1\ \text{kHz})$	1.219×10^3	V_{out}^-	-12.7 V
		P_d	59.4 mW

EXAMPLE 12.12   Evaluate the parameters of the operational amplifier macromodel for the amplifier described in Table 12.2.

Solution.   From Eq. (12.141) $I_{C1} = 30\times10^{-12}\times0.90\times10^6/2 = 13.5\ \mu$A. The positive slew-rate performance is degraded by the capacitor C_S as given by Eq. (12.142). With $I \approx I_{C1}$

$$C_S = \frac{2\times13.5\times10^{-6}}{0.72\times10^6} - 30\times10^{-12} = 7.5\ \text{pF}.$$

The base currents of the input transistors are related to the bias and offset currents by Eq. (12.144), from which we find $I_{B1} = 256.35$ nA and $I_{B2} = 255.65$ nA. Knowing I_{C1} and I_{B1} permits calculation of $\beta_1 = I_{C1}/I_{B1} = 52.66$. The required voltage offset is provided by mismatching the input transistor pair. From Eq. (12.143) at room temperature and assuming that $I_{s1}^{(1)} = 8.0\times10^{-16}$ A,

$$I_{s1}^{(2)} = 8.0\times10^{-16}\exp\left[\frac{0.299}{26}\right] = 8.0925\times10^{-16}\ \text{A}.$$

Also,

$$I_{C2} = I_{C1}\exp\left[\frac{V_{os}}{V_T}\right] = 13.5\times10^{-6}\exp\left[\frac{0.299}{26}\right] = 13.656\ \mu\text{A}.$$

Therefore, $\beta_2 = I_{C2}/I_{B2} = 53.42$. We now compute the value of the input-stage emitter current source from Eq. (12.146a) as

$$I = \left[\frac{1+52.66}{52.66} + \frac{1+53.42}{53.42}\right] \times 13.5 \ \mu A = 27.509 \ \mu A$$

and therefore from Eq. (12.146b) with $V_A = 200$ V, $R_E = 7.27$ MΩ.

The unity-gain frequency of a fully compensated operational amplifier that falls off at 6 dB/octave can be determined from the product of the gain at some high frequency and the frequency of measurement, that is,

$$\omega_u = A_{VD} \times \omega = 1.219 \times 10^3 \times 2\pi \times 10^3 = 7.659 \times 10^6 \ \text{rad/s}.$$

Thus, from Eq. (12.145) $R_{C1} = 1/(7.659 \times 10^6 \times 30 \times 10^{-12}) = 4.35$ kΩ. The final resistor of the input stage is R_{e1} which is determined from Eq. (12.147)

$$R_{e1} = \frac{52.66+53.42}{2+52.66+53.42}\left[4.35 \times 10^3 - \frac{0.026}{13.5 \times 10^{-6}}\right] = 2.38 \ \text{k}\Omega.$$

Capacitor C_1 is chosen to provide the required excess phase according to Eq. (12.148) with the result $C_1 = 4.529$ pF.

The macromodel has more components whose value must be specified than there are constraints in Table 12.2, and therefore some component values can be specified arbitrarily. The interstage network design is begun by arbitrarily setting $R_2 = 100$ kΩ and $g_d = 1/R_{C1} = 229.89 \ \mu$mhos. The CMR requirement of 106 dB determines the value of g_{cm} via Eq. (12.150). The result is $g_{cm} = 1.1522$ nmhos. The dc and ac output impedances are given in Table 12.2 as $R_{out} = 566 \ \Omega$ and $R_{o-ac} = 76.8 \ \Omega$. Therefore, from Eq. (12.151), $R_{o2} = 566 - 76.8 = 489.2 \ \Omega$. It is also seen that at high frequencies R_{o2} is bypassed by the Miller-effect multiplied capacitance due to C_F. Therefore, $R_d = R_{o-ac} = 76.8 \ \Omega$. The differential mode voltage gain of 4.17×10^5, together with previously specified component values, determines $g_b = 37.08$ mhos.

Resistor R_p across the power supply provides the required power dissipation of 59.4 mW. Assuming that ± 15-V power supplies are used, $R_p = 15.363 \ k\Omega$.

Parameter values for the output stage components are determined from Eqs. (12.154) to (12.156). The leakage current for the diodes that limit the short-circuit output current is found from Eq. (12.154):

$$I_{sD1} = (2 \times 13.5 \times 10^{-6} \times 10^5 \times 37.08 - 25.9 \times 10^{-3})\exp\left[-\frac{76.8 \times 25.9 \times 10^{-3}}{26 \times 10^{-3}}\right]$$

$$= 5.9547 \times 10^{-32} \ \text{A} = I_{sD2}.$$

From Eq. (12.155), $R_c = 0.01 \ \Omega$ and because $g_c R_c = 1$, $g_c = 100$ mhos. Finally, the leakage current for diodes D_3 and D_4 are arbitrarily chosen to be 8×10^{-16} A, and therefore $V_c = 1.604$ V and $V_E = 3.104$ V.

The performance of the macromodel was compared to that of a device-level model of the operational amplifier circuit using the SPICE circuit simulation program.[9, 10] Figure 12.51 shows the slew-rate performance of a voltage follower subjected to a 10-V step input. According to simple theory the initial jump in the voltage-follower response is $\Delta v_o = (C_F/C_S)\Delta v_{in} = 2.5$ V. The device-level model provides a value of 3.4 V while the macromodel shows 3.6 V. From the curve of Fig. 12.51 it is seen that there is close agreement between the device-level and macromodel

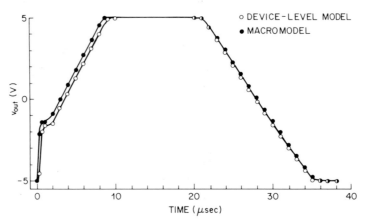

Fig. 12.51 Slew-rate performance of a voltage follower subjected to a 10-V step input. (From Coyle et al.[12])

results. However, the macromodel has only eight *p-n* junctions compared with 52 for the actual circuit. The total processor time to perform the simulation using the macromodel is less by almost a factor of 10 from that of the device-level models. This last fact, coupled with the excellent agreement between the responses, illustrates the utility of the macromodel approach for simulating complex analog circuits and systems.

REFERENCES

1. J. G. Greame, G. E. Tobey, and L. P. Huelsman, editors. *Operational Amplifiers—Design and Applications.* New York: McGraw-Hill, 1971.

2. J. Millman and C. C. Halkias. *Electronic Devices and Circuits.* New York: McGraw-Hill, 1967.

3. G. Meyer-Brotz and A. Kley. "The common mode rejection of transistor differential amplifiers." *IEEE Trans. on Circuit Theory,* **CT-13**: 171-175 (June 1966).

4. L. J. Giacoletto. *Differential Amplifiers.* New York: Wiley-Interscience, 1970.

5. R. J. Widlar. "Super-gain transistors for IC's." *IEEE J. of Solid-State Circuits* **SC-4**: 249-251 (August 1969).

6. R. C. Jaeger and G. A. Hellworth. "On the performance of the differential cascode amplifier." *IEEE J. of Solid-State Circuits* **SC-8**: 169-174 (April 1973).

7. J. E. Solomon. "The monolithic op-amp: a tutorial study." *IEEE J. of Solid-State Circuits* **SC-9**: 314-334 (December 1974).

8. J. E. Solomon, *op.cit.,* pp. 316 and 319.

9. R. J. Widlar. "Design techniques for monolithic operational amplifiers." *IEEE J. of Solid-State Circuits* **SC-4**: 184-191 (August 1969).

10. W. E. Hearn. "Fast slewing monolithic operational amplifier." *IEEE J. of Solid-State Circuits* **SC-6**: 20-24 (February 1971).

11. P. C. Davis, S. F. Mayer, and W. R. Saari. "High slew rate monolithic operational amplifier using compatible complementary P-N-P's." *IEEE J. of Solid-State Circuits* **SC-9**: 340-347 (December 1974).

12. G. R. Coyle, B. M. Cohn, D. O. Pederson, and J. E. Solomon. "Macromodeling of integrated circuit operational amplifiers." *IEEE J. of Solid-State Circuits* **SC-9**: 353-363 (December 1974).

13. R. J. Widlar. "New developments in IC voltage regulators." *IEEE J. of Solid-State Circuits* **SC-6**: 2-7 (February 1971).

14. A. P. Brokaw. "A simple three-terminal IC bandgap reference." *IEEE J. of Solid-State Circuits* **SC-9**: 388-393 (December 1974).

PROBLEMS

12.1 Design a simple bipolar difference amplifier to have a difference mode voltage gain, $A_{dd} = 40$, a common-mode rejection ratio of 100, and an input bias current of less than $10 \, \mu A$. Assume that $\beta = 250$, $r_{bb'} = 700 \, \Omega$ and $g_m r_0 = 10^4$.

12.2 What is the input offset voltage and input offset current for a bipolar difference amplifier with a $100 \, \mu A$ bias current source if $\Delta V_{BE} = 3$ mV, the tolerance on resistor ratios is 5% and $\beta = 100 \pm 20\%$? What is the drift in these parameters?

12.3 An ideal difference amplifier with current source bias is subject to a large common-mode input signal. Two power supplies, whose values are V_+ and V_-, are used to enable the output to be nominally at 0 V. For *n-p-n* transistors, the differential-mode gain is limited by the maximum positive value V_{CM}, of the common-mode voltage. Derive an expression for the maximum value of A_{dd}.

12.4 Find the differential-mode input resistance of a difference amplifier with Darlington-connected input transistors.

12.5 The input bias current of a bipolar difference amplifier can be reduced by using controlled current sources to inject a current proportional to the base current into the input transistors as shown in Fig. P12.5. Do a simple analysis of this circuit to determine the input bias current in terms of the current gain of the transistors. Does this current cancellation scheme affect the input offset current?

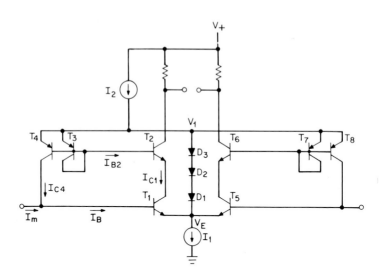

Fig. P12.5

12.6   Show that n related current sources can be obtained by connecting n source transistors across the reference transistor similar to the arrangement shown in Fig. 12.14.

Derive expressions for the output currents I_j, I_k of the jth and kth transistors and show how they are related to each other and also to the reference current I_{ref} of the reference transistor. How do the transistor current gains affect these currents? What is a practical limit to the value of n?

12.7   Show that the output resistance of the three-transistor current source shown in Fig. 12.17 is

$$R_{out} = \frac{\beta_2}{2} r_o$$

where r_o is the common-emitter output resistance of T_2 and β_2 is its gain.

12.8   A multiple current source of the type shown in Fig. P12.8 with $I_3 = 2I_2 = 4I_1$ is useful in the realization of D/A converters. The emitters are all of identical area.

a)   Analyze the circuit to obtain relationships between the controlled currents I_1, I_2, I_3, and the reference current I.

b)   Obtain expressions for the ratios of controlled currents.

c)   Design a binary weighted current source with $I_3 = 1$ mA.

d)   Are there any limits to the number of controlled currents that can be derived from one reference?

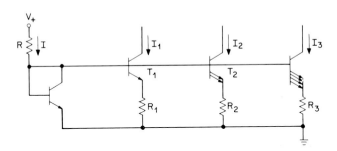

Fig. P12.8

12.9 Treat the cascode amplifier of Fig. 12.34 as a compound transistor and compute its small-signal hybrid-*pi* parameters. Use the following transistor parameters: Ordinary *n-p-n* transistors $\beta = 100$, $g_o = \eta g_m$, $\eta = 1.5 \times 10^{-4}$; superbeta transistors, $\beta = 1000$, $\eta = 1 \times 10^{-4}$. All collector currents are $I_C = 5 \times 10^{-6}$ A.

12.10 A Class A emitter follower has a peak value of output voltage $\hat{V}_o$, where $\hat{V}_o < |V_-|$. Let the peak value of load current corresponding to the most negative value of output voltage, namely $V_o = -\hat{V}_o$, be denoted by $\hat{I}_o$. Show that the quiescent value of base current, I_{BQ}, must obey the inequality

$$I_{BQ} > V_- \hat{I}_o / (\hat{V}_o + V_-).$$

12.11 Design the Class AB output stage shown in Fig. 12.37 to be capable of developing 20 V peak to peak across a 1 kΩ load. Calculate also the maximum power that can be dissipated in the output transistor. Assume that the source resistance of the driving stage is 1 kΩ. The stabilization factor $S = \partial I_C / \partial I_{co}$ is given by

$$S = \frac{(1+\beta)[1+(R_S/R_E)]}{1 + \beta + (R_S/R_E)}.$$

Assume that $\beta = 100$ and that the stabilization factor is to be less than 10. Determine a suitable value of R_E and supply voltage V_+ and V_-.

12.12 An operational amplifier in a TO-5 type metal can with $K_T = 0.3°C/W$ dissipates 200 mW. The output voltage is 10 V peak and the supply voltages are ±12 V. What is the maximum usable gain for this operational amplifier?

12.13 Show that the composite *n-p-n p-n-*p transistor shown in Fig. P12.13 behaves like a *p-n-*p transistor. Derive a set of small-signal parameters (such as the hybrid-*pi* or *h*-parameters) to show that the current gain for the structure is $\beta_1 \alpha_2$ and discuss the frequency response.

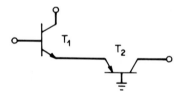

Fig. P12.13

12.14 Analyze the "composite transistor" difference amplifier shown in Fig. P12.14. In particular, show that:

a)   The bias current source I_o and the p-n-p current gain set the current level in the amplifier.

b)   The n-p-n current gain determines the input bias current and any mismatch determines the input offset current.

c)   Mismatched emitter junctions produce an input offset voltage.

d)   Mismatched p-n-p current gains produce an unbalanced collector current and therefore an input offset voltage.

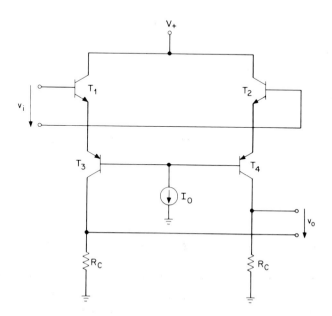

Fig. P12.14

12.15 Determine the macromodel parameters for the operational amplifier whose performance characteristics are listed in the following table.

C_C	5 pF	A_v (1 kHz)	1.6×10^4
S_R^+	100 V/μs	$\Delta\phi$	40°
S_R^-	71 V/μs	CMR	100 dB
I_B	120 nA	R_{out}	75 Ω
I_{os}	6 nA	I_{SC}^+	25 mA
V_{os}	2 mV	I_{SC}^-	25 mA
$A_v(0)$ 2×10^5	V^+	13 V	
		V^-	-13 V

13

Analog Systems

This chapter deals with applications of operational amplifiers and other analog integrated circuits to the realization of analog system functions. Topics such as signal conditioning, filtering, analog-to-digital, and digital-to-analog conversion, and the phase-locked loop (which can be used for FM modulation and detection) will be discussed.

13.1 SIGNAL CONDITIONING WITH THE OPERATIONAL AMPLIFIER [1]

The simplest application of the operational amplifier is the inverting amplifier with gain $-G$ shown in Fig. 13.1. For the time being consider that the amplifier has infinite bandwidth, infinite input impedance, and zero output impedance. The expression for closed-loop gain is found by applying

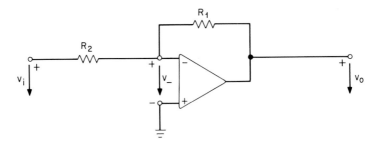

Fig. 13.1 An inverting amplifier with gain $G = -R_1/R_2$.

Kirchhoff's current law to the inverting input node of the operational amplifier and assuming that the input current of the operational amplifier is zero. Thus

$$\frac{v_i - v_-}{R_2} + \frac{v_o - v_-}{R_1} = 0. \tag{13.1a}$$

After collecting terms we find that

$$v_i - \left[1 + \frac{R_2}{R_1}\right]v_- = -\frac{R_2}{R_1}v_o. \tag{13.1b}$$

The constraint relation for the operational amplifier is

$$v_o = A(v_+ - v_-) \tag{13.2}$$

where v_+ is the voltage from the noninverting input terminal of the operational amplifier to ground, which is zero in this particular case.

Substituting Eq. (13.2) in Eq. (13.1b) and solving for v_o/v_i yields

$$G = \frac{v_o}{v_i} = -\frac{AR_1}{R_1 + (1+A)R_2} = \frac{1 - (1/\beta)}{1 + (1/A\beta)} \tag{13.3}$$

where the *feedback factor* $\beta = R_2/(R_1 + R_2)$ and where $A\beta$ is called the *loop gain*. Note that when the *open-loop gain* A of the operational amplifier becomes infinite the *closed-loop gain* G becomes

$$G = \frac{v_o}{v_i} = -\frac{R_1}{R_2} \tag{13.3a}$$

and $(v_+ - v_-) \rightarrow 0$, by virtue of v_o being finite and A infinite. In the present configuration this implies that $v_- \rightarrow 0$ and that input is said to be a *virtual ground*. Because of the negative feedback, the closed-loop gain G depends only on the ratio of passive components R_1 and R_2. When the gain is high but finite, it can be shown that

$$G \approx -\frac{R_1}{R_2}\left\{1 - \frac{1}{A}\left[1 + \frac{R_1}{R_2}\right]\right\} = \left\{1 - \frac{1}{\beta}\right\}\left\{1 - \frac{1}{A\beta}\right\} \tag{13.3b}$$

which is not independent of the open-loop gain A. However, the correction term becomes very small as A increases.

Negative feedback reduces the sensitivity of the closed-loop gain G to variations in the open-loop gain A of the amplifier. This beneficial effect can be seen by differentiating Eq. (13.3) with respect to A to yield (after some algebra)

$$\frac{dG}{G} = \frac{R_1 + R_2}{R_1 + (1+A)R_2}\frac{dA}{A} = \frac{1}{1+\beta A}\frac{dA}{A}. \tag{13.4}$$

EXAMPLE 13.1   Find the closed-loop gain and gain sensitivity of an inverting amplifier with $R_1 = 25$ kΩ and $R_2 = 2.5$ kΩ if the open-loop gain of the operational amplifier is (a) 100; (b) 1000; (c) 10,000; and (d) 100,000.

Solution.   The feedback factor $\beta = R_2/(R_1+R_2) = 2.5/(25+2.5) = 9.091\times10^{-2}$. From Eq. (13.3), $G = -10/[1 + (11/A)]$, and the gain sensitivity is obtained from Eq. (13.4). Evaluating these expressions yields

a) $G = -9.009$; $dG/G = 9.91\times10^{-2}$ dA/A

b) $G = -9.891$; $dG/G = 1.088\times10^{-2}$ dA/A

c) $G = -9.989$; $dG/G = 1.099\times10^{-3}$ dA/A

d) $G = -9.9989$; $dG/G = 1.099\times10^{-4}$ dA/A

Thus, we see that as the open-loop gain of the operational amplifier becomes very large, the closed-loop gain approaches the ideal value of -10 and it becomes more stable with respect to changes in open-loop gain. From part (d) if the open-loop gain doubles, $dA/A = 1$ and dG/G is about 0.01%. In this case the negative feedback caused a ten-thousandfold decrease in both gain and sensitivity to variations in the *active elements.*

13.1.1 Effect of Finite Input and Nonzero Output Impedance

In the above derivations it was assumed that the operational amplifier had infinite input impedance and zero output impedance. The effect of nonideal impedances at the amplifier ports will now be examined. The situation is modeled by the equivalent circuit of Fig. 13.2. Defining R_I as the parallel combination of r_i, R_1, and R_2, and R_O as the parallel combination of r_o, R_1, and R_L, and writing the Kirchhoff current law equations at the input and output nodes of the amplifier yields

$$G = \frac{-\dfrac{R_1}{R_2}\left[A - \dfrac{r_o}{R_1}\right]}{A + \dfrac{R_1 r_o}{R_I R_O} - \dfrac{r_o}{R_1}} . \qquad (13.5)$$

Note that if $r_o \rightarrow 0$ and $r_i \rightarrow \infty$, Eq. (13.5) reduces to Eq. (13.3). To put the effect of r_o and r_i in evidence write

$$\frac{R_1}{R_I} = \frac{1}{\beta} + \frac{R_1}{r_i} \qquad (13.6a)$$

and define R_L' as $R_L \| R_1$, so that

$$\frac{R_O}{r_o} = \frac{R_L'}{R_L' + r_o} \triangleq \alpha. \qquad (13.6b)$$

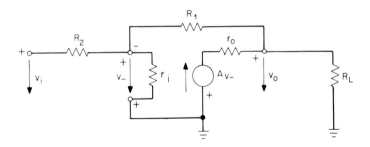

Fig. 13.2 Equivalent circuit of the inverting amplifier shown in Fig. 13.1.

Substituting Eq. (13.6) into Eq. (13.5) yields

$$G \approx \frac{1 - (1/\beta)}{1 + \dfrac{1}{A\alpha}\left[\dfrac{r_i+\beta R_1}{\beta r_i}\right]} . \qquad (13.7)$$

Comparing Eq. (13.7) to Eq. (13.3) it is seen that the nonzero output impedance of the operational amplifier effectively reduces the amplifier gain by the factor α. This factor can be neglected when $r_o \ll R_L'$. Similarly, it is seen that if r_i is finite and not much greater than βR_1 the loop gain is reduced by the divider action between r_i and βR_1. Thus nonzero output impedance and finite input impedance both reduce the loop gain and result in a decrease in closed-loop gain G.

It can be shown (Problem 13.1) that the closed-loop output impedance is

$$Z_o \approx \frac{r_o}{A\beta-1} \qquad (13.8)$$

and the input impedance is $Z_i \approx R_2$.

EXAMPLE 13.2   (a) For the inverting amplifier discussed in Example 13.1 driving a load resistance $R_L = 1\ \mathrm{k\Omega}$, find the closed loop gain if the amplifier has a gain $A = 1000$, and output impedance $r_o = 500\ \Omega$, and an input impedance $r_i = 10\ \mathrm{k\Omega}$. (b) Find the output impedance.

Solution.

a) $R_I = r_i \| R_1 \| R_2 = 1.852\ \mathrm{k\Omega}; \quad R_O = r_o \| R_1 \| R_L = 329\ \Omega.$ From   Eq. (13.6), $\alpha = 0.658$, and from Eq. (13.7),

$$G = \frac{1-11}{1 + \dfrac{1}{0.658\times1000}\left[\dfrac{10^4 + 1.852\times10^3/11}{10^4/11}\right]} = -9.833.$$

b)   From Eq. (13.8), $Z_o = 500/(90.9-1) = 5.56\ \Omega$. Note that the nonzero output impedance and the finite input impedance resulted in a further decrease in closed-loop gain. Also, the negative feedback greatly reduced the output impedance.

13.1.2 Frequency Response of Amplifier

Actual operational amplifiers do not have infinite bandwidth. It can be shown that if the closed-loop response is stable the gain of the operational amplifier cannot fall by more than 12 dB/octave until the gain is less than unity. In this section the amplifier frequency response is assumed to be governed by a single time constant

$$A(f) = \frac{A}{1+jf/f_c}\ , \tag{13.9}$$

where f_c is the cutoff frequency, i.e., the frequency at which the amplitude has fallen 3 dB. Substituting this expression in Eq. (13.3) yields

$$G(f) \approx \frac{G}{1 + j\ \dfrac{f}{(1+A\beta)f_c}} \tag{13.10}$$

which shows that the 3-dB point of the closed-loop gain G is $(1+A\beta)$ times the 3-dB point of the operational amplifier gain A. Thus, the gain-bandwidth product of the amplifier is constant but, through the use of negative feedback, gain is traded for increased bandwidth and reduced sensitivity to active element variation.

13.1.3 Noninverting and Differential Amplifiers

The circuit of a noninverting feedback amplifier is shown in Fig. 13.3. The

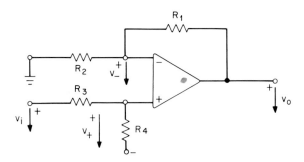

Fig. 13.3 A noninverting feedback amplifier.

constraint equations are:

$$v_+ = \frac{R_4}{R_3 + R_4} v_i$$

$$v_- = \frac{R_2}{R_1 + R_2} v_o$$

$$v_o = A(v_+ - v_-). \tag{13.11}$$

From these equations it immediately follows that

$$G = \frac{AR_4/(R_3 + R_4)}{1 + [AR_2/(R_1 + R_2)]} = \frac{\beta'/\beta}{1 + (1/A\beta)} \tag{13.12}$$

where $\beta' = R_4/(R_3 + R_4)$.

Note that the loop gain $A\beta$, is the same as for the inverting amplifier and therefore the stability is determined only by the elements in the feedback loop. The scaling feature β'/β is a function of all four impedances.

An interesting special case of the noninverting amplifier is the so-called positive follower for which $R_1 = R_3 = 0$, $R_2 = R_4 = \infty$. Under these conditions the gain becomes

$$G = \frac{1}{1 + (1/A)} \approx 1 - \frac{1}{A} \to 1. \tag{13.12a}$$

Other interesting features of this configuration include wide bandwidth, high-input impedance, and low-output impedance.

A differential amplifier using only one operational amplifier is shown in Fig. 13.4. Using the same analysis techniques as for the previous cases, it can be shown that the output voltage is given by

$$v_o = \frac{R_4(R_1 + R_2) - R_1(R_3 + R_4)}{R_2(R_3 + R_4)} v_{cm}$$

$$- \frac{R_1}{R_2} v_1 + \frac{R_4}{R_3} \cdot \frac{1 + (R_1/R_2)}{1 + (R_4/R_3)} v_2. \tag{13.13}$$

If $(R_1/R_2) = (R_4/R_3)$, this equation reduces to $v_o = (R_1/R_2)(v_2 - v_1)$. The value of these resistor ratios determines the amplifier gain, whereas the accuracy of the match between the ratios determines the common mode rejection. It must also be noted that the effective value of the input resistors R_2 and R_3 includes the internal impedance of the driving sources. The resistors R_2 and R_3 should be much larger than the nominal value of source impedance, and the variability in the source impedance will affect the common mode rejection of this amplifier configuration. In addition, the finite input impedance of the operational amplifier and its common mode rejection ratio also impose limits on the overall common mode rejection. The accu-

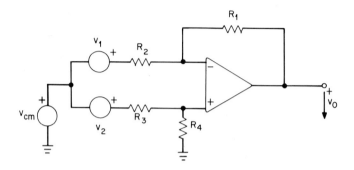

Fig. 13.4 A differential amplifier using only one operational amplifier.

racy of the closed-loop gain is also limited by the open-loop gain of the operational amplifier.

More elaborate, multiamplifier configurations are available which exhibit better common mode performance, improved gain accuracy, and higher input impedance.[2]

13.1.4 Integrators

An inverting integrator which includes the effect of operational amplifier offset voltage V_{os} and bias current I_B, is shown in Fig. 13.5. The analysis follows directly from the Kirchhoff current law equation at node a, namely, $i_1 = i_2 + I_B$. The output voltage becomes

$$v_o = -\frac{1}{RC} \int_{-\infty}^{T} v_i \, dt + \frac{1}{RC} \int_{-\infty}^{T} V_{os} \, dt + \frac{1}{C} \int_{-\infty}^{T} I_B \, dt + V_{os}, \qquad (13.14)$$

where $V_{os} \approx v_a$.

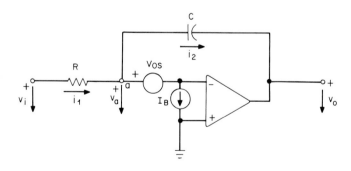

Fig. 13.5 An inverting integrator.

The first term is the desired result whereas the remaining terms are errors. The integral error terms are the most troublesome because they increase continually with time. They set an upper limit on the permissible integration time because eventually they will drive the operational amplifier to saturation. The bias current error can be reduced by increasing the integrator capacitance but this requires a reduced resistance R for the same time constant. Reducing R implies an increased input current from the driving circuitry. Balancing the input circuit by inserting a resistance R from the positive amplifier input to ground reduces the error current from the total bias current to the amplifier offset current.

When the finite gain and bandwidth $\omega_c = 1/\tau_0$ of the operational amplifier are considered, the following transfer function is obtained

$$\frac{v_o(s)}{v_i(s)} = - \frac{A_o/RC\tau_0}{\left[s + \dfrac{1}{A_oRC}\right]\left[s + \dfrac{A_o}{\tau_0}\right]} , \tag{13.15}$$

where the dc gain of the operational amplifier $A_o \gg 1$ and $A_oRC \gg \tau_0$. An ideal integrator has a single pole at the origin whereas this function has two poles on the negative real axis. At low frequency the real integrator deviates from the ideal because of the finite gain A_o. The deviation at high frequency is due to the finite bandwidth of the operational amplifier given by the amplifier pole at $1/\tau_0$.

The time domain response of the integrator to a step voltage excitation $Vu(t)$ is:

$$v_o(t) = - A_o V \left\{ 1 - \frac{\exp(-t/A_oRC)}{1 - (\tau_0/A_o^2RC)} + \frac{\exp(-tA_o/\tau_0)}{(A_o^2RC/\tau_0) - 1} \right\} u(t). \tag{13.16a}$$

An approximate expression for $v_o(t)$ valid for small values of t is found by expanding the exponentials in a Maclaurin series

$$v_o(t) \approx - \frac{V}{RC} \left\{ t + \frac{\tau_0}{A_o} \left[-1 + \exp\left(- \frac{tA_o}{\tau_0} \right) \right] \right\} u(t). \tag{13.16b}$$

For small values of time the principal error is a time-delay term due to the finite bandwidth of the operational amplifier. For large values of time the output voltage is approximately

$$v_o(t) \approx - A_o V \left[1 - \exp\left(- \frac{t}{A_oRC} \right) \right] u(t). \tag{13.16c}$$

The time response of the integrator is shown in Fig. 13.6. For accurate integration the process should be terminated at a time much less than A_oRC and at an amplitude much less than A_oV.

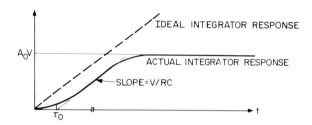

Fig. 13.6 Time response of the inverting integrator shown in Fig. 13.5.

EXAMPLE 13.3   An operational amplifier with an open loop gain $A_o = 1000$ and a cutoff frequency $f_c = 159$ Hz is used in an integrator whose RC time constant is 100 ms. What is the response of this integrator to an input voltage $v_i = 5u(t)$?

Solution. $\tau_o = 1/2\pi f_c = 1$ ms.

$$v_o = -1000 \times 5 \left[1 - \frac{\exp(-t/10^3 \times 10^{-1})}{1 - (10^{-3}/10^6 \times 10^{-1})} + \frac{\exp(10^3 t/10^{-3})}{(10^6 \times 10^{-1}/10^{-3}) - 1} \right] u(t)$$

$$= -5000[1 - \exp(-10^{-2}t) + 10^{-8}\exp(-10^6 t)]u(t)$$

For small values of time

$$v_o \approx -50\{t + 10^{-6}[\exp(-10^6 t) - 1]\}u(t)$$

13.2 FREQUENCY SELECTIVITY—THE BIQUAD [3]

In many communications applications it is necessary to contain signals within prescribed parts of the frequency spectrum. If this filtering process is accomplished with lumped finite circuit elements, the resulting filter transfer function is the ratio of two polynomials in the complex frequency variable s. High-order transfer functions are rarely synthesized directly because of extreme sensitivity of the response to variations in the values of circuit elements.[4] A way of realizing a high-order transfer function with low sensitivity to element variations is to decompose the transfer function into a product (or sum) of biquadratic factors

$$\frac{V_{\text{out}}(s)}{V_{\text{in}}(s)} = \frac{ms^2 + cs + d}{s^2 + as + b}. \tag{13.17}$$

An active *RC* structure which can realize any voltage transfer function of the above kind will now be described. The structure can realize complex poles and zeros without the presence of inductors because it contains active elements, in this case operational amplifiers. These operational amplifiers also make the individual sections self-isolating, so that a high-order transfer function can be realized as a cascade of these second-order sections. Furthermore, the design of each of these second-order sections has the important property for integrated circuit implementation that the coefficients a, b, c, d, and m can be independently adjusted by selecting the values of certain resistors. The topology of the network remains constant in every case. This is highly desirable in an integrated circuit realization, because the same structure will implement a wide range of transfer functions. The particular structure we refer to is shown in Fig. 13.7.

The circuitry enclosed by the dotted lines is the dynamic part of the system. The overall feedback loop encloses a lossy integrator as stage one, a simple inverting integrator in the second stage, and a unity gain inverting amplifier as stage three. Analysis of this structure shows that the voltage transfer function V_1/V_{in} is given by

$$\frac{V_1(s)}{V_{in}(s)} = -\frac{cs}{s^2+as+b} \tag{13.18}$$

which is a simple bandpass characteristic and that V_2/V_{in} is

$$\frac{V_2(s)}{V_{in}(s)} = -\frac{d}{s^2+as+b} \tag{13.19}$$

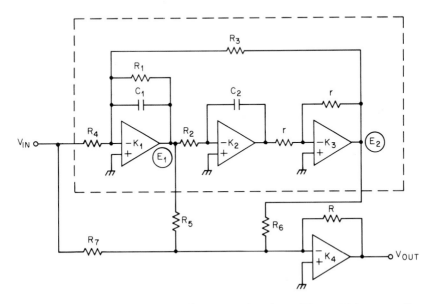

Fig. 13.7 A biquad structure using four operational amplifiers which can realize any biquadratic transfer function.

which is a simple lowpass characteristic. The fourth and final stage is a scaling and summing amplifier, whose output is a linear combination of the input $V_{in}(s)$ and the above two outputs $V_1(s)$ and $V_2(s)$. It is left as an exercise for the reader (Problem 13.5) to derive the above transfer functions and to show that the sum is actually a biquadratic.

Complex pole (and zero) pairs are characterized by an undamped resonant frequency, $\omega_p(\omega_z)$ and a quality factor $Q_{po}(Q_{zo})$. The original biquadratic function, Eq. (13.17), can therefore be rewritten as

$$\frac{V_{out}(s)}{V_{in}(s)} = m \, \frac{s^2 + \left(\dfrac{\omega_z}{Q_{zo}}\right)s + \omega_z^2}{s^2 + \left(\dfrac{\omega_p}{Q_{po}}\right)s + \omega_p^2} \, . \tag{13.20}$$

Expressions are now presented for determining the network resistor values in terms of each of these parameters.

$$\omega_p^2 = \frac{1}{R_2 R_3 C_1 C_2} \tag{13.21}$$

For minimum sensitivity of ω_p with respect to passive element variations choose

$$\omega_p = \frac{1}{R_3 C_1} = \frac{1}{R_2 C_2} \, . \tag{13.22}$$

It is reasonable to let $C_1 = C_2 = C$ and then $R_2 = R_3$. Small adjustments of the pole frequency ω_p are carried out by varying R_3, which takes the operating point slightly away from the sensitivity minima. For the choice of equal time constants [Eq. (13.22)]

$$Q_{po} = \frac{R_1}{R_3} \, . \tag{13.23}$$

Since R_3 is already used to determine ω_p, R_1 must be used to independently adjust Q_{po}.

For the case when $\omega_z < \omega_p$ and $\omega_z/Q_{zo} < \omega_p/Q_{po}$,

$$\frac{\omega_z}{Q_{zo}} \frac{Q_{po}}{\omega_p} = 1 - \frac{R_1 R_7}{R_4 R_5} \tag{13.24a}$$

and

$$\left(\frac{\omega_z}{\omega_p}\right)^2 = 1 - \frac{R_3}{R_4} \frac{R_7}{R_6} \, . \tag{13.24b}$$

If either $\omega_z \geqslant \omega_p$ or $\omega_z/Q_{zo} \geqslant \omega_p/Q_{po}$, then the summing amplifier K_4 must be provided with a noninverting input to which R_5 and/or R_6 is connected.

Recall that R_1 and R_3 have already been adjusted and observe that R_4 and R_7 are common to both these expressions. Hence the zero or notch frequency ω_z is adjusted by means of R_6 and the notch attenuation factor Q_{zo} is adjusted by means of R_5. The overall gain is adjusted by means of R_8. In the first two tuning steps R_1 and R_3 are adjusted by observing the response at V_1, and in the second two tuning steps V_{out} is observed. When carried out in the stated sequence R_3, R_1, R_6, and R_5, these tuning operations are mutually independent.

EXAMPLE 13.4   Determine element values for a biquad section with the transfer function

$$\frac{v_{out}(s)}{v_{in}(s)} = \frac{s^2+50s+10^6}{s^2+80s+4\times10^6} .$$

Solution. Comparing the given transfer function with Eq. (13.20) we identify $\omega_z = 10^3$; $Q_{zo} = 10^3/50 = 20$, $\omega_p = 2\times10^3$; $Q_{po} = 2\times10^3/80 = 25$. From Eq. (13.22) $\omega_p = 1/RC = 2\times10^3$. Arbitrarily choosing $R_8 = R_2 = R_3 = 1\,\text{k}\Omega$ yields the value $0.5\,\mu\text{F}$ for C, C_1, and C_2. We now set $Q_{po} = 25 = R_1/R_3$ and solving for $R_1 = 25\,\text{k}\Omega$. Now the conditions $\omega_z < \omega_p$ and $\omega_z/Q_{zo} < \omega_p/Q_{po}$ are satisfied so the circuit of Fig. 13.7 can be used unaltered. From Eqs. (13.24a) and (13.24b) we find that $R_6 = 1.33R_7/R_4$ and $R_5 = 2.67R_7/R_4 = 2R_6$. Without any other constraints set $R_4 = R_7 = 1\,\text{k}\Omega$, and therefore $R_5 = 2.67\,\text{k}\Omega$ and $R_6 = 1.33\,\text{k}\Omega$.

The above derivations have assumed an ideal operational amplifier. In actuality, the gain A of an operational amplifier is frequency dependent and is usually assumed to obey a single time constant relationship

$$A(\omega) = \frac{A}{1+(s/\omega_c)}$$

where ω_c is the radian cutoff frequency of the operational amplifier and A is its dc gain. When this type of amplifier is used in the biquad the pole Q may be shown to be

$$Q_p = \frac{Q_{po}}{1 + \dfrac{2Q_{po}}{A\omega_c}(\omega_c-2\omega_p)} \qquad (13.25)$$

where Q_{po} is the low frequency value of the pole Q.

The upper frequency limits for the pole pair realized by the biquad of Fig. 13.7 occurs when the denominator of this expression vanishes causing Q_p to become infinite. This upper frequency limit is

$$f_p\bigg|_{max} = \frac{f_c}{2}\left[1 + \frac{A}{2Q_{po}}\right]. \qquad (13.26)$$

Typically for a $\mu A741$ amplifier $A = 33,000$ and f_c is 1.2 kHz. Assuming a desired pole Q of $Q_{po} = 100$ results in $f_{pmax} \approx 100$ kHz. Note that at this frequency Q_p becomes infinite and the circuit becomes unstable. Hence it is concluded that with presently available operational amplifiers the use of the biquad in realizing high Q pole pairs is restricted to the frequency range below 100 kHz. In practice there are simpler structures that will realize pole pairs with Q's of less than 15, while the structure of Fig. 13.7 is useful in the Q range from 15 to 150. For higher Q's it is convenient to use crystal filters. It has been shown[3] that the sensitivity of the natural modes of this biquad structure to its passive elements is comparable to the passive LC case. The operational amplifiers are silicon monolithic integrated circuits, while the resistors and capacitors are made using either thick- or thin-film circuits. It is practical to put the necessary thin-film resistors on a silicon chip containing several operational amplifiers. For very low-frequency versions, it may be necessary to use appliqued capacitors in a hybrid film circuit realization, since there is a physical limitation to the size of capacitor that can be integrated.

13.3 NONLINEAR AMPLIFICATION [1,19]

There are many applications in which it is desirable to alter the dynamic range of a signal by means of an amplitude-dependent gain function. Some examples are the automatic gain controls in radio receivers, companding (i.e., amplitude compression and expansion) used in telephone transmission, function generation for analog computation, and instrumentation applications.

A logarithmic amplifier is presented in this section as indicative of a simple nonlinear function. The circuit for this amplifier is shown in Fig. 13.8.

Sah[20] did a detailed evaluation of the logarithmic relationship between collector current and forward bias base-emitter voltage for silicon planar

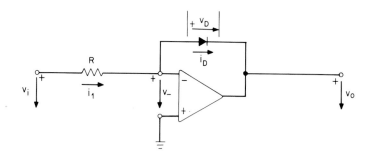

Fig. 13.8 A logarithmic amplifier based on the nonlinear voltage-current characteristic of a *p-n* junction.

bipolar transistors. He found that when the base-collector voltage is zero

$$I_c = I_s\left[\exp\left(\frac{V_{BE}}{V_T}\right) - 1\right]$$ (13.27a)

where I_s is the saturation current of the collector-base junction and is in the 10^{-14} A range. This relation holds for about nine to ten decades from 10^{-12} A to about 10^{-3} A. Thus for $V_{BE} \geqslant 100$ mV

$$V_{BE} = V_T\ln\left(\frac{I_c}{I_s}\right).$$ (13.27b)

Assume an ideal operational amplifier of gain A,

$$i_1 = \frac{v_i - v_-}{R} = I_c.$$ (13.27c)

In this circuit $v_o = -Av_-$ so that

$$v_D = v_- - v_o = (1+A)v_-.$$ (13.27d)

If the amplifier gain is very high $i_1 \approx v_i/R$, and therefore from Eqs. (13.27)

$$v_o \approx - V_T\left[\ln\left(\frac{v_i}{R}\right) - \ln I_s\right].$$ (13.28)

Apart from the temperature-sensitive scale factor V_T and offset term $V_T\ln I_s$, the output is the logarithm of the input over about ten orders of magnitude for which the collector current follows the exponential law. The temperature dependence can be compensated by means of a differential amplifier connection and a temperature-sensitive resistor network in the feedback loop.[19]

The dynamic range of this type of logarithmic amplifier is limited by the offset voltage of the operational amplifier. This can be seen by noting that if the input signal is restricted to 10 V and if the maximum feedback diode current is to be 1 mA, then $R = 10$ kΩ. If the amplifier offset voltage is 1 mV, then for 1% accuracy the input signal must be greater than 100 mV. Hence the dynamic range is just 40 dB.

13.4 COMPARATORS [5]

A comparator is a nonlinear analog system element that provides a digital (true-false) output indication when a prescribed amplitude condition is satisfied by the analog input signal. The transfer characteristic of an ideal zero crossing detector is discontinuous at $v_i = 0$ because infinite gain is required to change the output indication with no change in input voltage. Actual comparators only approach the abrupt transition of the ideal characteristic. When the transition region of the transfer characteristic has a finite slope, the comparator has a range of input values for which the output is uncertain, i.e., not in one of the two distinct output states. This uncertainty in the input value at the output-state transition limits the accuracy of the comparator.

It will now be shown that the accuracy of a bipolar transistor comparator can be improved by placing an amplifier ahead of the nonlinear element. Assume that the comparator consists of an amplifier with gain A followed by a zero crossing detector as shown in Fig. 13.9(a). The transfer characteristic of the detector is shown in Fig. 13.9(b) and it has a region of uncertainty of

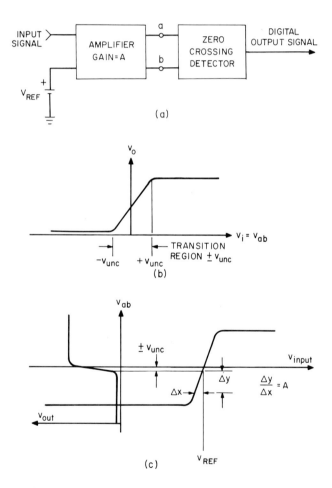

Fig. 13.9 (a) Comparator consisting of an amplifier with gain A followed by a zero crossing detector; (b) transfer characteristic of the zero crossing detector; (c) derivation of the overall transfer characteristic.

width $\pm V_{unc}$ with respect to the input variable v_i. The added amplifier serves two purposes, namely:

1.   Its gain A translates a change in its output of $\pm V_{unc}$ to a change of $\pm V_{unc}/A$ at its input.

2.   The comparison point of V_{ref} at its input is translated to zero at its output.

Figure 13.9(c) shows on the right-hand side the characteristic curve of the amplifier, whose gain is A. Note, this curve is offset by V_{ref}. On the left-hand side is the characteristic curve of the zero crossing detector, together with its region of uncertainty in input variable v_{ab} of $\pm V_{unc}$. Note that $A = 2V_{unc}/v_{in}$ or $v_{in} = 2V_{unc}/A$, showing the desired reduction in input uncertainty.

Note further that the amplifier should be a differential amplifier with a high common mode rejection ratio, so that its output will be independent of the value of the reference voltage V_{ref}. Clearly V_{ref} must be within the specified common mode range of the amplifier.

13.5 ANALOG-TO-DIGITAL CONVERSION [6]

There are many methods for performing analog-to-digital conversion. Two methods are described in this section. They are the dual-slope integration method and delta modulation.

13.5.1 Dual-Slope Integration Analog-Digital Converter

The dual-slope integration method involves integrating the analog input for a fixed time T_1 followed by integration toward a fixed reference voltage V_R for an unknown time T_2. It will be shown that the time T_2 is proportional to the analog input voltage to be converted and that the result is independent of the component values R and C.

Neglecting offset voltage and current, the output voltage of an inverting operational integrator is given by [see Eq. (13.14)]

$$v_o = -\frac{1}{RC} \int_0^{T_1} v \, dt = -\frac{\bar{v}}{RC} T_1, \qquad (13.29)$$

where T_1 is the fixed integration time for the input signal, and $\bar{v}$ is the average value for the input signal over the interval $0 \leqslant t < T_1$. During the time interval $T_1 \leqslant t < T_2$ the input is a constant V_R. The time T_2 is chosen such that $v_o(T_2) \equiv 0$. Thus

$$v_o(T_2) \equiv 0 = -\frac{1}{RC} \int_{T_1}^{T_1+T_2} V_R \, dt + v_o(T_1)$$

$$= -\frac{\bar{v}}{RC} T_1 - \frac{V_R}{RC} T_2. \qquad (13.30a)$$

Solving this equation for T_2 yields

$$T_2 = -\frac{T_1}{V_R}\, \bar{v}_{\text{in}}, \tag{13.30b}$$

which is greater than zero, for $\bar{v}_{\text{in}} < 0$.

A circuit that performs these operations is shown in Fig. 13.10. The comparator is used as a zero crossing detector on the output of the integra-

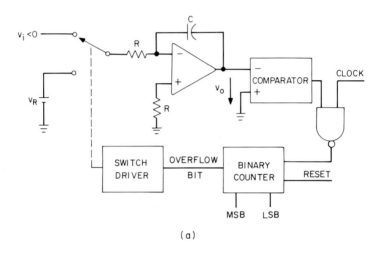

(a)

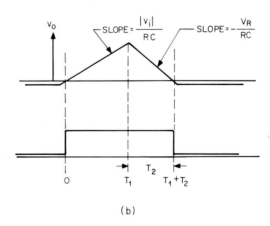

(b)

Fig. 13.10 (a) Block diagram of a dual-slope integration A/D converter; (b) integrator and comparator outputs.

tor. When the integrator output goes positive the comparator enables the counter and the interval $(0, T_1)$ begins. During this interval the counter counts clock pulses until it overflows at a time defined as T_1. At time T_1 the input of the integrator is switched to the reference voltage V_R and the second time interval from T_1 to $(T_1 + T_2)$ begins. The number stored in the counter when it is disabled by the comparator as the integrator output passes through zero, is a digital code representation of the input signal, which has an error in the range -1 to 0 times the least significant bit weight. The error can be made to have a zero mean value by inhibiting the counter for one-half of a clock pulse period at the start of the second time interval. The counter is reset to zero at the start of an A/D conversion cycle.

Note that long-term variation in the clock frequency does not affect the accuracy because Eq. (13.30b) is in terms of the ratio of time intervals and voltages. Similarly any error in the comparison point is compensated. Unfortunately, the effect of offset voltage and current in the operational integrator is cumulative in both of the time intervals and is the fundamental limitation on the accuracy of this converter.

It is left as an exercise to derive relations between clock frequency, comparator settling time, number of bits in a digital word, and conversion time.

Integrated circuit implementations of this type of A/D converter are described in References 22 and 23.

13.5.2 Delta Modulator A/D Converter [7,8]

A delta modulator A/D converter is shown in Fig. 13.11. The two major sections are a delta modulator which converts the analog input waveform to a one-bit digital code, and a digital-code converter that converts that one-bit coding to a multibit code. The delta modulator consists of a comparator that compares the input signal, $f(t)$, with a piecewise constant approximation, $\hat{f}(t)$. If the input signal, $f(t)$, is larger than the approximation, $\hat{f}(t)$, the comparator output is positive. The clock triggers a pulse generator to produce an impulse train $k\delta(t - T_C)$ where T_C is the clock time. The pulse generator is a modulator that multiplies the clock impulse train by $+1$ if the comparator output is 1 and by -1 if the comparator output is 0. This modulated impulse train, $\Delta(t)$, is the delta modulated output signal. The piecewise constant approximation to the input signal, $\hat{f}(t)$ is formed by integrating $\Delta(t)$.

The response of the delta modulator, like all other encoders, is restricted in amplitude by the finite dynamic range of the active devices. In addition its speed of response is limited by a phenomena known as *slope overload* which arises if the input waveform $f(t)$ changes, during a clock time, by more than the step height in the approximating signal. If the input signal

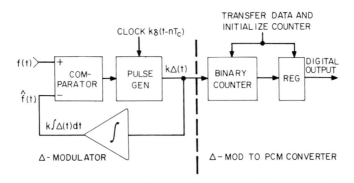

Fig. 13.11 An A/D converter based on Δ-modulation and code conversion.

is a sine wave of amplitude A and frequency f_1, its maximum slope is $2\pi f_1 A$ which occurs at the zero crossing. Equating this to the maximum rate of rise of the approximating signal which is k/T_C, where k is the step size, yields the following relation between signal amplitude, input frequency, delta modulation rate, and step size.

$$A = \frac{k}{2\pi f_1 T_C} \qquad (13.31)$$

Thus large dynamic range A implies high clock rates $(1/T_C)$.

The code converter is a binary counter. The value corresponding to zero input signal is preloaded into the counter at the beginning of each digitizing period. Each positive impulse in the output of the delta modulator increases the count and each negative impulse decreases it. Thus the number stored in the counter is characteristic of the level of the stepped approximation to the input signal. An M-bit digital code word required 2^M impulses of the delta modulator output stream to completely determine the count. Thus the sampling frequency, F_S, and the delta modulator clock rate, $F_C = 1/T_C$, are related by

$$F_C = 2^M F_S . \qquad (13.32)$$

After the required number of pulses have been counted, the result is transferred to the register and the counter is reinitialized.

EXAMPLE 13.5   A delta modulator PCM encoder is to be used to quantize a signal whose maximum frequency is 4 kHz. The quantized signal is to have 256 levels. What clock frequency is required?

Solution. The upper frequency limit of 4 kHz means that 8000 samples/s are required assuming an ideal low-pass sampling filter. The 256 levels in the discrete approximation correspond to an 8-bit code. Therefore, from Eq. (13.32) the clock frequency $F_C = 2.048$ MHz.

13.6 ANALOG MULTIPLIERS [1]

13.6.1 Variable Transconductance Multiplier

A variable transconductance analog multiplier circuit is shown in Fig. 13.12. It is basically a difference amplifier with a current source in the emitter circuit of the differential pair. One of the signals to be multiplied is applied differentially to the bases of the difference amplifier while the other signal is used to vary the magnitude of the total emitter current about a quiescent value. The output is a differential signal at the collectors of T_1 and T_2. This circuit is also known as a balanced modulator.

Applying Kirchhoff's current law to the common emitter terminal yields,

$$I_3(t) = I_{s1}\exp\left(\frac{V_{BE1}}{V_T}\right)\left[1 + \exp\left(\frac{v_x + a}{V_t}\right)\right] \tag{13.33}$$

where it has been assumed that $I_{E1} \approx I_{s1}\exp(V_{BE1}/V_T)$, $I_{E2} \approx I_{s2}\exp(V_{BE2}/V_T)$, and that $I_{s2}/I_{s1} = \exp(a/V_T)$. Since a is determined by the ratio of transistor leakage currents, it is almost zero for a well-matched monolithic transistor pair. In addition, note that $I_3(t) = I_{dc} + I(t)$ where $|I(t)|$ must be restricted so as to keep the difference amplifier transistors in their active region. Neglecting any small differences in the IR drop in the base circuit of T_1 and T_2, $v_x = V_{BE2} - V_{BE1}$. It therefore follows that

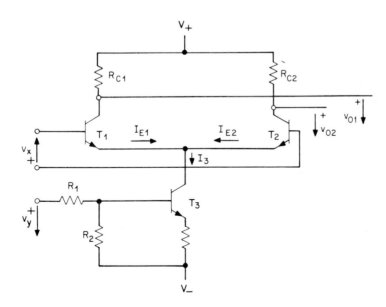

Fig. 13.12 Variable transconductance analog multiplier.

the differential output voltage is

$$v_o(t) = I_{C2}R_{C2} - I_{C1}R_{C1} = I_{E2}\alpha_2 R_{C2} - I_{E1}\alpha_1 R_{C1}$$

$$= -\frac{I_3(t)\alpha_1 R_{C1}}{1 + \exp\left[\dfrac{v_x+a}{V_T}\right]}\left[1 - \exp\left(\frac{v_x+a+b}{V_T}\right)\right], \qquad (13.34)$$

where total current $I_3(t)$ is given by Eq. (13.33) and where $\alpha_2 R_{C2}/\alpha_1 R_{C1} = \exp(b/V_T)$. When the differential pair is perfectly balanced $(a = 0, \ b = 0)$

$$v_o(t) \approx \frac{\alpha_1 R_{C1}}{2V_T}\, I_3(t)v_x(t). \qquad (13.35)$$

Note that this approximate form of the output voltage is valid when v_x is small so that the higher terms involving powers of v_x can be neglected. It may be shown that if the signal v_y causes $I_3(t)$ to have small variations $i_3(t)$ about its quiescent value, then

$$i_3(t) \approx \frac{R_2}{R_1+R_2}\ \frac{1}{R + (V_T/I_E)}\ v_y(t).$$

Hence the product term becomes

$$v_o(t) = \frac{\alpha_1 R_{C1}}{2V_T}\ \frac{R_2}{R_1+R_2}\ \frac{1}{R + (V_T/I_E)}\ v_x(t)v_y(t),$$

which is clearly a function of the product of $v_x(t)$ and $v_y(t)$.

A disadvantage of this multiplier is that both the scale factor and the dc level are functions of temperature through V_T. Because of these difficulties this circuit is not used as a precision multiplier, but it is used as an electronically variable attenuator and balanced modulator. It forms the basis for the phase detector that is needed in the phase-locked loop to be described shortly, and for the current ratioing multiplier.[9]

13.6.2 Analog Multiplier as a Phase Detector

The multiplier used as a phase detector for the phase locked loop is sometimes known as the doubly balanced modulator. It is shown in Fig. 13.13.

The input stage, containing transistors T_1, T_2, is a differential amplifier whose differential gain when it is balanced is given by

$$A_d = \frac{R_C}{r_e} = \frac{R_C I_3}{2V_T}. \qquad (13.36)$$

In the phase detector application which is considered here, v_x is a square wave signal of sufficient amplitude to change the state of transistor pairs T_3, T_4 and T_5, T_6 which are used as current mode switches (see Section 14.1).

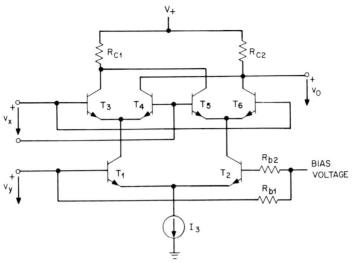

Fig. 13.13 A doubly balanced modulator.

The effect of these cross connected current mode switches is to multiply the differential amplifier output by ± 1 in response to signal v_x. The Fourier series expansion of this square-wave multiplying factor is assumed to be a summation of sine terms

$$m(t) = \frac{4}{\pi} \sum_{n=0}^{\infty} \frac{1}{2n+1} \sin[(2n+1)\omega_o t] \qquad (13.37)$$

where ω_o is the fundamental frequency of v_x.

Suppose that the second signal, v_y, is

$$v_y(t) = V_i \sin(\omega_i t + \theta_i) \qquad (13.38)$$

where θ_i is the phase of the components of v_i, measured with respect to $m(t)$. Then from Eqs. (13.36), (13.37), and (13.38), the output v_o is given by

$$v_o = \sum_{n=0}^{\infty} \frac{2A_d V_i}{(2n+1)\pi} \Big[\cos[(2n+1)\omega_o t - \omega_i t - \theta_i]$$
$$- \cos[(2n+1)\omega_o t + \omega_i t + \theta_i]\Big]. \qquad (13.39)$$

The dc component is proportional to the phase difference θ_i

$$v_o' = \frac{2A_d V_i}{\pi} \cos \theta_i . \qquad (13.40)$$

If the signal $v_x(t)$ which generated the square-wave multiplying factor $m(t)$ is expanded in a Fourier cosine series, the output voltage v_o' in Eq. (13.40) will be proportional to $\sin \theta_i$.

Up to this point it has been shown that the doubly balanced modulator can function as a phase detector. These results will be used in the next section on the phase locked loop.

EXAMPLE 13.6   What is the amplitude of the output voltage of an analog multi-plier phase detector with a load resistance of $2 \text{ k}\Omega$, a 1 mA current source, a 100 mV rms input signal, and a phase difference of $35°$ between the input signal and reference signal?

Solution.   From Eq. (13.36) the gain $A_d = (2\times10^3\times1\times10^{-3})/(2\times26\times10^{-3}) = 38.5$. The output voltage is given by Eq. (13.40)

$$v_o' = \frac{2\times38.5\times0.1}{\pi} \cos 35° = 2.45 \cos 35° = 2.01 \text{ volts.}$$

13.6.3 Current Ratioing Multiplier [9]

In this multiplier, shown in Fig. 13.14, an input voltage signal v_y, applied to the difference amplifier T_7, T_8 varies the current (I_7, I_8) through diodes D_7, D_8. These diodes are connected to difference amplifiers T_3, T_4 and T_5, T_6. As a result of this connection, the following current ratios are preserved in the circuit:

$$\frac{I_7}{I_8} = \frac{I_3}{I_4} = \frac{I_6}{I_5}. \qquad (13.41)$$

A proof of this relationship is given later in this section.

The second voltage signal, v_x, is applied to difference amplifier T_1, T_2 and changes currents

$$I_2 = I_3 + I_4; \quad I_1 = I_5 + I_6 \qquad (13.42)$$

By virtue of the cross-connection of transistors T_3, T_4, T_5 and T_6 with diodes D_7 and D_8, the output voltage V may be shown to be proportional to the product of input voltages v_x and v_y. A detailed analysis of this circuit now follows.

The heart of the circuit is the gain cell shown in Fig. 13.15 which consists of diodes D_1, D_2 and transistors T_3, T_4. Applying Kirchhoff's laws to this configuration yields

$$V_{D7} - V_{D8} = V_{BE3} - V_{BE4}. \qquad (13.43)$$

For identical diodes $I_7 \approx I_{sD}\exp(V_{D7}/V_T)$ and $I_8 \approx I_{sD}\exp(V_{D8}/V_T)$. Hence

$$\ln\left(\frac{I_7}{I_8}\right) = V_{D7} - V_{D8}. \qquad (13.44)$$

Similarly for identical transistors T_3, T_4 one has $I_3 \approx I_{sT}\exp(V_{BE3}/V_T)$ and $I_4 \approx I_{sT}\exp(V_{BE4}/V_T)$ and so

$$\ln\left(\frac{I_3}{I_4}\right) = V_{BE3} - V_{BE4}. \qquad (13.45)$$

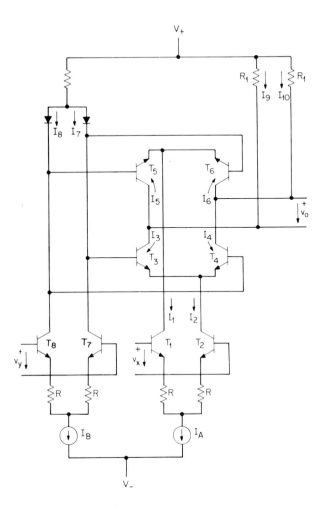

Fig. 13.14 Current ratioing multiplier.

By virtue of Eqs. (13.43), (13.44), and (13.45) now follows the first part of Eq. (13.41). The second part of this equation is established similarly.

By applying Kirchhoff's laws to certain nodes of the network one obtains also

$$I_A = I_1 + I_2; \quad I_B = I_7 + I_8$$

$$(13.46)$$

$$I_9 = I_3 + I_5; \quad I_{10} = I_4 + I_6.$$

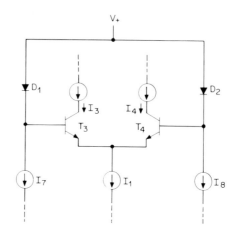

Fig. 13.15 Basic gain cell.

Furthermore

$$v_x \approx R(I_1 - I_2); \quad v_y \approx R(I_8 - I_7) \tag{13.47}$$

where the small additional terms due to the differences in transistor base-emitter voltages have been neglected.

Finally

$$V = R_1(I_9 - I_{10}); \quad V_o = GV = GR_1(I_9 - I_{10}). \tag{13.48}$$

Combining Eqs. (13.41) through (13.48) yields after some algebra

$$V_o = -\frac{GR_1}{I_B R} v_x v_y, \tag{13.49}$$

where G is the gain of the operational amplifier. From Eq. (13.12),

$$G = \frac{\beta'/\beta}{1 + (1/A\beta)}. \tag{13.12}$$

Here A is the open-loop gain of the operational amplifier. Equation (13.49) establishes that the output voltage is proportional to the product of the two input voltages. Multiplication demands matched diodes and transistors and so a monolithic bipolar integrated circuit configuration is a natural choice. Advantages of this circuit are differential input facility, good linearity, wide bandwidth with low ac feedthrough, and excellent temperature stability.

A monolithic integrated circuit version of the current-ratioing multiplier including additional circuitry to compensate for errors is described in Reference 21. The accuracy of this circuit is about 0.5%.

13.7 THE PHASE-LOCKED LOOP [10-14]

The phase-locked loop is a feedback control system consisting of a phase comparator, low-pass filter, dc amplifier, and voltage-controlled oscillator, as shown in Fig. 13.16. It can be used as a tuned amplifier and FM demodulator, frequency synthesizer, tone detector, and in AM detection.

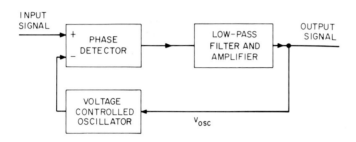

Fig. 13.16 Block diagram of a phase-locked loop.

The voltage-controlled oscillator (VCO) driven by the processed output of the phase comparator generates a signal whose phase is compared to that of the input signal. The control action of the feedback loop tries to drive the VCO output into phase correspondence with the input signal. Because the loop has a limited response speed, the VCO output does not reproduce all of the noise that is superimposed on the input signal.

Before analyzing the behavior of the loop, consider the transfer function of the voltage-controlled oscillator. It has a normal frequency of oscillation denoted by ω_n. The deviation of the instantaneous oscillation frequency from this rest value is a single valued, generally nonlinear, function of control voltage. In most practical cases the percentage frequency deviation is small enough that the control law can be taken as linear, and therefore the deviation from the rest frequency is given by

$$\omega_0(t) - \omega_n = k_1 v_{osc}(t). \qquad (13.50)$$

Because frequency is the derivative of phase $\omega_0(t) - \omega_n = d\theta_o(t)/dt$, and therefore the VCO output phase is

$$\theta_o(t) = \int_{-\infty}^{t} k_1 v_{osc}(x)\,dx \qquad (13.51)$$

which is a periodic function with period 2π. The output signal from the VCO is $B \cos \omega_0 t$, and is applied to one input of the phase detector. The other input to the phase detector is the input signal which is taken as equal to $A \sin \omega_i t$, where $\omega_i(t) - \omega_n = d\theta_i(t)/dt$.

The phase detector analyzed in the previous section had an output proportional to the sine of the input phase difference. If the low-pass filter has an impulse response $h(t)$ and the dc amplifier has gain k_a, then the oscillator control signal is

$$v_{osc}(t) = \frac{ABk_dk_a}{2} \sin \lambda(t) * h(t) \tag{13.52}$$

where k_d is the phase-detector gain constant, $\lambda(t) = \theta_i(t) - \theta_o(t)$ is the phase-detector input phase difference, A and B are amplitudes of the detector input signals, and $*$ denotes the convolution operation. Hence the phase difference $\lambda(t)$ is given by

$$\lambda(t) = \theta_i(t) - k_1 \int_{-\infty}^{t} v_{osc}(x)\,dx. \tag{13.53}$$

Differentiation of the above equation and transposition now yields the differential equation

$$\frac{d\lambda(t)}{dt} + k_1 v_{osc} = \frac{d\theta_i}{dt}. \tag{13.54}$$

Substitution of Eq. (13.52) into Eq. (13.54) finally gives

$$\frac{d\lambda(t)}{dt} + \frac{ABk_1k_dk_a}{2} \sin \lambda(t) * h(t) = \frac{d\theta_i}{dt}. \tag{13.55}$$

This equation describes the time behavior of the phase error in terms of the deviation of the instantaneous input frequency from the rest frequency of the VCO.

When the low-pass filter is a simple one-pole filter, this equation is a second-order nonlinear differential equation which cannot be solved in closed form. The transient behavior can be studied for particular cases by using phase-plane techniques. However, the important properties of the phase-locked loop can be deduced without knowledge of the actual phase trajectory. Only the specific details of the lock-up mechanism cannot be determined.

The parameters characterizing the behavior of the phase-locked loop are now defined.

Lock range   The voltage-controlled oscillator in the phase-locked loop tries to track the frequency of the incoming signal. It can do this as long as it has sufficient control range and as long as the steady-state phase-detector operation is restricted to a single branch of its transfer characteristic. If the low-pass filter has unity dc gain, then the steady-state solution $d\lambda(t)/dt = 0$ of Eq. (13.55) yields

$$\sin \lambda = \frac{2\Delta\omega_i}{ABk_1k_dk_a} \tag{13.56}$$

where $\Delta\omega_i = d\theta_i(t)/dt$ is the input frequency deviation from the rest frequency of the VCO. The loop can remain locked to the input frequency as long as the sine of the steady-state phase error is less than unity. Hence the *lock range* is

$$LR = \max|\Delta\omega_i| = \frac{ABk_1k_ak_d}{2},\qquad(13.57)$$

that is, the lock range is proportional to the product of the gain constants. This is the maximum steady-state frequency deviation.

Capture range The *capture range* is that range of input frequencies in which an initially unlocked loop can acquire an input signal and reach a locked state. The capture range depends upon the gain constant of the system and the frequency response of the low-pass filter.

When the phase-locked loop is in an unlocked state, the output of the phase detector is a time-varying function. If the low-pass filter has an ideal characteristic that passes the low-frequency beat components of the phase comparator output and rejects all the high-frequency components, then the capture range and the lock range are identical. For any realizable filter transfer function, $H(j\omega) = |H(\omega)|\exp[j\phi(\omega)]$, the loop gain at any nonzero frequency is less than the dc gain and therefore the capture range is less than the lock range.

An approximate expression for the capture range can be found by opening the loop at the VCO input terminals. The VCO control voltage in the steady state for an input frequency deviation $\Delta\omega_i$ is given by

$$v_{osc}(t) = \frac{ABk_dk_a}{2}\,|H(\Delta\omega_i)|\sin(\Delta\omega_i t).\qquad(13.58)$$

A bound on the control (error) voltage at the input frequency deviation $\Delta\omega_i$ is $\hat{v}_{osc}(t) = ABk_ak_d|H(\Delta\omega_i)|/2$. This control voltage produces an output frequency deviation $\Delta\omega_{0c} = ABk_1k_dk_a|H(\Delta\omega_i)|/2$. If the loop is to achieve lock when it is closed, the VCO frequency deviation at the capture frequency, $\Delta\omega_{0c}$, must equal the input frequency deviation at the capture frequency, $\Delta\omega_{ic}$. Hence an approximate expression for the capture frequency $\Delta\omega_{ic}$ in terms of the low-pass filter characteristic follows immediately by equating $\Delta\omega_{ic}$ and $\Delta\omega_{0c}$.

$$\Delta\omega_{ic} \approx \frac{ABk_1k_dk_a|H(\Delta\omega_{ic})|}{2}\qquad(13.59)$$

The voltage necessary to shift the VCO frequency by $\Delta\omega_{0c}$ has a steady-state value $v_{osc}(t)$ equal to $\hat{v}_{osc}(t)$ with $\Delta\omega_i$ replaced by $\Delta\omega_{ic}$ of Eq. (13.59). This yields

$$v_{osc}(t) = \hat{v}_{osc}(t) = \frac{ABk_a k_d |H(\Delta\omega_{ic})|}{2} = \frac{\Delta\omega_{ic}}{k_1}. \qquad (13.60)$$

Note that for an ideal low-pass filter $|H(\omega)| = 1$ and therefore the lock range and capture range are equal.

Capture time   The *capture time*, T_c, is the time required for the phase-locked loop to produce the dc steady-state control voltage equivalent to the peak ac control voltage that was necessary to shift the VCO output frequency by $\Delta\omega_{0c}$. This time is obviously a function of the particular low-pass filter transfer function that is used. An important special case is the single pole RC low-pass filter whose attenuation function, $|H(\omega)|$, is given by

$$|H(\omega)| = \frac{1}{\sqrt{1 + \omega^2 T_c^2}} \qquad (13.61)$$

where $T_c = RC = 1/\omega_c$, and ω_c is the 3 dB cutoff frequency of the filter. Substituting this expression into Eq. (13.59) yields

$$\left(\frac{\Delta\omega_{ic}}{k}\right)^2 \left[\frac{\Delta\omega_{ic}/k}{\sqrt{\omega_c/k}}\right]^4 = 1, \qquad (13.62)$$

where $k = ABk_1 k_d k_a/2$ is related to the loop gain. A plot of normalized capture range $\Delta\omega_{ic}/k$ versus filter bandwidth $\sqrt{\omega_c/k}$ is shown in Fig. 13.17. The capture range is approximately proportional to the square root of the filter cutoff frequency for small frequencies.

Wideband low-pass filters increase the capture range and reduce the capture time at the expense of reduced noise immunity.

Small-signal behavior of the phase-locked loop   The phase-locked loop behaves like a linear dynamic system if the operating range is suitably restricted. This occurs either by intentionally designing the phase comparator to operate in its linear range, or for "small" perturbations about the operating point when lock is achieved. For these reasons the small-signal behavior of the phase-locked loop is of interest and will now be investigated.

The linearized servo block diagram for the phase-locked loop is shown in Fig. 13.18. In this analysis the input and feedback variables Ω_i and Ω_0 are frequencies, and therefore the comparator includes the operator $1/s$ because it integrates the frequency difference $\Delta\Omega(s)$. Using straightforward analysis techniques, the closed-loop transfer function, $T(s)$, relating the output voltage, $V_c(s)$, to the *input frequency shift*, $\Omega_i(s)$, is found to be

$$T(s) = \frac{V_c}{\Omega_i} = \frac{1}{k_1} \frac{k_1 k_d k_a H(s)}{s + k_1 k_d k_a H(s)}. \qquad (13.63)$$

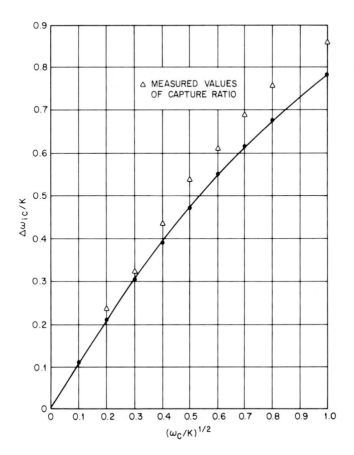

Fig. 13.17 Normalized capture range of a phase-locked loop with a single pole RC low-pass filter versus filter bandwidth. (Reprinted from *The Bell System Technical Journal*, Copyright 1965, The American Telephone and Telegraph Co.)

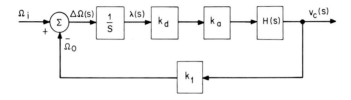

Fig. 13.18 Block diagram of linearized phase-locked loop.

Note from this expression that the phase-locked loop has an inherent low-pass transfer characteristic which is modified by the form of the low-pass filter function $H(s)$. However, this is not a conventional transfer function in that the excitation is the frequency modulation of the input carrier frequency. Thus the phase-locked loop acts as a frequency noise or jitter filter with respect to the input signal.

If $H(s)$ in Eq. (13.63) is a constant, then $T(s)$ is of the form $\alpha/(s+\alpha)$ and the loop is known as a *first-order phase-locked loop*. When $H(s)$ is the simple RC low-pass filter used above, the denominator of Eq. (13.63) becomes a quadratic and the system is known as a *second-order loop*. The second-order phase-locked loop using an RC low-pass filter has a bandpass transfer function. In order to ensure loop stability, and to permit independent adjustment of the static phase error and the 3-dB closed loop or noise bandwidth that is defined as

$$B = \int_{-\infty}^{\infty} |T(j\omega)|^2 d\omega, \tag{13.64}$$

a simple lag network is often used as the low-pass filter. For further details the reader is referred to the specialized literature.[13]

13.8 PHASE-LOCKED LOOP APPLICATIONS

FM demodulation   A frequency modulated carrier described by

$$v_i(t) = A \sin[\omega_n t + \theta_i(t)] \tag{13.65}$$

is applied to the input of the phase-locked loop shown in Fig. 13.16. In this expression ω_n is the carrier frequency and

$$\theta_i(t) = k \int_{-\infty}^{t} m(x)\,dx$$

is the phase modulation where k is a constant and $m(t)$ is the baseband modulating signal. The VCO is designed to have a rest frequency equal to the carrier frequency ω_n, when the applied control voltage is zero. When both the input signal and the VCO output are at frequency ω_n, the phase-detector output is zero. Any shift in input frequency, $\omega_i = d\theta_i/dt$, produces a control voltage, $v_{osc}(t)$, given by Eq. (13.52) which drives the VCO frequency toward the instantaneous-input frequency. The instantaneous-input phase is

$$\theta_i(t) = k \int_{-\infty}^{t} m(x)\,dx$$

while from Eq. (13.51) the instantaneous VCO phase is

$$\theta_o(t) = k_1 \int_{-\infty}^{t} v_{osc}(x)\,dx.$$

Thus, to within a multiplying constant, $v_{osc}(t)$ is a replica of the baseband modulating signal and the phase-locked loop is an FM demodulator.

Because the phase comparator transfer function k_d is a function of the input signal amplitude there is a limiting value for the input signal amplitude for which the lock range of the phase-locked loop, LR, is equal to the frequency deviation, $\Delta\omega_i$, of the FM signal. This limiting threshold, V_L, is given by virtue of Eq. (13.57) as

$$V_L = \frac{2\omega_i}{k_a k_1 k_d V_0} , \qquad (13.66)$$

where V_0 is the VCO output amplitude. At input amplitudes greater than V_L the lock range is greater than the FM frequency deviation. Above the limiting threshold it is immaterial whether or not the phase-locked loop is preceded by a limiter to remove input amplitude variations. Below threshold it is advantageous not to use a limiter because the amplitude dependence of the lock range automatically restricts the range of the phase-locked loop during a noise spike.

Figure 13.19 is the circuit for a monolithic integrated circuit phase-locked loop designed for demodulation of commercial FM IF signals.[15] The VCO is a Wien bridge oscillator using an external capacitor, C. The frequency is controlled by changing the amplifier gain by means of the junction field-effect transistor, F_1, which determines the position of the poles in the right-half plane. The phase comparator is a balanced multiplier using a junction field-effect transistor F_2 as a variable coupling element.

The measured lock and capture ranges as a function of input amplitude at the 10.7 MHz IF frequency is shown in Fig. 13.20(a). The interference rejection characteristics of the phase-locked loop when the loop is initially in lock is shown in Fig. 13.20(b). The performance characteristics tabulated in Table 13.1 are comparable to those of a conventional three-stage IF amplifier and detector of a high quality FM receiver.

Table 13.1 Performance Characteristics for Integrated Phase-Locked FM Amplifier/Demodulator—Measuring Conditions:
$\Delta f = 75$ kHz; $f_{mod} = 1$ KHz; $V_{CC} = V_{EE} = 9.0$ volts

	$f_0 = 10.7$ MHz	$f_0 = 20.0$ MHz
Recovered Audio	90 mV rms	75 mV rms
Limiting Threshold	0.8 mV rms	1.5 mV rms
AM Rejection	>40 dB	>40 dB
Dynamic Range	>40 dB	>40 dB
Output Distortion	<1 percent	<1 percent

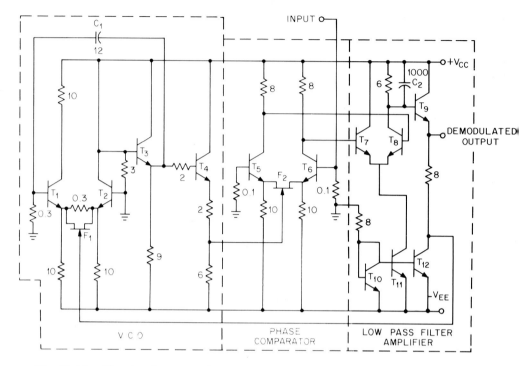

Fig. 13.19   Circuit diagram of an integrated circuit phase-locked loop FM amplifier/demodulator. Resistor values are in kilohms; capacitor values are in picofarads. (From Grabene and Camenzind.[15])

AM demodulation   An AM demodulator using a phase-locked loop is shown in Fig. 13.21. The phase-locked loop is used to lock on to the carrier of the amplitude-modulated input signal. The VCO output of the phase-locked loop is used as the local oscillator signal for the synchronous detector. The 90°-phase shift is necessary so that the VCO output and the amplitude-modulated input will be either in-phase or 180° out-of-phase which is the condition for maximum output from the product detector. The filter at the output removes modulation products at twice the carrier frequency.

13.9 ISOLATED TWO-PORT AMPLIFIERS [16]

The amplifier structures discussed in previous sections of this chapter did not have isolation between the input and output ports. However, there are many applications where such isolation is required. Traditionally isolation has been provided by means of transformers, which are not amenable to integrated circuit implementation. In this section is described an isolated two-port

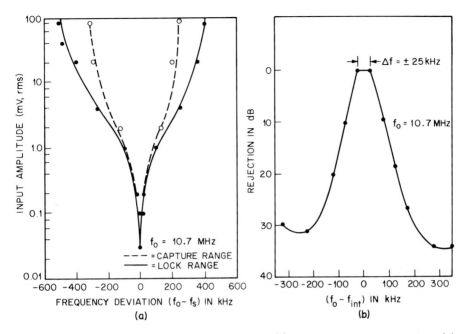

Fig. 13.20 Performance of FM demodulator. (a) Lock and capture range; (b) interference rejection characteristics. (From Grabene and Camenzind.[15])

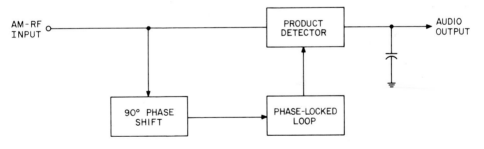

Fig. 13.21 The phase-locked loop as an AM demodulator.

structure which can be implemented using hybrid integrated circuit techniques. This structure, shown in Fig. 13.22, consists of a floating amplifier associated with each port which is optically coupled in a unilateral manner. The fundamental element of this structure is the double detector optocoupler consisting of a light-emitting diode (LED) which equally illuminates two isolated identical photodetecting diodes. The conversion of light to diode

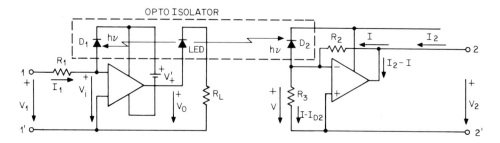

Fig. 13.22 An isolated two-port amplifier.

current is a basic physical phenomenon, involving the generation of one electron-hole pair for every absorbed photon. Hence the two photodetector currents are approximately equal and track together. The light-emitting diode is forward biased and so its current is temperature sensitive and a non-linear function of the junction voltage. In addition, the current transfer ratio

$$\Delta = \frac{I_L}{I_D} , \tag{13.67}$$

where I_L is the LED current and I_D is the photodetector current is typically 1000. As a consequence, gain must be provided to compensate the opto-coupler transfer loss. In addition, the current-voltage characteristic must be linearized and made independent of temperature. This may be achieved by putting the LED and one photodetecting diode, D_1, into a negative feedback loop, as is shown on the input side of Fig. 13.22. Figure 13.23 shows the

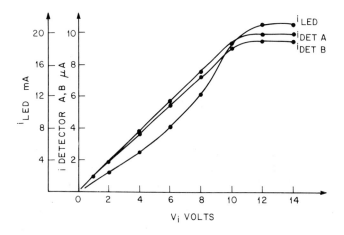

Fig. 13.23 Static transfer characteristics of the optocoupler feedback amplifier.

measured static transfer characteristic and clearly there is a considerable linear range of detector current versus input voltage. In fact, gain stability of about 1% with respect to temperature is obtained in the range of $-50°C$ to $+100°C$.

Applying Kirchhoff's laws to the input side of the structure yields after some algebra

$$I_{D1} = \left[\frac{V_1 - (V_0/A_1)}{R_1} \right] \left[1 + \frac{R_L \Delta}{R_1 A_1} \right]^{-1}, \tag{13.68}$$

where V_1 is the LED voltage, A_1 is the input operational amplifier gain, and Δ is defined in Eq. (13.67). For typical values of these parameters $V_0/A_1 \ll V_1$ and $R_L \Delta / R_1 A_1 \ll 1$, and so

$$I_{D1} \approx V_1/R_1. \tag{13.69}$$

Similar treatment of the output side of the structure yields

$$V_2 = I_{D2} R_2 \left[1 - \frac{1}{A_2} \left(1 + \frac{R_2}{R_3} \right) \right]^{-1} \tag{13.70}$$

where A_2 is the gain of the output operational amplifier. For typical parameter values $[1 + (R_2/R_3)]/A_2 \ll 1$, and so

$$V_2 \approx I_{D2} R_2. \tag{13.71}$$

Combining Eqs. (13.69) and (13.71) gives the overall voltage gain G_v

$$G_v = \frac{V_2}{V_1} \approx \frac{R_2}{R_1}. \tag{13.72}$$

When the output port is terminated in resistance R, then $V_2 = -I_2 R$ and so the overall current gain G_I becomes

$$G_I = \frac{I_2}{I_1} \approx -\frac{R_2}{R}. \tag{13.73}$$

The total signal power into the two port is given by $P = I_1 V_1 + I_2 V_2$, and so

$$P = I_2^2 R \left[\left(RR_1/R_2^2 \right) - 1 \right]. \tag{13.74}$$

Clearly, when $RR_1 = R_2^2$ the network is lossless. When $RR_1 > R_2^2$, then $P > 0$ and the network attenuates, and when $RR_1 < R_2^2$, the network has a power gain.

EXAMPLE 13.7 Determine the voltage gain and internal signal levels for the isolated two-port amplifier of Fig. 13.22 with $\Delta = 1000$, $R_1 = 10\ k\Omega$, $R_2 = R_3 = 100\ k\Omega$, $R_L = 1\ k\Omega$, and a peak input voltage of 100 mV.

Solution. The current in the input port photodiode $I_{D1} \approx V_1/R_1 = 10\ \mu A$. A photodiode current of $10\ \mu A$ requires an LED current of $10\ mA$ since $\Delta = I_L/I_D = 1000$. This LED current develops a 10 V signal across R_L, and allowing another volt for the forward bias of the LED and 1 V for the drop across each output transistor of the op-amp, the floating power supplies for the port 1 op-amp must be at least ± 12 V. The photodiodes are well matched and therefore $I_{D2} = I_{D1}$ which means $V_2 \approx 10^{-5} \times 10^5 = 1$ V. Thus, the voltage gain of the two-port structure is 10.

The gallium phosphide LED has a flat response for light generation up to approximately 100 kHz. Combined with the 20 dB/decade fall-off of the operational amplifier, this leads to the gain characteristic shown in Fig. 13.24, which clearly depends on the LED feedback resistor R_L. As may be seen, a wide response with a small transmission peak may be obtained for a suitable value of R_L.

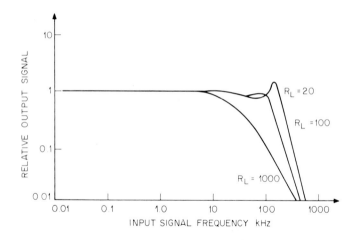

Fig. 13.24 Frequency response of the optical feedback amplifier.

Figure 13.25 shows how a pair of back-to-back isolated two-ports may be arranged as a hybrid for bilateral transmission of signals.

13.10 FILTERING USING CHARGE TRANSFER DEVICE AS TIME-DELAY ELEMENTS [17]

Charge transfer devices (CTD's) are either charge-coupled devices (CCD's) or bucket-brigade devices (BBD's). The operation of these devices was explained in Section 3.7. These devices are naturally realized as silicon integrated circuits and are well suited to the time delay of analog signals.

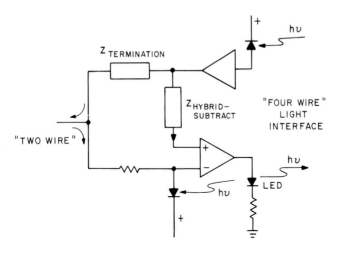

Fig. 13.25 Optically isolated hybrid. Only the left side of the symmetric circuit is shown.

Two filter structures utilizing these devices will be discussed. One structure is a resonator and the other a transversal filter.

13.10.1 Second-Order Resonator [24]

A block diagram of a second-order sampled data resonator incorporating a delay line is shown in Fig. 13.26. If the delay elements are replaced by capacitors and the summers and scalars are implemented using operational amplifiers and resistors, a form of the biquad described in Section 13.2 is obtained.

Some advantages of this sampled data resonator over the biquad are that the present structure is completely compatible with silicon integrated circuit technology, that the resonant frequency can be varied electrically by changing the clock rate, and multiplexing is possible because it is a sampled-data filter. Among the disadvantages of the structure are a limit on the dynamic range imposed by the size of the charge packets and a minimum clock rate which implies very long delay lines in order to obtain low resonant frequencies.

For the case illustrated by Fig. 13.26, the input to the filter is a continuous signal, which appears at the filter output during the off interval between sampling. The feed-through signal must be eliminated by sampling the output in synchronism with the clocking of the delay line. The sampling operations can be arranged to provide signal multiplexing if desired.

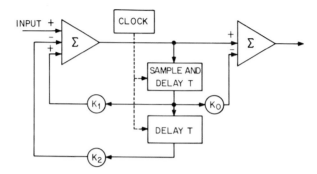

Fig. 13.26 Second-order resonator using a sampled-data delay line.

A straightforward analysis of this sampled data system in terms of z-transform notation yields a system function

$$H(z) = \frac{z^2 - K_0 z}{z^2 - K_1 z + K_2} \qquad (13.75)$$

with the filter poles located at

$$z_1, z_2 = \frac{K_1}{2} \pm \left[\frac{K_1^2}{4} - K_2 \right]^{1/2}, \qquad (13.76)$$

where $z = \exp(sT)$ and T is the sampling period.

For resonances to occur, $0 < K_1^2/4 < K_2$. Coefficients K_1 and K_0 may be positive or negative, but stability conditions require that the poles lie in the interior of the unit circle in the z-plane.

$$|z_1, z_2| = \left| \frac{K_1}{2} \pm j \left[K_2 - \frac{K_1^2}{4} \right]^{1/2} \right| = |K_2| < 1, \qquad (13.77)$$

and so K_2 must be positive and less than unity.

The poles z_1, z_2 of $H(z)$ in polar form can be written

$$z_1, z_2 = \sqrt{K_2} \exp \left[\pm j \cos^{-1} \left(\frac{K_1}{2\sqrt{K_2}} \right) \right]. \qquad (13.78)$$

As K_2 approaches unity z_1 and z_2 approach the unit circle, and the Q of the resonator, which may be shown to be

$$Q = \cos^{-1} \frac{(K_1/2\sqrt{K_2})}{|\ln K_2|}, \qquad (13.79)$$

increases. The frequencies of resonance, f_0, may be shown to be

$$f_0 = \frac{\cos^{-1}(K_1/2\sqrt{K_2})}{2\pi T} .$$ (13.80a)

The resonant frequency f_0 is a function of the coefficients K_1 and K_2, and depends inversely on the delay time T which is the reciprocal of the sampling frequency. In some implementations the sampling frequency is a submultiple of the clocking frequency which transfers information down the delay line. If

$$T = kT_c$$

where k is an integer and T_c is the clock period, then from the last two relations

$$f_0 = \frac{[\cos^{-1}(K_1/2\sqrt{K_2})]f_c}{2\pi k} .$$ (13.80b)

Hence the resonant frequency is a linear function of the clock frequency and is therefore easily varied electrically.

The bandwidth B of a high Q system is approximately given by

$$B \approx \frac{\ln\sqrt{K_2}}{\pi T} .$$ (13.81)

Some experimental results[17] of the resonator used as an $f_0 = 400$ Hz bandpass filter are now presented. The delay line is a 20-stage bucket-brigade device with taps after the tenth and twentieth stages. Hence the clock frequency f_c is ten times the sampling frequency, or $k = 10$. The transfer function $|H(z)|^2$ evaluated on the unit circle $|z| = 1$ is shown in Fig. 13.27, for two values of Q. The effect on Q of changing coefficient K_2 is shown in Fig. 13.28a. The variation of resonant frequency as a function of clock frequency is shown in Fig. 13.28b. The deviation from linearity of the curve at high clock frequencies is due to frequency limitations of the devices.

13.10.2 Transversal Filter [18]

A block diagram of a transversal filter is shown in Fig. 13.29. The continuous signal $v(t)$ is first sampled by sampler S, which is followed by $(M-1)$ stages of delay D, each of which delays the signal by an integral number of clock periods kT_c. The signal is accessed at M tap points between delay stages. At each tap, say the jth, the signal is multiplied (or weighted) by parameter h_j ($j = 1, 2, ..., M$) and the weighted signals are summed together by a summing amplifier Σ to give the filter output. Note that in contrast to the resonator described in the previous section this structure has no feedback

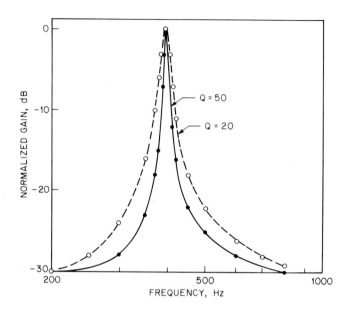

Fig. 13.27 Frequency response of 400 Hz bandpass filter with $Q = 20$ and $Q = 50$.

loops and is therefore called nonrecursive. As a consequence, filters of this type are always stable. When an arbitrary input signal $v_{in}(t)$ is applied,

$$v_{\text{out}}(nT_c) = \sum_{j=1}^{M} h_j v_{in}[(n-j+1)\,T_c].$$ (13.82)

The total time delay T_d of the filter is

$$T_d = MT_c.$$ (13.83)

The above output is approximately equal to the convolution of the input signal with the impulse response $h(t)$ of the filter. That is,

$$v_{\text{out}}(nT_c) \approx \int_0^{T_d} h(\tau)\,v_{in}(nT_c-\tau)\,d\tau.$$ (13.84)

A transversal filter can have an arbitrary impulse response $h(t)$, by changing weighting coefficients h_j. On the other hand, because of the lack of feedback very many taps may be required, which may be impractical to implement. Transversal filters find particular application as linear phase filters.

A particular implementation of the transversal filter, using a three-phase CCD as a delay line is shown in Fig. 13.30. Clock signals Φ_1, Φ_2, and

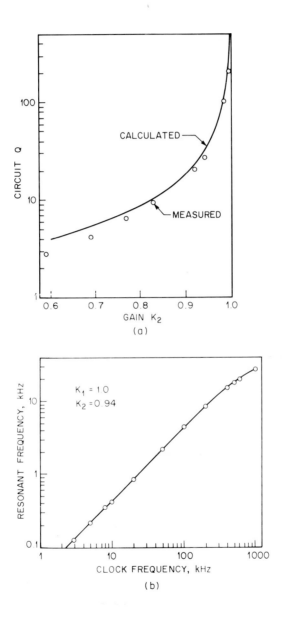

Fig. 13.28 (a) Variation of Q with changes in K_2; (b) resonant frequency versus clock frequency.

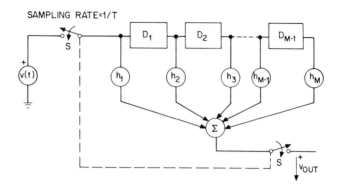

Fig. 13.29 Block diagram of a transversal filter.

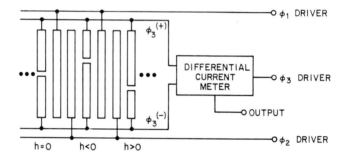

Fig. 13.30 Split electrode, three-phase CCD transversal filter.

Φ_3 are phase related, and in one delay period a charge packet is transferred through all three phases of the clock. Note that the phase three electrodes are split. Signals are tapped by measuring the differential current induced in the $\Phi_3^{(+)}$ and $\Phi_3^{(-)}$ clock lines as the charge is transferred. Tap weight is provided by the ratio of split, with $h_j = 0$ when the split is in the middle and the weights restricted to $-1 \leqslant h_j \leqslant 1$. This is no hardship and merely constitutes a scaling as is evident from Eq. (13.82). The dynamic range of the transversal filter is limited by the size of the charge packet and the amount of residual charge, which is taken as the zero level. The impulse response $h(t)$ of this transversal filter is a periodic function with frequencies lying between dc and $1/2T_c$. The time duration T_d of signals that can be processed using CCD is ultimately limited by the storage time, that is, the time it takes a stored charge to leak away. This time is of the order of several seconds at room temperature. A further limitation affects the signal bandwidth which is limited to less than half the clock frequency of the filter. For CCD's this limitation is about 20 MHz. A final limitation refers to the accuracy with which weighting coefficients h_j can be determined. Errors in these coefficients are due to photolithographic tolerances of the metal pattern.

REFERENCES

1. J. A. Graeme. *Applications of Operational Amplifiers*. New York: McGraw-Hill, 1973.

2. *Ibid.*, pp. 205-207.

3. L. C. Thomas. "The biquad, parts I and II." *IEEE Trans. on Circuit Theory* **CT-18**: 350-361 (May 1971).

4. S. K. Mitra. *Analysis and Synthesis of Linear Active Networks*. New York: John Wiley & Sons, 1969, Chapter 5.

5. J. Millman and H. Taub. *Pulse, Digital, and Switching Waveforms*. New York: McGraw-Hill, 1965.

6. D. F. Hoeschel, Jr. *Analog-to-Digital/Digital-to-Analog Conversion Techniques*. New York: John Wiley & Sons, 1968.

7. D. Goodman. "The application of delta modulation to analog-to-PCM encoding." *Bell Syst. Tech. J.* **48**: 321-343 (February 1969).

8 P. Z. Peebles, Jr. *Communication Systems Principles*. Reading, Massachusetts: Addison-Wesley Publishing Co., 1976, Chapter 7.

9. B. Gilbert, "A precise four-quadrant multiplier with subnanosecond response," *IEEE J. Solid-State Circuits* **SC-3**: 365-373 (December 1968).

10. H. Taub and D. Schilling. *Principles of Communications Systems*. New York: McGraw-Hill, 1971.

11. G. S. Moschytz. "Miniaturized RC filters using phase-locked loop." *Bell Syst. Tech. J.* **44**: 823-870 (May-June 1965).

12. K. K. Clarke and D. T. Hess. *Communication Circuits*. Reading, Massachusetts: Addison-Wesley Publishing Company, 1971, pp. 623-628.

13. J. Klapper and J. T. Frankle. *Phase-Locked and Frequency-Feedback Systems*. New York: Academic Press, 1972.

14. R. R. Cordell and W. G. Garrett. "A highly stable VCO for application in monolithic phase-locked loops." *IEEE J. of Solid-State Circuits* **SC-10**: 480-485 (December 1975).

15. A. Grabene and H. Camenzind. "Frequency selective integrated circuits." *IEEE J. of Solid-State Circuits* **4**: 216-225 (August 1969).

16. S. Waaben. "High performance opto-electronic circuits." *Proc. of ISSCC*: 30-31 (February 1975).

17.   C. H. Séquin and M. F. Tompsett. *Charge Transfer Devices.* New York: Academic Press, 1975, Chapter 6.

18.   D. D. Buss, D. R. Collins, W. H. Bailey, and C. R. Reeves. "Transversal filtering using charge-transfer devices." *IEEE J. of Solid-State Circuits.* **SC-8**: 138-146 (April 1973).

19.   E. J. Kennedy. "Low current measurements using transistor logarithmic dc electrometers." *IEEE Trans. on Nuclear Science* **NS-17**: 326-334 (February 1970).

20.   C. T. Sah. "Effect of surface recombination and channel on P-N junction and transistor characteristics." *IRE Trans. on Electron Devices* **ED-9**: 94-108 (January 1962).

21.   B. Gilbert. "A high-performance monolithic multiplier using active feedback." *IEEE J. Solid-State Circuits* **SC-9**: 364-373 (December 1974).

22.   F. H. Musa and R. C. Huntington. "A CMOS Monolithic 3½-digit A/D converter," *Proc. of ISSCC*, Feburary 1976, pp. 144-145.

23.   A. G. E. Dingwall and B. D. Rosenthal. "Low-power monolithic COS/MOS dual slope 11-bit A/D converter." *Proc. of ISSCC*, February 1976, pp. 146-147.

24.   D. H. Smith, C. M. Puckette, and W. J. Butler. "Active bandpass filtering with bucket-brigade delay lines." *IEEE J. of Solid-State Circuits.* **SC-7**: 421-425 (October 1972).

PROBLEMS

13.1   An inverting operational amplifier has an open-loop gain A, which is large but finite. Show:

a)   That the closed-loop gain

$$G \approx \frac{1 - (1/\beta)}{1 + (1/A\beta)}$$

where $\beta = R_2/(R_1+R_2)$.

b)   That the input resistance

$$R_{in} \approx (R_1+R_2)\,\frac{1+A\beta}{1+A} \approx R_2.$$

c)   That the closed-loop output resistance $R_o \approx r_o/(A\beta-1)$ where r_o is a nonzero output resistance (see Fig. 13.2).

d)   If the amplifier is to have a 40-dB closed-loop gain and a sensitivity to variation in open-loop gain of 0.1%, find the required open-loop gain.

13.2   Discuss the effect of offset voltage on the performance of an inverting integrator. Discuss also the effect of operational amplifier bandwidth and derive Eq. (13.15) for the transfer function.

13.3 A constant amplitude phase shifter is shown in Fig. P13.3. Calculate the amplitude and phase of the transfer voltage gain. How is the gain modified when the open-loop gain A is finite?

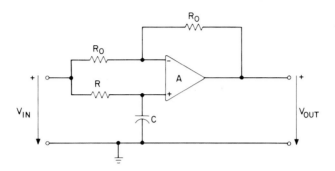

Fig. P13.3

13.4 A summing amplifier is shown in Fig. P13.4. Discuss the performance of this circuit when both the open-loop gain A and the input resistance r_i are finite.

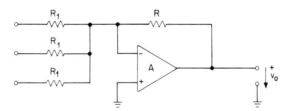

Fig. P13.4

13.5 Analyze the circuit of Fig. 13.7 to derive the transfer functions given in Eq. (13.18) and Eq. (13.19). Show that the transfer function $V_{out}(s)/V_{in}(s)$ is a biquadratic function.

13.6 The circuit shown in Fig. P13.6 is a filter. Find its voltage transfer function and state whether it is a low-pass, high-pass, or bandpass filter. If the denominator of the transfer function is given by $s^2 + (\omega_p/Q_{po})s + \omega_p^2$, state how you could change the undamped resonant pole frequency ω_p and the pole quality factor Q_{po} so that the change achieved in each parameter is mutually independent.

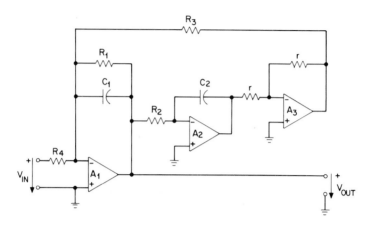

Fig. P13.6

13.7   Consider the dual slope integration A/D converter of Fig. 13.10 and assume a clock frequency f_c and a comparator settling time of τ seconds. Find:

a)   How many clock cycles are required to convert an input to an N-bit digital word?

b)   What is the minimum conversion time for a full-scale input, when the effect of offset voltage and offset current in the integrator are taken into account?

13.8   Consider a part of the variable transconductance multiplier shown in Fig. P13.8. Show that the output voltage V_o is approximately given by

$$V_o = \frac{R}{2} \frac{I_o}{V_T} e$$

where $V_T = kT/q$, the thermal voltage. Discuss the variation of V_o with temperature T, assuming that R is a diffused resistor with $R = R_o(1 + \alpha[T - T_o])$ and that $I_o \approx I_{so}\exp(V_B/V_T)$, where V_B is a fixed bias reference voltage.

13.9   A first-order phase-locked loop has the phase-comparator characteristic shown in Fig. P13.9. The phase-locked loop is adjusted so that $\lambda = 0$ if the carrier alone is present. The time constant of the phase-locked loop $\tau = 1/k_1 k_d k_a$ equals 2×10^{-5} s. At time $t = 0$ the carrier abruptly changes frequency by 6 kHz. Plot λ as a function of time for the range $0 \leqslant \lambda \leqslant 3\pi/2$.

13.10   A phase-locked loop, whose VCO is initially operating at frequency ω_n is used to demodulate the signal $\cos(\omega nt + \pi t/4\tau - \theta_s(t))$.

Find $v_{osc}(t)$ when $\theta_s(t) = 0$.

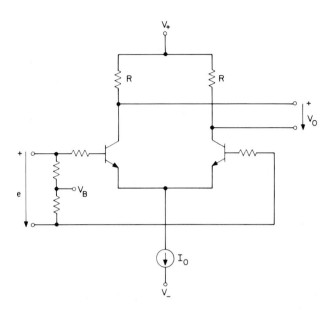

Fig. P13.8

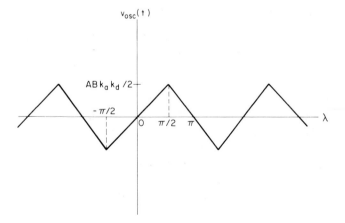

Fig. P13.9

When the phase-locked loop has reached equilibrium, it is found that
$\theta_s(t) = -\alpha t$ for $0 \leqslant t \leqslant 2\pi/\alpha$ and $\theta_s(t) = 0$ elsewhere. What is the
minimum value of the product $\alpha\tau$ needed to avoid a spike?

14

Digital Circuits

The most complex digital system can be constructed from circuits which implement a small number of basic operations such as AND, OR, and INVERT. These circuits are known as logic gates and their design and characterization is the subject of this chapter. More complex functional blocks and some system applications are considered in the next chapter.

14.1 CHARACTERIZATION OF LOGIC CIRCUITS

Digital systems often require a large number of logic elements. Hence, it is desirable to be able to describe the behavior of the entire digital system in terms of the terminal behavior of the logic elements rather than in terms of the constituent parameters of the individual circuit components. In addition, it is difficult to gain access to many points internal to an integrated circuit. Fortunately, the presence of transistors in the logic elements makes them essentially unilateral and self-isolating. The difficulty in obtaining a terminal characterization is the nonlinearity of the circuits.

As a vehicle to illustrate the terminal characteristics of a logic circuit and to motivate the introduction of terminology, we choose the current mode switch which is a difference amplifier used in a large signal mode.

Analysis of the Current Mode Switch   The current mode switch, shown in Fig. 14.1, is the decision element in an emitter-coupled logic (ECL) gate. It is a difference amplifier used in a large-signal mode. Kirchhoff's Current Law (KCL) applied at the common emitter node, and Kirchhoff's Voltage Law (KVL) around the base-emitter loop yields

$$I = I_{E1} + I_{E2} \tag{14.1}$$

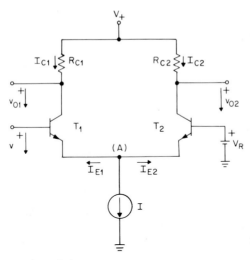

Fig. 14.1 A current mode switch.

where the current I is supplied by a constant current source, and the signal is

$$v_i = V_R + V_{BE1} - V_{BE2}$$

$$= V_R + V_T \ln\left(\frac{I_{E1}}{I_{E2}}\right). \qquad (14.2)$$

In writing Eq. (14.2), it is assumed that the current gain of the transistors is large so that $I_C \approx I_E$ and that the transistors are not in saturation so that interaction with the substrate is negligible, and therefore from Eq. (2.30)

$$V_{BE} \approx V_T \ln\left[\frac{(1-\alpha_I)I_E}{(1-\alpha_N \alpha_I)I_{s1}}\right].$$

Equation (14.2) is solved for I_{E1} to obtain an explicit relation for the output voltages v_{o1} and v_{o2} in terms of v_i. Substituting the result into Eq. (14.1) and using $I_C \approx \alpha_N I_E$ yields

$$v_{o2} = V_+ - I_{C2}R_{C2} = V_+ - \frac{\alpha_N I R_{C2}}{1 + \exp\left[\dfrac{v_i - V_R}{V_T}\right]} \qquad (14.3a)$$

$$v_{o1} = V_+ - \frac{\alpha_N I R_{C1}\exp\left[\dfrac{v_i - V_R}{V_T}\right]}{1 + \exp\left[\dfrac{v_i - V_R}{V_T}\right]} \qquad (14.3b)$$

When $v_i \ll V_R$, the exponential function is approximately zero and the outputs assymptotically approach

$$v_{o2} = V_+ - \alpha_N I R_{C2} = V_L$$

and

$$v_{o1} = V_+ = V_U$$

where the upper logic level, V_U, is the logic 1 level in a positive logic system, and the lower logic level, V_L, is a logic 0. The difference between these two levels is called the *logic swing*, v_l

$$v_l \triangleq V_U - V_L. \tag{14.4}$$

Transistor T_1 enters its active region as v_i is increased, causing I_{C2} to decrease and v_{o2} to increase because I is constant. The exponentials are unity when $v_i = V_R$ and if $R_{C1} = R_{C2} = R_C$ as is usually the case,

$$v_{o1} = v_{o2} = V_{th} = V_+ - \frac{\alpha_N I R_C}{2},$$

where V_{th} is the output voltage of the current mode switch at the *threshold point* to be defined subsequently. Note that the threshold point for this type of decision element is equidistant from the two assymptotic output levels and that the input voltage swing is also symmetric about this point, as shown in Fig. 14.2. Since in this case the logic swing is $v_l = \alpha_N I R_C$, we can write

$$V_{th} = V_+ - \frac{v_l}{2}.$$

The output voltages can be written in a more compact form in terms of the threshold voltage and logic swing.

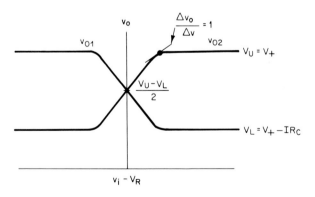

Fig. 14.2 Transfer characteristic of current mode switch.

$$v_{o1} = V_{th} + \frac{v_l}{2} \tanh\left(\frac{v_i - V_R}{2V_T}\right)$$

(14.5)

$$v_{o2} = V_{th} - \frac{v_l}{2} \tanh\left(\frac{v_i - V_R}{2V_T}\right)$$

The logic swings can differ on opposite sides of the switch if the load resistors are not of equal value. To account for this possibility, we will add a subscript to the logic swing to signify an association with a particular side of the switch. Usually $v_{l1} = v_{l2} = v_l$.

When $v_i \approx V_R$ both transistors are in their active region so that the gain is large and slight changes in v_i result in large changes in v_o. We will now compute the incremental gain dv_o/dv_i and show that the input voltage swing required to switch the current I is of the order of several times the thermal voltage V_T. Differentiating v_{o2} as given by Eq. (14.3a) with respect to v_i yields

$$\frac{dv_{o2}}{dv_i} = \frac{v_{l2}}{V_T} \frac{\exp\left(\dfrac{v_i - V_R}{V_T}\right)}{\left[1 + \exp\left(\dfrac{v_i - V_R}{V_T}\right)\right]^2},$$

(14.6)

where $v_{l2} = \alpha_N I R_{c2}$ is the logic swing at output 2. Observe that when v_i is very negative or very positive the incremental gain approaches zero. Intuitively, when the gain is less than unity the circuit is passive, whereas when the gain exceeds unity the circuit is active. Therefore the location of the *unity gain points* is of great interest. Since the slope of the curve for v_{o2} in Fig. 14.2 is always positive, Eq. (14.6) is set equal to $+1$ to find the location of the unity gain points. The result is

$$z = \left(\frac{v_{l2}}{2V_T} - 1\right) \pm \frac{v_{l2}}{2V_T}\left(1 - \frac{4V_T}{v_{l2}}\right)^{1/2}$$

(14.7a)

where $z = \exp[(v_i - V_R)/V_T]$. Replacing the square root by the first three terms of its binomial expansion, and retaining the physically meaningful root yields

$$z = \exp\left(\frac{v_i - V_R}{V_T}\right) = \frac{V_T}{v_{l2}} \quad \text{for} \quad 4V_T \ll v_{l2}$$

(14.7b)

and therefore the input voltage corresponding to unity gain is

$$v_{ug} = V_R - V_T \ln\left(\frac{v_{l2}}{V_T}\right).$$

(14.8)

This locates the unity gain point denoted by A in Fig. 14.2. To locate unity gain point B note that the circuit is symmetric about the vertical axis. Therefore the circuit does not care which side is the input and which side is the reference. This is equivalent to saying that the transfer characteristic has odd function symmetry about an input voltage V_R. Hence the exponent in Eq. (14.6) can be changed to $(V_R - v_i)/V_T$. Evaluating the derivative dv_{o2}/dv_i of the resultant expression yields

$$v_{ug} = V_R - V_T \ln\left(\frac{V_T}{v_{I2}}\right). \qquad (14.9)$$

The region between the two unity gain points is known as the *transition region*. The difference in input voltage between the unity gain points is the *transition width*. From Eqs. (14.8) and (14.9) the transition width is found to be

$$v_w = 2 V_T \ln\left(\frac{v_{I2}}{V_T}\right). \qquad (14.10)$$

EXAMPLE 14.1   Find the lower logic level V_L, the logic swing v_l, the unity-gain input voltage points and the transition width for a current mode switch with $V_+ = 5$ V, $R_{C1} = R_{C2} = 1$ kΩ, $I = 1$ mA, $\alpha_N = 0.95$, and operating at a temperature of 300 K.

Solution.   The low-level output voltage is

$$V_L = V_+ - \alpha_N I R_C = 5 - 0.95 = 4.05 \text{ V}.$$

The upper logic level corresponds to no current on that side of the switch and therefore $V_U = V_+ = 5$ V. The logic swing is $v_l = 950$ mV, and from Eqs. (14.8) and (14.10),

a) $v_{ug} = V_R \pm 0.026 \ln\left(\dfrac{950}{26}\right) = V_R \pm 0.094$, and

b) $v_w = 188$ mV.

   Up to this point we have considered the transfer characteristics of an isolated current mode switch. In any useful system the switching elements will be driven by, and in turn drive, other switching elements. The behavior of a cascade of two identical switching elements shown in Fig. 14.3(a) is now investigated. When the input voltage applied to the driving gate is either V_U or V_L, the output voltage of both the driving gate and the driven gate must be either V_U and V_L because the gates are identical and because no external voltage level translators are used to couple them. This is known as *input-output compatibility. The* operating point of the driven stage depends upon that of the driving stage since the interconnection implies that $v_{o1} = v_{i2}$. This is used in the construction of Fig. 14.3(b) to determine the operating

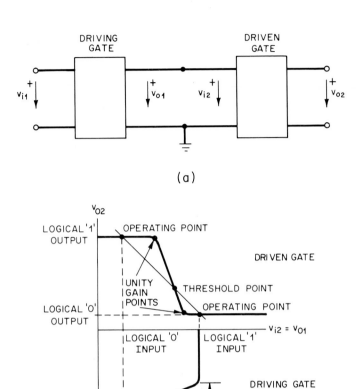

Fig. 14.3 (a) Cascade of two logic elements used to examine the consequences of input-output compatibility, and (b) construction to determine the operating and transition points of the driven element.

points of the driven gate in that the output axis of the driving gate is aligned with the input axis of the driven gate. If the output voltage of the driving element is low corresponding to a logic 0, then the input of the driven element is also low. Therefore the output voltage of the driven element is that value corresponding to a logic 0 input excitation denoted as point *A*. The operating point for a logic 1 input is determined in a similar manner and is denoted by point *B*.

Since both axes have equal scales, a line joining the operating points has a slope of -1. This line intersects the transfer characteristic at a point in the transition region called the *threshold point.*

Up to this point, we have considered voltage levels in a logic circuit to have unique, deterministic values. In a practical system, this is not the case for a number of reasons:

1. Variability of components in a logic gate from circuit to circuit and as a function of operating conditions will cause these levels to shift. As a result, worst-case conditions under which a system will still perform the desired logical operation must be considered.

2. Desired voltage levels are continually being perturbed by undesired and perhaps unknown signals, which are termed noise. These noise signals might be caused by inductive or capacitive coupling between signal lines in the system, or they might be induced from external high-frequency, high-power sources. In any event, such noise may cause a logic element to indicate an erroneous output.

The magnitude of the noise voltage that the element can tolerate before changing states can be determined from the transfer characteristic.

Noise margin and noise sensitivity   The *noise margin, NM,* of a logic gate is defined as the input voltage difference between the operating point and the nearest unity gain point.[1] (What we have defined as noise margin is sometimes called the *minimum noise margin.)* Since there are two quiescent operating points on the transfer characteristic, there are two noise margins denoted NM^0 for a logic 0 input state and NM^1 for a logic 1 input state. These are indicated in Fig. 14.4.

For an input perturbation somewhat less than the noise margin, the incremental gain of the logic element is zero. Hence, the perturbation is not observable at the output. For an input in excess of the noise margin, the element is in the transition region where the incremental gain is larger than unity. Since the element may be enclosed in a feedback loop, such an input could cause regeneration which could result in a permanent transition in logic state. Note that the noise margins NM^0 and NM^1 are not necessarily equal. They are equal in the present case because of the symmetry of the transfer characteristic and placement of the operating points.

The noise margin is not the only noise figure of merit which can be defined. An alternative is the *noise sensitivity, NS,* which is the input voltage difference between either operating point and the threshold point. (This definition of noise sensitivity is sometimes called noise margin.) Since there are two quiescent operating points, two noise sensitivities are defined. They are denoted by NS^0 for a logic 0 input state and by NS^1 for a logic 1 input state, as shown in Fig. 14.4.

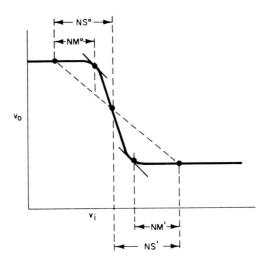

Fig. 14.4 Definition of noise margin and noise sensitivity.

To appraise the noise performance of a gate, it is also necessary to know the input impedance of the element corresponding to its two quiescent operating states, the output impedance of the driving source and some information of how the noise is coupled into the logic circuit.

Another convenient figure of merit is known as the *noise immunity, NI.* It is defined as the ratio of noise sensitivity to logic swing,

$$NI \triangleq \frac{NS}{v_l} , \qquad (14.11)$$

that is, noise immunity is noise sensitivity normalized to the logic swing.

Fan-in and fan-out   The basic logic element, in this case the current mode switch, is incorporated into more complicated structures in order to perform some useful operation. Logic gates are analyzed in the next sections. A logic gate has an output that is a function of several input variables. The number of inputs to a logic gate is often termed the *fan-in.* The logic circuit is called upon to drive another logic circuit which may load the circuit to varying degrees. A measure of the drive capability is the *fan-out,* which is the number of unit loads that the circuit can drive. A unit load is often chosen to be the smallest load presented by any member of a particular logic family. Specially designed elements with a large fan-out are used when a large number of gates must be driven. An example is a clock driver.

Only the static characterization of a logic gate was considered above. A complete characterization of the gate must also describe its time response during the time interval when it is undergoing a transition. A quantitative transient analysis is extremely difficult because of the nonlinear behavior of these circuits. Sufficient information for practical purposes can be obtained by characterizing the transient behavior in terms of propagation delay, rise time, and fall time that will now be defined.

Propagation delay, rise time, and fall time   Consider, for example, an inverting gate excited by a rectangular pulse, as shown in Fig. 14.5. The *rise time, t_r,* is the time required for the response to rise from its 10% to 90% points, although other definitions are possible. Similarly, the *fall time, t_f,* is taken to be the time required for the response to fall from the 90% to the

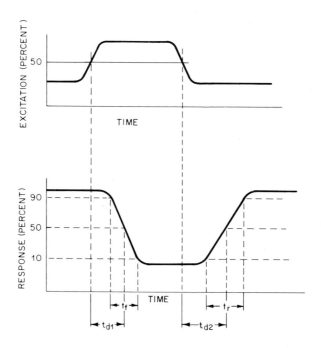

Fig. 14.5 Definition of rise time, fall time, and delay time.

10% point. The rise and fall times are usually different because of nonlinearities of the gate. It is not uncommon to find a single time specified which is the average of the rise and fall times. Moreover the rise and fall times observed at the output of a gate are due solely to the response of the gate only if the gate has no external load and if the excitation waveform has negligible rise and fall times. If this is not the case and the circuit is linear then the rise and fall times are the convolution of the excitation waveform and the impulse response of the circuit. For nonlinear circuits no general statement can be made. However, in practice it is usually possible to use an excitation whose rise and fall times are practically negligible.

The leading edge *delay time,* t_{d1}, is the time between 50% points of the leading edges of the excitation and response waveforms. (Delay times are often defined between voltage levels rather than between 50% points.) The leading edge of the input waveform is taken as the reference. Similarly the trailing edge delay time, t_{d2}, is the time difference between 50% points of the trailing edges of the response and excitation. The leading and trailing edge delay times are often different because of storage time associated with the active devices that are driven into saturation. The *average delay time,* $t_d = (t_{d1}+t_{d2})/2$ is often specified for simplicity. Note that all these times arise physically from the necessity of charging dynamic elements in the logic circuit and its external load. Hence, their exact values depend on fan-in and fan-out conditions, together with the response curves which must be given to correctly interpret the results.

14.2 EMITTER-COUPLED LOGIC GATE [2]

Emitter-coupled logic (ECL) gates use the current mode switch described in the previous section as the decision making element. It is a fast form of bipolar transistor logic because the transistors are never allowed to saturate. Consequently there is no excess charge to be removed from the base region before the turn-off process can begin.

A multiple (N) input ECL gate is shown in Fig. 14.6. The basic current mode switch of Fig. 14.1 is modified to include N transistors in parallel on the input side of the switch. The base of the transistor on the right side of the switch is connected to reference voltage V_R. The ideal current source in Fig. 14.1 is replaced by a transistor current source in Fig. 14.6. Emitter followers are used to increase the fan-out of the gate by isolating the current mode switch from the loads by virtue of their impedance-transforming capability, and to provide a dc level translation which makes the input and output levels compatible. The ECL gate simultaneously implements both the OR and NOR functions because if any of the inputs is a logic 1, the current is diverted to the input side of the switch. Under this condition the output voltage on the reference side of the gate is V_U. Hence, the reference side of the gate is also called the *OR side.* Since all of the current is drawn through the load resistor on the input side of the switch, the corresponding output voltage is V_L. Therefore the input side of the switch is also called the *NOR side.*

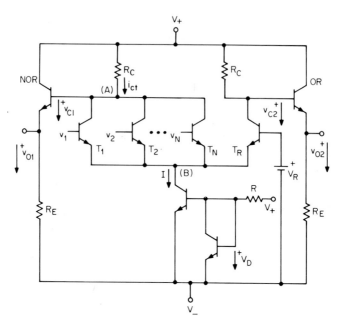

Fig. 14.6 An N-input ECL gate.

14.2.1 DC Analysis of ECL Gate

We consider an N-input ECL gate in which m of the N inputs are at voltage V_U corresponding to a logic 1 and the remaining $(N\text{-}m)$ inputs are at V_L corresponding to a logic 0. Applying Kirchhoff's current law at node E yields

$$I = I_s m \left[\exp\left(\frac{V_U - V_E}{V_T} \right) - 1 \right]$$

$$+ (N-m) \left[\exp\left(\frac{V_L - V_E}{V_T} \right) - 1 \right]$$

$$+ \left[\exp\left(\frac{V_R - V_E}{V_T} \right) - 1 \right] \tag{14.12}$$

where V_E is the voltage at node E. Only the first term in the Ebers-Moll equations describing the transistors need be considered because the values of current and collector-load resistors are chosen such that the transistors are not permitted to saturate. Also note that the second term in Eq. (14.12) is negligible because those transistors are always cut off.

The first step in determining the transfer characteristic is to replace V_U by a general input voltage v_i in the reduced form of Eq. (14.12). Solving the resultant equation for the common emitter-node voltage V_E yields

$$V_E = V_R - V'_{BE} + V_T \ln\left[1 + m \exp\left(\frac{v_i - V_R}{V_T}\right)\right], \qquad (14.13)$$

where

$$V'_{BE} = V_T \ln\left[\frac{I}{I_{s1}} + m + 1\right] \approx V_T \ln\left(\frac{I}{I_{s1}}\right), \qquad (14.14)$$

because I/I_s is much larger than practical values of $m + 1$. From Eq. (14.14) it is seen that V'_{BE} is the base-emitter voltage of the reference transistor when it conducts the full current I.

Next an expression for the NOR side output voltage is found by applying KCL at node A and using the Ebers-Moll equations to describe the active transistors.

$$i_{CT} = \sum_{k=1}^{m} i_{ck} = m\alpha_N \exp\left(\frac{v_i - V_E}{V_T}\right) \qquad (14.15)$$

Using V_E given by Eq. (14.13) together with $v_{c1} = V_+ - i_{CT}R_C$ yields

$$v_{c1} = V_+ - \frac{\alpha_N I R_C}{1 + \frac{1}{m}\exp\left(\frac{v_i - V_R}{V_T}\right)} . \qquad (14.16)$$

Unity has been neglected in comparison with the exponential term of Eq. (14.15) in obtaining this result. Note that this equation is similar to Eq. (14.3a) and that the effect of the m conducting input transistors is reflected in the multiplying factor of the exponential term in the denominator. The logic swing of the ECL gate is $v_i = \alpha_N I R_C$ as in the case of the current mode switch.

For dc considerations the emitter followers are merely level shifters. Therefore the output voltage v_{o1} of the gate will be the collector voltage v_{c1} minus the base-emitter voltage V_{BEO} of the emitter follower. Hence,

$$V_{o1} = V_+ - V_{BEO} - \frac{v_l}{1 + \frac{1}{m}\exp\left(\frac{v_i - V_R}{V_T}\right)} . \qquad (14.17)$$

Note that this equation defines a family of curves; one for each value of m as shown in Fig. 14.7. Observe from Fig. 14.2 that the transfer characteristics of the current mode switch are nearly symmetrical about V_R. Thus V_R may be regarded as the average input voltage, and since the gate must be input-output compatible V_R is also the average value of the output voltage. Hence the average current in the emitter follower resistor R_E is V_R/R_E. Therefore

$$V_{BEO} = V_T \ln \left[\frac{V_R}{I_{sf} R_E} \right] \tag{14.18}$$

where I_{sf} is I_{s1} for the emitter follower transistor. Equation (14.17) is the desired relation for the NOR side transfer function.

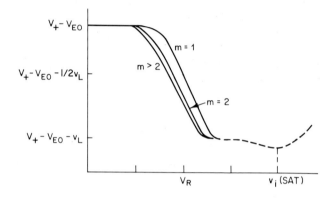

Fig. 14.7 Transfer function of the NOR side of an ECL gate for several values of m.

A similar analysis for the OR side of the gate yields

$$v_{o2} = V_+ - V_{BEO} - \frac{v_I}{1 + m \exp\left[\dfrac{v_i - V_R}{V_T}\right]} \tag{14.19}$$

When $m = 1$ Eqs. (14.17) and (14.19) reduce to Eqs. (14.3a) and (14.3b), and both outputs have the value $V_{th} = V_+ - V_{BEO} - v_i/2$ when $v_i = V_R$. If the transfer characteristic is to be perfectly symmetrical for this case, then V_R should be chosen equal to $V_+ - V_{BEO} - v_i/2$ because of input-output compatibility. Under these conditions the noise sensitivities NS^0 and NS^1 are both equal to $v_i/2$ because of the symmetry.

To find what constraints on the input voltage levels are necessary to prevent the input transistors from saturating, note that when v_i is at the logic 1 level (denoted by $V^{(1)}$), the collector voltage of the input transistor falls to $V^{(0)} + V_{BEO}$, where $V^{(0)}$ is the logic 0 voltage level. Under this condition the collector-base junction must be reverse biased. Since $v_{CB} = 0$ is satisfactory,

$$v_{CB}(T_1) = V_L + V_{BEO} - V_U$$
$$= V_{BEO} - v_l \geqslant 0, \qquad (14.20a)$$

and hence

$$v_l \leqslant V_{BEO}. \qquad (14.20b)$$

Thus, the logic swing of an ECL gate must be less than the drop across the base-emitter junction of the output-emitter follower if the input stage is to remain out of saturation.

The location of the unity gain points of the current mode switch are given by Eqs. (14.8) and (14.9). This result also holds for the ECL gate with $m = 1$, and therefore the transition width is given by Eq. (14.10). The derivation of the unity gain points for other values of m is left as an exercise. The noise margins are

$$NM^0 = \frac{v_l}{2} - V_T \ln\left(\frac{mv_l}{V_T}\right) \qquad (14.21a)$$

$$NM^1 = \frac{v_l}{2} - V_T \ln\left(\frac{v_l}{mV_T}\right). \qquad (14.21b)$$

The logic swing is

$$v_l = \frac{\alpha_{NC}\alpha_N}{2-\alpha_{NC}} \frac{R_C}{R} (V_+ - V_D) \qquad (14.22)$$

where α_{NC} is the forward (normal) alpha of the current source transistor and V_D is the voltage developed across the current source reference diode. From this equation it is evident that the logic swing is dependent upon the *ratio* of the load resistor to the reference resistor and the *product* of the current gains. The output levels also depend upon these parameters. Since the upper output voltage level is essentially fixed at $V_+ - V_{BEO}$, the lower output level, the logic 1 state noise margin, NM^1, and the logic 1 noise sensitivity, NS^1 will be affected by variations in resistance ratio and current gain.

EXAMPLE 14.2   The ECL gate of Fig. 14.6 has a current $I = 1$ mA, $R_E = 500$ Ω, $\alpha_N = 0.95$, and $I_{s1} = I_{sf} = 2 \times 10^{-16}$ A. Choose values for R_C and V_R such that the transistors do not saturate and the input and output levels are compatible.

Solution.   The CMS transistors do not saturate if $v_l \leqslant V_{BEO}$ [Eq. (14.20b)]. The upper logic level of the CMS is $V_U = V_+$ and the lower logic level is $V_L = V_+ - \alpha_N I R_C$. Therefore the logic swing is $v_l = \alpha_N I R_C \leqslant V_{BEO}$. Input-output compatibility requires the output voltage corresponding to the upper logic level equal the input voltage corresponding to that logic level, and a similar requirement for the lower logic level. Thus $v_{oU} = V_+ - V_{BEO}$ and $v_{oL} = V_+ - V_{BEO} - v_l$. Furthermore

V_R is the average input voltage and is therefore $V_R \approx V_+ - 1.5\,V_{BEO}$, where the limiting value is used for v_l. Note from Eq. (14.18) that V_{BEO} is a function of V_R which is presently unknown, and must be determined iteratively. Assume for an initial guess that $V_{BEO} = 0.8$ V and therefore at room temperature ($V_T = 0.026$ V) we find from Eq. (14.18) that $V_{BEO}^{(1)} = 0.813$ V. For the second iteration use $V_{BEO} = V_{BEO}^{(1)}$ in Eq. (14.18) to obtain the result that $V_{BEO} = 0.813$ V.

Therefore $V_R = V_+ - 1.5\,V_{BEO} = 3.78$ V. Finally $v_l = \alpha_N I R_C \leqslant V_{BEO}$ from which we find that $R_C \approx 850\ \Omega$.

EXAMPLE 14.3 (a) Find the noise margins for the ECL gate of the previous example if $m = 1$ and if $m = 3$. (b) Find the transition width.

Solution. (a) From Eqs. (14.21a) and (14.21b),

$$NM^0 = \frac{v_l}{2} - V_T \ln\left(\frac{mv_l}{V_T}\right)$$

and

$$NM^1 = \frac{v_l}{2} - V_T \ln\left(\frac{v_l}{mV_T}\right).$$

Using the results of the previous example, $v_l = 813$ mV. Therefore for $m = 1$, $NM^0 = NM^1 = 316$ mV. When $m = 3$, the transfer characteristics are no longer symmetrical and $NM^0 = 288$ mV and $NM^1 = 345$ mV. (b) From Fig. 14.4 it is evident that $v_l = NM^0 + NM^1 + v_w$. Thus for $m = 1$, $v_w = 181$ mV and for $m = 3$, $v_w = 180$ mV.

Sometimes the current source is replaced by a resistor. In that case the current I does not remain constant throughout the switching process and a small correction term must be added to the current to account for the finite resistance of the current source.[1] This correction term results in the negative slope of the low section of the NOR transfer characteristic. Furthermore, if the input transistors do saturate, additional terms of the Ebers-Moll equations must be considered or the transistor must be described by the Gummel-Poon model which provides a more precise description in the saturation region. The resultant behavior of the gate is described by the broken curve; where $v_{i\,\text{sat}}$ is the input voltage corresponding to the edge of saturation.

Input and output characteristics The input characteristic is the current-voltage relationship at the input to any one of the N transistors. On the assumption that the relevant input transistor is not driven into saturation, only the first term of the Ebers-Moll equations need be considered. The input current of the driven transistor, say T_1, is the difference between the emitter and collector currents $i_{b1} = (1 - \alpha_N)i_{e1}$. This expression for i_{b1} is a function of the emitter voltage V_E at the common node of the current mode switch as given by Eq. (14.13) with $m = 1$, because we are assuming that only one transistor is conducting. If we also assume that the transistor is conducting sufficiently that $\exp[(v_i-V_E)/V_T] \gg 1$, it follows that

$$i_{b1} = \frac{(1-\alpha_N)I}{1 + \exp\left[\dfrac{V_R-v_i}{V_T}\right]} . \tag{14.23}$$

A plot of this equation is shown in Fig. 14.8(a). When the effects of the nonideal current source and saturation are considered, a modified characteristic shown dotted in Fig. 14.8(a) is obtained. The zero and one operating points are also shown. Note that the dynamic resistance of the gate is infinite at the logic 0 input level. In the transition region the input resistance of the gate may be obtained by differentiating Eq. (14.23). At the point $v_i = V_R$ the dynamic input resistance is approximately equal to $4V_T/(1-\alpha_N)I$ which is about 2 or 3 kΩ. At operating point (1) the input resistance is about ten times this value. In any event, the base-spreading resistance of the transistor is always much less than the input resistance and it may be neglected.

The output characteristic of the gate is a relation between the output current and output voltage. There are two branches depending upon whether the output level is a logic 1 or a logic 0. Applying KCL to the output node of the emitter follower and substituting the appropriate Ebers-Moll expression for the emitter current yields

$$i_{o1} = \frac{v_{o1}}{R_E} + I_{s1}\left[\exp\left(\frac{v_{c1}-v_{o1}}{V_T}\right)\right]. \tag{14.24}$$

Plots of this equation for the logic 1 output $(v_{c1} = V_+)$ and the logic 0 output $(v_{c1} = V_+ - v_l)$ are shown in Fig. 14.8(b).

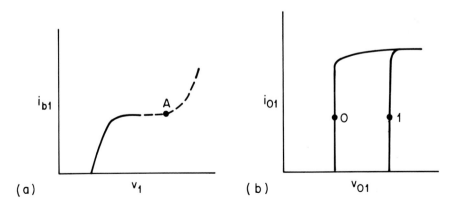

(a)   (b)

Fig. 14.8 (a) Input and (b) output characteristics of an ECL gate.

14.2.2 Transient Behavior of the ECL Gate

The analysis of nonlinear dynamic systems is extremely difficult in practice. Even in the present case where the nonlinearities are due to p-n junctions and are well behaved, the solutions often cannot be obtained in closed form and it is necessary to employ numerical techniques and a digital computer to obtain the solution. Many sophisticated computer programs are available for setting up and solving the equations that arise in the simulation of circuit behavior.[3,4]

It was noted above that for most purposes the transient behavior of the gate could be characterized by a delay time between the excitation and response waveform, and the rise and fall times of the response waveform. The formulas for these characteristic times are obtained by replacing each nonlinear capacitor by a linear capacitor that is equivalent in the sense that both capacitors store the same amount of charge when subjected to the same voltage excursion.[5]

The base circuit response of the current mode switch will now be considered. The response at the common emitter node is usually much faster than the collector circuit response because with at least one transistor conducting at all times, the impedance between that node and ground is small. Recall that there is a depletion capacitance associated with each junction. The effect of the collector-base capacitor of a transistor is multiplied by the nonlinear equivalent of the Miller effect. Assuming that the emitter junction is abrupt, the collector junction is linearly graded and the logic swing is about 0.8 V, the equivalent linear junction capacitance, $\overline{C}_j$, is

$$\overline{C}_j \approx 1.41 C_{BEO} + 1.62 C_{BCO} \qquad (14.25)$$

where C_{BEO} is the zero bias value of the emitter-base junction capacitance and C_{BCO} is the zero bias value of the collector-base capacitance.

The input capacitance has a diffusion capacitance component in addition to the depletion capacitance. The average value of the base-current dependent diffusion capacitance, $\overline{C}_d$, is given by

$$\overline{C}_d \approx \frac{1}{m \alpha_o R_C \omega_T} , \qquad (14.26)$$

where m is the number of transistors on the NOR side of the gate whose inputs rise to the logic 1 level at $t = 0$ and where ω_T is the gain-bandwidth product of the transistor. In deriving this equation the frequency dependence of α was described by a single pole.

We assume that the voltage at the emitter terminal of the current mode switch is constant so that the stray capacitance from the emitter terminal to ground has no effect. The equivalent base circuit is a linear, time-invariant single-pole RC circuit with time constant

$$\tau_{b1} = r_b(\overline{C}_j + \overline{C}_d), \tag{14.27}$$

where r_b is the dynamic resistance of the base-emitter junction seen from the base terminal. The delay time required for the response of a linear RC circuit to reach its 50% point after the application of a unit step is $t_d \approx \tau \ln 2$, and the 10% to 90% rise time is $t_r \approx \tau \ln 9$ where τ is the time constant. For the linearized base circuit

$$t_{db} = 0.7\tau_{b1}, \tag{14.28a}$$

and

$$t_{rb} = 2.2\tau_{b1}. \tag{14.28b}$$

The response of this linearized base circuit is not completely identical to that of the actual base circuit. There is good agreement between the actual delay and rise time and these estimates because of the way in which the parameters of the linearized circuits are chosen.

The capacitance loading the collector terminal has components due to the collector-base capacitance and the collector-substrate capacitance of the current mode switch transistors, the input capacitance of the output emitter follower, and the parasitic capacitance associated with the collector-load resistor. Assuming that the logic swing $v_l = 0.8$ V, that the supply voltage $V_+ = 5$ V, that the collector-base junction is linearly graded, and that the feedthrough coupling is included in the first term yields the following approximate expression for the collector-node capacitance.

$$\overline{C}_1 = 1.55NC_{CBO} + 0.88C_{ef} + C_{Rc} + 0.4A_hC_h + 0.55A_vC_v \tag{14.29}$$

In this expression N is the number of input transistors, C_{ef} is the zero bias value of the emitter-follower collector-base capacitance, A_h and A_v are the horizontal and sidewall areas of the input transistor array, and C_h and C_v are the corresponding zero bias values of parasitic capacitance.

The collector-circuit time constant is $\tau_{c1} = R_c\overline{C}_1$. Under the very reasonable assumption that the input swing is large enough to cause the current mode switch to regenerate the waveform ($v > 10V_T$), the 50% delay time of the current mode switch is the sum of the base- and collector-circuit delay times.

$$t_{dg} = t_{db} + t_{dc} = 0.7(\tau_{b1} + \tau_{c1}) \tag{14.30}$$

The ECL gate has signal reshaping properties by virtue of the gain because the output circuit is excited by the input circuit only during the time interval in which the input signal is in the transition region. If the input signal amplitude is larger than about $10V_T$, the rise and fall times of the collector waveform are effectively limited by the collector-circuit time constant and are given by

$$t_{rg} = t_{fg} = 2.2\tau_{c1}. \tag{14.31}$$

To obtain the total response time of the gate we must consider the effect of the output emitter follower. Two cases must be considered because when the output voltage is rising, the emitter follower is active, whereas when the CMS output voltage falls, the emitter follower is driven toward cutoff which increases the response time. The output emitter follower of the gate is assumed to be driving a load capacitance C_L. When the emitter follower is active, it loads the output of the current mode switch with a capacitor whose value is $C_L/(1+\beta_{Nf})$ where β_{Nf} is the common-emitter current gain of the emitter-follower transistor. Therefore the delay time and rise time of the ECL gate, from Eqs. (14.29) through (14.31), is

$$t_r = 2.2\tau'_{c1} \tag{14.32a}$$

$$t_{dr} = 0.7(\tau_{b1} + \tau'_{c1}) \tag{14.32b}$$

where $\tau'_{c1} = R_{C1}[\overline{C}_1 + C_L/(1+\beta_{Nf})]$.

The fall time for a waveform at the output of a linear amplifier is given by $\sqrt{t_w^2 + t_a^2}$ where t_a is the fall time of the amplifier and t_w is the fall time for its excitation waveform.[6] Thus, for the ECL gate with an output emitter follower

$$t_f \approx \sqrt{t_{fg}^2 + t_{fef}^2}$$
$$\approx \sqrt{(2.2\tau_{c1})^2 + (0.8R_EC_Lv_l/V_R)^2}. \tag{14.33}$$

The first term in this equation is the RC time constant in the collector circuit of the current mode switch as given by Eq. (14.31). The second term of Eq. (14.33) describes the time required to remove sufficient charge from C_L to change the voltage across it by v_l volts. The actual relations are nonlinear because the charge will be removed both by the resistor R_E and the transistor as it reenters the active region. Since the reference voltage is in the center of the transition region, V_R is the average value of the output voltage and V_R/R_E is the average current through the load resistor. Note that if the circuit were linear, $v_l = It_f/C_L$.

The delay time for the ECL gate on the falling waveform is the sum of the gate delay time, t_{dg}, given by Eq. (14.30) and the delay time of the emitter follower with a negative going excitation.

$$t_{df} \approx 0.7(\tau_{b1}+\tau_{c1}) + \frac{0.5R_EC_Lv_l}{V_R} \tag{14.34}$$

We have just calculated the response times of the NOR side of the ECL gate. The response times of the OR side are different because there is only one transistor on the OR side of the current mode switch and because there is no feedthrough coupling between the input and the OR side output of the current mode switch. The OR side response times can be obtained from the above results by setting $N = 1/2$ in Eq. (14.29).

EXAMPLE 14.4 Find the rise time, fall time, and delay time for the ECL gate designed in Example 14.2 when it drives a 20-pF load capacitance. Assume that the gate has three input transistors, that the parasitic capacitance of the input transistor array to the substrate is 4.5 pF, that the parasitic capacitance of the collector load resistor is 0.75 pF, and that the emitter follower collector-base capacitance is 1 pF. The parameters of the transistor are $\omega_T = 9.5 \times 10^8$, $\alpha_0 = 0.95$, $r_b = 200 \ \Omega$, and $C_{BEO} = C_{CBO} = 0.33$ pF.

Solution. From Eqs. (14.25) through (14.27) $\bar{C}_d = 0.439$ pF, $\bar{C}_j = 1$ pF, and base time constant $\tau_{b1} = 0.3$ ns. The collector node capacitance is found from Eq. (14.29) with $N = 3$, and the sum of the last two terms being 4.5 pF. Thus $\bar{C}_1 = 7.5$ pF.

We are now in a position to evaluate the required times from Eqs. (14.32) through (14.34). First

$$t_{c1}' = R_{C1}\left(\bar{C}_1 + \frac{C_L}{1+\beta_{Nf}}\right) = 7.23 \text{ ns.}$$

From Eq. (14.32b) the delay time on the rising edge of the waveform is

$$t_{dr} = 0.7(\tau_{b1} + \tau_{c1}') = 5.27 \text{ ns,}$$

and the rise time obtained from Eq. (14.32a) is $t_r = 15.9$ ns. From Eq. (14.33) the fall time $t_f = 14.1$ ns. Finally, from Eq. (14.34) the delay time on the falling waveform is $t_{df} = 6.33$ ns.

The response times for a gate depend upon the value of the collector-load resistor which enters into the RC time constants and on the magnitude of the current because it affects the transistor time constants. If the switching time depended only on RC products, the product of switching time and power dissipation would be a constant. The implication being that it takes a certain amount of energy to change the logic state of the gate and the actual switching time is inversely proportional to the power consumed. The situation is complicated by the variability of transistor parameters with current which means that the time delay-power product is no longer constant. It turns out that there is usually a minimum delay time-power product beyond which an additional penalty must be paid for increased speed. The time-delay power products for different logic families are shown in Fig. 14.9. The energy range for modern ECL gates is 50 to 100 pJ.

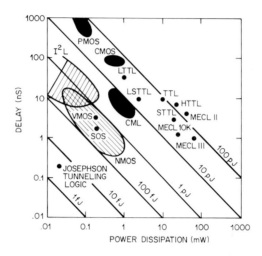

Fig. 14.9 Time delay versus power for different logic families.

14.2.3 Multilevel ECL Gating

The basic ECL gate implements the OR and NOR functions in positive logic. It is possible to extend the logic capability of each stage of gating by passing the same unit of current through several current mode switches in series. A specific example of multilevel gating is shown in Fig. 14.10.

In this circuit, if variable C is a logic 1, $i_c \approx I$ and current is available at the common emitter node of the upper current mode switch. The value of logic variables A and B determine the path of the current through the upper switch. If either A or B is a logic 1, implying that either $v(A)$ or $v(B)$ is greater than V_{R2}, then the current will take the path to the left. Hence $i_{CT} > 0$ if and only if the logic function $C(A+B)$ is true. Since

$$v_{o1} = V_+ - i_{CT}R_{C1} - V_{BEO} \tag{14.35}$$

the logical function at the output terminal is $F_1 = \overline{C(A+B)}$. In general, logical variables applied as inputs to a current mode switch on a particular level are combined by the OR operator, while variables applied to switches on different levels are combined by the AND operation. We now consider the function implemented at output v_{o2}.

If $v(C) < V_{R1}$, the current I is drawn through load resistor R_{C2} and therefore one term of F_2 is C. Furthermore, if $v(C) > V_{R1}$ and neither $v(A)$ nor $v(B)$ is greater than V_{R2}, the current I is once again drawn through

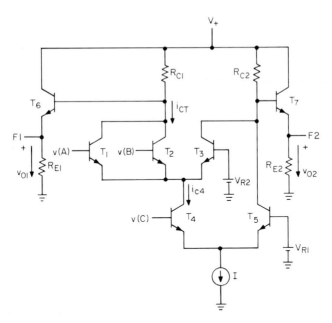

Fig. 14.10 A multilevel ECL gate.

the right-side load resistor. Hence,

$$F_2 = \overline{\overline{C} + C(\overline{A+B})} = C(A+B)$$

where DeMorgan's Laws[7] were used to simplify the last expression.

To illustrate the increase in logical capability afforded by multilevel gating, DeMorgan's Laws are applied to $F_1 = \overline{C(A+B)}$ so as to write it in a form suitable for implementation by OR/NOR gates. The result is:

$$F_2 = \overline{F}_1 = \overline{\overline{C(A+B)}} = \overline{C} + \overline{(A+B)}.$$

The mechanization of this equation is shown in Fig. 14.11. Three standard OR/NOR gates are used in place of one two-level gate. Certainly the two-level gate is more complex in that it requires two reference sources and a level shifter to translate the inputs to the lower gating level, but the additional circuitry only makes this two-level gate about twice as complex as a single-level gate. Multilevel gating is very popular in the implementation of MSI complexity functions in ECL.

14.2.4 Wired-AND and Wired-OR Capability

Wired function capability is something that we shall encounter again in other logic families. It is something for nothing because a metal connection is one

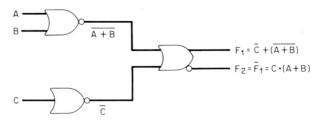

Fig. 14.11 Logical equivalent of the circuit shown in Fig. 14.10 using single level gates.

of the cheaper system components. Both the wired-OR and the wired-AND functions are possible with ECL logic. First consider two gates sharing a common emitter-follower load resistor as shown in Fig. 14.12.

Gate 1 implements the logical function F_1 whereas Gate 2 implements logical function F_2. The output voltage v_{out} will be high if F_1 is true because in this case transistor T_1 will conduct. If F_2 is false when F_1 is true, transistor T_2 will be cut off. If both F_1 and F_2 are true, both emitter followers will be active. Hence

$$F = F_1 + F_2.$$

This operation is called *wired-OR, OR-tied,* or *dotted-OR.*

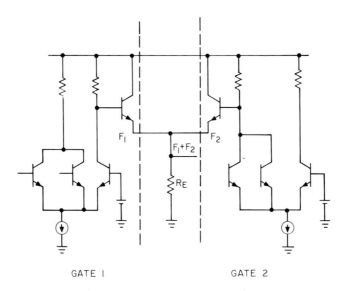

Fig. 14.12 The wired-OR function in ECL.

The *wired-AND* operation in ECL is performed at the common collector node of two current mode switches sharing the load resistor. The output function *F* is true only if both current mode switches divert current from the shared collector-load resistor. This function is known as the *wired-AND, dotted-AND,* or *tied-AND* function. Note that it was used in the circuit shown in Fig. 14.10.

14.3 COMPENSATED ECL [8]

We have seen that emitter-coupled logic is fast because the transistors are not driven into saturation and because of its relatively small logic swing. Although the absolute value of the noise margin is small, the noise behavior is relatively good because of the low impedance levels. Nevertheless, the low absolute value of the noise margin does influence system design because temperature and voltage variations from gate to gate further reduce the noise margin, and the relative effect of these variations is much greater because of the small logic swing. Plots of the transfer characteristics of ECL gates are shown in Fig. 14.13. From 14.13(a) and 14.13(c) it is seen that the transfer

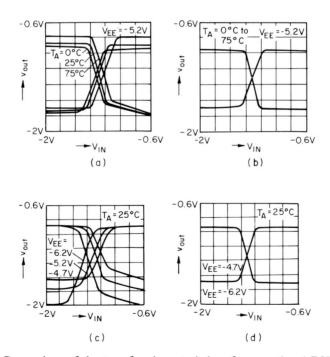

Fig. 14.13 Comparison of the transfer characteristics of conventional ECL and compensated ECL. Parts (a) and (b) compare temperature behavior. Parts (c) and (d) compare behavior with variations in supply voltage.

characteristics of conventional ECL varies markedly with changes in temperature and supply voltage. A large digital system must be designed to tolerate some variation in temperature and supply voltage between its component parts. The variation in the behavior of the individual gates influences the total system cost because it dictates, in large measure, power-supply voltage regulation and tolerable temperature variation. From the remaining parts of this figure, it is seen that transfer characteristics of fully compensated ECL gates, which are described in this section, are practically independent of temperature variation from 0°C to 75°C and power supply variation from 4.7 to 6.2 V.

The transfer characteristics of the ECL gate are a function of the threshold reference voltage, V_R, against which the input signal is compared and the magnitude of the current that is controlled by the current mode switch. The switch current, which might be a function of the supply voltage, affects the logic swing v_l, whereas the value of the reference voltage determines how symmetrically the actual input voltage excursion fits on the transfer characteristic. The voltage levels are also functions of temperature because the voltage drop across a p-n junction is temperature dependent.

The circuit diagram for a compensated ECL gate is shown in Fig. 14.14. The bias reference network may be common to many gates on the same chip.

Transistors T_7, T_8, and T_{12}, and resistors R_5 and R_6 form a bandgap voltage reference as described in Section 12.13. Transistor T_7 is operated at a higher current level than T_8. The voltage developed across R_5 is the differential base-emitter voltage $\Delta V_{BE} = V_{BE7} - V_{BE8}$ which has a positive temperature coefficient. Transistors T_7 through T_{11} operate at about the same current levels. Assuming high-gain transistors, the voltage at the emitter of T_{11} is $V_D + (R_6/R_5)\Delta V_{BE}$ and therefore the current source reference voltage is

$$V_{cs} \approx V_D + \left(1 + \frac{R_6}{R_5}\right)\Delta V_{BE} \qquad (14.36)$$

where $V_{BE9} - V_{BE11} \approx \Delta V_{BE}$ was assumed. Increasing temperature causes V_D to decrease and ΔV_{BE} to increase. The ratio R_6/R_5 is temperature independent to a first approximation and correct choice of this ratio will render V_{cs} temperature independent. Transistor T_{13} is a substrate p-n-p transistor which is used as a shunt regulator to render V_D independent of variations in the supply voltage V_+. Its presence ensures that the current through R_{10} remains constant with supply voltage. This current is also the collector current of T_{12} and so $V_{BE12} = V_D$ will remain constant with changes in supply voltage. It is shown in Section 12.13, where this bandgap reference is discussed in more detail, that the voltage V_{cs} is independent of tempera-

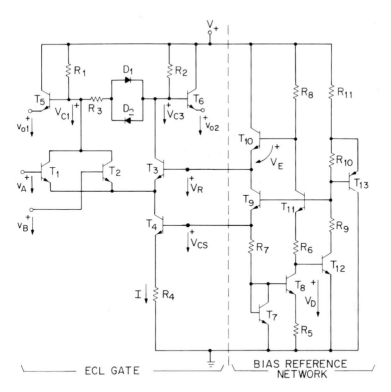

Fig. 14.14 Circuit of a compensated ECL gate. Note that the bias reference network can be common to several gates.

ture or supply voltage when it is approximately equal to the bandgap voltage of silicon. Also note that with reference to transistor T_{10},

$$V_+ - V_R = V_E + \frac{R_8 \Delta V_{BE}}{R_4} ; \tag{14.37}$$

as temperature increases, V_E decreases and ΔV_{BE} increases. Hence $V_+ - V_R$ can be made constant by a proper choice of resistor R_8. Recall from Eqs. (14.17) and (14.19) that the output voltage of the gate is also a function of V_+ and therefore, according to Eq. (14.37), the reference voltage can be maintained in the center of the transition region independent of temperature.

The ECL gate itself differs from the conventional gate discussed above in that there is a diode-resistor network between the collector nodes of the current mode switch and that the current source contains an emitter resistor. Since V_{cs} is constant and the base-emitter voltage of a transistor decreases with increasing temperature, T_4 provides a current I to the switch that increases as a function of temperature. The combination of the temperature-dependent current I and the temperature-dependent currents, I_{D1} and I_{D2}, in the diode-resistor collector-coupling network compensates for the temperature dependence of the output emitter followers T_5 and T_6.

Depending on the state of the gate, either D_1 or D_2 will be conducting. Figure 14.15 is the equivalent circuit when T_3 is conducting. The current mode switch and its current source are represented by the current source $\alpha^2 I_E$. Note that V_{c1}, the voltage at the collector node on the off-side of the gate, does not rise to V_+ because of the diode coupling network. The analysis of this circuit to obtain the output voltage as a function of temperature is straightforward. The starting point is the Kirchhoff current law equations at nodes A and B, together with the observation that the logic swing is given by

$$v_l = (I' - I_x - I_{B4})R = I_x R + V_{dx} \tag{14.38}$$

where it is assumed that $R_1 = R_2 = R_3 = R$. Solving for I_x and substituting the result into the current law equation at node B yields a relation for I' in terms of the emitter current. The final result for the low-state output voltage (v_{o2} in this case) is

$$v_{o2} = V_{OL} = V_+ - \frac{2}{3}\left\{\frac{\alpha^2 R}{R_4}\left[V_{cs} - V_{BE4}(T)\right]\right.$$

$$\left. + \frac{1}{2}V_{DX} + \left[I_{BL} + \frac{1}{2}I_{BU}\right]R\right\} - V_{BEOL}$$

$$\tag{14.39}$$

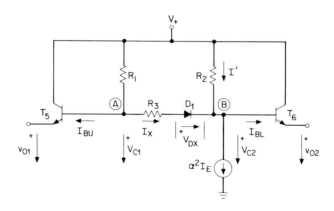

Fig. 14.15 Equivalent circuit of compensated ECL gate of Fig. 14.14 when transistor T_3 is conducting.

where V_{BEOL} is the base-emitter voltage of the emitter follower T_6 when the output is a logic 0. Similarly, the high state output from emitter follower T_5 is

$$v_{o1} = V_{OU} = V_+ - \frac{1}{3} \left\{ \frac{\alpha^2 R}{R_4} [V_{cs} - V_{BE4}(T)] \right.$$

$$\left. + V_{DX} + (I_{BL} + 2I_{BU})R \right\} - V_{BEOU}.$$

$$(14.40)$$

We now assume that base currents I_{BL} and I_{BU} are small so that they cause a negligible voltage drop across resistance R, that the ECL transfer characteristic is symmetric about the threshold voltage and that the gate is input-output compatible. Thus, from Eqs. (14.39) and (14.40),

$$\frac{1}{2}(V_{OU} + V_{OL}) = V_+ - \frac{1}{2}(V_{BEOU} + V_{BEOL}) - \frac{1}{3}V_{DX} - \frac{\alpha^2 R}{2R_4}(V_{cs} - V_{BE4})$$

$$= V_R. \qquad (14.41)$$

The reference voltage is a constant obtained from Eq. (14.37). Another expression for the logic swing v_l is obtained from Eqs. (14.39) and (14.40):

$$v_l = V_{BEOL} - V_{BEOU} + \frac{1}{3}\frac{\alpha^2 R}{R_4}(V_{cs} - V_{BE4}). \qquad (14.42)$$

The transfer characteristic is temperature independent if the threshold level and logic swing are independent of temperature. Taking the derivatives of Eqs. (14.41) and (14.42) yields

$$\frac{dV_{BEOU}}{dT} + \frac{dV_{BEOL}}{dT} + \frac{2}{3}\frac{dV_{DX}}{dT} - \frac{\alpha^2 R}{R_4}\frac{dV_{BE4}}{dT} = 0 \qquad (14.43a)$$

and

$$\frac{dv_l}{dT} = \frac{dV_{BEOL}}{dT} - \frac{dV_{BEOU}}{dT} - \frac{1}{3}\frac{\alpha^2 R}{R_4}\frac{dV_{BE4}}{dT} = 0. \qquad (14.43b)$$

From these two equations we obtain a relation for the ratio of R/R_4:

$$\frac{\alpha^2 R}{R_4}\frac{dV_{BE4}}{dT} = 3\frac{dV_{BEOU}}{dT} + \frac{dV_{DX}}{dT}.$$

Let us assume that all of the diodes have the same temperature variation, that the logic swing is 800 mV, and that $V_{BE4}(T_o) = 0.7$ V. From this last equation $(\alpha^2 R/R_4) = 4$, and from Eq. (14.42) with $V_{BEOL} \approx V_{BEOU}$, $V_{cs} = 1.3$ V.

14.4 TRANSISTOR-TRANSISTOR LOGIC

Transistor-transistor logic (T^2L) is probably the most popular type of integrated circuit logic. It is a type of saturating logic, so called because some of the transistors are allowed to saturate under certain conditions.

There is no discrete component counterpart for T^2L. It was developed in the course of simplifying the design and improving the performance of integrated circuit diode-transistor logic (DTL). Therefore we digress to describe DTL before beginning the discussion of T^2L.

14.4.1 Diode Transistor Logic (DTL)

A diode transistor logic (DTL) gate implements the NAND operation. It consists of a diode AND gate in cascade with a diode-level shifter and a transistor inverter, as shown in Fig. 14.16. The diode AND gate consists of diodes D_A, D_B, and D_C, and resistor R. If any of the input voltages is approximately 0 V, the corresponding diode conducts and point P is clamped

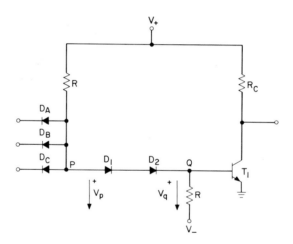

Fig. 14.16 A DTL gate.

about one diode drop above ground. Resistor R_1 provides bias current for level-shifting diodes D_1 and D_2 which are always forward biased. They provide the voltage translation to ensure that the circuit has compatible input and output voltage levels. Transistor T_1 and resistor R_C form the inverter stage, which is off when point P is clamped one diode drop above ground. Under that condition the output is a logic 1 using the positive logic convention.

Since the input and output levels are compatible, the logic 0 input voltage is $V_{CE\ \text{sat}}$. Hence point P is one diode drop above $V_{CE\ \text{sat}}$.

$$V_p^{(0)} = V_{CE\ \text{sat}} + V_D. \tag{14.44}$$

The base voltage of transistor T_1 is two diode drops lower than V_p because the level-shifting diodes D_1 and D_2 are always forward biased by current through resistor R and R_1. Hence the voltage at point Q is given by

$$V_q = V_p^{(0)} - 2V_D' = V_{CE\ \text{sat}} - V_D < 0, \tag{14.45}$$

where V_D' is the drop across each level-shifting diode. Thus $V_q < 0$ and transistor T_1 is always cut off whenever any of the inputs is at the logic 0 level. In this state there is only a small leakage current in the collector of T_1 and the current in the collector load resistor is determined primarily by the external load. For a negligible load the logic 1 output level is approximately V_+.

At coincidence, when all of the inputs are at the logic 1 level (all input diodes are not conducting), V_p can rise no higher than

$$V_p^{(1)} = 2V_D + V_{BE\ \text{sat}}. \tag{14.46}$$

Therefore the lowest input voltage that will be interpreted as a logic 1 level is approximately $3V_D$ because all of the input diodes must be off.

To show that transistor T_1 is in saturation under these conditions, assume that $V_q = V_{BE\ \text{sat}}$. Hence, the current I_{R1} through resistor R_1 is $[V_{BE\ \text{sat}} - V_-]/R_1$. The base current I_B of transistor T_1 is

$$I_B = I_R - I_{R1} = \frac{V_+ - V_p^{(1)}}{R} - \frac{V_{BE\ \text{sat}} - V_-}{R_1} \tag{14.47}$$

and the output transistor will be in saturation if $I_C < \beta I_B$. This condition can always be satisfied by proper selection of R and R_1. It is left as an exercise to show that the fan-out capability of the DTL gate, neglecting current diverted to the substrate, is given by

$$n < \beta_F \frac{V_+ - 2V_D - V_{BE\ \text{sat}}}{V_+ - V_D - V_{CE\ \text{sat}}} - \frac{V_+ - V_{CE\ \text{sat}}}{V_+ - V_D - V_{CE\ \text{sat}}} \frac{R}{R_C} \tag{14.48}$$

where β_F is the effective or forced β of the transistor in saturation. The upper limit on fan-out is set by the current gain of the output transistor.

The logic swing v_l is given by

$$v_l = [V_+ - V_{CE\ \text{sat}}]\left[1 - \frac{r_c}{R_C} - \frac{N r_c}{R}\right] + \frac{N r_c}{R} V_D \tag{14.49}$$

where the output transistor in saturation has its collector circuit modeled by a voltage source in series with a small saturation resistance r_c.

EXAMPLE 14.5 The DTL gate of Fig. 14.16 has $R = 8 \text{ k}\Omega$, $R_C = 1 \text{ k}\Omega$, and $R_1 = 10 \text{ k}\Omega$. It is operated with $V_+ = 5$ V and $V_- = -5$ V. Assume that the diode leakage current $I_s = 10^{-14}$ A, $\beta = 50$, $\beta_F = 30$, $V_{CE \text{ sat}} = 0.3$ V, $r_c = 50 \ \Omega$, and $V_{BE \text{ sat}} = 0.8$ V. Find the logic swing and fan-out.

Solution: Assume that the gate is driven by a similar gate so that the low logic level input is $V_{CE \text{ sat}} = 0.3$ V. If only one input is held low all of the current through resistor R goes through the one diode. Again using an iterative technique, assume that $V_D = 0.8$ V, so that

$$I_R = \frac{V_+ - V_p}{R} = \frac{5 - 0.3 - 0.8}{8 \times 10^3} = 0.488 \text{ mA}$$

which implies that $V_D = 640$ mV at 300 K. Recomputing I_R yields a value of 0.508 mA and a new value for V_D of 641 mV. The variation of V_p will actually be somewhat different than this because of the current through the level shifter, but the correction will be small judging by the convergence of the above iterative procedure.

The fan-out is given by Eq. (14.48),

$$N < 30 \left[\frac{5 - 2(0.640) - 0.8}{5 - 0.64 - 0.3} \right] - \frac{5 - 0.3}{5 - 0.640 - 0.3} \frac{8}{1} = 12.32,$$

but since fan-out must be an integer, $N < 12$. Finally the logic swing is

$$v_l = (5 - 0.3) \left[1 - \frac{50}{1000} - \frac{12 \times 50}{8000} \right] + \frac{12 \times 50}{8000} (0.640) = 4.07 \text{ V}.$$

14.4.2 Transistor-Transistor Logic Gate [9]

A three-input T^2L gate is shown in Fig. 14.17. In this circuit the base-emitter junctions of the input transistor T_1 are equivalent to the input diodes of the DTL gate. Similarly, the collector-base junction of T_1 performs the level shifting function. The advantages of T^2L over DTL are that it requires less silicon surface area than the integrated DTL circuit, and that it is faster than DTL because the capacitances are smaller and the transistor action of T_1 provides active turn-off of the inverter transistor T_2.

The operation of the T^2L gate has some similarity to that of the DTL gate. When an input is held low, the corresponding base-emitter junction conducts. We will now show that the input transistor is in saturation and the inverter transistor is effectively cut off under this condition.

Recall that a transistor is in saturation if both junctions are forward biased, or if $I_C < \beta I_B$. The current into the base of T_1 is

$$I_1 = \frac{V_+ - V_p}{R}. \tag{14.50}$$

The voltage at node p corresponding to a logic 0 level $V_i^{(0)}$ at any input is

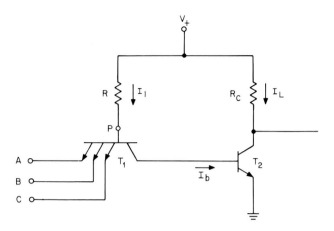

Fig. 14.17 A 3-input TTL gate.

$$V_p^{(0)} = V_i^{(0)} + V_{BE1}^{(0)} = V_i^{(0)} + V_{BE \text{ sat}}. \tag{14.51}$$

This voltage is also developed across the series-connected base-collector junction of T_1 and the base-emitter junction of T_2, which are in the direction to be forward biased by a positive V_p. However, $V_p^{(0)}$ is less than the voltage necessary to heavily forward bias these two diodes. Hence T_1 is in saturation because both of its junctions are forward biased, and also because the collector current of T_1 is almost zero since the base of T_2 cannot supply the current required for operation in the active mode.

Using the four-layer transistor model described in Chapter 2, the collector-base junction voltage of T_1 corresponding to zero collector current is found to be

$$\exp\left(\frac{V_{BC}}{V_T}\right) \approx \frac{\alpha_N I_{s1}\left[\exp\left(\dfrac{V_{BE}}{V_T}\right) - 1\right]}{(1-\alpha_S)I_{s2}} \tag{14.52a}$$

where it was assumed that the collector-substrate bias is zero. If it is assumed that the exponentials are very large compared to unity, $I_{s2} = 10I_{s1}$, $\alpha_N \approx 1$, and $\alpha_S = 0.2$, this equation reduces to

$$V_{BC} \approx V_{BE} - 2.1 V_T. \tag{14.52b}$$

The inverter is not truly cut off because the base-emitter junction is slightly forward biased. We define the transistor to be effectively cut off when the collector current is less than 1% of its value at the edge of saturation. The voltage across the intrinsic base-emitter junction when the transistor is at the edge of saturation is $V_{BE}' \approx V_{BE \text{ sat}}$ and the corresponding collector current is

$$I_C' \approx \alpha_N I_{s1} \exp\left(\frac{V_{BE}'}{nV_T}\right). \tag{14.53a}$$

Now

$$I_C(\text{cutoff}) \triangleq 0.01 I_C' \approx \alpha_N I_{s1} \exp\left(\frac{V_{BE}^*}{nV_T}\right) \tag{14.53b}$$

where V_{BE}^* is the intrinsic base-emitter voltage when the transistor is effectively cutoff, and n is the emission coefficient, $1 \leqslant n \leqslant 2$, which approaches two for small currents. The approximations in Eqs. (14.53) are valid for $V_{BE} > \approx 3nV_T$. Taking the ratio of Eqs. (14.53a) to (14.53b) and solving for V_{BE}^* yields

$$V_{BE}^* = V_{BE}' - nV_T \ln(100)$$
$$\approx V_{BE}' - 9.2V_T \tag{14.54}$$

where $n = 2$ was assumed. At room temperature V_{BE}^* is about 240 mV less than V_{BE}'. From Eq. (14.51), Eq. (14.52b), and

$$V_p^{(0)} = V_{BC1}^{(0)} + V_{BE2}^{(0)},$$

it follows the $V_{BE2}^{(0)}$ is about $2V_T$ larger than $V_i^{(0)}$. Combining this result with Eq. (14.54) yields an upper bound on the logic 0 output level $V_i^{(0)}$, namely

$$V_i^{(0)} < V_{BE}' - 11V_T.$$

With $V_{BE2}' = 0.7$ V which is a typical value for $V_{BE\,\text{sat}}$, $V_i^{(0)}$ must be less than 410 mV at room temperature. The worst-case logic 0 output level for commercial T^{2}L integrated circuits is 400 mV.

A single input T^{2}L gate, including the parasitic substrate p-n-p transistors, is shown in Fig. 14.18. We have just shown that transistor T_1 is in saturation when the input is held low. Parasitic transistor T_3 is therefore in its active region. We will now show that the parasitic transistor has a beneficial effect under this condition because it reduces the input current I_i which the driving gate must sink.

From the four-layer transistor model the following expressions are obtained for the base and emitter currents of the composite input transistor (T_1 and T_3):

$$I_B = (1-\alpha_N)I_{s1}\left[\exp\left(\frac{V_{BE}}{V_T}\right) - 1\right] + (1-\alpha_I)I_{s2}\left[\exp\left(\frac{V_{BC}}{V_T}\right) - 1\right]$$
$$- \alpha_{SI}I_{s3}\left[\exp\left(\frac{V_{SC}}{V_T}\right) - 1\right]$$

$$\tag{14.55a}$$

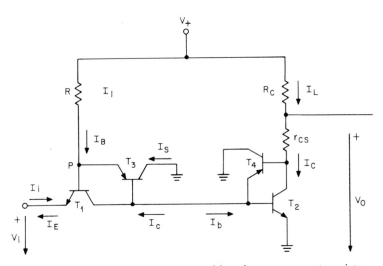

Fig. 14.18 A TTL inverter including the parasitic substrate p-n-p transistors.

$$I_E = I_{s1}\left[\exp\left(\frac{V_{BE}}{V_T}\right) - 1\right] - \alpha_I I_{s2}\left[\exp\left(\frac{V_{BC}}{V_T}\right) - 1\right]. \qquad (14.55b)$$

The last term of Eq. (14.55a) is negligible because the collector-substrate junction is reverse biased. Furthermore, from Eqs. (14.50) and (14.51)

$$I_B = I_1^{(0)} = \frac{V_+ - V_i^{(0)} - V_{BE\,\text{sat}}}{R}$$

is almost constant. The junction voltages V_{BE} and V_{BC} are further related by Eq. (14.52) because the collector current I_C of the composite input transistor is almost zero. Therefore, from Eqs. (14.52) and (14.55a)

$$\exp\left(\frac{V_{BE}}{V_T}\right) - 1 = \frac{I_1^{(0)}}{I_{s1}\left\{(1-\alpha_N) + \alpha_N\dfrac{1-\alpha_I}{1-\alpha_S}\right\}}, \qquad (14.56a)$$

$$\exp\left(\frac{V_{BC}}{V_T}\right) - 1 = \frac{I_1^{(0)}}{I_{s2}\left\{(1-\alpha_I) + \dfrac{(1-\alpha_N)(1-\alpha_s)}{\alpha_N}\right\}}. \qquad (14.56b)$$

Substituting these expressions into Eq. (14.55b) yields an expression for the input current in terms of I_1:

$$I_i^{(0)} = -\frac{I_1^{(0)}}{1 + \dfrac{\alpha_N\alpha_S}{1-\alpha_S-\alpha_N\alpha_I}}. \qquad (14.57)$$

The negative sign means the current is directed out of the input terminal and must be sinked by the driving gate. This current is divided among the m inputs of a multiple input gate that are held low. Note that if α_S were zero, $I_i = I_1$, whereas with $\alpha_S \neq 0$ some of the current I_1 is shunted to the substrate through the parasitic transistor T_3. Thus, the input current that must be sinked by the driving stage is less than $I_1^{(0)}$ which is the beneficial effect of the parastic transistor.

The input transistor is in the inverse active mode when all of the inputs to the gate are at the logic 1 level. This is because all of the base-emitter junctions are reverse biased and the base-collector junction is forward biased. Therefore $I_b^{(1)} > 0$, and

$$V_p^{(1)} = V_{BC1}^{(1)} + V_{BE2}^{(1)}$$
$$= V_{BC} + V_{BE\ sat} \approx 2V_D, \tag{14.58}$$

where V_D is the voltage across a saturated p-n junction, and in the absence of parasitic transistor effects

$$I_b^{(1)} = (1+\beta_I)I_1^{(1)}$$
$$= (1+\beta_I)\,\frac{V_+ - V_p^{(1)}}{R}\ . \tag{14.59}$$

However, the parasitic transistor T_3 is once again in the active mode. This time its shunting effect is undesirable since it reduces the base drive to the inverter, thus reducing the fan-out capability of the gate. The extent of this effect is determined by computing the current transfer ratio $I_b^{(1)}/I_1^{(1)}$.

When both the collector-substrate and base-emitter junctions are reverse biased, Eq. (14.55a) reduces to

$$I_B = I_1^{(1)} \approx (1-\alpha_I)I_{s2}\left[\exp\left(\frac{V_{BC}}{V_T}\right) - 1\right]. \tag{14.60}$$

The expression for collector current of the composite transistor is

$$I_C = \alpha_N I_{s1}\left[\exp\left(\frac{V_{BE}}{V_T}\right) - 1\right] - (1-\alpha_S)I_{s2}\left[\exp\left(\frac{V_{BC}}{V_T}\right) - 1\right]$$
$$- (1-\alpha_{SI})I_{s3}\left[\exp\left(\frac{V_{SC}}{V_T}\right) - 1\right] \tag{14.61a}$$

which, under the present conditions, reduces to

$$-I_C^{(1)} = I_b^{(1)} \approx (1-\alpha_S)I_{s2}\left[\exp\left(\frac{V_{BC}}{V_T}\right) - 1\right]. \tag{14.61b}$$

It was assumed in deriving Eqs. (14.60) and (14.61b) that the leakage currents are negligibly small. It follows from these two equations that

$$I_b^{(1)} \approx \frac{(1-\alpha_S)}{(1-\alpha_I)} I_1^{(1)} = (1-\alpha_S)(1+\beta_I) I_1^{(1)}. \tag{14.62}$$

Comparing this result with Eq. (14.59), we see that the parasitic *p-n-p* transistor reduces the base drive to the inverter transistor by the factor $(1-\alpha_S)$. Depending on the particular structure, this reduction in base drive might be quite significant.

The input current into each of the *m* inputs of the gate is

$$I_i^{(1)} = \frac{\alpha_I I_1^{(1)}}{(1-\alpha_I)m} = \frac{\beta_I}{m} I_1^{(1)}. \tag{14.63}$$

When all of the inputs are high, the inverter must be driven into saturation. In the absence of the parasitic *p-n-p* transistors T_3 and T_4, this requirement is satisfied by choosing resistor R such that

$$(1+\beta_I) I_1^{(1)} > \frac{I_C}{\beta_N}.$$

The base drive available for the inverter is reduced due to the shunting action of T_3 as given by Eq. (14.62). We now consider the effect of T_4.

Note that T_4 is in the active region when T_2 is in saturation. Thus, it too diverts current away from the base of T_2 and toward the substrate. The expressions for base and collector current obtained from the four-layer model, and given by Eqs. (14.55a) and (14.61), reduce to the following equations when the transistor is saturated.

$$I_C \approx \alpha_N I_{s1} \exp\left(\frac{V_{BE}}{V_T}\right) - (1-\alpha_S) I_{s2} \exp\left(\frac{V_{BC}}{V_T}\right)$$

$$I_B \approx (1-\alpha_N) I_{s1} \exp\left(\frac{V_{BE}}{V_T}\right) + (1-\alpha_I) I_{s2} \exp\left(\frac{V_{BC}}{V_T}\right)$$

The ratio I_C/I_B is the effective current gain of the composite transistor

$$\frac{I_C}{I_B} \triangleq \beta_F = \beta_N \left\{ \frac{\left|\left(\dfrac{I_{s1}}{I_{s2}}\right) \exp\left(\dfrac{V_{CE}}{V_T}\right) - \left(\dfrac{1-\alpha_S}{\alpha_N}\right)\right|}{\left|\left(\dfrac{I_{s1}}{I_{s2}}\right) \exp\left(\dfrac{V_{CE}}{V_T}\right) + \left(\dfrac{1-\alpha_I}{1-\alpha_N}\right)\right|} \right\},$$

$$\tag{14.64}$$

where we have used the relation $V_{CE} = V_{CB} + V_{BE}$ to simplify this result.

Recall that when a transistor is on the verge of saturation, $V_{BC} = 0$ and therefore $V_{CE} = V_{BE}$. The first term dominates in both the numerator and denominator because $V_{BE} \approx V_{BE\,sat}$ is many times V_T. Thus, the factor in braces is approximately unity. As the transistor is driven further into saturation, V_{CE} decreases and the other terms in the factor become significant. Since the factor in braces is at most equal to unity, $\beta_F \leqslant \beta_N$.

From Eqs. (14.62) and (14.64) it is seen that when the output current sinking capability is specified, resistor R must be chosen such that

$$
-I_1^{(1)} = \left[\frac{1}{(1+\beta_I)(1-\alpha_S)}\right]_1 \left[\frac{\left(\dfrac{I_{s1}}{I_{s2}}\right)\exp\left(\dfrac{V_{CE}}{V_T}\right) + \left(\dfrac{1-\alpha_I}{1-\alpha_N}\right)}{\left(\dfrac{I_{s1}}{I_{s2}}\right)\exp\left(\dfrac{V_{CE}}{V_T}\right) - \left(\dfrac{1-\alpha_S}{\alpha_N}\right)}\right]_2 \frac{I_C}{\beta_{N2}}
$$

(14.65)

where the subscripts on the factors in brackets indicate the transistor whose parameters are to be used in evaluating the factor.

Finally, the output voltage when all of the inputs are high is

$$
V_o^{(1)} = V_{CE\,sat} + I_C r_c \tag{14.66}
$$

where r_c is the collector saturation resistance of the output transistor. Transistor-transistor logic circuits are designed so that the output voltage is less than 0.4 V when all inputs are at logic 1 which places a requirement on the size of the output transistor to reduce r_c, and on the drive current $I_1^{(1)}$.

14.4.3 Fan-Out and DC Transfer Characteristics

Consider the situation of Fig. 14.19 in which a T^2L gate drives n identical T^2L gates. For simplicity it is also assumed that the driving gate is identical to the driven gates. We now calculate the fan-out of the driving gate.

The output stage of the driving gate must sink the input current from the n driven gates when the output is low. The worst case occurs when only one input from each of the driven gates is held low. For that case, the collector current of the driving gate is

$$
\begin{aligned}
I_C &= \frac{V_+ - V_o^{(g)}}{R_C} + nI_i^{(0)} \\
&\approx \left[\frac{1}{R_C} + \frac{n}{R}\right]\left(V_+ - V_o^{(g)}\right) - \frac{n}{R}V_D.
\end{aligned} \tag{14.67}
$$

Resistor R must be chosen so as to provide sufficient base drive to saturate the output transistor. The forced beta of the output transistor, $\beta_F = I_C/I_B$,

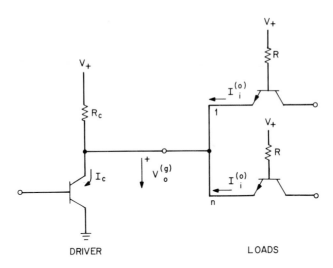

Fig. 14.19 A TTL gate driving n identical TTL gates.

given by Eq. (14.64), is a function of V_{CE}. Therefore the required base drive is dependent on the value of collector saturation resistance r_c which relates $V_o^{(g)}$ to V_{CE} as given by Eq. (14.66). From Eqs. (14.66) and (14.67)

$$V_{CE}^{(g)} = V_o^{(g)} - I_C r_c$$

$$= \left[1 + \frac{r_c}{R_C} + \frac{nr_c}{R_C}\right] V_o^{(g)} - \left[\frac{r_c}{R_C} + \frac{nr_c}{R}\right] V_+ + \frac{nr_c}{R} V_D$$

$$(14.68)$$

where once again, $V_{BE\ sat} \approx V_D$ the saturation voltage of a diode. Using this value of V_{CE} in Eq. (14.65), recalling that $I_1^{(1)} \approx (V_+ - 2V_D)/R$, and solving for the fan-out n yields

$$n = \frac{X\beta_N(1+\beta_I)(1-\alpha_S)(V_+ - 2V_D) - (R/R_C)\left[V_+ - V_o^{(g)}\right]}{V_+ - V_o^{(g)} - V_D},$$

$$(14.69a)$$

where

$$X = \frac{\dfrac{I_{s1}}{I_{s2}} \exp\left(\dfrac{V_{ce}^{(g)}}{V_T}\right) - \left(\dfrac{1-\alpha_S}{\alpha_N}\right)}{\dfrac{I_{s1}}{I_{s2}} \exp\left(\dfrac{V_{ce}^{(g)}}{V_T}\right) + \left(\dfrac{1-\alpha_I}{1-\alpha_N}\right)},$$

$$(14.69b)$$

and where both transistors of the gate are assumed to have the same parameters.

Another condition that must be satisfied is that the input emitters of the driven gates must be reverse biased when the output of the driving gate is high. The worst-case condition occurs when all of the inputs to each of the driven gates are high. Then the driving gates must supply a current $I_1^{(1)} \beta_I / m$ to each of the m inputs of each gate:

$$\frac{\beta_I}{m} I_1^{(1)} \approx \frac{\beta_I}{mR} (V_+ - 2V_D).$$

The input emitter junction is reverse biased whenever $V_o^{(g)} = V_i^{(0)} > \sim 2V_D$. Therefore, the high-level output imposes the following limitation on fan-out:

$$V_o^{(g)} = V_+ - \frac{n\beta_I}{m} \frac{R_C}{R} (V_+ - 2V_D) \geqslant 2V_D.$$

This equation is treated as a constraint on the ratio R_C/R:

$$\frac{R}{R_C} \geqslant \frac{n\beta_I}{m}.$$

Substituting the limiting case of this constraint into Eq. (14.69a) and solving for n yields the desired expression for fan-out:

$$n = \frac{X\beta_N(1+\beta_I)(1-\alpha_S)(V_+ - 2V_D)}{\left[\dfrac{1+\beta_I}{m}\right]\left[V_+ - V_o^{(g)}\right] - V_D}. \tag{14.70}$$

An input-output transfer characteristic for the T²L gate is shown in Fig. 14.20. Because the gates are input-output compatible, the logic swing is approximately

$$v_l \approx V_+ - V_{CE\ sat}. \tag{14.71}$$

The hump in the transfer characteristic is caused by a reduction in the gain of the driving gate when the input impedance of the driven gate decreases as its input transistor goes from the inverse active to the saturation region. Assuming that the transistor comes out of cutoff when the collector current is 1% of the current it carries just at the onset of saturation, the transition width $v_w \approx 4.6V_T$. Finally, if the power supply is more than a few volts and the gate output is a logic 0 level for half the time, the average power consumption is approximately

$$P_{av} \approx \left[\frac{1}{R} + \frac{1}{2R_C}\right] V_+^2 \tag{14.72}$$

and is typically 5 to 20 mW.

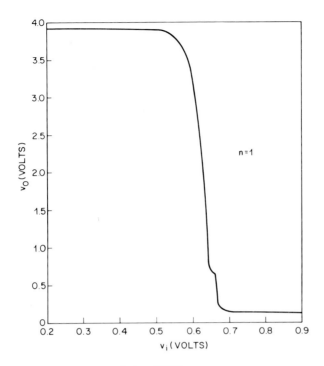

Fig. 14.20 Input-output characteristic of TTL gate.

14.4.4 Transient Analysis of the TTL inverter

We have already seen that the transient analysis of digital logic circuits is very difficult because of the nonlinearities and the complexity of the waveforms. The transient analysis of even a simple T^2L gate is extremely difficult and nonlinear circuit analysis routines on a digital computer must be used to get an accurate answer. However, under certain simplifying assumptions we can perform a piecewise linear analysis of a T^2L inverter to obtain an estimate of the delay, rise, and fall times.

Consider the inverter shown in Fig. 14.21(a) where the capacitors are *linear* and store the same average charge when subjected to the same voltage excursion as the nonlinear capacitors they replace.[5] Capacitor C_1 is the base-emitter capacitance of T_2 together with any other capacitance appearing from base to ground, C_2 is the collector-base capacitance plus any strays, and C_3 is the collector-substrate capacitance plus any strays from collector to ground. We saw in the previous section that the voltage across the collector-base junction of T_1 does not change much and for that reason the collector-base capacitance of T_1 is neglected. Furthermore its collector-

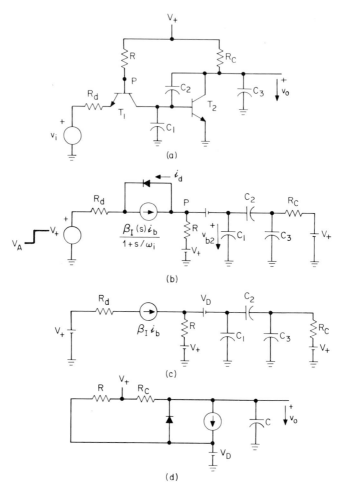

Fig. 14.21 (a) TTL gate with equivalent linear capacitors; (b) equivalent circuit for T_1 in saturation and T_2 cutoff—t_I; (c) equivalent circuit for T_1 in inverse active mode—t_{II}; (d) equivalent circuit for T_1 in inverse active mode and T_2 in active region—T_{III}.

substrate capacitance is lumped in with C_1. Finally, since V_p is almost constant, the base-emitter capacitance of T_1 is effectively from the emitter to ground and is neglected.

We begin the analysis by assuming that initially the input voltage $v_i = V_A \approx V_{CE\ sat}$, and at $t = 0$ it is switched to a value approximately V_+. Therefore transistor T_1 is in saturation and transistor T_2 is cut off leading to the equivalent circuit of Fig. 14.21(b) where a piecewise linear model has been used for T_1, and T_2 is not shown explicitly because it is cut off. Our first task is to calculate the time at which T_1 comes out of saturation. We find T_1 comes out of saturation when current $i_d = 0$. The current decay in the transistor is described by a first-order differential equation whose solution is

$$i_d(t) = i_d(\infty) + [i_d(0) - i_d(\infty)]\exp(-\omega_i t). \qquad (14.73)$$

The endpoint conditions $i_d(0)$ and $i_d(\infty)$ are obtained by applying KCL at the input node, namely

$$i_d(0) = \beta_I \frac{V_+ - 2V_D}{R} - \frac{V_A - 2V_D}{R_d}$$

$$i_d(\infty) = \beta_I \frac{V_+ - 2V_D}{R} - \frac{V_+ - 2V_D}{R_d}$$

$$= \frac{\beta_I}{R}\left[1 - \frac{R}{\beta_I R_d}\right](V_+ - 2V_D),$$

where it is assumed that $V_p \approx 2V_D$. Note that if $\beta_I R_D < R$, $i_d(\infty) < 0$ as is required. Solving Eq. (14.73) for that value of t at which $i_d = 0$ yields

$$t_1 = \frac{1}{\omega_i} \ln \frac{V_+ - V_A}{\left[1 - \beta_I \dfrac{R_d}{R}\right](V_+ - 2V_D)}. \qquad (14.74)$$

For $t > t_1$, transistor T_1 is in the inverse active mode resulting in the equivalent circuit shown in Fig. 14.21(c). However, in practical circuits β_I is small and the effect of the $\beta_I i_b$ source is neglected for simplicity. The equivalent circuit then reduces to three capacitors driven by two Thevenin voltage sources. This circuit is of second order but reduces to first order if $R_C = 0$ or $R_C = \infty$. The most pessimistic case occurs when $R_C = 0$. The time constant of this circuit is $T = R(C_1 + C_2)$, and the response $v_{b2}(t)$ is

$$v_{b2}(t) = (V_+ - V_D) + [V_A + 2.1V_T - V_+ + V_D]\exp\left(-\frac{t}{T}\right), \qquad (14.75)$$

where we have used

$$v_{b2}(0) = V_A + V_{BE1} - V_{BC1} \approx V_A + 2.1V_T.$$

We solve for the time at which $v_{b2} = V_D$ yielding

$$t_{II} = T \ln \frac{V_+ - V_D - V_A - 2.1V_T}{V_+ - 2V_D}. \qquad (14.76)$$

Once the base voltage v_{b2} reaches V_D, transistor T_2 is in the active region and there is no further change in v_{b2}. The equivalent circuit is shown in Fig. 14.21(d) where, once again, the contribution of the $\beta_I i_b$ source to the base drive of T_2 is neglected. In addition C_1 is neglected because v_{b2} is constant. Furthermore the value of C_2 is multiplied as a result of Miller effect and it appears in parallel with C_3.

Transistor T_2 is in the active region until $v_0 = V_D$. That time, denoted t_{III}, is easily calculated from this equivalent circuit:

$$t_{\mathrm{III}} = R_C C' \ln \frac{\beta_n (V_+ - V_D) R_C / R}{(V_D - V_+) + \beta_N (R_C/R)(V_+ - V_D)} . \tag{14.77}$$

EXAMPLE 14.6 Calculate the turn-on time until the output transistor reaches saturation for a TTL inverter with the following parameters: $R_d = R_C = 1\,\mathrm{k\Omega}$, $R = 10\,\mathrm{k\Omega}$, $V_+ = 5\,\mathrm{V}$, $V_A = 0.3\,\mathrm{V}$, $\beta_I = 1$, $\beta_N = 50$, $\omega_i = 10^9$, $C_1 = C_2 = 2\,\mathrm{pF}$ and $C_3 = 5\,\mathrm{pF}$.

Solution. From Eq. (14.74) the time required for the input transistor to come out of saturation is:

$$t_1 = 10^{-9}\ln \frac{5-0.3}{[1-0.1][5-2(0.8)]} = 0.43\ \mathrm{ns}.$$

The time it takes v_{b2} to reach the diode voltage $V_D = 0.8\,\mathrm{V}$ is obtained from Eq. (14.76):

$$t_{\mathrm{II}} = 10^3 \times 4 \times 10^{-12}\ln \frac{5-0.8-0.3-2.1(0.026)}{5-2(0.8)} = 0.493\ \mathrm{ns}.$$

The capacitance in the collector circuit of T_2 is $C' = 5 + 51(2) = 107\,\mathrm{pF}$. Therefore, from Eq. (14.77)

$$t_{\mathrm{III}} = 10^3 \times 107 \times 10^{-12}\ln \frac{50(5-0.8)(1/10)}{(0.8-5) + \dfrac{50}{10}(5-0.8)} = 23.9\ \mathrm{ns}.$$

The turn-on time is the sum of the above three components and is $t_{\mathrm{on}} = 24.8\ \mathrm{ns}$.

The turn-off transient begins when the input voltage v_i applied to the inverter is abruptly switched from a value close to V_+ to $v_i = V_A \approx V_{CE\,\mathrm{sat}}$. Prior to this change in input excitation, transistor T_2 was in saturation and T_1 was in the inverse active region so that both junctions of T_2 and the collector-base junction of T_1 were forward biased and storing excess charge. As the input voltage falls somewhat below $V_p \approx 2V_D$, the base-emitter junction of T_1 conducts[†] and the voltage V_p falls to $V_A + V_D$ as transistor T_1 enters saturation. The equivalent circuit is shown in Fig. 14.22(a) from which it is seen that there is a capacitive voltage divider from point P to

---

[†] By assumption of a step input voltage and a piecewise linear transistor characteristic, the base-emitter diode conducts abruptly and a voltage V_D is established across its terminals.

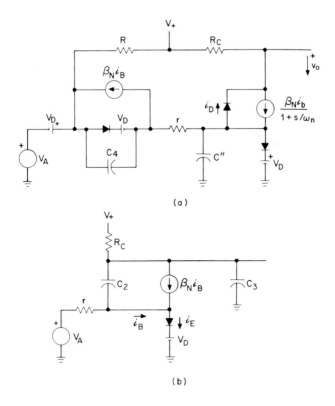

Fig. 14.22 (a) Equivalent circuit for turn-off transient $-T_{IV}$; (b) equivalent circuit for turn-off transient $-t_V$ and t_{VI}.

ground consisting of the Miller-effect multiplied collector-base capacitance of T_1 in series with the input capacitance of T_2. The capacitor values for the voltage divider are

$$C' = (1+\beta_N)C_4$$

and

$$C'' = C_1 + \left[1 + \frac{\beta_N}{1 + (s/\omega_n)}\right]C_2.$$

The capacitive voltage divider is excited by a step voltage of magnitude $V_D - V_A$ as the voltage V_p falls from about $2V_D$ to $V_D + V_A$. Most of this excitation is developed across C'' because T_2 is in saturation and there is no effective multiplication of C_2 as can be seen from the initial value theorem of Laplace transforms which yields

$$C''(t=0) = \lim_{s \to \infty} \left\{ C_1 + \left[1 + \frac{\beta_N}{1 + (s/\omega_n)}\right] C_2 \right\} = C_1 + C_2$$

$$\ll C'.$$

Therefore a turn-off signal is applied to the base-emitter junction of T_2 and the excess stored charge begins to recombine with a time constant equal to $1/\omega_n$. This current decay is described by an equation similar to Eq. (14.73):

$$i_D = i_D(\infty) + [i_D(0) - i_D(\infty)]\exp(-\omega_n t), \qquad (14.78)$$

where

$$i_D(0) = \frac{\beta_N(V_+ - V_D)}{R} - \frac{V_+ - V_D}{R_C}$$

and

$$i_D(\infty) = \frac{\beta_N(V_A - V_D)}{r} - \frac{V_+ - V_D}{R_C}.$$

In evaluating $i_D(\infty)$, note that it depends on the series resistance in the collector circuit of T_1. Furthermore $i_D(\infty)$ is always negative, which is required if T_2 is to come out of saturation. We substitute these endpoint conditions into Eq. (14.78) and solve for the time at which $i_D = 0$ yielding

$$t_{IV} = \frac{1}{\omega_n} \ln \frac{\beta_N\{1 + (r/R)[(V_+ - 2V_D)/(V_D - V_A)]\}}{\beta_N - (r/R_C)[(V_+ - V_D)/(V_D - V_A)]} . \qquad (14.79)$$

EXAMPLE 14.7   Consider the same gate as in Example 14.6 with $r = 200\ \Omega$, $\omega_n = 5 \times 10^7$ and $V_A = 0.4$ V. Calculate storage time t_{IV}.

Solution.   From Eq. (14.79)

$$t_{IV} = 2 \times 10^{-8} \ln \frac{50\{1 + (0.2/10)(3.4/0.4)\}}{50 - (0.2/1)(4.2/0.4)} = 4 \text{ ns.}$$

At the end of region IV, transistor T_2 enters the forward active region with the equivalent circuit shown in Fig. 14.22(b). This transistor will be cut off when its emitter current $i_E = 0$. From the equivalent circuit it is seen that

$$i_E(0) = \frac{V_+ - V_D}{R_C} + \frac{V_A - V_D}{r}$$

and

$$i_E(\infty) = \frac{(\beta_N+1)(V_A-V_D)}{r}.$$

The time t_V at which $i_E = 0$ is

$$t_V = R_C C' \ln \frac{[(V_+-V_D)/R_C] + [\beta_N(V_D-V_A)/r]}{(V_D-V_A)(\beta_N+1)/r}, \qquad (14.80)$$

where $C' = C_3 + (1+\beta_N)C_2$.

EXAMPLE 14.8   Evaluate t_V for the gate of Example 14.6.

Solution. $C' = 5 + (51\times2) = 107$ pF

$$t_V = 107\times10^{-9}\ln \frac{(4.2/1) + 50(0.4/0.2)}{0.4\times51/0.2} = 2.28 \text{ ns.}$$

At the end of region IV, the output transistor is cut off but the output voltage is still essentially V_D. In region IV the capacitance from the output node to ground is charged towards V_+ through the collector load resistor R_C. The endpoint conditions are $v_o(0) = V_D$ and $v_o(\infty) = V_+$ and therefore

$$v_o(t) = V_+ + [V_D-V_+]\exp\left(-\frac{t}{T}\right)$$

where $T = R_C(C_2+C_3)$. The time required for the output node to reach $0.9V_+$ is

$$t_{VI} = T \ln \frac{V_+-V_D}{0.1V_+}. \qquad (14.81)$$

Using the element values from the previous examples yields a value of 14.9 ns for t_{VI}. Thus the total turn-off time for the illustrative gate is $t_{off} \approx 21.2$ ns.

14.4.5 Active Pull-Up and Tristate Inverters

A modified T^2L gate with an active load pull-up is shown in Fig. 14.23. Transistors T_2 and T_3 are turned off when either or both of the inputs A and B are low. When T_2 and T_3 are off, the base of emitter follower T_4 is high and this transistor and the diode D conduct pulling the output high. The output voltage is given by

$$V_o^{(1)} = V_+ + V_{BE4} - V_D - I_{B4}R_2. \qquad (14.82)$$

Note that the maximum value attained by $V_o^{(1)}$ is two diode drops below V_+ as contrasted with the inverter using a passive load for which $V_o^{(1)} \rightarrow V_+$.

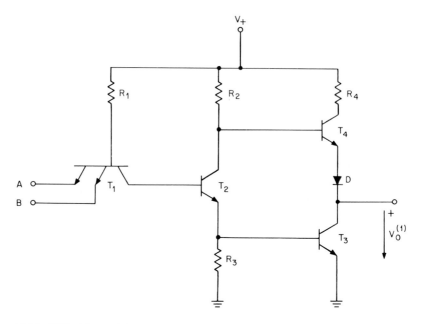

Fig. 14.23 TTL driver with an active load pull-up.

The output impedance of the gate is low because transistor T_4 is an emitter follower. Therefore, the high state output voltage is not affected by the input current of the driven gates.

When both the inputs A and B are high, T_1 will be in the inverse active region and T_2 and T_3 will be in saturation, while T_4 and diode D are off. The diode is required to ensure that T_4 will be off when T_2 is in saturation. The logic 0 output voltage is $V_o^{(0)} = V_{CE3\ sat}$, and transistor T_3 must be able to sink current from the driven gates. However, all of the collector current for T_3 is supplied by the driven gates because the emitter follower is off under this condition. Therefore the fan-out capability of this inverter is greater than that of the simple inverter considered previously. During the transition there is an interval of time in which transistors T_3 and T_4 and diode D conduct simultaneously. Current limiting resistor R_4 limits the size of the resulting current spike.

There are many occasions in the design of large digital systems where it is desirable to have both the advantage of an active pull-up and the capability to perform the wired-OR function. This situation often occurs in the implementation of memories whose capacity exceeds the number of words available on a single chip. The wired-OR function for T^2L is described in the next section. The active pull-up driver just described is not suitable for a wired-OR function because the power supply can be short circuited if one of the drivers tries to pull the output bus high while another driver tries to pull it low.

The tristate inverter shown in Fig. 14.24 employs a "disable" signal to put the driver in a "don't care" condition by turning off both output transistors. It is often possible to obtain the output disable signal from existing system inputs when this type of driver is used as the output of a circuit that implements a complex system function.

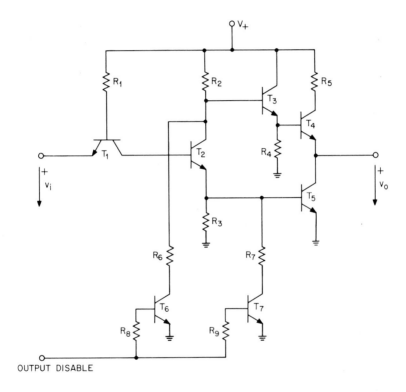

Fig. 14.24 A TTL inverter with tristate output.

14.4.6 Wired-OR Function and the AND-OR-INVERT Gate

We have seen how the wired-AND and wired-OR functions were used in ECL to increase the logical capability of a stage of gating. In this section we will consider the use of the wired-OR function in T^2L and DTL and examine, as an example, the AND-OR-INVERT (AOI) gate.

Consider two transistors T_1 and T_2 having a common collector terminal and a common emitter terminal connected to ground. If the voltage applied to either base corresponds to a logic 1, the corresponding transistor saturates and therefore the output voltage measured from collector to ground is small and interpreted as a logic 0. The only case for which the output is a logic 1 is when both A and B are logic 0 levels. Hence this circuit realizes the OR-INVERT (NOR) function:

$$F = \overline{A + B}.$$

Note that it differs from the usual T^2L or DTL output circuit only in that there are two transistors in parallel. Therefore we can get an OR function simply by providing an open collector output which can be wired to the output of a normal gate.

The basic logic function of DTL or T^2L is the NAND gate with which all combinational functions can be realized. However, it is often possible to effect a saving in the total number of gates required to implement a function by using the AOI function,

$$F = \overline{AB + CD}.$$

A particular implementation of the T^2L AOI gate is shown in Fig. 14.25.

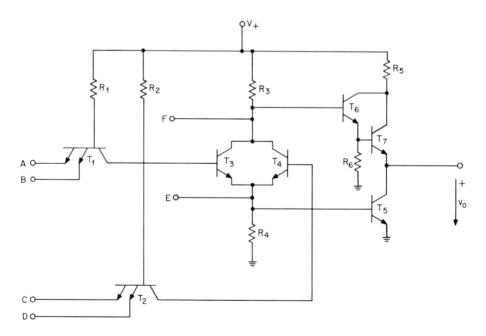

Fig. 14.25 A TTL AND-OR-INVERT gate with active load pull-up.

This circuit employs a variation of the active load pull-up circuit described in the previous section. The Darlington emitter follower performs the same function as the emitter follower and level shifting diode in the circuit of Fig. 14.23. An advantage of the Darlington connection is that only one isolation region is required because the transistors have a common collector. The wired-OR function is performed by the parallel connected transistors that form the phase splitter that drives the output stage. Additional transistors connected between points E and F provide more OR terms.

14.4.7 Schottky Diode Clamping [10]

We have seen before that the turn-off response time of a saturating logic gate is increased because of the time it takes to dissipate the excess charge stored in the saturated transistors. One very popular method of controlling the stored charge, and thus increasing the speed of the gate, is to put a Schottky barrier diode across the collector-base junction.

The Schottky barrier diode, described in Section 2.8, is a majority carrier device which does not store any charge. With a properly chosen metal, it has a lower forward voltage drop than a silicon p-n junction. Hence it can be used to control saturation of the transistor by diverting current from the transistor base and adding this current to the collector. The connection and a symbol for the composite device is shown in Fig. 14.26. We now derive expressions for the collector-emitter voltage, and storage time of the clamped transistor.

Kirchhoff's current law applied at nodes A and B, and inside the transistor, yield

$$I_B = I_1 - I_D, \tag{14.83a}$$

$$I_C = I_{C1} + I_D, \tag{14.83b}$$

$$I_E = I_B + I_C = I_1 + I_{C1}. \tag{14.83c}$$

The transistors are described by Ebers-Moll type equations and it is assumed that the junctions are sufficiently forward biased so that the exponential terms dominate. Therefore

$$I_B = (1-\alpha_N)I_{Ef} + (1-\alpha_I)I_{Cf}, \tag{14.84a}$$

$$I_C = \alpha_N I_{Ef} - I_{Cf} \tag{14.84b}$$

where $I_{Ef} = I_{s1}\exp(V_{BE}/V_T)$ and $I_{Cf} = I_{s2}\exp(V_{BC}/V_T)$.

The clamp diode limits the extent to which the transistor can saturate by diverting some of the input current away from the base of the transistor as is evident from Eq. (14.83a). This reduces the amount of stored charge due to the forward current of the collector-base junction I_{Cf}. It is this stored

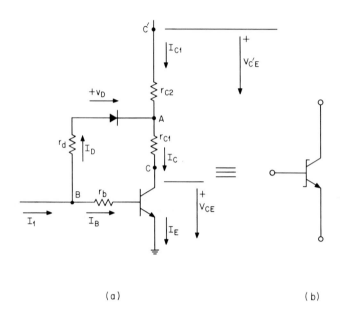

(a)   (b)

Fig. 14.26 (a) Equivalent circuit and (b) symbol for Schottky-diode-clamped bipolar transistor.

charge which must be removed before the transistor can come out of saturation. Clamping of the collector-base junction also increases the collector-emitter saturation voltage. From Eqs. (14.84a) and (14.84b) we obtain the following equation for the base current in terms of I_{Cf} and the collector current I_C:

$$I_B = \frac{1-\alpha_N}{\alpha_N} I_C + \frac{1-\alpha_N\alpha_I}{\alpha_N} I_{Cf}. \qquad (14.85)$$

The following relation between the clamp diode current I_D, the input current I_1, and the load current I_{C1} is obtained from Eqs. (14.83a), (14.83b), and (14.85).

$$I_D = \alpha_N I_1 - (1-\alpha_N) I_{C1} + (1-\alpha_N\alpha_I) I_{Cf} \qquad (14.86)$$

The terminal relation for the Schottky diode is $I_D \approx I_{DS}\exp(V_D/V_T)$ where I_{DS} is the saturation current of the diode. For a clamped transistor $I_{Cf} \ll I_D$ because the Schottky diode conducts at a smaller forward bias than a p-n junction. The voltage V_D across the Schottky diode is obtained from Eqs. (14.86), and neglecting I_{Cf} which is small because of the clamping action,

$$V_D = V_T \ln \left\{ \frac{\alpha_N I_1 - (1-\alpha_N) I_{C1}}{I_{DS}} \right\}. \tag{14.87}$$

Another constraint is imposed by Kirchhoff's voltage law applied to the collector-base loop:

$$V_D + I_D r_d = I_B r_b + V_{BC} - I_C r_{c1}.$$

This equation is solved for V_{BC} in terms of V_D and the currents because the Schottky diode conducts heavily and clamps the collector-base junction. Substituting from Eqs. (14.83c), (14.85), and (14.86) for the currents and neglecting the terms in I_{Cf} yields

$$V_{BC} \approx V_D + [\alpha_N (r_d + r_{c1}) - (1-\alpha_N) r_b] I_1$$
$$+ [\alpha_N r_{c1} - (1-\alpha_N)(r_d + r_b)] I_{c1}. \tag{14.88}$$

Knowing V_{BC} we can now calculate $I_{Cf} = I_{s2} \exp(V_{BC}/V_T)$. In connection with Eq. (14.85), we assumed that $I_{Cf} \ll I_C$, and therefore $I_{Ef} = I_E = I_{C1} + I_1$. Hence

$$V_{BE} = V_T \ln \left\{ \frac{I_{C1} + I_1}{I_{s1}} \right\}. \tag{14.89}$$

The saturation voltage from the externally accessible collector terminal to the emitter is

$$V_{C'E} = V_{BE} - V_{BC} + I_D r_{c1} + I_{C1}(r_{c1} + r_{c2})$$
$$\approx V_T \ln \left[\frac{I_{DS}}{I_{s1}} \frac{1 + \beta_F}{1 - (\beta_F/\beta_N)} \right]$$
$$- \left[\frac{\alpha_N r_d - (1-\alpha_N) r_b}{\beta_F} - (1-\alpha_N)(r_d + r_b) - r_{c2} \right] I_{c1}, \tag{14.90}$$

where $\beta_F = I_{C1}/I_1$ is the forced beta of the clamped transistor.

When the transistor is in saturation it stores excess charge which must be removed before the transistor can come out of saturation. The stored charge can be expressed in terms of the transistor currents as

$$Q_{BN} = \tau_{CN}^b I_C = \alpha_N \tau_{CN}^b I_{Ef}, \tag{14.91a}$$

where Q_{BN} is the charge stored in the base region due to injection in the normal operating mode,

$$Q_{BI} = \tau_{EI}^b I_E = \alpha_I \tau_{EI}^b I_{Cf} \tag{14.91b}$$

is the charge stored in the base region due to current injection when the transistor is in the inverted operating mode, and finally the charge stored in the collector region due to injection current from the base when the transistor is inverted is

$$Q_C = \tau_{EI}^c I_B = (1-\alpha_N)\tau_{EI}^c I_{Cf}. \tag{14.91c}$$

Both Q_{BI} and Q_C are zero when the transistor is operating in the normal mode. In saturation, the collector current is limited by the external circuit to a value $I_{C\ \text{sat}} < \beta_N I_B$. The external collector circuit current is limited to a value I_{C1}'. Hence the collector current becomes $I_C = I_{C1}' + I_D$, and $\Delta Q_{BN} = \tau_{CN}^b(I_{Cf} + I_D)$. Therefore the excess stored charge is

$$\Delta Q_{BN} + Q_{BI} + Q_C = \left[\alpha_I\tau_{EI}^b + (1-\alpha_N)\tau_{EI}^c + \tau_{CN}^b\right]I_{Cf} + \tau_{CN}^b I_D \tag{14.92}$$

It is customary to express the first term on the right side of Eq. (14.92) in terms of an equivalent relaxation time and the excess base current. Using Eq. (14.85), the excess charge is written as

$$Q_x = \tau_s\left[\frac{1-\alpha_N\alpha_I}{\alpha_N}\right]I_{Cf} + \tau_{CN}^b I_D, \tag{14.93a}$$

where

$$\tau_s = \frac{\alpha_N\tau_{CN}^b + \alpha_N\alpha_I\tau_{EI}^b + \alpha_N(1-\alpha_N)\tau_{EI}^c}{1-\alpha_N\alpha_I}. \tag{14.93b}$$

The differential equation describing the decay of charge is

$$-I_B = \frac{dQ_{BN}}{dt} + \frac{dQ_x}{dt} + \frac{Q_{BN}}{\tau_{BN}^b} + \frac{Q_x}{\tau_s}. \tag{14.94}$$

Consider the situation in which the transistor is saturated with a base current I_{B1}. The excess base current is $I_{BX} = I_{B1} - (I_{C\ \text{sat}}/\beta_N)$ gives rise to the excess stored charge Q_x. At time $t = 0$ the base current is reduced to a value I_{B2} so as to turn off the transistor. However, the transistor remains in saturation until $Q_x = 0$. During this storage time interval $dQ_{BN}/dt = 0$ and Q_{BN} remains constant at the value $\tau_{CN}^b I_{C\ \text{sat}}$. Hence, the differential equation becomes

$$\frac{dQ_x}{dt} = -I_{B2} - \frac{I_{C\ \text{sat}}}{\beta_N} - \frac{Q_x}{\tau_s}, \tag{14.95}$$

where $\beta_N = \tau_{BN}^b/\tau_{CN}^b$. Using the definition $\tau_s \triangleq dQ_x/dI_B$ to introduce dI_B

into this equation, and separating variables, yields

$$\int_0^{t_s} \frac{dt}{\tau_s} = -\int_{I_{BX}}^0 \frac{dI_B}{I_{B2} + \dfrac{I_C \text{ sat}}{\beta_N} + \dfrac{Q_x}{\tau_s}}, \tag{14.96}$$

where, once again, $I_{BX} = Q_x/\tau_s$ is the excess base current. The upper limit of integration of the right-side integral corresponds to the condition of zero-excess stored charge when the transistor comes out of saturation. The solution of this equation is[7]

$$t_s = \tau_s \ln\left\{ \frac{\left[\dfrac{1-\alpha_N\alpha_I}{\alpha_N}\right]I_{Cf} + \dfrac{\tau_{CN}^b}{\tau_s} I_D + \dfrac{I'_{C1}}{\beta_N} + I_{B2}}{\dfrac{I'_{C1}}{\beta_N} + I_{B2}} \right\} \tag{14.97a}$$

where we have approximated $I_{C \text{ sat}}$ by I'_{C1} which is the limiting value of current in the external collector circuit. Except for the possibility of a small circulating current in the Schottky diode-collector loop, $I_{B2} = I_{1\text{ off}}$ the turn-off value of the input current.

There are many cases in which $I_{B2} \gg I_{Cf}$. For example, there is an active turn-off for the inverter transistor in a T²L gate. When $I_{B2} \gg I_{Cf}$, Eq. (14.97a) is of the form $\tau_s\ln(1+x)$ where $x < 1$, and $\ln(1+x) = x - 1/2\,x^2 + \cdots$. Therefore the storage time is approximated by

$$t_s = \tau_s\left\{ \frac{\left[\dfrac{1-\alpha_N\alpha_I}{\alpha_N}\right]I_{Cf}}{(I'_{C1}/\beta_N) + I_{B2}} \right\} + \tau_{CN}^b\left\{ \frac{I_D}{(I'_{C1}/\beta_N) + I_{B2}} \right\}. \tag{14.97b}$$

Recall that $I_{Cf} = I_{s2}\exp(V_{BC}/V_T)$ where V_{BC} is given in terms of the Schottky diode characteristics and the resistors in the diode-collector-base loop by Eqs. (14.87) and (14.88). We can make I_{Cf} very small compared to $(I'_{C1}/\beta_N) + I_{B2}$ by a suitable choice of diode size for a given diode metal.

The reduction in storage time due to Schottky diode clamping is illustrated in Fig. 14.27 where the storage time of a clamped transistor inverter is compared to that of an inverter using an identical transistor without the Schottky diode clamp. The transistor parameters used in the calculation are $I_{s1} = 10^{-14}$ A, $I_{s2} = 10^{-13}$ A, $\alpha_N = 0.98$, $\alpha_I = 0.67$, $\tau_s = 20$ ns, and $\tau_{CN}^b = 1$ ns. The Schottky diode reverse saturation current $I_{DS} = 10^{-10}$ A. It

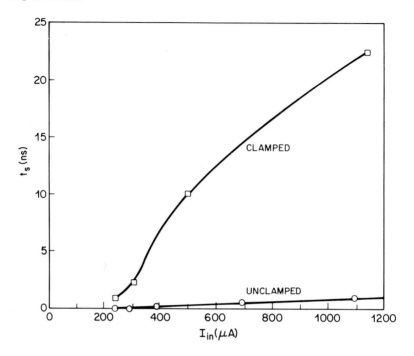

Fig. 14.27 Reduction in storage time due to Schottky diode clamping.

is assumed that resistors r_d, r_b, and R_{c1} are zero to simplify the calculation. The collector current is supplied from a 5 V source through a total resistance of 500 Ω and the base turn-off current $I_{B2} = 250$ μA. The storage time of the clamped transistor is more than an order of magnitude lower than that of the unclamped transistor. Similar results and experimental verification are given by Tarui, *et al.*[10] who, unfortunately, do not give the parameters of the transistor and diode.

The top view and cross section of a Schottky clamped transistor is shown in Fig. 14.28. The diode and transistor are fabricated as a single structure by enlarging the isolated region by an amount sufficient to provide the required area of *n*-type epitaxial material over the buried collector adjacent to the base region of the transistor. The base contact window is enlarged to include an opening equal to the Schottky diode area over the *n*-epitaxial material. The Schottky diode can, in principle, be formed and connected by the interconnection metallization. Often additional metal deposition steps are used to get low-leakage Schottky diodes.

14.4.8 T^{2}L Gates with Reduced Propagation Delay [11]

Another method of improving the turn-off delay of a saturating driver is to use a transistor in a feedback loop from collector-to-base to clamp the driver

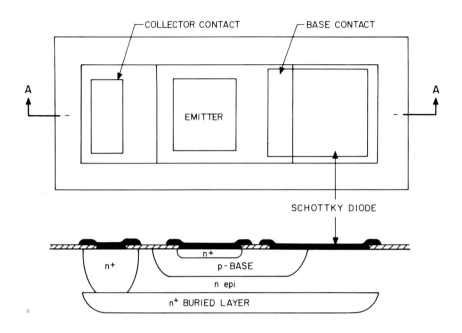

Fig. 14.28 Top view and cross section of Schottky diode clamped transistor.

transistor. This method requires more silicon area than the Schottky clamp but does not require any additional processing steps.

The original T^2L gate is modified by the addition of a feedback transistor, T_3, and associated bias resistors, R_3 and R_4, as shown in Fig. 14.29. Without the clamp transistor T_3, transistor T_2 saturates heavily because of the large base current supplied via resistor R_1. When the feedback transistor T_3 is included, voltage V_D is arranged by means of resistors R_3 and R_4 to exceed V_{BE2} by about 200 mV. Hence, in the ON state $V_D \approx 1$ V and transistor T_3 is saturated. From Kirchhoff's Current Law applied at node B and the output node, it is seen that the base current of T_2 is less than the collector current of T_1 and that I_{C2} is increased by the emitter current of T_3. Recall that one of the conditions for saturation is that $\beta_N I_B > I_C$. Since this feedback connection tends to decrease I_B and to increase I_C, the degree to which T_2 saturates can be easily controlled.

This saturation control leads to simple circuit layouts because T_1 and T_3 have a common collector, and is not critically dependent on absolute device parameters. Moreover, the output level V_o can be adjusted within a certain range because $V_o = V_D - V_{BE3}$.

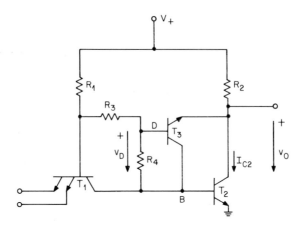

Fig. 14.29 TTL gate with a feedback transistor clamp on the output transistor to reduce turn-off time.

The improvement in turn-off switching time obtainable with this modified circuit is comparable with that obtainable by using Schottky diode clamping discussed in the previous section. This method is fully compatible with standard bipolar transistor technology and so the processing is simpler than that required for Schottky diode clamping where a special metal deposition is often required. However, the circuits require more silicon area.

14.5 INTEGRATED INJECTION LOGIC [12-14]

Integrated injection logic (I^2L), also known as merged transistor logic (MTL), is a bipolar transistor logic that is closely related to direct-coupled transistor logic (DCTL). Among its advantages are a good speed-power product, typically about 1 pJ, high packing density, and simple device processing. First generation I^2L circuits were processed using a standard bipolar processing sequence and therefore other analog and digital circuits could be made on the same chip. The I^2L gate uses a transistor operating in the inverse mode, that is with the normal collector region as the emitter and the normal emitter region as the collector. Additional processing steps are included in second generation I^2L processing sequences as described in Section 6.18 to improve the performance of these inverted transistors while still permitting other types of bipolar circuits to be made on the same chip. The ability to produce other types of circuits on the same chip is essential because I^2L gates operate at very low voltage and current levels, have logic swings that are not compatible with other forms of logic, and have limited capability to drive other types of logic gates. Level translators are required on-chip to provide buffering and drive capability. Most I^2L circuits are buffered to look like T^2L at the terminals. It is because of this requirement for buffering that I^2L is only used for LSI (>100 gates) applications.

The I^2L gate   An interconnection of DCTL gates is shown in Fig. 14.30. The basic logic operation in DCTL is the NOR. In implementations of DCTL each gate has n transistors whose bases form the n gate inputs and whose collectors are OR-tied to form the output. The output pull-up resistor is considered to be part of the collector circuit.

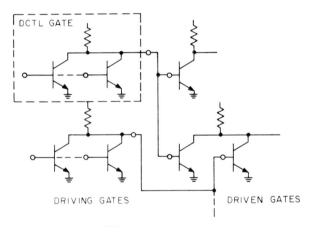

Fig. 14.30 Interconnection of DCTL gates.

Another way of looking at the interconnection of DCTL gates is that the pull-up resistor and the collector voltage supply form a current source that supplies sufficient base current to the driven stage to saturate the transistors. The base current of the driven stage can be diverted by saturating any of the transistors in the driving stage. The circuit is redrawn in Fig. 14.31 to emphasize this interpretation. The basic gate now consists of a single base, multiple collector transistor with a current source in the base.

Figure 14.32(a) is the symbol for an I^2L gate. The current source is a lateral *p-n-p* transistor consisting of a diffused *p-n* junction called the injector (which is the emitter-base junction of the lateral *p-n-p* transistor) and the base of the vertical *n-p-n* transistor. The multicollector *n-p-n* transistor is actually a multiemitter vertical transistor which is operated in the inverse mode, that is, with the epitaxial layer as the emitter and the shallow n^+ regions as collectors. Unless additional processing steps are used, the vertical *n-p-n* structure is basically a low-gain device. Note from Fig. 14.31 that all emitters of the gate transistor are grounded, and therefore the complete

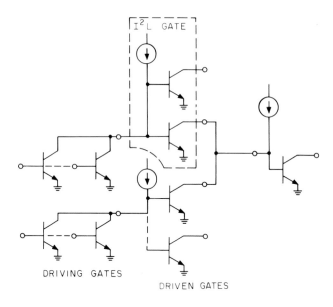

Fig. 14.31 Circuit of Fig. 14.30 redrawn to show I^2L gate configuration.

I^2L circuit can be built on a common buried layer which serves as the emitter. Each gate consists of a p-type base diffused region containing several n^+ regions which serve as collectors. The injector is a p-diffused region which is shared by many gates. The top and cross-sectional view of an I^2L structure is shown in Fig. 14.33. I^2L has a high packing density because the structure is self-isolating and therefore no isolation diffusion is required. (A pure I^2L circuit can be made with just five masking operations; namely, deep n^+ diffusion, p-type base diffusion, shallow n^+ diffusion, contact windows, and metallization.) Furthermore, ground is distributed through the common buried layer because all n-p-n transistors have a grounded emitter and V_+ is distributed by means of the p^+ diffusion that forms the injector. This leaves the metallization solely for logic connections between cells.

It is convenient to consider that the basic logic function of the I^2L gate is a wired-AND followed by an inversion.

Current gain of the I^2L composite transistor[16]   Before proceeding with the analysis of a loaded I^2L gate, it is useful to examine a simple dc model for the composite p-n-p n-p-n transistor that is the heart of the I^2L gate. This model is shown in Fig. 14.32(b). Currents across each junction of this multijunction structure are represented as a sum of injected and collected com-

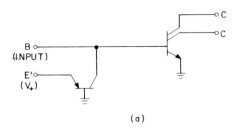

(a)

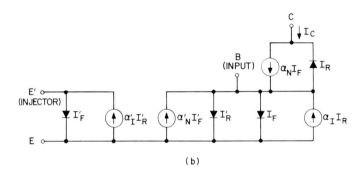

(b)

Fig. 14.32 (a) Symbol fo I^2L gate and (b) circuit model for the gate transistor.

ponents in the manner of the Ebers-Moll model. Note that all of the components to the left of the broken line form the Ebers-Moll model for the *p-n-p* injector transistor whose injected currents are $I_F' = I_{s1}' \lambda(V_{E'E})$ and $I_R' = I_{s2}' \lambda(V_{BE})$ where the prime relates to the injector transistor and we have used the terminology of Chapter 2 that $\lambda(x) = \exp(x/V_T) - 1$. Similarly, the injected currents for the *n-p-n* transistor are $I_F = I_{s1}\lambda(V_{BE})$ and $I_R = I_{s2}\lambda(V_{BC})$.

An important parameter of this composite transistor is the so-called upward current gain β_u defined as

$$\beta_u = \frac{I_C}{I_{BT}}\bigg|_{V_{E'E}=0, V_{BC}=0} \tag{14.98a}$$

which must be greater than unity at each output for the device to function as a gate. This can be seen by assuming that the gates are identical and observing that if a particular output of the test gate under consideration is connected to the inputs of N gates (fan-out of N), then the base current of the test gate is I and its collector current is NI in the worst case when the test

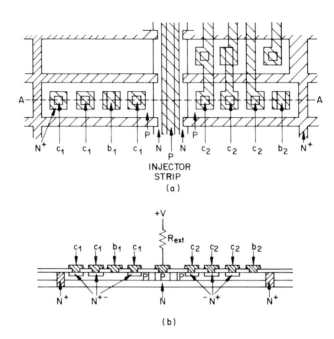

INJECTOR
STRIP
(a)

(b)

Fig. 14.33 (a) Top view and (b) cross section of multicollector transistors on both sides of an injector.

gate is ON. Under this condition the test gate transistor must saturate. A necessary and sufficient condition for saturation is that $\beta I_B > I_C$. Therefore, $\beta_u I > NI$ and the condition is $\beta_u > N$ in order for the I^2L gate to function. The total base current I_{BT} consists of the component I_R' which goes into the collector of the saturated p-n-p transistor, and I_{BI} which is the base current of the vertical n-p-n transistor. It is assumed that the p-n-p transistor is symmetric so that $\alpha_N' = \alpha_I'$. The effect of each transistor on the upward current gain can be put in evidence by rewriting Eq. (14.98a) as

$$\beta_u = \frac{I_C}{I_R' + I_{BI}} = \frac{\beta_I}{1 + \beta_I \Gamma} \qquad (14.98b)$$

where $\beta_I = I_C/I_{BI}$ is the common emitter current gain of the vertical transistor operating in the inverted mode and

$$\Gamma = \frac{I_R'}{I_C} = \frac{I_{s2}'}{\alpha_N I_{s1}} \qquad (14.98c)$$

is a loss factor which can be reduced by providing either a diffused n^+ collar or an oxide collar around the I^2L gate.

This method of modeling the I^2L gate can be extended to structures with more than one collector by adding more sections to represent the additional vertical *n-p-n* transistors. Usually, the collector contacts are arranged in a row with the injector at one end and the whole array surrounded by a collar. The value of β_u is a function of the position of the collector contact in the array, being greatest for the contact closest to the injector.

14.5.1 DC Analysis of I^2L Gate

Consider the interconnection of I^2L gates shown in Fig. 14.34. It consists of a pair of test gates G_1 and G_2, driven by driver gates DG_1 and DG_2, and loaded by gates LG_1 to LG_N. Qualitatively the operation is as follows.

The base current I_{BT} of test gate G_1 will be I if driver gate DG_1 is cut off. The gate G_1 is ON and it will be saturated if it can sink the combined current of the N load gates. This is the worst case for saturation because if driver gate DG_2 is also cut off, gate G_2 will also be ON and it will share the current with G_1. If both DG_1 and DG_2 are saturated, gates G_1 and G_2 are cut off and load gates LG_1 to LG_N are turned on. We now derive expressions for the logic swing and fan-out of an I^2L gate.

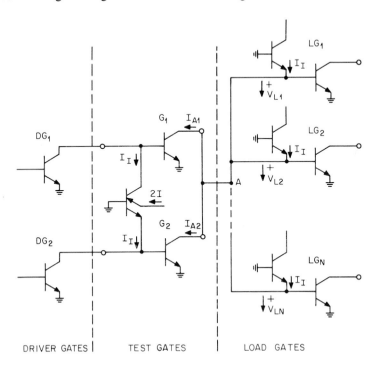

Fig. 14.34 An interconnection of I^2L gates to define driving and loading problems.

When test gate G_1 is ON, it must sink a current $I_{C1} = NI$ and the transistor must be saturated with a saturation voltage

$$V_{CE\ sat} = V_T \ln\left[\frac{I_{s2}}{\alpha_N I_{s1}} \frac{1 + (1-\alpha_I)N}{1 - (N/\beta_u)}\right]. \tag{14.99}$$

In deriving this equation from the Ebers-Moll equations we have used the result that β_I for the vertical n-p-n transistor is expressed in terms of the model parameters by $\beta_I = \alpha_N/(1-\alpha_N)$, because the transistor is operated in the inverse mode, and that the fan-out of the gate is $N = I_{C1}/I$. We note from this expression that the voltage across the saturated transistor is reduced if β_u is large and that the fan-out must be less than the upward gain, $N < \beta_u$ as previously mentioned. The output voltage of the saturated transistor differs from $V_{CE\ sat}$ by the drop across the parasitic resistance. However, this effect is much smaller than in conventional logic circuits and is neglected. Hence, $V_{CE\ sat} = V_L$ is the lower logic level in a positive logic system.

Point A assumes the logic 1 output level when both test gates are cut off. There is a potential current-hogging problem because the transistor is not a unilateral device. In particular, the input voltage of each load gate is a function of its output current. This means that it is possible for one load gate to turn on before the others and begin to rob the current so that the remaining gates do not turn on. The effect of this current feedback on the input voltage can be determined by solving the Ebers-Moll type equations describing the model for V_{BE} when the external input current $I_{BI} = 0$. The result is

$$V_{BE} = V_T \ln\left[\frac{I + (1-\alpha_I)I_{C1}}{(1-\alpha_N\alpha_I)I_{s1} + I_{s2}'}\right]. \tag{14.100a}$$

Fortunately V_{BE} is not a strong function of I_{C1} because α_I is close to unity (α_I is the normal downward current gain α_N for the vertical n-p-n transistor), and therefore current hogging is not a serious problem. When gate G_1 is ON, V_{BE} is at its maximum value which is the logic 1 level V_U for the preceding stage. Assume $\alpha_I \approx 1$, Eq. (14.100a) can be rewritten as

$$V_{BE} \approx V_T \ln\left[\beta_u \frac{I}{\alpha_N I_{s1}}\right]. \tag{14.100b}$$

The upper logic level V_U is V_{BE} evaluated at the particular value of injector current I.

The logic swing $v_l = V_U - V_L$ is obtained from Eqs. (14.99) and (14.100b) as

$$v_l = V_T \ln\left[\frac{(\beta_u - N)I}{I_{s2}}\right]. \tag{14.101}$$

Note that the logic swing decreases as the fan-out is increased because of the increased output staturation voltage $V_{CE\ sat}$.

The speed-power product of first generation I^2L is typically 1 pJ for gate delays greater than about 10 ns. At low power levels the speed-power product is a constant and the speed is determined by the ability to charge junction and stray capacitances. At high-power levels the propagation delay decreases, but not as rapidly as the power consumption increases.

14.5.2 Second Generation I^2L

Although I^2L circuits could be made by a standard bipolar processing sequence, it became evident that great improvements in performance would result from changes in the device structure. Three main types of changes are discussed here. They are the use of Schottky diodes, oxide isolation, and changes in the doping profile to improve the performance of the vertical *n-p-n* transistor. Details of the processing sequences are described in Section 6.18.

Schottky I^2L I^2L gates are usually used at power levels such that the speed-power product is determined by the charging of capacitors. Thus, it is proportional to Cv_l^2 where v_l is the logic swing. The speed-power product can be improved by limiting the logic swing.[17]

The logic swing of conventional I^2L gates is about 700 mV. It can be reduced to about 250 mV by forming a Schottky barrier diode in each collector contact, as shown in Fig. 15.35(a). The transfer characteristics of conventional and Schottky I^2L are shown in Fig. 15.35(b).

Schottky I^2L gates require a more complex structure than conventional I^2L because lightly doped *n* regions are required for the Schottky barrier diodes.

It has been reported[17] that Schottky I^2L using palladium silicide diodes yields a factor of two improvement in speed-power product.

The use of a Schottky diode as an antisaturation clamp between collector and base of the vertical *n-p-n* transistor has also been reported.[18] The clamp diode reduces the stored charge in the vertical transistor thereby increasing the speed of the gate. In addition, it also reduces the output voltage swing of the gate, thus reducing the power.

Other methods of storage control   When the *n-p-n* transistor is ON, the lateral *p-n-p* injector transistor is in saturation and the base drive to the vertical *n-p-n* transistor is reduced. The *p-n-p* transistor should be designed such that its current falls off rapidly in saturation, thus reducing the stored charge in the base of the vertical *n-p-n* transistor at high current levels. This will improve the turn-off delay of the gate. Improvement in the turn-on delay requires a reduction in the charging time of the base node capacitance. This

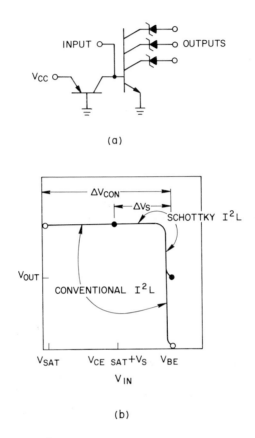

(a)

(b)

Fig. 14.35 (a) Schottky I^2L gate and (b) comparison of its transfer characteristics with those of conventional I^2L.

can be achieved by increasing the current gains of the lateral *p-n-p* transistor. It has been reported[19] that a 20% improvement in speed results from simple horizontal device geometry changes to increase the gain of the *p-n-p* transistor.

Another method of storage control involves tying one collector of the multicollector *n-p-n* transistor back to the base. Ideally, when β_u is very large, the feedback through this diode limits the effective upward gain in the cell to the ratio of the normal collector area to the diode area. This ratio is a more accurately controlled quantity than the gain of a transistor. This method is very effective in decreasing the stored charge and propagation delay at higher current levels. However, it results in longer propagation delays at low current levels because the devices are larger, and the current noise margin is reduced.[19]

Optimized impurity profiles   The standard bipolar transistor processing sequences produce doping profiles such as that shown in Fig. 14.36(a). A structure with this type of doping profile is not well suited for I^2L because the light doping of the *n*-epitaxial layer (which is now used as the emitter) relative to the base results in a low emitter injection efficiency (see Chapter 2). Also, the collector-base capacitance of the I^2L transistor is large and the breakdown voltage is low because of the n^+ region which is used as the collector.

A significant improvement in speed and power-delay product can be achieved by using an additional selective boron diffusion or ion implantation prior to epitaxial growth. During epitaxial growth, this boron will outdiffuse

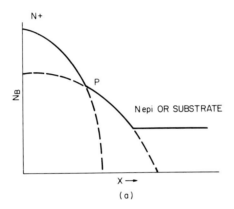

(a)

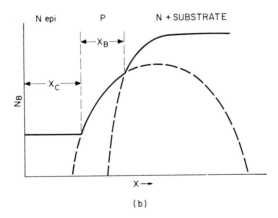

(b)

Fig. 14.36 (a) Doping profile of a standard epitaxial, double-diffused transistor and (b) doping profile of transistor produced by outdiffusion of *p*-type impurities from buried layer.

from the n^+ doped substrate into the epitaxial layer producing the doping profile shown in Fig. 14.36(b). The injector and a low resistance contact to the base region of the n-p-n transistor are formed by a subsequent p^+ diffusion. This doping profile results in an n-p-n transistor with (1) a high emitter injection efficiency, (2) an aiding diffusion field in the graded p-base region, (3) a low value of collector-base depletion capacitance, and (4) an improved collector-base breakdown voltage. A further advantage of this doping process is that the base width is independent of the epitaxial layer thickness. It is a function of the growth rate of the epitaxial layer and the diffusion rate of dopant atoms from the substrate, and is typically 0.4 μm. Other processing sequences use ion implantation to form the base region directly in the epitaxial film.

Experimental evidence[18] indicates that the use of the optimized doping profiles and Schottky diodes in the collector contacts as described previously, results in a threefold decrease in the minimum propagation delay time and a 40% decrease in the power-delay time product as compared to I^2L circuits made with standard device processing.

Isoplanar integrated injection logic[19] Isoplanar, integrated injection logic (I^3L) is a nice example of an advanced I^2L process. It uses oxide guard rings to prevent parasitic p-n-p action between adjacent gates (thus reducing the loss factor Γ) and ion implantation to produce doping profiles that are tailored to I^2L devices. Standard, high-performance vertical n-p-n transistors are also available. The performance is further improved by designing the p-n-p injector to have a high gain and by the use of antisaturation clamps. An interesting circuit technique called *graduated collectors* is used to overcome the decrease in β_u of the collectors far away from the injector in the "stick" geometry shown in Fig. 14.37. The problem arises because the distributed resistance of the base region causes transistors closer to the injector to be more heavily biased than those farther away. Therefore, at a fixed injector current, the collector current density of collector 4 is less than that of collec-

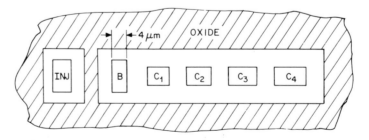

Fig. 14.37 Top view of an I^3L stick transistor with graduated collectors.

tor 1. If the collector areas are equal, the current gain at collector 4 is less than that of collector 1. In the graduated collector scheme, the collector areas are adjusted such that the upward gain of all collectors is equal at some reasonable value of injector current for LSI circuits, for example, 500 μA. In the I^3L structure this requires that $A_3 = 1.8A_1$, and $A_4 = 2.2A_1$, where A_i is the area of the ith collector.

Propagation delays of 4 ns/gate at an injector current of 1 mA have been reported for a 6-gate inverter chain.[20] The power-delay time product is less than 0.15 pJ/gate and packing density is in excess of 250 gates/mm^2.

14.6 MOS LOGIC [21,22]

The MOSFET is an attractive element for logic circuits for a number of reasons. First, its small size and self-isolating properties imply a high packing density on the chip. Second, its relatively simple processing results in high-chip yields at the 500 to 1000 equivalent gate level. Third, the interconnection of complex circuits is simplified because two interconnection levels are inherent in the structure. Finally, the high impedance levels of MOSFET's lead to low power consumption. In this section the basic MOSFET inverter and gate are described and analyzed.

14.6.1 The MOSFET Inverter with Active Load

The circuit shown in Fig. 14.38 is one type of MOSFET inverter circuit. It is a two-transistor circuit using p-channel enhancement mode transistors in which T_2 is the load. Many other combinations are possible because there are four types of MOSFET's; namely, p-channel enhancement or depletion mode, and n-channel enhancement or depletion mode. However, in digital circuits the driver is always an enhancement mode device because enhancement mode transistors require gate and drain bias sources of the same polarity which permits direct coupling between stages.

In this section we analyze and compare the performance of single channel (n or p) inverters using enhancement mode loads operating in either the triode region or the saturation region, single channel inverters using depletion mode loads, and complementary channel (CMOS) inverters.

The simplest inverter in both structure and circuit complexity is the inverter with a saturated, enhancement mode load. This configuration is obtained by connecting together the gate and drain of the load device T_2 (setting $V_{GG} = V_{DD}$ in Fig. 14.38). For the load transistor in the saturation region $|V_{GS} - V_t| < |V_{DS}|$. There is a square law relationship between the drain current and the voltage in the saturation region:

$$I_D = \frac{\beta}{2} (V_{GS} - V_t)^2 \qquad (14.102a)$$

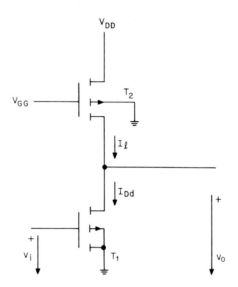

Fig. 14.38 An MOSFET inverter using *p*-channel enhancement mode transistors.

where V_t is the threshold voltage and

$$\beta = \frac{\kappa_{ox}\epsilon_o\mu\,W}{t_{ox}L}.$$ (14.102b)

For a load transistor in the saturation region $V_{GS} = V_{DD} - v_o$ and therefore the drain current

$$I_{Dl} = \frac{\beta_l}{2}\,(V_{DD} - v_o - V_t)^2$$ (14.103)

where β_1, the beta of the load transistor, is obtained from Eq. (14.102b) by setting $W = W_l$ and $L = L_l$.

Figure 14.39 shows the parabolic load line of the load transistor for several values of β_l drawn on the drain characteristics of the driver transistor. The logic 1 output $V_o^{(1)}$ occurs when the load current is zero. From Eq. (14.103) this value is

$$V_o^{(1)} = V_{DD} - V_t.$$ (14.104)

The logic 0 output level $V_o^{(0)}$ is a function of β_l which decreases with decreasing values of β_l.

Referring to Fig. 14.38, we see that for the driver transistor $v_i = V_{GS}$ and $v_o = V_{DS}$. Two distinct cases must be considered in deriving the transfer characteristics of the inverter. When the output voltage is somewhat greater than $V_o^{(0)}$ the driver transistor is in saturation. Therefore the drain current of the driver transistor is given by

$$I_{Dd} = \frac{\beta_d}{2} \ (v_i - V_t)^2 \tag{14.105}$$

where β_d is given by Eq. (14.102b) with the value of W and L appropriate to the driver device. The inverter is assumed to have no dc load, and therefore $I_{Dd} = I_l$. Substituting Eqs. (14.103) and (14.105) for I_l and I_{Dd}, solving for v_o in terms of v_i, and observing that the output voltage must always be less than $V_{DD} - V_t$ yields

$$v_o = V_{DD} - \left(1 - \sqrt{\beta_R}\right) V_t - \sqrt{\beta_R} \ v_i \tag{14.106}$$

where $\beta_R = \beta_d / \beta_l$ is called the beta ratio. This expression is valid whenever $|v_o| \geqslant |v_i - V_t|$ and $|v_i| \geqslant V_t$, that is, while the driver transistor is saturated. We have tacitly assumed that both the driver and load transistors are made by the same process, and therefore have identical mobilities and oxides. Hence, from Eq. (14.102b) the beta ratio, β_R, is a geometric constant with typical values of about 40.

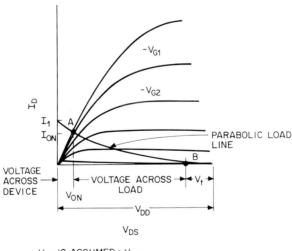

Fig. 14.39 Characteristic of driver with parabolic load line for several values of β_l.

The driver transistor is in the triode region when the output voltage $|v_o| \leqslant |v_i - V_t|$. The drain current of T_1 is then described by

$$I_{Dd} = \frac{\beta_d}{2} \left[2(V_{GS} - V_t) V_{DS} - V_{DS}^2 \right]$$

$$= \frac{\beta_d}{2} \left[2(v_i - V_t) v_o - v_o^2 \right]. \tag{14.107}$$

The drain current of the driver transistor given by this equation must equal the load current given by Eq. (14.103).

Noting that the maximum value of the output voltages is $V_o^{(1)} = V_{DD} - V_t$, we define normalized output and input voltages as

$$u = \frac{v_o}{V_o^{(1)}}$$

and

$$w = \frac{v_i - V_t}{V_o^{(1)}}.$$

Using Eqs. (14.103) and (14.107) and these normalized variables yields the following relation between the input and output voltages:

$$(1 + \beta_R) u^2 - 2(1 + \beta_R w) u + 1 = 0. \tag{14.108}$$

Note that only one root of this equation is physically acceptable. This implicit relation is valid when the driver transistor is not in saturation. The boundary between the region of validity of this equation and that of Eq. (14.106) is found most easily by setting $|v_o| = |v_i - V_t|$ in Eq. (14.106) and solving for v_i. The result is

$$v_i = \frac{V_{DD} + \sqrt{\beta_R} \, V_t}{1 + \sqrt{\beta_R}} , \tag{14.109a}$$

$$v_o = \frac{V_o^{(1)}}{1 + \sqrt{\beta_R}} . \tag{14.109b}$$

This pair of equations define the value of input and output voltage for which the driver transistor is on the border between the saturation region and the triode region. For v_o less than the value prescribed by Eq. (14.109b) the driver transistor is in the triode region.

14.6.2 Operating Points, Transition Width, and Noise Margins

We saw previously that the logic 1 output level of the inverter with an enhancement mode load in the saturation region is $V_o^{(1)} = V_{DD} - V_t$. The logic 0 output level of the inverter is obtained by invoking the input-output compatibility condition; namely, that the logic levels at the input of a gate are the same as those at the output. Thus, when the input to an inverter is $V_i^{(0)} = V_o^{(1)}$, its output is $V_o^{(0)}$. The logic 0 output level can therefore be found by solving Eq. (14.108) for that normalized output, $u^{(0)}$, corresponding to $w^{(0)} = (V_o^{(1)} - V_t)/V_o^{(1)}$.

An approximate value for $V_o^{(0)}$ can be found by considering the inverter to be a resistive voltage divider as shown in Fig. 14.40. Note that the driver, which is in the triode region, is modeled by a simple resistor, while the load which is in saturation is modeled by a current source whose Thevenin equivalent circuit is shown. The current in the load can be written as

$$I_l = I_{Dl} = \frac{g_{ml}}{2}\,(V_{DD} - V_t - v_o) \tag{14.110}$$

where $g_{ml} = \beta_l(V_{DD} - V_t - v_o)$. Note that r corresponds to the drain-source impedance of the driver when $I_{Dd} = I_l$. The drain conductance $g_D \triangleq \partial I_D/\partial V_{DS}$ is found from Eq. (14.107) to be

$$g_{Dd} = \beta_d[V_{GS} - V_t - V_{DS}]$$
$$= \beta_d[v_i - V_t - v_o]. \tag{14.111}$$

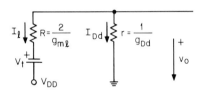

Fig. 14.40 Equivalent circuit for determining approximate value of the low state output voltage.

For small values of output voltage $v_o \approx V_o^{(0)}$ such that $v_o \ll v_i - V_t$, this conductance is independent of the output voltage. Therefore from Fig. 14.40 $v_o \approx I_{Dd}/g_{Dd} = I_l r$.

It follows therefore, that

$$V_o^{(0)} \approx (V_{DD} - V_t)\,\frac{g_{ml}}{g_{ml} + 2g_{Dd}}. \tag{14.112a}$$

Usually $g_{ml} \ll g_{Dd}$, so that this equation reduces to

$$V_o^{(0)} \approx (V_{DD} - V_t)\,\frac{g_{ml}}{2g_{Dd}}. \tag{14.112b}$$

This approximate expression for $V_o^{(0)}$ can be written in an interesting form in terms of β_R by observing that for $v_o \ll v_i - V_t$, $g_{ml} \approx \beta_l(V_{DD}-V_t)$ and $g_{Dd} \approx \beta_D(v_i-V_t)$. Substituting in Eq. (14.112b) yields

$$\frac{V_o^{(0)}}{V_o^{(1)}} = \frac{1}{2\beta_R} \frac{V_{DD}-V_t}{V_{DD}-2V_t} . \tag{14.113}$$

This equation shows that the logic 0 output level is inversely proportional to the beta ratio β_R.

The further importance of β_R in affecting the performance of the inverter is seen by taking the derivative of Eq. (14.106) to obtain the gain of the inverter in the transition region.

$$\frac{dv_o}{dv_i} = -\sqrt{\beta_R}. \tag{14.114}$$

This equation is valid when both the driver and load are in saturation. It shows that a large value of β_R is required to provide high gain in the transition region. Differentiation of Eq. (14.108) which describes the transfer characteristic when the driver transistor is in the triode region shows that the gain is a function of both β_R and the normalized output voltage.

$$\frac{dv_o}{dv_i} = -\frac{2\beta_R u^2}{1 - (1+\beta_R)u^2} \tag{14.115a}$$

This expression is valid in the region where $u = v_o/V_o^{(1)} \ll 1$. Therefore,

$$\frac{dv_o}{dv_i} \approx -2\beta_R u^2, \tag{14.115b}$$

which approaches zero as the output voltage decreases. Hence, we expect the transfer characteristic to flatten out as the operating point of the driver transistor moves into the triode region.

The normalized transfer function for an enhancement mode inverter load in saturation is plotted in Fig. 14.41 for several values of β_R. The transfer characteristic of this inverter for the case $\beta_R = 25$ is shown in Fig. 14.42 along with the transfer characteristic for an identical driving stage, which defines the location of the logic 0 and logic 1 operating points. The logic 1 input noise margin, NM^1, is defined in the conventional manner as the difference between the logic 1 input operating point and the corresponding unity gain point. The logic 0 noise margin, NM^0, is defined as the difference between the logic 0 operating point and the point at which the output begins to change because the transfer characteristic is discontinuous.

EXAMPLE 14.9 An n-channel MOSFET inverter with an enhancement load transistor in saturation has the following device parameters: $V_t = 3$ V, $W_d = 500$ μm, $W_l = 20$ μm, $L_l = L_d = 10$ μm. The power supply voltage is

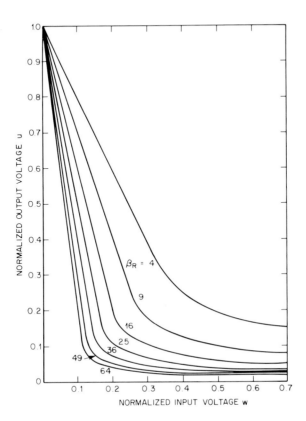

Fig. 14.41 Normalized transfer function of an inverter with an enhancement mode load in saturation for several values of β_R. (After Penney and Lau.[21])

$V_{DD} = 15$ V. Find $V_o^{(1)}$, $V_o^{(0)}$, the voltage gain in the transition region, and the boundary between the saturation and triode regions for the driver transistor.

Solution. First find the beta ratio $\beta_R = \beta_d/\beta_l = W_d L_l / W_l L_d = 25$. The logic 1 state output voltage $V_o^{(1)} = V_{DD} - V_t = 15 - 3 = 12$ V from Eq. (14.104). In this positive logic system $V_o^{(1)} = V_U$. An approximate value for the logic 0 state output voltage is most easily found from Eq. (14.113)

$$V_L = V_o^{(0)} = \frac{1}{2 \times 25} \cdot \frac{15-3}{15 - (2\times3)} \cdot 12 = 0.32 \text{ V}$$

From Eq. (14.114) the voltage gain in the transition region is $-\sqrt{\beta_R} = -5$. Finally, the boundary points determined from Eq. (14.109) are $v_i = 6$ V and $v_o = 2$ V.

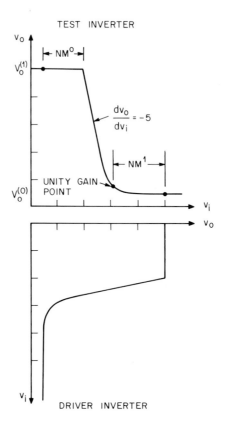

Fig. 14.42 Transfer characteristics of identical driver and test inverters with $\beta_R = 25$ to locate the operating points and define the noise margins.

14.6.3 Inverter with Enhancement Mode Load in Triode Region

Returning to the circuit of Fig. 14.38, if $|V_{GG} - V_t| > |V_{DD}|$ then the load transistor T_2 will always be in the triode region. This configuration requires a second power supply for the gate of the load transistor. The average dc power drain from the V_{GG} supply is negligible but considerable power can be consumed during transients. Offsetting the need for a second power supply is the larger logic swing because, as we shall soon see, the output can rise to V_{DD}. In addition, the triode load provides the fastest transient response of all single channel MOS inverters.

The load current for the nonsaturated load transistor is

$$I_l = \frac{\beta_l}{2} \left[2(V_{GG} - V_t - v_o)(V_{DD} - v_o) - (V_{DD} - v_o)^2 \right]. \qquad (14.116)$$

Recall from the saturated load case that the logic 1 output level occurs when $I_{Dd} = I_l = 0$. Therefore, Eq. (14.116) it is found that $V_o^{(1)} = V_{DD}$. In terms of the normalized output voltage

$$u = \frac{v_o}{V_o^{(1)}} = \frac{v_o}{V_{DD}}$$

and a biasing parameter

$$m = \frac{V_{DD}}{2(V_{GG} - V_t) - V_{DD}} ,$$

the load current can be written as

$$I_l = \frac{\beta_l}{2} \frac{V_{DD}^2}{m} (1-u)(1-mu). \tag{14.117}$$

Once again, two cases must be considered in the derivation of the transfer characteristic. When the driver transistor is in saturation its drain current is given by Eq. (14.105). Setting that equation equal to Eq. (14.117) yields

$$\beta_R w^2 = \frac{(1-u)(1-mu)}{m} , \tag{14.118}$$

where $w = (v_i - V_t)/V_o^{(1)}$ is the normalized input voltage. This equation is valid when $|v_i - V_t| \leq v_o$. The drain current for the driver transistor when it is in the triode region is given by Eq. (14.107). From that equation and Eq. (14.117) we find the following relation between the normalized voltages,

$$\beta_R[2wu - u^2] = \frac{(1-u)(1-mu)}{m} , \tag{14.119}$$

which is valid for $|v_i - V_t| > v_o$, or $|w| > |u|$. When values are specified for m and β_R, Eqs. (14.118) and (14.119) can be solved to give v_o for values of v_i. Curves of this normalized transfer function are plotted for various values of β_R in Fig. 14.43(a) for $m = 0.7$, and in Fig. 14.43(b) for $m = 0.3$.

Note that Eq. (14.103) which describes the load current when the load transistor is saturated can be written in the form of Eq. (14.117) with $m = 1$. Thus, the saturated load inverter is a special, limiting case of the inverter with a triode mode load. It is therefore interesting to plot the transfer function of the inverter for one value of β_R and several values of m on the same set of axes. This is done in Fig. 14.44 for $\beta_R = 25$ and $m = 0.3$, 0.7, and 1.0. From this figure it is seen that the transfer function for the inverter with a triode mode load is always above that of the inverter with saturated load. This is because for any value of output voltage the triode mode load provides at least as much current as the saturated load. Therefore the driver transistor must have a larger input voltage when the load is in the triode

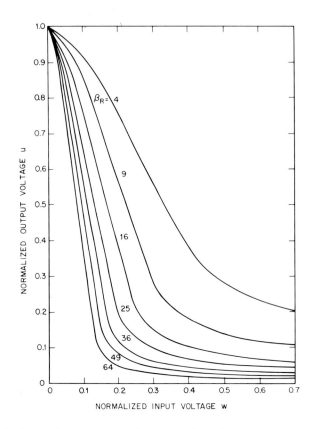

Fig. 14.43(a) Normalized transfer function of an inverter with an enhancement mode load in the triode region for several values of β_R and $m = 0.7$. (After Penney and Lau.[21])

region. It is the lower impedance of the triode mode load at intermediate points of the transfer characteristic that results in a faster rise time for the transient response.

14.6.4 Inverter with Depletion Mode Load

The circuit diagram for an MOS inverter using a depletion mode load is shown in Fig. 14.45. As we shall see, this inverter has higher gain in the transition region for a given β_R than either of the enhancement mode load inverters discussed above. It requires only one power supply and has gain even with $\beta_R = 1$. However, these advantages are achieved at the expense

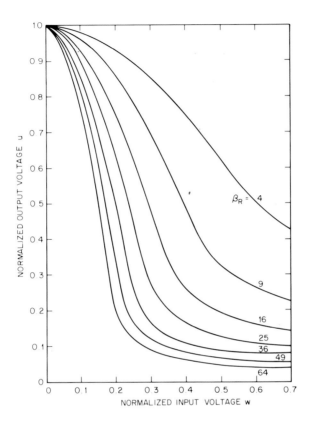

Fig. 14.43(b) Normalized transfer function of an inverter with an enhancement mode load in the triode region for several values of β_R and $m = 0.3$. (After Penney and Lau.[21])

of more complex processing because both enhancement and depletion mode devices must be made on the same chip. The processing is described in Section 6.16.

The depletion mode transistor has an inversion channel when $V_{GS} = 0$. It is described by the same equations as the enhancement mode transistor except that the threshold voltage is of the opposite polarity and corresponds to the input voltage at which conduction ceases. This is in contrast to the enhancement mode transistor for which the threshold voltage defines the onset of conduction. In the analysis of the depletion mode load inverter it is necessary to differentiate between the threshold voltage of the enhancement mode driver and that of the depletion mode load. We denote by V_t the

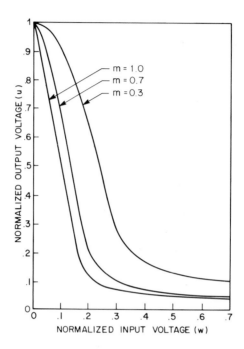

Fig. 14.44 Transfer function of MOSFET inverters with $\beta_R = 25$ and $m = 0.3$, 0.7, and 1.0.

enhancement mode threshold voltage, and by V_p (for pinch-off) the depletion mode threshold voltage.

The equations for the drain current of a depletion mode transistor are therefore

$$I_D = \frac{\beta}{2} (V_{GS} + V_p)^2 \tag{14.120a}$$

when the transistor is saturated, and

$$I_D = \frac{\beta}{2} \left[2 V_{DS}(V_{GS} + V_p) - V_{DS}^2 \right] \tag{14.120b}$$

when it is not saturated. The boundary between the saturated and nonsaturated regions is $V_{DS} = V_{GS} + V_p$. For the circuit of Fig. 14.45, $V_{DS} = V_{DD} - v_o$ and $V_{GS} = 0$. The load device is in saturation when $v_o < V_{DD} - V_p$. In saturation the load is a constant current source whose value is

$$I_l = \frac{\beta_l}{2} V_p^2. \tag{14.121a}$$

For $v_o > V_{DD} - V_p$ the load device is in the triode region, and the load current is

$$I_l = \frac{\beta_l}{2} \frac{V_{DD}^2}{m} (1-u)(1+mu) \tag{14.121b}$$

where, in this case, m is the pinch-off parameter defined by

$$m = \frac{V_{DD}}{2V_p - V_{DD}}. \tag{14.121c}$$

Note the similarity of Eq. (14.121b) to Eq. (14.117) for the load current of an enhancement mode load in the triode region.

The analysis of the inverter with a depletion mode load is similar to that of inverters with enhancement mode loads except that three cases must be considered because the load is in the triode region for some values of the output, and in the saturation region for other values.

Consider first the case in which the load is in the triode region and the driver is in the saturation region. Equating I_l given by Eq. (14.121b) to I_{Dd} given by Eq. (14.105) and collecting terms, yields the following expression involving the normalized input and output variables:

$$mu^2 + (1-m)u + (m\beta_R w^2 - 1) = 0. \tag{14.122}$$

The physically meaningful root is

$$v_o = (V_{DD} - V_p)\left\{1 + \left[\left(\frac{V_p}{(V_{DD} - V_p)}\right)^2 - \beta_r\left(\frac{v_i - V_t}{V_{DD} - V_p}\right)^2\right]^{1/2}\right\} \tag{14.123}$$

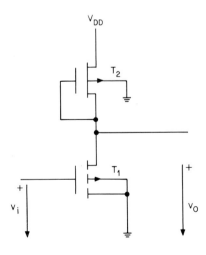

Fig. 14.45 MOSFET inverter with a depletion mode load.

which is a valid description of the output voltage in the range $V_{DD} \geqslant v_o \geqslant V_{DD} - V_p$. Setting the radical equal to zero and solving for v_i yields the input voltage that puts the load transistor on the border of the saturation and triode regions.

$$v_i = V_t + \frac{V_p}{\sqrt{\beta_R}} \tag{14.124}$$

For $v_o < V_{DD} - V_p$ the load transistor is in saturation. While the driver transistor is also in saturation, equating I_l of Eq. (14.121a) to I_{Dd} of Eq. (14.105) yields

$$\beta_R (v_i - V_t)^2 = V_p^2. \tag{14.125}$$

This equation is valid for $V_{DD} - V_p > v_o > v_i - V_t$ because the driver transistor is in the triode region when $|v_i - V_t| \geqslant v_o$. Note that only one value of input voltage satisfied Eq. (14.124) which is valid for a finite range of output voltage values. This means that the gain in the transition region is infinite if the transistors are ideal. Actually, there is a variation in threshold voltage with back-gate bias which we have neglected. This effect causes the transition gain of the depletion load inverter to be finite

When the driver is in the triode region and the load transistor is saturated, we find from Eqs. (14.107) and (14.122b) that

$$v_o^2 - 2(v_i - V_t)v_o + \frac{V_p^2}{\beta_R} = 0. \tag{14.126}$$

The solution of this quadratic equation is valid for $v_o < v_i - V_t$.

These approximate results agree reasonably well with transfer characteristics given by Carr and Mize.[22] Their results, shown in Fig. 14.46, were derived by numerical calculation with the aid of a digital computer, using a more exact transistor model including the effect of back-gate bias on the load transistor.

An approximate value for the logic 0 output voltage is obtained by considering the driver transistor to be a conductance g_{Dd} passing a current I_l. Therefore, from Eq. (14.111) and Eq. (14.121a),

$$v_o \approx \frac{I_l}{g_{Dd}} \approx \frac{V_p^2}{2\beta_R(v_i - V_t)} . \tag{14.127}$$

14.6.5 Complementary MOS Circuits [23]

The most sophisticated MOS structure is the complementary symmetry (CMOS) circuit which utilizes both n-channel and p-channel devices fabricated in the same substrate. Processing techniques are discussed in Section 6.15. The availability of complementary MOSFET's permits the implementation of logic circuits which are faster than single channel MOS circuits and whose static power consumption is essentially determined by the leakage current of the *p-n* junctions.

A CMOS inverter circuit is shown in Fig. 14.47. Both transistors are enhancement mode devices. Isolation between the transistors is obtained by using an *n*-type substrate which contains the *p*-channel transistors and *p*-type diffused regions containing *n*-type transistors. The *n*-type substrate is connected to V_{DD} while the *p*-tubs are connected to V_{SS}, thus reverse biasing the isolation junction. The input source controls both devices and the load capacitance is moved in either direction by an active device. Since most of the power is used to charge the load capacitance, we would expect that the power consumption would be approximately proportional to frequency.

The drain characteristics of both devices are plotted on the same set of axes, as shown in Fig. 14.48, as an aid in qualitatively describing the operation of this inverter. We now consider the case of the *p*-channel device being turned off by positive going input signal. Application of Kirchhoff's voltage law to the input and output loops yields the following expressions for the terminal voltages of the transistors:

$$V_{GSp} = v_i - V_{DD}, \tag{14.128a}$$

$$V_{DSp} = v_o - V_{DD}, \tag{14.128b}$$

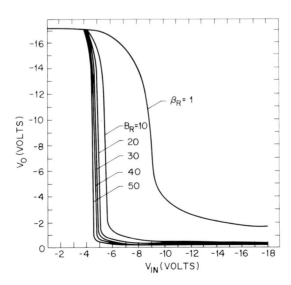

Fig. 14.46 Transfer characteristics of an MOSFET inverter with depletion mode load. (From Carr and Mize.[22])

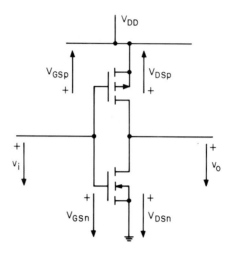

Fig. 14.47 CMOS inverter circuit.

$$V_{GSn} = v_i, \tag{14.128c}$$

$$V_{DSn} = v_o. \tag{14.128d}$$

Note that the drain terminals of the two devices are connected together, and since the drain current is defined as positive when directed into the drain terminal, $I_{Dn} = -I_{Dp}$. Initially the inverter output is at the logic "1" level and the input is at 0 V. Under these conditions the p-channel device is biased heavily ON and in the triode region while the n-channel device is cut off. As the input v_i increases, the operating point of the p-channel device shifts to characteristic curves corresponding to lower values of current. Until $v_i > V_{tn}$, the threshold voltage for the n-channel device, the n-channel dev-

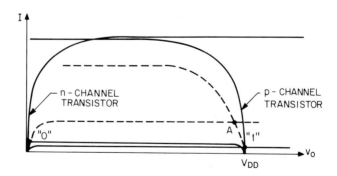

Fig. 14.48 Drain characteristics of CMOS inverter devices.

ice remains cut off, the p-channel device remains in the triode region, and the output voltage remains at $V_o^{(1)}$. When v_i exceeds V_{tn}, the n-channel device comes out of cutoff and goes into saturation. The drain current then rises and the output voltage begins to fall. Point A in Fig. 14.48 is a typical operating point in this regime. As v_i continues to increase, both transistors enter the saturation region. In this condition the gain becomes very large, the drain current reaches its maximum value, and the output voltage falls rapidly. The transition point is that input voltage v_i' corresponding to both transistors in the saturation region. It is found by equating the drain currents for both devices in the saturation region yielding

$$\beta_p \left[1 - v_i^* - v_p \right]^2 = \beta_n \left[v_i^* - v_n \right]^2, \tag{14.129}$$

where $v_p = |V_{tp}|/V_{DD}$, $v_n = V_{tn}/V_{DD}$, and $v_i^* = v_i'/V_{DD}$ is the normalized transition voltage. Solving for v_i^* yields

$$v_i^* = \frac{\sqrt{\beta_R}\,(1-v_p) + v_n}{1 + \sqrt{\beta_R}} \tag{14.130}$$

where in this case $\beta_R = \beta_p/\beta_n$.

Thus the normalized transition voltage depends upon the β ratio and the normalized threshold voltages of the devices. Finally, the n-channel transistor is driven into the triode region and the p-channel transistor is cut off.

The transfer characteristics for the CMOS inverter are shown in Fig. 14.49. In each of the quiescent output states $v_o = V_o^{(0)}$ and $v_o = V_o^{(1)}$, the steady-state current and the voltage developed across the ON device are very small by virtue of its low resistance. The steady-state current is determined by the OFF device. Since the circuit is symmetric, $V_o^{(0)} \approx 0$ and $V_o^{(1)} \approx V_{DD}$.

14.6.6 Power Dissipation of MOSFET Inverters

We have already seen from the composite drain characteristic curves of Fig. 14.48 for the CMOS inverter, that whenever the output is $V_o^{(0)}$ or $V_o^{(1)}$ the inverter consists of the series connection of a transistor in cut-off and a transistor heavily biased in the triode region. Under these conditions the current is junction-leakage current and the static power dissipation is approximately zero. In a CMOS inverter, power is only consumed during the transient. Usually the load is entirely capacitive. If the rise and fall times of the output waveform are small enough so that V_{DD} is developed across the load capacitance, C_L, the transient power is

$$P = C_L V_{DD}^2 f_o, \tag{14.131}$$

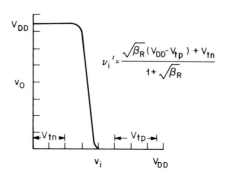

$$v_i' = \frac{\sqrt{\beta_R}(V_{DD}-V_{tp}) + V_{tn}}{1 + \sqrt{\beta_R}}$$

Fig. 14.49 Transfer characteristic of CMOS inverter.

where f_o is the frequency of a square wave excitation driving the inverter.

The power dissipation of a single channel MOSFET inverter when it is ON is proportional to the beta of the load device. The load current for an inverter with an enhancement mode load biased in the saturation region is

$$I_l \approx \frac{\beta_l}{2} (V_{DD}-V_t)^2$$

when we assume that the output voltage $V_o^{(0)} = 0$ in Eq. (14.103). Hence

$$P_{es} = V_{DD}I_l = \frac{\beta_l}{2} V_{DD}(V_{DD}-V_t)^2. \tag{14.132}$$

From Eq. (14.116) with $v_o = 0$, the load current for the inverter with an enhancement mode, triode-operated load is

$$I_l \approx \frac{\beta_l}{2} V_{DD}[2(V_{GG}-V_t) - V_{DD}]$$

and therefore

$$P_{et} = \frac{\beta_l}{2} V_{DD}^2[2(V_{GG}-V_t) - V_{DD}]. \tag{14.133}$$

Finally, for the inverter with a depletion mode load connected as shown in Fig. 14.45, when $v_o \approx 0$ the load is in saturation. Therefore, from Eq. (14.121a), the load current is

$$I_l = \frac{\beta_l}{2} V_p^2 \tag{14.121a}$$

and

$$P_{ds} = \frac{\beta_l}{2} V_{DD} V_p^2.$$ (14.134)

14.6.7 Transient Response of the MOSFET Inverter

We now consider the transient response of the various MOSFET inverters described above. The analysis is greatly simplified because the speed of the intrinsic device is much greater than that of the circuit, as will now be shown.

A figure of merit for a voltage-controlled current source is the ratio g_m/C. It can be shown that the maximum operating frequency of a MOSFET is

$$f_{max} = \frac{g_m}{2\pi C_g}$$ (14.135)

where C_g is the total gate capacitance. When the transistor is in the saturation region, $g_m = \beta(V_{GS}-V_t)$. The capacitance of the gate is that of a parallel plate capacitor,

$$C_g = \frac{\kappa_{ox}\epsilon_o WL}{t_{ox}}.$$

Substituting these relations into Eq. (14.135) and simplifying, yields an explicit expression for f_{max}:

$$f_{max} = \frac{\mu}{2\pi L^2} |V_{GS}-V_t|,$$ (14.136)

from which it is seen that f_{max} varies inversely as the square of the channel length. It is independent of the width because both g_m and C_g increase directly with the width.

Typically f_{max} is of the order of 1 GHz, but in a practical circuit the stray capacitance dominates. The maximum switching rate is limited by the rate at which the current source can charge the stray capacitance. This rate is usually about 1 to 10 MHz.

It can be shown that the pair delay for the CMOS inverter is given approximately by

$$t_D \approx 0.9\tau \left\{ \frac{1}{(1-\nu_n)^2} + \frac{1}{\beta_R(1-\nu_p)^2} \right\},$$ (14.137a)

which is valid for $\nu_n + \nu_p < 1$, $\beta_R \geqslant 0.2$. In this expression

$$\tau = \frac{C_L}{\beta_m V_{DD}} , \qquad (14.137b)$$

where C_L is the capacitance loading the inverter output. The rise and fall times are related in a less straightforward manner. However, an empirical condition for equal rise and fall times is

$$\beta_R(1-\nu_p)^2 \approx (1-\nu_n)^2. \qquad (14.138)$$

We consider that the MOSFET responds instantaneously to a change in input excitation and that the output response is limited by the external load capacitance. It is further assumed that all of the capacitances are lumped into the simple equivalent capacitance C_L. The equivalent circuit to be analyzed is shown in Fig. 14.50.

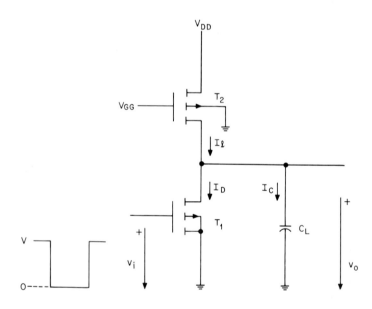

Fig. 14.50 Equivalent circuit for transient analysis of an MOSFET inverter.

Initially transistor T_1 is conducting and therefore the output voltage is at the logic 0 value which is taken here to be 0 V. At $t = 0$ a negative step is applied to the input which reduces the input voltage to zero. For $t > 0$, $I_D = 0$ and the capacitor C_L is charged by the load current I_l since transistor T_1 responds instantly. It was shown previously that the load current for an enhancement mode load in saturation is

$$I_l = \frac{\beta_l}{2} (V_{DD}-V_t)^2 \left\{ 1 - \frac{v_o}{V_{DD}-V_t} \right\}^2 \tag{14.103}$$

which can be written as

$$I_l = \frac{\beta_l}{2} \left[V_o^{(1)} \right]^2 [1-u]^2.$$

When the load is in triode region,

$$I_l = \frac{\beta_l}{2} \frac{V_{DD}^2}{m} [1-u][1-mu], \tag{14.117}$$

where $V_{DD} = V_o^{(1)}$. Both of these expressions are of the form

$$I_l = I_{SC}(1-u)(1-mu), \tag{14.139}$$

where I_{SC} is the short circuit current when the normalized output $u = 0$, and where $m = 1$ for the load in saturation. The differential equation describing the charging of capacitor C_L is found by setting $I_C = I_l$.

$$C_L \frac{du}{dt} = \frac{I_{SC}}{V_o^{(1)}} (1-u)(1-mu) \tag{14.140}$$

Note that $V_o^{(1)}/I_{SC}$ is the equivalent resistance of the load device at the instant the charging process begins. We therefore define a characteristic time constant:

$$\tau = \frac{C_L V_o^{(1)}}{I_{SC}} = \frac{2mC_L}{\beta_l V_o^{(1)}}. \tag{14.141}$$

Equation (14.140) can be integrated by separation of variables.

$$\frac{dt}{\tau} = \frac{du}{(1-u)(1-mu)} \tag{14.142}$$

The solution of this equation is

$$t = \frac{\tau}{1-m} \ln \left\{ \frac{1-mu}{1-u} \right\}. \tag{14.143}$$

Inverting this result to obtain an expression for u yields

$$u = \frac{v_o}{V_o^{(1)}} = \frac{1 - \exp[-(1-m)t/\tau]}{1 - m \exp[-(1-m)t/\tau]}. \tag{14.144}$$

Figure 14.51 is a plot of normalized output voltage, u, versus normalized time, t/τ, for several values of m. The case $m = 0$ corresponds to a simple time invariant RC circuit. Therefore, the MOSFET inverter with an active

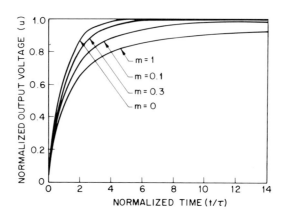

Fig. 14.51 Normalized turn-off transient response of an MOSFET inverter.

load is always slower in charging the load capacitance than a simple resistor. The triode mode load provides a faster rise time than does the saturation mode load.

The response to an excitation which turns on the driver transistor T_1 is more complex because the drain current of T_1 is the sum of the load current and the capacitive discharge current, and because the driver transistor is in the saturation region for part of the time and in the triode region for the balance of the time.

The equation to be considered is obtained by applying Kirchhoff's current law to the output node,

$$\frac{I_D}{I_{SC}} = -\tau \frac{du}{dt} + (1-u)(1-mu), \qquad (14.145)$$

where I_D is given by Eq. (14.105) during the initial phase when T_1 is in saturation and by Eq. (14.107) when the driver is in the triode region. In deriving Eq. (14.145) it was assumed, once again, that the response of the transistor is much faster than that of the circuit. Furthermore, we now assume the input excitation is a step so that transistor T_1 is in a stable condition at $t = 0^+$. With these assumptions the differential equation for the initial phase of the falling response when T_1 is in saturation is

$$\frac{dt}{\tau} = \frac{du}{mu^2 - (1+m)u + (1-m\beta_R w^2)} . \qquad (14.146)$$

This equation must be integrated subject to the initial condition that $u = 1$ at

$t = 0$. The result of the integration is

$$\frac{t}{\tau} = \frac{-2}{\sqrt{q}} \tanh^{-1} \left\{ \frac{2(1-m\beta_R w^2)u - (1+m)}{\sqrt{q}} \right\} - K$$

where K is the constant of integration and

$$q = (1+m)^2 - 4m(1-m\beta_R w^2).$$

Solving for the normalized output voltage u yields

$$u = \frac{\sqrt{q}}{2(1-m\beta_R w^2)} \tanh \left[\frac{\sqrt{q}}{2} \left(-\frac{t}{\tau} + K \right) \right] + \frac{1+m}{2(1-m\beta_R w^2)}$$

$$(14.147)$$

at $t = 0$, $u = 1$. Using these values in Eq. (14.147) and solving for K yields

$$K = \frac{2}{\sqrt{q}} \tanh^{-1} \left\{ \frac{1 - m(1+2\beta_R w^2)}{\sqrt{q}} \right\}. \qquad (14.148)$$

This equation for u is valid as long as T_1 is in saturation, i.e., for $u > w$. The time at which this equation is no longer valid is found by setting $w = u$ in Eq. (14.147) and solving for the time denoted by t'. The result is

$$t' = \left\{ K - \frac{2}{\sqrt{q}} \tanh^{-1} \left[\frac{2u(1-m\beta_R w^2) - (1+m)}{\sqrt{q}} \right] \right\} \tau. \qquad (14.149)$$

The differential equation applicable for $t > t'$ is obtained by substituting Eq. (14.117) for the drain current in the triode region into Eq. (14.145).

$$\frac{dt}{\tau} = \frac{du}{m(1+\beta_R)u^2 - [(1+m) + 2m\beta_R w]u + 1}. \qquad (14.150)$$

This differential equation is of the same form as Eq. (14.146). In this case

$$q = [(1+m) + 2\beta_R mw]^2 - 4m(1+\beta_R)$$

and the normalized output voltage, u, is

$$u = \frac{1 + m(1+2\beta_R w)}{2m(1+\beta_R)} + \frac{\sqrt{q}}{2m(1+\beta_R)} \tanh \left[-\frac{\sqrt{q}}{2} \left(\frac{t-t'}{\tau} \right) + K' \right],$$

$$(14.151)$$

where K' is the constant of integration which is specified by the initial condition $u = w$ at $t - t' = 0$.

$$K' = \tanh^{-1} \left\{ \frac{2mw - (1+m)}{\sqrt{q}} \right\}. \qquad (14.152)$$

Figure 14.52 is a plot of the response of the inverter to a turn-on excitation for the case $V_{DD} = 15$ V, $V_{GG} = 17$ V, $V_t = 5$ V, $v_i = 12$ V, and $\beta_R = 20$. The turn-on response for an inverter with an enhancement mode load is always faster than the turn-off response.

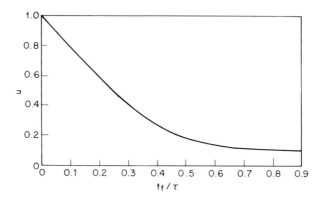

Fig. 14.52   Turn-on transient response of MOSFET inverter.

14.6.8 The NOR and NAND Gates

A simple NOR gate consisting of two identical MOS input devices, T_A and T_B, and an enhancement mode load device T_l operating in the saturation region is shown in Fig. 14.53(a).

If both inputs are at the logic 0 level, the current I will be zero and the output voltage v_o will be $V_o^{(1)} = V_{DD} - V_t$. If any of the inputs are high, i.e., at the logic 1 level, the output voltage v_o will be low and equal approximately to $V_o^{(0)}$ for the inverter. The logic operation performed by this gate is

$$F = \overline{A+B} = \overline{A} \cdot \overline{B}.$$

The gate input devices are identical and are equivalent to the driver transistor in the inverter. The size of these devices is determined by specifications on the low level output $V_o^{(0)}$ of the gate through the specification of the drain conductance given by Eq. (14.111). When p-channel transistors are used, $V_{DD} - V_t < 0$ and the NOR function is implemented in negative logic.

A NAND gate is formed by placing the driver devices T_A and T_B in series as shown in Fig. 14.53(b). In this circuit the drain current is nonzero only when both of the driver devices T_A and T_B conduct. This occurs only when both of the inputs are high. The number of inputs is limited in practical cases to about two or three because the resistance from the output termi-

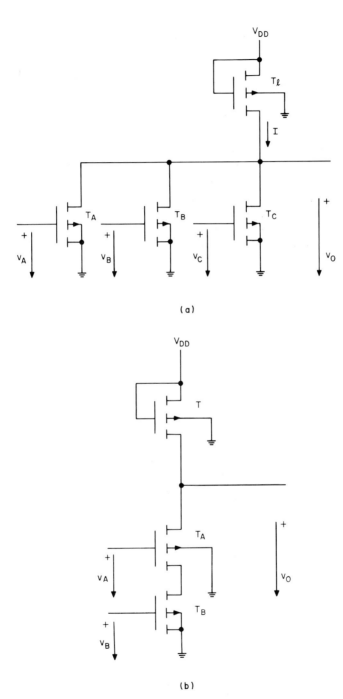

Fig. 14.53 (a) An MOS NOR gate; (b) an MOS NAND gate.

nal to ground when the gate is in the ON condition is determined by the $V_o^{(0)}$ specification. To obtain the required output resistance, the drain conductance of each of the input devices must be multiplied by the number of inputs. This rapidly increases the area of the NAND gate.

To see how to extend the CMOS inverter circuit described above to perform the NAND and NOR functions, recall the word description of these operations. The output of a NAND gate is low if and only if all of the inputs are high. The output is low when there is a path through ON devices to ground. Thus a CMOS NAND gate is formed by putting several n-channel devices in series from the output terminal to ground, and a like number of p-channel devices in parallel from the output terminal to V_{DD}. A two-input CMOS NAND gate is shown in Fig. 14.54(a).

A two-input NOR gate may be constructed as an inverter with two n-channel devices in parallel and two p-channel devices in series as shown in Fig. 14.54(b).

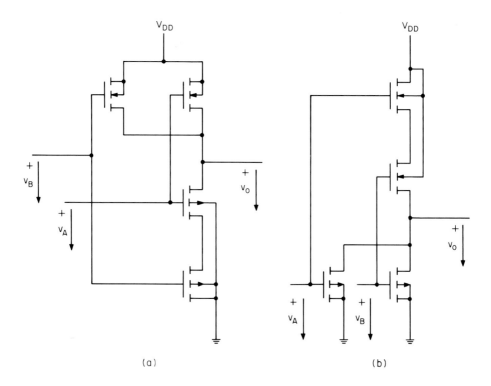

Fig. 14.54 CMOS (a) NAND and (b) NOR gates

14.7 EMITTER-FUNCTION LOGIC [24]

Emitter-function logic (EFL) is a high-performance bipolar logic based on noninverting gates and is a suitable for LSI implementation. The advantages of EFL are that the noninverting gate has a shorter propagation delay time than an inverting gate operating at the same power level, that it permits the use of low-power supply voltages thus reducing power dissipation, that it results in simpler circuit configurations for circuit functions such as flip-flops, and that the resultant chip layouts have a high gate density.

The basic emitter-function logic gate is shown in Fig. 14.55a. It consists of a multiple-input common-base voltage amplifier T_1 driving a multiple-output common-collector output stage T_2. A major reduction in propagation delay time is achieved because the voltage across the collector-base capacitance of the input stage T_1 must change only by the logic swing v_l instead of $2v_l$ as in an inverting gate. The reduction in swing reduces the delay time by about 30%. The multiple emitters of the input transistor perform the AND function on the input variables. A diode-connected clamp transistor T_3 is used to prevent saturation of T_1 when there is current from more than one emitter. Multiple emitters on the output transistor T_2 are used to perform the wired-OR operation. The symbol for the EFL gate is shown in Fig. 14.55(b).

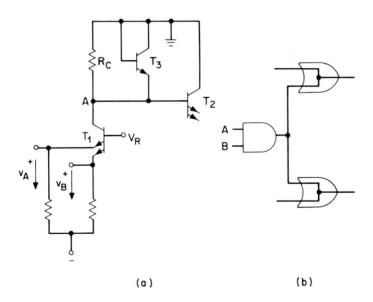

(a)   (b)

Fig. 14.55 Emitter function logic (a) circuit of basic gate and (b) symbol.

Input transistor T_1 is cut off when all of its inputs are high and the internal gate node, called node A in Fig. 14.55(a), tries to rise to ground. The output node cannot rise higher than V_{BE} below ground. Because of input-output compatibility

$$V_i^{(1)} = V_o^{(1)} = -V_{BE}.$$

The input transistor will conduct if one or more of the inputs is not held high. In the low state the voltage at the input emitter will be $V_i^{(0)} = -V_R - V_{BE}$, while the voltage at node A will be $-V_{BE}$. Thus, the logic swing at node A is $v_l = V_{BE}$. The output tries to follow node A except that it is translated downward by V_{BE}. However, the output node cannot fall below $V_i^{(0)}$ because an n-p-n emitter follower cannot pull down on a load. Therefore T_2 begins to turn off when V_A falls below $-V_R$. Furthermore, T_2 should be cut off when $V_A = -V_{BE}$. These considerations dictate setting $V_R = 0.5\,V_{BE}$.

Note that the voltage swing at the input and output is $0.5\,V_{BE}$ whereas the swing at node A is V_{BE}, thus providing the same noise margin as ECL. There is no requirement for constant current through the pull-down resistor and therefore a supply voltage as low as $-2\,V_{BE}$ is sufficient. This results in a 75% reduction in power consumption compared with an ECL gate with no degradation in speed or noise margin.

14.8 THRESHOLD LOGIC GATES [25,26]

A generalization of Boolean logic is *threshold logic* which offers the possibility of implementing a logic function using fewer gates because each gate can perform a more complex operation. Boolean logic gates are special cases of threshold gates as we shall soon see.

A threshold gate with inputs x_i is defined by a set of real-valued weights w_i and a real-valued threshold t such that

$$f(\mathbf{x}) = 1 \quad \text{when} \quad \sum w_i x_i \geq t,$$

(14.153)

$$f(\mathbf{x}) = 0 \quad \text{when} \quad \sum w_i x_i < t.$$

The output of the threshold gate can assume one of two values denoted by 0 and 1. It assumes the 1 or true level when the weighted sum of the input variables equals or exceeds the threshold value. The output is false when this condition is not satisfied.

It is easy to show that the threshold gate is a generalization of Boolean gates. For example, if all of the weights are unity and the threshold $t = N$, (where N is the number of inputs) the threshold gate becomes an AND gate. Similarly, if all of the weights and the threshold are unity, the threshold gate degenerates to an OR gate. It is left as an exercise to determine the threshold logic equivalents of the NAND, NOR, and INVERT gates.

Two mechanizations of the threshold gate are shown in Fig. 14.56. In the FET circuit of Fig. 14.56(a) the resistive divider forms the weighted sum of the input variables. The voltage source in series with the drain determines the threshold level. A bipolar transistor circuit using current sources and current mode switches is shown in Fig. 14.56(b). Each input variable controls a separate current mode switch. The weighting is provided by scaling the value of the current source. The controlled currents are summed at a node whose potential is sensed by a comparator. The reference voltage of the comparator is proportional to the threshold level.

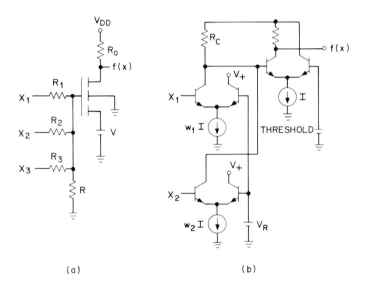

Fig. 14.56 Threshold logic gate using (a) FET transistors and (b) bipolar transistors.

All Boolean logic functions cannot be represented by nondegenerate threshold logic functions. The necessary and sufficient conditions for a logic function to be a threshold function are given by Muroga.[25]

Threshold logic should be considered as a means of reducing the amount of circuitry required to implement complex logic functions. The interested reader is referred to the cited references for a more complete exposition of this topic.

REFERENCES

1.   C. S. Meyer, D. K. Lynn, and D. J. Hamilton, Editors. *Analysis and Design of Integrated Circuits.* New York: McGraw-Hill, 1968, pp. 154-156.

2.   *Ibid.,* p. 112.

3.   L. W. Nagel and D. O. Pederson. "SPICE (simulation program with integrated circuit emphasis)." Berkeley, California, University of California, Electronics Research Laboratory, *Memorandum ERL-M382,* April 1973.

4.   L. O. Chua and Pen-Min Lin. *Computer-Aided Analysis of Electronic Circuits: Algorithms and Computational Techniques.* Englewood Cliffs, N. J.: Prentice-Hall, 1975.

5.   T. R. Bashkow. "Effect of nonlinear collector capacitance on collector current rise time." *IRE Trans. on Electron Devices* **ED-3**: 167-172 (October 1956).

6.   J. Millman and H. Taub. *Pulse, Digital, and Switching Waveforms.* New York: McGraw-Hill, 1965, chapter 4.

7.   F. J. Hill and G. R. Peterson. *Introduction to Switching Theory and Logical Design,* 2nd Edition. New York: John Wiley & Sons, 1974, chapters 3 and 4.

8.   H. H. Muller, W. K. Owens, and P. W. J. Verhofstadt. "Fully compensated emitter-coupled logic: Eliminating the drawbacks of conventional ECL." *IEEE J. of Solid-State Circuits* **SC-8**: 362-367 (October 1973).

9.   C. S. Meyer, *et al., op. cit.,* chapter 10.

10. Y. Tarui, Y. Hayashi, H. Teshima, and T. Sekigawa. "Transistor Schottky-barrier-diode integrated logic circuit." *IEEE J. of Solid-State Circuits* **SC-4**: 3-12 (February 1969).

11. S. K. Wiedmann. "A novel saturation control in TTL circuits." *IEEE J. of Solid-State Circuits,* **SC-7**: 243-251 (June 1972).

12. K. Hart and A. Slob. "Integrated injection logic: A new approach to LSI." *IEEE J. of Solid-State Circuits* **SC-7**: 346-351 (October 1972).

13. H. H. Berger and S. K. Wiedmann. "Merged-Transistor logic (MTL)—a low-cost bipolar logic concept." *IEEE J. of Solid-State Circuits* **SC-7**: 340-346 (October 1972).

14. N. C. de Troye. "Integrated injection logic—present and future." *IEEE J. of Solid-State Circuits* **SC-9**: 206-211 (October 1974).

15. "Special Issue on I^{2}L." *IEEE J. of Solid-State Circuits* **SC-12** (April 1977).

16. F. J. Hewlett, Jr. "I^{2}L current gain design." *IEEE J. of Solid State Circuits* **SC-12**: 206-208 (April 1977).

17. F. W. Hewlett, Jr. "Schottky I^{2}L." *IEEE J. of Solid-State Circuits* **SC-10**: 343-348 (October 1975).

18. B. B. Roesner and D. J. McGreivy. "A new high speed I^{2}L Structure." *IEEE J. of Solid-State Circuits* **SC-12**: 114-118 (April 1977).

19. F. Hennig, H. K. Hingarh, D. O'Brien, and P. W. J. Verhofstadt. "Isoplaner I^{2}L: A high performance bipolar technology." *IEEE J. of Solid-State Circuits* **SC-12**: 101-109 (April 1977).

20. R. E. Crippen, D. O'Brien, K. Rallapalli, and P. W. J. Verhofstadt. "Microprogram sequencer utilizing I³L technology." *Proc. International Solid-State Circuits Conference,* Philadelphia, Penn., pp. 98-99, February 1976.

21. W. M. Penney and L. Lau, Editors. *MOS Integrated Circuits.* New York: Van Nostrand Reinhold, 1972.

22. W. N. Carr and J. P. Mize. *MOS/LSI Design and Application.* New York: McGraw-Hill, 1972.

23. J. R. Burns. "Switching response of complementary-symmetry MOS transistor logic circuits." *RCA Review* **25**: 627-661 (December 1964).

24. Z. E. Skokan. "Emitter function logic—logic family for LSI." *IEEE J. of Solid-State Circuits* **SC-8**: 356-361 (October 1973).

25. S. Muroga. *Threshold Logic and its Applications.* New York: Wiley-Interscience, 1971.

26. F. J. Hill and G. R. Peterson, *op. cit.,* chapter 18.

PROBLEMS

14.1 Show that all combinational logic functions can be realized using either only NAND or NOR gates.

14.2 Plot the transfer characteristics of the current mode switch for input voltages in the range from $0 \leqslant v_i \leqslant V_+$.

14.3 A 3-input ECL gate has a bias current $I = 1$ mA. Calculate and plot the input impedance when $m = 1$ and when $m = 3$. The transistors have $\beta_N = 99$.

14.4 (a) The ECL gate performs the OR and NOR operations in a positive logic system. What operations does it perform if negative logic is used? (b) What operation does a TTL gate perform if negative logic is used rather than positive logic?

14.5 Derive the location of the unity gain points for an ECL gate as a function of m, the number of input transistors that are ON. Also, determine the corresponding noise margins.

14.6 Emitter-coupled logic circuits are usually operated with the collector resistors connected to ground and a negative voltage supply in the emitter circuit; that is, the ground is in the collector circuit instead of the emitter circuit as described in the text. Explain why the use of a negative supply provides better noise performance. Hint: Transistors are in the active region in ECL circuits.

14.7 Figure P14.7 shows the output circuit of an emitter-coupled logic gate driving the inputs of N ECL gates. Only one input circuit is explicitly shown.

    a) Derive an expression for the dc fan-out of the ECL gate.

    b) Evaluate the fan-out for the given element values if $NM^1 = 300$ mV and if $NM^1 = 100$ mV.

14.8 Consider an ECL gate with the NOR side collector load $R_{C1} = 290 \ \Omega$, the OR side collector load $R_{C2} = 300 \ \Omega$, a resistor of 1.18 kΩ in the emitter circuit of

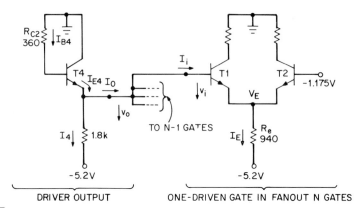

Fig. P14.7

the current mode switch to provide bias current and $1.5 \, k\Omega$ emitter follower load resistors. A $-V_- = -5.2$ V supply is used in the emitter circuit and $V_+ = 0$. The reference voltage $V_R = -1.175$ V. Assume $\beta_N = 99$.

a)   Plot the OR and NOR transfer characteristics at $-50°C$, $25°C$, and $100°C$.

b)   Determine the noise margins NM^0 and NM^1 and the logic swing at these temperatures.

c)   Is it necessary to make the reference voltage vary as a function of temperature? If yes, suggest a suitable circuit.

14.9   Two ECL gates of the type described in Problem 14.8 are interconnected in a digital system.

a)   Determine the operating points, noise margins, logic swing and transition point if the driving gate is at $-50°C$ and the driven gate is at $100°C$.

b)   Repeat part (a) when the driving gate is at $100°C$ and the driven gate is at $-50°C$.

14.10   Implement the exclusive-OR function

$$F = A\bar{B} + \bar{A}B$$

using

a)   ECL gates

b)   T²L gates.

14.11   Construct truth table for AND, OR, NAND, and NOR operations on three binary variables.

14.12   Show that if the current diverted to the substrate by the saturated output driver transistor is neglected, the fan-out capability of the DTL gate is given by Eq. (14.48). This assumption is valid if the circuit uses dielectric isolation or if discrete transistors are used.

14.13 Figure P14.13 is a possible DTL NOR gate. Discuss the operation of this circuit and determine the operating points.

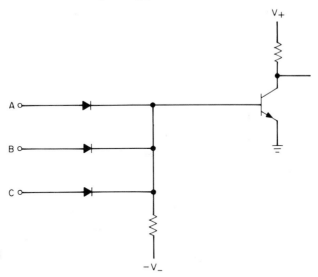

Fig. P14.13

14.14 Compute the input current of the T^2L gate shown in Fig. 14.17 with $R = 4\,k\Omega$, $R_c = 1\,k\Omega$, $\beta_N = 50$, $\beta_I = 0.1$ and $\alpha_S = 0.2$.

14.15 Consider the T^2L gate with active pull-up (sometimes called a "totem pole" output) shown in Fig. 14.23 with $R_1 = 4\,K\Omega$, $R_2 = 1.3\,K\Omega$, $R_3 = 1\,K\Omega$, $R_4 = 100\,\Omega$, and $\beta_N = 100$

a) Calculate the output voltage when both inputs are at 3 V.

b) Calculate the output voltage when at least one input is at 0.4 V.

14.16 Modify the T^2L gate described in Problem 14.15 to have tristate output capability using a scheme such as the one shown in Fig. 14.14. Determine suitable values for the resistors and state any requirements for the transistors.

14.17 For the T^2L inverter with active load pull-up shown in Fig. P14.17.

a) Derive an expression for V_o when the voltage at the base of T_2 is 0 V.

b) Perform a small-signal analysis on the phase splitter T_2 and show that the voltage gain to the emitter $A_{VE} \approx 1$ and that the voltage gain to the collector $A_{VC} \approx -R_2/R_3$. When is the small-signal analysis no longer valid?

c) What is the voltage gain A_{VC} when transistor T_3 goes ON?

d) What is the voltage gain from the base-to-collector of T_3? The load resistance for T_3 is the dynamic impedance of diode D in series with the impedance seen looking into the emitter of an active emitter follower.

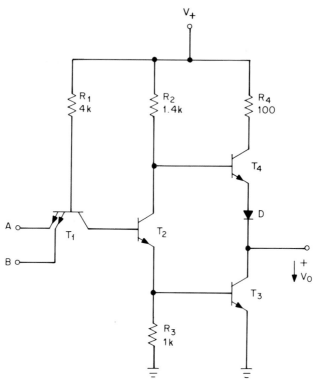

Fig. P14.17

e)   Plot the transfer characteristic from the base of the phase splitter T_2 to the output for the element values shown in Fig. P14.17.

f)   Plot the transfer characteristic from input to output for the element values shown in Fig. P14.17.

14.18 The parameters of an n-p-n bipolar transistor are $I_{s1} = 10^{-14}$ A, $I_{s2} = 10^{-13}$ A, $\alpha_N = 0.98$, $\alpha_I = 0.67$, $\tau_s = 20$ ns, and $\tau_{CN}^b = 1$ ns. The reverse saturation current of the Schottky diode is $I_{DS} = 10^{-10}$ A. This transistor is used as an inverter with a 500 Ω collector load and a 5 V supply. The inverter is driven by a constant current source. Calculate $V_{CE\,sat}$ and the storage time t_s with and without the clamp diode when the input current is 180 μA, 250 μA, and 1 mA. The base turn-off current is 250 μA in all cases. current is 250 μA in all cases.

14.19 Sketch the layout for a 3-input I^{2}L decoder. The decoder has as inputs A, B, and C, and produces all possible combinations of these variables as outputs.

14.20 What is the logic swing v_l and the upper logic level V_U for an I^{2}L gate at an injector current of 10 μA? Assume $\beta_u = 10$, $\alpha_N = 0.7$, $I_{s2} = 10^{-14}$ A.

14.21 An MOSFET inverter with an enhancement mode load transistor in saturation has the following device characteristics: $V_t = 2$ V, $W_d = 240$ μm, $W_l = 15$ μm, $L_d = L_l = 10$ μm. The power supply is $V_{DD} = 12$ V.

a) Find the unity gain points and the transition width.

b) Find the noise margins NM^0 and NM^1.

c) Find the gain in the transition region.

d) Find the logic levels $V_o^{(0)}$ and $V_o^{(1)}$.

14.22 Find the operating points and logic swing for an MOSFET inverter with $\beta_R = 1$ and thereby show that it is of no practical value.

14.23 Solve Eq. (14.18) and Eq. (14.119) to obtain expressions for the transfer characteristic of an MOSFET inverter with a triode mode enhancement load.

14.24 Find the unity gain points, logic levels, transition width, transition region gain, and noise margins for an inverter with a triode-mode enhancement load if $\beta_R = 36$, $V_{DD} = 5$ V, $V_t = 2$ V, $V_{GG} = 12$ V.

14.25 Repeat Problem 14.24 for an MOSFET inverter with depletion mode load if $\beta_R = 25$, $V_{DD} = 5$ V, $V_t = 2$ V, and $V_p = -2$ V.

15

Digital Systems

This chapter on digital systems is the last chapter on the topic of circuits and systems. The greatest impact of the silicon integrated circuit technology has probably been in the field of digital electronics. A visible example for the scientist and engineer of this impact is the popular scientific calculator that has all but replaced the slide rule. Some of these calculators are even programmable so that they can perform a sequence of calculations automatically. This is one of many examples which demonstrate the important results accruing from the power to pack the equivalent of many hundreds of logic gates on a small chip of silicon.

The availability of an extremely powerful logic capability on a chip has changed the architecture of many systems. For example, the microprocessor has created a trend toward using stored program concepts to implement operations rather than hard-wired logic. Array logic and microprocessors will be discussed later in the chapter. Furthermore, many analog signal-processing functions are now being performed by sampling and digitizing the analog signals, and operating on the resultant data stream with a digital signal processor. The digital filter is a modern example which illustrates such signal processing, but a detailed discussion of such filters is beyond the scope of this book.[1,2]

15.1 SYSTEM TERMINOLOGY AND SOME ALTERNATIVE REALIZATIONS [3,4]

A multi-input, multi-output digital system has a set of inputs $x_1, x_2, ..., x_M$ called the input vector $\mathbf{x}$ and a set of outputs $y_1, y_2, ..., y_p$ called the output vector $\mathbf{y}$. In the previous chapter we briefly considered *combinational systems* which are *memoryless systems whose present output is a function solely of the present input*

$$\mathbf{y}(k) = f[\mathbf{x}(k)].$$ (15.1)

This is a special case of the general digital system or *general state machine* shown in Fig. 15.1. The output of the general system is a function of the present input excitation as well as previous inputs and outputs. Knowledge of the previous history of the system is contained in the *state vector* $\mathbf{q}$ with elements q_1, q_2, ..., q_N. The state vector must contain sufficient memory about the system history to determine both the next state and the present output if the present inputs are known. This terminology for finite state systems is analogous to the familiar state variable representations of continuous or discrete state systems.

The equations describing the general system are

$$\mathbf{q}(k+1) = g[\mathbf{q}(k), \mathbf{x}(k)]$$

(15.2)

$$\mathbf{y} = f[\mathbf{q}(k), \mathbf{x}(k)]$$

where k indexes the states. Many systems contain periodic clocks which permit state transitions only at fixed instants of time. These systems are known as *synchronous systems*. *Asynchronous systems* can undergo state transitions at any time. We see from Fig. 15.1 that the general digital system contains a memory to store the state vector and two combinational logic blocks, one of which operates on the present state vector and present input vector to produce the present output vector. The second combinational block operates on the present state vector and the present input vector to produce the next state vector.

The operations of this general machine can be restricted by constraints on the structure so as to define five classes of operation. The simplest of these classes is the combinational, or Class 0, machine discussed previously. Class 0 machines are memoryless. A Class 1 machine has memory but no

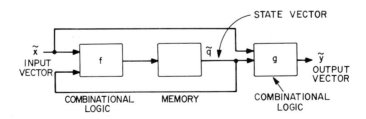

Fig. 15.1 A general state machine.

feedback or feedforward paths so that

$$\mathbf{y}(k+1) = fg[\mathbf{x}(k)], \tag{15.3}$$

that is, the next output is obtained by a combinational transformation on the previous input vector. This machine requires only one combinational network which can precede or follow the state vector storage. An example of a Class 1 machine is a *shift register.*

Next in complexity is the Class 2 machine in which the present output vector and the next state vector are determined solely by the present state vector. Thus, Class 2 machines have state vector output and the next state vector has *direct state transition.* Class 2 machines are described by Eqs. (15.2) with the input vector $\mathbf{x} = 0$. *Counters* are an example of Class 2 machines.

For the Class 3 machine the next state vector is a function of both the present state and the input, while the output depends solely on the state vector. Hence, it has *conditional state transition* and *state output,* and for a fixed input vector the next state is determined by the present state. Thus the Class 3 machine can be viewed as a Class 2 machine with several possible state sequences, one of which is selected by the particular input vector. Since Class 3 machines can choose between alternative state sequences, they can perform all algorithms. The Class 4 machine, which has both *conditional state transition* and *conditional output,* can only provide some simplifications.

The Class 2, 3, and 4 machines in which the next state is a function of the present state are *sequential machines.*

These five classes of digital machines have been discussed in general terms with no reference to the many possible physical implementations. Before the advent of large-scale integrated circuits, a digital system would have been built as an interconnection of discrete or small-scale integrated circuit gates assembled on printed circuit cards. Advances in integrated circuit technology have enabled machines of reasonable complexity to be built entirely on one chip. There are several ways in which this LSI technology can be utilized.

Custom or *random* logic makes maximum use of the available silicon area. The gates needed to implement the required system function are placed and interconnected on an LSI chip that is committed by design to perform a particular operation. Custom logic is a desirable method for circuits that will be in high-volume production because the yield improvement due to the reduction in chip area will result in a cost reduction that exceeds the increased design cost. However, the cost of each new integrated circuit design is typically many tens of thousands of dollars. There are many cases in the design of very large systems in which a custom logic circuit is the best design alternative, in spite of its high cost, because it permits the system to be more effectively partitioned into circuit boards and racks, and therefore results in an overall cost saving.

The high cost of an integrated circuit design stimulated the development of circuit configurations that can be adapted easily to perform a wide variety of system functions. The goal was to develop a circuit that can be produced in a large volume to meet the combined demand of many users. The circuit is customized to perform the specific-user function, either at a very late stage of production or by the user himself.

The *gate array* is an IC chip that consists entirely of unwired logic gates. It is customized for a particular application by means of a special metallization mask. Gate arrays containing more than 500 gates have been made.

Great advances have been made in memory technology in the past few years. It is now possible to obtain single memory chips that contain 16 kilobits of memory, and the day of the 64 kilobit memory chip is not too far away. Memory will be considered in a later section of this chapter. Two engineering considerations have made large memories possible, namely: (1) memory cells can be arranged in very orderly arrays which permit interconnecting the cells to each other and to the peripheral circuits; (2) reduction of cell size and the consequent small signal swings allow a high ratio of cells to relatively sophisticated sensitive detectors.

Memory technology can be applied to the implementation of system functions in two ways. First, read-only memories (ROM) may be used to implement combinational functions. An extension of this technique makes use of "generalized" ROM's and some flip-flops to mechanize a Class 3 or Class 4 state machine, known as programmable logic arrays (PLA). The PLA is discussed in more detail later in this chapter. Second, one may use a stored program computer to execute the steps of the algorithm that describes the system function. The functional behavior of the system is changed by altering the program instructions that are usually stored in ROM. This recent technique relies on the introduction of microprocessors which are central processing units (CPU) fabricated on one or a few LSI chips. Recently, single-chip computers that contain both the processor and memory have appeared in the market.

Both the PLA and microprocessor techniques rely on the fact that the character of the particular system is contained in the coding of ROM's and not in the basic chip structure. Consequently, one basic chip can be adapted to many system applications permitting it to be produced in very large quantities at low cost.

15.2 THE FLIP-FLOP AS A STORAGE ELEMENT

As previously mentioned, a requirement of sequential logic circuits is the ability to store information about the previous condition of the system. The most common bistable circuit is the bistable multivibrator, which has become known as the flip-flop. It consists of two inverting gates enclosed in a positive feedback loop as shown in Fig. 15.2(a). The same circuit is redrawn in Fig. 15.2(b) in the more conventional cross-coupled arrangement. NOR gates have been used in this realization, but there is also a NAND gate realization.

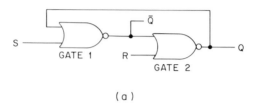

(a)

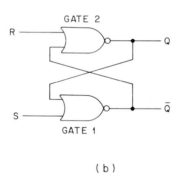

(b)

Fig. 15.2 (a) An R-S flip-flop drawn as two inverting gates in a positive feedback loop and (b) in the more conventional cross-coupled arrangement.

The outputs of the flip-flop are customarily labelled Q and $\bar{Q}$. The inputs to this type of flip-flop are denoted by S for set and R for reset. It will shortly be shown that when the set input is a logic 1, the Q output is driven to a logic 1, and similarly for the reset input and $\bar{Q}$ output. This circuit is a *static* flip-flop because once it is driven into a particular state it remains in that state as long as power is supplied to the circuit.

The truth table is expressed by the following equation:

$$Q_{N+1} = S_N + Q_N \bar{R}_N, \tag{15.4}$$

where $R_N S_N = 0$, and R_N, S_N are the reset and set inputs at time t_N. The truth table for the R-S flip-flop is given in Table 15.1.

Table 15.1 Truth Table for R-S Flip-Flop

S	R	Q_N	$\overline{Q}_N$	Q_{N+1}	$\overline{Q}_{N+1}$
1	0	0	1	1	0
0	1	0	1	0	1
0	0	0	1	Q_N	$\overline{Q}_N$
0	1	1	0	0	1
0	0	1	0	Q_N	$\overline{Q}_N$
1	1	ϕ	ϕ	undefined	

Observe that for nonzero inputs S and R, the outputs $Q_{N+1}, \overline{Q}_{N+1}$ follow the inputs except for the undefined case $S = R = 1$. On the other hand, for inputs $S = R = 0$, the outputs remain unchanged because of the feedback and so this circuit "remembers" the previous set or reset command.

EXAMPLE 15.1 Derive the truth table for an R-S flip-flop using NOR gates by an exhaustive argument.

Solution. Since the R-S flip-flop is a sequential circuit the outputs at any time depend on the inputs and outputs. The outputs at time t_N are denoted by Q_N and $\overline{Q}_N$, while at a subsequent time t_{N+1} they are denoted by Q_{N+1} and $\overline{Q}_{N+1}$. Assume that $Q_N = 0$ and $\overline{Q}_N = 1$. If the applied inputs are $S = 1$, $R = 0$, then it is easily seen from Fig. 15.2 that the presence of a logic 1 level on the input of gate 1 forces the output of gate 1 to the logic 0 level and thus $\overline{Q}_{N+1} = 0$. Both inputs to gate 2 are now zero and therefore $Q_{N+1} = 1$. This is the first line of the truth table shown in Table 15.1.

If the inputs are $S = 0$, $R = 1$, and $Q_N = 0$, $\overline{Q}_N = 1$, the output of gate 1 will remain at the logic 1 level and so $\overline{Q}_{N+1} = \overline{Q}_N = 1$. Both inputs to gate 2 are now high and hence $Q_{N+1} = Q_N = 0$.

When both inputs are low, $S = R = 0$, the output of gate 1 is the complement of the feedback signal. It therefore follows that $Q_{N+1} = Q_N$ and $\overline{Q}_{N+1} = \overline{Q}_N$. These results are given in the first three rows of Table 15.1.

Observe that the circuit of Fig. 15.2(b) is symmetric about a horizontal axis and the circuit of Fig. 15.2(a) is circularly symmetric. Hence, three more rows of the truth table are obtained by interchanging the designation of the R and Q terminals with the S and $\overline{Q}$ terminals.

The final case occurs when $R = S = 1$. Since one of the inputs to each of the gates is a logic 1 regardless of the output state, each of the outputs attempts to go to the logic 0 level. Therefore the output state is undefined for this input combination.

The flip-flop does not change states instantaneously. Consider once again the circuit shown in Fig. 15.2(a) where, this time, the gates are modeled as ideal gates in cascade with a propagation time delay, τ_p, as shown in Fig. 15.3(a). Assume that initially $Q = 0$, $\overline{Q} = 1$, $S = R = 0$, and that at time $t = 0$, an input is applied to set the flip-flop as shown in Fig. 15.3(b).

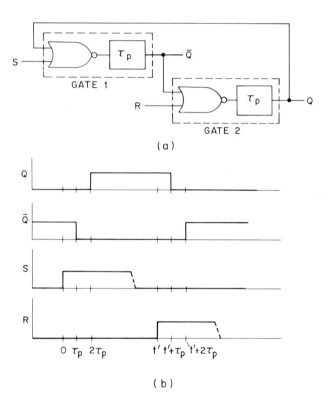

Fig. 15.3 (a) A flip-flop showing the time delays associated with each gate; (b) excitation and response waveforms.

There is no evidence at the output of the change in input excitation until $t = \tau_p$ when the $\bar{Q}$ output makes its transition to the logic 0 level. At this time both inputs to gate 2 are zeros. After another delay of τ_p seconds, the effect of this input change appears at the Q output. The feedback from the Q output to the input of gate 1 causes regeneration that holds the outputs at $Q = 1, \bar{Q} = 0$ for $t \geqslant 2\tau_p$ regardless of the level of the set input. At some time after $2\tau_p$ and before the reset input goes high, the set input must go to the logic 0 level. However, the input must remain stable for at least $2\tau_p$ in order for the outputs to stabilize. This time interval is sometimes called the *latching time*. Note that during part of the latching time both outputs are at the zero level.

Similar arguments show that when the flip-flop is reset, the Q output goes to the logic 0 level before the $\bar{Q}$ output goes to the logic 1 level. Therefore, if this flip-flop drives another similar flip-flop, the undefined state when both inputs are logic 1's is avoided.

It is left as an exercise to show that an R-S flip-flop can be made using NAND gates.

15.2.1 Clocked Flip-Flops

A state change for the R-S flip-flop described above can be initiated at any time. This may be undesirable because the flip-flop is then constantly subjected to the effect of noise on the input lines that could inadvertently switch its state. Furthermore, there is another cause of uncertainty because the set and reset signals are often derived from long chains of logic circuits and therefore the timing of these signals may depend critically on the propagation characteristics of the individual gates. The effect of these problems can be reduced by providing a level of gating in front of the flip-flop, as shown in Fig. 15.4, to enable the inputs for a limited time. This configuration is called a *gated,* or *clocked,* flip-flop because the control signal that enables the gates is often derived from a periodic clock. The period of the clock, T_c, and the pulse duration, t_p, must be chosen so that the circuit changes state only at the occurrence of a clock pulse and no more than once for each clock pulse. The signals at the R and S inputs are allowed to stabilize before the clock pulse is applied, and inputs must remain stable during the interval t_p. From the discussion of R-S flip-flops, $t_p \geqslant 2\tau_p$, so that the inputs are applied to the flip-flop for a time sufficient to enable it to change state. The pulse duration should be no longer than necessary so as to minimize noise problems.

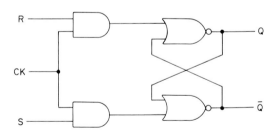

Fig. 15.4 A clocked R-S flip-flop.

These concepts can be quantified by considering the waveforms of the clock and input signal shown in Fig. 15.5. The waveforms have a nonzero rise time and therefore it is customary to use time measurements between the 50% levels to specify the relationship between the waveforms. Usually clocked flip-flops are designed to trigger on the edge of the clock waveform

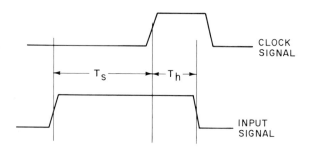

Fig. 15.5 Clock and input signals for the clocked R-S flip-flop.

rather than on a particular level. The *setup time*, T_s, is the time interval allowed for the inputs to stabilize before they are clocked into the flip-flop. The *hold time*, T_h, is the time interval after the clock transition has occurred during which the inputs must be held stable.

15.2.2 The J-K Flip-Flop

The *R-S* flip-flop has an undefined output state when $R = S = 1$ which can often be troublesome. This undefined state can be eliminated by means of some auxiliary input gating that compares the present inputs to the present outputs and causes the flip-flop to change state when both inputs are at the logic 1 level. The resultant configuration is known as a *J-K* flip-flop.' The input gating functions are

$$S = J\overline{Q}$$

$$(15.5)$$

$$R = KQ$$

where J and K are the externally accessible inputs as shown in Fig. 15.6(a). An *R-S* flip-flop with input gating connected according to Eq. (15.5) is shown in Fig. 15.6(b). Observe that this circuit is no more complex than the clocked *R-S* flip-flop shown in Fig. 15.4, and for this reason clocked *J-K* flip-flops are much more prevalent than clocked *R-S* flip-flops. The truth table for the *J-K* flip-flop is shown in Table 15.2 and is given by the equation

$$Q_{N+1} = J\overline{Q}_N + \overline{K}Q_N. \qquad (15.6)$$

For many applications it is desirable to have unclocked, direct set and direct reset inpu̇ts in addition to the J and K inputs. These direct inputs take precedence over the J and K inputs and are useful in initializing the flip-flops to a known state. It is left as an exercise to modify the circuit of Fig. 15.6 to obtain this feature.

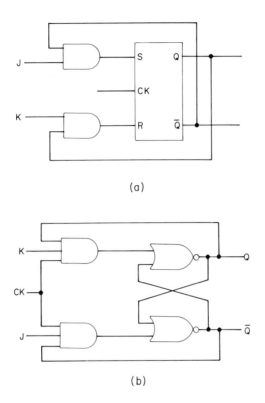

Fig. 15.6 (a) Input gating to convert an *R-S* flip-flop into a *J-K* flip-flop. (b) A *J-K* flip-flop using AND-OR-INVERT gates.

Table 15.2 Truth table for J-K Flip-flop

J	K	Q_N	Q_{N+1}
0	0	0	0
0	0	1	1
0	1	0	0
0	1	1	0
1	0	0	1
1	0	1	1
1	1	0	1
1	1	1	0

EXAMPLE 15.2 A toggle, or *T*, flip-flop is a single input device whose state changes at the occurrence of each input pulse. Suitably connect a *J-K* flip-flop to obtain this behavior.

Solution. The state of a *T* flip-flop is given by $Q_{N+1} = \overline{Q}_N$ which is a special case of Eq. (15.6) with $J = K = 1$. Hence, permanently connecting both the *J* and *K* inputs to the logic 1 level converts a J-K flip-flop into a T flip-flop. Note that the flip-flop is an example of a Class 2 machine because the output is the state and the next state is determined solely by the present state.

15.2.3 The D Flip-Flop

The D, or data, flip-flop is obtained by connecting an inverter between the J and K inputs of a J-K flip-flop as shown in Fig. 15.7. Direct set and reset are also provided to precondition the state of the flip-flop. The truth table is specified by

$$Q_{N+1} = D_N \qquad (15.7)$$

which is a special case of Eq. (15.6) with $J = \bar{K} = D$. The output is the data input delayed by one clock period and therefore the D flip-flop is an example of a Class 1 machine.

When a single-rank J-K flip-flop is used in Fig. 15.7, the output follows the input when the clock is high, and it holds that value of the input prior to the clock edge when the clock is low. Therefore, this implementation of the D-type flip-flop is sometimes called a *follow-and-hold circuit*. It is often used for temporary data storage.

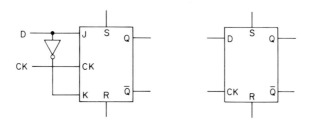

Fig. 15.7 A D flip-flop.

15.2.4 Master-Slave Flip-Flop

A master-slave flip-flop is a structure whose input and output terminals never communicate directly. One popular arrangement consists of two flip-flops in cascade which are clocked by complementary signals as shown in Fig. 15.8(a). The master section follows the inputs on one phase of the clock while it is decoupled from the slave flip-flop by the interstage gating. During this time interval the system observes the stable outputs of the slave section. On the other clock phase the inputs are locked out of the master section and the state of the master is transferred to the slave. The slave section of the master-slave flip-flop shown in Fig. 15.8(a) is a clocked R-S flip-flop while the master section is shown as an unclocked *R-S* flip-flop. The input gating is explicitly shown to indicate that the slave outputs are fed back to the inputs so as to turn the composite structure into a *J-K* flip-flop.

A NAND gate realization for the *J-K* master-slave flip-flop is shown in Fig. 15.8(b). It requires eight NAND gates and a clock inverter. Usually direct set and reset inputs are also provided, although only a direct reset input is shown.

The only operational problem inherent in this structure is known in switching theory as an "essential hazard." An essential hazard can occur if the clock inverter is slow enough to permit the inputs to the slave section to

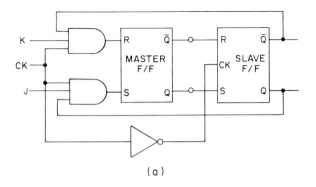

(a)

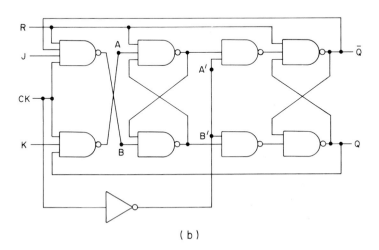

(b)

Fig. 15.8 (a) The master-slave flip-flop as a cascade of two flip-flops. (b) A J-K master-slave flip-flop using NAND gates.

remain enabled on the 0-to-1 clock transition when the master flip-flop has already begun to change state. This condition might result in a double transition of the output state during one clock pulse.

15.3 TIMING PROBLEMS AND CLOCK SKEW [6,7]

In this section are mentioned *clock skew* and *races,* which are problems that arise because of inevitable variations in the propagation time of signals through a dynamic system. These variations are due to differences in performance between nominally identical logic elements and the variation of properties of the transmission media through which these elements are interconnected. Recall that the next state and present output of a sequential machine are functions of the present state and present input. A *clock mode* or *synchronous* machine is one in which the state transition time is controlled by a periodic clock signal. Only one state transition is permitted per clock period. The clock period must be long enough to permit the effect of any input change to propagate to, and stabilize at, the inputs of the flip-flops before the occurrence of the clock pulse. The maximum value of propagation delay time, often equal to twice the typical value, must be used for each gate in the logic chain when calculating the minimum clock period. Synchronous systems are slower than *asynchronous,* or *level mode,* systems that permit state transitions to be initiated directly by the inputs.

Clock pulse distribution is a big problem in large, high-speed systems because logic elements are physically separated and because several clock drivers may be needed to drive all of the points that must be gated simultaneously. The result, called *clock skew,* is a time displacement between supposedly simultaneous clock pulses that may cause false operation of the circuit. This effect is illustrated by the cascade of two *J-K* master-slave flip-flops with time delay in the clock line, as shown in Fig. 15.9(a). When there is no clock skew, an input change, say from $x = 1$ to $x = 0$, is stored in the first flip-flop and the transition is observable at y_1 on the trailing edge of the first clock pulse. The change in state of y_1 is sensed by the second flip-flop on the next clock pulse, and y_2 goes to zero on the trailing edge of the second clock pulse, as illustrated in Fig. 15.9(b). When there is a nonzero delay in the clock line the first flip-flop still behaves correctly, but if the clock skew is large enough, the master section of the second flip-flop will still be enabled when y_1 makes its transition. If this happens, the transition is immediately sensed by the second stage master section and y_2 undergoes a state change on the falling edge of the first clock pulse instead of the second clock pulse as desired. This sequence of events is shown in Fig. 15.9(c).

The distribution of clock signals is a difficult problem often further complicated when multiphase clock signals with precise timing relationships must be derived from the master clock.

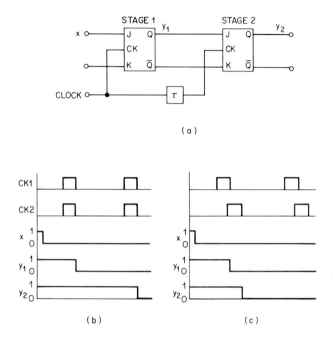

Fig. 15.9 Illustration of the effect of clock skew.

The master clock is buffered from the loads by at least one isolating amplifier to prevent performance from being affected by load variations. In very large systems the load on each phase of the clock may be large enough to require more than one driver. Finally the loads on the various clock drivers should be equalized and the characteristics of the drivers should be such that the delay through each of the paths is approximately equal so as to minimize the clock skew.

While asynchronous systems are faster than synchronous systems, they are troubled by a phenomena called a *race,* which could result in the machine assuming an erroneous state. A *critical race* occurs when a change in input causes the machine to leave a stable state for one or more other stable states, with the final result depending on the way in which signals propagate through logic chains internal to the system. Critical races can sometimes be eliminated by restricting the tolerable variation in gate propagation delay, and by tailoring the delays in the interconnections. This ensures that the proper signal always wins although such a modification might be expensive to do in practice. Alternatively, state variables may be assigned to ensure that critical races are not possible, even though more than the minimum number of state variables may now be required.

15.4 MSI COMPLEXITY CIRCUITS

In this section are described several system functions whose implementation requires 15 to 100 equivalent gates, and are therefore classified as *medium scale integration* (MSI). A general shift register, a presettable counter, a fast adder, and an *arithmetic logic unit* (ALU) are discussed.

15.4.1 Shift Register

A shift register is a digital memory which is used for temporary data storage, data manipulation such as the conversion from a parallel data stream on N wires to a serial bit stream on one wire at N times the word rate, and as a sequentially accessed memory. The emphasis in this section is on the small, static shift register that is used for temporary storage and manipulation. The large sequential access store using CCD's is treated in the next section.

The general shift-right, shift-left register shown in Fig. 15.10 is similar to devices manufactured by Fairchild and Texas Instruments. It consists of a D-type master-slave flip-flop to store each of the N bits of information and input AND-OR gates to steer the data to the proper storage cell and to control the mode of operation. Data can be entered into all of the storage cells in parallel and observed at the output of all the flip-flops simultaneously, or it can be observed sequentially at Q_D as the contents of the storage cells are shifted to the right under control of a clock. Data can also be entered in a serial mode under control of the clock. The choice between parallel and serial input modes is effected by the logic level on the mode control input which enables one of the two sets of input AND gates. The register length can be extended indefinitely by connecting the Q_3 and $\overline{Q}_3$ outputs of one register section to the J and $\overline{K}$ inputs of the succeeding section. The serial data input terminal for the leftmost section of the composite register is formed by tying together the J and $\overline{K}$ terminals of the leftmost register chip. The chip is wired internally for the shift-right function, that is, data shifting in the direction from Q_0 to Q_3, which is the more common shifting pattern with serial data inputs. This operating mode is selected by a logic-0 level signal on the mode control input. The shift-left function is obtained by wiring the output Q_i to the parallel input Q_{i-1}. It is enabled by a logic-1 level on the mode control which also enables the shift-left clock input. This latter input is useful in case the shift rate is not the same in both directions. Note that without external gating this circuit cannot be used for both parallel data input and shift-left operation. Finally, there may be a master reset line which activates the direct reset (direct clear) input of the flip-flops when it is necessary to clear the register.

This circuit has a complexity of 50 to 60 equivalent logic gates, which puts it at the middle of the MSI spectrum. Versions of this circuit are avail-

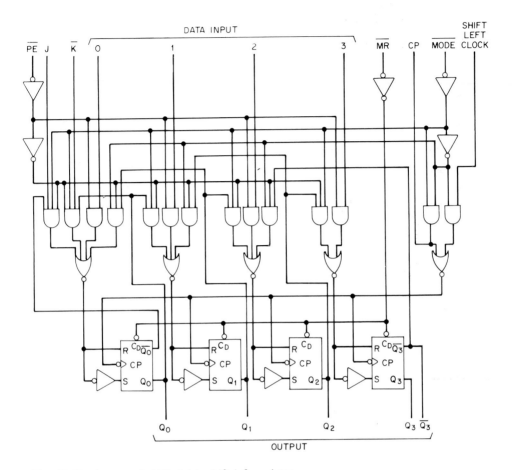

Fig. 15.10 A general shift-right, shift-left register.

able in a high speed, Schottky diode clamped TTL that can perform the shift-right function at a 100 MHz rate.

Sixteen-pin DIP packages are very popular because they are relatively inexpensive and permit a high density of packages on a printed circuit board. The shift register as described above requires 17 pins necessitating a larger, costlier package which would make it somewhat less attractive. Therefore, Texas Instruments eliminated the master reset line, whereas Fairchild chose to eliminate the separate clock inputs in order to reduce the pin count.

15.4.2 Counters

Counters were classified above as Class 2 state machines because the next state is completely determined by the present state. The 4-bit synchronous

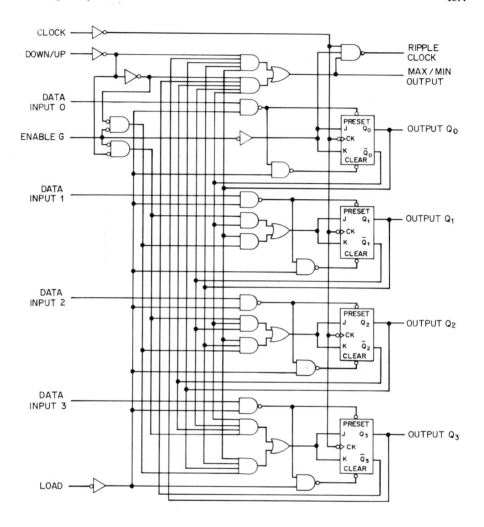

CLOCK

DOWN/UP

DATA INPUT 0

ENABLE G

DATA INPUT 1

DATA INPUT 2

DATA INPUT 3

LOAD

RIPPLE CLOCK

MAX / MIN OUTPUT

OUTPUT Q_0

OUTPUT Q_1

OUTPUT Q_2

OUTPUT Q_3

Fig. 15.11   A 4-bit synchronous counter.   (Courtesy of Texas Instruments.)

counter shown in Fig. 15.11 is a Class 3 machine because it has some additional combinational gating to permit up/down counting and presetting a
count in the flip-flops. The ability to preset a count permits this circuit to be
used as a modulo-N frequency divider or as a variable-length interval timer.

The state vector is stored in four, clocked J-K master-slave flip-flops
that are clocked synchronously to avoid the momentary occurrence of false
output states that could occur if the next state changes were allowed to ripple
through the flip-flop chain.

To see how the counter operates, recall that tying together the J and K
inputs of a J-K flip-flop yields a toggle flip-flop that undergoes a state transi-

tion on every clock pulse during the time that the toggle input is at the logic-1 level. The toggle inputs of Q_1, Q_2, and Q_3, are driven by two gates. Assuming for the moment that the counter is in the up-counting mode, the ith flip-flop toggles if all the Q outputs of the preceding stages are high in the presence of an enabling control signal. Toggle signals for the down counting mode are formed by ANDing the $\overline{Q}$ outputs of preceding gates. The enabling control signals that activate one of the two AND gates in each AND-OR gate are derived by performing the AND operation on the UP/DOWN mode control signal and the input signal that enables the counter. An additional AND-OR gate forms the function

$$Q_0 Q_1 Q_2 Q_3 \text{UP} + \overline{Q}_0 \overline{Q}_1 \overline{Q}_2 \overline{Q}_3 \text{DOWN}$$

where UP and DOWN is the UP/DOWN mode control. This function is a logic 1 when the maximum count is reached in the UP-counting regime and when the minimum count is reached in the DOWN-counting mode. The max/min count and the ripple clock outputs are used for cascading these counters to extend their capability.

The direct set and direct reset (preset and clear) capabilities of the flip-flops are utilized to permit the loading of an initial count. The four data input signals are steered to the flip-flops under control of a load signal via two NAND gates for each bit.

The circuit shown in Fig. 15.11 is one of several counters in the Texas Instruments catalog. It has a complexity of 58 equivalent gates and a typical maximum clock frequency of 25 MHz. Note also that once again the functional block has been configured to fit in a 16-pin package.

15.4.3 Adders

Binary addition is a basic arithmetic operation that is required in digital computation and signal-processing applications. Given the ith bit of the augend and addend words, A_i and B_i, and the carry from the adjacent bit of lesser significance, C_{i-1}, the carry and sum outputs are

$$C_i = A_i B_i + (A_i + B_i) C_{i-1}$$

$$(15.8)$$

$$S_i = (A_i + B_i + C_{i-1}) C_i + A_i B_i C_{i-1},$$

as may be easily verified from a truth table. A circuit which implements Eqs. (15.8) is called a *full adder*. There is no carry input in the least significant bit position and therefore a full adder is not required for that bit. A *half adder* is the mechanization of Eqs. (15.8) with $C_{i-1} = 0$. It is left as an exercise to verify these equations and to provide a NAND-gate implementation of a full adder.

The addition of two N-bit binary words can be performed in parallel by providing $N - 1$ full adders and a half adder, as shown in Fig. 15.12(a). It also can be performed serially with considerably less hardware at the expense of increased computation time using the circuit of Fig. 15.12(b). The two input words are presented as serial bit streams with the least significant bit first and with a time of T seconds between successive bits. A flip-flop provides one bit of storage for the carry output. The contents of the carry flip-flop are reset to zero when the least significant bits A_0 and B_0 are applied to the adder, which then generates the sum bit S_0 and the carry bit C_0 according to Eqs. (15.8). The clock is phased so that C_0 is stored in the master section of the flip-flop and presented at the output of the slave section when the inputs to the adder are A_1 and B_1. During this clock period S_1 and C_1 are generated and the process is repeated. One bit of the result is computed during each clock cycle and therefore, in general, $N + 1$ clock cycles are required to compute the sum of two N-bit numbers. The sum of two N-bit words may be $N + 1$ bits long, and therefore some restriction on the numbers is required because it is usually necessary to maintain synchronism between the input and output bit streams. An example of the use of serial adders in digital signal processing is given by Jackson, *et al.*[5]

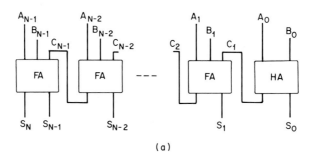

(a)

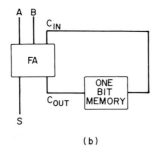

(b)

Fig. 15.12 (a) A parallel, ripple carry, N-bit binary adder; (b) a serial binary adder.

The parallel, *ripple-carry adder* shown in Fig. 15.12(a) does not produce the final result significantly faster than the serial adder just described because comparable times are required to compute sum and carry outputs at each bit position and because carries from bits of lesser significance must ripple through the more significant bit positions thereby affecting the carries for those bits. The ordinary ripple-carry adder is usually not used for long words.

A significant improvement in speed is obtained by using either carry look-ahead or fast-carry propagation techniques. Carry look-ahead is the more popular technique and therefore will be discussed first.

Observe from the first of Eqs. (15.8) that the carry output from the ith bit is a function of the inputs of all bits of lesser significance. Carry look-ahead adders have additional combinational circuitry with inputs A_j, B_j, $1 \leqslant j \leqslant i$ which generates the carry functions C_j, $j \leqslant i$. For a 4-bit adder the carry functions are

$$C_1 = A_1 B_1 + (A_1 + B_1) C_0$$

$$C_2 = A_2 B_2 + (A_2 + B_2) C_1$$

$$\quad = A_2 B_2 + (A_2 + B_2)[A_1 B_1 + (A_1 + B_1) C_0]$$

$$C_3 = A_3 B_3 + (A_3 + B_3) C_2$$

$$C_4 = A_4 B_4 + (A_4 + B_4) C_3 \qquad\qquad (15.9)$$

where C_0 is the carry input whose purpose will be apparent shortly and C_4 is the carry output. The expression for C_1 is substituted into that for C_2 which then becomes quite complex. When that expression is substituted into C_3, the resultant equation is significantly more complicated. The complexity of the carry functions increase very rapidly and therefore, from a practical viewpoint, total carry look-ahead can only be applied over a limited number of stages, typically four.

The circuit for a 4-bit carry look-ahead adder is shown in Fig. 15.13. It consists of the carry function generator described by Eqs. (15.9) and four identical circuits to produce the sum outputs given by

$$S_i = (A_i + B_i) C_i + (A_i B_i + C_i) C_{i-1}. \qquad\qquad (15.10)$$

This circuit has a complexity of 44 equivalent gates. It is left as an exercise to verify that this circuit implements Eqs. (15.9) and (15.10).

EXAMPLE 15.3   The look-ahead carry function extends over the four bits of the adder described above. Can this adder be used in the addition of words that are longer than four bits?

Solution.   A partial carry look-ahead adder for the addition of words longer than four bits can be made by connecting the carry output C_4 from the fourth bit of one block to the C_0 input of the next more significant block. The carry signal then rip-

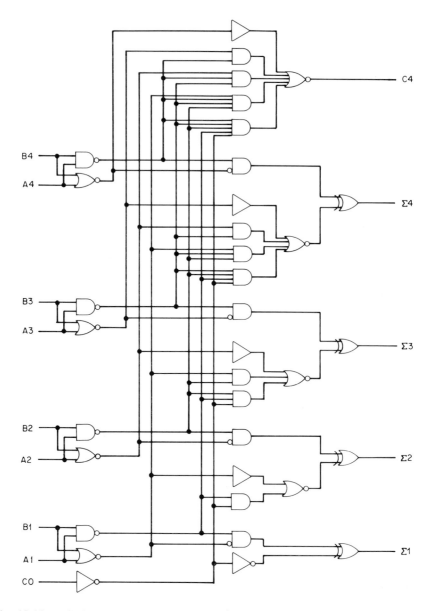

Fig. 15.13 A 4-bit carry look-ahead adder. (Courtesy of Texas Instruments.)

ples from block to block. An example of a TTL implementation of this type of adder is the TI SN7483A whose logic diagram is shown in Fig. 15.13. This circuit can form the sum of two 8-bit words in 23 ns and the sum of two 16-bit words in 43 ns.

Fast-carry propagation techniques are based on rewriting the carry expression to identify under what conditions a carry output is generated in a particular bit position, and under what conditions a carry is propagated through a bit position. Again referring to the carry expression of Eq. (15.8), one notes that the ith bit produces a carry if both A_i and B_i are one regard-

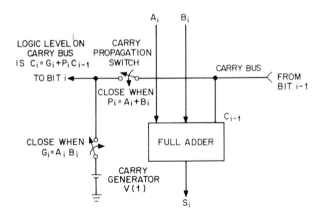

Fig. 15.14 A fast carry propagation adder.

less of the carry input. One therefore defines a *carry-generate* function

$$G_i = A_i B_i. \qquad (15.11)$$

A carry into the ith bit position is propagated if

$$P_i = A_i + B_i = 1 \qquad (15.12)$$

where P_i is the *carry-propagate* function. The carry function can be written in terms of G_i and P_i as

$$C_i = G_i + P_i C_{i-1}. \qquad (15.13)$$

This equation can be used to obtain a fast adder if the functions G_i and P_i are used to control the condition of switches on the carry-transmission line as shown in Fig. 15.14. This type of adder, which was first proposed by Kilburn, uses the inputs A_i and B_i at each bit position to *simultaneously* set up all of the carry-propagation paths between stages. The switches in the propagation path should be transmission gates rather than logic gates containing threshold elements. Two suitable switches are an emitter follower and level-shifting diode as shown in Fig. 15.15(a) and the CMOS transmission gate shown in Fig. 15.15(b). Carry signals are injected on the line at those stages for which $G_i = 1$.

To our knowledge the Kilburn type of adder has not been commercially produced as an integrated circuit. The concept of carry-generate and carry-propagate functions can be extended to fast carry propagation between multibit adder blocks of the type described above. This topic will not be considered further here.

15.5 LSI COMPLEXITY CIRCUITS: I. MEMORIES

There are two basic memory architectures, namely, random access and sequential access. A random-access structure consists of a two-dimensional

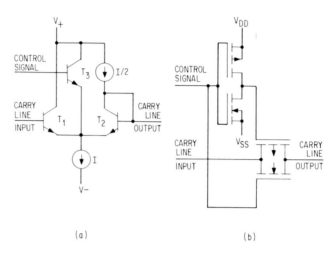

(a)   (b)

Fig. 15.15 Transmission switches for the carry propagation adder using (a) an emitter follower and level shifting diode and (b) a CMOS transmission gate.

array of storage cells with addressing circuitry for accessing any storage location. The access time is almost independent of storage location, and the locations can be accessed in random order. A sequential-access store is comparable to a pipe that is filled with information on one side and emptied on the other. Therefore, the time to access a particular piece of information depends on its position in the information stream. The time required for a piece of information to traverse the pipe is called the *latency time.* Memories are very repetitive and orderly and therefore quite amenable to LSI implementation.

   This section begins with a description of the read-only memory (ROM) which is a random-access structure whose contents can only be read.

15.5.1 Read-Only Memories

The block diagram for a *read-only memory* (ROM) is shown in Fig. 15.16. Information is stored in the rectangular array of crosspoint elements which are usually a single diode, bipolar transistor, or MOS transistor. The rows of the array are called *word lines* and the columns are called *bit lines*. During manufacture the crosspoints are selectively connected to the bit lines. A group of bits that are read simultaneously are called a *word.* Integrated circuit ROM's typically have 4- or 8-bit output words.

   The binary address of the desired output word is applied to the *address-input buffers* which provide an interface between the logic levels of the external system and those internal to the ROM, and convert from single-rail logic signals to dual-rail. The buffered address signals are applied to a *word-line decoder* which is a collection of AND gates connected so as to select one of the word lines. They are also applied to a *bit-line decoder* that selects one of the groups of bit lines. The *column selectors* route the bit line signals to the *detectors*. There is one detector and one output amplifier for each bit of the output word. A simplified circuit diagram showing the selection and detection circuitry associated with one row and column of a bipolar transistor ROM is shown in Fig. 15.17. The chip-select circuitry disables the ROM so that it can be used as part of a memory with a word capacity exceeding that of the individual ROM chips. If the ROM has tristate output circuits (see Section 14.4) the chip-select circuits can simply inhibit the output driver.

The two common methods of coding bipolar transistor or diode arrays are by the selective opening of contact windows or by the selective removal of metal links between the crosspoint element and the bit line. MOS arrays are coded by including or omitting the gate metallization.

Programmable read-only memory (PROM)   The requirement for committing to a particular code for the ROM during processing is a serious disadvantage for the small user who requires just a few pieces of each code. This disadvantage is overcome by a more complex relative of the ROM called the *programmable read-only memory,* or PROM, which uses polysilicon or nichrome fuse links to connect the crosspoint element to the bit lines. MOS PROM's use the trapping of charge at the interface of a dual-dielectric gate insulator to irreversibly shift the threshold voltage (see Chapter 3) of a

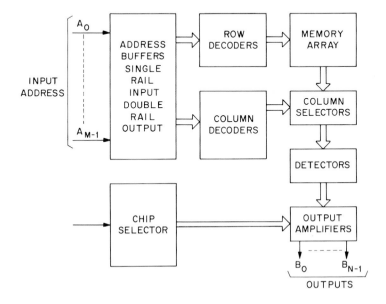

Fig. 15.16 Block diagram of a read-only memory.

selected crosspoint transistor, thus providing the second logic state. Additional circuitry is provided to permit an uncoded word to be selected by the addressing circuits and the desired code to be applied to the output terminals in such a manner as to selectively destroy the fuse links or otherwise write the memory. Once the memory is written it cannot be erased. There is a speed penalty incurred by using a PROM over a ROM of comparable design because of the additional circuitry for writing the array. However, the PROM concept does permit the user to make himself even one unit with a custom code at a price that is possible only by pooling the demand of many users.

Reprogrammable read-only memories (REPROM's)   There are many applications in which the nonvolatility of the ROM is desirable but where it is also occasionally necessary or desirable to rewrite the ROM. For example, when the ROM is the program store for a microcomputer, rewriting the ROM reprograms the computer to perform a new task.

REPROM's available today are MOS devices that use either the trapped charge at the interface of a dual-dielectric gate insulator or FAMOS (see Chapter 3) to provide a crosspoint element with two logic states. The difference between an MOS PROM and a REPROM or EAROM (erasable ROM) is that less energy is required to remove the trapped charge from a REPROM crosspoint, and therefore it is possible to bulk erase a REPROM using X-ray or ultraviolet irradiation. REPROM's that can be electrically block erased are currently under development.

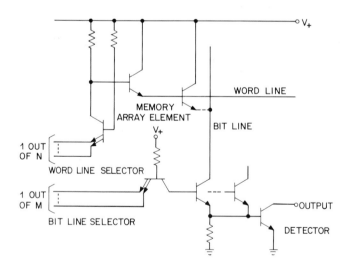

Fig. 15.17 Simplified circuit diagram of a bipolar ROM.

REPROM's could also be called *read mostly memories* because the read access time is much shorter than the write time and because only a limited number of write cycles are possible.

Applications   ROM's are used as fixed program stores, in particular the microprogram store, in digital computers. They are also used as translation stores or lookup tables for mathematical functions or code conversion.

ROM's can also be used to implement general combinational logic functions. The ROM is structurally a large AND-OR gate network whose input variables are the input address bits. Corresponding to every combination of the input variables there is a word in the ROM array which specifies the desired output response of the combinational network for that particular excitation. Address decoding is by means of an AND gate to select one of the word lines and another AND gate to select the appropriate group of bits for the output word. These gates are part of the row decoder and column decoder, respectively, in the block diagram of Fig. 15.16. The use of two decoders to select the output word is necessary in order to obtain a good aspect ratio for the crosspoint array. There are sufficient bits on each word line to provide typically 8 or 16 output words which are channeled through an OR gate at the detector input. Each word of the array corresponds to a cell of the Karnaugh map[6,7] of the logic function. The advantages of this approach to implementing combinational logic is that it is simple and it can be reasonably efficient if there is a large number of output functions. The inefficiency in this method is caused by the restricted form of the decoder which exhaustively enumerates all cells of the Karnaugh map. As a consequence, the realization of $F = AB = ABC + AB\overline{C}$ by means of a ROM with three inputs A, B, and C requires two words of the array because there is no efficient method to handle the "don't care" situation with the variable C. There is also a speed penalty for using the ROM instead of custom-wired logic.

15.5.2 Random Access Read-Write Memories

The architecture of random access read-write memories (RAM) is very similar to that of the ROM described in the last section. A block diagram of a RAM is shown in Fig. 15.18. The major differences between a RAM and a ROM are that the RAM has additional circuits for writing a data word into a storage cell, additional logic to control the read and write functions, and a more complex memory cell. Signal swings sensed by the bit detectors are much smaller in a RAM, and therefore more elaborate detectors must be used and noise is more troublesome.

Some terminology is defined before describing the memory cells in more detail. Once again, a memory is said to have *random access* when the

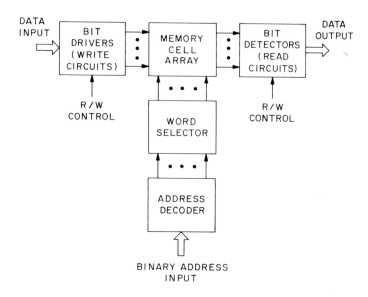

Fig. 15.18 Block diagram of a read/write memory.

cells can be addressed in random order and the time to operate on the contents of any cell is substantially independent of the cell location. A memory is *volatile* if its contents can be destroyed in the event of a power interruption. There are two types of semiconductor memory cells, namely, static and dynamic. *Static memory cells* are basically bistable flip-flops that can hold their state for as long as power is applied to the circuit. *Dynamic memory cells* store the information as charge on a capacitor which is constantly decaying because of loss mechanisms within the cell. Therefore the contents of a dynamic memory must be periodically refreshed. Certain memory cells can be read only by destroying their contents and are said to have *destructive read-out*. The dynamic semiconductor cells and the magnetic cores are in this class. When information is not destroyed during read-out, as with the static semiconductor cell, the process is called *nondestructive*.

The static memory cell[20]   A static memory cell using bipolar transistors is shown in Fig. 15.19(a). Ignoring the bit lines and their associated pull-up resistors, the cell is a direct-coupled bistable flip-flop with the common emitter terminal connected to an externally controlled voltage source which is the word-line driver.

When the *word-line* (WL) voltage is low, the flip-flop has a supply voltage of several volts and it is in a stable state holding previously stored information. This is the standby condition.

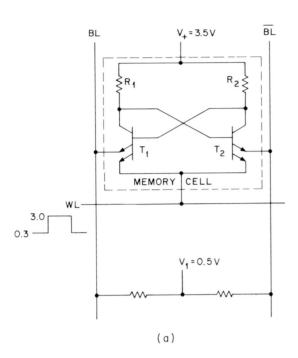

(a)

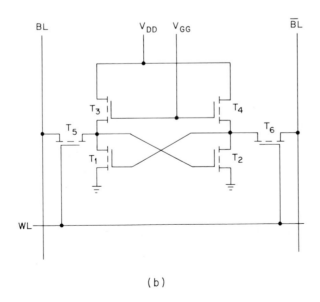

(b)

Fig. 15.19 (a) A bipolar transistor static memory cell; (b) a 6-transistor static MOS memory cell.

The second pair of emitters and the bit lines (BL and $\overline{BL}$) are used for the read and write operations. The cell is written by driving the bit lines with a differential signal. The read operation involves raising the word-line voltage to its high value (~ 3 V) which transfers the standby current from the word-line to the bit line of the saturated transistor. The cell is written by raising either BL or $\overline{BL}$ to a high level when the word line is high. When the word-line voltage is lowered, the flip-flop regeneratively switches to the desired new state.

Static MOS transistor cells are also based on the bistable multivibrator.[14] MOS transistors are also used as load devices and therefore a large number of transistors are required per cell. A 6-transistor cell is shown in Fig. 15.19(b). The word-select-line controls transistors T_5 and T_6 which couple the state of the flip-flop to the bit lines. The cell is written by establishing the desired state on bit lines BL and $\overline{BL}$, followed by raising the word-line signal to select the cell. The write circuitry maintains the desired state on the bit lines and forces the flip-flop to change state.

One-transistor cell dynamic RAM[18, 19]   Read/write memories using the 1-transistor dynamic cell are probably the most important LSI circuits. The industry standard at the present time is the 4-kilobit dynamic RAM and the 16-kilobit dynamic RAM is rapidly approaching the leadership position. Unlike the static cell just described, the 1-transistor dynamic cell has destructive read-out and must also be periodically refreshed because its contents decay with time. The operation of a 1-transistor cell dynamic RAM will now be described.

Most dynamic RAM's are MOS circuits in which each cell consists of an inversion layer capacitance in series with a selection transistor. The information is stored in the form of charge on the capacitor. An inversion layer capacitor is chosen because it has higher capacitance density than a junction capacitor, thus permitting a smaller cell size for the same voltage swing. The cells are arranged in an array with the gates of the selection transistors in each row connected to that word line, and the drain of the selection transistors in a column connected to the bit line as shown in Fig. 15.20. Also associated with the cell are parasitic gate-source and gate-drain capacitors of the selection transistor which will be neglected in this discussion, and the parasitic bit-line capacitance C_B. During the read operation, the selection transistor goes on and the charge on the cell storage capacitor C_S is redistributed between C_S and C_B. Denoting by V_S and V_B the voltages across C_S and C_B before the start of the read operation, the voltage change on the bit line is readily shown to be

$$\Delta V_B = \frac{V_S - V_B}{1 + (C_B/C_S)} . \qquad (15.14)$$

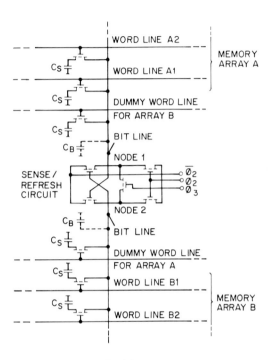

Fig. 15.20 A 1-transistor dynamic cell RAM.

This equation shows that the cell storage capacitance must be chosen large in comparison with the stray bit-line capacitance in order to have a large signal swing. A lower limit on the permissible bit-line swing is imposed by the sensitivity of the bit-line detector because $\Delta V_B > V_{tr}$ the transition width of the detector. Equation (15.14) is usually used to specify the storage capacitance C_S in terms of the detector sensitivity $V_{tr}/(V_S - V_B)$ and the bit-line capacitance C_B. From this result, one concludes that small cells require a sensitive detector.

EXAMPLE 15.4   The storage capacitance of a single-transistor dynamic RAM cell is $C_S = 0.5$ pF and the stray bit-line capacitance is $C_B = 2.0$ pF. Before the beginning of the read operation $V_S = 5$ V and $V_B = 2$ V. What is the bit-line voltage swing during the read operation?

Solution.   From Eq. (15.14)

$$\Delta V_B = \frac{5-2}{1 + (2/0.5)} = 600 \text{ mV}.$$

The detector is usually a gated flip-flop which is a symmetric 2-terminal amplifier. It is used in each bit line as the cell refresh amplifier and as the data detector for the READ operation. Its performance is relatively independent of processing variations because of the symmetry, and each of the two nodes can be used for a bit line thus effectively doubling the number of cells per detector. As shown in Fig. 15.20, the memory array is split into two sections to take advantage of the balanced detector. Each of these arrays contains an additional word line which is used as a dummy for the opposite array so that the detector can be operated in a quasi-balanced fashion.

The READ/REWRITE cycle begins by precharging the whole circuit prior to sensing. Control voltage ϕ_3 is brought high which turns on T_c, the cross transistor of the flip-flop, and selects the dummy word line in each array. This charges both bit lines and the storage capacitors in both dummy cells to the switching voltage of the detector. The flip-flop is then turned off by dropping control signal ϕ_2. During the time that the flip-flop is off, the location of the memory cell whose contents is to be read or refreshed is selected according to the address applied to the address decoders. The selected location is at the intersection of the chosen word line and bit line. In addition, the dummy word line in the other array must also be selected. This selection process establishes a voltage between the nodes of the flip-flop whose sense is determined by the voltage of the storage capacitor. The differential voltage is amplified by the flip-flop after it is again energized by raising the control signal ϕ_2. Positive feedback in the flip-flop regenerates the zero and one levels and forces the storage capacitors in the cells to charge to these voltages, thus performing the REWRITE function.

This READ/REWRITE or REFRESH cycle must be performed on each cell of the memory about once every 2 ms because leakage currents cause the cell voltage to decay. Circuits must be provided on the RAM to control the refresh operation and there might be a slight speed penalty at the system level because the memory might be in the midst of a refresh cycle and not immediately accessible. It is the overhead of the refresh operation that makes the dynamic RAM unattractive in small systems.

READ and WRITE operations from the external terminals require a sense amplifier, such as a gated flip-flop, and a write amplifier connected to the bit-lines through the bit line select transistors. Capacitive balance is obtained by connecting all of the even rows to one side of the flip-flop, and all of the odd rows to the other. Prior to sensing, the bit-line capacitors are precharged to about half of the reference voltage by precharging one line to the reference voltage while the other line is held at zero, followed by shorting the lines together with the cross transistor. After the cross transistor is turned off, the appropriate bit line is connected to one side of the flip-flop thereby unbalancing it. When the flip-flop is turned on, this differential voltage is amplified to produce the outputs D_o and $\overline{D}_o$.

The WRITE operation is performed by a single-ended to double-ended amplifier that establishes the correct voltage on the selected bit lines. The desired cell is chosen by raising the voltage on the word line. The output impedance of the write amplifier is low enough that it forces the storage capacitor of the selected cell to charge to the appropriate input voltage.

Static versus dynamic memories   There is no question that the dynamic memory cell is smaller than the static memory cell. The important consideration is the total complexity of the complete memory. Dynamic memories require circuitry to continually refresh the memory cells, a more involved set of control signals than does the static RAM, and additional time must be devoted to the refreshing operation which must be done about once every 2 ms. In large systems containing many memory chips, the prorated share of these overhead functions and the possible increase in the complexity of the program software to permit refreshing the memory is more than offset by the increased density and lower power consumption of the dynamic memory. It appears that, at least for the present, systems such as microprocessors that require small amounts of RAM are more efficiently served by static memory designs.

State-of-the-art static MOS memories are one size unit smaller than state-of-the-art dynamic MOS memories. At the present time both the 4-kilobit static RAM and the 16-kilobit dynamic RAM are approaching the position of industry standards. Several interesting techniques are employed in the design of static MOS RAM's.

Recall that a RAM is static if the contents of its cells do not have to be refreshed. Therefore one approach to achieving a static MOS RAM is to use a dynamic RAM with automatic on-chip generation of refresh signals.[8] This method requires two additional transistors in each cell to perform the refresh operation. The increase in area due to the transistors is partially offset by a reduction in the size of the storage capacitor. This reduction is possible because the cell is being constantly refreshed. The refresh signal is provided by an on-chip oscillator that generates a 100 kHz, 6 to 8 V p-to-p signal.

A new 4-kilobit static RAM[9] using the 6-transistor static cell achieves its very high speed by using on-chip circuitry to produce a substrate bias that reduces the parasitic junction capacitance and MOS transistor body effect, oxide isolation and depletion mode loads to increase circuit density and improve circuit performance, and a reasonably thin gate oxide to provide high gain transistors.

Other static RAM designs use dynamic logic in the access circuitry to reduce the size and power consumption of the peripheral circuits.

15.5.3 CCD Sequential Stores

The *charge-coupled device* (CCD) and *bucket-brigade device* (BBD) are structures which, under control of suitable clock signals, move packets of charge across the semiconductor surface. These *charge transfer devices* (CTD's) are similar to a ladder network whose series arms are switches and buffer amplifiers, and whose shunt arms are storage capacitors as shown in Fig. 15.21. When the odd-numbered switches are closed, the odd-numbered capacitors are charged to a voltage equal to that of the preceding even-numbered capacitor (or the input for Stage 1). The bucket-brigade stores the signal in capacitors whereas the CCD's store the signal in potential wells. These devices are analog memory devices which are used in digital applications by including a thresholding element at the output to differentiate between a "0" level and a "1" level. Applications of these devices to analog systems were described in Chapter 13. The operation of the CCD cell was described in Chapter 3.

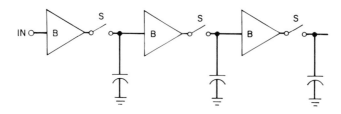

Fig. 15.21   A CCD sequential store.

CCD's appear to be very attractive for large sequential access stores. The organizations for a CCD sequential store are the serial-parallel-serial, the serpentine, and the loop, as shown in Fig. 15.22. The simplest is the serial-parallel-serial structure which has a single data access point and therefore the least peripheral circuitry and the highest bit-packing density. The information is clocked serially into the N-bit input shift register at the full clock rate f_c. After N clock cycles the input register is full and its contents are transferred in parallel into the top register in the array. At the same time the contents of each register in the array are shifted in parallel to the next lower register. The bottom register is the output register. It is loaded in parallel and emptied serially at the full clock rate. Séquin and Tompsett[10] estimated that a 4-kilobit array organized as sixty-four 64-bit registers operating at a 1 MHz clock rate would consume about 5 mW and have a mean access time of 2.05 μs.

The serpentine arrangement shown in Fig. 15.22 provides a shorter mean access time at the expense of more peripheral circuitry and increased power consumption. Electrically the serpentine consists of M, N-bit serial registers in cascade. Regenerating circuits are inserted at the interface of each pair of registers to threshold the binary signal so as to eliminate the

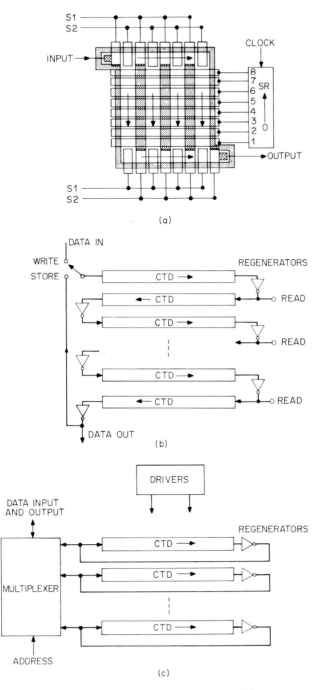

Fig. 15.22 (a) Serial-parallel-serial, (b) serpentine, and (c) loop organization for CCD sequential stores.

degrading effects of dark current generation and transfer inefficiency in the CCD registers. A secondary effect is to provide access to the bit stream at N-bit delay intervals along the structure for reading the data. Thus, the mean access time is only $(N/2)$-bit delay intervals. Because of the large number of regenerators that are required, each regenerator must be smaller than the detector for the serial-parallel-serial structure, and they still might be the controlling factor determining the line-to-line spacing of the arrays. A commercially available Fairchild Semiconductor 9216-bit CCD memory intended to replace a conventional shift register is organized as 1024 words by 9 bits. Each 1024-bit register block uses a serpentine organization with 128 bits between refresh amplifiers and with adjacent rows of the serpentine propagating the bit stream in opposite directions. The mean access time is 165 μs. The serpentine structure is a natural choice for applications such as digital filters that require a delay line tapped at uniformly spaced intervals.

The loop structure is similar to the serpentine in that it uses serial registers and a regenerator after each register. In the loop organization the regenerated bit stream is fed back to the input of the same loop through a data multiplexor. This permits read/write access to the data stream with a maximum of N-bit delay times at the expense of more complex addressing circuits. An Intel 16-kilobit CCD memory chip is organized as 64 loops, each containing 256 bits. At a clock frequency of 1 MHz the mean access time is 128 μs with an on-chip dissipation of 150 mW and additional 450 mW dissipation in the off-chip clock drivers.

15.6 LSI COMPLEXITY CIRCUITS: II. PROGRAMMABLE LOGIC ARRAYS

The term *programmable logic array* (PLA) is presently used for two somewhat different, but related, devices. The simpler device is a generalization of the ROM in which the restrictive address decoder is replaced by an AND-gate matrix whose coding is determined by the particular logic function to be mechanized. This eliminates the need for storing redundant words in the array in order to take care of the "don't care" conditions. Stated in a slightly different form, the PLA can implement *prime implicant factors,* whereas the ROM implements *minterms.*[6,7] The more elaborate devices include some flip-flops on the chip to provide storage for the state vector so that the PLA becomes a mechanization of a Class 3 state machine.

The structure of a combinational PLA made by Signetics[11] is shown in Fig. 15.23. Its $N = 16$ input variables can potentially address $2^N = 65,536$ minterms in the input space. However, the AND matrix is only capable of storing the address of 48 words in the OR matrix. Since each of the words in the AND matrix does not have to be a function of all of the input variables, the subset of input elements addressed by the AND matrix can be

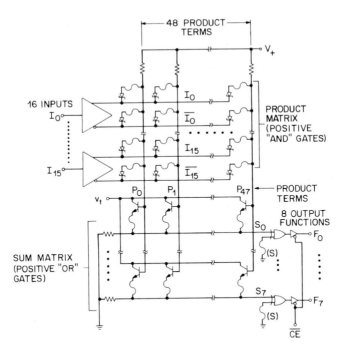

Fig. 15.23 Block diagram of a programmable logic array. (Courtesy of Signetics.)

greater than the 48 AND terms. For example, a particular AND gate which implements a prime implicant that is a function of 12 input variables performs the same function as 16 words of ROM.

This PLA has 8-bit output words and therefore the OR matrix contains forty-eight 8-bit words. Additional design flexibility is provided by coupling the OR matrix to the outputs through exclusive-OR gates. This permits each bit of the output function to be individually programmed as either the true or complement of the function in the array. This is a very useful feature because often the complement of a function requires fewer prime implicants than the function itself.

It is interesting to note that although this PLA is more versatile than a ROM it contains about the same number of crosspoints as a relatively small ROM. The AND matrix contains $32 \times 48 = 1536$ bits while the OR matrix contains $48 \times 8 = 384$ bits for a total of 1920 bits. A 1-kilobit ROM with 4-bit output words has 1024 crosspoints in the OR array and 512 crosspoints in the address decoder. The PLA has more peripheral circuits than the ROM because it has 16-input and 8-output variables. It is probably about twice as complex as a 1-kilobit ROM, yet it can implement any 8-output logic function of 16 variables that can be written as a sum of no more than 48 prime implicants.

The PLA described above accepts input variables which are converted into dual-rail (both true and complement) logic signals in what amounts to a 1-bit decoder before those signals are applied to the AND array which does the actual address decoding. Some reduction in the size of the AND array is possible if certain of the variables always appear as products, i.e., if the variables *A* and *B* always appear together. This reduction is obtained by forming the appropriate product in a 2-bit decoder whose output is then used to activate the AND array. Fleisher and Maissel[12] cite the example of a 2-bit adder which requires 11 rows in the AND array when 1-bit decoders are used versus 5 rows when 2-bit decoders are used. This concept is most effectively mechanized by means of an array decoder that can be customized.

A chip layout for a PLA with customized decoder and flip-flop registers which permit the array to realize sequential state machines is shown in Fig. 15.24. The register stores the state of the dynamic system. Any of the outputs of the OR array can be fed back to the input decoder through another array called the "feedback OR array." Clocked flip-flops are used to prevent race conditions. Several PLA's have been reported, at least experimentally, with about a dozen flip-flops, 16 to 20 input variables, 48 to 70 rows in the AND array, and 8 to 16 output variables.

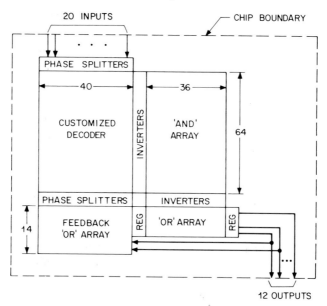

Fig. 15.24 Chip layout of a PLA with a customized decoder and flip-flop registers. The customized decoder is another AND array that can be used to address the rows of the principal AND array more efficiently. (After Fleisher and Maissel.[12])

As an example of the capability of PLA chips of this complexity, the logic for an IBM terminal control unit is implemented using seven of these PLA's.[13] A more specific example is the bit serial adder/subtractor for binary coded decimal (BCD) inputs shown in Fig. 15.25 which is given by Carr and Mize,[14] and is based on a design by Irwin.[15] The logic diagram contains 16 gates, two full adders each requiring 9 gates, and a total of 11 flip-flops. It can be mechanized by a PLA using 18 input variables, 32 prime implicant factors, 1 output variable, and 14 feedback variables. The arrays contain a total of 1632 crosspoints.

A programmable logic array is a viable economic alternative to a completely custom realization of a logic function in quantities up to several tens of thousands of chips because the basic chip can be produced in much larger quantities and customized for the specific application. Programmable logic arrays can also be made field programmable using the fusible link technology described in the section on PROM's. Furthermore, the coding for the array is relatively easy to design and check, and the procedure can be automated with computer-aided design techniques.

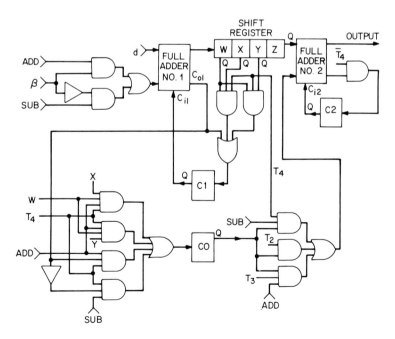

Fig. 15.25 A BCD bit serial adder/subtractor. (From Carr and Mize.[14])

15.7 LSI COMPLEXITY CIRCUITS: III. MICROPROCESSORS AND MICROCOMPUTERS

A *microprocessor* is defined as a computer central processing unit (CPU) built on one, or a few, MOS or bipolar LSI chips with a complexity of at least 2000 equivalent logic gates. Most single chip designs are MOS circuits because the MOS transistor is basically a surface device whose properties are controlled by the horizontal dimensions which are typically several micrometers. Defect densities for MOS processes are of the order 1 to 100 defects/cm^2. Bipolar transistor properties are determined by the vertical device structure, particularly those of the base region which has fractional micrometer thickness for high-speed LSI devices. The result is that the defect density in the base-emitter region of bipolar transistors is typically 10^5 defects/cm^2. As a result, the overall defect density in the active area of bipolar circuits is between 10 and 500 defects/cm^2. (See Chapter 16 for a discussion of yield.) A single LSI chip containing the CPU, ROM, and RAM is known as a *microcomputer*.

Microprocessors were first introduced by Intel in 1971 as a means of economically applying LSI technology to the realization of complex digital system functions. The idea was to provide one type of circuit that could be used in many applications, thus enabling it to be produced in large quantities. Microprocessors have literally revolutionized equipment design because they provide complex control and computational capability without the need to design a custom LSI circuit. The designer of a microprocessor-based system must design peripheral circuits which interface with the computer, and must write the software (computer program) that customizes the microprocessor for the particular application.

Before discussing some architectural details of the microprocessor and some applications, it is useful to philosophize about the role of microprocessors *vis-a'-vis* PLA's and custom logic. It is extremely difficult to set down any rules for choosing between these alternatives because the choice involves such factors as the complexity of the function to be realized, speed requirements, and production quantities. In general, a random logic design is superior for large-volume production, for high-speed operation, and for those functions which do not require many inputs and outputs and do not have multiple decision paths. On the other hand, microprocessors can handle large numbers of inputs and outputs very efficiently, and are well suited to complex control and computational algorithms. However, they must go through many overhead operations, such as moving data around in registers, in order to perform the desired operation, and therefore they can be quite inefficient for simple operations. It is not difficult to specify a reasonably simple function that will overtax a sophisticated microprocessor. In addition, the line of demarcation between custom logic and microprocessors is blurred because new and simpler microprocessor designs continually appear which require less support hardware, and are therefore competitive for many applications although they have excessive logical capability. A great advantage of the microprocessor is that the logical operation is specified in *firmware,* that is, software instruction stored in a ROM. It is easier and cheaper to code a new ROM or PROM than to redesign a custom LSI logic chip. In the midst of all this confusion is the PLA which is not well understood and is underutilized at this time. It is a simpler circuit than the microprocessor and therefore should provide faster response for those applications within its capability. The structural simplicity should lead to higher yields, and therefore, to cheaper devices. There has not been much incentive to define standard PLA configurations and classes of operations that these configurations could realize. This lack of incentive is probably due to the meteoric rise of the microprocessor. So, for the most part, the PLA seems to be a device whose time either has not yet come or has already passed, and the microprocessor seems presently to be the favored device.

Some guidelines for choosing between a microprocessor and alternative implementations are as follows:

1. A microprocessor is more efficient than custom logic if the production quantities are not very large. The crossover points between using a microprocessor, several LSI custom logic chips, or many commercial MSI integrated packages on a printed-circuit board are very difficult to determine. They depend on factors such as development cost of custom LSI circuits, the requirement for flexibility in the logic configuration late into the development cycle, and the value of board space in a large system. Some of these factors are examined in a idealized situation in Section 15.8.

2. Functional flexibility or expansion capability is required. This can be obtained as a trade-off between hardware and software so that the capability of the system can be extended without a system redesign.

3. Multiple inputs and outputs are required.

4. Complex control or computational algorithm with multiple decision paths is required.

5. Large memories are required.

6. High-speed operation is not required.

In the remainder of this section the basic architectural features of a microprocessor and some typical applications will be described.

15.7.1 Architecture

The essential parts of a digital computer are the central control unit which controls the operation of the machine according to instructions contained in the program, the arithmetic logic unit, the *input/output* (I/O) unit, and the

memory for storage of program instructions and data. Figure 15.26(a) is a
block diagram of a computer in which all communication between the I/O
unit and the memory is through the central controller. The I/O unit has its
own controller to handle communication with the peripheral circuits. This
architecture is used in some microprocessors, but the throughput is low
when many ports must be serviced because the central controller must
devote time to transferring data between the I/O unit and memory.

The architecture can be modified as shown in Fig. 15.26(b) to provide
an access path between the I/O unit and memory that is independent of the
central controller. In the *direct memory access* (DMA) structure, there is
direct communication between the I/O unit and memory under control of
the DMA controller. When the central controller wants to transfer data to or
from the memory, it signals the DMA controller which then handles all of

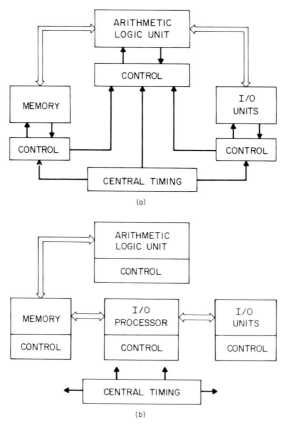

Fig. 15.26 (a) Block diagram of a computer having communication between I/O and
memory through the central controller; (b) a direct memory access computer architecture.

the details of the transfer. This architecture is most useful for applications such as communication systems that require high input/output rates.

Figure 15.27 is the block diagram of the Intel 8080 which is an 8-bit CPU with DMA capability, external program storage, program interrupt, and parallel data I/O. It is a sophisticated and fairly powerful computer with an ALU that can perform either binary or binary-coded decimal arithmetic. There is an on-chip scratch-pad memory consisting of three 16-bit registers organized as 8-bit subregisters which can be individually selected. Data instructions enter and leave the microprocessor chip through the I/O buffer latch via the 8-bit data and instruction bus. The destination for the incoming data is the accumulator or one of the scratch-pad registers, while incoming instructions are routed to the instruction register. The 8080 can handle up to 256 input ports and 256 output ports. Address words are 16 bits long so the 8080 can address up to 64K words of program and data memory. The *last-in, first-out* (LIFO) or *push-down stack* is used for servicing calls and returns for subroutines and I/O interrupts. The push-down stack is maintained in memory by the software. The stack pointer holds the location of the last entry into the stack. The possibility of a very large stack makes it easy to service multiple interrupts. Finally there is a 16-bit program counter with its associated incrementing and decrementing circuitry. The 8080 requires only 6 peripheral chips as compared to 20 for its predecessor, the Intel 8008. Its extensive instruction set of 74 instructions, good speed, and I/O capability permit it to solve a great variety of problems. A minimum microcomputer system using the Intel 8080 requires a total of nine IC packages. It is a more powerful microprocessor than is needed for many applications.

The successor to the 8080 is the Intel 8048 which is an 8-bit microcomputer that contains 1024 words of ROM memory and 64 words of RAM data storage on the same chip as the processor. The processor design is similar to that of the 8080. The 8048 has a larger basic instruction set (90 instructions). This means that more operations can be performed directly in one or two machine cycles, instead of via the execution of a multistep routine. The important observation for our purposes is that the present state of the SIC art is such that it is practical to produce a powerful, single-chip, 8-bit microcomputer with a 2.5 μs cycle time.

At the low end of the microprocessor scale are the 4-bit data word designs that are derived from calculator chips. One of the members of this class is the Texas Instruments TMS1000 series 1-chip microcomputer which uses 8-bit instruction words, and have either 1K-word or 2K-word program ROM's and either 64 or 128 words of data RAM on the same chip as the CPU. Data is transferred between the input or output port and the memory through the ALU. This machine has no system level interrupt capability since it is intended to be used in applications where it is dedicated to one input port and one output port. An interesting and novel feature of this

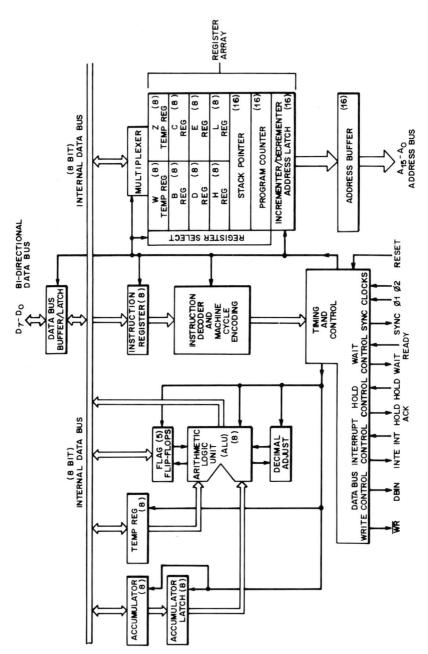

Fig. 15.27 Block diagram of Intel 8080 microprocessor. (Courtesy of Intel Corp.)

machine is that the output register signals are derived from a PLA that is driven by the outputs from the accumulator register and ALU. This permits a logical operation such as code conversion to be performed on the output of the accumulator.

15.7.2 Applications of Microprocessors

Point-of-sale terminals are the electronic successors to the traditional electromechanical cash register. They can provide many additional features such as accepting input data from electronic scales and optical price sensors, elaborate control algorithms to prevent the operator from committing certain errors, and inventory control and credit card verification when they are used as terminals in a store-wide or bank data network.

The simpler electronic cash registers are well within the capability of 4-bit microprocessors.[16, 17] In this application a keyboard is the only input device, and there are usually three output ports which control the cash drawer, numeric display, and the printer. Speed of the microprocessor is usually no problem because of the limitation to a keyboard entry. It may even be possible to use excess microprocessor capability to "debounce" the key switch contacts and to provide the logic for discrimination of 2-key rollover, i.e., two keys being struck accidently in very rapid succession. Several thousand words of program storage and several tens of words of data storage appear to be sufficient for the cash register task, putting it within the realm of the 1-chip microcomputers such as the TMS1000 series mentioned above.

Another point of sale application for the 4-bit microprocessor is the TRANSACTION telephone which is used for credit card verification. The credit card and another information card containing the telephone number of the appropriate data center and information about the merchant are inserted into the phone, and the amount of the proposed purchase is entered via the TOUCH-TONE® pad. The microprocessor assembles this information into a properly formatted message, establishes contact with the data center, and controls transmission of the data and reception of the reply which is then displayed visually.

A more elaborate supermarket checkout terminal incorporating multiple input ports to accept data from electronic scales, an optical reader, and a general communications interface in addition to the usual keyboard has been described.[21] It can control multiple cash drawers, several types of displays, an automatic coin dispenser, and the printer. Because of the communications interface it can act as a node in a store-wide data-gathering system. The terminal uses the National IMP-16, a powerful 16-bit microprocessor, plus a clock generator and priority interrupt controller, 6K words of ROM, and 1.2K words of RAM. The arithmetic capability of the computer is used to provide a detailed breakdown of each transaction, automatically taking into account such things as sales tax tables for appropriate categories of merchandise.

The final microprocessor application to be described is that of a controller for a Private Automatic Branch telephone Exchange (PABX).[22] A stored program controller is very desirable for telephone exchanges because it permits features to be added or modified by simply changing the program. This concept has been in use in the Bell System for over a decade in the form of Electronic Switching System (ESS) machines for central offices and the CSS201 Private Automatic Branch Exchange (PABX). A simplified diagram of a microprocessor-oriented PABX structure is shown in Fig. 15.28.

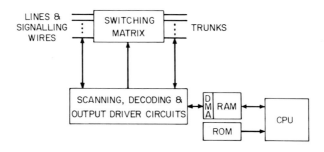

Fig. 15.28 A microprocessor-oriented PABX.

In this system the line-scanning circuits scan the lines and trunks at least once every 50 ms, and write the input status in memory using the DMA capability of the microprocessor. Besides updating the input status, the I/O circuits fetch the necessary output data from the status tables in the RAM. The CPU confines its activity to processing data in the RAM which places a much less stringent speed requirement on the microprocessor than trying to do the I/O function. There is still a considerable amount of processing involved in servicing each line, and it must be performed in an average of 250 μs if the PABX is to have a capacity of 200 lines. Processor speed is still of utmost importance since even a relatively simple task can require about 30 program instructions which take about 100 μs to execute, and it might be necessary to do several tasks during one 250 μs service period. Besides speed, the microprocessor needs a good addressing capability and a system interrupt capability. Call processing involves logical operations and testing. It does not require much arithmetic capability.

A PABX can be made with a microprocessor in the same class as the Intel 8080 using a minimum of about 8K words of ROM for the program and 3K words of RAM for the dynamic data. The full 64K-word addressing capability of the 8080 can be utilized in some applications.

15.8 ECONOMICS OF INTEGRATED CIRCUITS

The effect of silicon-integrated circuit technology in providing complex, low-cost digital system functions is very apparent in all modern electronic equipment. Almost any function can be implemented if the potential economic reward justifies the investment for circuit design and perhaps process development. As an example of how much a company would be willing to spend if the potential market is large enough, each company that entered the race to design a 4-kilobit dynamic random-access memory invested anywhere from several hundred thousand to several million dollars for design and development. There are only a very few examples of circuits that warrant such a large investment.

A problem which arises frequently is that of partitioning a complex system so as to take advantage of IC technology but in a way that "minimizes" the installed cost of the resultant part in the system. There is no unique answer to the problem because it depends on the exact properties of the process technology in hand, the circuit design cost as a function of circuit complexity, the potential demand for the circuit, and the worth of the circuit to the overall system. Obviously a general discussion is beyond the scope of this book.

Instead, consider the economics associated with several different methods for implementing a hypothetical digital-system function with a complexity of about 150 gates. It will be shown that the preferred method is a function of the quantity of systems to be produced. The following assumptions are made to simplify the analysis and make it tractable. In each case the effect of not making the assumption is explained.

1. Only mature processing technologies will be used so that no money need be allocated for specific process development. The only development expenses are for circuit design and simulation, development of tests to ascertain whether each chip is functional, layout and mask making, etc.

2. All realizations are equally acceptable at the system level. Differences such as the size of the resultant IC package, power consumption, and so on, may therefore be neglected. In a real system these factors probably would be considered and additional terms would have to be added to the appropriate equations, but the extension is almost self-evident.

3. The marginal cost of a particular realization remains constant regardless of quantity which means that there are no start-up costs if the total number of circuits is produced in several lots, and that the processing

facility has sufficient capacity to produce any number of circuits that are required.

4. The only difference between the realizations that are considered is the level of integration of the integrated circuits.

The alternative realizations to be considered are:

1. Total integration on one custom LSI chip in a DIP package.

2. Total integration on two custom MSI chips in a single, more complex DIP-type package.

3. The system function can be realized by using two existing MSI chips and a custom metallized MSI chip in a DIP-type package.

EXAMPLE 15.5 Determine the design and manufacturing costs for each of the alternative system realizations. Plot the total cost to produce N copies of the circuit as a function of N for each of these realizations and comment on the result.

Solution. For each of these alternative approaches one must determine, at least symbolically, the cost of developing all of the custom components and the cost of manufacturing an assembled part. The main tasks in developing an integrated circuit are circuit design and simulation, layout and mask preparation, and the preparation of a test sequence that is applied by the computer controlled test set to each chip to determine whether it is operating satisfactorily. Development costs are an increasing function of circuit complexity whose form is something like a staircase function with costs varying slightly over the range of complexity corresponding to SSI and MSI. There is probably a more rapid cost variation for LSI complexity circuits and the jump between classes is larger than in direct proportion to the complexity. Therefore

$$D_1 \geqslant D_2 > D_3 \qquad (15.15)$$

where D_i is the development cost associated with approach i.

The cost of making a chip is proportional to its area because a processed wafer costs a fixed amount. The chip cost is inversely proportional to its yield which is also a function of area. For chips at the MSI and LSI levels, the cost increases at a greater rate than the chip area. Thus the cost of making the two custom MSI chips for the second approach is probably less than that of the custom LSI chip. However, it is assumed that the LSI chip is mounted in a simple DIP package, whereas the dual MSI chip assembly requires a more elaborate and expensive package in addition to more assembly operations. Let C_i be the total cost of n circuit assemblies by approach i. The *marginal cost* M of producing one more circuit is

$$M_i = \frac{dC_i(n)}{dn} . \qquad (15.16)$$

Under the above assumptions the total cost is a linear function

$$C_i(n) = D_i + nM_i, \qquad (15.17)$$

and clearly with $D_1 \geqslant D_2$ if $M_1(n) \geqslant M_2(n)$, the LSI approach would never be preferred because $C_1(n) \geqslant C_2(n)$ for all values of n. The more interesting case occurs when $D_1 > D_2$ and $M_1(n) < M_2(n)$ yielding two straight lines which intersect at

$$N_1 = \frac{D_1 - D_2}{M_2 - M_1} \qquad (15.18)$$

showing that the LSI implementation is preferable for $n > N_1$.

The development cost D_3 for the third approach is lowest because only one custom metallized MSI chip and perhaps a custom metallized interconnection pattern for the ceramic substrate need to be developed. The marginal cost is the highest because of the three chips and the more complex package and assembly operations. For $D_3 < D_2$ and $M_3(n) > M_2(n)$ the third alternative is preferable when

$$n < N_2 = \frac{D_2 - D_3}{M_3 - M_2} . \qquad (15.19)$$

These results are plotted in Fig. 15.29.

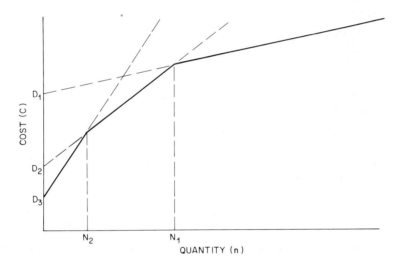

Fig. 15.29 Cost versus demand for three alternative integrated circuits.

It has thus been shown that the preferred system implementation is a function of the required number of systems and depends on the relative magnitudes of the development costs and the marginal costs.

REFERENCES

1. L. R. Rabiner and B. Gold. *Theory and Application of Digital Signal Processing.* Englewood Cliffs, N. J.: Prentice-Hall, 1975.

2.   A. V. Oppenheim and R. W. Schafer. *Digital Signal Processing.* Englewood Cliffs, N. J.; Prentice-Hall, 1975.

3.   C. R. Clare. *Designing Logic Systems Using State Machines.* New York: McGraw-Hill, 1973.

4.   L. A. Zadeh and C. A. Desoer. *Linear System Theory.* New York: McGraw-Hill, 1963.

5.   L. Jackson, J. F. Kaiser, and H. S. McDonald. "An approach to the implementation of digital filters." *IEEE Transactions on Audio and Electroacoustics* **AU-16**: 413-421 (September 1968).

6.   F. J. Hill and G. R. Peterson. *Introduction to Switching Theory and Logical Design*, 2nd Edition. New York: John Wiley & Sons, 1974.

7.   E. J. McCluskey. *Introduction to the Theory of Switching Circuits.* New York: McGraw-Hill, 1965.

8.   H. J. Boll, E. N. Fuls, J. T. Nelson, and L. D. Yau. "Automatic refresh dynamic memory." *Proc. ISSCC*, February 1976, pp. 132-133.

9.   R. D. Pashley, W. K. Owen III, K. R. Kokkonen, R. M. Jecman, A. V. Ebel, C. N. Ahlquist, and P. Schoen. "A high performance 4k static RAM fabricated with an advanced MOS technology," Proc. ISSCC, February 1977, pp. 22-23.

10. C. H. Séquin and M. F. Tompsett. *Charge Transfer Devices.* New York: Academic Press, 1975.

11. Signetics Corporation. "Field programmable logic arrays." Sunnyvale, California, February 1976.

12. H. Fleisher and L. I. Maissel. "An introduction to array logic." *IBM J. of Research and Development* **19**: 98-109 (March 1975).

13. J. C. Logue, N. F. Brickman, F. Howley, J. W. Jones, and W. W. Wu. "Hardware implementation of a small system in programmable logic arrays." *IBM J. of Research and Development* **19**: 110-119 (March 1975).

14. W. N. Carr and J. P. Mize. *MOS/LSI Design and Application.* New York: McGraw-Hill, 1972.

15. J. Irwin. "MOS shift registers in arithmetic operations." *Electronic Engineer* **29**: 71-73 (April 1970).

16. E. Sonn. "Four-bit chip set cuts cash register's cost and size." *Electronics* **49**: 154-157 (April 1976).

17. B. S. Franklin. "Long live the 4-bit mirco." *Mini-Micro Systems* **10**: 48-50 (April 1977).

18. K. V. Stein, A. Schling, and E. Doering. "Storage array and sense/refresh circuit for single-transistor memory cells." *IEEE J. of Solid-State Circuits* **SC-7**: 336-340 (October 1972).

19. R. C. Foss and R. Harland. "Peripheral circuits for one-transistor cell MOS RAM's." *IEEE J. of Solid-State Circuits* **SC-10**: 255-261 (October 1975).

20. D. J. Lynes and D. A. Hodges. "Memory using diode coupled bipolar transistor cells." *IEEE J. of Solid-State Circuits* **SC-5**: 185-191 (October 1970).

21. M. Schwartz. "Checkout terminal takes on many supermarket tasks." *Electronics* **49**: 157-161 (April 1976).

22. Z. G. Vranesic and S. G. Zaky. "Nonnumeric Applications of Microprocessors." *Proc. IEEE* **64**: 954-959 (June 1976).

PROBLEMS

15.1 Construct an *R-S* flip-flop using NAND gates and derive the truth table. What is the difference between this structure and a flip-flop made of NOR gates?

15.2 Verify that Eqs. (15.8) define the full adder function and then provide a NAND-gate implementation of the full adder.

15.3 Construct a *J-K* master-slave flip-flop using AND-OR-INVERT gates.

15.4 Construct a *J-K* master-slave flip-flop in which the storage for the master section is dynamic. Use capacitive charge storage to achieve this and discuss what advantages this type of circuit would possess. What technology would be particularly amenable to such an implementation?

15.5 Design a system to convert serial data arriving at bit rate F_1, to words which are outputted on an 8-bit wide data bus. You are provided with the system clock at frequency F_1. Any other required control signals must be derived internally. Signals on the parallel data bus are to remain stable while the word is being assembled.

15.6 a) Devise a logic system that will carry out binary subtraction, given as inputs the minuend, the subtrahend, and the borrow from a previous stage.

b) How can you combine this circuit with a full adder to make an adder/subtractor circuit?

15.7 Construct a circuit which will form the product of two positive binary numbers using a shift-and-add algorithm, that is, by repeatedly adding shifted versions of the multiplicand to the partial product.

15.8 a) A ROM is a combinational logic circuit. Flip-flops are constructed from combinational logic by means of feedback. Specify the coding of the ROM array and any external connections that must be made so that the completed circuit obeys the truth table of a *J-K* flip-flop.

b) Commercial memories usually have binary coded addressing. For this application would you prefer binary or linearly coded addressing? How big would the array have to be in each case?

15.9   Multiply two 4-bit numbers $A_3A_2A_1A_0$ and $B_3B_2B_1B_0$ using the decomposition

$$A_3A_2A_1A_0 = A_3A_200 + 00A_1A_0$$

and

$$B_3B_2B_1B_0 = B_3B_200 + 00B_1B_0.$$

Show how a ROM multiplier, consisting of four, 16-word, 4-bit ROM's and five, 2-bit adders may be constructed. What is the increase in storage capacity necessary, if the multiplier is to be realized by means of a single ROM?

15.10 A ROM is to be used as a look-up table for the function $\sin \theta$, in the range $0 \leqslant \theta \leqslant 90°$. The required resolution is to be in steps of $0.01°$. By writing $\theta = I + F$, where I represents the integral and F the fractional part, one may utilize the trigonometrical identity

$$\sin \theta = \sin(I+F) = \sin I \cos F + \cos I \sin F.$$

It is now possible to use four smaller ROM's for $\sin I$, $\cos I$, $\sin F$, and $\cos F$, respectively. If the input $X = \theta$ and the output $Y = \sin \theta$, correct to the nearest thousandth, specify the size of the required ROM's. Draw a system block diagram showing suitable interconnections.

15.11 Derive a digital system which behaves like a monostable multivibrator. The output pulse width, called the fall-back time for a monostable multivibrator, is to be specified as an integral number of clock pulses by means of signals applied to a B bit wide data input bus. The pulse commencement is to be triggered by an external trigger.

15.12 Figure P15.12 shows a modulo-5 ripple counter constructed from J-K flip-flops and NAND gates. The addition of gates to the counter allows the counter array to bypass some states. All J-K terminals not explicitly shown are fixed at logic 1 level. Check that Fig. P15.12 represents a modulo-5 counter and employ your ingenuity to construct a modulo-7 counter using only flip-flops and NAND gates.

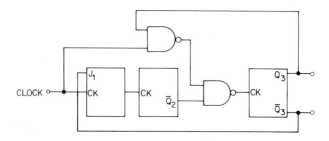

Fig. P15.12

15.13 A CMOS monostable multivibrator circuit is shown in Fig. P15.13(a). Assume each CMOS NOR gate has an ideal input-output transfer characteristic as shown in Fig. P15.13(b), where V_t is the transition voltage. Let the input, V_{is}, of gate 1 consist of a rectangular pulse of duration τ, and adequate height to carry the input to logic level 1. Moreover, let the pulse width $\tau > t_{d1} + t_{d2}$, where t_{d1}, t_{d2} are the propagation delays of gates 1 and 2, respectively.

Sketch the voltage waveforms V_{i1}, V_{01}, V_{i2}, and V_{02}, on the assumption that the pulse at terminal $i1$ commences at time $t = -T_{d1}$. Show that the circuit returns to its initial state after a time T and express T in terms of V_{DD}, V_t, R, and C.

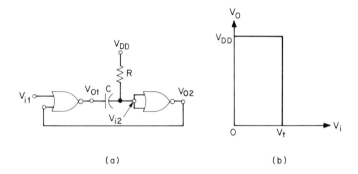

(a)   (b)

Fig. P15.13

15.14 A second-order digital filter is described by the difference equation

$$y(n) = a_0 x(n) + a_1 x(n-1) + a_2 x(n-2) - b_1 y(n-1) - b_2 y(n-2)$$

where x is the input signal and y is the output signal. Assume that all signals are bounded by ± 1 and are represented in two's-complement notation, using B bits

$$x(k) = -x^0(k) + \sum_{j=1}^{B-1} x^j(k) 2^{-j}, \quad x^j(k) = 0 \quad or \quad 1.$$

rewritten as

$$y(n) = \sum_{j=1}^{B-1} 2^{-j} \phi[x^j(n), x^j(n-1), x^j(n-2), y^j(n-1), y^j(n-2)]$$

$$- \phi[x^0(n), x^0(n-1), x^0(n-2), y^0(n-1), y^0(n-2)]$$

where ϕ is a function with five binary arguments defined as

$$\phi(u_1, u_2, u_3, u_4, u_5) = a_0 u_1 + a_1 u_2 + a_2 u_3 - b_1 u_4 - b_2 u_5.$$

b)   Show that the second-order section may be realized by means of the ROM, registers, and adder connected as shown in Fig. P15.14. Explain how the structure works and calculate the size of the ROM array.

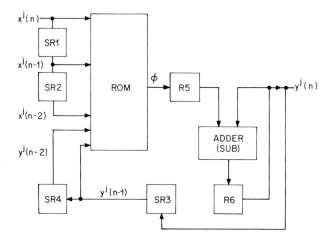

Fig. P15.14

15.15   A first-order digital filter is characterized by the first-order difference equation $y(n+1) = x(n) + Ky(n)$, which links an input x to an output y. Constant K is real, positive, and smaller than unity. All three quantities y, x, and K are to be represented by 8-bit words. The filter is to be implemented by means of a microprocessor, whose instruction set does not include "multiply." The machine is capable of executing one instruction per microsecond. How many input samples x can be processed by this digital filter per second? How big a ROM do you need to store the instructions? How much read/write memory is required?

15.16   A digital system, containing a microprocessor, is used to process analog signals. The microprocessor and the D/A convertor in the system are operating at less than full capacity. That is, they can perform all required operations and have more time left over to perform other tasks. Is it possible to perform the A/D conversion by means of the existing system hardware and an additional comparator? If it is possible, describe the program that is required. Make reasonable assumptions about the instructions available in the microprocessor instruction set and sketch a suitable program. Estimate also the number of machine cycles required to perform a B-bit conversion. Hint: Consider the method of successive approximations for A/D conversion, in which a D/A converter is incorporated in a feedback loop.

16

Failure, Reliability, and Yield of Integrated Circuits

The reliability of a circuit component is the probability of failure-free performance of a required function, under stated conditions for a given period of time. It is impossible to state categorically that a component or system will perform properly. The most that can be done is to calculate or measure a probability of successful operation.

In the first part of this chapter the failure mechanisms,[1] the reliability,[2-4] and the testing[5-9] for reliable operation of integrated circuits are discussed. In the second part we discuss yield[10-13] of integrated circuits. A knowledge of probability theory and statistics[14,15] on the part of the reader is assumed. Special topics such as component screening procedures or accelerated life testing methods will be introduced where required.

16.1 FAILURE RATE OF INTEGRATED CIRCUITS [16]

All electronic devices have a finite lifetime. Any integrated circuit component that is to be used in a large production run must be statistically characterized not only for its expected lifetime, but also for the value of other critical parameters, such as failure or hazard rate to be defined subsequently. Intuitively, a device that dies early can be said to have a high failure rate, while another device that lives to a great old age can be said to have a low failure rate. Hence the age at death, called the *time to failure* or *lifetime,* is a good figure of merit for defining the failure rate of a device. More specifically the *failure rate,* λ_i, of the ith device is defined as the reciprocal of its *time to failure,* t_i, that is $\lambda_i = 1/t_i$. Failure rate λ of a set of similar devices is usually measured in terms of failed devices divided by the total hours of successful operation, as will now be shown.

Assume next that a set of N devices is put on life test at time $t = 0$, and that devices die at times t_1, t_2, ..., respectively. Hence the total number of operating hours for the devices is

$$\sum_{r=1}^{N} t_r.$$

The average failure rate, λ_{av}, can now be defined as the number of dead devices divided by the total operating hours

$$\lambda_{av} = \frac{N}{\displaystyle\sum_{r=1}^{N} t_r} \, ,$$

The *mean time to failure* (MTF), t_{av}, is

$$t_{av} = \sum_{r=1}^{N} \frac{t_r}{N} \, .$$

From the above two equations it follows that $t_{av} = 1/\lambda_{av}$. It is convenient to introduce the following failure unit (FIT):

Definition   One device failure in 10^9 device operating hours is termed "one FIT."

EXAMPLE 16.1   A data processor, which incorporates 115,000 transistors of a certain type, is to experience no more than one transistor failure per year, while in use. What transistor failure rate is necessary to achieve this reliability?

Solution.   Assuming 365 days per year gives 8760 hours per year. Hence 115,000 transistors represent $115,000 \times 8760 = 1.0074 \times 10^9$ device-hours per year. If the transistors have a failure rate of f FIT's, then

$$\left(\frac{f}{10^9} \text{ failures/device-hour} \right) \times (1.0074 \times 10^9 \text{ device-hours/year}) = 1 \text{ failure/year}.$$

From this equation follows $f = 0.993$ FIT's and so a transistor failure rate of one FIT would be satisfactory.

However, a failure rate of one FIT or 10^9 failure-free operating hours is approximately equal to 125,000 years of failure-free operation for one device. It is impossible to wait that long to determine whether all of a large number of similar devices have failed by then and to record the time of individual demise. As a consequence, some sort of *accelerated-stress life testing*[17] is needed to determine reliability and will form the subject of Section 16.5. If the degradation mechanism is known, it is often possible to test a small sample from a large population of devices under higher than normal stresses for short, tractable periods of time. One then infers, or extrapolates, from these tests the device reliability under normal operation. It is next necessary to introduce certain terms used in reliability engineering.

Let there be N devices in the population, whose "reliability" one wishes to ascertain, and let these devices be placed on test at time $t = 0$. Further, it is ascertained that $N_f(t)$ of the devices have failed and that $N_s(t) = N - N_f(t)$ devices survive at time t. The ratio of surviving devices to original devices placed on test is defined as the *reliability, R(t)*,

$$R(t) = \frac{N_s(t)}{N} .$$ (16.1)

Similarly, the ratio $N_f(t)/N$ is a measure of the unreliability of the devices and is usually termed the *cumulative failure function, F(t)*, where

$$F(t) = \frac{N_f(t)}{N} = 1 - R(t).$$ (16.2)

The *instantaneous failure rate,* also called the *hazard rate,* $\lambda(t)$, may be obtained from Eq. (16.1) by differentiation

$$\lambda(t) = - \left[\frac{1}{N_s(t)} \right] \frac{dN_s(t)}{dt} = - \frac{1}{R(t)} \frac{dR(t)}{dt} ,$$ (16.3)

and is one of the most important functions of reliability theory. It gives the instantaneous rate of failure of devices per device surviving at time t. Defining the *failure density function* $f(t) = dF(t)/dt$ one then obtains from Eq. (16.2)

$$\lambda(t) = \frac{f(t)}{R(t)} .$$ (16.4)

EXAMPLE 16.2   The exponential distribution function is given by the National Bureau of Standards' *Handbook of Mathematical Functions,* as

$$f(t) = \beta^{-1}\exp\left[- \left| \frac{t-\alpha}{\beta} \right| \right],$$

where α and β are constants. Assume the failure of a device population is governed by this distribution, with parameter $\alpha = 0$. Find the reliability, $R(t)$ and the hazard rate, $\lambda(t)$, of the population.

Solution.   From Eq. (16.2), $R(t) = 1 - F(t)$ where

$$F(t) = \int_0^t f(\mu)\,d\mu = \beta^{-1}\int_0^t \exp(-\mu/\beta)\,d\mu = 1 - \exp(-t/\beta).$$

Therefore, $R(t) = \exp(-t/\beta)$. From Eq. (16.4), $\lambda(t) = f(t)/R(t) = \beta^{-1}$. Hence the exponential failure density possesses a constant hazard rate.

Note that $f(t)\,dt$ is the probability that the device will fail in the interval from t to $t + dt$. Furthermore, because all devices fail eventually,

$$\int_0^\infty f(t)\,dt = 1.$$

Any one of the four basic functions $R(t)$, $F(t)$, $\lambda(t)$, or $f(t)$ suffices to completely describe the reliability of the device. Functions $R(t)$ and $F(t)$ are dimensionless, whereas functions $\lambda(t)$ and $f(t)$ are inversely proportional to time. When only a sample $n < N$ of the devices is tested, the expected number of good components after time t equals $nR(t)$, while the expected number of failed components equals $nF(t)$. Moreover, the mean time to failure, t_{av}, becomes

$$t_{av} = \int_0^\infty tf(t)\,dt. \tag{16.5a}$$

After substitution from Eqs. (16.1) through (16.4) and integration by parts, it may also be shown that

$$t_{av} = \int_0^\infty R(t)\,dt. \tag{16.5b}$$

Similarly, the variance of lifetime, σ_t^2, may be obtained by noting that

$$\sigma_t^2 = \int_0^\infty (t - t_{av})^2 f(t)\,dt, \tag{16.6a}$$

and σ_t is called the standard deviation. After substitution and integration by parts,

$$\sigma_t^2 = 2 \int_0^\infty tR(t)\,dt - t_{av}^2. \tag{16.6b}$$

EXAMPLE 16.3   Find the mean time to failure (MTF) and the variance σ_t^2 for the exponential failure density of Example 16.2.

Solution.   From Eqs. (16.5), $t_{av} = \lambda \int_0^\infty t \exp(-\lambda t)\,dt$. But it is well known that

$$\int_0^\infty x^{\beta-1} \exp(\alpha x)\,dx = \frac{\Gamma(\beta)}{\alpha^\beta}$$

where $\Gamma(\beta) = (\beta-1)\Gamma(\beta-1)$ and where $\Gamma(n+1) = n!$ when n is an integer. For the above integral $\alpha = \lambda$ and $\beta = 2$ so $t_{av} = \lambda\Gamma(2)/\lambda^2 = 1!/\lambda = 1/\lambda$. To find the variance we use Eq. (16.6), and so

$$\sigma_t^2 = \lambda \int_0^\infty t^2 \exp(-\lambda t)\,dt - t_{av}^2.$$

Setting $\beta = 3$, $\alpha = \lambda$ in the above equation, yields

$$\sigma_t^2 = \lambda\Gamma(3)/\lambda^3 - 1/\lambda^2 = 2!/\lambda^2 - 1/\lambda^2 = 1/\lambda^2.$$

Hence $\sigma_t = 1/\lambda = t_{av}$, for this failure density.

With respect to the hazard rate, $\lambda(t)$, examination of failure data over many years has led to the recognition of several epochs of device failure, all illustrated in Fig. 16.1.

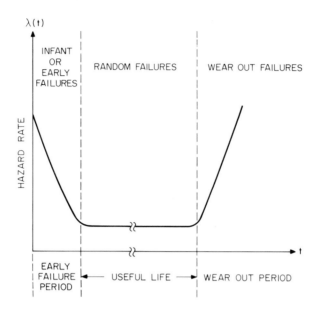

Fig. 16.1 Bathtub shaped curve of hazard rate showing the different reliability regions.

Infant mortality Early in the lifetime of a device there are a large number of failures, due to built-in weakness or defect. These early failures are called "infant mortality" and lead to decreasing hazard rate with respect to time. Stress tests and screens to eliminate the potential failures are discussed in Section 16.3.

Useful life During the middle period of the device lifetime fewer failures occur, but it is necessary to know which mechanism is the principal determinant of the hazard rate under the conditions of interest. In this region, sometimes called the "random failure region" of constant hazard rate (further discussed in Section 16.4.1), the device characteristics are essentially constant and when failure occurs it is usually catastrophic. For some integrated circuits, the hazard rate in this region is log-normal and falls slightly with time, as discussed in Section 16.4.2.

Wear-out region As a device reaches old age, it begins to deteriorate rapidly. Its instantaneous hazard rate increases monotonically and many failures occur. This failure region is called the "wear-out region" and is caused by material degradation due to high temperatures, effect of electric fields, and slow chemical reactions. Integrated circuits do not usually reach the wear-out region in normal operation. Exceptions occur when integrated circuits are exposed to ionizing radiation fields and when the hermeticity of integrated circuit packages is impaired by progressive corrosion. An "intrinsic" wear-out process leads to the ultimate death of every device.

Actual failure mechanisms for integrated circuits are considered next.

16.2 INTEGRATED CIRCUIT FAILURE MECHANISMS [2]

The physical or chemical process that causes devices to fail is termed the *failure mechanism*. The cause of rejection of any failed device is termed the *failure mode*. Thus electromigration is an example of a failure mechanism, which can lead to the failure mode of an open interconnection. A further example is given by excess charge near a silicon-oxide interface (failure mechanism) which causes drift of the parameters of an MOS transistor (failure mode).

Failure mechanisms for bipolar integrated circuits can be divided roughly into three groups, namely: (1) chip-related failures, such as oxide defects, metallization defects, and diffusion-related failures; (2) assembly-related problems such as chip mount, wire bonds, or package failures; and (3) miscellaneous, undetermined, or application-induced failures.

A typical diagram of failure-mode distributions is shown in Fig. 16.2. No attempt has been made to portray accurately actual percentage values of the failure modes, which will differ for SSI, MSI, and LSI circuits.

Diffusion-related failures   Nonuniform current-flow may occur within a device because of dopant diffusion-related causes. These may affect the base width, the emitter resistivity, the curvature of junctions, and other device parameters.

Oxide-related failures   Contamination of oxide, during or after its growth, directly affects its dielectric properties, particularly its breakdown strength. Presence of surface charge, Q_{ss} (see Chapter 3), at or near an oxide-silicon interface can affect the turn-on voltage, V_{TH}, of a device and other parameters, such as dc gain, and leakage current. Oxide-charge values of large magnitude can cause surface inversion. Other surface-related failures arise because of ion migration in the thermally grown oxide and along its surface, dipole polarization effects or charge trapping effects (see Chapter 3).

Metallization-related failures   The mass transport of metal atoms by momentum exchange with conducting electrons is called "electromigration." It occurs in metal lines at high current densities and elevated temperatures, and consists of the movement of metal atoms toward the positive end of the

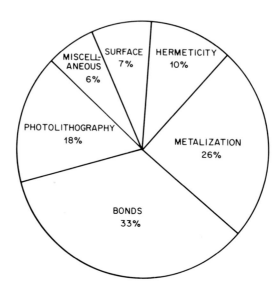

Fig. 16.2 Failure mode distributions of integrated circuits.

conductor, while voids move in the opposite direction. As a consequence, metal disappears from certain regions and ultimately an open-circuit occurs. The degradation of integrated circuits with aluminum metallization operating at high current densities (in excess of 10^5 A/cm^2) and at elevated temperatures (in excess of 100°C) is described by the Arrhenius model, which is discussed in Section 16.5. Electromigration occurs for many metals, including gold, silver, copper, and platinum. The current density at which reliability problems occur with gold films is substantially higher than that for aluminum films. This has led to the use of gold in circuits requiring high current densities.

Another significant cause of metallization failure is the formation of microcracks, where the metallization passes over an oxide step. It occurs frequently at the base-contact window where the oxide step is 6000 to 8000 Å high. Steeper steps lead to thinner metal deposits which have a greater probability of failure than normal deposits under high current stress. Microcracks are not usually detectable with optical microscopes, but may effectively be detected by scanning electron microscopes.

Contact alloying is used to form ohmic contacts between aluminum metallization and the silicon contact region. It is generally performed below the aluminum-silicon eutectic temperature of 577°C and therefore a liquid phase is not formed. Solid-state diffusion of silicon into aluminum occurs and the aluminum-silicon interface moves down into the silicon, particularly near the contact cut. Overalloying occurs when the process is conducted at a temperature that is too high, causing a discontinuity between the aluminum over SiO_2 and the aluminum at the edge of the contact region. This, in turn, restricts the cross-sectional area of the conductor stripe and leads to failure. Overalloying can also occur for other silicon-metal systems.

Metallization may also fail because of poor ohmic contacts with silicon, poor bondability to aluminum or gold wires, or poor adhesion to the silicon dioxide.

Chip mount failures   Chip-to-package alloy bond failures have been attributed to low-strength bonds caused by inadequate process control, to voids occurring in the chip-to-header bond, to cracked chips caused by improper handling during the bonding process, and to the presence of solder balls which form during chip-to-package bonding.

In the case of glass frit mounts, failures have been attributed to poor control being maintained over the glass composition. As a consequence brittle mounts are formed which, in addition, may transmit mechanical stresses to the semiconductor element. Epoxy mounts may fail under temperature stress because the thermal coefficient of expansion of most epoxies exceeds by ten or twenty times the coefficients of expansion of both the semiconductor chip or that of the substrate to which it is to be mounted. Moreover, if the epoxy is not properly thermally cured, its thermal properties are unstable and its expansion characteristics may be uncontrolled.

Wire-bond failures   Gold wire is usually thermocompression bonded to the chip metallization as described in Chapter 10. Failure of gold wire bonds to aluminum-metallized chips may be due to the formation of intermetallic compounds that lead to loss of strength and an increase in resistance. When one of the compounds formed is the intermetallic phase $AuAl_2$, the interaction is termed "purple plague." This undesirable gold-aluminum interaction causes degradation and failure and is sometimes accompanied by void formation.

Aluminum wires are ultrasonically bonded at room temperature using an aluminum alloy containing 1% of silicon to add hardness. In the case of aluminum wire, control of its diameter, tensile strength, and ductility is very important. Failure of such control leads to "over bonding," "under bonding," cracks at the heel of the bond, misaligned bonds, and contaminated bonds.

The incidence of chip-to-package-bond failures is usually much lower than that of wire-bond failures.

Package-related failures   The hermeticity of metal-can packages depends on a glass-to-metal seal that isolates the leads going to the device. The coefficient of expansion of glass is matched to that of the Kovar header. Failure may occur due to poor hermeticity or due to the corrosion of the Kovar header.

Ceramic packages usually consist of two halves, joined together by a glass seal whose coefficient of thermal expansion closely matches that of the ceramic. Failure of ceramic packages is usually due to loss of integrity of the seal, either because of a poorly controlled sealing process or during subsequent use.

Plastic packages, usually of epoxy material, are formed by molding the device in molten plastic. Failure of plastic packages is usually due to environmental factors, such as humidity and temperature stress. However, with proper design and testing, plastic packages can have a high degree of reliability.

16.3 EARLY FAILURES—RELIABILITY SCREENING PROCEDURES

Reliability screening selects from a collection of devices those having superior reliability and rejects those devices that are potential early failures (freaks). Assuming that all devices in a lot are initially within specification, screening is a test procedure that classifies a device as to longevity, based on time-zero or short-time measurements. Table 16.1 shows standard screening sequences documented in Notices 1 and 2 to Military Standard 883.[5-7] The highest level of reliability specified by the military is Class A which refers to devices intended for use where maintenance and replacement are extremely difficult or impossible, such as electronics in an unmanned spacecraft. Class A devices presently require a failure rate of approximate 0.001%/1000 hr. Class B devices, used, for example, in avionics have present failure rates of about 0.005%/1000 hr, while Class C devices have failure rates of about 0.05%/1000 hr. Devices for commercial uses are rated lowest by the military, requiring present failure rates of about 0.1%/1000 hr. Note that these quoted failure rates may have to be revised downwards with improvements of the technology.

Tests 1 to 8 shown in Table 16.1 are called short-time tests because they can be carried out in less than 48 hr, while Test 9 referred to as "burn-in" is an early life measurement that requires 168 hr. Marginal devices that are eliminated as a result of a burn-in test may actually fail during the test or they may exhibit performance characteristics after burn-in that indicate early failure in actual applications. Sufficient screening, consisting of specific tests that are designed to excite expected failure mechanisms, can remove most of those devices that cause a high initial failure rate.

The number and type of screens used in a particular application is a compromise between the economic benefit of the reduced component failure rate and the cost of the screens. It is often necessary to determine which parameters of a device should be measured as precursors to failure. Moreover, one must determine whether to measure the parameter, or its drift in some time increment, or the rate of such parameter drift. This requires a knowledge of the process acting on a given device type.

The design of screening tests for integrated circuits involves the determination of the probable failure mechanism for the device and its environment, a suitable stress, and measurable parameters required by the device application. The important principle here is to operate the devices under stress for a suitable time such that unreliable devices fail or exhibit undesirable parameter variations. However, the stress must not be so great that reliable devices are prematurely aged or that other failure mechanisms are excited. The usual burn-in screen for integrated circuits lasts approximately 168 to 250 hr, with stress levels of near-rated power dissipation, and temperature of 125°C or above for IC's whose maximum operating temperature is 75°C.

Figure 16.3 shows a possible distribution of parameter values Q in a set of devices after screening. Acceptable parameter values are those that lie within the range $Q_L < Q < Q_U$, where Q_L and Q_U represent the lower and

Table 16.1 Screening Sequence-Method 5004-MIL-STD-883

No. of Test	Screen	Reliability Classes		
		A	B	C
1	Internal visual	condition A	condition B	condition B
2	Stabilization bake	24 h	24 h	24 h
3	Thermal shock	15 cycles and	15 cycles or	15 cycles or
4	Temperature cycle	10 cycles	10 cycles	10 cycles
5	Mechanical shock	20 000 g	no	no
6	Centrifuge	30 000 g	30 000 g	30 000 g
7	Hermeticity	yes	yes	yes
8	Critical electrical parameters	yes	no	no
9	Burn-in	168+72 h	168 h	no
10	Final electrical	yes	yes	yes
11	X-ray radiograph	yes	no	no
12	External visual	yes	yes	yes

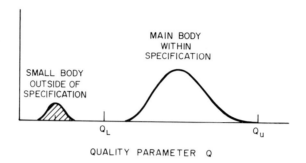

Fig. 16.3 Bimodal distribution of quality parameters.

upper tolerance limits respectively. The assumption here is that if a device parameter is outside a marginal tolerance limit (Q_L, Q_U) after the burn-in test, then there is a high probability that the parameter will be outside failure-tolerance limits at some early time of actual operation. As may be seen, the freaks tend to form a second distribution which lies outside the acceptable limits. From practical screening experience it may be concluded that screening provides, as a minimum, an order of magnitude improvement in device reliability by the elimination of freaks or early failures.

16.4 USEFUL LIFE REGION

Consider now the useful life of a device, which is depicted by the period of constant or almost-constant hazard rate in Fig. 16.1. Here the weak devices, depicted by the high rates of failure of the infant mortality region, have been weeded out. Let us first assume that the hazard rate is constant.

16.4.1 Constant Hazard Rate

Since $\lambda(t) = \lambda$, it follows from Eqs. (16.2), (16.3), and (16.4) that

$$f(t) = \lambda \exp(-\lambda t); \quad R(t) = \exp(-\lambda t); \quad F(t) = 1 - \exp(-\lambda t)$$

$$(16.7)$$

which are the exponential reliability functions. They are plotted in Fig. 16.4. It was shown in Example 16.3 that the mean time to failure, t_{av}, and the standard deviation, σ_t, are given by the reciprocal of the failure rate

$$t_{av} = 1/\lambda; \quad \sigma_t = 1/\lambda.$$

$$(16.8)$$

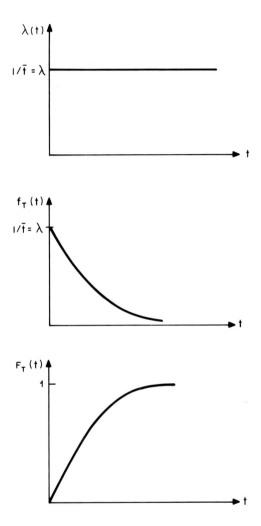

Fig. 16.4 Hazard rate, probability density, and cumulative distribution for the exponential reliability function.

From these considerations, it is reasonable to assume that in the middle region of the lifetime curve of Fig. 16.1 the occurrence of a failure is represented by exponential reliability functions.

16.4.2 Log-Normal Density Function [17,18]

The log-normal distribution is a two-parameter distribution that is particularly appropriate for accelerated stress life testing of integrated circuits.

Assume that members of some population display a normal distribution, when classified by some property x. Then

$$f_X(x) = \frac{1}{\sigma\sqrt{2\pi}} \exp\left[-\frac{(x-\mu)^2}{2\sigma^2}\right] = \frac{1}{\sigma}\,\phi\left(\frac{x-\mu}{\sigma}\right),$$

where $\phi(\cdot)$ is the normal probability density function. Assume further, that the measurable device lifetime, t, bears the relation $x = \ln t$ with the parameter x. In that case, the density function, $f(t)$, is log-normal and given by

$$f_T(t) = f_X(x=\ln t)\left|\frac{d \ln t}{dt}\right| = \frac{1}{\sigma t}\,\phi\left(\frac{\ln t-\mu}{\sigma}\right). \tag{16.9}$$

Parameter μ is called the *logarithmic mean value* and parameter σ is called the *logarithmic standard deviation*. These parameters should be carefully distinguished from the mean of the log-normal distribution, $E(t)$ (expected value), and from the standard deviation of the log-normal distribution, σ_t, to be obtained subsequently.

The cumulative density function, $F(t)$, is given by

$$F(t) = \int_0^t f_T(u)\,du = \Phi\left(\frac{\ln t-\mu}{\sigma}\right), \tag{16.10}$$

where $\Phi(\cdot)$ is the standard normal cumulative distribution function. Finally, the hazard rate $\lambda(t)$ may, by virtue of Eqs. (16.4), (16.9), and (16.10), be expressed as

$$\lambda(t) = \frac{1}{\sigma t}\,\frac{\phi\left(\dfrac{\ln t-\mu}{\sigma}\right)}{1 - \Phi\left(\dfrac{\ln t-\mu}{\sigma}\right)}. \tag{16.11}$$

The shape of these three curves, $F(t)$, $f(t)$, and $\lambda(t)$—shown in Fig. 16.5—is strongly dependent on the logarithmic standard deviation σ, while the logarithmic mean μ determines the spread. In the case of the hazard rate, $\lambda(t)$, a maximum is reached at time t_0, where $t_0 > \exp(\mu-\sigma^2)$, provided σ lies in a suitable range of values. Moreover, hazard rate often appears to be constant or slightly decreasing over a time interval of interest.

The median (or fiftieth percentile) of the log-normal distribution, $t_{.50}$, can be obtained from Eq. (16.10) and is

$$t_{.50} = \exp\{\mu + \sigma\Phi^{-1}(0.50)\} = \exp(\mu). \tag{16.12}$$

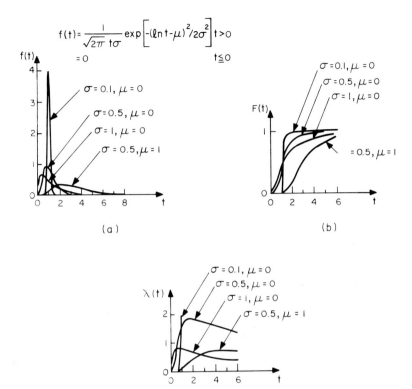

$$f(t) = \frac{1}{\sqrt{2\pi}\, t\sigma} \exp\left[-(\ln t - \mu)^2/2\sigma^2\right] \quad t > 0$$
$$= 0 \qquad t \le 0$$

Fig. 16.5 (a) Probability density; (b) cumulative distributions; (c) hazard rate for the log-normal reliability function.

The mean or expectation, $E(t)$, of the log-normal distribution is

$$E(t) = \int_0^\infty t f_T(t)\, dt = \exp\left[\mu + \frac{\sigma^2}{2}\right]. \tag{16.13}$$

The above equation gives the mean or average life of the population and clearly $E(t)$ is always greater than the median $t_{.50}$ for nonzero values of σ. For reasons of mathematical convenience the median life, $t_{.50}$, is generally emphasized, in preference over the mean life, $E(t)$.

The standard deviation, σ_t, of the log-normal distribution is given by

$$\sigma_t = \left\{\int_0^\infty [t - E(t)]^2 f_T(t)\, dt\right\}^{1/2} = E(t)[\exp(\sigma^2) - 1]^{1/2}. \tag{16.14}$$

Note carefully the distinction between σ_t and the logarithmic standard deviation σ.

EXAMPLE 16.4 A group of 500 submarine cable transistors was subjected to steady stress tests at 220, 250, and 280°C. From a plot of the cumulative failure function versus time in hours, it was found that the standard deviation $\sigma = 1.91$ hr. It was also found by graphical extrapolation that the median, $t_{50} = 10^{10}$ hr, at the service temperature of 100°C. Calculate numerically by the use of Eq. (16.11) the value of the failure rate corresponding to 10,000 hr, 20,000 hr and 100,000 hr.

Solution. The argument of ϕ and Φ of Eq. (16.11) was computed to be -7.233, -6.870, and -6.028, corresponding to 10^4, 2×10^4, and 10^5 hr. By the use of tables of ϕ and Φ it was then found that the corresponding failure rates were 9.11×10^{-17}, 5.89×10^{-16}, and 2.67×10^{-14} failures per hour. If the failure rate is expressed as %/1000 hours, as is frequently done, then these figures must be multiplied by 10^5. These results represent a rising failure rate with time.

16.5 ACCELERATED-STRESS LIFE TESTING [17,19-22]

The purpose of a life test is to be able to predict device reliability when the device is operating in a specified environment. Frequently, a device is so reliable under normal operating conditions that years of testing would be required in order to predict its reliability. Hence there exists a great difficulty, because the greater the reliability of a device, the more difficult it is to determine this reliability.

A solution to this dilemma is to design accelerated-stress life tests in which a device is run at a higher stress level than encountered in normal operation. As a consequence, the device has a shorter life than under normal conditions. Results obtained at more severe stress levels are then extrapolated to normal stress levels so as to obtain an estimate of the life distribution.

The technique of accelerated-stress life testing will be illustrated by examples taken from silicon-transistor and integrated-circuit device tests. The particular failure mechanism treated is that of charge accumulation at the silicon surface (near a transistor or diode junction), which is responsible for a degradation of characteristics. This degradation of parameters is a common failure mode for silicon transistors and integrated circuits with aluminum contacts and thermally grown SiO_2 surface layers.

For many semiconductor products undergoing accelerated stress life testing at selected elevated temperatures, the Arrhenius model[19-21] has proved satisfactory for explaining lifetime as a function of temperature. The following assumptions are made:

1. For any temperature, the lifetime distribution is log-normal, as described by Eqs. (16.9), (16.10), and (16.11).

2. The logarithmic standard deviation, σ, of the distribution is constant and therefore independent of temperature.

3. The logarithmic mean, μ, of the distribution depends on temperature according to the Arrhenius relation

$$\mu(T) = \alpha + \beta T^{-1}, \tag{16.15}$$

where α and β are parameters, characteristic of the particular device and of the stress test method. More particularly, parameter β is given by

$$\beta = \frac{E_a}{k} \tag{16.16}$$

where E_a is called the activation energy and where k is Boltzmann's constant. Recently published values[21] of apparent activation energies are given in Table 16.2.

Table 16.2 Apparent Activation Energies

Device Type	Apparent Activation Energy, E_a, in Electron Volts
Transistors:	
Germanium, ungettered	0.88
gettered with vycor or molecular sieve	1.24
Silicon bipolar with	
surface-inversion failures	1.02
with Au-Al bond failures	1.02~1.04
with metal penetration into Si	1.65
p-n-p-n	1.65
Diodes:	
Four-layer	1.41
Varactors	2.31~2.38
Others	1.13~2.77

The above model is based on experimental observations. The following discussion is a plausible reconciliation to its use in accelerated stress testing.

An electronic device may be thought of as a system consisting of very many particles. Experience indicates that most change processes (such as failure) in which transfer or rearrangement of particles occur require that an energy barrier be surmounted. In order to fail under stress the device has to receive a fixed amount of degradation. Let us assume that a device under the stress of normal use , T', fails after a time, t', and that the amount of degradation can be measured by the product of a power of time to failure, $(t')^n$, and the reaction rate, $r(t')$, of the degrading process. The latter rate, according to the Arrhenius model, is given by

$$r(t') = C \exp\left[-\frac{E_a}{kT}\right],$$ (16.17)

where C is a constant. Clearly, from the above equation, the higher the temperature stress, the larger the reaction rate. Let us further assume, that a similar device under higher temperature stress, T'', fails after a shorter time, t'', with the same amount of degradation. A generalized relationship is

$$(t')^n r(t') = (t'')^n r(t'')$$ (16.18)

and from Eq. (16.17) and Eq. (16.18) now follows

$$\ln\left(\frac{t'}{t''}\right) = \frac{E_a}{nk}\left[\frac{1}{T'} - \frac{1}{T''}\right].$$ (16.19a)

The apparent activation energy E_a/n is the activation energy of the actual mechanism causing the change in the device parameter scaled by the factor $1/n$. The actual value of E_a will depend on the type of device and on the failure mechanism (see Table 16.2).

Determination of the apparent activation energy   The threshold voltage of an MOSFET exhibits a shift in value as the device ages. This shift ultimately leads to device failure. It is observed that the shift ΔV_t in threshold voltage at constant test temperature varies as t^n where n is between one-third and one depending on the gate insulator structure. The activation energy for the degradation process can be determined by observing from Eq. (16.19a) that

$$\frac{E_a}{n} = \frac{k[\ln t' - \ln t'']}{\left[\frac{1}{T'} - \frac{1}{T''}\right]}.$$ (16.19b)

Under the assumption that no additional failure mechanisms are excited by operating the device at high stress, the apparent activation energy is found as the slope of an so-called Arrhenius plot, which is a plot of median lifetime on a logarithmic scale verus reciprocal temperature. If the Arrhenius model holds for the particular device, the apparent activation energy can be determined by measuring the median time to failure at two different high-stress levels. This value will hold for a broad class of devices.

Knowledge of the apparent activation energy and the median time to failure t'' at one high-stress level permits the determination of the time to failure t' under normal stress. The ratio t'/t'' is known as the *acceleration factor*.

Life distributions The life distributions for different temperatures can be plotted as $\Phi[(\ln t - \mu)/\sigma]$ versus $\ln t$. If log-normal probability paper is used, the ordinates have a normal probability scale and the abscissae have a logarithmic time scale or vice-versa. Parallel straight lines representing the distributions at different temperatures are obtained if the probability distribution is log-normal with constant σ. If σ is not constant, two lines corresponding to different test temperatures T' and T'' would intersect at a point corresponding to a time t_1. This implies that for times greater than t_1 fewer devices fail under the higher stress excitation which is contrary to intuition and experience.

Testing methods At this point a word about the testing method is in order. *Constant-stress testing* involves holding every test unit at a constant elevated stress until failure occurs. Several different constant stresses are generally used, with a number of test units tested at each stress level. *Step-stress testing* subjects every test unit to successively higher stresses until it fails. Each test unit is exposed to a specific constant stress for a specific length of time. If the unit does not fail in the time period of exposure, then it is exposed to a specified higher stress for a specific length of time, and so on. Stress on a unit is therefore increased step by step until it fails.

As a simple example of the preparation of a life distribution plot, consider Table 16.3 giving life test data of unit failures at stated test times. The total number of devices n under test is 20. Cumulative failures i would ordinarily correspond to a plotting position $(100i/n)\%$ failures. In practice, plotting position $100i/(n+1)$ provides a statistically unbiased estimate, besides avoiding the 100% point, when the whole sample fails. A straight line has been fitted to the data points, which are plotted on Fig. 16.6. The median life $\exp(\mu)$ for the particular stress is the age at which 50% of the devices have failed. From Fig. 16.6, $\exp(\mu) = 430$ hr. More formally, if P represents the cumulative fraction failed and if t_p is the corresponding life, then for the log-normal distribution

$$\ln t_p = \mu + \Phi^{-1}(P)\sigma$$

where $\Phi^{-1}(\cdot)$ is the inverse of the standard normal cumulative distribution function. In particular Table 16.4 gives some values of $\Phi^{-1}(\cdot)$. It may then

Table 16.3 Life Test Sample Failures at Stated Test Times

Test Time Hours	Number of Failures	Number of Cumulative Failures, i	Plotting Position $100i/(N+1)$
10	0	0	0
20	1	1	4.76
30	1	2	9.52
50	1	3	14.28
100	2	5	23.80
150	1	6	28.56
200	1	7	33.32
300	0	7	33.32
500	3	10	47.60
800	2	12	57.12
1500	4	16	76.16
2500	3	19	90.48
3500	0	19	90.48
6000	0	19	90.48

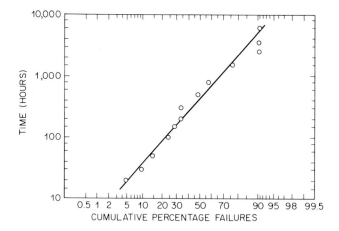

Fig. 16.6 Test time in hours versus cumulative percentage of failures for the data of Table 16.3.

TAble 16.4 Some values for the Inverse
of the Standard Normal Cumulative Distribution

P	$\Phi^{-1}(P)$	P	$\Phi^{-1}(P)$
0.01	−2.326	0.84	+1
0.10	−1.282	0.90	+1.282
0.16	−1 (approx.)	0.99	+2.326
0.50	0		

be shown from the above equation that

$$\ln t_{0.16} = \mu - \sigma; \quad \ln t_{0.84} = \mu + \sigma;$$

(16.20)

$$\ln t_{0.50} = \mu.$$

As a consequence of Eq. (16.20) the standard deviation σ may be approximated by any of the following expressions

$$\sigma \approx \ln\left(\frac{t_{0.50}}{t_{0.16}}\right) = \ln\left(\frac{t_{0.84}}{t_{0.50}}\right) = \frac{1}{2}\ln\left(\frac{t_{0.84}}{t_{0.16}}\right). \qquad (16.21)$$

EXAMPLE 16.5   The following data of junction temperature versus median life was obtained from an accelerated stress life test of a certain transistor type.

Junction Temperature (K)	538	515	488
Median Life (hr)	20	150	2000

Estimate the activation energy, E_a, characteristic of this transistor.

Solution. Applying Eq. (16.19b) to the above data in pairs yields estimates of activation energies of 2.079 eV ($t_1 = 2000$, $t_2 = 150$) and 2.095 ($t_2 = 150$, $t_3 = 20$). The average value of E_a is therefore approximately 2.09 eV.

EXAMPLE 16.6   Analyze the three distributions plotted in Fig. 16.7.

Solution. In a practical situation, life distributions for several stress levels are displayed on a plot like Fig. 16.7. This figure shows the operating life of a semiconductor device versus cumulative percent failures for junction temperatures $T_j = 20°C$, $250°C$, and $300°C$. The curves are slightly S-shaped because of the presence of freaks. Hence the total distribution shown in Fig. 16.7 is really the sum of the desired main distribution plus the freak distribution. The freak distribution may be removed from the combined distribution by noting that the inflection point gives the percentage d_f of freaks present in the total distribution. For junction temperature $T_j = 200°C$, $d_f = 16\%$, for $T_j = 250°C$, $d_f = 26\%$. The new abscissa, d_m, for the main distribution is now given by the empirical formula $d_m = 100(d_t - d_f)/(100 - d_f)$ where d_t is the abscissa for the total distribution. Recalculation of abscissa for the three curves of Fig. 16.7 then gives the main distributions of Fig. 16.8, which have been approximated to be a straight line in each case. The slope of each line is proportional to the logarithmic standard deviation σ. With the use of Eq. (16.21), one obtains $\sigma_{200°C} = 0.44$, $\sigma_{250°C} = 0.38$, and $\sigma_{300°C} = 0.42$. The average value of σ is 0.41. The fact that the standard deviation does not change significantly indicates that the distribution is being shifted in lifetime by stress, without distortion, and therefore satisfies the second requirement of the Arrhenius model listed previously.

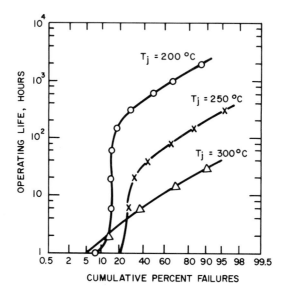

Fig. 16.7 Experimental plot of life test data showing the effect of freak distributions.

A graphical estimate of the Arrhenius regression line [see Eq. (16.15)] for the median life, $\exp(\mu)$, can be made on Arrhenius plotting paper. Time to failure of each test unit is plotted horizontally and corresponds to a stress temperature on the vertical axis. Figure 16.9 shows points obtained from the data of Fig. 16.8 with respect to median life, and the corresponding regression line fitted by eye to pass through the data points at each temperature stress.

To obtain the regression line for the distribution percentiles, one enters the log-normal plot of Fig. 16.8 on the probability scale at each percentage of interest, then one goes vertically to the selected temperature stress line and then horizontally to read off the life in hours. Thus the estimate of the tenth percentile at 200°C, 250°C, and 300°C is 270, 23, and 3.2 hr, respectively. These three points are plotted as a dashed line on Fig. 16.9 and give the regression line for 10% failures. The line for 1% failures is also shown.

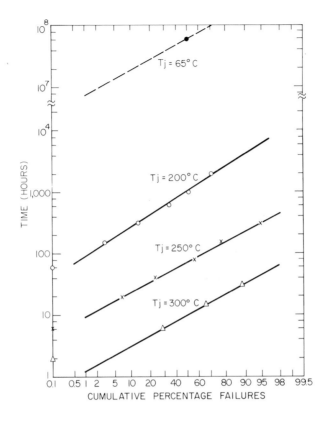

Fig. 16.8 Experimental plot of main life test data exluding the effect of freak distributions.

Regression parameters α and β [see Eq. (16.15)] may be graphically obtained as follows. For two widely spaced absolute temperatures T' and T'', the graphical estimate of corresponding median life t' and t'' is obtained from the Arrhenius plot. Then the estimated beta, $\hat{\beta}$, estimated alpha, $\hat{\alpha}$, and activation energy $\hat{E}_a$ are

$$\hat{\beta} = \frac{T'T''}{T''-T'} \ln\left(\frac{t'}{t''}\right); \quad \hat{\alpha} = \ln(t'') - \hat{\beta}/T''; \quad \hat{E}_a = K\hat{\beta}.$$

$$(16.22)$$

EXAMPLE 16.7   Estimate parameters α, β and E_a for the data of Fig. 16.9.

Solution.   For the regression line of median life of Fig. 16.9, $T' = 473.2$ K (200°C) and $T'' = 573.2$ K (300°C). Corresponding lifetimes are $t' = 1000$ hr and $t'' = 10$ hr. Hence $\hat{\beta} = 5425$ and $\hat{\alpha} = -8.46$ and $\hat{E}_a = 1.12$ eV. Note that the

1.12 eV activation energy is representative of failures in silicon transistors (see Table 16.2). As $\hat{E}_a$ is proportional to $\hat{\beta}$, the slope of the regression plot of Fig. 16.9 represents the activation energy associated with the failure mechanism of the devices under test.

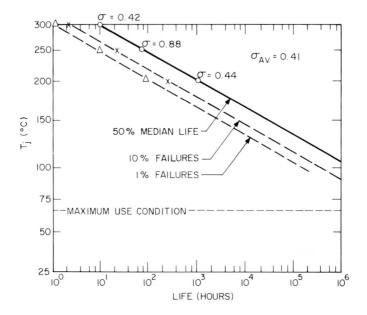

Fig. 16.9 Plot of Arrhenius regression line for the data of Fig. 16.8.

EXAMPLE 16.8 Extrapolate the failure distribution shown in Fig. 16.8 to the lower stress level of 65°C.

Solution. If the maximum-use condition for the transistors is 65°C, then extrapolation of the median regression line on Fig. 16.9 indicates a life of 6×10^7 hr. The distribution for a usage temperature of $T = 65°C$ can be extrapolated and added to the log-normal plot of Fig. 16.8 by drawing a line with the average slope corresponding to $\sigma_{av} = 0.41$ and passing through the point ($t = 6 \times 10^7$, 50%). This completes the extrapolation of failure distribution to a low stress level, from measurements carried out at high stress levels.

Reliability studies enable many transistor improvements to be suggested.[21] Transistor modifications undertaken have raised the activation energy and reduced the maximum failure rate for special types of silicon transistors. Similar improvements should be possible in the case of integrated circuits.

Accelerated aging tests for transistors and integrated circuits under temperature-humidity stress have also been devised.[21] Such tests are particularly valuable for communication equipment exposed to the natural environment.

16.6 YIELD OF INTEGRATED CIRCUITS

A basic concept affecting the economics of integrated circuit manufacture is that of *yield*. Suppose a circular silicon wafer of radius R contains N chips each of total area $A = ab$. Let each chip contain a complete electronic circuit, which itself consists of many identifiable subcircuits or cells such as logic gates, memory cells, etc. These subcircuits are placed into the active area of the chip, denoted by $A_c \leqslant A$, which is considered sensitive to the presence of defects. Area A_c is surrounded by a periphery which may contain bonding pads or other relatively defect-insensitive circuit components.

The wafer is subjected to a sequence of processing steps, after which each chip is electrically tested according to some specification and pronounced good if it meets this specification and bad otherwise. This procedure is purely experimental and will result in a number, $N(0) \leqslant N$, of chips with zero defects, that have passed the test. Then the experimental yield of good chips, Y, for the wafer is given by

$$Y = \frac{\text{number of good chips on wafer}}{\text{total number of chips on wafer}} = \frac{N(0)}{N}. \qquad (16.23)$$

Clearly the larger the number Y, the more good chips are available for sale to consumers and the lower the cost per good chip. The value of Y just computed is an *a posteriori* yield for a chip of specified geometry and design. It is desirable to be able to predict the yield for a design *a priori* so that the benefits of alternative designs can be weighed before committing to manufacture. Experiments or prior experience are used to statistically characterize the yield determining parameters of the process technology and the design rules. These parameters are used in statistical models to be described in the remainder of this chpater to predict the yield of the proposed design. In addition to A, A_c, and N defined previously, other quantities that enter into the yield calculations are the active or effective wafer area A_{we}, the total wafer area A_w, and the total number of defects, M, in the active or effective wafer area A_{we}. Consequently,

$$A_c = \frac{A A_{we}}{A_w}; \quad N = \frac{A_w}{A}. \qquad (16.24)$$

From these quantities derive the important concepts of average defects, d, per chip and average defect density, D, per effective area

$$d = \frac{M}{N} = DA_c; \quad D = \frac{M}{A_{we}}. \tag{16.25}$$

EXAMPLE 16.9   Find the average defect densities, when $M = 96$, $N = 52$, and $A_c = 2.2 \times 10^{-2}$ cm^2.

Solution.   From Eq. (16.25), $d = 96/52 \approx 1.84$. Further

$$D = d/A_c = \frac{1.84}{2.2 \times 10^{-2}} \times 10^{-2} = 84.$$

The chip defect density, d, depends directly on both A_c and D. However, D is a function of the technology and so depends on minimum tolerances in some way. Hence density D may also be a function of A_c, and so the overall dependence of d on A_c may be quite complex. The functional dependence of d on A_c may have to be investigated experimentally, perhaps by mapping the location of defects on many wafers and then calculating values for d as hypothetical chip size is varied. This matter will not, however, be discussed further. Most subsequent equations will be in terms of d, the average defect density per chip and will not show an explicit area dependence.

It has not yet been explained what is meant by a "defect." Furthermore, the average chip defect density, d, may not be a known number. Consider in fact a collection of P wafers each of which may have been processed slightly differently due to normal process variations. This lot P may display somewhat different material properties because the wafers may come from different ingots or different parts of the same ingot. As a consequence, M, the number of defects per wafer is, in fact, a random variable. Because of Eqs. (16.24) and (16.25), defect densities d and D will then also be random variables. Finally, the yield, Y, which depends on the defect density will also be a random variable. Hence the use of probability theory and statistics[14-15] is necessary for the study of yield.

With regard to the definition of a chip, it could be the total wafer, in which case $A = A_w$ and $d = M$. For present-day technology the number of defects on this chip, M, may be a number running into tens of thousands and so the yield would be almost zero. Nevertheless, improvements in processing technology cause M to decrease with time, thus permitting larger chip size. In present-day technology, a chip is a small part of a wafer.

At the other end of the spectrum a chip could refer to the area that contains a single electronic subcircuit or cell, where C is the number of such cells on each chip. In that case the "chip" area is approximately A_{we}/NC and the defect density is $d = M/NC$. Clearly, for sufficiently large N and C the chip area is very small and the number of defects on the "chip" is likely to be zero if the cell is good and will at most be one if the cell is bad. In any case, if the processing is reasonably good, the cell yield will be close to unity because the good cells greatly outnumber the bad cells.

In experimental work on yield, one often plots a *histogram* of the results of the fabrication process, that is, the discrete plot of the number $N(k)$ of chips with k defects versus the number k of defects per chip. The total number of all chips, N', in a collection of P wafers is

$$N' = \sum_k N(k). \tag{16.26}$$

A *theoretical value* for $N(k)$ is given in Section 16.8 for distinguishable defects and in Section 16.9 for indistinguishable defects. Furthermore the notation $N_E(k)$ for the experimental histogram and $N_T(k)$ for the theoretical histogram will be used and the assumption is made that

$$N_E(k) \approx N_T(k)$$

The number of good (defect-free) chips is $N(0)$, since it is assumed that the presence of even one defect renders a chip "bad." In that case, the experimental process yield, Y_{av}, is given by

$$Y_{av} = \frac{N_E(0)}{\sum_k N_E(k)}. \tag{16.27}$$

The total number M' of defects in a collection of P wafers is given by

$$M' = \sum_{k=1} kN_E(k), \tag{16.28}$$

and so the experimental average defect density, d, is given by

$$d = \frac{M'}{N'} = \frac{\sum\limits_{k=1} kN_E(k)}{\sum\limits_{k=0} N_E(k)}. \tag{16.29}$$

Knowledge of the histogram $N_E(k)$ allows the independent computation of experimentally determined values for the yield Y_{av} and the average defect density d.

EXAMPLE 16.10   Tests on a collection of digital logic chips gave rise to the following experimental histogram:

k	N(k)	K	N(k)	k	N(k)
0	188	5	32	10	13
1	138	6	36	11	6
2	97	7	22	12	7
3	48	8	15	13	6
4	41	9	10	14	5
				15	2

Find the defect density d.

Solution. From the above data, the sum $\sum_k N(k) = 666$ and the sum $\sum_k kN(k) = 1838$. Hence, from Eq. (16.29)

$$d = \sum_k kN(k)/\sum N(k) = 1838/666 \approx 2.73.$$

Note that, in a modern, well-controlled process, it would be unlikely to find so many chips with more than about two or three defects.

The nature of the defects will next be investigated in order to statistically model the yield.

16.7 DEFECT DENSITY [10-13]

It is assumed that the wafer material either contains local imperfections before the start of the fabrication process or acquires such imperfections during fabrication. Imperfections may further be classified as point or spot defects, line defects, or area defects. Oxide pinholes, pits, dislocations in crystal structure, and imbedded abrasive grains are all examples of point defects. Scratch marks, slip lines caused by contaminants during mechanical polishing, and slip lines originating from chipped edges of the wafer after exposure to heat, are all examples of line defects. New polishing techniques and more careful handling has eliminated most line defects. Mounds created during epitaxy are an example of an area defect.

For further analysis of the nature of yield, only the presence of *point defects* occurring *randomly* on the wafer surface will be assumed. This is equivalent to stating that defects have no tendency to *cluster* in certain parts of the wafer, and so the occurrence of a defect in one place does not affect the occurrence of another defect in another place on the wafer.

One important question now presents itself, which later will prove of probabilistic importance: Are defects distinguishable or indistinguishable from one another?

The view adopted in this chapter is that the distinguishability of defects does not depend on their exact location on the wafer. Nor is it assumed that one can conclude from measurements on a test set (which merely decides whether a circuit is electrically good or bad) whether or not defects are distinguishable.

On the other hand, it is assumed that defects can be examined by other techniques to determine and classify their physical nature. Some possible categories for defects are metal pinholes, metal shorts, oxide pinholes, and diffusion pipes.

Finally, it is assumed that all defects within one of these classes are *indistinguishable*. This is stated as follows: Classes of point defects are distinguishable but defects within the same class are considered to be indistinguishable.

A helpful analogy is that of distinguishable or indistinguishable balls (defects) that can be placed into urns (chips). M balls are *distinguishable* if they can be labeled with the recognizable numbers 1, 2, $\cdots$, M. Urns are used with *replacement,* if more than one ball may be placed into any one urn. M *indistinguishable defects* may be placed onto N chips (with replacement) in N_1 ways, where

$$N_1 = \binom{N+M-1}{M} = \frac{(N+M-1)!}{M!(N-1)!} . \tag{16.30}$$

The above placement leads to the so-called *Bose-Einstein* statistics. On the other hand, M *distinguishable defects* may be placed onto N chips (with replacement) in N_2 ways, where

$$N_2 = N^M \tag{16.31}$$

This placement leads to the so-called *Maxwell-Boltzmann* statistics. The above two equations are basic tools needed to enumerate yield formulae as will be shown next.

16.8 DISTINGUISHABLE POINT DEFECTS

Assume M distinguishable defects are distributed on a wafer of N chips and define a random variable X_i such that $X_i = 1$ if the ith chip contains k defects and $X_i = 0$ otherwise. Hence the number of chips with k defects on the wafer is given by

$$\sum_{i=1}^{N} X_i$$

and the expected number of such chips $NP(z=k)$ is given by

$$\sum_{i=1}^{N} E\{X_i\}$$

where z denotes the random variable "number of defects" on chips. Furthermore, this expected number is defined as

$$E\{X_i\} = \sum_{m=0}^{1} mP\{X_i=m\} = P\{X_i=1\}.$$

From these remarks now follows

$$P(z=k) = \frac{1}{N} \sum_{i=1}^{N} P\{X_i=1\}. \tag{16.32}$$

EXAMPLE 16.11   Compute the probability, $P(z=k)$, that k distinguishable defects are on a chip.

Solution.   The "sample space" or total number of ways for distributing M distinguishable defects on N chips is given by Eq. (16.31) as N^M. Consider $P\{X_i=1\}$, the probability that the event "k defects on the ith chip" has taken place. A subset of k distinguishable defects may be chosen from M distinguishable defects in $\binom{M}{k}$ ways. Then k defects may be placed in the ith chip in only one way and the remaining $(M-k)$ defects may be placed on $(N-1)$ chips according to Eq. (16.31) in $(N-1)^{M-k}$ ways. The ratio of favorable events to the sample space N^M is the required probability.

$$P\{X_i=1\} = \binom{M}{k} \frac{(N-1)^{M-k}}{N^M} = \binom{M}{k}\left[\frac{1}{N}\right]^k\left[1 - \frac{1}{N}\right]^{M-k} \qquad (16.33)$$

But this probability is the same for all chips and so from Eq. (16.32)

$$P(z=k) = \binom{M}{k}\left[\frac{1}{N}\right]^k\left[1 - \frac{1}{N}\right]^{M-k} \qquad (16.34)$$

The above probability $P(z=k)$ is a special case of the binomial law and has an associated mean value of $E\{z\} = M/N = d_0$ and variance $\sigma_z^2 = d_0[1 - (1/N)]$, where M is a function of the technology and $1/N$ is an index of the chip size. The quantity d_0 is perhaps the most important index of the yield. The defect density for MOS technologies is typically 1 to 100 defects/cm². The defect density for bipolar technologies is between 10 and 500 defects/cm², primarily because of the high defect density associated with the thin base region of high-speed LSI devices. Three separate cases may be distinguished:

Case 1   Let M be large with $d_0 \gg 1$. Taking the variance, $\sigma_z^2 = d_0[1 - (1/N)]$, one may assume that the actual number of defects per chip, k, lies in the vicinity of the mean value d_0. That is, $k \approx d_0 \pm \sigma_z$. For this case it may be shown[15] that Eq. (16.34) can be approximated by the Gaussian density

$$P(z=k) \approx \frac{1}{\sqrt{2\pi}\,\sigma_z} \exp\left\{-\frac{1}{2}\left[\frac{k-d_0}{\sigma_z}\right]^2\right\}. \qquad (16.35)$$

This situation represents poor yield, due to a large chip size.

Case 2   A fixed large M will be considered with $d_0 \approx 1$. As a consequence, $N \approx M$ and for this case it may be shown that Eq. (16.34) can be approximated by a Poisson density, namely,

$$P(z=k) \approx \frac{d_0^k\exp(-d_0)}{k!}. \qquad (16.36)$$

For this density the variance is known to equal the mean value and so $\sigma_z^2 = d_0$. This situation represents good yield for normal sized chips and will be emphasized in the subsequent discussions.

Case 3   A fixed large M will be considered with $d_0 \ll 1$. As a consequence of Eq. (16.36), $P(z=k) < P(z=k-1)$. Ultimately, most chips will be defect-free and a few chips will have one defect. That is $P(z=0) \approx 1$ and $P(z=1) \approx 0$. This situation represents a large yield and very small chips. As an extreme limit, N may be made so large that the chip size approaches the size of the spot defect, which previously was considered negligible. Since the average yield Y_{av} is given by $P(z=0)$ one obtains from Eq. (16.34)

$$Y_{av} = \left(1 - \frac{1}{N}\right)^M = \left(1 - \frac{d_0}{M}\right)^M .$$

For the limit of the Poisson approximation, with M large,

$$Y_{av} = \exp(-d_0) \tag{16.37}$$

A plot of the exponential yield relationship is given as the bottom curve of Fig. 16.10. The exponential relationship, representative of a Poisson process, is deemed important and also appropriate because it stands for the occurrence of *rare events*. With improvements in processing technology, the occurrence of defects should ultimately become a rare event and therefore be characterizable by a Poisson process. In actuality, the exponential yield formula proves to be a pessimistic estimate of the observed yield.

The variance, σ_Y^2, for the yield $Y = P\{z=0\} = (1-1/N)^M$ may be shown to equal

$$\sigma_Y^2 = \frac{1}{N}\left[\left(1 - \frac{1}{N}\right)^M - \left(1 - \frac{2}{N}\right)^M\right] + \left[\left(1 - \frac{2}{N}\right)^M - \left(1 - \frac{1}{N}\right)^{2M}\right].$$

$$\tag{16.38}$$

EXAMPLE 16.12   Compute, σ_Y^2, for $Y = [1 - (1/N)]^M$ and thereby show that $\sigma_y^2 = \exp(-d_0)[1 - \exp(-d_0)]/N$.

Solution.   Define the random variable $X_i = 1$ if the ith chip on a wafer is good and let $X_i = 0$ otherwise. Then

$$Y = \sum_{i=1}^{N} \frac{X_i}{N}$$

and

$$\sigma_y^2 = E\left\{Y - E\{Y\}\right\}^2 .$$

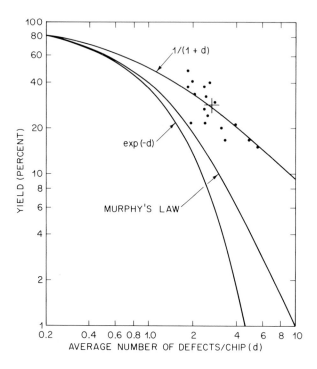

Fig. 16.10 Integrated circuit yield curves.

This may be written

$$\sigma_y^2 = E\left\{ \frac{1}{N} \sum_{i=1}^{N} X_i - \frac{1}{N} \sum_{i=1}^{N} E\{X_i\} \right\}^2$$

$$= \frac{1}{N^2} \left\{ \left[\sum_{i=1}^{N} P\{X_i=1\} \right] \left[1 - \sum_{i=1}^{N} P\{X_i=1\} \right] + \sum_{\substack{i,j \\ i \neq j}} P\{X_i=1, X_j=1\} \right\}.$$

The second term in the above equation sums the probability $P\{X_i=1, X_j=1\}$, that both the ith chip and the jth chip are defect-free. Now for distinguishable defects $P\{X_i=1\}$ is given by Eq. (16.33). To find the probability that chip i and chip j are defect-free one takes the sample space of Eq. (16.31). Cases favorable to the event $X_i=1, X_j=1$ are computed as follows: Chip i and chip j may be removed from N chips in only one way. The remaining M defects can then be distributed over the $(N-2)$ remaining chips in $(N-2)^M$ ways. As a consequence

$P\{X_i=1;\ X_j=1\} = [1 - (2/N)]^M$. After some algebra one finds that

$$\sigma_Y^2 = \frac{1}{N}\left[\left(1 - \frac{1}{N}\right)^M - \left(1 - \frac{2}{N}\right)^M\right] + \left[\left(1 - \frac{2}{N}\right)^M - \left(1 - \frac{1}{N}\right)^{2M}\right].$$

For the limit of the Poisson approximation

$$\sigma_Y^2 = \frac{1}{N}\exp(-d_0)[1 - \exp(-d_0)]. \tag{16.39}$$

These results can also be derived in a slightly different, but informative way. It is well known[15] that the probability $P\{NY'=n\}$, that n chips (urns) (from a total of N) are free of distinguishable defects (balls), is given by

$$P\{NY'=n\} = \binom{N}{n}\sum_{k=0}^{N-n}(-1)^k\binom{N-n}{k}\left[1 - \frac{n+k}{N}\right]^M. \tag{16.40a}$$

This formula may be simplified if M is allowed to become large, while $M/N = d_0$ remains finite. In that case

$$\left[1 - \frac{n+k}{N}\right]^M = \left[1 - \frac{[n+k]d_0}{M}\right]^M \rightarrow \exp[-(n+k)d_0] \qquad (M\ \text{large})$$

and so Eq. (16.40a) may be written

$$P\{NY'=n\} = \binom{N}{n}[\exp(-d_0)]^n[1 - \exp(-d_0)]^{N-n} \tag{16.40b}$$

The above equation for yield is recognized as the *binomial distribution,* with $p = \exp(-d_0)$ and $q = 1 - p$. The expected value or yield is given by $Np = N\exp(-d_0)$ and so $Y_{av} = \exp(-d_0)$ as was shown in Eq. (16.37). The variance $N^2\sigma_y^2$ is given by

$$Npq = N\exp(-d_0)[1 - \exp(-d_0)]$$

which is equivalent to Eq. (16.39).

The important conclusion of this section is that k distinguishable defects on a chip follow a *binomial law* as shown in Eq. (16.34), while the yield follows the law of Eq. (16.40a). In the limit of a large value of M and a value $d_0 \approx 1$, the number k of defects on a chip follow a *Poisson* law, as shown in Eq. (16.36), while the yield follows a *binomial law* as shown in Eq. (16.40b). In the whole of this discussion it was assumed that the number of defects per wafer, M, was not a random variable. This, in fact, is not true and will lead to a modification of certain of the derived equations. A discussion of the randomness of M is given in Section 16.11.

16.9 INDISTINGUISHABLE POINT DEFECTS [23]

Assume all M defects on a wafer of N chips belong to the same defect class and so are indistinguishable among themselves. Then proceeding as in the previous section, Eq. (16.32) is obtained and one must evaluate $P\{X_i=1\}$, the probability that the event "k indistinguishable defects on the ith chip," has taken place. The evaluation proceeds as in Section 16.8, except that now Eq. (16.30) must be substituted. The result is

$$P(z=k) = \frac{\binom{N+M-k-2}{M-k}}{\binom{N+M-1}{M}} . \tag{16.41}$$

EXAMPLE 16.13   Derive the yield equation (16.41) for indistinguishable defects.

Solution.   From Eq. (16.32),

$$P(z=k) = \frac{1}{N} \sum_{i=1}^{N} P\{X_i=1\},$$

where the random variable $X_i = 1$ if the ith chip contains k indistinguishable defects. The total number of ways of distributing M indistinguishable defects on N chips is $\binom{N+M-1}{M}$. A subset of k indistinguishable defects may be chosen from M indistinguishable defects in only one way. After that choice has been made, k defects may be placed on the ith chip according to Eq. (16.30) in $\binom{1+k-1}{k} = 1$ way. The remaining $(M-k)$-defects may be placed on the remaining $(N-1)$-chips in $\binom{N+M-k-2}{M-k}$ ways. As a consequence of these remarks,

$$P\{X_i=1\} = \frac{\binom{N+M-k-2}{M-k}}{\binom{N+M-1}{M}} .$$

The probability $P(z=k)$ may be written in the form[15]

$$P(z=k) = \frac{P(z=0)\left\{d_0 - \dfrac{k-1}{N}\right\}\left\{d_0 - \dfrac{k-2}{N}\right\}}{\left\{1 + d_0 - \left[\dfrac{k+1}{N}\right]\right\}\left\{1 + d_0 - \left[\dfrac{k}{N}\right]\right\}} \cdots \frac{d_0}{1 + d_0 - \left[\dfrac{2}{N}\right]} \tag{16.42}$$

with

$$P(z=0) = \frac{1 - \dfrac{1}{N}}{1 + d_0 - (1/N)} .$$

It is easy to show that for small values of k, and large, increasing values of N, $P(z=k)$ asymptotically approaches

$$P(z=k) = p(1-p)^k; \quad p = 1/(1+d_0); \quad q = 1 - p. \tag{16.43}$$

This is the well-known *geometric distribution* and characterizes the limiting probability for good yield for moderately sized chips in the presence of indistinguishable defects.

The theoretical value of yield Y_{av} is obtained from Eq. (16.41) as

$$Y_{av} = P\{z=0\} = \frac{N-1}{N+M-1} .$$ (16.44)

For the limiting situation as N becomes large, the yield Y_{av} asymptotically approaches

$$Y_{av} = \frac{1}{1+d_0}$$ (16.45)

The above relation is called the Price[23] yield law.

EXAMPLE 16.14   An IC process produces chips having an average defect density of two defects per chip. Compare the exponential yield law of Eq. (16.37) and the Price yield law of Eq. (16.45). Which yield law is more optimistic?

Solution.   According to the exponential yield law $Y_{av} = \exp(-2) = 0.135$ or 13.5%. According to the Price yield law $Y_{av} = 1/3 = 0.333$ or 33.3%. Hence the Price yield law is a more optimistic predictor.

Also of interest is the variance, σ_y^2, or the standard deviation σ_y of the yield Y. In practice, yield curves differ most markedly at low yields. Results should be stated in terms of the expected value of the yield $E\{Y\}$ and the resolution or spread of values about the mean,

$$E\{Y\} = Y_{av}; \quad Y_{av} - \sigma_y \leqslant Y \leqslant Y_{av} + \sigma_y$$ (16.46)

To compute σ_y^2 one proceeds similarly as in Section 16.8. It can be established after some algebra that

$$\sigma_y^2 = \frac{M(M-1)(N-1)}{N(M+N-1)^2(M+N-2)} .$$ (16.47)

In the limit when M is large, σ_y^2 is approximately

$$\sigma_y^2 = \frac{(d_0)^2}{N(1+d_0)^3} .$$ (16.48)

An alternative derivation uses the well-known probability[15] that exactly n of N specified chips (urns) will be empty of indistinguishable defects (balls), namely,

$$P\{NY'=n\} = \frac{\binom{N}{n}\binom{M-1}{N-n-1}}{\binom{N+M-1}{M}} .$$ (16.49)

The expected value of the yield Y', namely $Y_{av} = E\{NY'\}/N$, is then

$$Y'_{av} = \frac{1}{N} \sum_{n=0}^{N} nP\{NY'=n\} = \frac{N-1}{N+M-1} \, . \tag{16.50}$$

and the variance may be obtained similarly,[14, 15] again giving Eq. (16.47).

Equation (16.49) for the yield is a *hypergeometric probability* law that characterizes an integrated circuit process when the defects are considered indistinguishable, while Eq. (16.41) for the number of indistinguishable defects per chip is, in the limit, a *geometric probability* law.

The yield model for indistinguishable defects, culminating in Eqs. (16.44), (16.45) or (16.47), (16.48) is essentially that proposed by Price and it is believed by many people to give the most realistic interpretation of the yield behavior of present-day integrated circuits. A plot of $1/(1+d_0)$ versus d_0 is given as the top curve of Fig. 16.10.

16.10 SOME EXPERIMENTAL RESULTS [24]

Several years ago, yield studies were carried out on a memory chip. A flip-flop storage cell was considered "good" if it could store information and change state, and was considered bad otherwise. A sample consisting of 663 chips, which displayed only flip-flop failures, was studied.

Table 16.5 lists the wafers from 1 to 20, the number of defects per wafer M_i and the number of chips per wafer N_i. Also shown are the number of defect-free chips $N_i(0)$. The last two columns of the table show computed values for the average defect density $d_i = M_i/N_i$ and for the wafer yield $Y_i = N_i(0)/N_i$.

Due to some broken wafers the number of defects per wafer, M_i, varies from a low value of 26 to a high value of 149 and similar behavior is displayed by the number of chips N_i, which go from a low of 6 per wafer to a high of 52.

EXAMPLE 16.16   Examine the histogram for the 663 logic chips of Fig. 16.11 and, if possible, determine the distribution.

Solution.   A logarithmic plot of the histogram is shown in Fig. 16.12. As may be seen, a straight line can be made to fit the data and so log $N(n) = A - n$ log B. This relationship suggests [see Eq. (16.43)] a geometric distribution and hence indistinguishable defects. Multiplying Eq. (16.43) by N, taking the logarithm and comparing with the above result yields $A = \log N + \log p$ and log $B = \log[1/(1-p)]$. From Fig. 16.12, $\log[1/(1-p)] = (\log 200 - \log 2)/14.5$ and so $p = 0.27$. But from Eq. (16.43), $p = 1/(1+d_0)$ and so $d_0 = 2.7$ defects/chip.

A value of d_0 could also have been obtained from the histogram as follows: The total number of defects M is given by

$$M = \sum_{k=1}^{14} kN(k) = 1792.$$

Table 16.5 Data from Yield Experiment

Wafer	M_i	N_i	$N_i(0)$	$d_i = M_i/N_i$	$Y_i = \dfrac{N_i(0)}{N_i}$
1	149	36	3	4.14	0.083
2	93	47	19	1.98	0.404
3	84	32	9	2.63	0.281
4	106	22	2	4.82	0.091
5	88	37	14	2.38	0.378
6	96	52	25	1.85	0.481
7	148	38	8	3.90	0.211
8	79	25	5	3.16	0.200
9	26	10	4	2.60	0.400
10	113	46	15	2.46	0.326
11	80	24	4	3.33	0.167
12	72	37	8	1.95	0.216
13	117	49	13	2.39	0.265
14	29	6	1	4.83	0.167
15	124	44	13	2.82	0.296
16	83	33	8	2.52	0.242
17	71	13	2	5.46	0.154
18	55	27	9	2.04	0.333
19	90	37	8	2.43	0.216
20	89	48	18	1.85	0.375
	1792	663	188		

The total number of chips, N, is given by

$$N = \sum_{k=0}^{14} N(k) = 663.$$

Hence the average defect density $d_0 = M/N = 1792/663 = 2.7$. The yield now follows as

$$Y_{av} = \frac{N(0)}{N} = \frac{188}{663} = 0.284.$$

A scatter plot of the yield Y_i for wafer i versus d_i is also shown on Fig. 16.11 together with the Price relation $Y_{av} = 1/(1+d_0)$, the Poisson exponential relation $exp(-d_0)$ and Murphy's[25] law to be given later. Note that these curves favor the Price hypothesis. The mean point corresponding to a yield of 28.4% and an average defect density d of 2.7 is also indicated by a cross on this curve.

It is also of interest to establish whether the actual experimental yields for each wafer, $N_i(0)/N_i$, lie in a 95% theoretical confidence interval given by $(Y-2\sigma_y, Y+2\sigma_y)$. Here, by virtue of Eqs. (16.44) and (16.47), one has

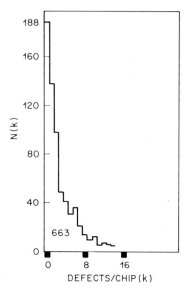

Fig. 16.11 Histogram of experimental data.

$$Y_{av} = (N_i-1)/(M_i+N_i-1) \text{ and}$$

$$\sigma_y^2 = [M_i(M_i-1)(N_i-1)]/[N_i(M_i+N_i-2)(M_i+N_i-1)^2].$$

The original formula has been used here because neither the values of M_i nor N_i are very large. Note that index i runs from 1 to 20. Table 16.6 shows results of this computation. As may be seen, with one exception, all experimentally established values of Y_{av} lie within the 95% confidence interval and this result strongly reinforces the hypothesis that the Price yield law is operative.

Taking the average defect density d_0 as 2.7 defect/chip, and the active chip area A_c as 2.2×10^{-2} cm^2, gives the average defect density D per unit area as

$$D = \frac{d}{A} = \frac{2.7}{2.2\times10^{-2}} = 122 \text{ defects/cm}^2.$$

16.11 ON THE RANDOMNESS OF M [25]

The number M of defects on each wafer is not a deterministic quantity, as explained above. It is assumed that the number of defects M (over a collection of P wafers) is a random variable that can be characterized by a probability density function $f_M(M)$. Because $M = dN$, then d can also be characterized by a defect density function $f_d(d)$. Hence the equation for $P(z=k)$

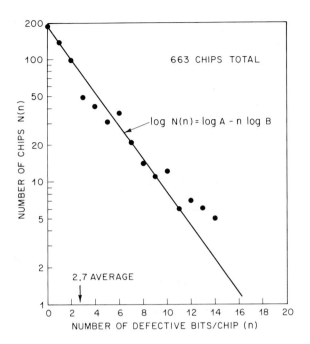

Fig. 16.12 Logarithmic plot of histogram.

should be modified to

$$P(z=k) = \int_0^\infty P(z=k\,|\,M)f_M(M)\,dM \tag{16.51}$$

where $P(z=k\,|\,M)$ is called the *conditional probability*.

Alternatively one may write

$$P(z=k) = \int_0^\infty P(z=k\,|\,d)f_d(d)\,d(d). \tag{16.52}$$

This form, in terms of the variable d, will be favored throughout the rest of the chapter. These equations presuppose that $f_M(M)$ and $f_d(d)$ can be determined in some way. The trouble, however, is that these densities can only be ascertained indirectly, because the only information available is usually the experimentally obtained histogram $N_E(k)$.

$$N_E(k) \approx N_T(k) = N'P(z=k) = N'\int_0^\infty P(z=k\,|\,d)f_d(d)\,d(d). \tag{16.53}$$

Neither $P(z=k\,|\,d)$ or $f_d(d)$ are known, other than theoretically, and they must somehow be inferred from the experimental knowledge of $N_E(k)$. Finally, a variety of functions $P(z=k\,|\,d)$ and $f_d(d)$ may produce the same $N_E(k)$ via Eq. (16.53).

Table 16.6 Results of Yield Calculation

Wafer	Experimental $Y_{av} = N_i(0)/N_i$	Theoretical $Y_{av} = \dfrac{N_i-1}{M_i+N_i-1}$	$2\sigma_y$	$Y-2\sigma_y$	$Y+2\sigma_y$	Result
1	0.083	0.190	0.118	0.072	0.308	Yes
2	0.404	0.331	0.112	0.219	0.443	Yes
3	0.281	0.270	0.134	0.136	0.404	Yes
4	0.091	0.165	0.145	0.020	0.310	Yes
5	0.378	0.290	0.126	0.164	0.416	Yes
6	0.481	0.347	0.106	0.241	0.453	No
7	0.211	0.200	0.116	0.084	0.316	Yes
8	0.200	0.233	0.148	0.085	0.381	Yes
9	0.400	0.257	0.237	0.020	0.494	Yes
10	0.326	0.285	0.112	0.173	0.397	Yes
11	0.167	0.223	0.150	0.073	0.373	Yes
12	0.216	0.333	0.126	0.207	0.459	Yes
13	0.265	0.291	0.109	0.182	0.400	Yes
14	0.167	0.147	0.266	0.000	0.413	Yes
15	0.296	0.257	0.113	0.144	0.370	Yes
16	0.242	0.278	0.132	0.146	0.410	Yes
17	0.154	0.146	0.180	0.000	0.326	Yes
18	0.333	0.321	0.148	0.173	0.469	Yes
19	0.216	0.286	0.125	0.161	0.411	Yes
20	0.375	0.346	0.110	0.236	0.456	Yes

Seeds,[26] possibly from experimental study of defects introduced during a single processing step, conjectured an exponential density function

$$f_d(d) = \frac{1}{d_0} \exp\left[-\frac{d}{d_0}\right] \qquad (16.54)$$

where d_0 is the average number of defects per chip. He observed that this density, together with the Poisson law (distinguishable defects) $P(z=k\,|\,d_0) = d_0^k exp(-d_0)/d_0!$ of Eq. (16.36), produces via Eq. (16.52)

$$P(z=k) = \left[\frac{d_0}{1+d_0}\right]^k \frac{1}{1+d_0} . \qquad (16.55)$$

Equation (16.55) with $k = 0$ is the Price yield law obtained after a single processing step by applying the so-called Poisson-Seeds hypothesis. Wadsack[24] has raised some very good critical points as to why the above hypothesis may be inferior to the direct Price law of Eq. (16.45). He observed that the standard deviation σ_y of the Poisson-Seeds model, namely,

$$\sigma_y = \frac{1}{1+d_0} \frac{d_0}{\sqrt{1+2d_0}} , \tag{16.56}$$

has much less "resolution" than the standard deviation of the Price model, namely

$$\sigma_y = \frac{d_0}{1+d_0} \frac{d_0}{\sqrt{1+d_0}} \frac{1}{N} . \tag{16.57}$$

This is obvious from the presence of the factor $1/N$. It is clearly demonstrated by the data of Table 16.6. The original value of σ_y given in Eq. (16.47) has been used, because the number of chips N, was not large enough to justify the limiting value as given by Eq. (16.48). Note further, that chips from all but one wafer lie in the 95% confidence interval.

Actually, one might logically expect that M defects on a wafer (which is part of a sufficiently large collection of P wafers) would exhibit a Gaussian density function rather than an exponential one. It is assumed here that all wafers in the lot are cut from the same ingot and are processed at the same time under identical fabrication conditions. Such a density, in terms of average defects per chip, would be given by

$$f_d(d) = \frac{1}{\sqrt{2\pi}\,\sigma_d} \exp\left[-\frac{1}{2}\left[\frac{d-d_0}{\sigma_d}\right]^2\right]. \tag{16.58}$$

This density would account for the small differences existing among otherwise identical members of a population.

Murphy[25] made a practical attempt at settling the nature of $f_d(d)$. The result, know as Murphy's law, has found widespread rule-of-thumb application in integrated circuit fabrication facilities and it is here mentioned for historical reasons rather than for its theoretical basis, which is doubtful.

Murphy assumed distinguishable defects, but he favored a Gaussian distribution for $f_d(d)$ as given by Eq. (16.58). Rather than compute the cumbersome integral of Eq. (16.52) using these functions, he approximated the Gaussian density by means of the triangular density $f(d)$, lying in the range $0 \leqslant d \leqslant 2d_0$:

$$f(d) = \begin{cases} d/d_0^2 & 0 \leqslant d \leqslant d_0 \\ 2/d_0 - d/d_0^2 & d_0 \leqslant d \end{cases} .$$

Using this density in Eq. (16.52), together with the Poisson density of Eq. (16.36), yields Murphy's law

$$Y_{av} = \left[\frac{1 - \exp(-d_0)}{d_0}\right]^2$$

$$\sigma_y / Y_{av} = \left[\left(\frac{d_0}{2} \right)^2 \left[\frac{1 + \exp(-d_0)}{1 - \exp(-d_0)} \right] - 1 \right]^{1/2}. \tag{16.59}$$

A plot of Murphy's yield formula is added to Fig. 16.11. It is more optimistic than the Poisson yield law and less optimistic than the Price yield law. Another density function, that has been suggested,[27-29] based on the examination of yield data, is the gamma density

$$f(d) = \frac{A_c d^{\alpha - 1}}{\Gamma(\alpha)(A\beta)^{\alpha}} \exp(-d/A_c\beta),$$

where A_c is the effective chip area and α and β are parameters that can be established experimentally. When the above density is used together with the Poisson probability of Eq. (16.36) and substituted into the integral of Eq. (16.52), the probability

$$P\{z=k\} = \frac{\Gamma(k+\alpha)(A\beta)^k}{k!\Gamma(\alpha)(1+A\beta)^{k+\alpha}}$$

is obtained where Γ stands for the gamma function. The above probability is known as the negative binomial distribution and gives acceptable explanations of some experimental yield data.[29, 30]

   Introduction of the randomness of M, coupled with the difficulty of establishing the form of $f_M(M)$, has a smearing effect because the variance σ_y^2 obtained after application of Eq. (16.52) invariably seems to have less resolution than that of the ordinary equations. Moreover, there may be cases where information as to the form of $f_M(M)$ is unobtainable or where this information has been lost. In that case, it seems desirable to ignore the randomness of M and simply use the ordinary equations, unmodified by application of Eq. (16.52). This procedure may be stated as an hypothesis.

Hypothesis   When the number of defects per wafer M (in a fabrication process involving a collection of P wafers, with N chips per wafer) is a random variable whose pdf, $f_M(M)$, is doubtful or otherwise undeterminable, then the lot should be considered as a "superwafer" having PM defects and PN chips. The yield for this superwafer should then be determined by the ordinary formulae, neglecting the possible randomness of M.

16.12 YIELD FOR MULTISTEP PROCESS [31]

A slightly modified version of Eq. (16.45) is obtained from the following considerations.

   Assume distinguishable defects are observed by an inspector after each processing step and that the inspector can determine whether or not a defect is fatal to the chip. Only the presence of a fatal defect is considered to render that chip bad. Moreover, let the probability that a defect is nonfatal be p and that it is fatal be $q = 1 - p$. In that case, the probability of having j fatal defects in k observed defects is

$$\binom{k}{j}p^{k-j}q^j.$$

This probability, when multiplied by $P(z=k)$ of Eq. (16.55), yields after summation from $k = j$ to $k = \infty$ the probability $P(z_f=j)$,

$$P(z_f=j) = \sum_{k=j}^{\infty} \frac{d_0^k\binom{k}{j}p^{k-j}q^j}{(1+d_0)^{k+1}}$$

$$= \frac{(d_0q)^j}{(1+d_0)^{j+1}} \sum_{r=0}^{\infty} \binom{r-j}{j}\left[\frac{d_0p}{1+d_0}\right]^r = p'(q')^j, \tag{16.60}$$

where $p' = 1/(1+qd_0)$, $q' = 1 - p'$, and z_f stands for the random variable "number of fatal defects." Equation (16.60) is clearly a geometric distribution with $E\{z_f\} = qd_0$ and $\sigma_{z_f}^2 = qd_0(1+qd_0)$. It is therefore identical with the geometric distribution for ordinary distinguishable defects as given by Eq. (16.55), except that d_0 has been replaced by qd_0. One can therefore state the following result: Fatal spot defects have the same statistical distribution as have ordinary spot defects and so statistically they behave exactly alike. Yield laws for fatal and ordinary spot defects are identical in form.

To obtain the yield after one processing step, one may put $j = 0$ in Eq. (16.60) yielding

$$Y_{av} = P(z_f=0) = \frac{1}{1+qd_0}. \tag{16.61}$$

Equation (16.61) is a slightly modified version of the Price yield law, Eq. (16.45), and it includes the factor q, which depends on the relative presence of fatal defects introduced by the processing step. Denoting the average defect density for the ith processing step as d_{0i} and the probability of fatal defects for that step as q_i, it then follows that the yield after k processing steps is given by

$$Y_{av} = \prod_{i=1}^{k} \frac{1}{1+q_id_{0i}}. \tag{16.62}$$

Equation (16.62) clearly shows that the yield can be increased by:

a)   Keeping the number of processing steps, k, as small as possible, because each factor $1/(1+q_id_{0i}) < 1$ and therefore decreases yield.

b)   Achieving the lowest average defect density, d_{0i}, by improvement of production masks and wafer processing at every processing step $i = 1, 2, \cdots, k.$

c)   Lowering the probability of fatal defects, q_i, by improved circuit design for every processing step $i = 1, 2, \cdots, k.$

The utilization of Eq. (16.62) for a process presupposes the presence of inspectors able to identify and record fatal and nonfatal defects at every processing step, which may be unrealistic. Also, Eq. (16.62) presupposes the independence of every processing step; that is, the defects introduced at the ith step are independent of those produced at the $i-1$st step.

The joint defect density after two processing steps for *distinguishable spot defects* is given by the convolution integral

$$f_{d_1+d_2}(d) = \int_0^d f_{d_1}(d') f_{d_2}(d-d') d(d'),\qquad(16.63)$$

on the assumption of an exponential density function for the industrial process, that is,

$$f_{d_i}(d) = \frac{1}{d_{0i}} \exp\left(-\frac{d}{d_{0i}}\right).\qquad(16.64)$$

Quite generally, after k processing steps, the joint density $f_{d_1+d_2+\cdots+d_k}(d)$ is given by

$$f_{d_1+d_2+\cdots+d_k}(d) = \int_0^d f_{d_1+d_2+\cdots+d_{k-1}}(d') f_{d_k}(d-d') d(d')$$

$$= \sum_{i=1}^{k} \frac{d_{0k}^{(k-2)} \exp\left(-\dfrac{d}{d_{0i}}\right)}{\displaystyle\prod_{r-1}^{k}(d_{0i}-d_{0r})}\qquad(16.65)$$

where $d_{0i}-d_{0i} \triangleq 1.$

The probability of n defects after k processing steps is then

$$P_k(z=n) = \int_0^\infty P(z=n|d) f_{d_1+d_2+\cdots+d_k}(d) d(d),$$

where $P(z=n|d)$ is given by Eq. (16.36) for normal industrial processing. Computation of the above integral then yields

$$P_k(z=n) = \sum_{i-1}^{k} \frac{d_{0i}^{k-1}}{\displaystyle\prod_{r=1}^{i}(d_{0i}-d_{0r})} \frac{d_{0i}^n}{(1+d_{0i})^{n+1}}.\qquad(16.66)$$

The histogram for distinguishable spot defects for a k-step process should therefore follow $N_E(n) \approx N_T(n) = N' P_k(z=n)$, where $N_T(n)$ is the theoretical number of chips containing n defects, and N' is the total number of chips in the P wafer lot.

The number of fatal defects, j, after k processing steps is obtained from Eq. (16.66) by merely substituting $q_i d_{0i}$ for d_{0i}. This yields

$$P_k(z_f{=}j) = \sum_{i=1}^{k} \frac{(q_d d_{0i})^{k-1}}{\prod_{i=1}^{k}(q_i d_{0i}-q_r d_{0r})} \frac{(q_i d_{0i})^{j}}{(1+q_i d_{0i})^{j+1}} . \qquad (16.67)$$

The histogram for fatal defects for a k step process should therefore follow

$$N_E(j) \approx N_T(j) = N'P_k(z_f{=}j).$$

When $j = 0$, $P_k(z_f{=}0)$ gives once more the yield Eq. (16.62), although some algebra is involved in showing this.

The expected total number n_f of fatal defects after k processing steps is by virtue of Eq. (16.60) equal to

$$n_f = E\{z_f\} = \sum_{j} jP(z_f{=}j) = \sum_{i=1}^{k} \frac{(q_i d_{0i})^{k}}{\prod_{r=1}^{k}(q_i d_{0i}-q_r d_{0r})}$$

$$= \sum_{i=1}^{k} q_i d_{0i} \qquad (16.68)$$

and indicates that the expected value of fatal defects at each processing step adds to give the total expected value.

If the sound theoretical basis for the Poisson-Seeds hypothesis, and therefore the validity of the exponential density Eq. (16.54), is questioned, another point of view is possible: The yield of Eq. (16.62) may be considered due to k classes of indistinguishable defects, each class being present on the wafer simultaneously. Each class has an average defect density of d_{0i} of which $q_i d_{0i}$ is ascribed to fatal defects. Hence

$$Y_{av} = \prod_{i=1}^{k} \frac{1}{(1+q_i d_{0i})}$$

can be considered the "Price" yield for a wafer which contains k classes of indistinguishable defects.

From this point of view a k-step process involving distinguishable defects may be considered equivalent to a one-step process involving k classes of indistinguishable defects, all classes of defects being thrown on the wafer simultaneously, so to speak.

16.13 A PROPOSED YIELD MODEL

The probability and statistics of yield as well as the physical nature of defects have been discussed at length in previous sections. This section will attempt to summarize past conclusions and arrive at a comprehensive, usable, yield model. Several facts are pertinent to the final choice of model:

1. Specific assumptions as to the nature of defects must be made. *Point defects* which are considered to belong to *distinguishable classes* of *indistinguishable defects are assumed.* This choice favors a Price type yield law.

2. All previous discussion has dealt with random point defects. In practice, defects also tend to cluster, frequently on or near the wafer periphery. The area of these defect clusters is a total loss or waste and must not be included in the area available to random point defects. Hence, yield formulae must be modified to take into account possible loss of area (and yield) through clustering.

3. Yield is reduced with increase in processing steps. Each such step may generate a distinguishable class of indistinguishable defects. Hence yield formulae must include a parameter that reflects this fact.

4. Changes in design rules will, under "relaxed design," *not cause* fatal defects to be generated and thereby reduce yield. However, when feature sizes or distances between features are reduced beyond a certain threshold point, large numbers of fatal defects may be created which will tend to *decrease* yield. Hence feature size under "tight design" rules, and the probability of occurrence of fatal defects, are linked and yield formula must reflect this fact.

5. Experimental yield must be established for different integrated circuit lots under identical test conditions. If the severity of the test, or the condition under which it is carried out, is allowed to vary significantly from lot to lot, then no useful yield comparison may be possible. This principle is assumed to be operative.

Consideration of the above facts and results of the previous sections led to the comprehensive yield formula now proposed

$$ Y_{av} = \frac{C}{\left[1 + q\left(\dfrac{s}{s_0}\right)d\right]^k} = \frac{C}{\left[1 + q\left(\dfrac{s}{s_0}\right)DA_c\right]^k} . \qquad (16.69) $$

Here parameter $q(s/s_0)$ is representative of the design rules and depends on the minimum spacing s and some empirical threshold value s_0. Parameter k is representative of the number of processing steps, and therefore of the number of different classes of identical defects on the chip. Parameter C is indicative of yield lost through other than random point defects, such as clusters of defects.

The nature of the parameters is now discussed in more detail.

As may be seen, the form of the yield equation strongly reflects a Price-type yield law. When $C = k = q(s/s_0) = 1$, the Price yield law is obtained.

If portion $(1-C)A_{we}$ of total wafer area is wasted by defect clusters, then the maximum yield for chips on this wafer has been reduced by a factor C. The remaining area CA_{we} of the wafer, is then subject to random point defects, whose average defect density is given by D. A plot of $\ln C$ versus chip area A_c will yield the value $\ln C$ as the intercept of the yield curve with the ordinate at $A_c = 0$. Parameter C lies in the range $0 \leqslant C \leqslant 1$ and, in the absence of area wastage through clustering or other causes, $C = 1$. Inclusion of this constant takes cognizance of fact 2.

Parameter $k \geqslant 1$ reflects the number of processing steps, and therefore the number of distinct defect classes as mentioned in fact 3. Ideally, parameter k is an integer, but in order to give the greatest possible flexibility to the proposed yield formula, k might be allowed to assume noninteger values greater than unity. However, this concession is probably unnecessary.

Consider a wafer divided into two equal parts. Let one part of the wafer contain chips of area A_1 designed with minimum spacing s_1. Let the other part of the wafer contain the identical type of chips of area A_2 designed with minimum spacing s_2. Let the yields Y_1 and Y_2 for these differently sized chips be obtained experimentally for many lots and then plotted in a scatter plot of Y_1 versus Y_2 as shown in Fig. 16.13. Such a scatter plot should behave according to two distinct situations:

Case 1 Relaxed design rules are deemed operative for both minimum spacing s_1 and s_2. This implies that these spacings are relaxed enough, so as not to decrease the yield by the creation of fatal defects. In that case parameter $q(s/s_o)$ may be taken constant and equal to q for both chips of area A_1 and A_2. Hence

$$Y_1 = \left[\frac{1+qDA_2}{1+qDA_1}\right]^k Y_2; \quad 0 \leqslant q \leqslant 1; \quad A_2 > A_1 \qquad (16.70)$$

As a consequence the scatter plot of Y_1 versus Y_2 should fall about a line of slope $[(1+qDA_2)/(1+qDA_1)]^k > 1$ as suggested in Fig. 16.13.

Case 2 Let relaxed design rules apply to chips with minimum spacing s_2, but let tight design rules apply to chips with minimum spacing s_1. By tight design rules we imply the increased creation of fatal defects and the consequent reduction in yield. If s_1 is made comparable with some threshold

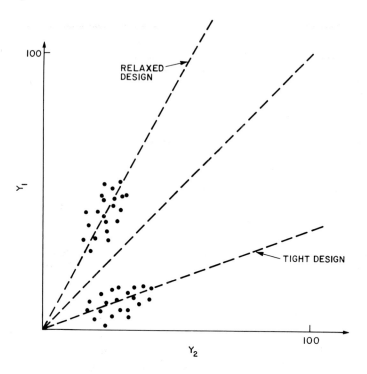

Fig. 16.13 Yield scatter plot.

value s_o, the number of fatal defects becomes large and the yield decreases appreciably, rapidly approaching zero for values of $s < s_o$. To accommodate this observed effect, $q(s/s_o)$ should be allowed to increase rapidly and it seems desirable therefore to allow it to exceed the value of unity. The exact form that $q(s/s_o)$ should take must depend on study of further yield data. However, a suggested simple empirical form is

$$q\left(\frac{s}{s_o}\right) = \frac{q}{1 - \exp\left[-\left[\frac{s}{s_o}\right]^m\right]} , \quad 0 \leqslant q \leqslant 1 \qquad (16.71)$$

where m is an integer greater than unity. For values of s greater than about $4s_o$, $q(s/s_o) \approx q < 1$. For $s \ll s_o$, $q(s/s_o) \approx qs_o^m/s^m$ and may assume large values. This in turn would decrease the yield in the catastrophic case $s \sim s_o$. For the case $m = 1$

$$Y_1 \approx \left[\frac{1+qDA_2}{1 + q \dfrac{s_o}{s} DA_1}\right]^k Y_2; \quad s \ll s_o, \quad A_2 > A_1 \qquad (16.72)$$

and the scatter plot will fall about a line of slope less than unity when $s_o A_1 > s A_2$. This case is also depicted in Fig. 16.13 and essentially answers the requirements of fact 4.

The effect of changes in design rules on a chip like the 64-bit memory whose yield was experimentally obtained in Section 16.10 may be examined in light of the empirical modification proposed in Eq. (16.71) with constant m taken as unity and three. In this connection Eq. (16.69) will be used with $C = k = q = 1$. The active chip size for the 64-bit memory chip is $A_c = 22 \times 10^{-3}$ cm^2 and will be kept *fixed,* while the minimum spacing s is allowed to vary. If s could be indefinitely decreased without loss of yield through the creation of fatal defects, then clearly more than 64 memory cells could be placed into the same area A_c. The 64-bit chip was designed with $s = 8$ μm design rules and it will be assumed that under this regime design was just still relaxed. Hence, the threshold value s_o will be taken as 2 μm. The average defect density for this chip was found in Section 16.10 to be $D = 120$ or $d = DA_c = 2.7$. Computing Eq. (16.69) with modification of Eq. (16.71) for these data then gives the yield degradation curves shown in Fig. 16.14. These curves are probably typical and should be examined in light of further experimental data.

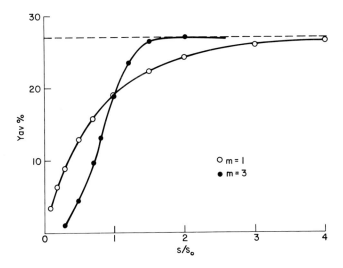

Fig. 16.14 Average yield versus spacing s.

When both s and A_c are simultaneously decreased, the yield will still ultimately fall to zero. If A_c is decreased as the square of s^2, a value of $m \geqslant 3$ will reflect the yield degradation.

When $k = 1/q$, Eq. (16.69) reduces to Stapper's formula given in Section 16.11. This formula has also previously been considered by Wadsack.[24] Values $k = q = 1$ reproduce the Price yield law. A value of k in the region of 5 (five processing steps) together with a suitable value of q will closely match Murphy's practical law, given in Eq. (16.59). When $k = 1/q \rightarrow \infty$ the exponential law is reobtained.

REFERENCES

1. G. L. Schnable and R. S. Keen. "On failure mechanisms in large scale integrated circuits." *Advances in Electronics and Electron Physics* **30**: 80-138 (1970). New York: Academic Press.

2. C. G. Peattie, J. D. Adams, S. L. Carrell, T. D. George, and M. H. Valek. "Elements of semiconductor-device reliability." *Proc. IEEE* **62**: 149-168 (February 1974).

3. M. L. Shooman. *Probabilistic Reliability—An Engineering Approach.* New York: McGraw-Hill, 1968.

4. B. V. Guedenko. *Mathematical Methods of Reliability Theory.* New York: Academic Press, 1969.

5. "Test methods and procedures for microelectronics." *MIL-STD 883.* May 1, 1968.

6. *Notice 1 to MIL-STD 883* (this provides tentative methods for screening, lot qualification, limit testing, and failure analysis of microcircuits), May 20, 1968.

7. *Notice 2 to MIL-STD 883* (this provides tentative methods for screening, lot qualification, and limit testing of microcircuits), November 20, 1969.

8. J. Vaccaro. "Semiconductor reliability within the U.S. department of defense." *Proc. IEEE* **62**: 149-168 (February 1974).

9. J. B. Brauer. "Microcircuit testing—matching the value with the cost: part 1 and 2." *Microelectronics* **2**: 23-31 (June and July 1969).

10. A. Gupta and J. W. Lathrop. "Yield analysis of large integrated-circuit chips." *IEEE J. Solid-State Circuits* **SC-7**: 389-395 (October 1972).

11. T. R. Lawson, Jr. "A prediction of the photoresist influence on integrated circuit yield." *SCP and Solid State Technology* **9**: 22-25 (July 1966).

12. T. Yanazawa. "Yield degradation of integrated circuits," *IEEE Trans. on Electron Devices* **ED-19**: 190-197 (February 1972).

13. A. Gupta, W. A. Porter, and J. W. Lathrop. "Defect analysis and yield degradation of integrated circuits." *IEEE J. of Solid-State Circuits* **SC-9**: 96-102 (June 1974).

14. M. G. Kendall and A. Stuart, *The Advanced Theory of Statistics*, Volumes 1-3. New York: Hafner Publishing Company, 1958.

15. E. Parzen. *Modern Probability Theory and Its Applications.* New York: John Wiley & Sons, 1960.

16. R. G. Stewart. "A causal definition of failure rate, theorems stress dependence and applications to devices and distributions." *IEEE Trans. on Reliability* **R15**: 95-113 (December 1966).

17. F. H. Reynolds. "Thermally accelerated aging of semiconductor components," *Proc. IEEE* **62**: 212-222 (February 1974).

18. L. R. Goldthwaite. "Failure rate study for the log-normal lifetime model." *Proc. 7-th, National Symposium on Reliability and Quality Control*, pp. 208-213, January. 1961.

19. W. Nelson. "Analysis of Accelerated life test data—part I: the Arrhenius model and graphical methods." *Trans. on Electrical Insulation* **EI 6**: 165-181 (December 1971).

20. D. S. Peck. "The analysis of data from accelerated stress tests," *9-th Reliability Physics Symposium*, pp. 69-76 (March 1971), Las Vegas.

21. D. S. Peck and C. H. Zierdt, Jr. "The reliability of semiconductor devices in the Bell System." *Proc. IEEE* **62**: 185-211 (February 1974).

22. L. E. Miller. "Reliability of semiconductor devices for submarine-cable systems." *Proc. IEEE* **62**: 230-243 (February 1974).

23. J. E. Price. "A new look at yield of integrated circuits," *Proc. IEEE* **58**: 1290-1291 (August 1970).

24. R. L. Wadsack, private communication.

25. B. T. Murphy. "Cost-size optima of monolithic integrated circuits." *Proc. IEEE* **52**: 1537-1545 (December 1964).

26. R. B. Seeds. "Yield, economic and logistic models for complex digital arrays." *1967 IEEE International Convention Record*, Part 6, pp. 60-61.

27. C. H. Stapper, Jr. "On a composite model of the IC yield problem." *IEEE J. of Solid-State Circuits.* **SC-10**: 537-539 (December 1975).

28. C. H. Stapper, Jr. "Defect density distribution for LSI calculations." *IEEE Trans. on Electron Devices* **ED-20**: 655-656 (July 1973).

29. C. H. Stapper. "LSI modeling and process monitoring." *IBM J. of Research and Development* **20**: 228-234 (May 1976).

30. R. M. Warner. "Applying a composite model of the IC yield problem." *IEEE J. of Solid-State Circuits* **SC-9**: 86-95 (June 1974).

31. D. A. McGillis, private communication.

PROBLEMS

16.1 The failure of a semiconductor device population is governed by the log-normal failure density function. Write expressions for the reliability function and the hazard rate. Show that the hazard rate has a maximum at time $t > \exp(\mu - \sigma^2)$, where $\mu = \ln t_{.50}$.

If the argument of Φ can be taken as a measure of degradation with $\mu(T) = \ln t_{.50}$, show that the mean of the distribution can be expressed as $\mu(T) = \alpha + \beta/T$, where α and β are called "regression parameters."

16.2   Using Arrhenius plotting paper, show that in a graphical estimate of the regression parameters α and β mentioned in Section 16.5,

$$\hat{\beta} = \frac{T_1 T_2}{T_2 - T_1} \log\left[\frac{t_1}{t_2}\right]$$

$$\hat{\alpha} = \log t_2 - \hat{\beta}/T_2,$$

where T_1, T_2 are two absolute temperatures and where t_1, t_2 are corresponding mean lifetimes. Show also that an estimate for the activation energy is given by $\hat{E}_a = k\hat{\beta}$.

16.3   The Arrhenius line for median life for a certain transistor type yielded the following data:

Junction Temperature (°C)	270	240	180
Median Life (hr)	20	150	2000

Estimate the activation energy, E_a, characterizing the transistor type. Compute also the median life at a maximum use condition of 65°C.

16.4   An accelerated life test is carried out by means of a step-stress test involving r steps at a temperature T_i $(i = 1, 2, \cdots, r)$. Let the steps be so arranged that the increments of reciprocal temperature are all equal to ΔT^{-1}. If the value of

$$\Phi\left[\frac{\ln t - \mu(T)}{\sigma}\right]$$

at the rth step is the same as would be obtained by a single step-stress test at temperature T_r', show that

$$(T_r')^{-1} = (T_r)^{-1} - \frac{k}{E_a} \ln\left[\frac{1 - \exp\left[\dfrac{E_a r}{k \Delta T}\right]}{1 - \exp\left[-\dfrac{E_a}{kT}\right]}\right].$$

16.5 M distinguishable defects may be placed onto N integrated circuit chips in N^M ways. Show that the probability $P(z=k)$ of having k defects on one chip is given by

$$P(z=k) = \binom{M}{k}\left[\frac{1}{N}\right]^k\left[1 - \frac{1}{N}\right]^{M-k}.$$

Hence show that the yield, $\bar{Y}$, is given by

$$\bar{Y} = \left(1 - \frac{1}{N}\right)^{M}$$

Discuss the limiting form of this yield expression.

16.6 An integrated circuit wafer containing M distinguishable defects is divided into n chips. The probability that the random variable Y, has a value n/N where $0 \leqslant n \leqslant N$ is given by

$$P\{NY=n\} = \binom{N}{n} \sum_{k=0}^{N-n} (-1)^{k} \binom{N-n}{k} \left[1 - \frac{n+k}{N}\right]^{M}.$$

Find the expected value, Y_{av}, of the yield, the variance σ_{y}^{2}, and discuss the limiting form of the above probability.

16.7 M indistinguishable defects may be placed onto N integrated circuit chips in $\binom{N+M-1}{M}$ ways. Show that the probability $P(z=k)$ of having k defects on one chip is given by

$$P(z=k) = \frac{\binom{N+M-k-2}{M-k}}{\binom{N+M-1}{M}}.$$

Hence show that the yield, Y_{av}, is given by

$$Y_{av} = \frac{1 - (1/N)}{1 - d_{0} - (1/N)}$$

and discuss the limiting form of this yield expression.

16.8 An integrated circuit wafer containing M indistinguishable defects is divided into N chips. The probability that the yield has a value n/N $(0 \leqslant x \leqslant N)$ is given by

$$P(NY=n) = \frac{\binom{N}{n}\binom{M-1}{N-n-1}}{\binom{N+M-1}{M}}.$$

Find the expected values of yield, its variance, and discuss the limiting form of the above probability.

16.9 An integrated circuit wafer contains M point defects and is divided into N chips. It is established experimentally that the number of chips containing k defects each is given by $N(k)$ $(0 \leqslant k \leqslant M)$.

Show that the experimental value of yield is given by

$$\bar{Y} = \frac{N(0)}{\sum_{k=0}^{M} N(k)}$$

and the experimental value for defect density d_0 is given by

$$d_0 = \frac{\sum_{k=1}^{M} kN(k)}{\sum_{k=0}^{M} N(k)} .$$

If the histogram follows the law

$$N(k) = \frac{Nd_0^k}{(1+d_0)^{k+1}} ,$$

how might you graphically obtain a value for the average defect density d_0?

16.10 The conditional probability of finding n distinguishable defects per chip is given by

$$P(z=n/d) = \frac{d_0^n \exp(-d_0)}{n!} .$$

If the joint defect density after k processing steps is given by

$$f_{d_1, d_2, \cdots, d_k}(d) = \sum_{i=1}^{k} \frac{d_{0k}^{(k-2)} \exp\left[-\dfrac{d}{d_{0i}}\right]}{(d_{0i}-d_{01})(d_{0i}-d_{02}) \cdots (d_{0i}-d_{0k})}$$

where $d_{0i} - d_{0i} \triangleq 1$, evaluate the probability $P_k(z=n)$ after k processing steps, where

$$P_k(z=n) = \int_0^\infty P(z=n/d) f_{d_1, d_2, \cdots, d_k}(d) \, d(d).$$

Show also that the yield of a k step process is given by

$$\bar{Y} = \frac{1}{\displaystyle\prod_{r=1}^{k} (1+d_{0r})} .$$

16.11 An empirical yield formula which reflects the number of processing steps, k, and also the circuit design rules is given by

$$\bar{Y} = \frac{1}{(1+qd_0)^k}$$

where q is a measure of the number of fatal defects, and considered to reflect the design rules. Plot a family of curves taking $k = 1, 2, 3, 4$, and 5 for $q = 1, 0.7$, and 0.5. Note that $q = 1$ and $k = 1$ corresponds to the Price law. Compare plots obtained both to the Price and the exponential yield laws.

16.12 A hypothetical lot of five integrated circuit wafers containing digital logic circuit chips each with 256 cells, gave the following yield and defect data:

No. of Wafer	M_i	N_i	$N_i(0)$
1	96	104	54
2	89	96	50
3	93	94	46
4	72	74	37
5	55	54	27

Furthermore, the total number $N(k)$ of chips containing k defects was $N(0) = 217$, $N(1) = 104$, $N(2) = 52$, $N(3) = 24$, $N(4) = 12$, $N(5) = 6$, $N(6) = 3$, $N(7) = 2$, $N(8) = 1$, $N(9) = 0$, $N(10) = 1$. The active chip size, $A_c = 20 \times 10^{-3}$ cm^2, and the chip was designed for $s = 5$ μm design rules and fabricated using four processing steps. Write a thorough yield report on this logic family fabrication process.

16.13 Two similar logic chips were produced in two different manufacturing facilities under slightly different conditions.

Process A required three processing steps and produced defect densities $d_{01} = 0.9$, $d_{02} = 1.0$, and $d_{03} = 1.1$ at each step. Moreover, the corresponding fatal defect coefficients were $q_1 = 0.25$, $q_2 = 0.45$, and $q_3 = 0.15$.

Process B required four processing steps and produced defect densities $d_{01} = 1$, $d_{02} = 1.1$, $d_{03} = 1.2$, and $d_{04} = 0.8$. The corresponding fatal defect coefficients were $q_1 = 0.2$, $q_2 = 0.15$, $q_3 = 0.35$, and $q_4 = 0.1$.

Assume both processes followed a modified Price law and compare the yield according to the two processing techniques. How can the above data be reconciled with the proposed yield model given by Eq. (16.69)?

16.14 Similar LSI logic chips were fabricated using four processing steps.

Chip A was produced under "relaxed" design rules. The average defect density $d = 1.2$, the fatal defect coefficient $q = 0.7$ and the minimum spacing, $s = 8$ μm. The threshold spacing $s_0 = 1.5$ μm.

Chip B was produced under "tight" design rules, by reducing the minimum spacing s toward the threshold value s_0. If the product of q and d can be modeled by the empirical formula

$$qd = \frac{d}{\left[1 - \exp\left(-\frac{s}{s_0}\right)^4\right]},$$

compare the yield under relaxed design procedures with the yield obtained as s is reduced toward the threshold value s_0. Use the proposed yield model of Eq. (16.69).

Index